Natural and Anthropogenic Disasters
Vulnerability, Preparedness and Mitigation

Natural and Anthropogenic Disasters
Vulnerability, Preparedness and Mitigation

Edited by

Madan Kumar Jha

Indian Institute of Technology Kharagpur
West Bengal, India

A C.I.P. Catalogue record for this book is available from the Library of Congress.

ISBN 978-90-481-2497-8 (HB)
ISBN 978-90-481-2498-5 (e-book)

Copublished by Springer,
P.O. Box 17, 3300 AA Dordrecht, The Netherlands
with Capital Publishing Company, New Delhi, India.

Sold and distributed in North, Central and South America by Springer,
233 Spring Street, New York 10013, USA.

In all other countries, except India, sold and distributed by Springer,
Haberstrasse 7, D-69126 Heidelberg, Germany.

In India, sold and distributed by Capital Publishing Company,
7/28, Mahaveer Street, Ansari Road, Daryaganj, New Delhi, 110 002, India.

www.springer.com

Cover photos credit: Left: www.en.wikipedia.org/wiki/cyclone_leon_eline and right: Dr. Goamthinayagam, Structural Engineering Research Centre, CSIR, Government of India, Chennai.

Printed on acid-free paper

All Rights Reserved
© 2010 Capital Publishing Company
No part of this work may be reproduced, stored in a retrieval system, or transmitted in any form or by any means, electronic, mechanical, photocopying, microfilming, recording or otherwise, without written permission from the Publisher, with the exception of any material supplied specifically for the purpose of being entered and executed on a computer system, for exclusive use by the purchaser of the work.

Printed in India.

Preface

"Nature provides for everybody's need, but not for everybody's greed."
—Mahatma Gandhi

"Concern for man and his fate must always form the chief interest of all technical endeavors. Never forget this in the midst of your diagrams and equations."
—Albert Einstein

With the advancement in human knowledge through rapid progress in science and technology, human beings have incurred major changes in natural processes through their various activities. As a result, a myriad of problems have been created and continue to do so; which will challenge us for many decades into the future. Unfortunately, the intricate nexus between nature and human life has been completely neglected by the humans in their endless quest of luxurious life-style and supremacy, both in developed and developing nations, and hence the sustainability of human development is severely threatened at the dawn of the 21st century. One of the best ways to ensure sustainable socio-economic and environmental systems is to hamper as little as possible the proper functioning of natural life cycles (i.e., ecosystems).

The most important challenges of this century are: *water security, food security, energy security* and *environmental security*. Owing to enhanced natural and anthropogenic disasters worldwide, these crucial challenges become much more complicated and daunting, particularly for developing and low-income countries. It is thus the need of the hour to highlight the risk of different disasters, the emerging tools and techniques, and the modern approaches to minimize disaster frequency and losses which can help develop a disaster-resilient society. Disaster management being highly multidisciplinary in nature, a comprehensive book dealing with different aspects of disaster management in the face of global climate change and socio-economic changes, and encompassing important disasters faced by humankind is presently lacking. The existing books on disaster management, though limited, provide either cursory treatment of the topic or are confined to specific aspects of disaster management. The present book attempts to fulfill this gap. This book provides clear, comprehensive, and up-to-date information about different facets of disaster management including emerging issues. The contributors were selected from India and abroad based on their proven knowledge in the specific area of their contribution. All of them are leading experts in their fields of specialization and are vibrant researchers from reputed research and/or teaching organizations.

The book contains 26 chapters. *Chapter 1* presents an overview of natural and anthropogenic disasters, including the definition/description of salient disaster-related terms, differential impact of disasters, and ancient and modern approaches for disaster management. *Chapter 2* presents a review of some of the major, recent and important earthquakes in the Indian shield region, results of geodetic measurements in the Himalaya and Indian shield region, and the discussion on seismic gaps. The issues related to the earthquake disaster mitigation are also discussed in this chapter. *Chapter 3* highlights the damages caused by the 2004 Tsunami and the actions taken to help the people and solve technical problems in the Ampara district of eastern Sri Lanka. *Chapter 4* discusses the devastation caused by the 2004 Tsunami along the Indian coasts and the rehabilitation measures implemented by the governments and various agencies of the affected areas. *Chapter 5* presents the experiences and outcomes from a comprehensive and integrated approach to support the restoration of groundwater-based water supply and drinking water quality after the 2004 Tsunami in Sri Lanka. *Chapter 6* presents a brief review on the tsunamis that affected Indian sub-continent, current status of tsunami warning system for the Indian Ocean, and the methodology for constructing tsunami travel time charts for the Indian Ocean. *Chapters 7, 8* and *9* are dedicated to the flood disaster encompassing the flood hazards in India and management strategies, and the roles of modeling and flood forecasting in managing devastating floods which are recurrent phenomena in India. *Chapter 10* deals with the impacts of drought and the management of drought in the agricultural drought regions of India with an emphasis on in-season drought coping practices and permanent drought amelioration. It also addresses issues concerning meteorological and hydrological droughts, and discusses existing policies and new approaches to combat droughts. *Chapter 11* presents an overview of various internationally accepted drought indicators and their application to drought-prone arid Rajasthan of western India. In *Chapter 12*, the relationships between the IODM (Indian Ocean Dipole Mode) and the cyclone frequency in the Bay of Bengal during post-monsoon season (also known as 'storm season' in South Asia) are discussed. The probable impact of IODM on the frequency of Monsoon Depressions and Storms has also been examined. *Chapter 13* focuses on the changes in extremes over Indian region using objectively defined indices of observed temperature extremes. It also explores the relationship between El Niño-Southern Oscillation (ENSO) and temperature extremes over India. *Chapter 14* deals with the fluctuation characteristics of annual, seasonal, and monsoon monthly rainfalls for the 15 physiographic divisions and 49 subdivisions of India using the longest available instrumental rainfall records from 316 rainfall stations.

Moreover, *Chapter 15* focuses on the issues pertaining to mine safety and disaster management, and the recent developments in tools and technologies and their applications in mining industries for safe mining operations and improved productivity. *Chapter 16* highlights forest fire disaster and addresses forest fire management issues with experiences from the kingdom of Swaziland. The role of emerging technologies in forest fire monitoring and management, possible early warning, and the mitigation strategies for forest fires are also discussed. In *Chapter 17*, an overview of climate change and climate variability and its impact on the water resources of India has been presented, together with the role of simulation modeling in evaluating the impact of climate change on water resources. *Chapter 18* discusses simulation modeling using the CERES- and GRO-family of crop simulation models to examine the impact of climate change on the growth and yield of salient crops under the agro-climatic conditions of Punjab State in India. *Chapter 19* explores the susceptibility of livestock production to climate change and the adaptation strategies for reducing the vulnerability of smallholder dairy production to climate change. *Chapter 20* highlights the implications of climate change for disaster management, presents climate change in the context of disaster risk mitigation, offers some suggestions and identifies important challenges to be overcome. *Chapter 21* discusses the potential of geospatial technologies in managing land and water resources so as to minimize or avoid land and water related disasters and ensure sustainable management of these vital resources. In addition, the current and future missions of Indian remote sensing satellites related to hydrology are discussed. *Chapter 22* presents the concept of Integrated Water Resources Management (IWRM) and its implication for solving real-world water problems. The fundamentals of decision support systems (DSS) are also described, together with a critical review of DSS applications in water management. *Chapter 23* emphasizes the need for shifting towards sustainable forest management to enhance our capacities in disaster preparedness, response, and reporting mechanisms, and thereby reducing human vulnerability to disasters. It also addresses the issues and challenges related to sustainable forest and other natural resources management. *Chapter 24* suggests a paradigm shift in the disaster risk reduction by adopting a participatory approach for efficient information management and proposes a conceptual model for effective information flow among the stakeholders involved in disaster management. *Chapter 25* presents management strategies for reducing damage to the coast-based major industrial installations by severe cyclones. The cyclone emergency management plan presented in this chapter can serve as useful guidelines for other disasters as well. Finally, the concluding *Chapter 26* succinctly discusses the challenges, and research and development needs for sustainable disaster management by highlighting modern management approaches such as 'comprehensive disaster management', 'adaptive management', 'IWRM' and 'adaptive IWRM' as well as modern tools and techniques. It also emphasizes the need for holistic and multidisciplinary approaches for resource and disaster management, proper education and training as well as for improved coordination among different agencies, and concerted and committed efforts by the policy makers, practitioners, and scientists to ensure sustainable disaster management and sustainable development on the earth.

    The information contained in this book will not only serve as general guidelines for efficient disaster management but will also help educate engineers and scientists about various natural and anthropogenic disasters as well as modern tools/techniques and approaches for their efficient management. This book will be very useful to the teachers, students and researchers of environmental, civil, agricultural, ocean, and land and water resources engineering fields as well as to the practitioners, planners and decision makers, especially of developing nations.

IIT Kharagpur, India
July 2009

Madan Kumar Jha

About the Editor

Madan Kumar Jha obtained his M.E. from the Asian Institute of Technology, Bangkok in 1992 availing full postgraduate scholarship. Thereafter, he obtained his PhD availing the 'Monbusho Scholarship' from United Graduate School of Agricultural Sciences, Ehime University, Japan in 1996. Between December 1996 and November 1997, he worked as a Water Resources Engineer at Panya Consultants Co. Ltd., Bangkok, and from December 1997 to November 1999 he was a Postdoctoral Fellow at Kochi University, Japan with 'JSPS Research Fellowship'. In 1999, he joined the Department of Agricultural and Food Engineering, Indian Institute of Technology Kharagpur, West Bengal, India and, presently, he is Associate Professor in the same department. Dr. Jha has been honored with salient scholastic awards such as the 'AMA-Shin-Norinsha-AAAE Young Researcher Award' in 2005 by the Asian Association for Agricultural Engineering, Bangkok, 'Shankar Memorial Award' in 2009 and 'Distinguished Services Award 1999-2000' by the Indian Society of Agricultural Engineers, New Delhi. Dr. Jha is also a recipient of the 'Alexander von Humboldt Fellowship' (2004-2005). Recently, he has been awarded 'JSPS Invitation Fellowship (Long-term)' by the Japan Society for the Promotion of Science, Tokyo and 'Fellow' of the Indian Water Resources Society, Roorkee, India.

Dr. Jha has to his credit 44 papers in international refereed journals and a couple of papers in national refereed journals, four technical reports and more than 50 papers in international and national conference proceedings. He has authored/co-authored two books, viz., "Applications of Remote Sensing and GIS Technologies in Groundwater Hydrology: Past, Present and Future" and "Land and Water Management Engineering" as well as has contributed five book chapters. He has also filed one patent and obtained seven copyrights. He is 'Associate Editor' of *International Agricultural Engineering Journal* published by the Asian Association for Agricultural Engineering, Bangkok and 'Co-Editor' of *Research Journal of Chemistry and Environment* published by the International Congress of Chemistry and Environment, India. He is also serving as a reviewer of several international journals related to water resources engineering.

Dr. Jha's research interest includes basin-wide simulation-optimization modeling of groundwater systems, inverse modeling for determining hydraulic parameters of aquifer and well systems using conventional and non-conventional optimization techniques, river-aquifer and tide-aquifer interactions, artificial recharge, rainwater harvesting, flow and solute transport in the vadose zone, and the application of RS, GIS and MCDM techniques for groundwater development and management.

Acknowledgements

The preparation of a book presenting comprehensive and thoroughly up-to-date information about different facets of disaster management cannot be possible without the invaluable help and cooperation from others. This book is part of my efforts to contribute to the solution of real-world problems concerning land and water. I am very grateful to the contributors who accepted my invitation and extended full cooperation during revision of the manuscripts according to my detailed comments/suggestions. Each chapter has been revised at least once. It was my pleasure and special experience to work with the leading scientists of India and abroad. I am also thankful to my PhD students Mr. Pramod Tiwary, Mr. Arbind Kumar Verma and Mr. Mukesh Tiwari as well as to my final year M.Tech. students Mr. Amanpreet Singh, Mr. Ganesh M. Bongane and Ms. Binitha V. Pai at the Indian Institute of Technology (IIT), Kharagpur, West Bengal, India for their help in correcting formats of some of the manuscripts and preparing subject index. Chapters 1 and 26 have been greatly benefited from Coppola (2007), UNESCO (2007), UNISDR (2005), UNDP (2004) and Wikipedia, which is gratefully acknowledged. Thanks are also due to other individuals who contributed indirectly to this project.

Many thanks to the manager and staff of Capital Publishing Company, New Delhi, India as well as to Springer, The Netherlands for their kind cooperation and keen interest in this project. Special thanks are due to the editorial team of the publishers for careful editing of the manuscripts.

To bring ideas into fruition require full hearts and bodies, for which I express my heartfelt thanks to my wife Manisha and son Piyush for their constant love, endless endurance and encouragement, and wholehearted support throughout this project.

Contributors

A. Sanjeewa P. Manamperi
International Water Management Institute
(IWMI), P O Box 2075, Colombo, Sri Lanka

Adlul Islam
ICAR Research Complex for Eastern Region
P.O.: Bihar Vet. College, Patna - 800 014
Bihar, India
E-mail: adlul@rediffmail.com

Alok K. Sikka
National Rainfed Area Authority, NASC
Complex Pusa, New Delhi - 110 012, India

Amal Kar
Central Arid Zone Research Institute
Jodhpur - 342 003, Rajasthan, India
E-mail: amalkar@cazri.res.in

A.S. Rao
Central Arid Zone Research Institute
Jodhpur - 342 003, Rajasthan, India

B.K. Prasad
Indian Institute of Forest Management
Bhopal - 462 003, M.P., India

B. Venkateswarlu
Central Research Institute for Dryland
Agriculture, Hyderabad - 500 059
Andhra Pradesh, India

Bhabagrahi Sahoo
Soil and Water Conservation Engineering
ICAR Research Complex for NEH Region
Nagaland Centre, Jharanapani
Medziphema - 797 106, India
E-mail: bsahoo2003@yahoo.com

Chris R. Panabokke
Water Resources Board, Sri Lanka

D. Dugaya
Indian Institute of Forest Management
Bhopal - 462 003, M.P., India

D. Nagesh Kumar
Department of Civil Engineering, Indian Institute
of Science, Bangalore - 560 012
Karnataka, India
E-mail: nagesh@civil.iisc.ernet.in

D.R. Kothawale
Climatology and Hydrometeorology Division
Indian Institute of Tropical Meteorology, Pashan
Pune - 411 008, India

D.V. Gopinath
Energy and Environment, 208, IV Cross, I Stage
Gangotri Layout, Mysore - 570 009, India
E-mail: dvgopinath@gmail.com

Dheeraj Kumar
Department of Mining Engineering, I.S.M.
University, Dhanbad - 826 004, Jharkhand, India
E-mail: dheeraj@dkumar.org

Dhrubajyoti Sen
Department of Civil Engineering, Indian Institute
of Technology, Kharagpur - 721 302
West Bengal, India
E-mail: djsen@iitkgp.ac.in

Falguni Baliarsingh
Department of Civil Engineering, College of
Engineering and Technology
Bhubaneswar - 751 003, Orissa, India

G.G.S.N. Rao
Central Research Institute for Dryland
Agriculture, Hyderabad - 500 059
Andhra Pradesh, India

G. Pratibha
Central Research Institute for Dryland
Agriculture, Hyderabad - 500 059
Andhra Pradesh, India

G. Rajeshwara Rao
Central Research Institute for Dryland
Agriculture, Hyderabad - 500 059
Andhra Pradesh, India

G. Ravindra Chary
Central Research Institute for Dryland
Agriculture, Hyderabad - 500 059
Andhra Pradesh, India
E-mail: gcravindra@gmail.com

H.N. Singh
Indian Institute of Tropical Meteorology
Pashan, Pune - 411 008, India

H.P. Borgaonkar
Climatology and Hydrometeorology Division
Indian Institute of Tropical Meteorology
Pashan, Pune - 411 008, India

J.V. Revadekar
Climatology and Hydrometeorology Division
Indian Institute of Tropical Meteorology
Pashan, Pune - 411 008, India
E-mail: jvrch@tropmet.res.in

K. Srinivasa Raju
Department of Civil Engineering, Birla Institute
of Technology and Science, Pilani, Hyderabad
Campus, Hyderabad - 500 078, A.P., India

K.L. Sharma
Central Research Institute for Dryland
Agriculture, Hyderabad - 500 059
Andhra Pradesh, India

K.P.R. Vittal
Central Arid Zone Research Institute
Jodhpur - 342 003, Rajasthan, India

K.V. Rao
Central Research Institute for Dryland
Agriculture, Hyderabad - 500 059
Andhra Pradesh, India

Karen G. Villholth
Geological Survey of Denmark and Greenland
(GEUS), Copenhagen, Denmark
International Water Management Institute
(IWMI), P O Box 2075, Colombo, Sri Lanka
E-mail: kgv@geus.dk

Leclerc Jean-Pierre
Laboratory of Chemical Engineering Sciences-
CNRS-Nancy-Université
B.P. 20451 54001 Nancy Cedex, France
E-mail: leclerc@ensic.inpl-nancy.fr

M.V. Venugopalan
NBSS&LUP, Amaravati Road, Nagpur - 440 010
Maharashtra, India

Madan Kumar Jha
AgFE Department, Indian Institute of
Technology, Kharagpur - 721 302
West Bengal, India
E-mail: madan@agfe.iitkgp.ernet.in

Md. Shahid Parwez
Asian Development Bank, Kathmandu, Nepal

Meththika Vithanage
Department of Geography and Geology
University of Copenhagen, Denmark

Muthiah Perumal
Department of Hydrology, Indian Institute of
Technology, Roorkee - 247 667, India
E-mail: p_erumal@yahoo.com

N.A. Sontakke
Indian Institute of Tropical Meteorology
Pashan, Pune - 411 008, India
E-mail: neelimasontakke@hotmail.com

Nify Benny
Department of Chemical Oceanography
Cochin University of Science and Technology
Kochi - 682 016, Kerala, India

Nityanand Singh
Indian Institute of Tropical Meteorology
Pashan, Pune - 411 008, India

O.P. Singh
India Meteorological Department, Mausam
Bhavan, Lodi Road, New Delhi - 110 003, India
E-mail: singh.op@imd.gov.in

P. Tiwary
AgFE Department, Indian Institute of
Technology, Kharagpur - 721 302
West Bengal, India
E-mail: ptiwary70@yahoo.co.in

P.C. Pandey
CORAL, Indian Institute of Technology
Kharagpur - 721 302, West Bengal, India

P.K. Mishra
Central Research Institute for Dryland
Agriculture, Hyderabad - 500 059
Andhra Pradesh, India

Paramsothy Jeyakumar
Eastern University of Sri Lanka, Batticaloa
Sri Lanka

Poonam Pandey
DCN Division, National Dairy Research Institute
Karnal - 132 001, Haryana, India

Prabhjyot-Kaur
Department of Agricultural Meteorology, Punjab
Agricultural University, Ludhiana - 141 004
Punjab, India
E-mail: ldh.aicrpam@gmail.com

Prasad K. Bhaskaran
OE&NA Department, Indian Institute of
Technology, Kharagpur - 721 302, India
E-mail: pkbhaskaran@naval.iitkgp.ernet.in

Priyanie H. Amerasinghe
International Water Management Institute
(IWMI), P O Box 2075, Colombo, Sri Lanka

Rohit R. Goswami
Department of Civil Engineering
Auburn University, Alabama, USA

Savita Bisht
Indian Institute of Forest Management
Bhopal - 462 003, M.P., India

S. Bandyopadhyay
Indian Space Research Organization (ISRO)
Bangalore - 560 097, India
E-mail: sbanerjee377@gmail.com

S.K. Sirohi
DCN Division, National Dairy Research Institute
Karnal - 132 001, Haryana, India

S.S. Hundal
Department of Agricultural Meteorology
Punjab Agricultural University
Ludhiana - 141 004, Punjab, India

S.V.R.K. Prabhakar
Climate Policy Project
Institute for Global Environmental Strategies
2108-11, Kamiyamaguchi, Hayama
Kanagawa, 240-0115, Japan
E-mail: sivapuram.prabhakar@gmail.com

Smita Sirohi
National Centre for Agricultural Economics and
Policy Research, New Delhi - 110 012, India
E-mail: smitasirohi@yahoo.com

Solidarités Staff
Solidarités, 50 rue Klock, 92110
Clichy La Garenne, France

Sujatha C.H.
Department of Chemical Oceanography, Cochin
University of Science and Technology
Kochi - 682 016, Kerala, India
E-mail: drchsujatha@yahoo.co.in

Tajbar S. Rawat
Indian Institute of Forest Management
Bhopal - 462 003, M.P., India
E-mail: tajbar.india@gmail.com

Vineet K. Gahalaut
National Geophysical Research Institute, Uppal
Road, Hyderabad - 500 007, A.P., India
E-mail: vkgahalaut@ngri.res.in

Wisdom M.D. Dlamini
Swaziland National Trust Commission
Lobamba, H107, Swaziland
E-mail: mwdlamini@gmail.com

Contents

Preface v
About the Editor viii
Acknowledgements ix
Contributors xi

1. Natural and Anthropogenic Disasters: An Overview 1
   Madan Kumar Jha
2. Earthquakes in India: Hazards, Genesis and Mitigation Measures 17
   Vineet K. Gahalaut
3. Impact of 2004 Tsunami on Housing, Sanitation, Water Supply and Wastes Management: The Case of Ampara District in Sri Lanka 44
   Leclerc Jean-Pierre and Solidarités Staff
4. Impact of December 2004 Tsunami on Indian Coasts and Mitigation Measures 60
   Sujatha C. H. and Nify Benny
5. Tsunami Impacts and Rehabilitation of Groundwater Supply: Lessons Learned from Eastern Sri Lanka 82
   Karen G. Villholth, Paramsothy Jeyakumar, Priyanie H. Amerasinghe, A. Sanjeewa P. Manamperi, Meththika Vithanage, Rohit R. Goswami and Chris R. Panabokke
6. Tsunami Early Warning System: An Indian Ocean Perspective 100
   Prasad K. Bhaskaran and P.C. Pandey
7. Flood Hazards in India and Management Strategies 126
   Dhrubajyoti Sen
8. Modeling for Flood Control and Management 147
   D. Nagesh Kumar, K. Srinivasa Raju and Falguni Baliarsingh
9. Real-time Flood Forecasting by a Hydrometric Data-Based Technique 169
   Muthiah Perumal and Bhabagrahi Sahoo
10. Drought Hazards and Mitigation Measures 197
    G. Ravindra Chary, K.P.R. Vittal, B. Venkateswarlu, P.K. Mishra, G.G.S.N. Rao, G. Pratibha, K.V. Rao, K.L. Sharma and G. Rajeshwara Rao
11. Indicators for Assessing Drought Hazard in Arid Regions of India 237
    K.P.R. Vittal, Amal Kar and A.S. Rao
12. Tropical Cyclones: Trends, Forecasting and Mitigation 256
    O.P. Singh

13. Temperature Extremes over India and their Relationship with
    El Niño-Southern Oscillation — 275
    J.V. Revadekar, H.P. Borgaonkar and D.R. Kothawale
14. Monitoring Physiographic Rainfall Variation for Sustainable Management of
    Water Bodies in India — 293
    N.A. Sontakke, H.N. Singh and Nityanand Singh
15. Emerging Tools and Techniques for Mine Safety and Disaster Management — 332
    Dheeraj Kumar
16. Management of Forest Fire Disaster: Perspectives from Swaziland — 366
    Wisdom M.D. Dlamini
17. Climate Change and Water Resources in India: Impact Assessment and
    Adaptation Strategies — 386
    Adlul Islam and Alok K. Sikka
18. Global Climate Change vis-à-vis Crop Productivity — 413
    Prabhjyot-Kaur and S.S. Hundal
19. Adapting Smallholder Dairy Production System to Climate Change — 432
    Smita Sirohi, S.K. Sirohi and Poonam Pandey
20. Climate Change-Proof Disaster Risk Reduction: Prospects and Challenges
    for Developing Countries — 449
    S.V.R.K. Prabhakar
21. Potential of Geospatial Technologies for Mitigating Land and Water Related
    Disasters — 469
    S. Bandyopadhyay and Madan Kumar Jha
22. Decision Support System: Concept and Potential for Integrated Water
    Resources Management — 503
    P. Tiwary, Madan Kumar Jha and M.V. Venugopalan
23. Sustainable Forest Management: Key to Disaster Preparedness and Mitigation — 536
    Tajbar S. Rawat, D. Dugaya, B.K. Prasad and Savita Bisht
24. Participatory Information Management for Sustainable Disaster Risk Reduction — 563
    S.V.R.K. Prabhakar and Md. Shahid Parwez
25. Cyclone Emergency Preparedness in DAE Coastal Installations, India — 578
    D.V. Gopinath
26. Sustainable Management of Disasters: Challenges and Prospects — 598
    Madan Kumar Jha

Index — *611*

Natural and Anthropogenic Disasters: An Overview

Madan Kumar Jha

1. INTRODUCTION

Disasters are undesirable and often sudden events causing human, material, economic and/or environmental losses, which exceed the coping capability of the affected community or society. They are caused either by natural forces/processes (known as '*natural disasters*') or by human actions, negligence, or errors (known as '*anthropogenic disasters*'). Natural disasters are generally classified into three major groups (CRED, 2009): (i) 'geophysical disasters' (e.g., earthquake, volcanic eruption, rockfall, landslide, avalanche, and subsidence); (ii) 'hydro-meteorological disasters' (e.g., flood, drought, storm, extreme temperature, wildfire, and wet mass movement); and (iii) 'biological disasters' (e.g., epidemic, insect infestation, and animal stampede). Similarly, anthropogenic disasters are broadly classified into two major groups (http://en.wikipedia.org/wiki/Disaster): (i) 'technological disasters' (e.g., disasters due to engineering failures, transport disasters, and environmental disasters); and (ii) 'sociological disasters' (e.g., criminal acts, riots, war, stampedes, etc.).

In the *anthropocene* (most recent period in the earth's history when human activities have a significant global impact on the earth's climate and ecosystems), most natural disasters are aggravated by anthropogenic activities. Thus, hazards, which are part of nature, often turn into disasters due to human actions or inactions (UNESCO, 2007). For example, severe flooding is often aggravated by deforestation, urbanization, silting, and buildings in floodplains. Destruction of nature's natural defenses such as wetlands and coastal swamps is responsible for the severe damage by tropical cyclones, apart from ecological damage and loss of biodiversity. Failure to impose building codes and implement earthquake-resistant techniques is mainly responsible for the collapse of buildings and human fatalities even by relatively low intensity earthquakes. Human activities are also responsible for the threat of global climate change and rising sea levels as a result of global warming caused by increased greenhouse gases (GHGs) concentrations in the atmosphere (IPCC, 2007). Today, the world is experiencing a dramatic increase of suffering from the effects of various disasters, ranging from extreme droughts to severe floods caused by the poor land and water management and probably by climate change. More than 2,200 major and minor water-related disasters occurred in the world during 1990-2001 period, of which floods accounted for 50%, waterborne and vector diseases accounted for 28%, and droughts accounted for 11% of the total disasters (UNESCO Water Portal, 2008).

Disasters result in a serious disruption in the functioning of a society and widespread damages to life, property, infrastructure, and environment. They have adversely affected mankind since the beginning of our existence. About 94% of natural disasters occur due to four major causes — earthquakes, tropical cyclones (also known as typhoon, hurricane, cyclone, tropical storm, and tropical depression depending on their location and strength), floods and droughts (UNDP, 2004; UNESCO, 2007). About 75% of the global population reside in the areas which were affected by one or more of these catastrophic phenomena at least once between 1980 and 2000. More than 184 deaths per day are recorded in different parts of the world due to disasters triggered by earthquake, tropical cyclone, flood or drought (UNESCO Water

Portal, 2008). It has been found that the earthquakes and volcanic eruptions cause almost half of all natural disaster casualties, whereas the storms and other hydro-meteorological disasters cause the most economic damage (UNESCO, 2007). Disasters not only trigger direct losses, but also aggravate other stresses and distress such as a financial crisis, social/political conflict, disease, and environmental degradation.

The United Nations had declared the 1990s as the "International Decade for Natural Disaster Reduction". However, it should be aptly called the 'decade of natural disasters' with a record of flooding, earthquakes, landslides and droughts rarely seen in such profusion (UNESCO, 2007). The loss from natural disasters in the 1990s exceeded the combined losses of the four previous decades. This pattern has continued into the 21st century with spectacular catastrophes. Ever-increasing population (world's population doubled in the past 40 years), rapid urbanization (urban population quintupled in the past 40 years), and growing techno-economic development have resulted in widespread unsustainable development which is responsible for an increasing burden of risk at the dawn of the 21st century. On the top of it, global climate change and extremes of weather increase the domain and cost of disasters. They are responsible for increasing calamities in terms of frequency, scale, severity (social, economic and environmental losses), and complexity (Coppola, 2007; UNESCO, 2007). For example, during a recent two-year period (December 2004 to November 2006), the world community faced an extremely severe wave of disasters. The August 2006 report of the UN Secretary-General revealed that in the period from June 2005 to May 2006, there were 404 disasters with nationwide consequences in 115 countries, including the death of 93,000 people and economic losses totaling 173 billion US dollars. Another UN report issued in November 2006 estimated that 91 million people had their lives devastated by natural disasters in the first eight months of 2006 (UNESCO, 2007). Table 1 summarizes major disasters experienced in different parts of the world between December 2004 and June 2009. Thus, the spectacular catastrophes of the 21st century such as the Indian Ocean tsunami (December 2004), powerful earthquakes in northern Pakistan and India (October 2005) as well as in China (May 2008), devastating Hurricanes Katrina and Rita in the U.S.A. (August-September 2005), Cyclone Nargis in Myanmar (May 2008), flooding of New Orleans in Louisiana, U.S.A. (August 2005), and recurrent floods of increasing magnitude in India and Bangladesh, just to name a few, already shattered any assumptions that some nations have solved the disaster risk problem. Through their overwhelming and destructive fury, they have proved that all the nations, whether rich or poor, have much to learn about preparation and mitigation for, response to, and recovery from various types of disasters that continue to afflict us (UNESCO, 2007; CRED, 2009). Indeed, despite even the best efforts taken by the governments and non-governmental organizations involved in disaster management, the fury of nature and/or the folly of human beings frequently result in catastrophic incidents in several parts of the world (Coppola, 2007). These catastrophes overwhelm not only the local response capacity but also the response capacity of entire region or nation.

In spite of the ample availability of scientific knowledge and expertise about natural/anthropogenic hazards and their mitigation, the global vulnerability is growing because of unsustainable development; increasingly large populations are at risk mostly in developing and low-income countries (O'Brien et al., 2006; MunichRe, 2007; UNESCO, 2007; CRED, 2009). There are obvious trends in disaster occurrence on the earth (Coppola, 2007; CRED, 2009): (i) there is an upward global trend in natural disaster occurrence observed over the last decade, (ii) almost all the countries face increased risk from known or unknown disasters, and (iii) the disasters have larger adverse impacts on humanity and environment. According to the World Bank as cited in UNESCO (2007), "Accelerated changes in demographic and economic trends have disturbed the balance between ecosystems, increasing the risk of human suffering and losses. Today's populated areas – cities and agricultural zones – constitute an increasingly valuable asset base. Potential human, social and economic losses from natural disasters grow year by year, independently of nature's

Table 1. Major natural and anthropogenic disasters experienced in different parts of the world between December 2004 and June 2009 (modified after Coppola, 2007)

Sl. No.	Type of disaster	Affected countries
1	Strong Earthquake	Peru, Pakistan, Japan, Chile, South Africa, China.
2	Tsunami	Thailand, Indonesia, Sri Lanka, India, Somalia, Maldives, Myanmar, Malaysia, Kenya, Tanzania, South Africa, Madagascar.
3	Volcanic eruption	El Salvador, Ecuador.
4	Landslide	United States, El Salvador, Nicaragua, India, Honduras, Indonesia, Guatemala, China.
5	Avalanche	India, Pakistan, Nepal, Greece, United States, Austria.
6	Severe flood	Vietnam, India, Bangladesh, China, Canada, Austria, Bulgaria, Germany, Romania, Switzerland, United States.
7	Famine	Malawi, Niger, North Korea, West Africa, Lesotho.
8	Severe drought	Paraguay, United States, Australia, Botswana, Bolivia, Cambodia, Djibouti, Somalia, Ethiopia, Eritrea, Tajikistan, Thailand.
9	Extreme heat	United States, Bangladesh, Nepal, Pakistan, India.
10	Hurricane	United States, Guatemala, Mexico, Iran, El Salvador, Indonesia, Haiti, Cuba, Jamaica, Nicaragua.
11	Tornado	United States, Bangladesh.
12	Typhoon/Cyclone	Japan, Myanmar, China, Philippines, Thailand, India, Nepal, Germany, Austria, Poland, Czech Republic.
13	Windstorm	Pakistan, United States, South Africa.
14	Wildfire	Spain, Portugal, United States, India, Russian Federation, Brazil, South Africa, Peru, Ukraine, Madagascar, Argentina.
15	Structural fire	France, Egypt, Argentina.
16	Marine disaster	Ecuador, United States, Bangladesh, India, Nepal, Pakistan.
17	Mine accident	China, United States.
18	Structural Failure	Bangladesh, India, Russian Federation, Australia, China, Spain, France.

forces. Increased vulnerability requires that natural disaster management be at the heart of economic and social development policy of disaster-prone countries". Therefore, disaster reduction has been on the top of global and national policy agenda in recent years. For instance, it is an important part of United Nations' Millennium Development Goals (MDGs) for poverty reduction. The United Nations Decade of Education for Sustainable Development (2005-2014) is a second strategic instrument for mitigating natural and anthropogenic disasters. UNESCO is closely involved in creating **public awareness** and **improving education** about disasters, which are two effective coping strategies to reduce the vulnerability of a disaster-prone area, region, or nation.

This chapter focuses on the fundamentals of natural and anthropogenic disasters and their trend, definition/description of salient disaster-related terms, differential impact of disasters on communities across the world, a succinct history of disaster management, global recognition of disaster management, and modern approaches for disaster management.

2. DEFINING DISASTER-RELATED TERMS

In this section, the definition/description of salient terms related to disaster management is presented.

2.1 Disaster and Its Characteristics

The term '*disaster*' is derived from the Greek pejorative prefix *dus-* plus *aster*, which mean "bad star". Thus, the root of the word 'disaster' comes from an astrological theme in which the ancients used to refer to the destruction or deconstruction of a star as a disaster (http://en.wikipedia.org/wiki/Disaster). In simple words, 'disaster' is the tragedy of a natural or anthropogenic hazard which negatively affects society and/or environment. Not all the adverse events (i.e., hazards) are disasters, rather only those adverse events which overpower the response capacity of a society, community, or nation are called disasters. According to UN (1992), 'disaster' is defined as "a serious disruption of the functioning of society, causing widespread human, material, or environmental losses which exceed the ability of the affected society to cope using only its own resources". CRED (2009) defines 'disaster' as "a situation or event which overwhelms local capacity, necessitating a request to a national or international level for external assistance; an unforeseen and often sudden event that causes great damage, destruction and human suffering". Disasters are measured in terms of losses of lives, losses or damage of property, infrastructure and national assets, and environmental degradation. The impacts of disasters could be direct or indirect as well as can be tangible or intangible.

Based on the nature of disaster occurrence, disasters can be characterized as *rapid-onset disasters* and *slow-onset* (or *creeping*) *disasters* (UNDP, 2004; Coppola, 2007). Rapid-onset disasters are triggered by an instantaneous shock and often occur with no or little warning. The impact of such disasters may unfold over the medium- or long-term (UNDP, 2004), though most of their damaging effects prolong within hours or days. The examples of rapid-onset disasters are earthquakes, tsunamis, volcanic eruptions, landslides, tornadoes, tropical cyclones, and floods. Slow-onset disasters, on the other hand, occur when the ability of concerned agencies to support people's needs degrades over weeks or months. They emerge along with and within development processes. The hazard can be felt as an ongoing stress for many days, months or even years (UNDP, 2004; Coppola, 2007). The examples of slow-onset disasters are drought, famine, soil salinisation, soil erosion, and AIDS epidemic. Furthermore, based on the intensity and extent of disasters, they can be characterized as *local*, *national* and *international* (Coppola, 2007). If a local/state government can manage the consequences of a disaster, it is known as a *local disaster*. If the intervention of a national government is required to manage the consequences of a disaster, it is known as a *national disaster*. If one or more national governments are unable to manage the consequences of disaster, it is known as an *international disaster*. International disasters necessitate intervention by a variety of international response and relief agencies/organizations.

2.2 Hazard

Although there is a dispute about the origin of the term '*hazard*', it likely came from either the Old French word *hasard*, a game of dice predating craps, or from the Arabic word *al-zahr*, which means "the dice" (Coppola, 2007). Clearly, the term 'hazard' is rooted in the concept of chance. In practice, the term 'hazard' is used to denote a danger or potentially harmful situation. Thus, 'hazard' is a situation which poses a level of risk to life, health, property, or environment (http://en.wikipedia.org/wiki/Hazard). Hazards are defined as "events or physical conditions that have potential to fatalities, injuries, property damage, infrastructure damage, agricultural loss, damage to the environment, interruption of business, or other types of harm or loss" (FEMA, 1997). According to UNISDR (2004), 'hazard' is defined as: "A potentially damaging physical event, phenomenon or human activity that may cause the loss of life or injury, property damage, social and economic disruption or environmental degradation".

Hazards can include latent conditions that may represent future threats. Most hazards are dormant or potential, with only a theoretical risk of harm. However, once a hazard becomes active (i.e., the hazard is certain to cause harm because no intervention is possible before the incident occurs), it can create an emergency situation (http://en.wikipedia.org/wiki/Hazard). There are many causes of hazards which can be classified into two major categories: (a) natural (geophysical, hydro-meteorological, and biological hazards), and (b) man-made or anthropogenic (environmental degradation, global warming, and technological hazards) (UNISDR, 2004). The level of risk/threat and vulnerability associated with a hazard determines whether the hazard will be called a disaster. Generally, severe and large-scale hazards are known as disasters.

2.3 Risk

The term '*risk*' is a concept which denotes the precise probability of specific eventualities. Technically, the notion of risk is independent from the notion of value and, as such, eventualities may have both positive (favorable) and negative (undesirable) consequences (http://en.wikipedia.org/wiki/Risk). Thus, the concept of '*risk*' can have varying meanings depending on the context. It is sometimes used in a positive manner to denote "venture" or "opportunity" (Jardine and Hrudley, 1997). Such variance in use comes from the Arabic word *risq* meaning "anything that has been given to you by God and from which you draw profit" (Kedar, 1970). In contrast, the Latin word *risicum* describes 'a specific scenario faced by sailors attempting to circumvent the danger posed by a barrier reef', which seems to be more appropriate for use in disaster management where the term 'risk' is always used in negative sense. Unfortunately, even among the risk managers, there is no single accepted definition for this term (Coppola, 2007). However, in general usage, the term 'risk' is used to focus only on potential negative impact to some characteristic of value which may arise from a future event (http://en.wikipedia.org/wiki/Risk). One of the simplest and most common definitions of risk used by many disaster managers is as follows: "Risk is the likelihood of an event's occurrence multiplied by the consequence of that event, if it occurs" (Ansell and Wharton, 1992; UNDP, 2004). That is,

$$\text{Risk} = \text{Likelihood } (L) \times \text{Consequence } (C) \tag{1}$$

Likelihood is expressed in terms of probability (e.g., 0.15 or 15%) or frequency (e.g., 1 in 1000 or five times per year) of an event whichever is appropriate for the analysis. Consequence is a measure of the effect of hazard on human, property, and/or environment (i.e., impact of the event). It is clear from Eqn. (1) that by reducing either the likelihood of a hazard (L) or the potential consequences which might result (C), risk can be effectively reduced. Similarly, any action which increases the likelihood or consequences of a hazard increases risk.

2.4 Vulnerability

The term '*vulnerability*' has been derived from the Latin word *vulnerabilis,* which means "to wound" (Coppola, 2007). It denotes the susceptibility of a person or community to physical or emotional injury, or attack (http://en.wikipedia.org/wiki/Vulnerability). Vulnerability is also the extent to which a community or system can be affected by the impact of a hazard. It is defined as: "The conditions determined by physical, social, economic, and environmental factors or processes, which increase the susceptibility of a community to the impact of hazards" (UNISDR, 2004). In the global warming context, 'vulnerability' is defined as "the degree to which a system is susceptible to or unable to cope with adverse effects of climate change, including climate variability and extremes" (http://en.wikipedia.org/wiki/Vulnerability). Thus, vulnerability is a complex and multifaceted concept which has physical, social, economic, and environmental dimensions. Vulnerability can be decreased through the actions which lower the susceptibility to changes that can harm a system, or it can be increased through the actions which enhance that susceptibility. For example, constructing resilient buildings to withstand an earthquake in earthquake-prone areas will decrease the buildings' vulnerability to this natural hazard, thereby lowering earthquake risk. People's vulnerabilities can be increased or decreased according to their practices, beliefs, and economic status (Coppola, 2007).

Moreover, from the hazards and disasters point of view, 'vulnerability' is a concept which links the people's relationship with their environment to social forces and institutions and the cultural values that sustain and contest them (http://en.wikipedia.org/wiki/Vulnerability). Thus, "the concept of vulnerability expresses the multidimensionality of disasters by focusing attention on the totality of relationships in a given social situation which constitute a condition that, in combination with environmental forces, produces a disaster" (Bankoff et al., 2004). Further, an emerging paradigm called *comprehensive vulnerability management* is defined as "holistic and integrated activities directed to the reduction of potential disasters by diminishing risk and susceptibility, and building resistance and resilience" (Simonovic, 2009).

2.5 Resilience

In relation to hazards and disasters, the term *'resilience'* is opposite of vulnerability and is a measure of susceptibility to avoid losses. It is defined as (UNISDR, 2004): "The capacity of a system, community or society potentially exposed to hazards to adapt, by resisting or changing in order to reach and maintain an acceptable level of functioning and structure". Resilience is determined by the degree to which a social system is capable of organizing itself to increase this capacity by learning from past catastrophic events for better future protection and to improve risk reduction measures (UNISDR, 2004).

2.6 Safe

Many people think that the word "*safe*" indicates complete elimination of all risks. However, such an absolute level of safety is virtually unattainable in the real world (Coppola, 2007). In fact, all the aspects of life involve a certain degree of risk. Disaster managers may determine the threshold of risk that defines a frequency of hazard occurrence below which the societies should not worry about a hazard. According to Derby and Keeney (1981), "a risk becomes safe or acceptable, if it is associated with the best of the available alternatives, not with the best of the alternatives which we would hope to have available".

2.7 Composite Disaster

Sometimes two or more completely independent disasters occur at the same time, which are termed *'composite disasters'* or *'compound disasters'* (Coppola, 2007). For examples, an earthquake strikes during flood event, cyclone and associated flooding, flood occurs in one part of a country and drought in another part, and epidemic outbreaks after a severe earthquake or flooding. Thus, composite disasters can occur either sequentially or simultaneously with one or more disasters. However, more commonly, one disaster triggers secondary hazards. Some secondary hazards only occur as a result of a primary hazard such as tsunami from earthquakes, volcanoes or landslides, whereas others can occur either due to or independent of other disasters such as landslides which can be triggered by heavy rains, earthquakes, volcanic eruptions, or other reasons, or can occur purely on their own (Coppola, 2007). Composite disasters usually worsen consequences and increase victims' stress and isolation. They can make search and rescue, and other response and recovery operations more difficult, and can significantly increase the risk of harm to victims and responders equally (Coppola, 2007).

2.8 Humanitarian Crisis

"A *'humanitarian crisis'* is an event or series of events which represents a critical threat to the health, safety, security or wellbeing of a community or other large group of people, usually over a wide area" (http://en.wikipedia.org/wiki/Humanitarian_crisis). According to Coppola (2007), humanitarian crisis is "a special situation that results from a combination of the realized consequences of a hazard and the severely diminished coping mechanisms of an affected population". In such situations, the health and life of a vast number of people are seriously threatened, and hence it is sometimes also known as a *'humanitarian disaster'*. The characteristics of a humanitarian crisis usually include mass occurrence of (Coppola, 2007): (i) starvation/malnutrition, (ii) disease, (iii) insecurity, (iv) lack of shelter, and (v) a gradually increasing number of victims. Armed conflicts, wars, epidemics, famine, natural disasters, and other major emergencies may all involve or lead to a humanitarian crisis. Humanitarian crises are likely to worsen without outside intervention. The examples of recent humanitarian crises are (http://en.wikipedia.org/wiki/Humanitarian_crisis): Cyclone Nargis in May 2008, Hurricane Katrina in August 2005, the 2005 Kashmir earthquake, the 2004 Indian Ocean tsunami, the 1994 Rwanda genocide, the Sri Lankan civil war, the Israeli-Palestinian conflict, the Afghan civil war, the Darfur conflict, and the Iraq war.

2.9 Disaster Risk Management and Coping Capacity

'Disaster risk management' is defined as: "The systematic management of administrative decisions, organization, operational skills and abilities to implement policies, strategies and coping capacities of the society or individuals to lessen the impacts of natural and related environmental and technological hazards" (UNDP, 2004).

'Coping capacity' is defined as: "The manner in which people and organizations use existing resources to achieve various beneficial ends during unusual, abnormal and adverse conditions of a disaster phenomenon or process" (UNDP, 2004).

2.10 Disaster Risk Reduction and Sustainable Resilient Community

The term *'disaster risk reduction'* is defined as "conceptual framework of elements considered with the possibilities to minimize vulnerabilities and disaster risks throughout a society, to avoid (prevention) or to limit (mitigation and preparedness) the adverse impacts of hazards, within the broad context of sustainable development" (UNISDR, 2004).

The goal of building resilient communities shares much with the principle of intergenerational equity inherent in the principle of 'sustainable development'. *'Sustainable resilient communities'* can be defined as "societies which are structurally organized to minimize the effects of abrupt change and at the same time, have the ability to recover quickly by restoring the socio-economic vitality of the community" (Simonovic, 2009).

2.11 Structural and Non-Structural Measures

The term *'structural measures'* means hardware measures, which are based on technological solutions. They denote "any physical construction to reduce or avoid possible impacts of hazards, which include engineering measures and construction of hazard-resistant and protective structures and infrastructure" (UNISDR, 2004). On the other hand, the term *"non-structural measures"* means software measures which include "legislation, policies, insurance, awareness, knowledge development, public commitment, and methods and operating practices, including participatory mechanisms and the provision of information, which can reduce risk and related impacts" (UNISDR, 2004).

3. IMPACT OF DISASTERS: RICH VERSUS POOR NATIONS

Every country has its own hazard profile, vulnerability level, and rise or fall of disaster management systems as well as has unique socio-economic, cultural, and political characteristics (Coppola, 2007). As a result, the disaster impact and management scenario vary appreciably from one country to another. The disparity is clearly visible between rich and poor countries of the world. Disasters strike hardest at the world's poorest nations, which have limited capacity to respond to and recover from disasters (UNESCO, 2007; CRED, 2009). In general, developing and low-income nations are more vulnerable to disasters; for instance, of the more than 2,200 disasters occurred in the world during 1990-2001, 35% occurred in Asia, 29% in Africa, 20% in the Americas, 13% in Europe, and 3% in Oceania (UNESCO Water Portal, 2008). The burden of losses due to various disasters is also greatest in poor countries, where 13 times more people die in disasters than in rich countries (UNESCO Water Portal, 2008). The basic disparities in disaster impact between rich and poor countries are summarized in Table 2. The ways in which disasters harm poor nations beyond the initial human fatalities, injury and devastation are briefly described below (modified after Coppola, 2007):

Table 2. Disparities in disaster impact between rich and poor countries (modified after UNDP, 2004; Coppola, 2007)

Rich countries	Poor countries
• They tend to suffer higher economic losses, but have adequate resources and effective plans to tackle these financial impacts.	• They have relatively less at risk in terms of financial value, but they maintain little or no buffer to absorb even low financial impacts. Economic impact can be considerable which in turn severely affects the socio-economic development of communities or nation.
• They have adequate resources, skill, and committed plans to take advantage of advanced and modern technologies to reduce loss of life and property such as efficient early warning and communication systems, enforced building codes and zoning, etc.	• They usually lack the resources and skill necessary to take advantage of advanced and modern technologies. They have little ability to enforce building codes and zoning even if these mechanisms do exist.
• Generally, they have immediate emergency and medical care that increase survivability and contain the spread of disease. Human fatalities and property losses are relatively less.	• They sustain massive primary and secondary casualties, and enormous loss of assets, infrastructure, agriculture, and ecosystems. They usually don't have immediate and efficient emergency services, which results in greater human suffering and chaotic situation.
• They transfer much of personal, private and public risk to insurance and reinsurance providers.	• They usually do not participate in insurance mechanisms. They divert funds from development programs to emergency relief and recovery.
• Recovery from disaster impact is much faster, and complete recovery is generally achieved.	• Recovery from disaster impact is very slow and often prolongs much longer, even years or decades. Complete recovery is often not achieved.

- Vital infrastructure is badly damaged or destroyed such as roads, railways, bridges, airports, ports, communication systems, power generation and distribution facilities, water supply and sewerage plants, which require years to rebuild. These damages not only create hindrance to relief and rescue operations but also significantly retard the recovery process.
- Dwelling houses are devastated or badly damaged, leaving disaster victims homeless for several months and thereby increasing their vulnerability and agony.
- Schools and colleges are destroyed or damaged, leaving students without an adequate source of education for months or even years.
- Hospitals and health centers as well as water supply systems are destroyed or damaged, which result in an increase in the vulnerability of disaster-affected people to disease; sometimes even the outbreak of epidemics. Also, water supply problem becomes grim.
- Formal and informal businesses are destroyed, resulting in surges in unemployment and decreased economic stability and strength. In addition, the destruction of agriculture severely affects the livelihoods of rural people.
- Disaster victims are forced to leave the affected area, often they never return, and thereby extract institutional knowledge, cultural and social identity, and economic viability from the areas which cannot afford to spare such resources. This situation badly disrupts the socio-economic condition of such areas.
- Poverty, almost a chronic evil in developing nations, leads to a rapid rise in crime and insecurity in the disaster-affected areas, which create added nuisance and unrest in the community.
- Sizeable portions of GDP are often diverted from development projects, social programs, or debt repayment to manage disaster consequences and start recovery efforts.
- National and/or international development efforts are inhibited, erased, or even reversed due to enormous financial stress caused by heavy disaster losses.
- A general feeling of despair and insecurity afflicts the disaster victims, which leads to increased rates of depression and a lack of motivation to regain independence from external assistance.

Moreover, it is well documented that the poor countries of the world often experience either stagnant or negative rates of development due to disaster impacts. For examples, countries like Honduras, Guatemala and Nicaragua which lost more than their entire gross domestic product (GDP) in hurricane Mitch in 1998 are still struggling to recover. This single storm reversed the rate of development in these and other Central American countries by at least a decade (as much as 20 and 30 years in some areas) (Coppola, 2007). Similarly, the Zimbabwe drought of the early 1990s was associated with an 11% decline in its GDP and a 60% decline in the stock market, whereas the 2000 drought in Brazil led to a 50% decrease in the projected economic growth (UNESCO Water Portal, 2008). The same effect can also be seen in many of the areas affected by the 2004 tsunami and earthquake events in Southeast Asia. For instance, the December 2004 tsunami caused a 20-year setback in the economic development of the Republic of Maldives, and it may take many years before the Maldivians can regain the level of prosperity they enjoyed prior to the tsunami disaster (Coppola, 2007; UNESCO, 2007). Such disasters are among the biggest obstacles to achieving the UN's Millennium Development Goals for poverty reduction. Further, it has been estimated that half of the population of the developing world is exposed to polluted sources of water which enhance disease incidence. Also, between 1991 and 2000, over 665,000 people died in 2,557 natural disasters, of which 90% were water-related disasters and a vast majority of victims (97%) were from developing nations (IFRC, 2001). Thus, disasters exacerbate the causes of poverty in developing nations and, therefore, minimizing the risk of natural hazards has become an important development issue in the 21st century (UNDP, 2004; UNESCO, 2007).

4. DISASTER MANAGEMENT

After the end of the Cold War in the early 1990s, the widely used term *'civil defense'* has been replaced with the term *'emergency management'* or *'disaster management'* which denotes a broader and modern thinking of protecting the civilian population. However, the terms *'civil protection'* and *'crisis management'* are widely used within the European Union (http://en.wikipedia.org/wiki/Disaster). An academic trend is to use the term *'disaster risk reduction'* which focuses on the mitigation and preparedness aspects of the disaster/emergency management cycle described in Section 4.3; the definition of the term is given in Section 2.10. This term has been adopted by the United Nations, which has developed an international strategy on promoting disaster risk reduction (called 'United Nations International Strategy for Disaster Reduction') because it has been proved to be very effective and less expensive. According to Haddow and Bullock (2004), *disaster management* or *emergency management* is a discipline which deals with and avoids risks. It involves preparation for a disaster (natural or anthropogenic) prior to its incidence, disaster response (e.g., emergency evacuation, quarantine, mass decontamination, etc.) as well as supporting and rebuilding society after the disaster. In general, any disaster management is a continuous process by which all the individuals, groups, and communities manage hazards in an attempt to avoid or ameliorate the impact of disasters. Actions taken depend in part on the perceptions of risk of exposed people (Wisner et al., 2004).

Of all the global environment issues, natural hazards are in some ways the most manageable (UNESCO, 2007). In spite of increasing frequency and severity of the catastrophes, it is becoming increasingly possible to prevent and mitigate the effects of a disaster because of gradually improved understanding about disasters and technological advancement (modern tools and techniques). Now, disaster risks can easily be identified and effective mitigation measures could be formulated (UNESCO, 2007; http://en.wikipedia.org/wiki/Disaster). The nature of disaster management in practice, however, largely depends on local socio-economic and political conditions.

4.1 Historical Perspective

Prior to the existence of humans on the earth, the presence of hazards or disasters is negated because the tragic events are not hazards or disasters without human involvement. With the appearance of humans on the earth, the incidence of hazards and disasters followed. The archeological discovery has revealed that our prehistoric ancestors faced many of the same hazards that exist today such as starvation, hostile elements, dangerous wildlife, violence by humans, diseases, accidental injuries, and so on (Coppola, 2007). Evidence indicates that they took measures to minimize the impact of hazards/disasters. The mere fact that they opted to live in caves is a testimony to this. Furthermore, the histological records including religious epics suggest that natural disasters occurred throughout the history and many of history's great civilizations were destroyed by floods, famines, earthquakes, tsunamis, El Nino events, and other widespread natural disasters (e.g., Fagan, 1999). Selected prominent disasters which occurred throughout the human history are summarized in Table 3. It should be noted that many of the historical disasters are record-breaking and unique even compared to the recent major disasters such as December 2004 tsunami which killed more than 300,000 people in an instant besides the loss of property, infrastructure and national assets in the affected countries.

Various applications of disaster management appear throughout the historical record. The story of Noah's Ark from the Old Testament, for example, provides the importance of early warning, preparedness, and mitigation in disaster management (Coppola, 2007). The evidence of risk management practices can be found as early as 3200 BC. In what is now modern-day Iraq lived a social group known as the 'Asipu'

Table 3. Selected notable disasters in the history (modified after Coppola, 2007)

Disaster	Country	Year	Number of persons killed
1. Mediterranean Earthquake	Egypt and Syria	1201	1,100,000
2. Shaanzi Earthquake	China	1556	830,000
3. Calcutta Typhoon	India	1737	300,000
4. Bengal Famine	India	1770	15,000,000
5. Caribbean Hurricane	Martinique, St. Eustatius and Barbados	1780	22,000
6. Tamboro Volcano	Indonesia	1815	80,000
7. Kangra Earthquake	India	1905	19,000
8. Influenza Epidemic	World	1917	20,000,000
9. Yangtze River Flood	China	1931	3,000,000
10. Famine	Russia	1932	5,000,000
11. Bihar-Nepal Earthquake	India and Nepal	1934	50,000
12. Bengal Famine	India	1943	3,000,000
13. Bangladesh Cyclone	Bangladesh	1970	300,000
14. Tangshan Earthquake	China	1976	655,000

(Coppola, 2007). When the community members faced a difficult decision, especially one involving risk or danger, they could appeal to the 'Asipu' for advice. The 'Asipu', using a process similar to modern-day hazard risk management, would first analyze the problem at hand, then propose several alternatives, and finally give possible outcomes for each alternative (Covello and Mumpower, 1985). Today, this methodology is known as *decision analysis*, which is the key to any comprehensive risk management endeavor. Early history is also marked by the incidents of organized emergency response (Coppola, 2007).

The birth of modern disaster management lies in the Civil Defense era (a particular period in the recent history) which witnessed the greatest overall move toward a centralized safeguarding of civilians. However, modern disaster management, in terms of the emergence of global standards and organized efforts to address preparedness, mitigation, and response activities for a wide range of disasters, did not begin to emerge until the mid-20th century. In most countries, this change materialized as a response to specific disaster events. At the same time, it was further stimulated by a shift in social philosophy because of advances in warfare technology, wherein the government played an increasing role in the management of disasters (Coppola, 2007).

In response to the threat posed by air raids and the ever-present and dreadful prospect of nuclear attacks, many developed nations started to form elaborate systems of civil defense. These systems included detection systems, early warning systems, hardened shelters, search and rescue teams, and local and regional coordinators. Most nations' legislatures also developed a legal framework to guide both the creation and maintenance of these systems through the enforcement of laws, the creation of national-level civil defense organizations, and the allocation of funding and personnel. Despite these encouraging efforts, surprisingly few civil defense units evolved over time into more comprehensive disaster management organizations (Quarantelli, 1995). However, the legal framework developed to support them remained in place, which provided a basis for modern disaster management.

Today, disaster management structures vary considerably from one country to another. Many countries developed their disaster management capabilities out of necessity, while other countries formed their disaster management structures not for the civil defense, but after the criticism for poor management of natural disasters; for examples, Peru in 1970, Nicaragua in 1972, and Guatemala in 1976, following destructive earthquakes in each country (Coppola, 2007). On the other hand, some countries still have no real disaster management structures regardless of their disaster history (Coppola, 2007).

4.2 Global Recognition

At an international level, the recognition of disaster management can be traced back to 11 December 1987 when the United Nations General Assembly declared the 1990s as the "International Decade for Natural Disaster Reduction" (IDNDR). This action was taken to promote internationally coordinated efforts to reduce material losses, and social and economic disruption caused by natural disasters, especially in developing countries. The stated mission of IDNDR was to improve each United Nations member country's capacity to avoid or minimize adverse effects from natural disasters and to develop guidelines for applying existing tools and techniques to reduce the impact of natural disasters. Thereafter, in May 1994, the United Nations member states met at the first World Conference on Natural Disaster Reduction in Yokohama, Japan to assess the progress achieved by the IDNDR. At this meeting they developed the "Yokohama Strategy and Plan of Action for a Safer World" (ISDR, 1994), which provided landmark guidance on reducing disaster risk and impacts. The participating member states accepted the ten principles to be applied to disaster management within their own countries. The tenth principle formalized the requirement that each nation's government should accept the responsibility for protecting its people, infrastructure, and other national assets from the impact of various disasters (ISDR, 1994).

Moreover, immediately following the Indian Ocean Tsunami in December 2004, the United Nations General Assembly convened the second World Conference on Disaster Reduction from 18 to 22 January 2005 in Kobe, Hyogo, Japan. The goal of this conference was to review the progress in disaster risk reduction since the 1994 Yokohama Conference and to make plans for next ten years (UNISDR, 2005). It provided a unique opportunity to promote a strategic and systematic approach to reducing vulnerabilities and risks to hazards. The participating member states adopted what is known as the "Hyogo Declaration" and agreed upon a 'Framework for Action' for the decade 2005-2015 which aims at building the resilience of nations and communities to disasters (UNISDR, 2005). The Hyogo Framework for Action includes the following five commitments or priorities for the decade 2005-2015 (UNISDR, 2005):

(1) *Make disaster reduction a priority:* Governments should integrate disaster risk reduction into their laws, programs and plans, and ensure the participation of local communities in planning.

(2) *Know the risks and take action:* Countries should define and understand potential risks so that they can develop early warning systems adapted to the needs of each community.

(3) *Build understanding and awareness:* Governments should provide information, include disaster reduction in formal and informal education, and ensure that invaluable local knowledge about disaster risks is preserved and transmitted.

(4) *Reduce risk:* Countries should apply safety codes to ensure that schools, hospitals, homes and other buildings do not collapse in earthquakes; avoid allowing communities to settle in hazard-prone areas such as floodplains; and protect forests, wetlands, and coral reefs that act as natural barriers to storms and flooding.

(5) *Be prepared and ready to act:* Governments and regional or local authorities should conduct risk assessments; adopt contingency plans; test preparedness by such measures as evacuation drills; and ensure emergency services, response agencies, policy makers and development organizations are coordinated.

The above-mentioned five Hyogo commitments or priorities are the foundation of disaster management practice. Unfortunately, this foundation is still lacking in many countries of the world (Coppola, 2007). Therefore, it has been recommended that in their approach to disaster risk reduction, national, regional and international organizations and other concerned actors must take into consideration the key activities listed under each of the five commitments/priorities and should implement them, as appropriate, to their own circumstances and capacities. For the details about the key activities under each Hyogo commitment, the interested readers are referred to the 'Report of the World Conference on Disaster Reduction' (UNISDR, 2005). An integrated, multi-hazard, comprehensive approach to address vulnerability, risk, assessment and disaster management, including prevention, mitigation, preparedness, response and recovery, is an essential element of a safer world in the 21st century (UNISDR, 2005). The Hyogo Framework for Action 2005-2015 is linked both to the Millennium Development Goals (MDG) and to the UN Decade of Education for Sustainable Development (2005-2014), which is led by UNESCO. Thus, there is a global recognition that the efficient management of natural and anthropogenic disasters is indispensable in order to make world safer and ensure sustainable development on the earth.

4.3 Modern Approaches

Modern approach to ensure comprehensive disaster management is based on a management cycle comprising four major components or phases (Alexander, 2002; Haddow and Bullock, 2004; Coppola, 2007): *mitigation* and *preparedness (preparation)* before the occurrence of disasters, and *response* and *recovery* after the occurrence of disasters (Fig. 1). Various diagrams illustrate the cyclic nature of disaster management by which these and other related activities are performed over time, though disagreement exists concerning how a "disaster management cycle" is visualized (Coppola, 2007). In practice, all of these activities are intermixed and are performed to some degree before, during, and after the disaster. Response is often visualized as beginning immediately after the disaster impact, though it is not uncommon for the actual response to begin well before the disaster actually occurs (Coppola, 2007). The four phases of disaster management cycle are described in the subsequent sections.

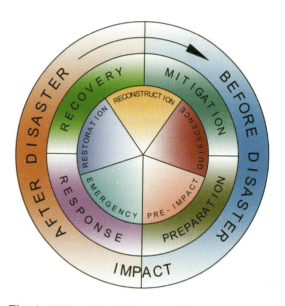

Fig. 1 Schematic of the comprehensive disaster management cycle (redrawn from Alexander, 2002).

4.3.1 Mitigation Phase

The goal of mitigation efforts is to prevent hazards from developing into disasters altogether, or to reduce the impact of disasters when they occur. The mitigation phase differs from the other phases because it focuses on the long-term measures for reducing or eliminating disaster risk (Haddow and Bullock, 2004). Thus, the identification of risks is a prerequisite to the mitigation process. Mitigation measures could be structural or non-structural. The implementation of mitigation measures/strategies can be considered as part of the recovery process if they are applied after the occurrence of a disaster (Haddow and Bullock, 2004).

4.3.2 Preparedness Phase

In the preparedness phase, disaster managers develop plans of action for the disasters which may occur in the future. It involves equipping people who may be impacted by a disaster or who may be able to help those impacted with the tools to increase their chance of survival and to minimize their economic and other losses. Another aspect of preparedness is 'casualty prediction', which is the study of number of deaths or injuries that may result from a disaster (http://en.wikipedia.org/wiki/Disaster). It gives planners an idea about the types of resources required to be ready in order to respond to a particular kind of disaster. Common preparedness measures include (http://en.wikipedia.org/wiki/Disaster): (i) communication plans with easily understandable terminology and methods; (ii) proper maintenance and training of emergency services, including mass human resources such as 'community emergency response teams'; (iii) development and exercise of emergency population warning methods along with emergency shelters and evacuation plans; (iv) stockpiling, inventory, and maintenance of disaster supplies and equipment; and (v) development of organizations of trained volunteers among civilian populations.

It is worth mentioning that preparedness and mitigation are the most cost-effective approach for reducing the impact of hazards/disasters. Therefore, it is said that a dollar invested in disaster preparedness and mitigation will prevent four to eight dollars in disaster losses (UNESCO, 2007).

4.3.3 Response Phase

The response phase involves taking action to reduce or eliminate the impact of disasters which have occurred or are currently occurring in order to prevent further suffering, financial loss, or a combination of both (Coppola, 2007). It includes the mobilization of necessary emergency services and first responders in the disaster-affected area. This is likely to include a first wave of core emergency services (e.g., firefighters, police, and ambulance crews), which may be supported by a number of secondary emergency services such as specialist rescue teams (http://en.wikipedia.org/wiki/Disaster). When the response process is conducted as a military operation, it is termed 'Disaster Relief Operation', but the commonly used term 'relief' is usually considered as one component of the response process. It should be noted that a well rehearsed emergency plan developed as part of the preparedness phase enables efficient coordination of rescue teams as well as emergency service providers, and that search and rescue efforts should commence at an early stage where needed.

4.3.4 Recovery Phase

The aim of the recovery phase is to restore the victims' lives to the normal/previous condition following the impact of a disaster (Haddow and Bullock, 2004; Coppola, 2007). It differs from the response phase in its focus; the recovery efforts are concerned with the issues and decisions that must be made after immediate needs are addressed (Haddow and Bullock, 2004). The recovery process primarily involves the actions such as rebuilding destroyed property and service systems, re-employment, and the repair of other essential infrastructure. This phase generally starts after the urgent response has ended, and may continue for months or years thereafter.

The above-mentioned modern approach of disaster/emergency management (i.e., *'comprehensive disaster management'*) should be followed in practice by implementing another modern concept of management known as *'adaptive management'*. The concept of 'adaptive management' is based on the insight that our knowledge about natural systems is presently limited, and hence our ability to predict future key factors influencing an ecosystem as well as system's behavior and responses is inherently limited (Loucks and van Beek, 2005; Pahl-Wostl, 2007). Consequently, 'adaptive management' treats management strategies and actions as experiments, not as fixed policies (Ludwig et al., 1993). That is, management decisions (e.g., designs and operating policies) should be flexible and adaptable; they should be continually improved as experience expands, new information emerges, and priorities change over time. The improved knowledge or new understanding arises from the facts that (Loucks and van Beek, 2005; van der Keur et al., 2008): (i) uncertainty exists in defining operational targets for different management goals, (ii) conflicting interests among stakeholders require participatory goal setting and a clear recognition of uncertainties involved, and (iii) the system to be managed is subject to change due to environmental and socio-economic changes. Adaptive management can be defined as "a systematic process for continually improving management policies and practices, as appropriate, by learning from the outcomes of implemented management strategies and the improved knowledge" (Loucks and van Beek, 2005; van der Keur et al., 2008).

Furthermore, 'adaptive management' takes into account the limitations of our current knowledge and experience as well as those learned by experiments. It helps us move toward meeting our changing goals over time in the face of this incomplete knowledge and uncertainty. Thus, adaptive management and decision-making is a challenging blend of scientific research, monitoring, and practical management which provides opportunities to *act*, *observe* and *learn*, and then *react* (Loucks and van Beek, 2005). Both monitoring and management actions need to be adaptive, continually responding to an improved understanding that comes from the analysis of monitored data in comparison to model predictions and scientific research (Loucks and van Beek, 2005). Loucks and van Beek (2005) emphasize that when predictions are highly unreliable, responsible managers should favor robust (i.e., good under a wide range of situations) actions, gain knowledge through research and experimentation, monitor results to provide feedback for the next decision, update assessments and modify operating policies in the light of new information, and avoid irreversible actions and commitments. Adaptive management in essence links science, values, and the experience of stakeholders and managers to the art of making management decisions (Maimone, 2004). For the details about 'adaptive management', the interested readers are referred to Holling (1978) and Walters (1986).

REFERENCES

Alexander, D. (2002). Principles of Emergency Planning and Management. Oxford University Press, New York.
Ansell, J. and Wharton, F. (1992). Risk: Analysis, Assessment, and Management. John Wiley and Sons, Chichester, U.K.
Bankoff, G., Frerks, G. and Hilhorst, D. (2004). Mapping Vulnerability: Disasters, Development, and People. Earthscan, London, U.K.
Coppola, D.P. (2007). Introduction to International Disaster Management. Elsevier, Amsterdam, The Netherlands.
Covello, V.T. and Mumpower, J. (1985). Risk analysis and risk management: A historical perspective. *Risk Analysis*, **5(2)**: 103-118.
CRED (2009). Annual Disaster Statistical Review 2008: The Numbers and Trends. Center for Research on the Epidemiology of Disasters (CRED), Brussels, Belgium, 25 pp.
Derby, S.L. and Keeney, R.L. (1981). Risk analysis: Understanding how safe is safe enough? *Risk Analysis*, **1(3)**: 217-224.
Fagan, B. (1999). Floods, Famines, and Empires. Basic Books, New York.
FEMA (1997). Multi Hazard Identification and Assessment. Federal Emergency Management Agency (FEMA), Washington D.C.
Haddow, G.D. and Bullock, J.A. (2004). Introduction to Emergency Management. Butterworth-Heinemann, Amsterdam, The Netherlands.
Holling, C.S. (editor) (1978). Adaptive Environmental Assessment and Management. Wiley, New York.

IFRC (2001). World Disasters Report 2001. International Federation of Red Cross and Red Crescent Societies, Geneva, Switzerland.

IPCC (2007). Climate Change 2007: The Physical Science Basis, Summary for Policymakers. Contribution of Working Group I to the Fourth Assessment Report of the Intergovernmental Panel on Climate Change (IPCC), Cambridge University Press, Cambridge, U.K. and New York, USA.

ISDR (1994). Yokohama Strategy and Plan of Action for a Safer World. UN World Conference on Natural Disaster Reduction, 23-27 May 1994, Yokohama, Japan.

Jardine, C. and Hrudley, S. (1997). Mixed messages in risk communication. *Risk Analysis*, **17(4)**: 489-498.

Kedar, B.Z. (1970). Again: Arabic Risq, Medieval Latin Risicum, Studi Medievali. Centro Italiano Di Studi Sull Alto Medioevo, Spoleto.

Loucks, D.P. and van Beek, E. (2005). Water Resources Systems Planning and Management: An Introduction to Methods, Models and Applications. Studies and Reports in Hydrology, UNESCO Publishing, UNESCO, Paris.

Ludwig, D., Hilborn, R. and Walters, C. (1993). Uncertainty, resource exploitation, and conservation: Lessons from history. *Science*, **260**: 17-18.

Maimone, M. (2004). Defining and managing sustainable yield. *Ground Water*, **42(6)**: 809-814.

MunichRe (2007). Natural Catastrophes 2006: Analyses, Assessments, Positions. Knowledge Series, MunichRe, Germany, 50 pp.

O'Brien, G., OKeefe, P., Rose, J. and Wisner, B. (2006). Climate change and disaster management. *Disasters*, **30(1)**: 64-80.

Pahl-Wostl, C. (2007). Transitions towards adaptive management of water facing climate and global change. *Water Resources Management*, **21(1)**: 49-62.

Quarantelli, E.L. (1995). Disaster Planning, Emergency Management, and Civil Protection: The Historical Development and Current Characteristics of Organized Efforts to Prevent and Respond to Disasters. University of Delaware Disaster Research Center, Newark, DE.

Simonovic, S.P. (2009). Managing Water Resources: Methods and Tools for a Systems Approach. UNESCO, Paris and Earthscan, London, 640 pp.

UN (1992). Internationally Agreed Glossary of Basic Terms Related to Disaster Management (DNA/93/36). Department of Humanitarian Affairs, United Nations, Geneva, Switzerland.

UNDP (2004). Reducing Disaster Risk: A Challenge for Development. United Nations Development Program (UNDP), Bureau for Crisis Prevention and Recovery, United Nations Plaza, New York, 146 pp.

UNESCO (2007). Disaster Preparedness and Mitigation: UNESCO's Role. Section for Disaster Reduction, Natural Sciences Sector, the United Nations Educational, Scientific and Cultural Organization (UNESCO), Paris, France, 48 pp.

UNESCO Water Portal (2008). Water and natural disasters in celebration of International Day for Natural Disaster Reduction. *UNESCO Water Portal Bi-monthly Newsletter*, **209**, 14 October 2008.

UNISDR (2004). Living with Risk: A Global Review of Disaster Reduction Initiatives. United Nations International Strategy for Disaster Reduction (UNISDR), Geneva, Switzerland.

UNISDR (2005). Report of the World Conference on Disaster Reduction. Kobe, Hyogo Prefecture, Japan, 18-22 January 2005, United Nations International Strategy for Disaster Reduction (UNISDR), Geneva, Switzerland, 42 pp.

van der Keur, P., Henriksen, H.J., Refsgaard, J.C., Brugnach, M., Pahl-Wostl, C., Dewulf, A. and Buiteveld, H. (2008). Identification of major sources of uncertainty in current IWRM practice: Illustrated for the Rhine basin. *Water Resources Management*, **22**: 1677-1708.

Walters, C. (1986). Adaptive Management of Renewable Resources. McMillan, New York.

Wisner, B., Blaikie, P., Cannon, T. and Davis, I. (2004). At Risk: Natural Hazards, People's Vulnerability and Disasters. Routledge, Wiltshire.

2 Earthquakes in India: Hazards, Genesis and Mitigation Measures

Vineet K. Gahalaut

1. INTRODUCTION

Earthquake, as its name suggests, means shaking of the ground which is caused by a sudden release of stored elastic energy in the rock mass that had accumulated as strain over time along faults. Seismic waves are generated due to sudden release of energy which extend outward from the point of origin (called "epicenter") like water ripples. The speed of these waves depends on the geologic composition of the materials through which they pass. Earthquakes can occur at a range of depths, and the focal depths (distance below the earth's surface at which accumulated energy is released) from 0 to 70 km are considered *shallow*, from 70 to 300 km are considered *intermediate* and greater than 300 km are considered *deep* (Richter, 1958). Some 50,000 earthquakes occur on an average every year as the earth's tectonic plates shift and adjust, including some of potentially devastating magnitude releasing enormous amounts of energy. Approximately 75% of the world's population live in the areas that were affected at least once by natural disasters namely earthquake, tropical cyclone, flood or drought between 1980 and 2000 (UNDP, 2004). Potential earthquakes often cause considerable causalities and economic damage, coupled with significant hydrologic/hydrogeologic changes (e.g., UNDP, 2004; Allen, 2007; Manga and Wang, 2007). In addition, many secondary hazards/disasters such as landslides, rockfalls, avalanches, tsunamis, etc. are known to occur in the aftermath of an earthquake.

The Himalayan mountain ranges are considered to be the world's youngest fold mountain ranges. The subterranean Himalayas are geologically very active. In a span of 53 years, four earthquakes of magnitude about 8 or more have occurred in this region. The peninsular part of India comprises stable continental crust. Although these regions were considered seismically least active, occasionally they were affected by the earthquakes, which caused great havoc as in the case of Latur earthquake in Maharashtra during 1993 (Seeber et al., 1996). The escalation in population has no effect on the frequency of occurrence of major and great earthquakes, but it does have effect on the fatalities caused by the earthquakes. The total deaths per earthquake appear to be on rise (Fig. 1), and this is particularly true with many developing Asian nations, including India. In developed nations, though the number of deaths per earthquake has come down, it appears to be on the rise in developing nations due to exponential urban growth and poor construction. It is difficult to believe but in the Indian subcontinent, on an average, every year more than 600 people lost their lives due to earthquakes in past 100 years. Thus, it is important to understand the regions of high earthquake risk and dense population and the regions where they overlap. With this view, a brief review of Indian seismicity, genesis and possible mitigation measures is presented in this chapter.

In this review, earthquake occurrence processes and earthquakes that occurred between the eastern and western Syntaxial bends in Assam and Kashmir, respectively have been considered. Thus, in the western region, the regions and the earthquakes that occurred in the Indo-Kohistan Seismic Zone (IKSZ), e.g., the recent 2005 Kashmir earthquake, which appears to be different from the Himalayan detachment earthquakes (Gahalaut, 2008), Hazara arc, Salt Ranges and Chaman fault region are excluded. In the

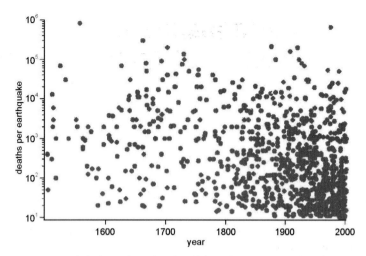

Fig. 1 Global earthquake fatalities since sixteenth century (Bilham, 2004).

eastern region, the regions and earthquakes that occurred in the Indo-Burmese arc and Sagaing fault regions are excluded. Since the focus is on the thrust belt of the Himalayan frontal arc, the regions of Tibet, Hindukush and Pamir are not considered. Thus, a review of some of the major, recent and important earthquakes in the Indian shield region, results of geodetic measurements in the Himalaya and India shield region about the rate and mode of convergence, and discussion on seismic gaps in the Himalayan arc region have been provided in this chapter. Finally, some issues related to the mitigation of seismic hazard and the feasibility of early warning systems are discussed.

2. GENESIS OF EARTHQUAKES

Occurrence of tectonic earthquakes is best explained by the theory of plate tectonics. According to this theory, entire lithosphere (i.e., consisting of crust and upper mantle with total thickness of 100-150 km) around the globe is divided into plates which are in continuous motion with respect to each other. It is the interaction of these plates along their edges which causes earthquakes. These earthquakes are referred to as interplate earthquakes. A few earthquakes occur within the plate interiors due to crustal heterogeneities and internal deformation and they are referred as intraplate earthquakes. Indian plate moves in the northeast direction at a rate of about 5 cm/year (about 1 mm/week) and its interaction with the Eurasian plate has led to the highest Himalayan mountain chain in the world and the highest plateau, i.e., the Tibetan plateau (Fig. 2). It has also led to the occurrence of some great earthquakes in the Himalayan and its contiguous regions. The earthquakes in the Himalayan arc occur due to the underthrusting of Indian plate beneath the Eurasian plate. According to the most acceptable and widely applicable model of underthrusting and earthquake occurrence, the convergence of the Indian and Eurasian plates is accommodated through slip on the detachment (Seeber and Armbruster, 1981). The detachment (also referred to as decollement or the Main Himalayan Thrust, MHT) is the surface between the underthrusting Indian shield rocks and the overlying Himalayan rocks (Fig. 3). The part of the detachment that lies under the Outer and Lesser Himalaya is seismogenic and slips episodically in a stick and slip manner. It accumulates strain during the interseismic period when it is locked, which is released during the infrequent earthquakes through sudden slip on the detachment. The detachment that lies under the Higher and

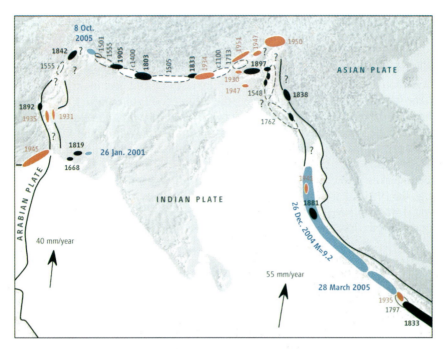

Fig. 2 Great and major earthquakes in the Indian subcontinent (Bilham, 2006). Question marks show the region where either the earthquakes have not occurred in past 200 years or there are no data to support or reject the occurrence of earthquakes.

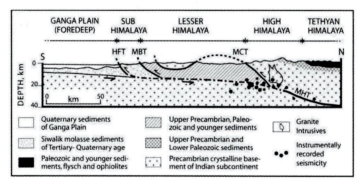

Fig. 3 Generalized north-south section across the Himalaya for the central portion of the Himalayan arc (Seeber and Armbruster, 1981). Figure after Kumar et al. (2006).

Tethys Himalaya slips aseismically and does not contribute to strain accumulation. In this model the major thrusts, namely, the Main Frontal Thrust (MFT), Main Boundary Thrust (MBT), Main Central Thrust (MCT) and Southern Tibet Detachment (STD) are assumed to be listric to the detachment.

The great thrust earthquakes in the Himalaya occur on the seismogenic detachment under the Outer and Lesser Himalaya, whereas the small and moderate earthquakes of the Himalayan seismic belt occur on the downdip part of the seismogenic detachment or on the mid-crustal ramp (Seeber and Armbruster,

1981; Ni and Barazangi, 1986; Molnar, 1990; Pandey et al., 1995, 1999; Gahalaut and Kalpna, 2001). Majority of the earthquakes of the Himalayan seismic belt are of thrust type with slip vectors perpendicular to the Himalayan arc. Further north of the Higher Himalaya, majority of the earthquakes exhibit normal type of motion on north-south oriented planes, while along the major faults, e.g., the Altyn Tagh and Kun-Lun, strike slip motion dominates. Those earthquakes are not included here in this review. Majority of the earthquakes in the Indian shield region occur along the well defined zones, which are referred as the failed rift regions. These are the regions along which rifting within the India plate developed but was immediately aborted as the India plate started moving northward after rifting from Africa and Madagascar. Thus, these regions became weak with heterogeneous intrusive material where many earthquakes of Indian shield region have occurred. Important of such failed rifts are Narmada-Son, Godavari and Kachchh failed rifts. Some of the earthquakes in the Indian shield region are referred as Stable Continental Region (SCR) earthquakes, e.g., the 1993 Killari earthquake. The 1967 Koyna earthquake also falls into that category with a distinction that it was triggered (but not caused) by the reservoir. Some small and moderate magnitude earthquakes occur along the Eastern and Western Ghats, which are linked to deformation in the rift shoulders.

3. REVIEW OF MAJOR EARTHQUAKES IN THE HIMALAYA AND INDIAN SHIELD REGION

The earthquakes in India and worldwide caused an enormous amount of damage to life and property (Table 1). The swiftness with which an earthquake unleashes its energy and the destruction that is left behind in its wake make earthquakes a hazard for mankind. To reduce the hazard from earthquakes, it is imperative that we understand the physics of the processes that occur during earthquakes and to review what happened during the past earthquakes. An earthquake happens when the accumulated strain energy in the earth is released suddenly. A part of the energy released, called fracture energy, is used in mechanical processes other than frictional heating on the fault zone as the rupture propagates; a part of the energy, frictional energy, is dissipated as heat on the fault surface and yet another part, wave energy, moves the particles on the fault generating seismic waves that are felt by people and recorded by instruments all over the world. The only part of the energy released in an earthquake that we have direct access to is the wave energy. Here we review interplate earthquakes of Himalayan region and intraplate earthquakes of Indian shield region.

Table 1. Top ten Indian earthquakes

Date	Earthquake	Magnitude	Deaths in India
16 Jun 1819	Kuchchh	8	1,500
12 Jun 1897	Shillong Plateau	8.7	1,500
04 Apr 1905	Kangra	8	19,000
15 Jan 1934	Nepal-Bihar	8.3	11,000
26 Jun 1941	Andaman	8.1	Thousands
15 Aug 1950	Assam	8.6	1,530
21 Aug 1988	Nepal-Bihar	6.6	1,004
30 Sep 1993	Killari	6.2	7,928
26 Jan 2001	Bhuj	7.7	13,805
26 Dec 2004	Sumatra-Andaman	9.2	10,749
08 Oct 2005	Kashmir	7.4	1,308

India Meteorological Department, www.imd.ernet.in

3.1 Seismicity of the Himalayan Region

The Himalaya is one of the most seismically active regions of the world (Fig. 2). Earthquakes of small, medium and large magnitudes have occurred in the Himalaya since earliest records (Guttenberg and Richter, 1954; Chandra, 1978). The seismicity in Himalaya appears to be nonuniform although major trends have been recognized which consist a belt of events beneath the lesser Himalaya between MCT and MBT (Seeber et al., 1981), which occurred along a huge circular arc, analogous to interplate earthquakes in island arcs.

The occurrence of earthquakes in the Himalayan arc and adjoining regions due to the collision process may be referred as the earthquakes of the Himalayan Continental Plate Margin (HCPM). The HCPM has produced three great earthquakes (1897 Shillong Plateau, 1934 Bihar-Nepal and 1950 Assam) with magnitude larger than 8.0 since the end of the 19th century and nine earthquakes magnitude larger than 7.5 since 1500 including five prehistorical earthquakes. All the great earthquakes of the Himalayan arc are considered to have occurred on the seismogenic detachment under the Outer and Lesser Himalaya (Fig. 3). The moderate sized earthquakes are reported to have occurred at a depth of 10 to 20 km (Seeber et al., 1981; Molnar, 1990) on the Basement Thrust (BT), a more steeply dipping thrust that juxtaposes basement of the Indian shield with the Tethyan slab, the pre-collisional leading edge of the Indian shield (Powell, 1979; Seeber and Armbruster, 1981; Ni and Barazangi, 1986; Molnar, 1990; Pandey et al., 1995, 1999; Gahalaut and Kalpna, 2001). We refer these earthquakes as the earthquakes of Himalayan Seismic Belt (HSB). In recent years, such moderate sized earthquakes which caught attention are 1991 Uttarkashi earthquake (Mw 6.8), and 1999 Chamoli earthquake (Mw 6.6). Both earthquakes occurred on the detachment under the Himalaya. The October 20, 1991 Uttarkashi earthquake affected region in the Garhwal Himalaya lies in the western part of the seismic gap between the rupture zones of 1905 Kangra earthquake and 1934 Bihar-Nepal earthquake. The last strong earthquake of Himalayan seismic belt occurred on March 29, 1999 (Ms 6.6) in the Kumaun-Garhwal Himalaya, and is known as the Chamoli earthquake.

3.1.1 Historical Earthquakes of Himalaya
(1) 1897 Shillong Plateau Earthquake
The great June 12, 1897 earthquake (Ms 8.0) in the Shillong Plateau in north-eastern India having epicenter at 25.7°N and 91.1°E is the largest well-documented intraplate earthquake in India, and probably one of the largest known anywhere (Fig. 1). The Shillong plateau is the only high ground between the Himalaya and the Bay of Bengal of about 250 km long and 80 km wide and about 1500 m above the plains of the Brahamaputra River. This earthquake has not only large magnitude but also caused heavy damage in surrounding district due to extensive liquefaction of the ground. The main event occurred at 17:15 hrs local time (11:09 GMT) and was recorded by 12 primitive seismograph in Europe. Many aftershocks were felt at and around the fault over a wide area through the end of 1898 (Oldham, 1899). Three large aftershocks occurred on successive days which when combined with the main event cause severe destruction in the surrounding region. No precise estimate of loss of lives is available but in spite of its large magnitude and high population density in some parts of epicentral area, it experienced less destruction as compared to the other earthquakes.

The highest intensity areas for this earthquake are Shillong Plateau and northern extension of this unit below the plains of western Assam. According to Seeber and Armbruster (1981), 1897 earthquake rupture was 550 km long with east-west strike and 300 km wide with no significant surface rupture (Oldham, 1899), which was meant to be the strong evidence for a shallow-dipping, detachment-like fault source. Molnar (1987) inferred an east-west extent of the rupture of 200±40 km from the western margin of the Shillong Plateau. A 170×100 km² rupture was estimated to have occurred in a predominantly thrust fault dipping north at about 5° having a depth of about 15 km and 23 km below the southern and northern margin of the rupture zone respectively by Gahalaut and Chander (1992). For more than a century it was believed that it occurred on a thrust fault dipping gently on north or some considered it as a Himalayan Basal thrust (Seeber et al., 1981; Molnar, 1987; Molnar and Pandey, 1989; Gahalaut and Chander, 1992) but according to recent analysis of geodetic data acquired in 1860, 1897 and 1936, Bilham and England (2001) recognize that the earthquake occurred on a ESE striking for 110 km, SSW-dipping reverse fault at 57° beneath the northern edge of the central Shillong Plateau with 16±5 m of slip on a fault between 9 and 45 km depth with a rake of 76° and named as Oldham Fault (Figs 4 and 5). They found that the northern edge of Shillong Plateau rose violently more than 11 m during rupture of a buried, 110 km long, reverse fault dipping steeply away from the Himalaya resulting in the destruction of structures over much of Plateau and surrounding areas. This solution is well constrained and consistent with an interpretation advanced by Auden (1949), and by gravity data, which suggest that Shillong Plateau is a horst of Peninsular India thrust up between the Himalaya and the Naga Hills (Auden, 1949). Ambrasys and Bilham (2003) confirmed that the rupture is inferred to have extended from 9 km to more than 30 km depth on a 100 km long SSE steeply-dipping reverse fault that slipped 15 m. They proposed a moment magnitude of this earthquake as 8.1. Surface faulting was found by Oldham (1899) in northwest part of the Shillong Plateau running for a distance of at least 19 km along the Chedrang fault, a north-south secondary fault, ruptured the surface at the western end of and above the main rupture and showing

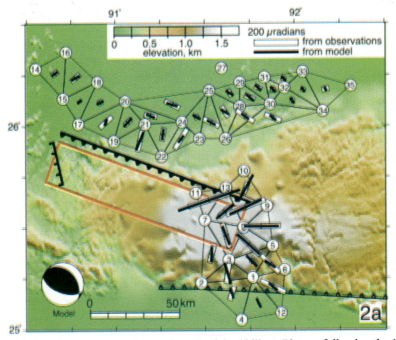

Fig. 4 Trigonometrical stations remeasured on and north of the Shillong Plateau following the 1897 earthquake. White rectangles are calculated from the triangulation observations, and black bars show the strains calculated for the best-fitting planar dislocation. Red rectangle indicates subsurface location of this SW dipping dislocation; thick black line with teeth shows the surface intersection of the continuation of this plane to the land surface (slip terminated 9 km below the surface). Short black line at western edge of fault plane indicates location of Chedrang fault. Line with open teeth to south of the Plateau shows location of the Dauki fault (Bilham and England, 2001).

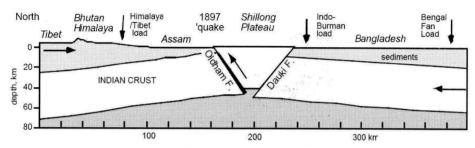

Fig. 5 N-S section from Tibet to the Bay of Bengal showing schematic geometry of Plateau pop-up (Bilham and England, 2001).

a vertical uplift of about 9.5 m with the east side up, on an approximately vertical exposed fault plane. Other fractures were also found at southwest of Chedrang fault and northern part of Plateau. In a recent study, Rajendran et al. (2004) opinioned that the earthquake rupture occurred near the northern limit of the Shillong Plateau. However, they could not precisely locate the rupture dimensions.

(2) 1905 Kangra Earthquake

The 4th April, 1905 Kangra earthquake occurred in the foothills of northwest Himalaya at 33°N and 76°E. The highest intensity, X on Rossi-Forel scale, was felt near the towns of Kangra and Dharamsala. Although this earthquake was assigned M = 8.4 by Richter (1958) and Ms = 8.0 by Kanamori (1977), Ambraseys and Bilham (2000) suggested a surface wave magnitude of Ms = 7.8 by reappraisal of the instrumental data with station corrections available for the event which is consistent with the intensity survey of this earthquake coordinated and compiled by Middlemiss (1910). Seven foreshocks were felt in just 24 hours prior to mainshock. In this earthquake more than 20,000 lives and 100,000 buildings were destroyed. The distribution of foreshocks suggests that 1905 large rupture is comparable to other great events of Himalaya except 1897 earthquake. Although extent of the rupture area is poorly constrained, several different possible rupture zones seem to be correlated with the observations. One is that the rupture occurred only beneath the area delimited by intensity VIII isoseismal, a zone of 100 km in length surrounding Kangra and Dharmsala. Second, its epicentral intensity distribution shows the maximum intensities ≥VIII around two regions Kangra and Dehradun which are separated by about 200 km, hence rupture was said to be the 280 km long fault zone along NW-SE trending boundary of Himalaya or third is that it is divided into two smaller segments that broke sequentially (Seeber and Armbruster, 1981).

The interpretation of levelling data of maximum seismic intensity reveals that 1905 rupture extends for about 250-300 km from the highest intensity area through the southeastern zone of high intensity where the coseismic elevation change was measured. Leveling data surveyed in 1904 and resurveyed in 1906 provide additional evidences that the rupture was not on the MBT but was on detachment that extends below the Himalayan front (Seeber et al., 1981; Chander, 1989; Gahalaut et al., 1992). The observed elevation changes were mostly coseismic, indicate less uplift on the northeast side than on the southwest side of MBT which would suggest normal movement during 1905 shock if slip occurred predominantly on the MBT. Ambraseys and Bilham (2000) calculated rupture area appropriate for a 7.8 magnitude earthquake in the range of 100×120 km^2 to 80×50 km^2 with 3-8 m of average slip. Assuming rupture between the zones of Himalayan Frontal thrust and moderate earthquakes bordering the southern edge of the Tibetan Plateau, the longest dimension of slip normal to the Himalayan arc is 80-100 km. Hence the greatest along-strike dimension of 120 km is significantly less than the proposed 280 km by Seeber and Armbruster (1981) but similar to the interpretation by Molnar (1990). This suggests that the main rupture was not extended continuously from Kangra through Dehradun. A reevaluation of the raw

24 Natural and Anthropogenic Disasters: Vulnerability, Preparedness and Mitigation

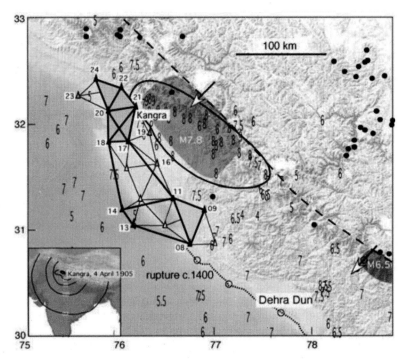

Fig. 6 Triangulation near the Kangra earthquake. Filled triangles indicate recovered GTS points: shaded triangles were used in shear strain analysis, and bold lines were used in linear strain analysis. The preferred rupture area is shaded. The MSK VIII contour (solid line) is interpolated from Ambraseys and Douglas (2004), with observations indicated by MSK number. Closed circles indicate M > 5.5 earthquakes since 1960; open circles indicate trench locations that define the c. 1400 rupture (dotted line). Dashed line approximates the 3.5 km elevation contour (Wallace et al., 2005).

leveling data reveals some systematic error in Middlemiss intensity distribution (Bilham, 2001) and shows that there was probably no or little uplift in that region. The absence of significant uplift or horizontal deformation restrict the rupture length less than 180 km (Fig. 6) which is consistent with the revised magnitude of M_s=7.8 with the probable rupture width of 50-70 km (Wallace et al., 2005).

(3) 1934 Bihar-Nepal Earthquake
The 1934 Bihar-Nepal earthquake is the most recent large historical event in the densely populated area of the Himalayan Front with an assigned Ms = 8.3 (Guttenberg and Richter, 1954) and moment Mo = 1.6×10^{28} dyne-cm (Chen and Molnar, 1977). Dunn et al. (1939) suggested that Bihar-Nepal earthquake did not result from movement along the MBT since no coseismic surface rupture was observed on this fault. He also observed that the largest region of intensity X is closely associated with a "slump belt" and a zone of soil liquefaction. The meizoseismal zone lies primarily on south of MBT and the region of intensity greater or equal to VIII extends about 300 km along the strike of the Himalaya and 250 km perpendicular to the strike (Seeber et al., 1981). In transverse direction it covers entire lesser Himalaya, the sub-Himalaya and the foredeep. The northern and southern boundaries of intensities fall in the transition zone of lesser and higher Himalaya and implies that seismic source extends under most of the area of about 75×10^3 km² of VIII intensity (Seeber et al., 1981). According to Chen and Molnar (1977), the epicenter was probably located in the Lesser Himalaya east of Kathmandu at 27.6°N and 87.1°E.

Macroseismic intensities and subsidence of the foreland revealed from leveling data suggest that the earthquake ruptured a 250–300 km along-strike segment of the arc (Bilham et al., 1998). The rupture area may have extended up to the MFT but probably not farther to the south (Chander, 1989). The northward extent of the rupture is not constrained at all. Three distinct areas of high intensities (≥VIII) have been discovered viz the large alluvial plains associated with ground failure, liquefaction, and slumping, the narrow belt near northern limit of outcropping belt suffered by high acceleration and Kathmandu Basin (Dunn, 1939). The 1833 event might have ruptured about the same arc segment as the 1934 earthquake (Bilham, 1995).

(4) 1950 Assam-Tibet Earthquake
The August 15, 1950 eastern Assam earthquake is the most recent great Himalayan Earthquake located in the extreme eastern and most remote portion of Himalayan front although epicenter of the earthquake probably lie in China. Numerous aftershocks beneath Himalaya in eastern Assam (Chen and Molnar, 1977) indicate part of rupture zone underlies Himalaya. The earthquake occurred at 28.38° N and 96.76° E having surface wave magnitude Ms = 8.4 (Guttenberg and Richter, 1954), moment magnitude Mw=8.6, and Moment Mo=10^{29} dyne-cm (Kanamori, 1977). It had the best instrumental coverage but not the best set of data on surface effect. On the basis of instrumental data, epicenter was located beyond the surface termination of Himalayan arc, in the Mishmi Mountains that bound the Assam basin towards the east-northeast and trend northwest (Seeber and Armbruster, 1981).

This event was extensively studied by Ben-Mehanem et al. (1974) who revised and added data to determine a fault plane solution with fault strike N26°W and dip of 60°E with nearly pure strike slip motion. Chen and Molnar (1977) found the P-wave first motion used by Ben-Mehanem et al. (1974) consistent with the thrust faulting on either a gently north-northwest or a steeply south-southeast dipping plane. They calculated seismic moment by assuming both low-angle thrust and Ben-Menahem et al.'s (1974) strike-slip fault plane solution and obtained a seismic moment of 7.8×10^{27} dyne-cm which is much smaller than the seismic moment of 2.5×10^{29} dyne-cm given by Ben-Menahem et al. (1974). A relocation of aftershocks of 1950 earthquake by Molnar and Pandey (1989) confirms that all the aftershocks lie beneath the Himalaya in a zone extending about 250±50 km west of the epicenter of the mainshock in east-west direction with 100 km width in north-south direction and thus the rupture occurred on a gently NNE dipping thrust fault.

3.1.2 Pre-Historical Earthquakes of Himalaya
Early earthquakes described in mythical terms include extracts in the Mahabharata (~1500 BC) during Kurukshetra battle (Iyengar and Sharma, 1999) and several semi-religious texts. Due to unavailability of adequate literature for Pre-historical earthquakes, little is known about Himalayan earthquakes in the 18th century and before. The significant large earthquakes which occurred in the period of 1500 to 1900 along Himalayan Seismic Belt (HSB) are 1505 Lo Mustang Earthquake, 1555 Srinagar earthquake, 1720 Uttar Pradesh earthquake, 1803 Uttar Pradesh earthquake, and 1833 Nepal earthquake.

(1) 1505 Lo Mustang Earthquake
The event of June 6, 1505 which occurred at 29.5°N and 83°E in southwestern Tibet was reported to be a great earthquake that was strongly felt in northern part of the Great Himalaya, along a distance of about 700 km from Guge in the northwest to Lo Mustang and Kyirong in the southeast (Ambraseys and Jackson, 2003; Bilham and Ambraseys, 2004). Bilham and Ambrasyes (2005) explicate that if slow earthquake occur, or if a substantial component of an historic great earthquake is caused by slip that does not radiate seismic energy, would result in the underestimation of seismic moment of that earthquake. Slow earthquakes imply the reduced frictional sliding as gently-dipping ruptures may be associated by modes

of failure that do not permit seismic radiation to escape into the body of the earth. Considering the 1505 earthquake as one of such type, they assigned the magnitude 8.6<Mw<8.8 to this earthquake. Rajendran and Rajendran (2005) examined the available documents and suggested three possibilities for occurrence of this earthquake. In the first possibility they found that the 1505 earthquake is smaller in size than 1803 event of magnitude 8.1 and occurred in Garhwal Kumaun Himalaya, the second possibility states that the magnitude of 1505 earthquake is comparable to the 1803 event and occurred in the Tibet-Nepal border and third possibility is that it was indeed a great earthquake but occurred in Tibetan Plateau, and not associated with the frontal thrust. The first possibility has been ruled out since most of the reports show the extensive destruction in northern Nepal and southern Tibet due to this earthquake. However, shaking from long ruptures is likely to increase the duration of shaking, causing much more damage to the structure than estimated one and increasing the severity of liquifaction. Hence, the other two possibilities are also ambiguous and research is still going on for better understanding of this earthquake. On the basis of reported destruction and intensities in area of radius 250 km Ambraseys and Jackson (2003) suggested a magnitude of 8.2 for this earthquake having a rupture length of about 400-700 km with a downdip width of 70-90 km associated with 7-15 m of slip (Bilham and Ambraseys, 2005).

(2) 1555 Srinagar Earthquake
The September 2, 1555 Kashmir earthquake occurred at 33.5°N and 75.5°E and is the western-most significant and large earthquake of Himalaya. Many authors suggest that it was a shallow, large magnitude earthquake of Ms=7.6 on the basis of the very long duration of aftershocks, intensity distribution and damaging reports which extended for more than 100 km southeast from Srinagar (Ambraseys and Jackson, 2003). Bilham and Ambraseys (2004) assigned the moment of the order of $2.69'10^{27}$ dyne-cm to this earthquake which may produce the earthquake of rupture area of length 100 km and width 80 km with a reverse slip of about 2 m.

(3) 1720 Uttar Pradesh Earthquake
The July 15, 1720 Uttar Pradesh earthquake occurred at 30°N and 80°E. This earthquake occurred near Delhi causing damage and apparent liquefaction but little else is known of this event (Kahn, 1874; Oldham, 1883). This event, from its location, could have been a normal faulting event, but because of the absence of damage accounts from the Himalaya it may have been a Himalayan earthquake (Bilham, 2004). Bilham and Ambraseys (2005) calculated the seismic moment 1.91×10^{27} dyne-cm for this earthquake which gives an idea of rupture dimension of about 100 km long and 80 km wide assuming a reverse slip of about 1.8 m.

(4) 1803 Uttar Pradesh Earthquake
The September 1, 1803 Uttar Pradesh earthquake occurred at 31.5°N and 79°E which caused massive damage and loss of life in Central Himalaya and Gangetic plains. Seeber and Armbruster (1981) suggested this to be decollement earthquake, which implies the earthquake having magnitude greater than 8.0. Khattri (1992) estimated its magnitude between 6.0 and 7.6. Ambrasyes and Jackson (2003) studied the earthquake and assigned the magnitude Ms = 7.5 estimating the size of the area over which the shock was clearly felt. A later study based on more complete data re-assessed its magnitude as Mw = 8.1 (Ambraseys and Douglass, 2004). Bilham and Ambraseys (2005) assigned the moment magnitude of the order of 1.51×10^{28} dyne-cm to this earthquake. Rajendran and Rajendran (2005) calculated the size of this event using Frankel's equation and obtained the moment magnitude Mw=7.7 for this earthquake. They concluded that this earthquake occurred on a subsidiary thrust of the MCT within Garhwal-Kumaun Himalaya and it cannot be characterized as great plate boundary earthquake.

(5) 1833 Nepal Earthquake

The August 26, 1833 earthquake occurred at 27.7°N and 85.7°E near Kathmandu within or close to the rupture of 1934 Bihar-Nepal earthquake (Bilham, 1995). It was felt over a large part of northern India and was located at 25.1°N and 85.3°E near Patna south of river Ganga (Bilham, 1995). Seeber and Armbruster (1981) proposed its location west of Kathmandu and suggested that it may have occurred in Central Himalayan Gap. Khattri and Tyagi (1983) placed the earthquake approximately 130 km west of the epicenter of 1934 Bihar-Nepal earthquake on the edges of Central Himalayan Gap and assign the event M = 7.6. Bilham (1995) explained that the earthquake consisted of three shocks, the first caused alarm, the second, five hours later brought most of people out of their home and the third, 15 min later was the main shock which caused widespread structural damage in India and Nepal. Based on the reported intensities, he estimated its moment magnitude Mw = 7.7 and locate it at 50 km north or northeast of Kathmandu. The slip associated with the event may have been 1-2 m on a thrust fault, which may have ruptured a region adjoining or overlapping the rupture zone of the great 1934 earthquake (Bilham, 1995).

3.1.3 Earthquakes Reported from Paleoseismological Investigations in the Himalaya

A few earthquakes have been reported on the basis of paleoseismological investigations in the Himalaya and Shillong Plateau. Sukhija et al. (1999) reported their results from the meizoseismal area of the 1897 earthquake which revealed well-preserved liquefaction and deformed syndepositional features at 10 selected sites in the alluvial deposits along two north flowing tributaries of the Brahmaputra river. In addition to the 1897 event, they provided evidence for at least three large seismic events. Two of them occurred during 1450–1650 and 700–1050 AD, the third predates 600 AD. Their analysis suggests a return period of about 400-600 years for the large earthquakes in the Shillong Plateau. Sukhija et al. (2002) reported paleoseismological evidence of occurrence of 1934 Nepal Bihar and 1833 Nepal earthquakes as well as evidence of occurrence of two prehistoric seismic events dated during 1700 to 5300 years BP and earlier than 25,000 years BP. Kumar et al. (2006) reported results of their paleoseismological investigations at sites along the Himalayan frontal Thrust between Chandigarh and Ramnagar (Nainital). Radiocarbon ages of samples obtained from the displaced sediments indicate that surface rupture at each site took place after ~A.D. 1200 and before ~A.D. 1700. Trench exposures and vertical separations measured across scarps in the eastern part of their region, are interpreted to indicate single-event displacements of ~11–38 m. Lave et al. (2005) presented paleoseismological evidence of occurrence of a great earthquake in the east central Nepal. They estimated that the earthquake occurred at ~1100 AD with a surface displacement of ~17 m and lateral extent and size that could have exceeded 240 km and Mw 8.8. Another major conclusion of this work was the absence of evidence of surface rupture during the 1934 Nepal-Bihar earthquake.

3.2 Indian Shield Region Earthquakes

Even today, earthquake occurrence in stable continental regions is not fully understandable. In most of the cases, the relationship between intraplate earthquakes and the subsurface seismogenic structure is not known. In order to improve our understanding of intraplate earthquakes, a considerable amount of work has been done in the past two decades. Several characteristics of intraplate earthquakes are now well known and more features appear to be emerging. For instance, intraplate earthquakes, especially those in stable continental interiors, usually have relatively long or very long recurrence intervals. Intraplate events are commonly characterized by very large areas of perceptibility compared with most shocks of similar magnitudes along plate boundaries. The interiors of any Lithospheric plates have also been found to be characterized by large horizontal compressive stress.

The Indian Peninsula is known as stable continental region dominated by Precambrian rocks. The seismicity of peninsular India was appeared to be very slow (Gutenberg and Richter, 1954; Chandra, 1977) prior to occurring of three damaging earthquakes in last 12 years. These earthquakes occurred in peninsular India, which is characterized by slow deformation and low seismic productivity (Johnston, 1989). Occurrence of three earthquakes during a brief interval of ten years naturally triggered a lot of discussion on the mechanism and pattern of recurrence of earthquakes in SCR-India. Other significant earthquakes in Indian Shield regions are 1819 Rann of Kachchh earthquake, 1969 Bhadrachalam earthquake and 1967 Koyna earthquake.

3.2.1 1819 Rann of Kachchh Earthquake

The June 16, 1819 Allah Bund (or Great Rann of Kachchh) earthquake is among one of the largest global intraplate earthquakes (Johnston and Kanter, 1990). The 1819 earthquake produced an about 80-90-km-long, 6-km-wide and 3-to-6m-high uplift known as the Allah Bund (Oldham, 1926; Bilham, 1999; Rajendran and Rajendran, 2001) across the Kori branch of Indus river (Fig. 7). A geometric moment magnitude of M=7.7±0.2 is obtained from the surface deformation and the inferred slip parameters, consistent with a magnitude estimated empirically from the intensity distribution (Bilham, 1999). Bilham (1999) suggested a shallow (from 10 km to near the surface) reverse-slip rupture on a 90-km-long 50–70° N-dipping fault plane to match the measured elevation changes from the event. Bilham et al. (2003) take the great depth and short lateral fault length of the 2001 Bhuj earthquake rupture into consideration and incorporate new topographic and remote sensing observations of the morphology of the Allah Bund fault scarp to obtain updated fault parameters. The 1819 event is estimated to have a

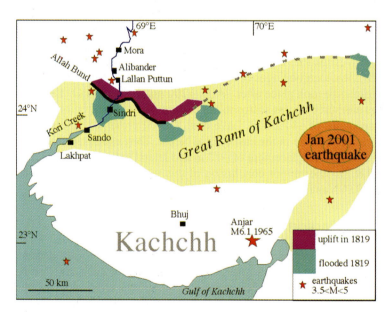

Fig. 7 Uplift during the 1819 Kachchh earthquake dammed the Kori River north of a zone of uplift termed the Allah Bund, and submerged the region to its south surrounding the fort at Sindri. On the basis of morphological changes recorded by Survey of India maps Oldham (1926) suggests that faulting may have extended a further 100 km to the east (dashed). (Bilham, 1999)

50-km-long rupture dipping 45° to the north with 3–8 m slip. The slip is set to 5.5 m in this study, consistent with a Mw = 7.7 earthquake for a rupture extending to 30-km depth.

3.2.2 1969 Bhadrachalam Earthquake

The April 13, 1969 Bhadrachalam earthquake is one of the major intraplate earthquake associated with paleorift zones in the stable continent region of Indian peninsula. It occurred on an 800 km long narrow NW trending Mesozoic rift of Godavari valley or Garben. Hence it is also known as Godavari valley earthquake. In northwest, it joins Narmada-Son lineament in the middle of the Indian peninsula and meets the east coast of India at southeastern end (Burke et al., 1978). It might be the first and only significant earthquake occurred in Godavari valley as only minor earthquakes have been reported in this region during past 200 years (Gupta et al., 1970). Indian seismological centre located the mainshock epicenter at 17.9°N and 80.6°E with body wave magnitude of m_b = 5.7 and moment magnitude Mw = 5.3 at a depth of 25 km. On the basis of body waveform analysis, Chung (1993) found strike slip dominant focal mechanism for this earthquake. The strike-slip motion represents a reactivated fault motion under the present-day stress field. The 1970 Broach earthquake of India is also an example of paleorift zone earthquake. It accompanied thrust motion with a small strike-slip component (Chung, 1993). These cases indicate that both thrust faulting or strike-slip faulting can occur on paleorift grabens reactivated by the present-day tectonic stresses depending on the orientation of the fault with respect to the applied stress field.

3.2.3 1967 Koyna Earthquake

The Koyna-Warna region of relatively stable peninsular India is a unique site in the world where the seismicity that reportedly began soon after the impoundment of the Koyna reservoir in 1961 has continued for over 40 years (Gupta, 1992, 2002; Talwani, 1997). The main Koyna earthquake of December 10, 1967 (M 6.3) is the largest earthquake near a reservoir, ever recorded globally, and the ongoing earthquake occurrences in the Koyna-Warna region have been considered as the reservoir triggered earthquakes (Chander and Kalpna, 1997; Rastogi et al., 1997; Talwani, 1997; Mandal et al., 1998; Gupta, 2002). The Koyna-Warna region in southwestern part of the Deccan volcanic province lies to the east of the west facing N–S escarpment parallel to the west coast from the gulf of Cambay to the peninsular tip. Our knowledge about the seismicity of the region prior to 1962 (i.e., prior to the reservoir impoundment) is very limited due to the absence of seismic stations in the area. Since 1963, more than 100,000 earthquakes, including about 170 of M ≥ 4, about 17 of M ≥ 5, have been reported from the Koyna-Warna region (Gupta, 2002) and frequency of the earthquakes of past 30 years is almost steady. Such frequency of earthquakes, especially those exceeding M 5.0, in a short span of time, is a rarity in the stable shield regions. Hence seismicity associated with Koyna earthquake is unique and of interest for many researchers since decades.

3.2.4 1993 Killari Earthquake

The 1993 Killari earthquake, also known as Latur earthquake (m_b 6.3, Ms 6.4, Mw 6.2), of September 29, 1993 occurred on the previously unmapped fault with no surface expression in the Deccan traps of peninsular India is considered to be a stable continental region, and caused destruction in an about 15 km wide area (MM intensity VIII). USGS estimated the epicenter of the mainshock at about 12 km west of Killari at a focal depth of 6.8 km. Several estimates of the fault plane solutions for this earthquake are available and all of them indicate predominantly reverse slip on the ESE trending nodal planes. The SSW dipping plane has been considered as the fault plane (Seeber et al., 1996; Baumbach et al., 1996; Gupta et al., 1998) on the basis of aftershock distribution, damage pattern and surface faulting. Despite its low magnitude, the damage due to the earthquake was extensive, because of extremely poor quality of buildings.

3.2.5 1997 Jabalpur Earthquake

The Jabalpur earthquake of May 21, 1997 is considered as the best-monitored Indian earthquake as it occurred in the central part of Indian shield with a very good azimuthal coverage of Indian seismic stations. The earthquake occurred in the lower crust of an ENE-WSW trending failed rift zone, known as Narmada-Son-Tapti failed rift zone which transects the Indian peninsular shield area into the northern and southern blocks. It is the most active paleorift zone, which evolved during the Archean and Proterozoic periods with magmatism of Cretaceous and Tertiary ages, which is similar to the New Madrid seismic zone (Liu and Zoback, 1997). Episodic reactivation of the failed rift is evident by the presence of varied rock formations ranging in ages from the late Archean to early Proterozoic. Two prominent deep faults termed as the Narmada South and Narmada North faults (NSF and NNF) with ENE-WSW strike have been mapped extensively in the region (Fermor, 1936; Nair et al., 1985; Roy and Bandyopadhyay, 1990). In the past, earthquakes with magnitude more than six have occurred in the failed rift zone and most of them are considered to be associated with the NSF.

The maximum intensity of shaking during the Jabalpur earthquake (m_b 6.0, Ms 5.6, Mw 5.8) was VIII on the MSK scale. The earthquake had a focal depth of about 36 km and is assumed to have occurred on the downdip extension of the NSF (Kayal, 2000). Fault plane solution of the earthquake suggest reverse slip on the SSE dipping steep plane. Field investigations related to damage survey indicate that the damage was maximum in a zone of 15×35 km^2 (Acharya et al., 1998) which lies to the north of reported earthquake epicenter by the India Meteorological Department (IMD) and USGS.

3.2.6 2001 Bhuj Earthquake

The Bhuj earthquake of January 26, 2001 was the largest intra-continental earthquake of the modern era of seismology, which occurred in the failed rift region on a steeply south-dipping reverse fault in the Rann of Kachchh region. The geologic structures in the epicentral region evolved during a long history of tectonic activity that began in the Proterozoic and involved several major tectonic episodes that fragmented Gondwanaland during Mesozoic and Paleogene periods (Biswas, 1987). In the late Triassic or early Jurassic, a number of smaller rift systems developed deep sedimentary basins, including the Kachchh basin. These structures were reactivated as a result of regional compression arising due to Indian plate movement in the Cenozoic which is evident from the fold and thrust belt along the Kachchh mainland fault system and the Allah Bund fault and also by the occurrence of major reverse earthquakes in the past 200 years, namely, the 1819 Allah Bund (Bilham, 1998), the 1956 Anjar and the 2001 Bhuj earthquakes. The main shock (m_b 6.9, Ms 7.9, Mw 7.6) occurred at an estimated depth of 25 km (IMD). Focal mechanism solutions suggest predominantly reverse motion on a steeply south dipping (51°-66°) fault. Thus the earthquake occurred in the lower crust on a steeply dipping reverse fault and the rupture did not extend up to the surface, or project towards a mapped surface fault. The earthquake caused large damage in the nearby cities. The main reason for this is the poor engineering, and poor quality of construction with poor building material.

3.3 December 2004 Sumatra-Andaman Earthquake

The December 26, 2004 Sumatra-Andaman earthquake (Mw 9.2) nucleated off the western coast of northern Sumatra and propagated north-northwest (Stein and Okal, 2005). The earthquake occurred along the eastern margin of the Indian plate where Indian plate obliquely subducts under the Sunda plate. The extent of aftershocks suggests that the rupture length of the earthquake was about 1400 km (Fig. 8). It is considered as the overall fourth largest earthquake since 1900 and is the largest since the 1964 Prince William Sound, Alaska earthquake. The tsunami generated by this earthquake caused more casualties than any other in the recorded history. In total, more than 200,000 people were killed in the countries surrounding Indian Ocean. With the rupture length of about 1400 km, it is the earthquake with a maximum rupture length. Another important aspect of this earthquake is that fast slip occurred in the southern part with a magnitude of slip reaching 15 m, which extended to the north-northwest direction at a velocity of 2.5 km/s, rupturing the 1300 km long plate boundary in about 8-10 minutes (Ammon et al., 2005; Ishii et al., 2005). The seismological data do not constrain either the coseismic surface displacements or coseismic slip on the rupture under Andaman-Nicobar islands, as the slip in this part occurred at a time scale beyond the seismic band (Ammon et al., 2005; Lay et al., 2005). In the subsequent one hour period, additional slow slip occurred in the Andaman-Nicobar region (Bilham, 2005; Banerjee et al., 2005). However, Vigny et al. (2005) argued against slow slip and suggested that the entire displacement at GPS sites in the northern Thailand occurred in less than 10 minutes after the earthquakes. Using far-field GPS sites located at about 400-3000 km from the rupture, they derived a slip

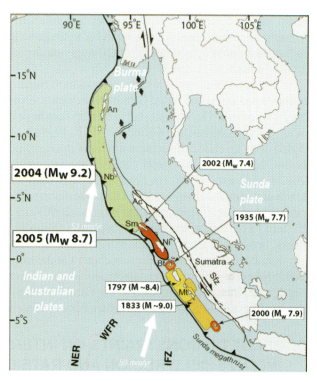

Fig. 8 The India Sunda subduction zone and rupture of the 2004 Sumatra Andaman earthquake (Briggs et al., 2006).

model for this earthquake. However, even these data do not provide a reliable estimate of coseismic surface displacements. In addition to the far-field GPS data (Vigny et al., 2005; Banerjee et al., 2005; Catherine et al., 2005), estimates of surface displacements have been reported from near-field GPS sites in Andaman-Nicobar Islands (Gahalaut et al., 2006). These data suggest that due to this earthquake the Andaman-Nicobar moved by 3-6 m horizontally in the SW direction and experienced a general subsidence of about 1-2 m.

4. GEODETIC CONSTRAINTS ON INTERSEISMIC DEFORMATION IN THE HIMALAYA AND RATE OF CONVERGENCE

In India, mainly the Survey of India undertakes the precise leveling work. However, data along a very few leveling lines are available in the public domain. Leveling observations along the Saharanpur-Mussoorie line have extensively been used to understand the effect of the 1905 Kangra earthquake (Chander, 1989; Gahalaut and Chander, 1992; Bilham, 2001) to assess the status of strain accumulation in the region (Gahalaut and Chander, 1994) and to understand the crustal deformation mechanism during earthquake cycle (Gahalaut and Chander, 1997a). The elevation changes along a leveling line from Pathankot to Dalhousie in Punjab Himalaya reveal uplift rate of 4-6 mm/year in the Lesser Himalaya and are consistent with strain accumulation on the detachment at the rate of 10-12 mm/year (Gahalaut and Chander, 1999). In central Nepal, the leveling data along a line from Birganj to Kodari via Kathmandu during the interseismic period reveal low uplift rate (<2 mm/year) in the Outer Himalaya while high uplift rate (6-8 mm/year) in the Lesser and southern Higher Himalaya (Jackson and Bilham, 1994). These data are consistent with the model of strain accumulation on the detachment at the rate corresponding to the plate convergence rate of 18-20 mm/year during the interseismic phase (Gahalaut and Chander, 1997b; Bilham et al., 1997).

In past two decades, the conventional land based geodetic techniques have been replaced by more accurate and fast space based GPS measurements. Extensive measurements have been undertaken in the Nepal Himalaya. The leveling, GPS, DORIS data have been analysed (Jackson and Bilham, 1994; Bilham et al., 1997; Gahalaut and Chander, 1997b, 1999; Jouanne et al., 1999; Avouac, 2003; Bettinelli et al., 2006) using an elastic dislocation model of interseismic strain and taking into account the uncertainty on India plate motion. The mean convergence rate across Central and Eastern Nepal is estimated to be 19 ± 2.5 mm/year (Fig. 9). The detachment was found to be locked from the surface to a depth of about 20 km over a width of about 115 km. The slight discrepancy between the geologically estimated deformation rate of 21 ± 1.5 mm/year (Lave and Avouac, 2000) and the 19 ± 2.5 mm/year geodetic rate in Central and Eastern Nepal can be explained by possible temporal variations of the pattern and rate of strain in the period between large earthquakes in this region. GPS measurements in the Garhwal and Punjab Himalaya show strain accumulation at the detachment at the rate of 18 and 14 mm/year (Banerjee and Burgmann, 2002; Jade et al., 2004). If the detachment is assumed to be fully locked then this estimate corresponds to the convergence accommodated in the Himalaya. It may be noted here that the estimates derived from the leveling and GPS measurements are very consistent.

5. GEODETIC CONSTRAINTS ON INTERSEISMIC DEFORMATION IN THE INDIAN SHIELD REGION AND RATE OF PLATE MOTION

GPS measurements have given a big boost in estimating the plate motion, to verify the models on plate motion and to estimate the crustal deformation through earthquake cycle in the plate boundary as well as plate interior regions. At present, in India, there are two permanent International GNSS (formerly GPS) System (IGS) stations located at Bangalore and Hyderabad. GPS measurements at these two sites along with several other permanent sites across the country have provided a robust estimate of Indian plate motion which is 5.1 cm/year towards N51°E in the central Indian region. This estimate is consistent with the models of plate motion. These measurements have also provided the evidence of extremely slow rate of strain accumulation within the India, which is consistent with the low frequency of earthquake occurrence, as compared to the plate boundary regions, e.g., Himalayan and Andaman region. It is estimated that about 2 mm/year of strain accumulation occurs across the N-S length of Indian peninsular region (Fig. 10). Large part of it is presumably accommodated across the Narmada-Son failed rift region (Banerjee et al., 2008).

6. HIMALAYAN SEISMIC GAP AND ASSOCIATED SEISMIC HAZARD

Within the framework of the seismotectonic models, the seismic gap hypothesis has been applied on the great earthquakes, which are caused by the rupturing of the convergent plate boundaries (Fedotov, 1965; Sykes, 1971). This hypothesis states that major earthquake are likely to occur along the sections of the detachment of convergent plate margins which have ruptured in past but have not experienced any great earthquake at least in past few decades. The possibilities of experiencing a great earthquake in such regions are directly proportional with time elapsed since the occurrence of last great earthquake because strain accumulates over decades or centuries due to consequence of slow plate motions.

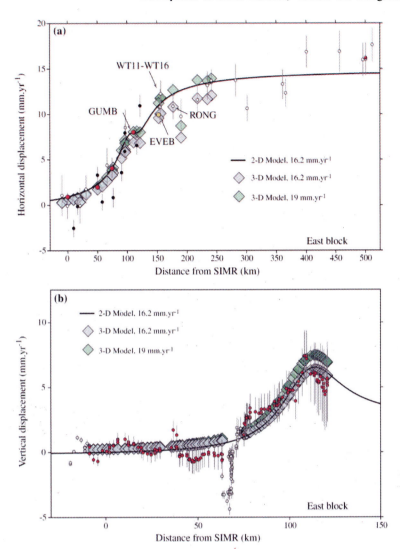

Fig. 9 (**a**) Horizontal velocities derived from GPS and DORIS stations across the Himalaya of Central and Eastern Nepal projected on a north-northeast cross-section (Bettinelli et al., 2006). The continuous black line shows prediction from a model of interseismic strain computed from a creeping dislocation embedded in an elastic half-space. Blue and green diamonds show, respectively, prediction of a 3D point-source dislocation model for a slip rate of 16.2 and 19 mm/year; (**b**) Observed (red dots) and modelled vertical displacements along the leveling profile across Central Nepal projected along the Kathmandu section (Jackson and Bilham, 1994). Grey dots show data not included in our determination of the best model. These data include some leveling data clearly affected by subsidence in Kathmandu valley and some points in the lowlands.

Seeber and Armbruster (1981) interpreted that the last four Himalayan great earthquakes have ruptured off about 1400 km of the Himalayan detachment. They estimated the rupture length 300 km for 1905 Kangra, 1934 Bihar Nepal and 1950 Assam earthquakes whereas for the 1897 earthquake, they suggested a rupture length of about 500 km. They identified two seismic gaps unruptured since 1800: first, the

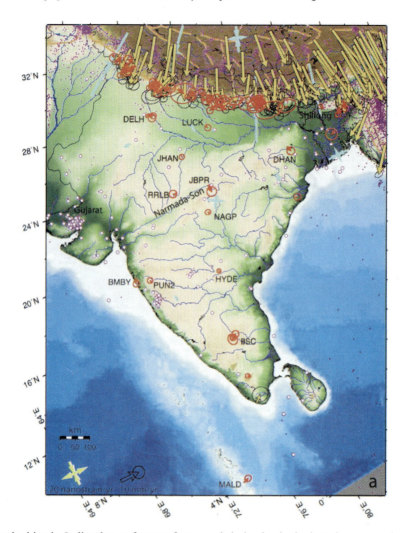

Fig. 10 GPS velocities in India plate reference frame and derived principal strain axes and magnitudes. Red arrows are GPS velocities of <4 mm/year magnitude, orange arrows have larger rates with respect to India. Magenta dots and circles indicate earthquakes. Note that stations lying south of the Narmada-Son region show almost zero velocity while those north of it but south of Himalayan region, show southward velocity of about 2 mm/year (Banerjee et al., 2008).

segment to the west of 1905 Kangra earthquake rupture located in Kashmir region and lies between 1885 and 1905 earthquakes and second, the segment in Uttaranchal between 1803 and 1833 ruptures. However, the tectonics of Kashmir Himalaya is poorly understood. They were not sure whether the detachment under this region is able to produce great earthquake. Hence according to Seeber and Armbruster (1981), a gap in Uttaranchal would be the most possible location of future great earthquake. They also suggested that 100 km gap between the ruptures of 1897 and 1950 earthquakes probably have been ruptured by 1943 and 1947 earthquakes and thus leaving no seismic gaps between these ruptures. They have suggested the repeat time of a great Himalayan earthquake to be about 200-270 years.

Khattri and Tyagi (1983) and Khattri (1987) analyze the space-time patterns of seismicity in the Himalaya plate boundary and recognize the existence of three seismic gaps: (1) the Kashmir Gap, an unruptured area to the west of the rupture of 1905 Kangra earthquake, (2) the Central Gap lying between the ruptures of 1905 Kangra earthquake and the 1934 Bihar-Nepal earthquake and (3) the Assam Gap between the ruptures of 1897 Shillong Plateau earthquake and the 1950 Assam earthquake. They found the reduced level of seismic activity in Central Gap from 1960 onwards and a similar seismic drop was noted in the Assam Gap from 1967 onwards; hence they characterized these two gaps as the region of relatively high level of seismicity from the point of view of future great earthquakes contrary to the view of Seeber and Armbruster (1981). Further they proposed the recurrence interval for these earthquakes between 200-500 years with a likely value of 300 years.

Molnar (1987), Molnar and Pandey (1989) and Molnar (1990) identified rupture lengths of great Himalayan earthquakes of 1905, 1934, 1897 and 1950 to be about 280 km, 200±40 km, 200±100 km and 250 km respectively. Thus, according to them only 30-35% of the plate margin has ruptured since 1897. Chander (1988) observed ground level changes due to 1905 Kangra earthquake and 1934 Bihar-Nepal earthquake and identified the susceptible seismic gap of nearly 700 km between the rupture zones of these earthquakes.

Bilham and Wallace (2005) and Bilham et al. (2001) analysed the seismic gap (Fig. 11) and estimated slip potential in each Himalayan segments (Fig. 12). They suggested that most of the Himalaya segments, except the 1934 and 1950 earthquake regions, can produce great earthquakes.

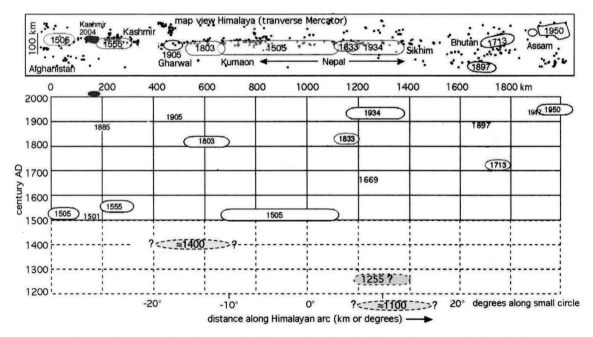

Fig. 11 The Himalayan arc plotted as angular distance from the Thakola graben in Nepal with the time-space history of the inferred ruptured zones of known major earthquakes. The rupture zones of none of these earthquakes are known precisely. Inferred rupture length and identified seismic gap in the Himalayan arc (Bilham and Wallace, 2005).

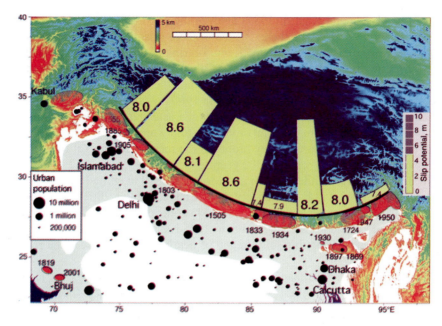

Fig. 12 Seismic potential of Himalayan arc (Bilham and Wallace, 2005). Urban population is shown by filled black circles. The height of each trapezoid is proportional to the current slip potential in meters, and the numbers refer to the potential size of Mw should the same segment length slip as is currently believed to have occurred in the last earthquake. The slip potential in the eastern Himalaya is tentative since the effects of the 1897 Shillong earthquake are uncertain and we know of no great historical earthquakes in Bhutan with the exception of a possible event in 1713 (Ambraseys and Jackson, 2003).

7. SEISMIC HAZARD AND ITS MITIGATION

The scientific challenge is to learn more about these giant earthquakes. From GPS geodesy we know how much closer India approaches Tibet each year (2 cm/year). From geological excavations of the faulted frontal range of the Himalaya we know how much slip occurs in a few giant earthquakes (8-10 m). From seismological studies we know how much slip occurred in recent great earthquakes (2-6 m). Simple arithmetic yields the astounding possibility that almost 3/4 of the Himalaya could have a magnitude 8 earthquake today, and that in parts of the west-central Himalaya we could have an earthquake as big as M = 8.2. After the 1993 Latur earthquake in peninsular India, seismological network for earthquake monitoring has been expanded and modernized. The India Meteorological Department (IMD), the primary agency for monitoring, maintains a network of more than 50 seismological observatories.

The biggest problem with future Himalayan earthquakes is not the determination of their timing; it is the problem that they will inevitably occur. The longer we wait, the larger they will be. Since population densities have increased tenfold or more since the last of the great Medieval earthquakes, and since building styles have in many cases become more vulnerable to seismic shaking, the next great Himalayan event will be much more disastrous than those in recorded history (Bilham, 2004). Various attempts have been made to quantify the seismic hazard in the Himalaya and plate interiors (Khattri et al., 1984; Kumar and Bhatia, 1999), and all of them reflect high seismic hazard in the Himalayan and NE India region.

Seismic hazard assessment of SCRs is complicated also because of the incompleteness in data (Jaiswal and Sinha, 2007). In the absence of high quality earthquake data from each of the SCRs, lack of attenuation relations of acceleration, and poor historical records of earthquake occurrence, assessment of the seismogenic processes may have to rely also on data from analogous settings elsewhere. As more data are generated from various source regions, it may be possible to identify analogies in their characteristic sizes, patterns of recurrence and deformation mechanisms, taking us closer to realistic hazard assessment. Nevertheless, the comparison of the probabilistic seismic-hazard map developed with the hazard map specified in the Indian Standards shows that the design parameters in the Indian Standards may significantly underestimate the seismic hazard in some regions of peninsular India.

Since earthquake prediction is not possible as yet, it is important to learn to live with earthquakes. Earth Scientists will continue to learn more about the earthquakes, their occurrence processes, potential regions of future earthquakes and the expected damage scenarios in a region in case such an earthquake occurs. It is important to follow the long-term and short-term mitigation schemes. Under the long-term scheme, it is necessary that a strict region-based building code is implemented in the country. This will definitely reduce the damage and loss of lives during the next great earthquake. Worldwide, significant development has taken place in this direction (Allen, 2007), which includes, for example, quantification in terms of probability of earthquake occurrence of the next big earthquake in a region on a given fault, attenuation laws using strong ground motion arrays etc. Under the short-term mitigation scheme, seismic microzonation, shake map, mapping and quantification of geological structures influencing the ground shaking, deployment of earthquake early warning systems etc. may be taken up.

In India, micro-zonation studies for major cities have been taken up by various agencies which are sponsored by the Ministry of Earth Sciences and the Department of Science and Technology (DST), Government of India, New Delhi. However, a unified and holistic approach in taking up such studies is required. The National Disaster Mitigation Authority (NDMA), under the Ministry of Home Affairs, has realised that in order to move towards safer and sustainable national development, development projects should be sensitive towards disaster mitigation. A few but significant steps towards vulnerability reduction, putting in place prevention and mitigation measures and preparedness for a rapid and professional response have been taken. NDMA has launched a massive awareness generation campaign and building up of capabilities as well as institutionalization of the entire mechanism through a techno legal and techno financial framework to gradually move in the direction of sustainable development. The various prevention and mitigation measures are aimed at building up the capabilities of the communities, voluntary organizations and government functionaries at all levels. Particular stress is being laid on ensuring that these measures are institutionalized considering the vast population and the geographical area of the country.

8. EARTHQUAKE EARLY WARNING SYSTEM

The idea of earthquake early warning is simple and is based on the principle that (i) electromagnetic waves travel faster than the elastic waves, and (ii) P waves travel faster than the more damaging elastic S waves in an elastic medium (Nakamura, 1984; Wu and Kanamori, 2005). P and S waves are generated simultaneously during the earthquake, but with distance they get segregated. Presently, two types of early warning systems (EWS) are in operation around the world. One is a front-detection EWS: seismometers installed in the earthquake source area give early warnings to more distant urban areas. The other is an on-site EWS, which determines the earthquake parameters from the initial portion of the P wave and predicts the more severe ground shakings of the following S-wave trains. Some crucial facilities, like nuclear power plant, high speed trains, gas pipelines, etc. may be shut down and people may get a few seconds to come out of the buildings. Such warning systems, also referred as Earthquake Alarms Systems (ElarmS) have been implemented in a few countries such as Japan, Taiwan, Mexico,

Turkey and Romania, and have been tested offline in California, Italy, Alaska (http://www.elarms.org/). Mexico is the classic example where EWS can help tremendously. Though Mexico city is not very prone to earthquakes, but a distant earthquake along the western Mexican coast can cause extensive damage due to its soil condition. Thus, a warning of few seconds at Mexico city about the earthquake along the coast can help in reducing the damage. A similar case exists in India. Though major cities in the north, like New Delhi, Lucknow, Kanpur, etc. are not in the earthquake-prone region, they are located close to the central seismic gap in the Himalayan region where great earthquake is expected. Hence, a warning of few seconds can reduce the damage. However, such systems are yet to be implemented in India.

One of the problems in implementing and installing earthquake early warning systems is the accurate estimation of the size of earthquake in real time and the assessment of extent of damage at the target site due to that earthquake (Allen, 2007). A starting and crucial point in this regard could be nuclear power plant facilities which are located in the Indo-Gangetic plains, e.g., Narora power plant in Uttar Pradesh. To implement EWS for safeguarding other facilities, we need to increase automation and public awareness about it. Also, there should be proper evacuation plans for the public in the residence, school and office complexes so that an early warning of few seconds may be utilized to save lives rather than increase in loss of lives due to chaos and stampede.

9. CONCLUDING REMARKS

Earthquake occurrence in the Himalaya and plate interiors is attributed to the ongoing plate motion and convergence with the neighbouring plates. Great earthquakes in the Himalaya occur through episodic slip on the detachment under the Outer and Lesser Himalaya, while small and moderate size earthquakes of the Himalayan seismic belt continue to occur near the downdip part of the seismogenically active detachment. Occurrence of great and major earthquakes along the Himalayan arc, e.g., 1934 Nepal-Bihar, 1950 Assam, 1905 Kangra in the twentieth century, 1803 and 1833 in nineteenth century and earlier earthquakes like 1505 and other events inferred from paleoseismological investigations attest that majority of the length of the Himalayan arc has potential to generate great or major earthquakes. Evidence that no great or even major earthquake has occurred in the Himalayan arc since 1950 and strain accumulation is underway in the central seismic gap region where great earthquakes have occurred in the past, suggest that a great earthquake may occur in this region. In other seismic gap regions (i.e., Kashmir and Assam), lack of evidence of strain accumulation coupled with poor historical earthquake records do not guarantee that great earthquakes did not and cannot occur in these regions. In-depth investigations are required to rule that out.

In the peninsular India, a few earthquakes that occurred in past 200 years have been reviewed in this chapter. All these earthquakes, except for the 1997 Jabalpur and 2001 Bhuj earthquakes, had very shallow focal depth (less than 10 km) and their focal mechanism is consistent with the approximately N-S maximum stress direction and on-going plate motion. Both 1997 Jabalpur and 2001 Bhuj earthquakes, along with a few other moderate magnitude earthquakes, occurred along the failed rift regions. Thus, these regions pose a serious hazard due to future major or moderate magnitude earthquakes in the peninsular India.

It is necessary to take up additional scientific studies, involving GPS measurements of crustal strains, crustal structure identification using various seismological and seismic methods, earthquake source parameters, ground motion attenuation relations, etc. At the same time, it is important to learn more about the previous earthquakes, their occurrence processes and to identify potential regions of future earthquakes. Based on the findings of such studies, it is desirable to develop damage scenarios and micro-zonation maps of some selected cities. Depending upon the feasibility, earthquake early warning systems may also be installed at some selected cities. The rapid growth of IT and communication technologies can help in disseminating the earthquake related information quickly and implementing it appropriately. Implementation of building codes and public awareness is the key to reduce the loss of lives and damage to civil structures, as it is said *"Earthquakes don't kill people, but buildings do"*.

REFERENCES

Acharya, S.K., Kayal, J.R., Roy, A. and Chaturvedi, R.K. (1988). Jabalpur earthquake of 22 May 1997. *Journal of the Geological Society of India*, **51**: 295-304.

Allen, R.M. (2007). Earthquake hazard mitigation: New directions and opportunities. *In:* G. Schubert (editor), Treatise on Geophysics, Vol. 4, Elsevier, pp. 607-647.

Ambraseys, N. and Bilham, R. (2000). A note on the Kangra Ms = 7.8 earthquake of 4 April 1905. *Current Science*, **79**: 45-50.

Ambraseys, N. and Bilham, R. (2003). Earthquakes and crustal deformation in northern Baluchistan. *Bulletin Seismological Society of America*, **93**: 1573-1605.

Ambraseys, N. and Jackson, D. (2003). A note on early earthquakes in northern India and southern Tibet. *Current Science*, **84**: 571-582.

Ambraseys, N.N. and Douglas, J. (2004). Magnitude calibration of north Indian earthquakes. *Geophysical Journal International*, **158**: 1-42.

Ammon, C.J., Ji, C., Thio, H.K., Robinson, D., Ni, S., Hjorleifsdottir, V., Kanamori, H., Lay, T., Das, S., Helmberger, D., Ichinose, G., Polet, J. and Wald, D. (2005). Rupture Process of the 2004 Sumatra-Andaman Earthquake. *Science*, **308**: 1133-1139.

Auden, J.B. (1949). A geological discussion on the Satpura hypothesis and Garo Rajmahal gaps. Proceedings of the National Institute Science of India, Vol. 15, pp. 315-340.

Avouac, J.-P. (2003). Mountain Building, erosion and the seismic cycle in the Nepal Himalaya. *Advances in Geophysics*, **46**: 1-80.

Banerjee, P. and Burgmann, R. (2002). Convergence across the northwest Himalaya from GPS measurements. *Geophysical Research Letters*, **29**: doi:10.1029/2002 GL015184.

Banerjee, P., Burgmann, R., Nagarajan, B. and Apel, E. (2008). Intraplate deformation of the Indian subcontinent. *Geophysical Research Letters*, **35**: L18301, doi:10.1029/2008GL035468.

Banerjee, P., Pollitz, F.F. and Bürgmann, R. (2005). The size and duration of the Sumatra-Andaman earthquakes from far-field static offsets. *Science*, **308**: 1769-1772.

Baumbach, M.H., Grosser, H.G., Schmidt, A., Paulat, A., Rietbrock, C.V., Rao, R., Raju, P.S., Sarkar, D. and Mohan, I. (1996). Study of the foreshocks and aftershocks of the intraplate Latur earthquake of 30 September 1993, India. *Memoir Geological Society of India*, **35**: 33-63.

Ben-Menahem, A., Aboudi, E. and Schild, R. (1974). The source of the great Assam earthquake: An intraplate wedge motion. *Physics of the Earth and Planetary Interiors*, **16**: 109-131.

Bettinelli, P., Avouac, J.P., Flouzat, M., Jouanne, F., Bollinger, L., Willis, P., and Chitrakar, G.R. (2006). Plate motion of India and interseismic strain in the Nepal Himalaya from GPS and DORIS measurements. *Journal of Geodesy*, DOI 10.1007/s00190-006-0030-3.

Bilham, R. (1995). Location and magnitude of the 1833 Nepal earthquake and its relation to the rupture zones of contiguous great Himalayan earthquakes. *Current Science*, **69**: 101-128.

Bilham, R. (1999). Slip parameters for the Rann of Kachchh, India, 16 June 1819 earthquake, qualified from contemporary accounts. *In:* I.S. Stewart and C. Vita-Finzi (editors), Coastal Tectonics, Geological Society of London, Vol. 146, pp. 295-318.

Bilham, R. (2001). Slow tilt reversal of the Lesser Himalaya between 1862 and 1992 at 78°E, and bounds to the southeast rupture of the 1905 Kangra earthquake. *Geophysical Journal International*, **144**: 713-728.

Bilham, R. (2004). Earthquakes in India and the Himalaya: Tectonics, geodesy and history. *Annals of Geophysics*, **47**: 839-858.

Bilham, R. (2004). Urban earthquake fatalities: A safer world or worse to come? *Seismological Research Letters*, **76**: 706-712.
Bilham, R. (2005). A flying start, then a slow slip. *Science*, **308**: 1126-1127.
Bilham, R. (2006). Moving Mountains in Himalaya. National Geographic Society, Boulder, USA, pp. 132-137.
Bilham, R. and Ambraseys, N. (2005). Apparent Himalayan slip deficit from the summation of seismic moments for Himalayan earthquakes, 1500-2000. *Current Science*, **88**: 1658-1663.
Bilham, R. and England, P. (2001). Plateau pop-up in the 1897 Assam earthquake. *Nature*, **410**: 806-809.
Bilham, R. and Wallace, K. (2005). Future Mw>8 earthquakes in the Himalaya: Implications from the 26 Dec 2004 Mw = 9.0 earthquake on India's eastern plate margin. Geological Survey of India, Special Publication, Vol. 8, pp. 1-14.
Bilham, R., Bendick, R. and Wallace, K. (2003). Flexture of the Indian plate and intraplate earthquakes. *Proceedings Indian Academy of Sciences, Earth Planet Sciences*, **112**: 1-14.
Bilham, R., Blume, F., Bendick, R. and Gaur, V.K. (1998). Geodetic constraints on the translation and deformation of India, implication for future great Himalayan earthquakes. *Current Science*, **74**: 213-229.
Bilham, R., Gaur, V.K. and Molnar, P. (2001). Himalayan seismic hazard. *Science*, **293**: 1442-1444.
Bilham, R., Larson, K., Freymueller, J. and Project Idylhim members (1997). GPS measurements of present-day convergence across the Nepal Himalaya. *Nature*, **386**: 61-64.
Briggs, R., Sieh, K., Meltzner, A.J., Natawidjaja, D., Galetzka, J., Suwargadi, B., Hsu, Y-J., Simons, M., Hananto, N., Suprihanto, I., Prayudi, D., Avouac, J.P., Prawirodirdjo, L. and Bock, Y. (2006). Deformation and slip along the Sunda Megathrust in the Great 2005 Nias-Simeulue Earthquake. *Science*, **311**: 1897-1901.
Catherine, J.K., Gahalaut, V.K. and Sahu, V.K. (2005). Constraints on rupture of the December 26, 2004, Sumatra earthquake from far-field GPS observations. *Earth and Planetary Science Letters*, **237**: 673-679.
Chander, R. (1988). Interpretation of observed ground level changes due to the Kangra earthquake, northwest Himalaya. *Tectonophysics*, **149**: 289-298.
Chander, R. (1989). On applying the concept of rupture propagation to deduce the location of the 1905 Kangra earthquake epicenter. *Journal of the Geological Society of India*, **33**: 150-158.
Chander, R. and Gahalaut, V.K. (1994). Preparations for great earthquakes seen in levelling observations along two lines across the Outer Himalaya. *Current Science*, **67**: 531-534.
Chander, R. and Gahalaut, V.K. (1999). On the cyclic nature of active crustal deformation in the Dehradun region. *Himalayan Geology*, **20**: 87-92.
Chander, R. and Kalpna (1997). On categorising induced and natural tectonic earthquakes near new reservoirs. *Engineering Geology*, **46**: 81-92.
Chandra, U. (1978). Seismicity, earthquake mechanisms and tectonics along the Himalayan mountain range and vicinity. *Physics of the Earth and Planetary Interiors*, **16**: 8-92.
Chen, W.P. and Molnar, P. (1977). Seismic moments of major earthquakes and the average rate of slip in central Asia. *Journal of Geophysical Research*, **82**: 2945-2969.
Chung, W.Y. (1993). Source parameters of two rift-associated intraplate earthquakes in peninsular India: the Bhadrachalam earthquake of April 13, 1969 and the Broach earthquake of March 23, 1970. *Tectonophysics*, **225**: 219-230.
Dunn, J.A. (1939). Seismological observations by the geological survey. *Memoir Geological Survey of India*, **73**: 76-87.
Dunn, J.A., Auden, J.B. and Ghosh, A.M.N. (1939). Earthquake effects. *Memoir Geological Survey of India*, **73**: 27-48.
Fedotov, S.A. (1965). Regularities of the Distribution of Strong Earthquakes in Kamchatka, the Kurile Islands and Northeastern Japan. *Tr. Inst. Fiz. Zemli Akad. Nauk SSSR*, **36**: 1-66.
Gahalaut, V.K. and Chander, R. (1992). A rupture model for the great earthquake of 1897, northeast India. *Tectonophysics*, **204**: 163-174.
Gahalaut, V.K. (2008). Coulomb stress changes due to 2005 Kashmir earthquake and implications on future seismic hazard. *Journal of Seismology*, DOI 10.1007/s10950-008-9092-4.
Gahalaut, V.K. and Chander, R. (1992). On the active tectonics of the Dehra Dun region from observations of ground elevation changes. *Journal of Geological Society of India*, **39**: 61-68.

Gahalaut, V.K. and Chander, R. (1997a). Evidence for an earthquake cycle in the NW outer Himalaya near 78°E longitude from precision levelling operations. *Geophysical Research Letters*, **24**: 225-228.

Gahalaut, V.K. and Chander, R. (1997b). On interseismic changes and strain accumulation for great thrust earthquakes in the Nepal Himalaya. *Geophysical Research Letters*, **24**: 1011-1014.

Gahalaut, V.K. and Chander, R. (1999). Geodetic evidence for accumulation of earthquake generating strains in the NW Himalaya near 75.5°E longitude. *Bulletin Seismological Society of America*, **89**: 837-843.

Gahalaut, V.K. and Kalpna (2001). Himalayan mid crustal ramp. *Current Science*, **81**: 1641-1646.

Gahalaut, V.K., Gupta, P.K., Chander, R. and Gaur, V.K. (1994). Minimum norm inversion of elevation change data for slips on the causative faults during the 1905 Kangra earthquake. *Proceedings of the Indian Academy of Sciences (Earth and Planetary Sciences)*, **103**: 401-411.

Gahalaut, V.K., Nagarajan, B., Catherine, J.K. and Kumar, S. (2006). Constraints on 2004 Sumatra-Andaman earthquake rupture from GPS measurements in Andaman-Nicobar Islands. *Earth Planetary Science Letters*, **242**: 365-374.

Gupta, H.K. (1992). Reservoir Induced Earthquake. Development in Geotechnical Engineering, Vol. 64, Elsevier, the Netherlands, 320 pp.

Gupta, H.K. (2002). A review of recent studies of triggered earthquakes by artificial water reservoirs with special emphasis on earthquakes in Koyna, India. *Earth-Science Reviews*, **58**: 279-310.

Gupta, H.K., Rastogi, B.K., Mohan, I., Rao, C.V.R.K., Sarma, S.V.S. and Rao, R.U.M. (1998). An investigation into the Latur earthquake of 29 September 1993 in southern India. *Tectonophysics*, **287**: 299-318.

Gutenberg, B. (1956). Great earthquakes 1896-1903. *Transaction American Geophysical Union*, **37**: 608-614.

Gutenberg, B. and Richter, C.F. (1954). Seismicity of the Earth and Associated Phenomena. Princeton University Press Princeton, NJ, 273 pp.

Ishii, M., Shearer, P.M., Houston, H. and Vidale, J.E. (2005). Extent, duration and speed of the 2004 Sumatra-Andaman earthquake imaged by Hi-Net array. *Nature*, **435**: 933-936.

Iyengar, R.N. and Sharma, D. (1999). Some earthquakes of the Himalayan region from historical sources. *Himalayan Geology*, **20**: 81-85.

Jackson, M. and Bilham, R. (1994). Constraints on Himalayan deformation inferred from vertical velocity fields in Nepal and Tibet. *Journal of Geophysical Research*, **99**: 13897-13912.

Jade, S., Bhatt, B.C., Yang, Z., Bendick, R., Gaur, V.K., Molnar, P., Anand, M.B. and Kumar, D. (2004). Preliminary tests of plate-like or continuous deformation in Tibet. *Geological Society of America Bulletin*, **116**: 1385-1391.

Jaiswal, K. and Sinha, R. (2007). Probabilistic seismic-hazard estimation for Peninsular India. *Bulletin of the Seismological Society of America*, **97**: 318-330.

Johnston, A.C. and Kanter, L.R. (1990). Earthquakes in stable continental crust. *Scientific America*, **262**: 68-75.

Jouanne, F., Mugnier, M., Pandey, M., Gamond, J., leFort, P., Surruier, L., Vigny, C. and Avouac, J.P. (1999). Oblique convergence in the Himalaya of western Nepal deduced from preliminary results. *Geophysical Research Letters*, **26**: 1933-1936.

Kanamori, H. (1977). The energy release in great earthquakes. *Journal of Geophysical Research*, **82**: 2981-2987.

Kayal, J.R. (2000). Seismotectonic study of the two recent SCR earthquakes in Central India. *Journal of Geological Society of India*, **55**: 123-138.

Khattri, K.N. (1987). Great earthquakes, seismicity gaps and potential for earthquake disaster along the Himalaya Plate boundary. *Tectonophysics*, **138**: 79-92.

Khattri, K.N. (1992). Seismic hazard in Indian Region. *Current Science*, **62**: 109-116.

Khattri, K.N. and Tyagi, A.K. (1983). Seismicity patterns in the Himalayan plate boundary and identification of the areas of high seismic potential. *Tectonophysics*, **96**: 281-297.

Khattri, K.N., Rogers, A.M., Perkins, D.M. and Algermissen, S.T. (1984). A seismic hazard map of India and adjacent areas. *Tectonophysics*, **108**: 93-108, 111-134.

Kumar, R.M. and Bhatia, S.C. (1999). A new seismic hazard map for the Indian plate region under the global seismic hazard assessment programme. *Current Science*, **77**: 447-453.

Kumar, S., Wesnousky, S.G., Rockwell, T.K., Briggs, R.W., Thakur, V.C. and Jayangondaperumal, R. (2006). Paleoseismic evidence of great surface rupture earthquakes along the Indian Himalaya. *Journal of Geophysical Research*, **111**, doi:10.1029/2004JB003309.

Lavé, J. and Avouac, J.P. (2000). Active folding of Fluvial terraces across the Siwalik Hills, Himalayas of central Nepal. *Journal of Geophysical Research*, **105**: 5735-5770.

Lay, T., Kanamori, H., Ammon, C.J., Nettles, M., Ward, S.N., Aster, R.C., Beck, S.L., Bilek, S.L., Brudzinski, M.R., Butler, R., DeShon, H.R., Ekström, G., Satake, K. and Sipkin, S. (2005). The great Sumatra-Andaman earthquake of 26 December 2004. *Science*, **308**: 1127-1133.

Liu, L. and Zoback, M. (1997). Lithospheric strength and intraplate seismicity in the New Madrid seismic zone. *Tectonics*, **16**: 585-595.

Mandal, P., Rastogi, B.K. and Sarma, C.S.P. (1998). Source parameters of Koyna earthquakes. *Bulletin of Seismological Society of America*, **88**: 833-842.

Manga, M. and Wang, C.-Y. (2007). Earthquake hydrology. *In:* G. Schubert (editor), *Treatise on Geophysics*, Vol. 4, Elsevier, pp. 293-320.

Middlemiss, C.S. (1910). The Kangra Earthquake of 4 April 1905. *Memoir Geological Survey of India*, **38**: 1-405.

Molnar, P. (1987). The distribution of Intensity Associated with the 1905 Kangra earthquake and Bounds on the Extent of Rupture. *Journal of the Geological Society of India*, **29**: 221-229.

Molnar, P. (1990). A review of the seismicity and the rates of active underthrusting and deformation at the Himalaya. *Journal of Himalayan Geology*, **1**: 131-154.

Molnar, P. and Pandey, M.R. (1989). Rupture zones of great earthquakes in the Himalaya region. *Proceedings of Indian Academy of Science, Earth Planetary Science*, **98**: 61-70.

Nair, K.K.K., Jain, S.C. and Yedekar, D.B. (1985). Geology, structure and tctonics of the Son-Narmada-Tapti lineament zone. *Records of Geological Survey of India*, **117**: 138-147.

Nakamura, Y. (1984). Development of the earthquake early-warning system for the Shinkansen, some recent earthquake engineering research and practical in Japan. The Japanese National Committee of the International Association for Earthquake Engineering, Tokyo, Japan, pp. 224-238.

Ni, J. and Barazangi, M. (1984). Seismotectonics of the Himalayan collision zone: Geometry of the underthrusting Indian plate beneath the Himalaya. *Journal of Geophysical Research*, **89**: 1147-1163.

Oldham, R.D. (1899). Report on the great earthquake of June 12, 1897. *Memoir of Geological Survey of India*, **29**: 1-379.

Oldham, R.D. (1926). The Kutch earthquake of 16th June 1819 with a revision of the great earthquake of the 12th June 1897. *Memoir of Geological Survey of India, Calcutta*, **46**: 80-147.

Pandey, M.R. and Molnar, P. (1988). The distribution of intensity of the Bihar-Nepal earthquake of 15 January 1934 and bounds on the extent of the rupture zone. *Journal of Geological Society of Nepal*, **5**: 22-44.

Pandey, M.R., Tandukar, R.P., Avouac, J.P., Lave, J. and Massot, J.P. (1995). Interseismic strain accumulation on the Himalayan Crustal Ramp (Nepal). *Geophysical Research Letters*, **22**: 751-754.

Pandey, M.R., Tandukar, R.P., Avouac, J.P., Vergne, J. and Heritier, Th. (1999). Seismotectonics of the Nepal Himalaya from a local seismic Network. *Journal of Asian Earth Sciences*, **17**: 703-712.

Powell, C.M. (1979). A speculative tectonic history of Pakistan and surroundings: Some constraints from the Indian ocean. Geodynamics of Pakistan, Geological Society of Pakistan, pp. 5-24.

Rajendran, C.P. and Rajendran, K. (2005). The status of central seismic gap: A perspective based on the spatial and temporal aspects of the large Himalayan earthquakes. *Tectonophysics*, **395**: 19-39.

Rajendran, C.P., Rajendran, K., Duarah, B.P., Baruah, S. and Earnest, A. (2004). Interpreting the style of faulting and paleoseismicity associated with the 1897 Shillong, northeast India, earthquake: Implications for regional tectonism. *Tectonics*, **23**: TC4009, doi:10.1029/2003TC001605.

Rajendran, K., Rajendran, C.P., Thakker, M. and Tuttle, M.P. (2001). The 2001 Kutch (Bhuj) earthquake: Coseismic surface features and their significance. *Current Science*, **80**: 1397-1405.

Rastogi, B.K., Chadha, R.K., Sarma, C.S.P., Mandal, P., Satyanarayana, H.V.S., Raju, I.P., Kumar, N., Satyamurthy, C. and Rao, A.N. (1997). Seismicity at Warna reservoir (near Koyna) through 1995. *Bulletin of Seismological Society of America*, **87**: 1484-1494.

Richter, C.F. (1958). Elementary Seismology. Freeman, San Francisco, California, 768 pp.

Seeber, L. and Armbruster, J. (1981). Great detachment earthquakes along the Himalayan Arc and long-term forecasting. *In:* D.W. Simpson and P.G. Richards (editors), Earthquake Prediction: An International Review, Maurice Ewing Series. American Geophysical Union, Vol. 4, pp. 259-277.

Seeber, L., Armbruster, J.G. and Quittmeyer, R.C. (1981). Seismicity and continental subduction in the Himalayan arc. Inter-Union Commission on Geodynamics, Working Group, Vol. 6, pp. 215-242.

Seeber, L., Ekstrom, G., Jain, S.K., Murty, C.V.R., Chandak, N. and Armbruster, J.G. (1996). The 1993 Killari earthquake in central India: A new fault in Mesozoic basalt flows? *Journal of Geophysical Research*, **101**: 8543-8560.

Stein, S. and Okal, E.A. (2005). Speed and size of the Sumatra earthquake. *Nature*, **434**: 581-582.

Sukhija, B.S., Rao, M.N., Reddy, D.V., Nagabhushanam, P., Hussain, S., Chadha, R.K. and Gupta, H.K. (1999). Paleoliquefaction evidence and periodicity of large prehistoric earthquakes in Shillong Plateau, India. *Earth and Planetary Science Letters*, **167**: 269-282.

Sukhija, B.S., Rao, M.N., Reddy, D.V., Nagabhushanam, P., Kumar, D., Lakshmi, B.V. and Sharma, P. (2002). Palaeoliquefaction evidence of prehistoric large/great earthquakes in North Bihar, India. *Current Science*, **83**: 1019-1024.

Sykes, L.R. (1971). Aftershock zones of great earthquakes, seismicity gaps, earthquake prediction for Alaska and the Aleutians. *Journal of Geophysical Research*, **76**: 8021-8041.

Talwani, P., Kumarswamy, S.V. and Sawalwede, C.B. (1996). The Re-evaluation of Seismicity Data in the Koyna-Warna Area. Report of the University of South Carolina, South Carolina, USA.

UNDP (2004). Reducing Disaster Risk: A Challenge for Development. United Nations Development Program (UNDP), Bureau for Crisis Prevention and Recovery, United Nations Plaza, New York.

Vigny, C., Simons, W.J.F., Abu, S., Bamphenyu, R., Satirapod, C., Choosakul, N., Subarya, C., Socquet, A., Omar, K., Abidin, H.Z. and Ambrosius, B.A.C. (2005). Insight into the 2004 Sumatra-Andaman earthquake from GPS measurements in Southeast Asia. *Nature*, **436**: 201-206.

Wallace, K., Bilham, R., Blume, F., Gaur, V.K. and Gahalaut, V. (2005). Geodetic constraint of the 1905 Kangra earthquake and interseismic deformation 1846-2001. *Geophysical Research Letters*, **32**: L15307, doi:10.1029/2005GL022906.

Wu, Y.-M. and Kanamori, H. (2005). Experiment on an onsite early warning method for the Taiwan Early Warning System. *Bulletin of the Seismological Society of America*, **95**: 347-353.

Impact of 2004 Tsunami on Housing, Sanitation, Water Supply and Wastes Management: The Case of Ampara District in Sri Lanka

Leclerc Jean-Pierre[*] and Solidarités Staff

1. INTRODUCTION

On 26th December 2004, a massive earthquake registering 9.0 on the Richter scale struck off the coast of Sumatra, Indonesia. This was followed by one of the most deadly tsunamis of the history that radiated through the Bay of Bengal at a rate of more than 500 km/h and directly impacted the coastal areas of Bangladesh, India, Indonesia, Kenya, Malaysia, Maldives, Mauritius, Myanmar, Reunion, Seychelles, Somalia, Sri Lanka, Tanzania and Thailand (Pilapitiya et al., 2006). The wave was around 10 m high, which led to considerable property damages and loss of life. It was not one of the highest waves like the wave of the 1960 tsunami (estimated at 25 m height) in Chile, but killed 1500 persons (Pascual et al., 2009). Sri Lanka was one of the most affected countries, followed by Indonesia, wherein 32,000 people lost their lives or were missing, about 100,000 houses were destroyed and nearly 50,000 houses were suffering damages (Pilapitiya et al., 2006). At the first stage, the population affected by the 2004 Tsunami has been temporarily lodged by their families, neighbors or in special camps built by government or NGOs.

One of the reasons for the extent of the damages by the 2004 Tsunami is due to the fact that the necessary infrastructure for disseminating a warning was not in place and the bulletins sent by the Pacific Tsunami Warning Centre officials to contact officials around the Indian Ocean region was inefficient. Moreover, in some areas, these public officials heard of the impending tsunami but dismissed any risk as there was no prior record of such event in their jurisdiction (Aitchison, 2005).

The objective of this chapter is to present an overview of the damages caused by the 2004 Tsunami, the actions taken to help the population and the measures taken to solve technical problems in the Ampara district of eastern Sri Lanka taking into account the specific characteristics of the area under investigation. The subject under discussion covers a wide social and technical aspect and the purpose of this chapter is not to answer the entire question or to cover all the problems of such situation. However, it provides relevant references related to different aspects which are not fully covered in this chapter such as psychological impact or mathematical modeling of wave propagation. This chapter focuses mainly on water supply, sanitation and waste management problems. However, the objective is not to recall technical procedures for water and wastewater management in the tropics since the literature is abundant on these topics (e.g., Lonhold, 2005; Action contre la faim, 2005).

---

[*]Corresponding Author

2. SIMULATION OF TSUNAMI OCCURRENCE

Tsunamis are waves of water generated by underwater earthquakes of shallow focus during which normal deformations of the sea bed occur. When these waves approach coastal areas where water depth decreases rapidly, the wave energy is concentrated by refraction which, combined with shoaling and local resonance effects, may result in a significant increase of the wave amplitude (Abou-Dina and Hassan, 2006). A lot of work has been done to simulate tsunami generation and propagation. They are mainly based on the linear theory of wave's motion in a fluid domain with a free surface and a horizontal bottom. Several models have been developed using a large range of assumptions from simple linear one-dimensional models to nonlinear multi-dimensional models. All the models require complex mathematical equations and large computing time. Due to the complexity and scarcity of this phenomenon, it is difficult to model and validate it. The 2004 tsunami has been recorded by numerous seismology centers around the world. These data are expected to improve the existing models and to calibrate the free parameters of the model. The propagation of the wave has been simulated using a finite element discretisation scheme to solve Navier-Stokes equations (CEA, 2004). Moreover, these simulations will help to refine the theoretical relation between magnitude, landslide, break characteristics and so on. Nevertheless, the ability to produce fast and accurate forecasts of tsunami is critical and therefore the tradeoff between high speed model runs and numerical errors that are introduced due to finite grid parameters must be evaluated (Burwell et al., 2007). It is important to underline that an efficient detection of the preliminary earthquake signs associated with an accurate online prediction of tsunami propagation and an efficient communication network might save lives and minimize damages to a greater extent. An efficient disaster preparedness project and an accurate awareness of the population about this tsunami disaster would probably efficiently mitigate the impact.

3. OVERVIEW OF AMPARA DISTRICT

A rapport about the Ampara district profile has been established by Consortium of Humanitarian Agencies (2004) and the data presented below are directly taken from this report. The Ampara district has an area of 4431.4 square kilometers and a population of 605,553; Sinhala, Tamil and Muslim are the three major communities living in the district and the ethnic ratio is 41.59% Muslim, 39.33% Sinhala, 18.76% Tamils and 0.32% others. The district consists of 20 Divisional Secretariat areas, 505 Grama Niladhari Divisions and 828 villages. The district is dominated by agriculture, but there is potential for industrial development. The main livelihood in the district is rice cultivation. The Ampara district is in the dry zone and agricultural production is made possible over two seasons per year with the help of irrigation facilities. The Snanayaka Samudraya is the main irrigation tank and several minor tanks are also there. The paddy area is 555,000 hectares and its average production is 250,000 metric tonnes per season, which is 1/5 of the country's requirement. Inland and sea fishing is the next important livelihood and about 17,000 families are involved in fishing.

The water supply system includes supply networks in the urban areas and shallow-wells in rural areas. However, the population living in the area along the coast is mainly composed of fishermen who use the water from shallow wells because they don't have enough money to finance the private connection and water cost. Most of the wells in this area are shallow wells, dug and cased with concrete casings of 1 to 1.5 m diameter and having a depth of 3 to 7 m. They are mainly private wells used for drinking, bathing, washing or other household uses. After the tsunami, it was observed that the salinity of the water inside these wells was very high. Therefore, the people stopped using them which resulted in an increased use of non-affected wells.

4. DAMAGES IN AMPARA DISTRICT

As mentioned earlier, Sri Lanka was one of most affected countries by the 2004 Tsunami: 32,000 people lost their lives or were missing, about 100,000 houses were destroyed, and nearly 50,000 houses suffered damages. The water supply system has been considerably damaged. The number of damaged wells is not accurately known. According to different sources, the value ranges from 12,000 to 100,000 in the worst case (ADB et al., 2005). This situation is serious for the population in Sri Lanka because 80% of the rural domestic water supply needs is provided from dug wells and tube wells. Ampara district was particularly affected severely; there were a total of 8,000 dead people, 2,300 missing and 127,500 displaced persons. In addition, 18,000 wells in this district were unusable, since they had been contaminated by seawater, "dead body", hazardous solid wastes and debris. The houses, toilets, shallow wells and water supply networks were fully destroyed in a large area along the ocean. The groundwater was polluted by solid wastes, fecal coliforms issues from the flooded septic tanks and salt. It led to a migration of the affected population toward inland. This resulted in a lack of housing, water supply and sanitation in the non-affected areas and a need of camps for the displaced people.

5. INVOLVEMENT OF NGOS, LOCAL AUTHORITIES AND RESCUE ORGANIZATION

The number of NGOs involved in the rescue after the 2004 Tsunami was the most important record since the history of humanitarian help. The specific time between Christmas and New Year's eve, the fact that Sri Lanka and Thailand are also known for tourism, and the incredibly high media coverage contributed to the public mobilization that was much larger than during Somalia or Yugoslavia wars. The amplitude of the seism and the lack of coordination due to the specific situation were responsible for a delay in the rescue, hopefully without dramatic consequences. One year after the Tsunami, the whole donations were estimated at more than 10 million dollars. This level has never been recorded before in the humanitarian history (Pascual et al., 2009). Before talking about technical aspects, it should be pointed out that the most important part of the rescue work was focused on physiological support to the victims and the identification of the recovered bodies. The fast and extensive decomposition of DNA under the environmental conditions in Southeast Asia made this identification difficult. The decomposition was even sped up when the bodies have stayed long time in salty water. The DNA analysis was also time-consuming and labor intensive task which explain the poor level of identification based on the DNA analysis (Lessig et al., 2006). A total of 1724 victims of the tsunami, including 669 females, 424 males and 631 children, were given counseling. Eight psychologists carried out psychotherapy and counseling. They identified that 8% of the clients were suffering from severe post-traumatic stress disorder and 7% were diagnosed as vulnerable to severe post-traumatic stress disorder. Several therapies have been used successfully to improve the situation, but a pressing need for long-term psychological care has been pointed out (Ranawaka and Dewaraja, 2006).

6. SOLID WASTE CLEANING

The extent of debris created as a result of the tsunami, particularly from destroyed houses, was very significant. Based on the assumption that the average weight of debris (mainly brick, concrete, wood and metal sheets) per destroyed house is in the range of 3000 kg for the 100,000 destroyed houses, there would be about 300 million kg of debris without making allowances for lost household goods, furnishings,

Fig. 1 Typical situation after the tsunami where people tried to save material for future rebuilding.

contents of shops, thousands of vehicles and boats, trees, destroyed road and bridges (Pilapitiya et al., 2006).

In addition to the difficulty due to the quantity of solid wastes, one of the major problems was that people tried to save material which represents an insignificant wealth as shown in Fig. 1. It was also necessary to pick out different types of material. One of the solutions to limit the amount of debris was to use the broken bricks and concrete as a foundation material for road rebuilding.

However, the specificity of the tsunami compared to an earthquake disaster is that large amounts of solid wastes were located under the sea in the fishing area. The evacuation of these underwater wastes necessitated developing specific programs involving boats, scuba divers, and lifting bags to remove trees, bricks, concrete blocks, damaged boats from the sea. This action was the first step to restart the fishing activity that was one of the major economic activities of the affected area. Unfortunately, as these types of activities are not part of the "usual" programs of the NGOs, only few NGOs including Solidarités implemented this type of project. This project included the following steps:

- Scanning of the area using a long rope sliding on the bottom of the sea dragged by boats; scuba divers follow the rope to detect and mark the debris.
- Once debris is found (tree, concrete well, fishing canoe, etc.), the divers fix it to the lifting bag, and then fill the bag with air in order to lift the debris up, and to move it out of the fishing area. This step is illustrated in Fig. 2. The lifting bags are filled with air—thanks to a hose, compressor and generator system installed on the boat.
- Debris is dragged onto the shore with the help of local fishermen.
- After cleaning, the work is validated through a check fishing done by the local fishermen.

This project necessitates training the local people to scuba diving methods and safety rules. The debris are thus collected and transported towards a solid waste disposal.

Fig. 2 A typical example of tree removal from the sea.

7. SANITATION STATUS

During the 2004 Tsunami, all the toilets located in the affected areas were destroyed and septic-tanks were filled with brackish water (i.e., seawater). This condition resulted in three major problems:

- Pollution of shallow wells by high level of fecal coliforms,
- Increase of diseases, and
- Physiological impact on the population due to the difficulties to find toilets and to respect privacy of population.

The impact of sanitation project after such a disaster is thus of the primary importance to reduce fecal pollution and diseases and to restore dignity. The first step is to empty and clean septic tanks; this labor raises three major problems. First, it is necessary to find a large number of tank trucks, second this is a time consuming and labor intensive task, and third it requires finding proper waste disposal sites.

In a few days after the 2004 Tsunami, the lack of tank trucks was already a crucial problem even if the local authorities made lots of efforts to place trucks to the NGO's disposal. All the tank trucks belonging to private companies had been rented in a few weeks. Some NGOs imported tank trucks from abroad, but that was time, money and administrative formalities consuming. This task monopolized numerous unskilled workers and drivers for a long time. Finally, wastewater disposal sites were selected by the local authorities as a priority. Here, it should be pointed out that the Sri Lankan authorities took all the precautions to select disposal sites far from habitation and with a limited risk of groundwater contamination.

The second step was to rebuild toilets for the inhabitants. Two strategies are poles apart: the first one consists of building numerous temporary toilets and the second one consists of building less permanent toilets. The first solution allows procuring in little time with low-cost and simple technical materials available everywhere, and sufficient toilets for a large population density as it is often required when there are many displaced persons. Unfortunately, the experience has shown that the maintenance essential for such solution is rarely assumed by the population, local authorities and NGOs. This solution is certainly the only one for large density refugees' camps but not for the present case. In Ampara district, the population was displaced from few hundred meters to few kilometers from its original housing. Moreover, the quality of the toilets is something important for the tsunami victims who are physiologically weak. Permanent toilets are preferable in this case though they are costly and their building requires more time. The construction of permanent toilets requires numerous administrative, socio-cultural (religious, cultural, etc.) and technical precautions. All these aspects are described in detail in the book published by "Action contre la faim" (2005). Figure 3 shows the septic tank design and Fig. 4 shows an example of a typical permanent toilet for four-person families.

Fig. 3 Internal structure of a septic tank including solid and liquid compartments.

In the present case, the main problem concerned the toilets in the "temporary camps" because their duration was not well defined; it depended on the time required to rebuild housing. This duration was initially strongly underestimated as equal to a few months due to only natural "communication and politic reasons" (the government wanted to reduce as much as possible the official time duration to rehouse the population), but the camps were finally installed for an expected duration of few years. This duration was more realistic for the situation and it justified the choice of permanent toilets. These

Fig. 4 Typical permanent toilets built for a private housing for a 4-rescue family.

toilets have also been installed in medical centers, schools and private housing. It should be pointed out that the positive impact on the mind and dignity has been claimed by the people. In all the cases, a sufficient water supply system should be installed to ensure high level of hygiene and to prevent seats of diseases.

8. WATER SUPPLIES

Water networks were severely damaged by the 2004 Tsunami, whereas the shallow water wells were polluted by "dead bodies", hazardous solid wastes, debris and salt from the seawater. All the solid materials inside the well can be manually removed and pathogens can be destroyed by chlorine treatment, but the presence of salt inside the water is a more difficult problem to evaluate and solve. Therefore, the water supply rehabilitation was preceded by water distribution and a detailed study of salt pollution of the wells.

Figure 5 illustrates the situation faced by the NGOs during well cleaning project. Even the accessibility for water analysis was problematic, while the cleaning of the surrounding area of the well needed heavy labor due to a large quantity of solid wastes.

8.1 Water Trucking Supply

Because of the complete destruction of water supply system in the tsunami affected area and the increasing of water need in other zones due to displaced persons, it was necessary to supply water to the population by water trucking. Since only a few tank trucks were available, bladder tanks were installed on each truck under the supervision of a technician. The equipment onboard included one bladder tank of 5 or 10

Fig. 5 Typical water well in the area: accessibility for water analysis was difficult due to the large amount of solid wastes surrounding the well.

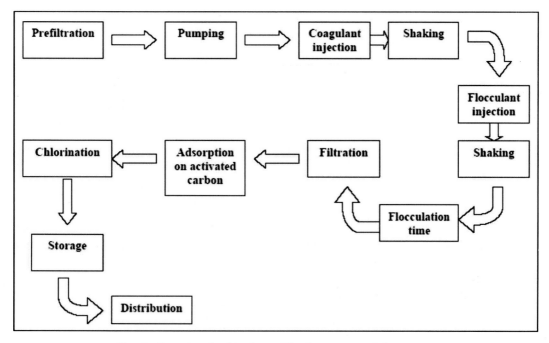

Fig. 6 Steps involved in the purification process of river water.

m^3, one motor pump and fittings, one pool tester for chlorine, one chlorination kit and a set of monitoring sheets. Several water reservoirs of capacities between 500 and 2000 liters were also installed in the strategic places, viz., schools, medical centers, displaced camps, and high density population areas. These reservoirs were regularly filled by purified water pumped from a river. The purification process used is described in Fig. 6. The purification process should be as simple as possible and should ensure robustness, easy maintenance and limited consumable compounds. This activity was run until the quality of underground water got improved. Sixty cubic meters of drinking water were delivered every day, seven days a week. Residual chlorine was checked before loading. In the case of chlorine content less than 0.5 mg/L, complementary chlorination was done directly in the tank before transportation. The average distance from the water source to the delivery area was about 25 km. The overall cost per cubic meter was ∈ 1.67 including the following:

- Fuel costs for the trucks and pumps,
- Trucks rental,
- Rental or depreciation of other equipment that are fitted on the truck (e.g., pumps and fittings),
- Truck and equipment maintenance costs, and
- Trucking personnel charge (excluding staff at the water source).

The first intervention allowed the population to be provided with a sufficient amount of water, but it was a temporary solution which required heavy logistic and hard labor. The main problem was to reduce progressively the distribution when the rehabilitation of the water network started or when shallow wells became utilizable. This should be done through a progressive awareness of the local population to reuse the rehabilitated shallow wells.

8.2 Salt Pollution of Groundwater

If all the solid materials inside the well can be manually removed and pathogens destroyed by chlorine treatment, the presence of salt in the water was more difficult problem to be solved. Preliminary transport analysis showed that salty water coming from the wave has been vertically mixed in the aquifers because of both forced and free convection (Illangasekare et al., 2006). In order to determine the boundaries of the wells' area affected by salt and the long-term evolution of salt concentration in the aquifer, several studies on groundwater salinity have been carried out (Leclerc et al., 2008; Villholth et al., 2005). The mapping of the electrical conductivity (EC) of groundwater showed that EC is high only in the area directly affected by the wave or just close to the limit of the impact. This observation has been confirmed in several places and it is quite probable that the structure of the water table has been deeply damaged. Data analysis revealed that there is a slow natural recovery process, which decreases groundwater EC. The rain seems to have a beneficial effect on this process, but the amplitude of the effect varied from one well to another. The natural recovery process is a combination of downward gravity flow of salty water and lateral flow of fresh water toward the sea.

After classifying the wells according to their behavior, the groundwater EC evolution remains a local phenomenon because two wells located very close to each other may have a different behavior (Leclerc et al., 2008). Therefore, a larger field investigation has been done during nearly two years to extend the study to 200 wells. From the beginning of 2005 to September 2006, the electrical conductivity (EC) of these wells has been monitored daily to analyze the evolution of salt concentration. The evolution of groundwater salinity has been plotted in terms of percentage of variation between the EC values at end of September 2006 and the EC values recorded just after the 2004 Tsunami. Figures 7 and 8 show the percentage of variation of the salinity of several wells located in Neelavanai (Fig. 7) and Karaithivu (Fig. 8). The dotted line in these figures indicates the limit of the tsunami wave impact determined by witness interviews and mud marks. The variation of groundwater EC varied considerably from one well to another. It decreased more than 25% for some wells, but also increased in the same order of magnitude for other wells. The wells having same behavior are sometimes located in the same stream line or in the same limited area, but these observations are too random and subjective to lead to definitive conclusions. Unfortunately, the location of the well doesn't have a clear influence on the groundwater salinity evolution.

The modeling of the salt transfer processes to evaluate the time needed to recover a normal salt concentration is complex because of the numerous ways of transfer (Leclerc et al., 2008). Two methods are poles apart. The first one is based on a detailed description of the elementary transfer processes at the laboratory scale as developed by Illangasekare et al. (2006). The advantage of this method takes into account the fundamental processes, but the inconvenience is that its scaling up at the field scale is difficult. The second method is derived from the chemical engineering approach by dividing the field in elementary cells of macro-scale (unlike detailed meshing using Computational Fluid Dynamics codes) and including only the main trends of salt transfers. Figure 9 shows a possible simple model to simulate the salt transfer process. The zone of fresh water polluted by the salt and the salty water zone are simulated by a cascade of elementary cells, each containing either only salt water (close to the sea), either fresh water (far inland) or two distinct salted zones (salty water zone and fresh water zone polluted by salt issued from the tsunami wave). In each zone of each cell, the salt concentration is assumed to be homogenous. For each cell, representing a given area, the pumping rate (Q_p) can be estimated from the water consumption by human in the area. The evaporated water flow rate (Q_+) can be estimated by correlation from the literature. It is proportional to the surface area and it depends on the temperature and water table depth. The fresh water flow rate (Q_i^F) of the groundwater and the salty water flow rate from the sea (Q_i^S) are approximately known to the hydrogeologists. The water and salt mass balance can be written for each

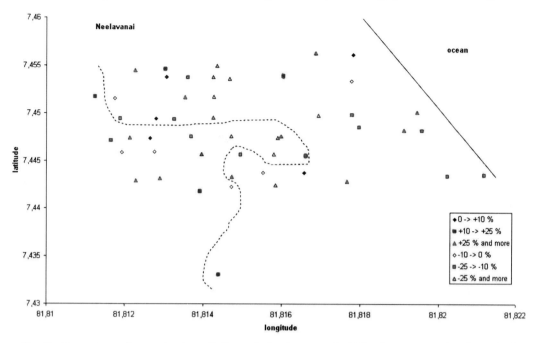

Fig. 7 Percentage of groundwater salinity evolution between the EC values measured at the end of September 2006 and the EC values measured just after the 2004 Tsunami in Neelavanai area.

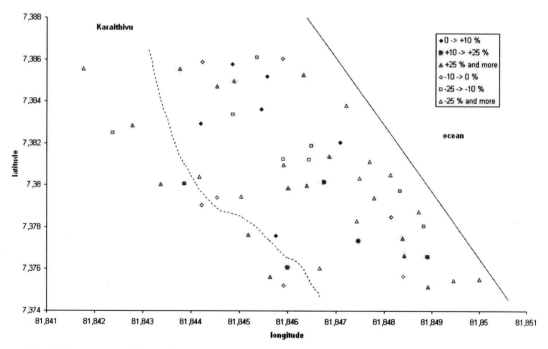

Fig. 8 Percentage of groundwater salinity evolution between the EC values measured at the end of September 2006 and the EC values measured just after the 2004 Tsunami in Karaithivu area.

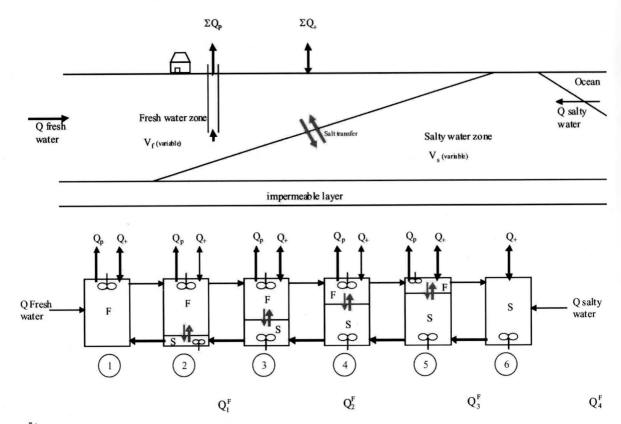

Fig. 9 Simple model for simulating salt transfer in groundwater.

cell. The number of cells should be large enough to limit the numerical dispersion but small enough to obtain reliable values of the parameters. However, even with such a simplified model, it is necessary to have accurate information about the soil properties (e.g., intrinsic permeability, local salt concentration, diffusivity, etc.).

Moreover, the salt was also responsible for negative impacts on agriculture. Some research work has been done to find suitable options to reduce the impact of salt on crop production with the specific objectives to tackle salt contamination and to assess the relationship between salt and crop production. It seems that there is a tendency for the adsorption of sodium in humus complex leading to the toxicity of crops which can be reduced by a high concentration of potassium provided by nutrients rich in potassium (Tchouaffe Tchiadje, 2006).

8.3 Water Supply Network

Rehabilitation of the water supply network does not present a complex technical labor. However, a preliminary solid waste management was required and several kilometers of pipes to rehabilitate the existing infrastructure were necessary. The major problem was to respect the local organization and to rebuild the network in agreement with the cadastral map. The 2004 tsunami created a debate about the minimum distance to the ocean for housing. The answer to this question is fully justified but time consuming. A large distance would create a lack of available land for the population leading to displace location and activities since the concerned people are fishermen. A short distance, on the other hand, would increase the risk in case of a new tsunami event. The rehabilitation has been done with a long-term vision because flow meters have been installed for each house in agreement with the local water board. Although the water was freely distributed after the tsunami, it should be considered for future that each family will pay for the water in order to ensure efficient and economic water supply on a long-term basis.

8.4 Well Cleaning and Rehabilitation

Several guidelines have been proposed to clean wells in a proper manner to prevent mixing between salty water and fresh water [see Appendix A, guidelines developed by UNICEF (2005) or Solidarités]. These guidelines suggest a systematic approach to clean the wells affected by the tsunami. The attempts are to provide water with turbidity low enough for bathing, to minimize the collapse or destruction of wells and also to minimize the potential for irreversible damage to the vital coastal aquifer by pollution from nearby latrines. Only the wells flooded by the 2004 Tsunami have been cleaned and a priority will be given to the wells used as common wells (e.g., wells in public lands, in refugees' camps, etc.). The protocol to clean the well is of primary importance in order to avoid irreversible mixing between fresh water and deep salty water. Nevertheless, the strict respect of this protocol does not ensure the complete success of well cleaning and rehabilitation, because several wells have been damaged despite these precautions.

9. CONCLUSIONS

The 2004 Tsunami was a unique natural event from several points of view. Even if several disasters of the same order of magnitude occur time to time, the inherent nature of the tsunami, the fact that numerous people disappeared and the number of countries directly affected by the tsunami or indirectly affected by death of tourists, have been responsible for an enormous psychological impact on the population. For these reasons, the 2004 Tsunami disaster led to the largest amount of donation ever recorded during the humanitarian history.

The scientific analysis of the impacts of the 2004 Tsunami and of the solutions to solve problems is not straightforward because damages were very important concerning both human and material losses. The rescue programs involved a large spectrum of activities from psychological aid to rebuilding action. The water supply network and the sanitation systems were fully destroyed along the coast. The rehabilitation required first to evacuate wastes generated from the destruction. Even if the procedures for rehabilitation after a disaster are well known and described in several books, two problems were considerably unusual: firstly, the groundwater was contaminated by salt and the salt concentration evolution was complex; the recovery process to lower salt concentration seemed to follow a random process or at least a process not straightforward to understand. Secondly, a large amount of the solid wastes were located under the sea. Since fishing was the major economic activity of the affected area, it was necessary to organize a special waste cleaning program based on the scuba diving and transportation boats management.

As it is better to prevent than to cure, the simplest but most efficient step is to enhance the efforts to promulgate earth science education in schools and colleges. This was well proved by the admirable reaction of a 10-year-old British girl, "Tilly Smith", who was holidaying at Maikhao Beach in Phuket, Thailand two weeks after learning about tsunamis from her Geography teacher "Andrew Kearney" at Danes Hill Preparatory School in Surrey, England (http://en.wikipedia.org/wiki/Tilly_Smith). Her knowledge about the signs of an impending tsunami (i.e., receding shoreline and frothing bubbles on the sea surface) and her quick reaction, coupled with considerable presence of mind, saved nearly one hundred foreign tourists. Because of this warning, the beach was evacuated before the 2004 Tsunami reached the shore, and was one of the few beaches on the Phuket Island with no reported casualties. Furthermore, proper monitoring of tsunami process and their impacts, together with an efficient early warning system can help mitigate this dreadful disaster. The data recorded during tsunamis will be useful for the scientists to validate and improve the numerical models for simulating tsunami wave propagation, which in turn could improve tsunami prediction as well as our knowledge about the tsunami phenomenon. Given an improved communication network, this should appreciably reduce the number of victims of probable tsunamis in the future.

ACKNOWLEDGEMENTS

This book chapter has been prepared with the help of the "Solidarités" members involved in the Sri Lanka humanitarian mission and the "Solidarités" head office. Sodidarités is a humanitary organization which provides aid and assistance to the victims of war or natural disasters in water, sanitation and food security. Solidarités would like to thank all the donors who supported the Sri Lanka mission: Agence de l'Eau Seine-Normandie, European Commission Humanitarian Aid Office (ECHO), Fondation de France, Croix-Rouge Française, Coordination pour le Tsunami, and USAID/OTI as well as all public authorities, private companies and local staff of Ampara district in Sri Lanka for their courage, motivation and hard labor.

REFERENCES

Abou-Dina, M.S. and Hassan, F.M. (2006). Generation and propagation of nonlinear tsunamis in shallow water by a moving topography. *Applied Mathematics and Computation*, **177**: 785-806.

Action contre la faim (2005). Water, Sanitation and Hygiene for Populations at Risk. Herman-Editeurs des Sciences et des Arts, Paris, France.

ADB, JBIC and WB (2005). Preliminary Damage and Needs assessment. Asian Development Bank, the Japan Bank for International Cooperation and the World Bank, Colombo, Sri Lanka.

Aitchison, J.C. (2005). The great Indian ocean tsunami disaster: Guest editorial. *Gondwana Research*, **8(2)**: 107-108.

Burwell, D., Tolkova, E. and Chawla, A. (2007). Diffusion and dispersion characterization of a numerical tsunami model. *Ocean Modelling*, **19**: 10-30.

CEA (2004). Seisme et tsunami de Sumatra 26 décembre 2004. http//www.dase.cea.fr/actu/dossiers_scientifiques/2004-12-26/index.html (accessed in January 2009)

Consortium of Humanitarian Agencies (2004). District Profile—Ampara. http://www.iicp.at/projects/slanka/reconstruction/ampara_profile.pdf (accessed in January 2009)

Illangasekare, T., Tyler, S.W., Clement, T.P., Villholth, K.G., Perera, A.P.G.R.L., Obeysekera, J., Gunatilaka, A., Panabokke, C.R., Hyndman, D.W., Cunningham, K.J., Kaluarachchi, J.J., Yeh, W.W.-G., Genuchten, M.T.V.

and Jensen, K. (2006). Impacts of the 2004 tsunami on groundwater resources in Sri Lanka. *Water Resources Research*, **42**, W05201, doi:10.1029/2006WR004876.

Leclerc, J.P., Berger, C., Foulon, A., Sarraute, R. and Gabet, L. (2008). Tsunami impacts on shallow groundwater on Ampara district in Eastern Sri Lanka: Conductivity measurements and qualitative interpretations. *Desalination*, **219(1-3)**: 126-136.

Lessig, R., Thiele, K. and Edelmann, J. (2006). Tsunami 2004: Experiences, challenges and strategies. *International Congress Series*, **1288**: 747-749.

Lonhold, J. (2005). Water and Wastewater Management in the Tropics. IWA Publishing, International Water Association (IWA), London, U.K.

Panabokke, C.R. and Perera, A.P.G.R.L. (2005). Groundwater Resources of Sri Lanka. Water Resources Board Report, Colombo, Sri Lanka.

Pascual, A., Dagot, L., Vallée, B. and Guéguen, N. (2009). Compliance without pressure, mediatization of a Tsunami and charitable donation: Compared effectiveness of the door-in-the-face and "you are free of…" techniques. *Revue Européenne de Psychologie appliqué*, **59(1)**: 79-84.

Pilapitiya, S., Vidanaarachchi, C. and Yuen, S. (2006). Effects of the tsunami on waste management in Sri Lanka: Guest Editorial. *Waste Management*, **26**: 107-109.

Ranawaka, D.S. and Dewaraja, R. (2006). Tsunami counseling project of the Sri Lanka National Institute of Professional Counselors. *International Congress Series*, **1287**: 79-81.

Tchouaffe Tchiadje, N.F. (2006). Strategies to reduce the impact of salt on crops (rice, cotton and chili) production: A case study of the tsunami affected area of India. *Desalination*, **206**: 524-530.

UNICEF (2005). Guidelines for the Rehabilitation of Tsunami Affected Wells. http://www.humanitarianinfo.org/srilanka/docs/watsan/Guidelines_on_well_rehabilitation-nov05.pdf (accessed in January 2009).

Villholth, K.G., Amerasinghe, P.H., Jeyakumar, P., Panabokke, C.R., Weerasinghe, M.D., Amalraj, N., Prathepaan, S., Bürgi, N., Lionelrathne, D.M.D.S., Indrajith, N.G. and Pathirana, S.R.K. (2005). Tsunami Impacts on Shallow Groundwater and Associated Water Supply on the East Coast of Sri Lanka. IWMI Report, International Water Management Institute (IWMI), Colombo, Sri Lanka, 68 pp.

APPENDIX A

Typical Cleaning Guidelines for Shallow Wells (UNICEF, 2005 or Solidarités)

STEP 1: Fill on the report sheet the general data including:

- DS Division
- Name of the village
- Implementing agency
- Person in charge
- Well's code
- Name/Location address
- Coordinates (use the GPS provided)
- Distance from the closest latrine
- Private(1)/Public (2) well
- Has been flooded by the tsunami (yes/no)
- Covered (yes/no)
- Presence of apron/drainage canal (yes/no)
- Water lifting system : pump (1), bucket (2)
- Date

STEP 2: Clear the area surrounding the well. Remove the debris, roots, etc. Fill in any holes and low spots around the well and compact the fill. This step should implicate the beneficiaries and to inform them that debris and dirty items around the well give direct contamination to the water by penetrating the soil during rainy weather.

STEP 3: Clean out the floating debris using a sieve or bucket. A hole will be dug at least 15 m far from the well to bury the debris.

STEP 4: Measure the well's details as shown below and fill the report sheet.

- Height of ring above ground (h)
- Inner diameter of the well (d)
- Total depth of the well (D)
- Depth of water in the well (H)
- Water column ($D - H$)

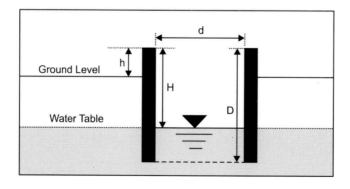

STEP 5: Measure the initial electrical conductivity (in µS/cm) of groundwater at the top and on the bottom, pH and turbidity.

STEP 6: Estimate the volume of water in the well by knowing the water level and inlet diameter.

STEP 7: Pump the water with pump intake placed close to the water surface to avoid sand intrusions and turbidity increase until a two feet high water column in the well is reached (never empty the well). The pump discharge line should be at least 15 meters away from the well, on the seaward side in order to reduce the potential of contaminated water recharging the well.

STEP 8: Remove the bottom debris manually. If the well recovers too fast, get out of the well and pump again until a minimum of two feet high water column is maintained so that all the debris can be removed during this time.

STEP 9: Clean the inside faces of the well, using scrubbing brush. Remove the floating debris, if necessary. Pump one more time if the turbidity has increased a lot and if the well has recovered partly.

STEP 10: Do chlorination shock: the chlorination concentration must be at 50 mg/L and dosed according to the volume of water in the well (V_{total}). The chlorine should be mixed properly inside the well.

STEP 11: Once cleaned and chlorinated, the well must be marked with the marking approved by the NGOs coordinating the well cleaning activity. Conductivity (EC), pH, chlorine concentration and turbidity of the groundwater need to be controlled after two weeks and the well cleaning procedure needs to be repeated until the values of these water quality parameters are acceptable.

Impact of December 2004 Tsunami on Indian Coasts and Mitigation Measures

Sujatha C.H.* and Nify Benny

1. INTRODUCTION

Tsunamis are among the most terrifying natural hazards known to man and have been responsible for tremendous loss of life and property. For example, the Indian coastline experienced the most devastating tsunami in recorded history on 26 December 2004. The tsunami was triggered by an earthquake of magnitude 9.3 M_w at 3.316° N, 95.854° E off the coast of Sumatra in the Indonesian Archipelago at 06:29 hours making it most powerful in the world in the past 40 years (DOD, 2005; Rossetto et al., 2007). Though various natural hazards, viz., droughts, flash floods, cyclones, landslides, and snowstorms had caused great threats to the Indian subcontinent, tsunamis were rather unknown. Other dangers include frequent summer dust storms, which usually track from north to south and cause extensive property damage in North India. Many powerful cyclones, including the 1737 Calcutta cyclone, the 1970 Bhola cyclone and the 1991 Bangladesh cyclone have led to widespread devastation along parts of the eastern coast of India and neighboring Bangladesh (http://en.wikipedia.org/wiki/Natural_disasters_in_India). Widespread death and property destruction were reported each year in exposed coastal states such as Andhra Pradesh, Orissa, Tamil Nadu, and West Bengal. India's western coast, bordering the more placid Arabian Sea, experiences cyclones only rarely; cyclones mainly strike Gujarat and, less frequently, Kerala (http://en.wikipedia.org/wiki/Natural_disasters_in_India). Floods are the most common natural disaster in India. The heavy southwest monsoon rains cause rivers to distend their banks, flooding surrounding areas. Many of the above hazards are related to the climate of India and cause massive losses of Indian life and property.

Due to the destructive potential of tsunamis, they have notable impact on the social and economic sectors of our societies. Tsunamis can spawn when the sea floor abruptly deforms and vertically displaces the overlying water (DOD, 2005). There are three mechanisms responsible for such a deformation: (a) earthquakes with epicenters located below the ocean floor which can make the floor vibrate; (b) mudslides on the ocean floor, particularly on the continental slope, can suddenly change the shape of the ocean floor; and (c) volcanic explosions, either on the ocean floor or on the nearby continent, can lead to shaking of the floor, or during explosion huge quantities of ash that accompanies rapidly on the ocean floor. The above-mentioned perturbations of the ocean floor create a disturbance of the ocean surface, which then propagates as a shallow-water wave (Shetye, 2005). A wave becomes a shallow-water wave when the ratio between the water depth and its wavelength tends to be very small and the rate at which a wave loses its energy is inversely related to its wavelength.

---

*Corresponding Author

The intent of this chapter is to highlight the devastation caused by 2004 Tsunami along the Indian coasts and the major mitigation measures adopted thereafter. Firstly, an overview of various studies and surveys conducted related to 2004 Tsunami is presented followed by a brief discussion of the destructions caused by this tsunami in the shorelines of Kerala, Tamil Nadu, Andhra Pradesh, Pondicherry and Andaman and Nicobar Islands. Lastly, the rehabilitation measures implemented by the governments and various agencies of each state are discussed.

2. TSUNAMI: MECHANISMS AND HISTORICAL PERSPECTIVE

Tsunami is a series of waves created when a body of water, such as an ocean, is rapidly displaced. Earthquakes, mass movements above or below water, some volcanic eruptions and other underwater explosions, landslides, underwater earthquakes, large asteroid impacts and detonation of nuclear weapons at sea—all have the potential to generate a tsunami (http://en.wikipedia.org/wiki/Tsunami). "Tsunami" is originally a Japanese word, "*Tsu*" means *Harbour*, and "*Nami*" means *wave*. It is widely known only in the Pacific Ocean and occasionally seems to have occurred in the past in the Indian and Atlantic Oceans (Ramaswamy and Kumanan, 2005). These ferocious tsunami waves are formed as the displaced water mass, which acts under the influence of gravity, attempts to regain its equilibrium and the size of the resulting tsunami waves is determined by the quantum of the deformation of the sea floor. More the vertical displacement, greater will be the size of the waves. As a rule, all earthquakes do not produce tsunamis. When large areas of the sea floor elevate or subside, a tsunami can be created. Tsunami and tides both produce waves of water that move inland, but in the case of tsunami the inland movement of water is much greater and lasts for a longer period, giving the impression of an incredibly high tide. Tsunami has much smaller amplitude (wave height) offshore, and a very long wavelength (often hundreds of kilometers long), forming only a slight swell usually about 300 mm above the normal sea surface (http://en.wikipedia.org/wiki/Tsunami). A tsunami can occur at any state of the tide and even at low tide—will still inundate coastal areas if the incoming waves surge high enough. As the tsunami approaches the coast and the waters become shallow, the wave is compressed due to wave shoaling and its forward travel slows below 80 km/hour (http://en.wikipedia.org/wiki/Tsunami). Thus, tsunami can be defined as a catastrophic ocean wave, usually caused by a submarine earthquake, by an underwater or coastal landslide or by the eruption of a volcano (http://www.britannica.com/EBchecked/topic/607892/tsunami).

In the Pacific Ocean, 79 tsunamis were observed with 117 casualties but with extensive damages to properties. According to the Tsunami Laboratory in Novosibirsk, during the 101-year period from 1900 to 2001 (DOD, 2005), 796 tsunamis were observed or recorded. Nearly nine tsunamis caused widespread destruction throughout the Pacific. In the Indian Ocean, overall 63 tsunamis seem to have occurred between 1797 and 1977 AD and amongst which eight tsunamis have been triggered by earthquakes of magnitude more than 8 (Kumanan, 2005). Tsunamis generated in the Indian Ocean pose a great threat to all the countries of the region, viz., Indonesia, Thailand, India, Sri Lanka, Pakistan, Iran, Malaysia, Myanmar, Maldives, Somalia, Bangladesh, Kenya, Madagascar, Mauritius, Oman, Reunion Island (France), Seychelles, South Africa and Australia (DOD, 2005). The earthquake of magnitude 8.25 Richter Scale Units which occurred on 28 November 1945 near Karachi created large waves of height 11 to 11.5 m in the Kutch region. At 00:58 GMT on 26 December 2004, a massive earthquake of magnitude between 9.1 and 9.3 struck the coastal area off northern Sumatra in Indonesia. A number of aftershocks also occurred, some of magnitude 7.1. These earthquakes triggered tsunamis that affected Indonesia and neighboring countries in Asia (including India, Malaysia, Maldives, Sri Lanka, and Thailand) and the east coasts of Africa (including Somalia and Yemen) (Fig. 1), causing serious damage to the coastal areas and small islands (e.g., DOD, 2005; Rossetto et al., 2007; Srinivas and Nakagawa, 2007). Rossetto et al.

Fig. 1 Countries most affected (shown in red dots) by the 2004 Indian Ocean Tsunami (*Source:* Wikipedia, http://en.wikipedia.org/wiki/2004_Indian_Ocean_earthquake#Countries_affected).

(2007) summarize the observations of lifeline performance, building damage and its distribution, and the social and economic impacts of the tsunami made by the Earthquake Engineering Field Investigation Team (EEFIT) in Thailand and Sri Lanka (EEFIT, 2006). Australia and the countries of Europe had large numbers of citizens traveling in the region on holiday. Both Sweden and Germany lost over 500 citizens each in the disaster. The U.S. Geological Survey (USGS) initially recorded the toll as 283,100 killed, 14,100 missing, and 1,126,900 people displaced. However, more recent analysis compiled by the United Nations lists a total of 229,866 people lost, including 186,983 dead and 42,883 missing (USGS, 2005). The figure excludes 400 to 600 people who are believed to have perished in Myanmar, which is more than their government's official figure of only 61 dead. Measured in lives lost, this is one of the ten worst earthquakes in recorded history, as well as the single worst tsunami in history. The tsunami caused serious damage and deaths as far as the east coast of Africa, with the furthest recorded death due to the tsunami occurring at Rooi Els in South Africa, 8,000 km (4,971 miles) away from the epicentre (http://www.geocities.com/unsasdsu/africatsunami.html). In total, eight people in South Africa died due to abnormally high sea levels and waves.

In the case of India, the affected areas were limited in the geographical area as well as the extent of socioeconomic and environmental damage. The waves seriously affected the coastal districts of three states, Tamil Nadu, Andhra Pradesh and Kerala (http://en.wikipedia.org/wiki/Effect_of_the_2004_Indian_Ocean_earthquake_on_India). Among these states Tamil Nadu was the worst affected. The impact was huge in terms of human loss, natural resources and basic livelihood assets. Approximately 2,260 km of the coastal area (besides the entire Nicobar Islands) was affected and the fisheries sector bore the brunt of the tsunamis accounting for 85% of the damages. A tsunami run-up survey was taken up by the National Institute of Ocean Technology (NIOT) and ICMAM Project Directorate of Department of Ocean

Development, Chennai and the National Geophysical Research Institute, Hyderabad along the severely affected Tamil Nadu coast and the Andaman and Nicobar Islands (GSI, 2005). The survey concentrated on the maximum run-up heights and distances, average run-up heights and areas of inundation, flow patterns of run-up and run down, and recording of eyewitness accounts. The maximum tsunami run-up height is defined as the vertical water surface elevation reached by the tsunamis above sea level. It was measured using the standard surveying instrumentation. The survey results conclude that tsunami run up heights along Tamil Nadu, Kerala and Andaman & Nicobar Islands show variation of tsunami heights between 1.5 and 7 m. Most of the loss of lives and damage to property and infrastructure were generally observed within 100 m distance from the seashore. Large variations in tsunami inundation and associated loss of life and property were seen even within small differences in local run-up and coastal topography.

3. AN OVERVIEW OF THE STUDIES ON 2004 TSUNAMI

3.1 Studies Conducted in Foreign Countries

Using an integrated approach, tsunami affected land, vegetation and inhabitants were assessed by Mattsson et al. (2009) to evaluate the potential to restore and protect coastal land in the context of Kyoto Protocol's Clean Development Mechanism in Hambantota district in the south-eastern part of Sri Lanka. Firstly, assessments of the status of the tsunami-affected area were carried out by collecting soil and well water samplings for carbon and salinity analysis. Secondly, identification of potential tree species for carbon sequestration and sustainable development was conducted to determine carbon stock and suitability to grow under the prevailing conditions. In addition, interviews to understand the local people's perception of forest plantations and land use were conducted. The results showed that the resilience process of salt intruded lands from the 2004 Asian tsunami has progressed rapidly with low salinity level in the soils 14 months after the event, while the well water showed evidence of salinity contamination. Studies were conducted by Leclerc et al. (2008) for a better understanding of the mechanism of the aquifer contamination in the Ampara district of eastern Sri Lanka after the tsunami and to forecast evolution of groundwater conductivity with time. After the tsunami, it was observed that the salinity of the water inside the private wells used for drinking, bathing, washing or any household use was very high.

An article by Srinivas and Nakagawa (2007) emphasizes the cyclical interrelations between environments and disasters, by studying the findings and assessments of the Indian Ocean earthquake and tsunami of 26 December 2004. It specifically looks at four key affected countries namely Maldives, Sri Lanka, Indonesia, and Thailand and found that many of the activities associated with tourism, fisheries and agriculture sector of these countries will take many years to recover from the tsunami impacts. Tanaka et al. (2007) explored the effects of coastal vegetation on tsunami damage based on field observations carried out after the Indian Ocean tsunami on 26 December 2004. Study locations covered about 250 km (19 locations) on the southern coast of Sri Lanka and about 200 km (29 locations) on the Andaman coast of Thailand. Imamura (2004) introduced a new information system called TIMING (Tsunami Integrated Media Information Guide) for monitoring tsunamis in Japan for advanced and real time information. Papathoma et al. (2003) estimated the tsunami vulnerability for the Herakleio coast in Crete. Bandibas et al. (2003) developed new software – Geohazard View – for the interactive management of geological hazard maps. Jaffe and Gelfenbaum (2002) created geological records not only of a single tsunami but also the effects of past tsunamis. Walsh et al. (2000) compiled a tsunami hazard map based on the modeled inundation for the Southern Washington Coast. Minoura et al. (1994) studied the sediment deposited by tsunamis in the lacustrine sequence of the Sanriku Coast, Japan. Dawson (1994) found that the geomorphologic processes associated with tsunami run-up and backwashes are highly complex.

Tsunami hazard along the Morocco coast was evaluated by Alami and Tinti (1991) by comparing tsunami data with the set of available earthquake data.

3.2 Studies Conducted along Indian Coasts

A study conducted by Chidambaram et al. (2008) indicated that tsunami waves encroached the coastal ecosystem and changed the geomorphology, sediment characteristics and water quality of Tamil Nadu region. The area from Parangipettai to Pumpuhar was included in the study to assess the salinity variation due to the impact of tsunami. Significant variations were observed in apparent resistivity values, due to percolation of sea water into shallow aquifers. Changes in the formation resistivity and formation factor have also been noticed, which indicates salinity increase in aquifers. Babu et al. (2007) investigated the impact of tsunami on texture and mineralogy of a major placer deposit in southwest coast of India. Laluraj et al. (2007) made an attempt to assess the impacts of tsunami on the hydrochemistry of Cochin estuary and observed drastic changes in estuarine water quality (high saline, turbid, cool and nutrient-rich water mass) generated by the tsunami. Chandrasekar et al. (2006a) applied geospatial technologies to evaluate the impact of tsunamis on certain beaches in South India and compiled a hazard map. Kurian et al. (2006) investigated the inundation characteristics and geomorphologic changes resulting from the 26th December 2004 tsunami along the Kerala Coast, India. They noted that river inlets had been conducive to inundation; the devastation was extremely severe there, as the tsunami had coincided with the high tide. Chandrasekar et al. (2006b) reported that the extent of inundation depends mainly upon the nature of the coastal geomorphology. Sheth et al. (2006) found that the tsunami effects varied greatly across different parts of the coast according to the number of waves experienced, inundation distance and height of the waves, and density of the area; noted that the topological and geographical features were more vulnerable than others. The number of lives lost was also influenced by the proximity of habitats to the coastline, exposure to previous disasters, and the local disaster management capability.

Studies made by Srinivasulu et al. (2005) on the effect of tsunami on the sediments along Tamil Nadu coast indicated that the characteristics of Tsunami sediment deposits depend on the shelf sediments of the area. The maximum sediment inundation was observed in the Nagapattinam district of Tamil Nadu coast. Impact of the tsunami of 26th December 2004 on the coral reef environment of Gulf of Mannar and Pak bay in the southeast coast of India was studied by Kumaraguru et al. (2005). Raval (2005) reported severe destruction along the coast of Nagapattinam, South India, primarily because of its geographic setting, which has favoured much inundation. Narayan et al. (2005a) found that the elevated landmass (medu) played a key role in preventing inundation and minimising damage along the coast of Tamil Nadu. It was concluded that the width of continental shelf has played a major character in the pattern of tsunami damage. It was inferred that the width of the continental shelf and the interference of reflected waves from Sri Lanka and Maldives Islands with direct waves and receding waves were responsible for the intense damage in Nagapattinam and Kanyakumari districts, respectively. During the damage survey, it was also noted that there was almost no damage or much lesser damage to houses situated on or behind the medu. Presence of medu at Killinjal village, Nabiyarnagar and Nagapattinam Port reduced the damaging impact of tsunami on the built environment. The place of local increase of damage was dependent on river orientation and the path of tsunami. The southern part of Tamil Nadu from Kanyakumari to Pudukkottai District suffered least damage due to the presence of Sri Lanka in the passageway of tsunami.

Moreover, the level of damage in the Gulf of Mannar was more than that in the Palk Strait, since only diffracted waves were able to enter into the Palk Strait. Chandrasekar et al. (2005) and Mujabar et al. (2007) found that beach morphology and sediment volume along the southern Tamil Nadu coast were modified by the tsunami inundation. In recent years, advanced scientific techniques like GIS and remote sensing have been applied to comprehend and model various aspects of tsunamis. Mohan (2005) reported that the elevated coastal dunes and beach ridges along the coastline could act as barriers to minimize the rate of inundation along the northern parts of the Tamil Nadu coast. Narayan et al. (2005b) found that the degree of inundation was strongly scattered in direct relationship to the morphology of the seashore and run-up level on the Kerala coast.

Being one of the world's productive ecosystems, coral reefs are vital to the coastal populations. The post-tsunami status assessment of the islands and coral reefs around them in the Gulf of Mannar indicated that the margins and peripheral landscape of the islands were altered to certain extent (Kumaraguru et al., 2005). Corals of Shingle, Mulli, Van, Karaichilli and Vilanguchalli islands have been disturbed by the tsunami. Hence, it is imperative that their socio-economic role is taken into account and ecological measures implemented accordingly. The study proposes that developing a barrier reef ecosystem at least in the gulf of Mannar and Pak Bay regions would help boost the environmental, social and economic conditions of the coastal people. Further, Chandrasekar and Immanuel (2005) categorized the tsunami hazard for certain beaches of South India using GIS techniques. Keating et al. (2004) modeled the inundation of the Kealakipapa Valley, Hawaii, using a 10m-tsunami wave model.

4. IMPACT OF 2004 TSUNAMI ON INDIAN COASTS

The effects of 2004 Tsunami on Indian coasts are summarized in Table 1 and the details of the impacts are described in the subsequent sections by dividing the Indian coasts into three parts.

Table 1. Details of tsunami inundation on Indian coasts (Rasheed et al., 2006)

Details	Kerala	Andhra Pradesh	Tamil Nadu	Pondicherry	A& N Island	Total
Coastal length affected (km)	985	250	1000	25	NR	2260
Penetration of water into main land (km)	0.50-2.0	1-2	1-1.5	0.30-3.0	NR	-
Average height of the tidal wave (m)	5	3-5	7-10	10	NR	-
No. of villages affected	301	187	367	26	14 Islands	895
Dwelling units	1557	6280	85,878	6,403	NR	100318
Population affected (lakh)	2.11	24.70	6.72	0.43	1.83	35.79
Cattles lost	195	NR	5.58.2	3445	NR	9222
Cropped area (ha)	790	NR	2487	790	NR	4067

Note: NR = Not reported; 1 lakh = 100,000.

4.1 Southwest Coast

This covers Kerala, and the Kanyakumari district (Tamil Nadu), which suffered heavy loss of life due to human-created local topographical features. In Kerala, there is no documented evidence of any such events, although there are reports of some previous tsunamis (1881, 1833, 1941, to mention a few) generated by earthquakes in the Andaman-Sumatra region. A 1945 earthquake of M 8.0 in the Mekran coast is believed to have generated significant tsunami run-up in some parts of Gujarat, the only documented report of any tsunami affecting the west coast. To the best of our knowledge, the 2004 Tsunami is the first of its kind to have affected the Kerala coast. Although 2004 Tsunami affected the

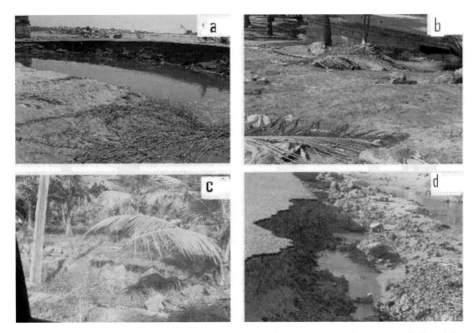

Fig. 2 Devastation caused by tsunami at Valiazhikkal region, north of Kayamkulam inlet and Edavanakkad, north of Cochin: **(a)** Deep trench formed close to the coastal road immediately to the north of Kayamkulam inlet; **(b)** Roof of a newly constructed house grounded due to wall collapse; **(c)** Boulders of sea walls thrown inland at Edavanakkad; and **(d)** Deep trench formed close to the coastal road at Edavanakkad (Kurian et al., 2006).

whole coastline of Kerala (Figs 2, 3 and 4), it devastated the low-lying coastal areas of Kollam, Alleppey and Ernakulam districts leading to the loss of life and property. In these districts, height of Tsunami waves varied from 1.9 to 5 m. The southernmost district of Thiruvananthpuram, however, escaped damage. This was possibly due to the wide turn of the diffracted waves at the peninsular tip, thereby missing Thiruvananthpuram (Sheth et al., 2006). Depending on the geographical orientation of the coast, geomorphology of the land mass, shallow water bathymetry and orientation of approaching waves etc., tsunami inundation and damage varies. Nearly 200 people were killed and hundreds injured in addition to the loss of houses and properties worth several crores of rupees. The coastal belt from Thottapally in Kerala to Kanyakumari in Tamil Nadu was monitored during 2005 January. Transects were selected on the basis of the intense impact of 2004 Tsunami. At each transect, stations were chosen at 5 km intervals, up to a distance of 25 km from shoreline. The water quality data after the 2004 Tsunami showed a slight deterioration at some of these transects.

4.1.1 Kerala

Ernakulam District: In Ernakulam district, maximum damage was suffered by Edavanakkad village, Kochi [Fig. 4 (a)], which is a prominent fishing center in Vypeen Island (about 20 km long and 3 km wide). It reports a loss of five lives, destruction of fifty houses and damaging of 350 houses. Damage due to the tsunami was more intense in the areas that were not protected by the seawall. At Edavanakkad, the seawall was about 100 m from the shoreline and was made of random loose rubble masonry comprising large boulders, some of which were displaced by the tsunami. The houses in Edavanakkad were built about 15 m from the seawall (Sheth et al., 2006).

Fig. 3 Examples of tsunami inundation from northern Kerala: **(a)** Collapsed embankment at Choottad, Cannannore; **(b)** Country boats thrown inland at Kadalur point, Calicut; **(c)** Damaged railway footbridge at Thalassery; and **(d)** Collapsed laterite banks at Kadalur point, Calicut (Kurian et al., 2006).

Fig. 4 **(a)** Run-up height of tsunami at Edavanakkadu; **(b)** Tsunami inundation at PuthuVypeen.

Allapuzha and Kollam Districts: The other strip most affected in Kerala is a shoestring cape south of the city of Allapuzha. It extends to 40 km long from Trikunnapuzha in the Allapuzha district to Karunagapalli in the Kollam district. The strip has a maximum width of less than 1 km and is bound by the open coast on the west and Kayanakulam Lake and backwaters on the east. A small opening in the strip connects the lake to the sea. The width of the land strip is less than 0.5 km at many places. As a result, the tsunami waves lashed over the entire strip of land, traveled across the backwaters, and rolled onto the opposite bank. Two areas that bore the brunt of the tsunami in the region were Arattupuzha in the Allapuzha district and the Alappad *panchayat* in the Kollam district. These areas are the gatekeepers of the opening where the lake disgorges into the sea. The villages are densely populated (with a density of 2,652 per km^2, versus the state average of 762 per km^2) on both sides of the coastal road that barely

Table 2. Comparison of water quality of selected coastal transects in Kerala before and after the 2004 Tsunami (Kurian et al., 2006)

Parameter	Kayamkulam		Alleppey		Vizhinjam	
	Before	After	Before	After	Before	After
Temp (°C)	28.26	28.26	28.53	28.50	28.26	28.26
Salinity (ppt)	33.24	33.62	33.54	33.75	33.24	33.82
pH	8.30	8.21	8.32	8.16	8.30	8.16
DO (mg/L)	5.67	4.84	5.61	5.85	3.16	4.69
NO_2 (µM)	1.69	0.18	0.13	0.38	0.04	1.69
NO_3 (µM)	3.22	3,25	-	3.88	1.29	4.25
SiO_4 (µM)	4.01	2.34	1.90	2.72	1.30	2.52
PO_4 (µM)	0.30	1.58	0.60	1.96	0.19	1.86

separates the lake from the sea. The isthmus has no intermediate bridge connected to the mainland and is linked to it only at the two ends. The area thus has to depend on boats for safe exit in times of emergency. The largest number of casualties (130) in Kerala was reported from the densely populated Alappad *panchayat* (including the villages of Cheriya Azhikkal and Azhikkal). In the village of Arattupuzha in the Allapuzha district, the death toll was 28. The seawall, where it existed, was dislodged and did not exist in many places along the north Kollam coast. A wave height of 4 m was reported by the survivors. The area has a history of floods that occur during the monsoon, especially when heavy rains coincide with high tide. The villagers are thus well trained in quick and safe evacuations, but there was no time for evacuation in these densely populated villages when the tsunami arrived. There was, however, no loss of life in villages such as Walia Azhikkal, where rapid evacuation was carried out successfully. After observations of abnormal behavior of the sea, the Kerala police had sounded an alert warning at 11:30 a.m. to caution people, but it was too late for some villages. The tsunami struck almost immediately afterward (Sheth et al., 2006).

The post-Tsunami results indicated that it affected the south west coast between Thottapally (Alleppey District) and Muttam (Ernakulam District). The concentration of nutrients, primary productivity and species diversity of plankton is lowered to a considerable extent. The population density of fish and microbes are reduced. The texture of the sediments changed into coarse sand with heavy minerals due to the high-energy backwash. The impact of Tsunami was maximum at Vizhinjam due to the geomorphic feature resembling inland basin. Construction of sea walls as a protection against tsunami does not seem to have any apparent merit. Edavanakkad (Ernakulam District), north of Cochin is a classical example where the sea wall, in spite of being well-built, was completely damaged with huge constituent boulders thrown far away inland (Kurian et al., 2006). Damage due to the tsunami was maximum in sectors adjoining Kayamkulam inlet in southern Kerala.

Measurements were made of land elevation, beach slope, tsunami flow direction and distance, maximum run-up height and duration of inundation, shoreline position and status of beach vegetation. Maximum intensity of the tsunami was observed in Kollam district followed by the districts of Alleppey and Ernakulam. The sectors adjacent to the Kayamkulam inlet between Kollam and Alleppey districts, recorded the maximum run-up height during the flood. The run-up level recorded at the northern side of inlet (Valiazheekkal) was 4.4 m. There was a drastic increase in the run-up level at Cheriazheekkal (southern side of inlet) and reached up to 5 m.

4.1.2 Tamil Nadu

The State of Tamil Nadu has been the worst affected on the main land, with a death toll of 7,793. Kanyakumari forms the southwest district of the State. Unlike the Kerala coast, the west coast of Tamil Nadu has flat land along the north coast in some areas such as Kolachel, where the soil is enriched with alluvium, but in other areas including further south, the land is mountainous. It slopes steeply upward from the coastline and is rocky. This was observed at Muttom Beach and at the tip of Kanyakumari. On some beaches such as Sothavilai, there are sand dune formations. Chandrasekhar et al. (2006) studied the impact of tsunami on the southern and western parts of Tamil Nadu including the districts of Kanyakumari and Tirunelveli. The results showed that the presence of Sri Lanka saved the beaches along the east coast from direct waves of high rapidity. Hence the beaches along the east coast are under least viability to any such similar hazards in future whereas high vulnerability prevails along the west coast beaches as they are devoid of any natural blockade (Narayana et al., 2005; Raval, 2005). The impact of tsunami rush was high in the southernmost part of the study area as most of the high vulnerable beaches falls on that region. They are Chinnamuttom, Kanyakumari, Manakudy and Pallam and are very much exposed to the refracted and diverted waves from Sri Lanka. Colachel suffered maximum destruction in the northwestern coast as the inundation has been encouraged by the river mouth. The impact was more in the beaches of low-lying flat topography as in the case of Manakudy and Colachel, etc.

Kanyakumari District: One of the most affected areas in the district was Kolachel and the surrounding villages, located 30 km west of Nagercoil, capital of the Kanyakumari district. The coastal area is flat at the sea level without any sea wall. Waves traveled inland by more than 300 m, which carved out new streams and estuaries. The town and neighboring villages recorded more than 500 fatalities, including half of them children. The wave height was reported to be 5 m, and the run up height was 2.6 m. A large number of deaths were triggered by the human created topology of this town. Harbour Road experienced the maximum number of casualties. A long (2.5 m deep and 6 m wide), open, dry channel called the Ananda Victoria Marthandam (AVM) Canal situated in the Harbour Road was meant to bring freshwater to this region. Besides, numerous open trenches were laid parallel to each other that catered to the special needs of the coir-making industry, which is the chief means of income in this region other than fishing. Tsunami waves made the canal and these trenches into death traps (Fig. 5). More than 300 bodies were recovered from the slush of the trenches and channel. The masonry retaining wall of the jetty at Kolachel was badly damaged. Muttom Beach, a popular tourist spot in Tamil Nadu frequently used as a romantic background by the Tamil film industry, did not suffer much damage, because it is more than 50 m above mean sea level. Tsunami wave, which arose at 10:30 a.m., washed away the tourists who came to see the beach. The Mannakudi *panchayat* near Kolachel, including Melamanakudi, was badly affected. A 160 m long bridge which connected the villages of Melamanakudi and Kelamanakudi on opposite banks of the Pazhyar River went missing after the tsunami. It was originally built with four spans out of which, the two end spans traveled upstream by 100 m and were beached on the banks; the central spans sunk into the waters (Fig. 6). Melamanakudi, a prosperous and scenic hilly village built on the waterfront, suffered complete damage. Structures 250 m from the shoreline suffered little or no damage, because the land sloped upward. About 150 people died at Melamanakudi. Sothavilai has a beautiful beach, which slopes upward from the shoreline and folds into sand dunes about 8-10 m high along parts of the beach. A beach resort has been constructed inland beyond the sand dunes. The tsunami waves lashed numerous teashops and other food stalls in the path leading to the beach causing loss of life and much scouring. An estimated 200 people, mainly stall owners and visitors, were victims. Kanyakumari experienced tsunami waves that were over 4.5 m high. Because Kanyakumari is on high land, the damage was limited—street furniture and compound walls were destroyed. About 1,200 people trapped at the rock memorial were successfully evacuated within a few hours of the incident.

Fig. 5 The AVM channel in Kolachel, which became a death trap; more than 300 bodies were found in the slush of the AVM channel and open trenches (Photo: Sheth et al., 2006).

Fig. 6 All spans of the Melamanakudi Bridge came off their bearings. The end spans drifted upstream, while the middle spans sank into the river (Photo: Sheth et al., 2006).

4.2 Southeast Coast

This includes the rich alluvial delta region of the Tamil Nadu coast and Pondicherry, which recorded the maximum loss of life and damage in mainland India.

4.2.1 Southeast Coast of Tamil Nadu and Pondicherry

The southeastern coastline of Tamil Nadu (800 km long) includes the Coromandel Coast in the north and the Fisheries Coast in the south. The coast north of Point Calimere up to Pondicherry was the worst affected region. It consists of the coastal districts of Nagapattinam, Cuddalore, Villipuram, Kanchipuram, and Chennai. The arrival of the first tsunami wave along the east coast coincided with high tide and amplified its effect. The coastline in this region is flat except for gentle slopes. From local residents, it was inferred that the beaches previously had a steeper slope than the present state. Many of the slopes and much of the vegetation have been cut during the last two decades to make way for coastal hamlets. The coastal districts of Tirunelveli, Tuticorin, Ramanathapuram, and Pudukkotai suffered relatively minor effects from tsunami waves. The tide gauge at Tuticorin Port, measured a wave height of nearly 1.8 m, which was only marginally higher than high-tide levels, partly because the tsunami waves were out of phase with the high tide.

In Chennai, besides seawater intrusion and deaths of numerous people, the impact also covers morphological changes along the coastline, where sea intruded in certain places and receded in others. In Adyar, the mouth of the river creek cut-off from the sea due to sandbar formation was cleared by big waves. The water quality of the area was affected due to increase in salinity and total dissolved solids along the tsunami-affected coast. A study was carried out in the tsunami-affected areas with respect to groundwater quality in the coastal areas of Adyar. The study area included Adayar and the surrounding areas in south Chennai that lie between latitudes 12°58′N and 13°08′N, and longitudes 80°16′E and 80°28′E (Palanivelu et al., 2006). Poor groundwater quality was evidenced in areas like Kottivakkam beach, Kuppam, Oorurolkot Kuppam (seashore), Raja Rangasamy Avenue (Thiruvanmaiyur), Foreshore Estate, Dhidir Nagar, Nochikuppam, Anna-MGR Memorial, which lie in close proximity to the sea and where sea water inundated during the tsunami. Apart from seawater-inundated areas, other areas showing poor water quality include R.A. Puram and Krishnamoorthy School, Adyar. Total Dissolved Solids (TDS) levels observed after the occurrence of the tsunami are within the range as observed during September 2004. Thus the recorded TDS values over time indicate that there is no major impact of the tsunami on water quality. As seepage of seawater is very less due to the short period of transgression during the tsunami, the aquifer has not been affected. Though seawater percolation into the ground through small pockets of waterlogged areas is possible, the effect would be less considering the short period of inundation of seawater during the tsunami. Thus it is clear that the groundwater quality has deteriorated due to lack of sufficient rainfall leading to seawater intrusion that is reflected in high TDS and chloride content of the analyzed samples.

Nagapattinam District: Nagapattinam District in Tamil Nadu was the most dreadfully affected territory within the entire mainland of India, where 6,051 fatalities were reported. Sujatha et al. (2008) studied the impact of tsunami waves on the sediments along the coastal belt of Nagapattinam. Sediments of the Nagapattinam beach, Tarangambadi temple, Tarangambadi temple site and Danish fort were analyzed for trace metals, total organic carbon, percentage carbon, hydrogen and nitrogen and nutrients (total phosphate, orthophosphate and exchangeable nitrate). The results show that the event has changed the chemical composition of the beach sediments and is threatening fishing grounds even in trace concentrations. Chromium showed the highest concentration at the Tarangambadi temple area compared to other stations. The Tarangambadi beach showed elevated concentrations of Nickel, Lead and Copper compared to other three stations. The variation of metal concentrations was drastic, due to the churning of the bottom sediments and their deposition along the shore by the waves, bringing changes in the beach profile and chemical concentrations. This remobilization contributes to environmental disturbance and is likely to have a far reaching effect on the life and activities of marine biota in general and pelagic fishery in particular.

Nagendra et al. (2005) studied the assemblages of foraminifera and its distribution pattern in the tsunami inundation areas of Nagappattinam coast. Samples were systematically collected from four locations (Nagappattinam (N), Kameshwaram (K), Vilundamavadi (V) and Velankanni (VK)) during the first week of January 2005. These locations experienced maximum inundation by tsunami waves. The foraminiferal assemblages recorded in the sediments brought by the tsunami waves were inferred to be derived from the shallow neritic zone along the Nagappattinam coast. Q-mode cluster analysis reveals that spatial variation of foraminiferal assemblage distribution and recognizes five biotopes along Nagappattinam coast. These changes in foraminiferal assemblage may be attributed to micro niches that control faunal distribution and also current circulation patterns along the coastal zones. The 200 m bathymetry of the continental shelf offshore east coast of Sri Lanka extends to Cuddalore-Nagapattinam sector along east coast of India, which might have channelized these waves to cause maximum damage in this part.

Velankanni: Velankanni, 12 km from Nagapattinam, where the Church of Mary, Mother of Good Health, a much admired pilgrimage center among believers for the healing powers is located. The central axis of the church is aligned with the main road leading to the beach. Thousands of worshippers had gathered in this town for Christmas, and many of them were at the beach on the morning of 26th December after attending mass. A tsunami wave approximately 5 m high first struck the shore at 9:20 a.m., followed by four more waves between 9:20 a.m. and 10 a.m. The waves inundated a distance of 900 m. The waves crashed onto the shore and continued on their rampage along the main road leading to the town. Though the water did not enter the church, it gushed up to the bus stand, inundating several shops and houses. A large number of bodies were found along this road. The inundation distance was more than 1,000 m (Sheth et al., 2006).

Seruthur: Seruthur is a coastal village adjacent to Velankanni. The height of Tsunami wave was the same as that of Velankanni. The number of deaths was lower due to: (i) lower population density, (ii) the base of the houses in this village was 1.5 m above ground level, and (iii) elevation of the ground was 1.5 m from sea level. The run-up height was 2 m above the plinth level, or 5 m above mean sea level. About 1000 boats and fishing trawlers were destroyed in Akkaraipettai, an important fishing port in the Nagapattinam district. Naliyanthottam, a godown storing gas cylinders was destroyed, with all cylinders being swept away by the tsunami. The godown was more than 800 m from the shoreline.

Nagore: Due to high elevation (more than 3 m from sea level), no serious damage was reported at the oil jetty at Chennai Petrol Chemicals Ltd. at Nagore. However, 1,200 houses were destroyed or damaged. Water entered almost 1,000 m inland. The rail link to Nagore, which is a pilgrim town, was cut off as a 7-km meter-gauge track between Nagapattinam and Nagore was completely damaged (Fig. 7).

Cuddalore District: Cuddalore was the second-most affected district and it borders Nagapattinam in the north. Silver Beach is a small amusement park developed for tourism and recreation activities, such as boating. Two rivers discharge into the sea at this scenic location. The first tsunami wave arrived at 8:35 a.m., with two subsequent waves within the next 15 minutes (Sheth et al., 2006). Almost three times as many women lost their life than men, with 391 female casualties, compared with 146 men. In Devanampattinam village in Cuddalore, for example, 42 women died compared with 21 men. In Pachaankuppam village, the only people to die were women (Sheth et al., 2006).

Kancheepuram District: Though the nuclear power plant at Kalpakkam in the Kancheepuram district was not affected by tsunami, the housing colony was very badly affected. The waves with a run up height of 1.5 m rushed into the colony damaging almost 1000 houses and the sea wall of 2 m high and 4 km

Fig. 7 Damaged railway line between Nagore and Nagapattinam in Tamil Nadu (Photo: Sheth et al., 2006).

long. Almost all courtyard walls of the houses for 1 km were collapsed. A school's compound wall was damaged, and electric poles were uprooted near the school grounds. The tsunami waves deposited a large quantity of sand near a pedestrian bridge.

Pondicherry: Pondicherry recorded 599 deaths. The damage was in two regions of Pondicherry—Karaikal and the city of Pondicherry. In Karaikal, the first tsunami wave struck the city at 8:45 a.m. and was followed by two more waves. The wave height was about 10 m and the run up height was 2.4 m. The damage was colossal in Karaikal- Nagapattinam stretch due to the flat topography of the region. In Kottuchery village located 5 km north of Karaikal town, the inundation was more than 1000 m. The compound wall of a self-financing Dental College was flattened. In Karaikal beach, which was situated in the estuary of Arasalar River, the damage was extensive. The road has been washed away. Several houses in fishing hamlet located were damaged (Ram Mohan, 2005). The run up was estimated to vary between 4 and 5 m. A settlement of 50 families of a fishing community called Karukalacheri was badly damaged. Pondicherry city, which lies north of Cuddalore district, experienced waves about 9 m but escaped relatively unscathed due to the stone seawall (height about 9 m above sea level) constructed almost three centuries ago. Although 107 bodies were recovered from Pondicherry and the surrounding hamlets, 56 of these bodies had been washed away from Tamil Nadu. About 25 people died on the Pondicherry promenade. They were fishermen living beyond the seawall boundary such as Kucchikupam, Ariyankuppam, Ponniyarpala, and Ganagachettikulam. At Kucchikupam, three waves washed out thatch huts, killed three people and damaged buildings with masonry brick walls. 16 people were reported dead at Ganagachettikulam, where people managed to build temporary houses within 10 days of the event.

4.2.2 Coast of Andhra Pradesh

The tsunami waves which struck the Andhra coast at about 9:05 a.m. brought about destruction to houses and loss of fertile land and employment. The waves penetrated about 0.2-1.0 km into the districts of Machalipatanam, Ongole and Kavali with an average wave height of 1.2-2.2 m. Comparatively less damage was reported at Srikakulam, Vishakhapatnam, Kakinada and Yanam (Pondicherry). Fishing and agriculture are the two dominant occupations in the villages along the coastline in Andhra Pradesh. The tsunami washed away most of the fertile topsoil and deposited heaps of sand, which ultimately left people with the choice of either changing the land use or changing the crop pattern.

Vishakhapatnam District: Vishakhapatnam, a very busy harbor and naval base, was protected by concrete tetrapods laid along the coastline. Vishakhapatnam Sea port and Vishakhapatnam Container terminal berths were empty on 26th December, so they suffered no serious damage. Vishakhapatnam Port had minor damage that created difficulty in navigational movements, cargo loading, and unloading. Fishing harbor boats and trawlers drifted away; 179 mechanized boats were partially damaged. The port suffered one full day's operational loss. Tsunami disrupted the fishing industry, many fishermen lost their boats and homes. It affected the sale of fish, which further destroyed the fishing industry.

East Godavari District: Sheth et al. (2006) have shown that East Godavari coastline suffered the least damage. The Kakinada Port Trust reported that some disruption in activities had disorganized its functions. The port was saved due to its peculiar shape and Hope Island acted like a shield. In this delta area, water channels or in some cases backwaters are extensively used to sail fishing boats right up to the edge of settlements. These channels helped to reduce the impact of tsunami water entering into the mainland. At Uppada village, however, the stone pitching of the embankment was damaged in some places. Studies by Mruthyunjaya Reddy et al. (2005) have shown that four villages were affected in the district. The seawater rose to 1-2 m, many boats and nets were damaged, but there was no damage to the houses.

Yanam and West Godavari Districts: Yanam has rich, black fertile soil deposited by the Godavari River. Paddy cultivation, high-quality brick manufacturing, ceramic tile molding, and fishing boat construction are some of the main occupations in Yanam. After the tsunami devastation, the two cottage industries of brick manufacturing and boat construction expanded rapidly. The soil is deep friable, well-drained sandy loam in the West Godavari district coastal belt. The land is a green mass of mangroves and a dense human-created forest of cash crops such as cashews, jackfruit, and coffee grown in the shade of coconut farms. The tsunami ruined the crops and replaced the standing cashew trees with sand deposits. Fresh ingress of water was seen in areas with dense mangroves. Part of the mangrove belt, in the path of the tsunami waves, was completely washed away, leaving another part partially or fully submerged in water for more than 48 hours. Silt had accumulated on the pneumatophores—the breathing roots of mangroves—choking the respiration of the mangroves. The tsunami left the area prone to further erosion of soil on seashores after the removal of vegetation cover. Slowly growing species such as mangroves, even if replanted, will take a long time to cover the entire land (Sheth et al., 2006).

4.3 Southwest Coast

Andaman and Nicobar Islands: Ramachandran et al. (2005) studied tsunami-induced damage to coastal ecosystems in four Nicobar Islands, viz. Camorta, Katchal, Nancowry and Trinkat. Tsunami-induced ecological damages are evident in all the four islands. There are damages to the structure and function of all the coastal ecosystems such as coral reefs, mangroves, sea grasses, estuarine mudflats, etc. The biological structure of the ecosystem could be easily disrupted as various species at different trophic levels were differentially removed, and with the structure altered, ecosystem functions could also be altered. Being a low-lying island, Trinkat has suffered maximum damage and has been cut into three pieces. The mangrove areas were affected to the extent of 335.70 ha (51%) in Camorta, 339.03 ha (69%) in Katchal, 152.53 ha (100%) in Nancowry and 240.06 ha (68%) in Trinkat. Such a major damage in mangrove area will severely affect the coastal productivity and destabilize coastal areas, which will accelerate shoreline erosion and increasingly affect the forest area due to salt-water intrusion into the forests. Extensive damage to coral reef is seen in all the four islands. The extent of reef area affected is

Table 3. Extent of area affected by the tsunami in the Andaman & Nicobar Islands (Ramachandran et al., 2005)

Class	21st December 2004	4th January 2005	Change in area (ha)	Percentage change in area
Camorta Island				
Mangroves	651.94	316.24	335.70	51
Plantation	4509.44	4147.02	-362.42	-8
Reserved Forest	8311.73	7771.42	-540.31	-7
Sand	356.49	725.21	+368.72	+103
Settlements	32.82	27.04	-5.78	-17
Mud over reef	733.24	896.93	+163.69	+22
Reef area	1775.12	1035.97	-739.15	-41
Water logged area	-	283.12	-	-
Katchan Island				
Mangroves	576.37	177.34	-399.03	-69
Reserved Forest	9801.72	8014.72	-1787.00	-18
Sand	473.82	1715.84	+1242.02	+262
Reef area	548.48	-	-	-
Water logged area	-	1640.60	-	-
Settlements	340.52	222.85	-117.67	-35
Nancowry Island				
Mangroves	152.53	0.00	-152.53	-100
Plantation	244.32	176.26	-68.06	-27
Reserved Forest	4212.49	4064.36	-148.13	-3
Settlements	25.42	-	-	-
Sand	254.32	175.34	-78.98	-31
Reef area	829.13	381.98	-447.14	-53
Mud flat	106.39	-	-	-
Water logged area	-	91.07	-	-
Trinkat Island				
Mangroves	352.93	112.33	-240.60	-68
Plantation	649.10	644.57	-4.53	-1
Reserved Forest	561.58	419.09	-142.49	-25
Sand	195.38	214.08	+18.70	+10
Settlements	71.52	30.09	-41.43	-58
Reef area	2432.12	986.47	-1445.65	-59
Mud over reef	-	440.01	-	-
Turbid water	-	552.44	-	-
Water logged area	-	126.89	-	-

41% at Camorta, 49% at Katchal, 53% at Nancowry and 59% at Trinkat. The sandy areas have considerably increased in two of the four islands. The increase in the extent of sandy beaches after the tsunami was 18.7 ha in Trinkat and 1242.02 ha in Katchal, whereas decrease in sand cover is witnessed in Camorta 368.72 ha (103.43%) and Nancowry 78.98 ha (31%).

5. REHABILITATION AND MITIGATION MEASURES ADOPTED BY THE LOCAL GOVERNMENT AGENCIES AND NGOs

In Kerala, the cooperative federation called Matsyafed undertook the rehabilitation measures in the coastal areas devastated by tsunami (Salagrama, 2006). The state government provided Rs. 134.86 crores for replacing and repairing damaged fishing implements. In the NGO sector, many agencies provided boat engines, and net to the fishers, mostly on a group ownership basis. Other mitigation measures included introducing self-help groups among women, giving assistance for small business in several areas and helping the youth in taking up alternate employment in construction and other sectors. Kurian et al. (2006) observed that shore protection systems (such as sea walls) and coastal engineering structures had variable impact on inundation along the Kerala coast. For example, the sea walls could control tsunami waves to some extent at Thangassery and Neendakara, as it was not so at Edavanakkad coast. Also, the presence of backwater at the Kayamkulam inlet increased the inundation along this coast. From these observations, it can be concluded that no single measure or uniform strategy is viable in mitigating tsunami-caused coastal hazards.

Foodstuff, garments and other provisions were brought to the camps in abundance from all quarters. The government offered medical assistance by the end of January 2005. All the affected people had moved to temporary shelters by the end of April. The government declared a grant of Rs. 7500.00 to all families moving into houses of their own choice and Rs. 3500 to all families moving into government built temporary shelters (SIFFS, 2005). Temporary shelters for tsunami victims who lost their houses were constructed by the government in four clusters in the tsunami affected wards of Alappad. Cheriazheekal, Srayikkadu, Aiyramthengu and the fishing harbor in the banks of the Kayamkulam lake estuary in Azheekal were the sites chosen for the shelters in Alappad. In Kollam district, 2969 houses were allotted to the affected people and 40 acres of land was acquired by the government for the relocation of affected families. Government has sanctioned Rs. 1.084 crores for the construction of sea wall in Alappad and Rs. 225.786 lakhs worth of free rations to the affected people. Kerala Water Authority spent Rs. 15.91 crores for various water supply works in Kollam district. Non-governmental organizations like Seva Bharati, the Quilon Social Service Society and Sahayi, working in the affected areas have built thatched (coconut palm frond) shelters with a small kitchen attached, for people (land owners) who opted for their own accommodation (SIFFS, 2005). Other non-governmental agencies involved in tsunami rehabilitation throughout the state include Kerala Voluntary Health Service, Kerala Vyapari Vyavasai Ekopana Samiti, Mata Amritanandamayi Math, Nehru Yuva Kendra, etc.

In Alappuzha district, Rs. 1.69 crores were spent for the relocation of those who lost their houses in the tsunami. Kerala Water Authority spent Rs. 8.28 crores for the construction of the sea wall, and Rs. 2.19 crores for the laying and repair of roads in the affected areas. The KSEB has done a line extension in Arattupuzha for Rs. 3.24 crores (SIFFS, 2005). Many fishermen have been given fishing boat, engine and nets by various agencies. Many stake nets and Chinese dip nets on the Kayamkulam estuary were washed away by the tsunami. The government has announced compensation for the Chinese dip nets but the environmentally harmful stake nets have not been compensated to date. Compensations for other indemnity suffered to fishing craft and gear in the inland sector have also been declared. The non-governmental agencies of the state organized different programs for the people, especially the children in the rehabilitation camps.

The main tsunami effects in Tamil Nadu related to marine fisheries included loss or damage to fishing boats, nets and engines; infrastructure for landing, fish processing and trade related activities, apart from loss of lives of active fishermen. The tragedy has invoked instant response from all government bodies, non-government organizations, celebrities, and corporate and international organizations (Salagrama, 2006). In Tamil Nadu, government introduced a package for replacing fishing nets for vallams and catamarans. Assistance was given to repair engines and damaged boats and catamarans and for obtaining bank loans at subsidized rates for the reconstruction. Government provided monetary assistance to the fishers, exempted payment of sales tax for purchase of goods necessary for reconstruction. Government also provided assistance to aquaculture owners, fish transporters and ice manufacturers and made arrangements for repairing fishing harbors, jetties and landing centers. According to Ghouse (2005), mitigation strategies in Tamil Nadu included agri-engineering based immediate, short-term and long-term strategies to bring back the soil, agriculture, horticulture, forestry, groundwater etc. to normalcy.

Andhra Pradesh government put up an institutional framework to offer support to affected people. The World Bank-aided Andhra Pradesh Rural Livelihood Program (Velugu) was the nodal agency for rehabilitation of the Tsunami affected fishing communities in the state. Velugu made arrangements for the repair of damaged boats, and the replacement of fully damaged boats by new boats, as well as provided loan for new boats, and replaced nets and other fishing gears (Salagrama, 2006). Sonak et al. (2007) documents several issues involved in the recovery of tsunami-affected areas and recommends the application of the ICZM (Integrated Coastal Zone Management) concept to the reconstruction efforts. The concept of ICZM has been effectively used in most parts of the world. This concept emphasizes the holistic assessment of coasts and a multidisciplinary analysis using participatory processes. It integrates anthropocentric and eco-centric approaches. Xue et al. (2004) notes that the World Bank (2002) views ICZM as a *"process of governance that consists of the legal and institutional framework necessary to ensure that development and management plans for costal zones are integrated with environmental and social goals, and are developed with the participation of those affected"*. ICZM is an accepted management framework to address coastal and marine environmental problems and conflicts, and to achieve sustainable use of coastal resources in developing countries.

6. TSUNAMI DISASTER REDUCTION AND PREPAREDNESS

India has witnessed some major disasters in recent memory and the lack of preparedness and inadequacy to face such calamities has been apparently visible. The Indian Ocean tsunami of 26 December 2004 was a disaster of unprecedented magnitude. Across the 12 affected countries in Asia and Africa, more than 230,000 people were reported dead or missing, over 2.1 million were displaced and left homeless, and millions of dollars of infrastructure was destroyed. The scale of the devastation presented enormous challenges for disaster response in the context of the evolving concept of disaster management (NDM, 2005). State governments have limited resources and infrastructure facilities to handle a major disaster. We know that natural hazards cannot be controlled. However, the vulnerability to these hazards can be reduced by planned mitigation and preparedness measures. There needs to be concerted and sustained steps towards reducing the vulnerability of communities to disasters. The past couple of years have witnessed a paradigm shift in the approach of central and state governments towards disaster management. The new approach proceeds from the conviction that development cannot be sustainable unless disaster mitigation is built into the development process. Another corner stone of the approach is that mitigation has to be multi-disciplinary spanning across all sectors of development.

Disaster management involves three phases—**pre-disaster**, **during the disaster**, and **post-disaster**. The pre-disaster phase consists of *risk identification*, *mitigation*, and *preparedness*. During the disaster, *emergency response* takes place, and in the post-disaster phase, *rehabilitation* and *reconstruction* are applied. The national government has set up a National Disaster Management Authority, with Prime Minister as its chairman. In the meanwhile, the Government has established a nodal agency (Delhi Disaster Management Authority) to facilitate, coordinate and monitor disaster management activities and promote good disaster management and mitigation practices in the state (http://www.ndmindia.nic.in/Mitigation/mitigationhome.html). Two Working Groups for (i) Prevention and mitigation and

(ii) Preparedness and response have been constituted and notified. These groups further consist of sub-groups to deal with awareness generation, capacity building, planning, techno-legal aspects, etc. In order to make the awareness generation activities more effective, the Government has developed a lot of resource material comprising leaflets, posters, planning documents, earthquake tips, etc. for wide dissemination. The IEC material serves as an agent to generate awareness and induce mitigation and preparedness measures for risk reduction. Multi-media CDs on disasters developed by various government agencies have been distributed in selected states of India. The efforts of the national government to introduce disaster management as a compulsory paper in all fields of study at graduation level in all the states of India, if and when realized, will ensure sustainable disaster awareness among the youth, leading to disaster risk reduction in due course.

7. CONCLUSIONS

In this chapter, a review of damages caused by the 2004 Tsunami disaster in the shorelines of India has been presented, together with the rehabilitation and mitigation measures taken by the local/national authorities. In addition, the plans and measures adopted by the Government of India for reducing tsunami disaster risk in the future are also highlighted in this chapter. In India, states of Tamil Nadu, Kerala, parts of Andhra Pradesh, Union Territory of Pondicherry and Andaman and Nicobar Islands witnessed the worst impacts of 2004 Tsunami on social, economic and ecological systems. It is evident that low-lying coasts were the prime victims of the tsunami surge, as they were unable to put up any resistance against the accelerating tsunami surge. Although the overall impact of the tsunami on the east coast was severe, damage to coastal protection structures was limited. Maximum damage observed along the coast of Nagapattinam district (Tamil Nadu) may be due to the lesser width of continental shelf and the interference of direct waves with the reflected waves from Sri Lanka. As an immediate measure, the state governments released funds for the treatment of those seriously injured in the tsunami and distributed free rations to the affected people. NGOs provided temporary shelters to all those who lost their houses and livelihoods. Governments and other social service societies made provision for replacing fishing equipments damaged in the tsunami and for the reconstruction of fishing harbors.

The 2004 Indian Ocean tsunami event allowed the assessment of technological approaches for disaster management. These approaches include GIS, GPS, RS, modeling, etc. Geographic Information System (GIS) facilitates common database operations with unique means of visualization and analysis. It assists users in statistical analysis and provides a base for interpreting how physical, social and economic factors interact in space. Global Positioning System (GPS) is a satellite-based positioning system for capturing locations of sample points, which can be used to reference satellite images or other spatial data layers. Further, Remote Sensing (RS) provides images of the earth's surface, which enable the classification of different types of land cover and the monitoring of land cover/land use change. High temporal resolution of some satellites has made remote sensing—in combination with GIS—an extremely useful tool for rapid mapping in support of disaster relief, as was the case during the tsunami in South-East Asia in December 2004. Furthermore, Tsunami Warning System is based on the concept that tsunamis travel at a much slower velocity (500 to 700 km/h) as compared to seismic waves (6 to 8 km/s). That is, the seismic waves move 30 to 40 times faster than the tsunami waves. Hence, after the occurrence of a damaging earthquake and quick determination of epicenter, warning time of a few minutes to 2 to 3 hours is available depending upon the distance from the epicenter to the coast line. This time can be utilized for warning the coastal community if quick detection and rapid communication systems are established. Thus, the geospatial technologies (GIS, GPS and RS), modeling techniques and Tsunami Early Warning System are promising tools for reducing the risk of tsunami disasters and developing effective disaster management plans.

Finally, actions related to reconstruction and recovery of tsunami hazards should ensure that the sustainability of coastal and marine ecosystems is not compromised, and is ideally enhanced as the goods and services they provide strengthen the livelihoods and immediate welfare of large coastal populations. A comprehensive coastal zone management strategy is required to reflect livelihood needs, reduce vulnerability to natural hazards, and the conservation of biodiversity and ecological services. Economic, environmental, social and cultural factors must all be taken into account when developing disaster risk mitigation strategies and solutions must be anchored in the prevailing circumstances of local situations.

REFERENCES

Alami, S.O.E. and Tinti, S. (1991). A preliminary evaluation of the tsunami hazards in the Moroccan coasts. *Science of Tsunami Hazards*, **9(1):** 31-38.

Babu, N., Suresh Babu, D.S. and Mohan Das, P.N. (2007). Impact of tsunami on texture and mineralogy of a major placer deposit in southwest coast of India. *Environmental Geology*, **52:** 71-80.

Bandibas, J., Warita, K. and Kato, H. (2003). Interactive presentation of geological hazardmaps using geohazardview. *Journal of Natural Disaster Science*, **25(2):** 75-83.

Chandrasekar, N., Immanuel, J.L., Sahayam, J.D., Park, J., Anderson, K., Aster, R., Butler, R., Lay, T. and Simpson, D. (2006). Global Seismographic Network records the great Sumatra-Andaman earthquake. *EOS Trans. AGU*, **86(6):** 57-64.

Chandrasekar, N. and Immanue, J.L. (2005). GIS supported categorisation of tsunami experienced beaches along the southern east coast of India: Usage in mitigation activities. Proceedings of the National Seminar on GIS Application in Rural Development, Hyderabad, India. pp. 349-362.

Chandrasekar, N., Immanuel, J.L., Rajamanickam, M., Sahayam, J.D., Singh, D.S.H. and Rajamanickam, G.V. (2006a). Geospatial assessment of Tsunami 2004 damages along the rocky coast of Kanyakumari, India. *In:* S.M. Ramasamy, C.J. Kumanan, R. Sivakumar and B. Singh (eds.), Geomatics in Tsunami, New India Publishers, New Delhi, pp. 135-146.

Chandrasekar, N., Immanuel, J.L., Sivasubramanian, P. and Rajamanickam, G.V. (2005). Impact of tsunami between the coast of Thoothukudi and Periyathalai. *In:* N. Chandrasekar (ed.), Quaternary Climatic Changes and Landforms, M.S. University Publishers, pp. 299-314.

Chandrasekar, N., Saravanan, S., Immanuel, J.L., Rajamanickam, M. and Rajamanickam, G.V. (2006b). Classification of tsunami hazard along the southern coast of India: An initiative to safeguard the coastal environment from similar debacle. *Science of Tsunami Hazards*, **24(1):** 3-24.

Chidambaram, S., Ramanathan, AL., Prasanna, M.V., Lognatan, D., Badri Narayanan, T.S., Srinivasamoorthy, K. and Anandhan, P. (2008). Study on the impact of tsunami on shallow groundwater from Portnova to Pumpuhar using geoelectrical technique: South east coast of India. Indian *Journal of Marine Sciences*, **37(2):** 121-131.

Dawson, A.G. (1994). Geomorphological effects of tsunami run-up and backwash. *Geomorphology*, **10(1-4):** 83-94.

DOD (2005). Preliminary Assessment of Impact of Tsunami in Selected Coastal Areas of India. Department of Ocean Development (DOD), Integrated Coastal and Marine Area Management Project Directorate, Chennai, India.

EEFIT (2006). The Indian Ocean tsunami of 26 December 2004: Mission findings in Sri Lanka and Thailand. Earthquake Engineering Field Investigation Team (EEFIT) Report, The Institution of Structural Engineers, London, U.K.

Ghouse, S.M. (2005). Tsunami Mitigation strategies in Coastal Areas of Tamil Nadu. *In:* S.M. Ramasamy and C.J. Kumanan (eds.), Tsunami: The Indian Context, Allied Publications, Chennai, India, pp. 287-291.

GSI (2005). Tsunami Impact. *In:* Great Tsunami of 26 December 2004 in Sumatra Region. Geological Society of India (GSI), Bangalore, India, pp. 33-38.

Imamura, F. (2004). Risk evaluation and real time information for tsunami disaster mitigation. Proceedings of the Wave, Tide Observations and Modeling in the Asian-Pacific Region, pp. 93-100.

Jaffe, B.E. and Gelfenbaum, G. (2002). Using tsunami deposits to improve assessment of tsunami risk. Proceedings of the Conference on Solutions to Coastal Disasters, ASCE, pp. 836-847.

Keating, B., Whelan, F. and Brock, J.B. (2004). Tsunami deposits at Queen's Beach, Hawaii: Initial results and wave modeling. *Science of Tsunami Hazards*, **22(1)**: 23-43.

Kumanan, C.J. (2005). Global scenario of tsunamis. *In:* S.M. Ramasamy and C.J. Kumanan (eds.), Tsunami: The Indian Context. Allied Publications, Chennai, India, pp. 15-26.

Kumaraguru, A.K., Jayakumar, K., Jerald, W.J. and Ramakritinan, C.M. (2005). Impact of the tsunami of 26 December 2004 on the coral reef environment of Gulf of Mannar and Palk Bay in the southeast coast of India. *Current Science*, **89**: 10.

Kurian, N.P., Pillai, A.P., Rajith, K., Krishnan, B.T.M. and Kalaiarasan, P. (2006). Inundation characteristics and geomorphological impacts of December 2004 tsunami on Kerala coast. *Current Science*, **90(2)**: 240-249.

Laluraj, C.M., Kesavadas, V., Balachandran, K.K., Gerson, V.J., Martin, G.D., Shaiju, P., Revichandran, C., Joseph, T. and Nair, M. (2007). Recovery of an estuary in the southwest coast of India from tsunami impacts. *Environmental Monitoring and Assessment*, **125(1-3)**: 41-45.

Leclerc, J.P., Berger, C., Foulon, A., Sarraute, R. and Gabet, L. (2008). Tsunami impact on shallow groundwater in the Ampara district in Eastern Sri Lanka: Conductivity measurements and qualitative interpretations. *Desalination*, **219**: 126-136.

Mattsson, E., Ostwald, M., Nissanka, S.P., Holmer, B. and Palm, M. (2009). Recovery and protection of coastal ecosystems after tsunami event and potential for participatory forestry CDM: Examples from Sri Lanka. *Ocean & Coastal Management*, **52**: 1-9.

Minoura, K., Nakaya, S. and Uchida, M. (1994). Tsunami deposits in a lacustrine sequence of the Sanriku coast, northeast Japan. *Sedimentary Geology*, **89(1/2)**: 25-31.

Mohan, V.R. (2005). December 26th, 2004 Tsunami: A field assessment in Tamil Nadu. *In:* S.M. Ramasamy and C.J. Kumanan (eds.), Tsunami: The Indian Context. Allied Publications, Chennai, India, pp. 139-153.

Mruthyunjaya Reddy, K., Nageswara, Rao, A. and Subba Rao, A.V. (2005). Recent Tsunami and its Impacts on Coastal Areas of Andhra Pradesh. *In:* S.M. Ramasamy and C.J. Kumanan (eds.), Tsunami: The Indian Context. Allied Publications, Chennai, India, pp. 129-138.

Mujabar, P., Chandrasekar, N., Saravanan, S. and Immanuel, J.L. (2007). Impact of the 26 December 2004 Tsunami on beach morphology and sediment volume along the coast between Ovari and Kanyakumari, Tamil Nadu, South India. *Shore & Beach*, **75(2)**: 22-29.

Nagendra, R., Kamalak Kannan, B.V., Sajith, C., Gargi Sen, Reddy, A.N. and Srinivasalu, S. (2005). A record of foraminiferal assemblage in tsunami sediments along Nagappattinam coast, Tamil Nadu. *Current Science*, **89**: 11.

Narayan, J.P., Sharma, M.L. and Maheswari, B.K. (2005a). Effects of medu and coastal topography on the damage pattern during the recent Indian Ocean tsunami along the coast of Tamil Nadu. *Science of Tsunami Hazards*, **23(2)**: 9-18.

Narayan, J.P., Sharma, M.L. and Maheswari, B.K. (2005b). Run-up and inundation pattern developed during the Indian Ocean tsunami of December 26, 2004 along the coast of Tamil Nadu (India). *Gondwana Research*, **8(4)**: 611-616.

Narayana, A.C., Tatavarti, R. and Shakdwipe, M. (2005). Tsunami of 26th December 2004: Observations on Kerala coast. *Journal of Geological Society of India*, **65(2)**: 239-246.

NDM (2005). Prevention/Protection and Mitigation from Risk of Tsunami Disasters. Strategy Paper, National Disaster Management, Ministry of Home Affairs, http://www.ndmindia.nic.in/ (accessed on 11 February 2009).

Palanivelu, K., Priya, M.N., Selvan, A.M. and Nateshan, U. (2006). Water quality assessment in the tsunami-affected coastal areas of Chennai. *Current Science*, **91(5)**: 583-584.

Papathoma, M., Dominey-Howes, D., Zong, Y. and Smith, D. (2003). Assessing tsunami vulnerability, an example from Herakleio, Crete. *Natural Hazards Earth System Science*, **3(5):** 377-389.

Ram Mohan, V. (2005). December 26, 2004 Tsunami: A field assessment in Tamil Nadu. *In:* S.M. Ramasamy and C.J. Kumanan (eds.), Tsunami: The Indian Context. Allied Publications, Chennai, India, pp. 139-153.

Ramachandran, S., Anitha, S., Balamurugan, V., Dharanirajan, K., Ezhil Vendhan, K., Marie Irene, P.D., Senthil, V.A., Sujjahad Hussain, I. and Udayaraj, A. (2005). Ecological impact of tsunami on Nicobar Islands (Camorta, Katchal, Nancowry and Trinkat). *Current Science*, **89(1):** 10.

Ramaswamy, S.M. and Kumanan, C.J. (2005). Tsunami: The Indian Context. Allied Publications, Chennai, India.

Rasheed. A.K.A., Das, V.K., Revichandran. C., Vijayan. P.R. and Thottam. T.J. (2006). Tsunami impacts on morphology of beaches along South Kerala Coast, West Coast of India. *Science of Tsunami Hazards*, **24(1):** 24-34.

Raval, U. (2005). Some factors responsible for the devastation in Nagapattinam region due to tsunami of 26th December 2004. *Journal of Geological Society of India*, **65(5):** 647-649.

Rossetto, T., Peiris, N., Pomonis, A., Wilkinson, S.M., Del Re, D., Koo, R. and Gallocher, S. (2007). The Indian Ocean tsunami of December 26, 2004: Observations in Sri Lanka and Thailand. *Natural Hazards*, **42:** 105-124.

Salagrama, V. (2006). Post-tsunami Rehabilitation of Fishing Communities and Fisheries-based Livelihoods in Tamil Nadu, Kerala and Andhra Pradesh, India. Proceedings of the ICSF Post-Tsunami Rehabilitation Workshop, pp. 159-210.

Sheth, A., Sanyal, S., Jaiswal, A. and Gandhi, P. (2006). Effects of the December 2004 Indian Ocean Tsunami on the Indian Mainland. *Earthquake Spectra*, **22(S3):** S435-S473.

Shetye, S.R. (2005). Tsunamis: A large-scale earth and ocean phenomena. *Resonance*, **10:** 8-19.

SIFFS (2005). The Tsunami in Kerala: An overview of tsunami rehabilitation in Alappad and Arattupuzha. South Indian Federation of Fishermen Societies (SIFFS), Kerala Information Centre, 16 pp.

Sonak S., Pangam, P. and Giriyan, A. (2007). Green reconstruction of the tsunami-affected areas in India using the integrated coastal zone management concept. *Journal of Environmental Management*, doi:10.1016/j.jenvman.2007.01.052.

Srinivas, H. and Nakagawa, Y. (2007). Environmental implications for disaster preparedness: Lessons Learnt from the Indian Ocean Tsunami. *Journal of Environmental Management*, doi:10.1016/j.jenvman.2007.01.054.

Srinivasulu, S., Nagendra, R., Rajalakshmi, P.R., Thangadurai, N., Arun Kumar, K. and Achyuthan, H. (2005). Geological signatures of sediments of M9 Tsunami event along Tamil Nadu coast. *In:* S.M. Ramasamy and C.J. Kumanan (eds.), Tsunami: The Indian Context. Allied Publications, Chennai, India, pp. 171-181.

Sujatha, C.H, Aneeshkumar, N. and Renjith, K.R. (2008). Chemical assessment of sediment along the coastal belt of Nagapattinam, Tamil Nadu, India after the 2004 tsunami. *Current Science*, **95(3):** 10.

Tanaka, N., Sasaki, Y., Jinadasa, K.B.S.N., Mowjood, M.I.M. and Homchuen, S. (2007). Coastal vegetation structures and their functions in tsunami protection: Experience of the recent Indian Ocean tsunami. *Landscape Ecological Engineering*, **3:** 33-45.

USGS (2005). Earthquake Hazards Program, http://earthquake.usgs.gov/eqinthenews/2004/usslav/ (accessed on 22 November 2008).

Walsh, T.J., Caruthers, Ch.G., Heinitz, A.C., Myers, E.P., Baptista, A.M., Erdakos, G.B. and Kamphaus, R.A. (2000). Tsunami hazard map of the Southern Washington Coast: Modelled tsunami inundation from a Cascadia Subduction Zone earthquake. Washington Division of Geological Earth Response Reports, GM-49, 12 pp.

World Bank (2002). Integrated coastal management. Coastal and Marine Management: Key Topics. The World Bank Group, Washington, USA, http://lnweb18.worldbank.org/ESSD/essdext.nsf/42ByDocName/KeyTopicsIntegratedCoastalManagementS (accessed on 22 November 2008).

Xue, X., Hong, H. and Charles, A.T. (2004). Cumulative environmental impacts and integrated coastal management: The case of Xiamen, China. *Journal of Environmental Management*, **71:** 271-283.

Tsunami Impacts and Rehabilitation of Groundwater Supply: Lessons Learned from Eastern Sri Lanka

Karen G. Villholth[*], Paramsothy Jeyakumar, Priyanie H. Amerasinghe, A. Sanjeewa P. Manamperi, Meththika Vithanage, Rohit R. Goswami and Chris R. Panabokke

1. INTRODUCTION

Huge devastation and human tragedy followed the December 26, 2004 tsunami in the Indian Ocean. The death toll from the earthquake, the tsunamis and the resultant floods totals to over 180,000 people in fourteen countries with tens of thousands reported missing, and over one million left homeless. After Indonesia, Sri Lanka was the second-hardest hit, with an estimated death toll of 35,000 people. Of immediate concern after the catastrophic event was the destruction of the traditional water supply system in the rural and semi-urban areas of the coastal belt, which for 80% of the population was based on groundwater, mostly drawn from private shallow open wells (Leclerc et al., 2008). Practically all wells within the reach of the flooding waves (up to a couple of kilometers inland) were inundated and filled with saltwater and contaminated with solid matter (sediment and waste), pathogens and other unknown chemicals, leaving the water unfit for drinking (Fig. 1a). Figure 1b shows the devastation of the 2004 Tsunami in a representative coastal town on the eastern coast of Sri Lanka. The density of wells in these areas is high, with practically each household having its own well (Fig. 2). Villholth et al. (2008) found a well density up to 600 per km^2. The local villagers reported that the water level reached close to the top of the coconut palm trees shown in the figure (approximately 5 m). Over one thousand lives were lost at this particular site. ADB et al. (2005) and UNEP (2005) estimated that the tsunami waves contaminated more than 50,000 wells in coastal Sri Lanka. This initial figure is highly underestimated, however, as the present research found, based on well statistics and flooding patterns, that approximately 18,000 wells were affected in an area representing only 3% of the affected coast line in Sri Lanka (Villholth et al., 2009). This suggests that the total number of affected wells is more in the range of half a million. The total number of people, mostly living in poor coastal rural areas, affected by the disruption of their water supply from wells could be in the range of 2.5 million. Approximately 75% of the coastline of Sri Lanka was impacted by the tsunami.

Though the detriment to the water quality was not immediately detectable, and little concerted effort was dedicated to analyzing contamination parameters and assessing various health-related contamination risks associated with the use of the flooded wells, it was obvious that salinity caused a pressing constraint for drinking, simply from a taste-based criterion. Pathogenic contamination was inevitable because of

---

*Corresponding Author

Fig. 1 Tsunami impact on open well and the general devastation of the coastal area in east Sri Lanka (*Courtesy* **a:** Scott Tylor and Jayantha Obeysekera; **b:** Karen Villholth).

Fig. 2 Tsunami-impacted site on the east coast of Sri Lanka (*Courtesy:* Scott Tylor and Jayantha Obeysekera).

the concomitant flooding of pit latrines, which often are located in close proximity to the wells (Fig. 2b), and because of a general mixing of waters during the inundation.

Intense efforts of cleaning of wells were initiated by the authorities, NGOs (non-governmental organizations), international aid organizations, incoming volunteers and individual well owners immediately after the event in an attempt to quickly restore pre-tsunami water quality conditions. However, due to the emergency character of the situation and the lack of technical knowledge, the methods applied were quite ad-hoc and haphazard, and purely documented and monitored. The special characteristics and vulnerability of the fresh, but shallow and unconfined groundwater systems in a coastal setting and the importance of the groundwater in the local water supply warranted a precautionary and informed approach to avoid further or prolonged salinity problems. The sparse data collected and word of mouth, however, seemed to agree on the fact that the cleaning and repeated emptying of wells did not have the expected positive impact on salinity (Leclerc et al., 2008; Lipscombe, 2007; Fesselet and Mulders, 2006; Clasen and Smith, 2005).

The approach to well cleaning can be defended by the circumstances of urgency. However, it was clear that significant resources went into these activities which were not grounded on proper understanding of best practices and which at best did little good to remediate the water quality problem and at worst may have exacerbated the salinity problem by prolonging the time to or prohibiting the recovery of freshwater conditions (Vithanage et al., 2009). In the Sri Lankan case, the water supply situation was ameliorated through a dedicated and immense effort of a host of relief and rehabilitation agencies and individuals, ensuring relatively safe drinking water following the catastrophe through intermediate alternative sources, especially water tankering[1], and hence the sub-optimal well cleaning procedures were not an immediate life-threatening problem. However, from a relief efficiency provision perspective and a groundwater sustainability perspective, recognizing that groundwater is the backbone of all water supplies in these areas, it was critical to optimize the procedures based on proper understanding of the scope of the problem and the best and most efficient practice to addressing it (Lytton, 2008).

The objective of this chapter is to present the experiences and outcomes from a comprehensive and integrated approach to supporting the restoration of the groundwater-based water supply and the drinking water quality after the 2004 Tsunami in Sri Lanka. This is achieved specifically from on-the-ground interactions with various actors and stakeholders, salinity well water monitoring initiated immediately after the tsunami, from technical research into the physical phenomenon of saltwater flooding, and finally from the derived guidelines for well cleaning and groundwater protection following seawater flooding. Based on this, recommendations for similar events in the future are proposed.

2. METHODOLOGY

The approach involved short-term support to the water supply relief efforts, initiated immediately after the tsunami as well as more long-term research to increase the understanding of the processes related to saltwater flooding and the implications for best practices for groundwater quality restoration and well water supply rehabilitation.

The following activities were initiated:

(1) On-the-ground guidance on well cleaning and awareness raising on water salinity issues.
(2) Monitoring of salinity and other water quality parameters in drinking water wells in affected areas.
(3) Household survey of perceptions of water quality aspects and remediation of water supply.
(4) Detailed studies of the physical processes related to saltwater flooding in coastal-near regions and the impacts of groundwater pumping and natural flushing on water quality restoration.
(5) Development of a set of internationally endorsed guidelines on well cleaning and groundwater protection after saltwater flooding.

2.1 On-the-Ground Guidance of Well Cleaning and Awareness Raising on Water Salinity Issues

It was clear from initial visits to the affected areas in late January 2005 and personal communication with the NGOs and local authorities that well cleaning was a major challenge. Procedures were carried out in an uncoordinated and haphazard manner with limited understanding of the physics of groundwater and

---

[1] Water provided in by various relief organizations. Supplied from fresh, unaffected surface or groundwater sources in mobile tankers and delivered to the road-side containers for general access.

wells and the possible risks associated with the cleaning procedures. A simple procedure of emptying the wells by use of electric pumps and discharging the saltwater next to the wells was the most common practice (Fig. 3). Wells were cleaned and chlorinated several times (Keba, 2006), up to five times during the 16 months following the tsunami (Villholth et al., 2009). Little coordinated monitoring and reporting of the efforts was performed. Excessive pumping and the recycling of the discharged saltwater likely retarded the rehabilitation of freshwater conditions in the groundwater (Vithanage et al., 2009). Although limited observations showed slow or no beneficial effects (Clasen and Smith, 2005; Fesselet and Mulders, 2006; Lipscombe, 2007; Leclerc et al., 2008) and concerns about over-pumping and seawater intrusion were raised, pumping continued unabated for a long time.

Partly a reason for the confusion related to well cleaning was the lack of easily available well cleaning guidelines (Clasen and Smith, 2005, Villholth et al., 2005; Illangasekare et al., 2006). No formalized guidelines existed that were relevant and appropriate for the situation encountered after a tsunami disaster. The existing emergency guidelines for the cleaning of open shallow wells either addressed the problem of wells being destructed and contaminated after flooding with dirty *freshwater*, and related more to the problems of microbiological contamination (e.g., WHO, 2005; CDC, 2005), or improperly addressed the problem of *saltwater* flooding (NGWA, 2005; OXFAM, 2006). All of these guidelines recommended the purging, or emptying, of the wells to remove contaminated water prior to and after a chlorination step and/or advised to continue pumping until the well water becomes clear and free of saltwater. Also, a report published after the tsunami addressing the role of groundwater in emergencies erroneously prescribed prolonged pumping after a tsunami to restore the water quality to drinking water standards (Vrba and Verhagen, 2006). The tsunami-dedicated guidelines from the National Groundwater Association (NGWA), USA were later removed from the internet. The risks of excessive pumping included physical instability of the well, ingress of contaminated water from other areas, e.g., from toilet pits, burial grounds or from accumulated salty surface water, saltwater upconing from below, and recycling of discharged saltwater. Furthermore, as mentioned, the repeated pumping of wells diverted important human resources from other emergency relief tasks.

As part of the comprehensive approach to rehabilitation support, the International Water Management Institute (IWMI), with headquarters in Colombo, Sri Lanka, initiated various on-the-ground initiatives to facilitate the well and water supply rehabilitation efforts:

Fig. 3 Cleaning a typical drinking water well after the 2004 Tsunami in Sri Lanka (*Courtesy:* Jannick M Christensen).

(1) Participation and contribution to various fora and meetings related to water and sanitation (WATSAN) rehabilitation, at local and national level, targeting various partners and stakeholders: the national WATSAN coordination unit of international donors and relief organizations, national and local authorities, and local and international NGOs. Contributions included inputs to discussions, delivery of information on the potential impacts of the tsunami from saltwater flooding, the physical processes involved in saltwater-freshwater interactions in coastal-near areas, and the best practices for well cleaning in terms of salinity remediation and potential risks and adverse impacts of faulty procedures.

(2) Information sharing and coordination of on-going activities related to the saltwater contamination and rehabilitation of water supply, to incoming and already present partners.

(3) Collaboration with international relief and environmental organizations on workshops for local health workers involved in the provision of safe drinking water and the development of joint guidelines for water pollution remediation, well cleaning and groundwater protection after the tsunami (IUCN, 2005; UNICEF, 2005).

(4) Development and dissemination of adapted guidelines for well cleaning and groundwater protection during the aftermath of the tsunami, initially advocating pumping, but later discouraging pumping as a means to ameliorate salinity (Villholth et al., 2005; Villholth, 2007).

2.2 Monitoring of Salinity in Drinking Water Wells in Affected Areas

In order to document the impacts of the tsunami on well water quality, especially salinity, and to follow the temporal changes and the process of recovery, a monitoring program was initiated in three representative impacted villages in the east coast of Sri Lanka (Fig. 4), with a rudimentary first sampling taken on January 26, 2005, just one month after the tsunami, followed by a systematic and more comprehensive monitoring program covering a total of approximately 150 wells, mostly shallow open drinking water wells (Villholth et al., 2005). The shallow unconfined aquifer system in these areas is developed in the coastal sediments dominated by Quaternary unconsolidated sand deposits intermixed with more fine materials, on the top of Precambrian metamorphic rocks (Cooray, 1985; Panabokke and Perera, 2004). The bedrock, consisting of granitic rock of the Precambrian metamorphic basement (the Vijayan complex), is found at about 15-25 m depth (Wickramaratne, 2004). The underground freshwater is delimited by a saltwater wedge entering from the sea side and thereby limiting the overall availability of freshwater in these systems, which in many cases are also limited by more or less brackish lagoons on the inland side (Fig. 4). Hence, these coastal groundwater systems are naturally vulnerable to salinity problems, even in the absence of a tsunami, and a general attention to long-term sustainability issues in

Fig. 4 Location map of the study area.

the perspective of increased population pressure and climate change is warranted. Average annual rainfall experienced in the area varies between 1000 mm and 1700 mm. Eighty percent of the annual rainfall occurs in the north-east monsoon period from November to April (Panabokke et al., 2002).

Repeated (on average at 2-month intervals) salinity monitoring in the same wells (concrete cylinder wells, average depth 3.4 m, and open at the bottom) over a period of two years after the tsunami was performed. The detailed methodology is described in Villholth et al. (2005).

2.3 Field Monitoring of the Impact of the First Rains after the Tsunami on Groundwater and Lagoon Water Quality

Water quality measurements were carried out in drinking water wells (same as in the salinity monitoring program) and in lagoon water just before and after the onset of the first full rainy season after the tsunami, i.e., September 2005 and November 2005, respectively (Villholth et al., 2006). The objective was to observe the effect of the first rains of the first rainy season after the tsunami on the well water quality. The parameters of pH, electrical conductivity (EC), Na, Cl, hardness, Ca, Mg, SO_4, NO_3, PO_4, K, Mn, Al, and Fe were monitored.

2.4 Household Survey

A household survey was conducted in the same areas as the salinity monitoring program in order to gain insight into the conditions prevailing in the villages with respect to the relief water supply and its rehabilitation, the water quality, the implication for daily routines related to water use, and the perception of the affected population regarding these matters. The survey was conducted in two rounds (in April/May of 2006 and January/February 2007, corresponding to 16 months and two years after the event, respectively) in the same 120 households—60 in each of the two most heavily impacted villages. The first survey took place during the phase of rehabilitation where tankered water constituted the major source of drinking water supply and alternative to shallow wells, and the second survey was conducted after the discontinuation of the emergency supply and after the population partially returned to use their wells (Villholth et al., 2009).

2.5 Detailed Studies of the Physical Processes Related to Saltwater Flooding

Saltwater entered the aquifer through the ground surface, from infiltration during the inundation (source 1 in Fig. 5), from ponds, drainage channels, and interim water accumulations in ground depressions after the event (source 2 in Fig. 5), and from flooded open wells (source 3 in Fig. 5). Finally, the saltwater-freshwater interface separating fresh and saltwater in the coast may have been disturbed or displaced from the impact of the tsunami (source 4 in Fig. 5). The problem of saltwater flooding is unique in the sense that in coastal areas, saltwater will naturally be underlying freshwater in the aquifer due to a higher density of the saltwater compared to freshwater. The contamination event, caused by the tsunami entailed an 'unnatural' and unstable situation with saltwater temporarily lying on top of freshwater. Due to the higher density of the saltwater it will tend to sink or overturn so that the natural balance of freshwater overlying saltwater is reestablished (Illangasekare et al., 2006).

The studies related to the physical processes of saltwater flooding and water and aquifer restoration in coastal areas included the following:

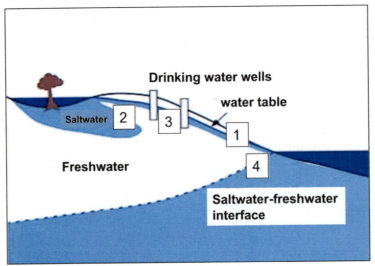

Fig. 5 Sources of saltwater ingress from the tsunami to the groundwater: 1. Infiltration through the soil surface during inundation; 2. Entry into open dug wells; 3. Percolation from depressions and ponds; and 4. Disturbance of the saltwater-freshwater interface.

- Field, laboratory and numerical modeling experiments to mimic the seawater flooding, saltwater movement, and the impacts on wells and of subsequent pumping; and
- Field investigations of the comparative process of recovery of the groundwater quality in natural, undisturbed areas and in areas disturbed by well cleaning and pumping for domestic use.

2.5.1 Field, Laboratory and Numerical Modeling Experiments

These experiments were conducted to understand fundamental processes affecting the mixing and movement of saltwater in a freshwater system after a saltwater flooding event. Results from the experiments were also used to support the development of guidelines for well cleaning related to seawater flooding events. The laboratory and numerical experiments were conducted using uniform homogeneous media and hydraulic conditions representing sandy aquifer settings in a flat terrain similar to the east coast of Sri Lanka, with a hydraulic conductivity of 1e–4 to 6e–3 m/s and a hydraulic gradient of 1% (Hogan et al., 2006; Goswami and Clement, 2007). A detailed model description and calibration was not attempted due to the lack of data and information about necessary boundary and initial conditions such as the initial conditions at the time of the tsunami (e.g., groundwater level and soil moisture content), and the stress function related to the tsunami waves (e.g., height and number of the waves, duration and extent of inundation(s), and exact mechanisms of saltwater entry). Therefore, experimental datasets were used to describe and visualize the general post-tsunami sub-surface flow, transport and mixing patterns and processes under representative and best knowledge conditions and to investigate some of the factors influencing the processes. Numerical modeling was conducted using the variable density groundwater model SEAWAT (Langevin and Guo, 2006). The model was validated by successfully simulating a concentration breakthrough curve from a saltwater tracer injection experiment in a three dimensional (3D) tank (Goswami et al., 2007). Laboratory experiments were conducted under controlled conditions in a quasi 2D and a 3D flow tank having dimensions of 53 cm (L) × 30.5 cm (H) × 2.7 cm (W), and 42 cm (L) × 27 cm (H) × 25 cm (W), respectively (Hogan et al., 2006; Goswami et al., 2007).

2.5.2 Field Investigations of the Comparative Process of Recovery of the Groundwater Quality in Undisturbed and Disturbed Areas

This part of the study investigated the chemical characteristics of the shallow groundwater post tsunami in two areas: an uninhabited area undisturbed by pumping and cleaning and in a village closeby where cleaning of wells and domestic use after the tsunami occurred. The two areas were located within 500 m distance in the same overall study area encompassing the monitoring program (Fig. 4). The water quality was monitored in twelve existing open wells (out of which four were abandoned, i.e. not in use after the tsunami) in the disturbed site while it was sampled from 12 shallow piezometers in the undisturbed site. The wells were located on two separate transects transversing the 2 km land strip from the sea to the inland lagoon (Vithanage et al., 2009). The monitoring of water quality was performed three times, or more, over a year, from October, 2005 to October, 2006.

2.6 Development of Internationally Endorsed Guidelines on Well Cleaning and Groundwater Protection after Seawater Flooding

Recognizing the need for practical guidelines dedicated to the emergency situation of cleaning wells after a seawater flooding event, this project developed, in consultation with experts from WEDC and WHO set of guidelines, which was integrated into their series of technical fact sheets. The guidelines supplement existing emergency well cleaning and disinfection guidelines (WHO, 2005) and focus on saltwater flooding. It intends to facilitate and accelerate provision of safe and contaminant-free water for drinking and other domestic purposes, protecting the coastal aquifer, minimizing the potential for saltwater intrusion (drawing saltwater into the well), and minimizing the physical destruction of the wells.

3. RESULTS AND DISCUSSION

3.1 Salinity in Drinking Water Wells in the Affected Areas

The results of the monitoring program showed that (Villholth et al., 2005; Villholth, 2007):
(1) The flooding waves had reached up to a distance of 1.5 km and 39% of the wells within a 2 km distance from the coast line were flooded by the tsunami waves.
(2) The impact varied between sites, probably depending on the specific initial force of the tsunami waves, and differences in topographic and coastline characteristics.
(3) Primarily, the groundwater was affected only in those areas flooded by the tsunami waves.
(4) The areas remained impacted by elevated salinity for up to 1.5 years after the tsunami, after which the impacted wells on average had the same salinity levels as the unaffected wells (Fig. 6).
(5) Reductions in average salinity levels are visually correlated with rainfall, indicating that rainfall was a primary agent for the restoration of the aquifers and the well water salinity (Fig. 6).

These findings are in line with those of Fesselet and Mulders (2006), Lytton (2008) and Leclerc et al. (2008). From these observations and by extension, it is clear that the scale and extent of flooding and salinization of the groundwater and the disruption of water supply systems were very severe. Furthermore, the impacts were long-lived. The prolonged elevated salinity levels in the wells indicated that the pumping and cleaning of wells did not have the overall intended impact of rapidly reverting the wells to pre-tsunami drinking water quality. Furthermore and very importantly, the data suggest that despite prolonged

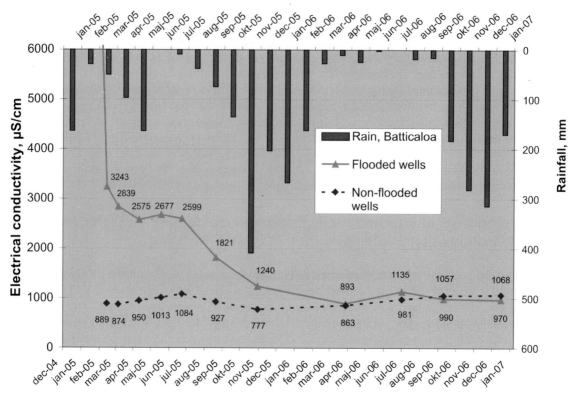

Fig. 6 Average well water salinity and monthly rainfall after the tsunami in the Batticaloa area in eastern Sri Lanka. Non-flooded wells are assumed to represent the background salinity level. (From Villholth, 2007)

effects, the aquifers have not suffered permanent and irreversible salinity impacts, irrespective of the non-optimal cleaning and pumping procedures performed.

3.2 Impact of the First Rains after the Tsunami on Groundwater and Lagoon Water Quality

From the water-quality monitoring before and after the onset of the first full rainy season after the tsunami it was found that chemical constituents originating from the tsunami influx (Na, Cl, Ca, Mg, SO_4, hardness) decreased as a result of the first rainwater flushing and dilution, while the parameters presumably derived from post-tsunami soil contamination (NO_3, PO_4) and intrinsic soil processes (Fe, Mn) increased as leaching processes initiated after a long dry season (Table 1) (Villholth et al., 2006). The same processes were inferred to occur with respect to the lagoon waters, though the dilution effect was more pronounced in the lagoon as opposed to the leaching effect (Table 1). Thus, the obtained results enhanced the understanding of overall water quality change processes occurring as a result of the tsunami flooding and the subsequent natural remediation processes and also the seasonal changes that occur irrespective of the tsunami. The findings support the recognition that these shallow aquifer systems are highly replenishable with short time scales for renewal, but at the same time highly vulnerable to surface leaching and contamination processes. Other monitoring has indicated a high risk of pathogenic

Table 1. Change of water quality before and after the onset of the first rainy season after the tsunami (Villholth et al., 2006)

	Wells				Lagoon	
	Sep-05		Nov-05		Sep-05	Nov-05
	Flooded	Non-flooded	Flooded	Non-flooded		
pH	7.82	7.79 / 7.77	7.67 / 7.70	7.65	7.7	7.62
EC mg/L	1821	1251 / 927	1240 / 948	777	28425	1498
Na mg/L	85.4	54.3 / 36.7	60.9 / 42.9	32.4	1847.5	86.8
Cl mg/L	410	248 / 156	278 / 180	123	11125	438
Hardness mmol/L	3.26	2.74 / 2.45	2.64 / 2.48	2.39	33.73	1.84
SO_4 mg/L	88.7	61.2 / 45.6	76.7 / 59.8	49.9	1700.0	65.5
NO_3 mg/L	7.28	7.74 / 8.00	11.1 / 17.0	20.4	0 / < 1[a]	2.00
PO_4 mg/L	0.11	0.54 / 0.78	0.44 / 0.77	0.97	0 / < 0.2	0.08
K mg/L	14.6	11.9 / 10.3	12.1 / 11.1	10.5	118.25	12.00
Mn μg/L	5.00	8.91 / 11.1	101.4 / 57.3	31.5	0.50	71.50
Al mg/L	0.20	0.10 / 0.04	0.07 / 0.06	0.06	0 / < 0.05	7.39
Fe μg/L	7.4	30.6 / 43.9	130.2 / 109.4	97.2	0 / < 20	4296.5

[a] Detection limit

contamination of well water, presumably from short-circuiting between domestic pit latrines and drinking water wells (Keba, 2006). ACF (2006) found all 122 wells sampled between July and November 2005 after the tsunami to be positive with respect to *E. coli* contamination. These findings illustrate that the pathogenic contamination was an essential part of the problem related to well water quality and drinking water provision, and in fact an issue that deserved relatively more focus.

3.3 Findings of Household Survey

The salient findings of the household survey, focusing on perceptions of the users and the problem areas emerging, showed that:

(1) Basically, all the households used their wells prior to the tsunami as their main/sole water supply and were quite satisfied with it (reliability and water quality-wise). Practically all wells of the households (94.2%) were flooded and salinized by the tsunami.
(2) Salinity was perceived as an over-riding immediate problem and constraint for the use of wells for drinking water provision.

(3) However, superstition and indefinite fear was also associated with the reluctance to using the wells just after the tsunami. Well cleaning, in addition to a perceived physical cleaning, had a psychological impact, supposedly rinsing the water from the evil content imprinted by the tsunami.

(4) No unsafe (unprotected) water sources were used just after the tsunami. This in combination with large focus on hygiene just after the tsunami meant that infectious diseases did not prevail in the immediate aftermath.

(5) Many people (approximately one quarter of respondents) expressed health problems after the tsunami related to the water supply, reporting diarrhea and skin rashes as the main problems, in the 1st and 2nd survey, respectively. Skin rashes were interpreted as caused by excessive and un-precautious well chlorination just after the tsunami and later diarrhea incidences due to diminished attention to hygiene after the immediate relief phase.

(6) People coped with the lack of traditional well water access by various means: saving water, sharing 'good' wells, improving hygienic practices to avoid water-related diseases, and using various water sources for different purposes, e.g., water for drinking and cooking from tankers and water for washing and cleaning from the wells.

(7) Alternative water sources, especially the tankered water did not provide a consistent and reliable source and acceptable quality of water also found by Keba (2006). Sometimes the chlorine taste was objectionable to the users and the supply short of demand.

(8) Two years after the tsunami, approximately half of the households (46%) still did not use their well for drinking but resorted to using neighbor wells or public stand posts, while they had reverted fully to using their own well for purposes of bathing and washing and partly for cooking. When asked for their reason for not using their well for drinking at this time, they still reported salinity as their main cause of concern. This result shows that a large fraction of the population continued to face problems or concerns regarding their drinking water supply after the major relief and rehabilitation efforts had terminated.

(9) The disproportionate focus on salinity as part of the rehabilitation efforts was exaggerated, purveying the erroneous impression to the users that salinity was a critical health problem, and this may in turn have retarded the acceptance of the people returning to the use of their wells for drinking when in fact salinity levels were not higher than the levels accepted pre-tsunami. This focus may also have diverted continued efforts of securing hygienic practices. In addition, people may have become accustomed to better quality water and hence were reluctant to use water of a quality which was acceptable to them before the tsunami.

(10) People were unclear about the process of rehabilitation of their water supply, e.g., the timing of the discontinuation of the tankering and the safety of returning to their wells [supported by Keba (2006)].

Summarizing and deriving some lessons-learned, there was a general need for more and balanced information to the affected people, in terms of the impacts on their wells, the water quality and health considerations, the implications for cleaning, disinfection, and use of wells and the provision of alternative sources, and the timing and process of the return to well use. These information activities, which could be made more interactive and participatory through integrated surveys extracting and applying the immediate perceptions of the people and using local champions as disseminators multiplying messages, would yield more confidence in the rehabilitation support and a smother transition from relief, to rehabilitation, to gradual recovery of livelihoods and 'normal life'. In addition, the results indicated that there was a general lack of competence in the more technical aspects and coordination of well cleaning, and chlorination of wells and tankered water. Prioritizing, rationalizing, and targeting efforts, some of the resources gone into repeated well cleaning could have been better spent in such activities.

3.4 Results of the Detailed Studies on the Physical Processes Related to Saltwater Flooding

3.4.1 Findings of Field, Laboratory and Numerical Modeling Experiments

Results from the combined numerical and laboratory experiments, relevant to the development of guidelines for well cleaning procedures, are as follows:

(1) The numerical model successfully reproduced the patterns of saltwater movement from the simultaneous sources—ground surface and well (sources 1 and 3 in Fig. 5), as observed in the laboratory (data not shown) (Goswami and Clement, 2007).
(2) The saltwater from the various sources of its entry migrated as separate plumes (Fig. 5), in the lateral and vertical direction towards the coast (data not shown) (Hogan et al., 2006).
(3) Stability of the saltwater plumes and mixing with contiguous freshwater depended on the ambient groundwater flow rate. Stability was found to increase with flow rate, resulting in less mixing of freshwater and saltwater (Hogan et al., 2006).
(4) Saltwater from the plumes migrated towards the coastal saltwater wedge and moved over the freshwater-saltwater interface without any noticeable mixing (Hogan et al., 2006).
(5) A major fraction of the saltwater that was injected into the wells rapidly and naturally descended as a large slug or plume from the bottom of the wells leaving only some residual saltwater in the well (Hogan et al., 2006).
(6) Early pumping of a contaminated well (within a week) and removal of discharged water significantly reduced well salinity. However, without pumping the salinity declined partially due to the sinking and dissipation of the saltwater from the well (Fig. 7). Pumping conducted at a later time would at best result in a temporary and small reduction of salinity as re-entry of saltwater from upstream sources or from the ground surface (source 1 in Fig. 5) caused a more long-term but less concentrated contamination of the well (Fig. 8) (Goswami and Clement, 2007).

The experiments simulated saltwater-flooding processes and the overall effects of pumping on well contamination and remediation. They also show the influence of natural cleaning on well salinity due to density-gradients and ambient groundwater flow. Rainfall enhances this process, though it was not directly included in the modeling effort. Deliberate cleaning of wells by pumping seemed to have a limited effect on salinity reduction, except when done very early where it may accelerate the innate process. However, re-contamination of wells is very likely from other upstream sources and from the saltwater front moving down from the soil surface. These findings support the observations in the monitoring program that wells were not rehabilitated to pre-tsunami salinity conditions from the large-scale, intensive cleaning efforts in the months after the tsunami. The idealized pattern of breakthrough in the wells as observed in the experiments, and the effect of single or few pumping events seen in Figs 7 and 8 was not captured in the continuous field monitoring study. This may be due to the low frequency of field monitoring which may not have sufficiently captured the temporal progress of salinity changes. Further complicating the analysis is the fact that the wells were cleaned repeatedly without maintaining any records. The well water, after cleaning, was discharged next to the wells, influencing and sustaining the local surface source next to the well. All these factors, in addition to the heterogeneity in the field and the occurrence

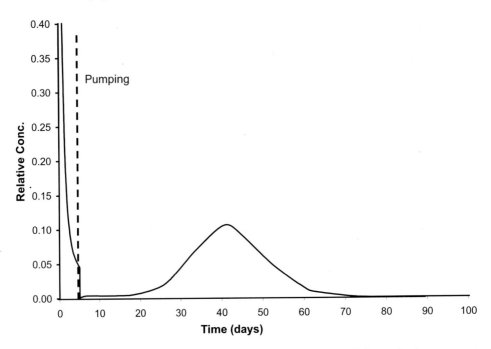

Fig. 7 Effect of early pumping, rapid natural dissipation of well salinity, and subsequent entry of saltwater from the surface source (Goswami and Clement, 2007).

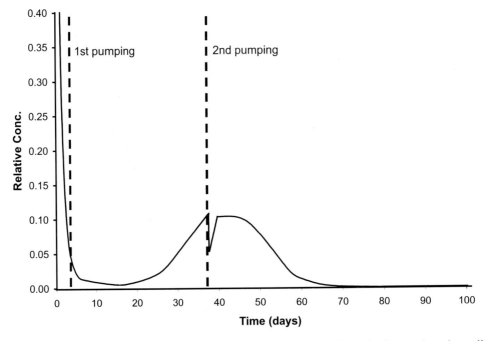

Fig. 8 Effect of a second late pumping. The re-entry of surface-derived saltwater into the well after pumping is almost immediate (Goswami and Clement, 2007).

of rainfall, make it difficult to compare breakthrough curves from the experiments and the field monitoring study.

The results presented above illustrate the processes pertinent to sandy, permeable unconfined aquifer systems which are representative of large parts of the Sri Lankan coast line (Panabokke and Perera, 2004) and typical coastal settings around the world. However, conditions different from those simulated here, e.g., hard rock or fractured systems, may present a context and problem different from the one addressed here.

3.4.2 Recovery of the Groundwater Quality in Undisturbed and Disturbed Areas

The results of the comparative study of the recovery of disturbed and undisturbed areas showed a discernible difference in the water quality and the recession of the contamination between the sites, focusing on parameters such as electrical conductivity (EC), concentration of Ca, Na, Cl, sulphate, and alkalinity. In general, the differences were in agreement with a larger degree of tsunami residual imprint on the wells in the disturbed area, e.g., showing higher levels and more small scale variability of EC and slower recovery in this area compared to the undisturbed area (Fig. 9). This suggests, assuming that the sites were very similar in other respects (e.g., rainfall conditions, and geological setting), that the pumping and cleaning of wells that went on at the disturbed site had the effect of disturbing the natural sinking and dissipation mechanisms of entered saltwater in the aquifer and prolonging the tsunami imprint by the continuous mixing of waters, by pumping, cleaning and recycling of waste water and discharged pumped-out well water locally. The finding was corroborated by the fact that the abandoned wells in the disturbed area similarly to the wells in the undisturbed area showed less of a tsunami imprint substantiating that cleaning and pumping of wells retards the natural cleaning process (Fig. 9).

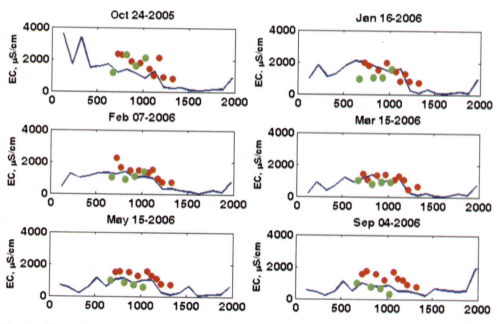

Fig. 9 Change in EC levels both in undisturbed (solid line) and disturbed sites (red circles: wells in use; green circles: abandoned wells) at six times during Oct. 2005 and Sep. 2006. The values on the x-axis indicate the distance of the observed sites from the sea (m).

Similar to EC, the patterns of Ca, Na, alkalinity and sulphate indicates relatively lower levels in the undisturbed and abandoned wells compared to the disturbed wells (Vithanage et al., 2009), supporting the increase of these parameters as an impact of the tsunami and the delayed recovery in areas subjected to post-tsunami well pumping and cleaning.

3.5 Guidelines on Well Cleaning and Groundwater Protection after Seawater Flooding

The guidelines, based on the findings of this study, emphasize that well pumping should be performed only once to remove debris, sludge and contaminated water standing in the well. Subsequent intensive pumping for cleaning, salinity removal, and disinfection should be avoided in order to limit the disturbance of natural ambient flushing and remediation processes. This advice is generally applicable for saltwater flooding in coastal areas where aquifer materials are dominated by loose sandy materials and the wells are open and shallow, typically the case in coastal villages in developing countries.

4. RECOMMENDATIONS

A disproportionate amount of efforts and attention went into well cleaning and raising concerns over well water salinity. This may be owing to a general lack of understanding among affected people and relief personnel of the impacts of saltwater flooding on groundwater quality and health, maybe to an extent where other equally pressing issues were not addressed adequately. In this chapter, we highlighted the lack of coordination and commensurate focus on information sharing, feedback, and participation from the affected people in the rehabilitation of wells and finding solutions to water supply. Also, there seemed to be a clear lack of focus and follow-up activities related to hygiene, pathogenic contamination and disinfection of wells, and safe water supply. These statements support the recommendation of developing more easily understandable information material related to saltwater flooding, and justifies the guidelines developed as part of this work. Also, approaches to enhance the involvement of the local affected population need to be developed and incorporated.

5. CONCLUSIONS

The analysis of the results of field investigations, and laboratory and modeling experiments presented in this chapter revealed that a tsunami can generate far-reaching and long-term impacts on groundwater resources, associated water supply systems, and affected people's access and perceptions to water. These impacts necessitate urgent and effective support to interim, effective and acceptable alternative water supply solutions, which may be required for an extended time and to ensure sustainable long-lasting solutions. The work presented in this chapter anticipated and responded to a need to incorporate informed knowledge, document experiences from the emergency response, to distill lessons learned, disseminate results, and develop guidelines to inform future actions (Clasen and Smith, 2005; Illangasekare et al., 2006; Lytton, 2008). The work was feasible due to the convenient location of IWMI headquarters in Colombo, Sri Lanka and its longstanding repute and working relations with a host of local, national and international partners. The guidelines and the associated support provided, in the form of workshops and meetings at various levels with stakeholders, NGOs, local and national authorities and donors where the problem of well and groundwater contamination and remediation approaches were discussed, were appreciated by the local actors (Lipscombe, 2007).

Evidence suggests that extensive and repeated well cleaning did not help and in fact could have aggravated and prolonged the salinity problem, and hence an overriding conclusion of this work is simple: avoid excessive groundwater pumping in connection with remediation, relief and rehabilitation efforts, both in relation to cleaning wells, groundwater restoration, and in connection with supplying alternative drinking water sources. These recommendations, along with associated advice on best practices for addressing the salinity problem and general drinking water provision primarily from groundwater have been distilled into a set of guidelines for well cleaning and groundwater protection for the use in similar future circumstances, supplementing existing guidelines. Existing guidelines would benefit from an amendment emphasizing that they do not apply to saltwater flooding conditions.

The work attempted to build a bridge between research and rehabilitation work, and in a sense supports a relief-rehabilitation-development nexus in a visionary approach. Indeed, the 2004 Tsunami presented a unique opportunity to gain fundamental insight into physical natural large-scale seawater flooding phenomena, a rare phenomenon normally studied only in small-scale artificially constructed systems. Nevertheless, the issue of saltwater ingress and saltwater flooding is likely to become increasingly important due to climate change and increased variability and extremes. Therefore, further research and support in terms of easily accessible information are needed with a view to preempt catastrophic events and prevent severe consequences, if they occur.

ACKNOWLEDGEMENTS

This chapter is based on the project funded by CARE and the Government of Canada (Agriculture and Agri-food Canada). All collaborating partners involved in this work are greatly appreciated for their individual contributions. Highlighted here are: T.H. Illangasekare, S.W. Tyler, T.P. Clement, J. Obeysekera, D.W. Hyndman, K.J. Cunningham, J.J. Kaluarachchi, W.W.-G. Yeh, M.-R. Van Genuchten, K.H. Jensen, P. Engesgaard, A.P.G.R.L. Perera, A. Gunatilaka, O. Woolley, M.D. Weerasinghe, N. Amalraj, S. Prathepaan, N. Bürgi, D.M.D.S. Lionelrathne, N.G. Indrajith, S.R.K. Pathirana, G. Jegan, M. Thilipan and V. Navalan.

REFERENCES

ACF (2006). Well hand over process: Water trucking phasing out, Manmunai North (Seelamunai/Dutchbar/New Dutchbar/Thiruchenthur). Internal paper, ACF, Batticaloa, Sri Lanka.

ADB, Japan Bank for International Cooperation and World Bank (2005). Sri Lanka – 2005 Post-Tsunami Recovery Program. Preliminary Damage and Needs Assessment, Colombo, Sri Lanka. January 10-28, 2005. http://www.adb.org/Tsunami/sri-lanka-assessment.asp (accessed on June 20, 2008).

CDC (2005). Disinfecting Wells Following an Emergency. Centers for Disease Control and Prevention (CDC), http://www.bt.cdc.gov/disasters/wellsdisinfect.asp (accessed on June 20, 2008).

Clasen, T. and Smith, L. (2005). The Drinking Water Response to the Indian Ocean Tsunami Including the Role of Household Water Treatment. Water, Sanitation and Health, Protection of the Human Environment, WHO, Geneva.

Cooray, P.G. (1985). Geology of Sri Lanka. National Museum of Sri Lanka, Colombo, Sri Lanka.

Fesselet, J.-F. and Mulders, R. (2006). Saline wells in Aceh. *Waterlines*, **24(3)**: 5-8.

Goswami, R.R. and Clement, T.P. (2007). Technical details of the SEAWAT model simulation results used to develop well cleaning guidelines. Technical Summary Report, Department of Civil Engineering, Auburn University, Aabama, USA.

Goswami, R.R., Villholth, K.G., Vithanage, M., Clement, T.P., Sakaki, T., Illangasekare, T.H. and Jensen, K.H. (2007). Variable density transport during well rehabilitation after a tsunami-induced saltwater contamination event. Paper presented at the ModelCARE Conference, 10-13 September 2007, Copenhagen, Denmark, pp. 402-403.

Hogan, M.B., Goswami, R.R., Villholth, K.G., Illangasekare, T.H. and Clement, T.P. (2006). Understanding the flow and mixing dynamics of saline water discharged into coastal freshwater aquifers. Selected peer-reviewed paper presented at the SWIM-SWICA meeting, 24-29 September 2006, Cagliari, Chia Laguna, Sardinia, Italy.

Illangasekare, T.H., Tyler, S.W., Clement, T.P., Villholth, K.G., Perera, A.P.G.R.L. Obeysekera, J. Gunatilaka, A., Panabokke, C.R., Hyndman, D.W., Cunningham, K.J., Kaluarachchi, J.J., Yeh, W. W.-G., Van Genuchten, M.-R. and Jensen, K.H. (2006). Impacts of the 2004 tsunami on groundwater resources in Sri Lanka. *Water Resources Research*, **42(5)**, W05201, doi:10.1029/2006WR004876.

IUCN (2005). After the Tsunami: Water Pollution. Series on Best Practice Guidelines (Sri Lanka). Information Paper No. 11.

Kar, A. and Gawri, C.P. (2006). Hurdles in framing groundwater management strategies in A&N Islands in the aftermath of tsunami. *In:* S. Romai, K.D. Sharma, N.C. Ghosh and Y.B. Kaushik (editors), Groundwater Governance: Ownership of Groundwater and its Pricing. Proceedings of the 12th National Symposium on Hydrology, 14-15 November 2006. New Delhi, pp. 181-202.

Keba, A.J. (2006). The role of beneficiary participation in the phase-out of post-emergency water tinkering, Ampara, Sri Lanka. Cranfield Univerity at Silsoe, U.K.

Langevin, C.D. and Guo, W. (2006). MODFLOW/MT3DMS-based simulation of variable density ground water flow and transport. *Ground Water*, **44(3)**: 339-351.

Leclerc, J.-P., Berger, C., Foulon, A., Sarraute, R. and Gabet, L. (2008). Tsunami impact on shallow groundwater in the Ampara district in eastern Sri Lanka: Conductivity measurements and qualitative interpretations. *Desalination*, **219**: 126-136.

Lipscombe, S. (2007). Groundwater salinity and handdug wells in Ampara, Sri Lanka. *Waterlines*, **26(1)**: 12-13.

Lytton, L. (2008). Deep impact: Why post-tsunami wells need a measured approach. *Civil Engineering*, **1612**: 42-48.

NGWA (2005). National Groundwater Association, USA. http://www.ngwa.org/pdf/welldisinfection.pdf (accessed on January 20, 2005).

OXFAM (2006). Technical Brief: Repairing, Cleaning and Disinfecting Hand Dug Wells. http://www.oxfam.org.uk/resources/downloads/emerg_manuals/draft_oxfam_tech_brief_wellcleaning.pdf (accessed on June 20, 2008).

Panabokke, C.R. and Perera, A.P.G.R.L. (2004). Groundwater Resources of Sri Lanka. World Water Assessment Program, Colombo, Sri Lanka.

Panabokke, C.R., Pathirana, S.R.K. and Wijekoon, D. (2002). Water quality of agro-wells in the coastal sand aquifer in Trincomalee District. Proceedings of the Symposium on Use of Groundwater for Agriculture in Sri Lanka, 30 September 2002, University of Peradeniya, Sri Lanka, pp. 85-98.

UNEP (2005). After the Tsunami: Rapid Environmental Assessment. http://www.unep.org/tsunami/tsunami_rpt.asp (accessed on June 20, 2008).

UNICEF (2005). Consequences of the Tsunami on the Coastal Aquifer in Eastern Sri Lanka. Guidelines on Wells Rehabilitation. UNICEF, Trincomalee, Sri Lanka.

Villholth, K.G. (2007). Tsunami impacts on groundwater and water supply in eastern Sri Lanka. *Waterlines*, **26(1)**: 8-11.

Villholth, K.G., Amerasinghe, P.H. and Jeyakumar, P. (2008). Tsunami impacts on shallow groundwater and associated water supplies on the east coast of Sri Lanka. *In:* P. Bhattacharya, AL. Ramanathan, A.B. Mukherjee and J. Bundschuh (editors), Groundwater for Sustainable Development: Problems, Perspectives and Challenges. A.A. Balkema (Taylor and Francis Group), London, pp. 211-222.

Villholth, K.G., Amerasinghe, P.H., Jeyakumar, P., Panabokke, C.R., Woolley, O., Weerasinghe, M.D., Amalraj, N., Prathepaan, S., Bürgi, N., Lionelrathne, D.M.D.S., Indrajith, N.G. and Pathirana, S.R.K. (2005). Tsunami Impacts on Shallow Groundwater and Associated Water Supply on the East Coast of Sri Lanka. International Water Management Institute (IWMI), Colombo, Sri Lanka, 68 pp.

Villholth, K.G., Jeyakumar, P., Manamperi, A.S.P., Jegan, G., Thilipan, M. and Navalan, V. (2009). Perceptions of water quality and water supply impacts and coping of the local population with the 2004 Sumatra-Andaman tsunami on the east coast of Sri Lanka. *Natural Hazards* (under review).

Villholth, K.G., Manamperi, A.S.P. and Bürgi, N. (2006). Chemical characteristics of tsunami-affected groundwater and lagoon on the east coast of Sri Lanka. Paper presented at the 32nd WEDC International Conference on Sustainable Development of Water Resources, Water Supply and Environmental Sanitation, 13-17 November 2006, Colombo, Sri Lanka.

Vithanage, M., Villholth, K.G., Mahatantila, K., Engesgaard, P. and Jensen, K.H. (2009). Effect of the Indian Ocean Tsunami on groundwater quality in coastal aquifers in eastern Sri Lanka. *Science of Tsunami Hazards*, **28(3)**: 218-231.

Vrba, J. and Verhagen, B.Th. (2006). Groundwater for Emergency Situations – A Framework Document. IHP VI, Series on Groundwater No. 12, UNESCO-IHP.

WHO (2005). Cleaning and Disinfecting Wells in Emergencies. WHO Technical Notes for Emergencies, Technical Note No. 1, http://www.who.int/water_sanitation_health/hygiene/envsan/technotes/en/ (accessed on June 20, 2008).

WHO (2008). Cleaning Wells after Seawater Flooding: Emergency Guidelines. WHO Technical Notes for Emergencies, Technical Note No. 3, http://www.who.int/water_sanitation_health/hygiene/envsan/technotes/en/ (accessed on June 20, 2008).

Wickramaratne, H.U. (2004). Use of Groundwater for Coastal Water Supply Schemes, Minimizing the Salinity Problem: A Case Study at Kattankudy in Batticaloa District, Sri Lanka. M.Sc. Thesis, Post Graduate Institute of Science, University of Peradeniya, Peradeniya, Sri Lanka.

Tsunami Early Warning System: An Indian Ocean Perspective

Prasad K. Bhaskaran[*] and P.C. Pandey

1. INTRODUCTION

Tsunamis are considered the most devastating natural hazard on coastal environments ever known. Densely populated cities on coastal belts are the engines of economic growth and the centers of innovation for global economy and hinterlands of respective nations. As we know most of global cities are located near the coast facilitating trade and commerce. They are also located near the mouths of major perennial rivers which serve as conduits for commerce connecting rest of the world. These locations place major cities at a greater risk of natural hazards viz., cyclones, flooding, sea-level rise, tsunamis, etc. With the increasing intensity of economic exploitation in coastal belts, there is also an increase in socio-economic consequences resulting from the hazardous action of tsunami waves generated from submarine seismic activity and other causes. On 26 December 2004, the countries within the vicinity of East Indian Ocean experienced and witnessed the most devastating tsunami in recorded history. This tsunami was triggered by an earthquake of magnitude 9.0 on the Richter scale at 3.4° N, 95.7° E off the coast of Sumatra in the Indonesian Archipelago at 06:29 hrs IST (00:59 hrs GMT).

Historical records of past tsunamis reveal that the most damaging world tsunamis generated by earthquakes during the past five decades are: (i) 1952 – Kamchatka Peninsula (Russian Far East): 18–19 m high (more than 2000 fatalities); (ii) 1960 – Chile: 25 m high (more than 500 fatalities); (iii) 1964 – Alaska: 67 m (more than 100 fatalities reported); and (iv) 26 December 2004 - Indian Ocean: up to 30 m high (more than 200,000 people dead) and 12 countries affected in three continents. The run-up levels associated with the past Indian Ocean tsunamis are summarized in Table 1.

While earthquakes could not be predicted in advance, once the signatures of an earthquake is detected it would have been possible to give warning of a potential Tsunami to the coastal stations. Such a warning system at present is in place across the Pacific Ocean. However, the tsunami warning system in the Indian Ocean had been set up quite recently after the 2004 event. In addition, coastal dwellers within the Pacific Ocean littoral belt are educated to get high ground quickly following waves. However, those in the Indian Ocean are quite unaware. In less than a day, tsunamis can travel from one side of the ocean to the other. People living near areas where large earthquakes occur may find that the tsunami waves will reach their shores within minutes of the earthquake. For these reasons, the tsunami threat for many areas, e.g., Indonesia, Philippines, Java, etc. can be immediate for tsunamis resulting from nearby earthquakes which take only few minutes to reach coastal areas, in comparison with sufficient response time for tsunamis from distant earthquakes which take approximately about 3 to 22 hours reaching other coastal destinations.

Tsunamis are rare in the Indian Ocean as the seismic activity is much less than what exist in the Pacific. Historical records state that there have been seven tsunamis set off by earthquakes near Indonesia, Pakistan and one at Bay of Bengal. Earthquakes occur due to collision of plates at their boundaries.

---

[*]Corresponding Author

Table 1. Summary of tsunami occurrences in the Indian Ocean during 1700-2007 period

Sl. No.	Affected location	Run-up height (m)	Date/Year	Earthquake magnitude at source	Source location
1	Tributaries of Ganges River (Bangladesh)	1.83	12 April, 1762	N.A.	Bay of Bengal
2	Port Blair, Andaman Islands	4.00	19 August, 1868	MW 7.5	Bay of Bengal
3	Car Nicobar Islands, Nicobar Islands	0.76	31 December, 1881	MS 7.9	Car Nicobar Islands, Andaman Sea
4	Dublat, India	0.30			
5	Nagapattinam, India	1.22			
6	Port Blair, Andaman Islands	1.22			
7	Chennai	1.5 (wave height)	26 August, 1883	Krakatao Volcanic Eruption	Islands of Java and Sumatra
8	Andaman & Nicobar Islands	N.A.	26 June, 1941	MW 7.7	Andaman Sea (12.5°N; 92.57°E)
9	Mumbai, India	1.98	27 November, 1945	MS 8.3	Arabian Sea (24.5°N; 63°E)
10	Karachi, Pakistan	1.37			
11	Ormara, Pakistan	13.0			
12	Pasni, Pakistan	13.0			
13	Victoria, Mahe Islands, Seychelles	0.30			
14	Not felt in India	-	19 August, 1977	MS 8.1	West of Sumba Islands, Indonesia
15	Cocos Islands, Australia	0.30	18 June, 2000	MS 7.8	Arabian Sea
16	13 countries surrounding Indian Ocean rim directly affected	34.90	26 December, 2004	MS 9.0	West coast of Northern Sumatra, Indonesian Archipelago
17	Indonesia	1.0	28 March, 2005	MS 8.6	Indonesia
18	Java	2.0	17 June, 2006	MS 7.7	Indonesia

Source: Information from the website of NGDC (National Geophysical Data Centre, NOAA, USA; http://www.ngdc.noaa.gov/hazard/tsu.shtml).

Scientists now believe that one plate that comprised the landmass from India to Australia has broken up into two (Orman et al., 1995). The earthquake location of recent 2004 Indian Ocean tsunami was near the meeting point of Australian, Indian and the Burmese plates. Scientists have advocated that this is a region of compression as the Australian plate is rotating counterclockwise into the Indian plate. The implication of this also means that a region of seismic activity has become active in the South-eastern Indian Ocean which has potential of triggering another deadly tsunami. Within the close vicinity of India, there are two potential tsunamigenic zones: Andaman-Sumatra trench (East India) and the Makran coast (West India).

Tsunamis are known as long gravity waves, and hence their travel time in the ocean depends only on the water depth and gravity, at least to the zeroth-order. As of today, no technology exists to predict a tsunami event well in advance (Synolakis, 1995). Contrary to popular belief, the tsunami travel times do not depend upon the magnitude of the under-water disturbance that generated the tsunami. For the Pacific Ocean, it has been clearly demonstrated that the computed tsunami travel times using the zero-order approximation are correct to plus or minus one minute for each hour of travel. The advantage of this zero-order approximation is that tsunami travel times to selected locations around the rim of the Indian Ocean as well as to selected island sites can all be pre-computed in advance once and for all. This set of information can be stored in the electronic format as well as a tsunami travel time atlas format and can be quickly accessed in real tsunami events with a minimum effort.

Isochrome charts on tsunami travel time were first made in 1947 by the American Coastal Service after the Aleutian disastrous tsunami in the Pacific coast (Zetler, 1947). Subsequently, in 1971 these charts were evaluated and about fifty such charts were used by the tsunami warning system (Holloway et al., 1986). In concurrence with the disastrous tsunami of Hawaii on 1 April 1946 (Okal et al., 2002) from an earthquake in the Aleutian Islands of USA, a Pacific Ocean tsunami warning system was established based in Ewa Beach, Oahu Island, Hawaii, USA. In the immediate vicinity of islands, the catastrophe resulting from a tsunami is advocated as enormous (Yeh et al., 1994).

The goal of this chapter is to present a brief review on tsunamis which affected Indian sub-continent, current status of tsunami warning system for the Indian Ocean, and methodology used in the construction of tsunami travel time charts for the Indian Ocean. Further, the importance of artificial neural network (ANN) to expedite warnings has been demonstrated, which is considered an essential pre-requisite for an early warning system in the Indian Ocean.

2. REVIEW ON TSUNAMIS AFFECTING THE INDIAN SUBCONTINENT

Though majority of reported tsunamis are from countries surrounding the Pacific Ocean rim, there are also few reported cases of tsunamis in the Indian Ocean. Considering the vast length of Indian coastline (7516 km), the threat arising from tsunami-genic event is potentially hazardous. From our past experiences, the tsunami-genic earthquakes occurred mostly at these three locations, viz.; (i) the Andaman Sea, (ii) geographical area about 400-500 kilometers South south-west off Sri Lanka, and (iii) in Arabian Sea about 70-100 km south off Pakistan coast. The oldest record of tsunami in the Indian Ocean dates back to November 326 B.C. earthquake near the Indus Delta (presently the Kutch region) where a major earthquake destroyed the Macedonian fleet (Lietzin, 1974). Historical records of tsunamis in the Indian Ocean reported 1.5 m at Chennai resulting from August 8, 1883 Krakatao volcanic eruption in Indonesia. Also, an earthquake of magnitude 8.25 occurred about 70 km south of Karachi, Pakistan on November 27, 1945. This resulted in a large tsunami magnitude of about 11.0 to 11.5 m high in the west coast of India (Kutch region) (Pendse, 1945). There are few more cases of earthquakes with a magnitude less than 8.0 which have given rise to some smaller tsunamis. Bapat et al. (1983) reported a few more earthquakes along the coast of Myanmar.

In the Andaman Sea, an earthquake of magnitude 8.1 occurred on June 26, 1941 resulting in a tsunami affecting the east coast of India. According to non-scientific sources, the heights of tsunami waves were in the order of 0.75 to 1.25 meters as no tide-gauges were in operation then. This was the strongest earthquake ever recorded in the Andaman & Nicobar Islands prior to which was the 1881 Nicobar Islands earthquake (magnitude 7.9 on the Richter scale). Hindcast studies employing mathematical calculations suggested that the height could be in the order of 1.0 meter. It is believed that nearly 5,000 people were killed by this tsunami in the east coast of India. Local newspaper sources are believed to have mistaken the deaths and damage to a storm surge; however, a search of meteorological records (Murty, 1984) does not show any storm surge on that day in the Coromandel Coast. Tremors from this earthquake were felt in cities along the Coromandel (eastern) Coast of India and even in Colombo,

Sri Lanka. This earthquake was followed by several powerful aftershocks (Tandon et al., 1974). Two events of magnitude 6.0 struck within 24 hours of the main shock on June 27, 1941. The first occurred at 07:32:47 UTC and was followed by another at 08:32:19 UTC. Consequently, these were then followed by 14 earthquakes of magnitude 6.0 until January 1942.

Considering the build-up of seismic activity in the Southeastern Indian Ocean and associated calamities which resulted from the past tsunami-genic events, a comprehensive tsunami travel time (TTT) atlas for the Indian Ocean was developed by Prasad et al. (2005) which can serve as an important tool for the early warning of tsunami. We discuss in subsequent sections how this information from the TTT atlas can be used in the context of ANN to reduce response time for an early warning system in the Indian Ocean.

3. CURRENT STATUS OF TSUNAMI WARNING SYSTEM FOR THE INDIAN OCEAN

Recognizing the imperative to put in place an Early Warning System for the mitigation of oceanogenic disasters that cause severe threat to nearly 400 million population living in the coastal belt with devastation of life and property, and further driven by the national calamity due to the Indian Ocean Tsunami of December 26, 2004, the Ministry of Earth Sciences (MoES), Government of India, has taken up the responsibility of establishing a National Tsunami Early Warning System (NTEWS). The Warning System has been established by MoES as the nodal ministry at a cost of Rs. 1,250 million in collaboration with the Department of Science and Technology (DST), Department of Space (DOS) and the Council of Scientific and Industrial Research (CSIR), Government of India. The National Tsunami Early Warning Centre has been set up at the Indian National Centre for Ocean Information Services (INCOIS), Hyderabad, Andhra Pradesh, India.

Tsunami-genic zones that threaten the Indian coast have been identified by considering the historical tsunamis, earthquakes, their magnitudes, location of the area relative to a fault, and also by tsunami modeling. The east and west coasts of India including the island regions are likely to be affected by tsunamis generated mainly by subduction zone related earthquakes from the two potential source regions, viz., the Andaman-Nicobar-Sumatra island arc and the Makran subduction zone north of Arabian Sea. The Indian Tsunami Early Warning System comprises a real-time network of seismic stations, Bottom Pressure Recorders (BPR) and tide gauges to detect tsunami-genic earthquakes and to monitor tsunamis.

The Early Warning Centre receives real-time seismic data from the national seismic network of the India Meteorological Department (IMD), New Delhi, and other international seismic networks. The system detects all the earthquake events of magnitude greater than 6 occurring in the Indian Ocean in less than 20 minutes of occurrence. BPRs installed in the deep ocean are the key sensors to confirm the triggering of a tsunami. The National Institute of Ocean Technology (NIOT) has installed four BPRs in the Bay of Bengal and two BPRs in the Arabian Sea. In addition, NIOT and Survey of India (SOI) have installed 30 Tide Gauges to monitor the progress of tsunami waves. Integrated Coastal and Marine Area Management (ICMAM) has customized the tsunami model for five historical earthquakes and the predicted inundation areas. The inundated areas are being overlaid on cadastral level maps of 1:5000 representative scales. These community-level inundation maps are extremely useful for assessing the population and infrastructure at risk. High-resolution coastal topography data required for modeling is generated by the National Remote Sensing Agency (NRSA), Government of India, using ALTM and CARTOSAT data. INCOIS has also generated a large database of model scenarios for different earthquakes that are being used for operational tsunami early warning.

Communication of real-time data from seismic stations, tide gauges and BPRs to the early warning centre is very critical for generating timely tsunami warnings. A host of communication methods are employed for timely reception of data from the sensors as well as for dissemination of alerts. Indian Space Research Organization (ISRO), Bangalore, has made an end-to-end communication plan using INSAT. A high level of efficiency is being built into the communication system to avoid single point failures.

A state-of-the-art early warning center is established at INCOIS (Indian National Centre for Ocean Information Services) located in Hyderabad, Andhra Pradesh, with all the necessary computational and communication infrastructure that enables reception of real-time data from all the sensors, analysis of the data, generation and dissemination of tsunami advisories following a standard operating procedure. Seismic and sea-level data are continuously monitored in the Early Warning Centre using a custom-built software application that generates alarms/alerts in the warning centre whenever a pre-set threshold is crossed. Tsunami warnings/watches are then generated based on pre-set decision support rules and disseminated to the concerned authorities for action, following a Standard Operating Procedure. The efficiency of the end-to-end system was proved during the large under-sea earthquake of 8.4 M that occurred on September 12, 2007 in the Indian Ocean.

The National Early Warning Centre will generate and disseminate timely advisories to the Control Room of the Ministry of Home Affairs (MHA) for further dissemination to the public. For the dissemination of alerts to MHA, a satellite-based virtual private network for disaster management support (VPN DMS) has been established. This network enables early warning center to disseminate warnings to the MHA, as well as to the State Emergency Operations Centers. In addition, alert messages will also be sent by Phone, Fax, SMS and e-mails to authorized officials. In case of confirmed warnings, the National Early Warning Centre is being equipped with necessary facilities to disseminate the advisories directly to the administrators, media and public through SMS, e-mail, fax, etc. The cyclone warning network of IMD and electronic ocean information boards of INCOIS could be effectively used for disseminating warnings directly to the public.

Periodic workshops will be organized in future for the user community by INCOIS to familiarize them with the use of tsunami and storm surge advisories as well as inundation maps. Easily understandable publicity material on earthquake, tsunami and storm surges will be generated by INCOIS and will be distributed in future to the general public for awareness.

4. ORGANIZATION OF TRAVEL TIME CHARTS

The locations for the present study have been chosen mainly for population centers around the Indian Ocean (Fig. 1). Also, any location deserving special consideration, even if it is not a major population center, has also been included. Database on countries surrounding the Indian Ocean rim was first identified. Accordingly, 35 countries surrounding the Indian Ocean has been selected listing major cities and population. Table 2 summarizes the country name and number of locations used in this study.

The study domain encompasses starting at the south-eastern end of the Indian Ocean (west coast of Australia) and proceeding northward to Bay of Bengal, then to the Arabian Sea. It should be noted that the Red Sea and the Persian Gulf (Arabian Gulf) are part of the Arabian Sea system which have experienced Tsunamis in the past, one notable example being the Arabian Sea tsunami of November 27, 1945. From the Red Sea, we proceed southward along the east coast of Africa until the Cape of Good Hope, and that limits the geographic domain for the TTT computation, as our interest here is only the Indian Ocean and not the Atlantic Ocean. Figure 2 depicts the earthquake locations for which TTT computations were conducted.

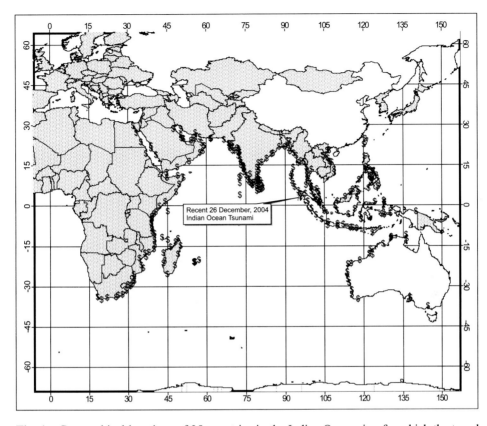

Fig. 1 Geographical locations of 35 countries in the Indian Ocean rim, for which the travel time of tsunami waves had been numerically computed by Tsunami Travel Time (TTT) model using the epicenter location (95.947°E; 3.307°N) of the recent (26 December, 2004) energetic event in the Indian Ocean.

Table 2. List of countries and corresponding number of coastal stations used for generating Tsunami Travel Time (TTT)

Country	Number of locations	Country	Number of locations	Country	Number of locations
Australia	19	Malaysia	12	Singapore	1
Bahrain	1	Maldives	1	Somalia	8
Bangladesh	3	Mauritius	1	South Africa	11
Brunei	1	Mozambique	9	Sri Lanka	15
Comoros	2	Myanmar	9	Sudan	2
Egypt	3	Oman	5	Taiwan	2
India	47	Pakistan	3	Tanzania	7
Indonesia	31	Philippines	11	Thailand	6
Iran	3	Qatar	2	U.A.E.	3
Kenya	3	Reunion	4	Vietnam	8
Kuwait	1	Saudi Arabia	3	Yemen	2
Madagascar	10	Seychelles	1		

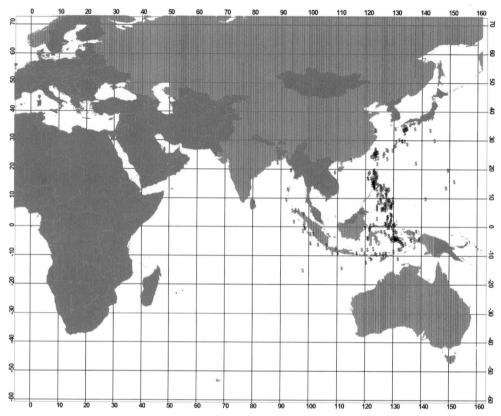

Fig. 2 Geographical coordinates of 250 past earthquake events in the Indian Ocean for which TTT were computed.

5. COMPUTATION AND DESCRIPTION OF TSUNAMI TRAVEL TIME CHARTS

The technique used to compute travel times over the entire grid is an application of Huygens principle which states that all points on a wave-front are point sources for secondary spherical waves. From the starting point, times are computed to all surrounding points. The grid point with minimum time is then taken as the next starting point and times are computed from there to all surrounding points. The starting point is continually moved to the point with minimum total travel time until all grid points have been evaluated. In the countries surrounding the Indian Ocean rim, 250 locations as mentioned above were selected for this study. The travel time of tsunami waves from the epicenter to various coastal regions has been evaluated for all the sample points identified.

On 26 December 2004, the countries in the Eastern Indian Ocean experienced the most devastating tsunami in recorded history (Bindra, 2005). This tsunami was triggered by an earthquake of magnitude 9.0 on the Richter scale off the coast of Sumatra in the Indonesian Archipelago at 06:29 hrs IST (00:59 hrs GMT). The extent of damage resulting from this tsunami has been cited in the post-tsunami field survey (Chapman, 2005). The dispersive signals of this energetic event were recorded by hydrophones and seismic stations in coastal locations around the Indian Ocean rim (Hanson and Bowman, 2005). The computational example for this Indian Ocean tsunami using our computational algorithm is shown in Fig. 3.

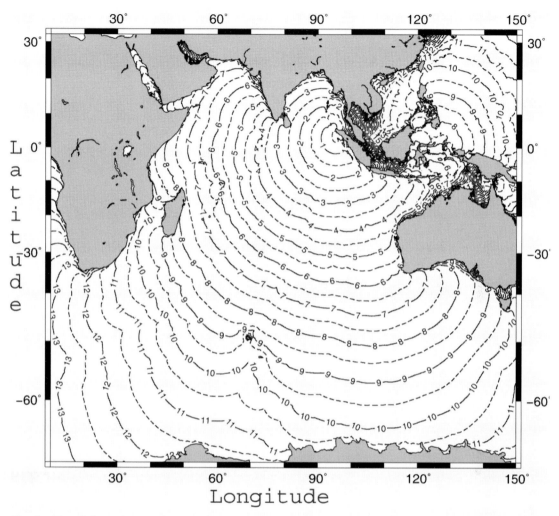

Fig. 3 Travel time of tsunami waves that resulted from the 26 December, 2004 event in the Indian Ocean. The contour intervals are sampled at every half an hour.

The most basic information a tsunami warning center requires is ETA (Expected Time of Arrival) of the first tsunami wave at selected coastal locations from the area of tsunami generation in the ocean. Almost always, the first wave in a tsunami event is not the wave with the greatest amplitude; nevertheless, tsunami travel time charts are generally constructed for the first wave, rather than the wave with the highest amplitude. Advanced knowledge of travel time for the first wave provides some additional valuable time for the evacuation of people, if and when evacuation is needed. In addition, tsunami travel times can be pre-computed, independent of the seismic moment magnitude of the earthquake, only for the first wave. The heights of the subsequent waves are not known until the event actually happens, and hence no pre-determination of the travel time of the highest wave can be made.

To the zeroth order, tsunami travel times are governed by the long gravity wave formula, which defines the speed of travel of the tsunami as equal to the square root of water depth, multiplied by the acceleration due to gravity (Murty et al., 1987). Of course there are higher order correction terms to the speed of tsunami travel, based upon dispersion characteristics. While these higher order terms are of scientific interest, nevertheless, for practical purposes, one can ignore the contributions from these terms, and use the simple long wave formula.

The calculation algorithm adopted by the authors for computing the isochromes tables is based on the well-known Huygens method and is actually a group of methods which obtain the time required for passing a given space interval with a given speed (Yuri et al., 1995). In this technique all the nodes of the calculated grid are partitioned into three sets, set M_1 containing the nodes with finally calculated tsunami travel times, set M_2 contains the nodes with preliminary estimated values that may later be refined and set M_3 the nodes with arrival time that has not been obtained yet. Each node (U) is associated with the notion of its domain of influence (S_u) which is aggregate of neighboring nodes satisfying criterion of proximity. It is considered that during one step of an algorithm the perturbation can cover the distance from the specified node only to its nearest pattern neighbors.

At the initial time all the nodes in the domain of the initial perturbation are assigned zero value of the arrival time (T_0) corresponding to the time of beginning of the earthquake and included in the set M_2^0. The remaining nodes of the calculation domain are included in M_3^0. Then the nodes of M_2^0 with patterns not influenced by the ones of the set M_3^0 are transferred to the set M_1^1 and excluded from further calculations. The algorithm afterwards acquires the regular character. Hence, at the nth step the exhaustive search of the nodes from the set M_2^n is performed in order of increasing tsunami travel times known at this point of time.

Let the node $A \in M_2^n$ (1)

and $T_A = \min\{T_{A_i}\}$ (where, $A_i \in M_2^n$); then the node for B such that $B \in S_A$ the tentative estimate (A) of arrival time is found from the relation:

$$T = T_A + T_{AB} \quad (2)$$

where $T_{AB} = \dfrac{2L_{AB}}{C_A + C_B}$ which is the time of perturbation propagation from the node 'A' to 'B'.

Here $L_{AB} = R \times \arccos(\sin\varphi_A \times \sin\varphi_B + \cos\varphi_A \times \cos\varphi_B \times \cos\Delta\psi)$ denotes the distance from node 'A' to 'B' through the great circle arc; 'R' stands for the radius of the earth; and $c = \sqrt{gh_i}$ is the local rate of perturbation propagation (h_i denotes the depth of the i-th node and 'g' the acceleration due to gravity).

In case where, $B \in M_3^n$, relation from (Equation 1) first yields the value of 'T' and the node 'B' itself is transferred to M_2^{n+1}, and if $B \in M_2^n$, the value of 'T' is refined from the minimizing relation:

$$T_B = \min\{T_A, T_A + T_{AB}\} \quad (\text{where } A \in S_B \cap M_2^n) \quad (3)$$

After the values of T, $\forall B : B \in S_A$ are obtained from Equations 1 and 2, the node 'A' is transferred to the set M_1^{n+1}. This procedure is repeated to the next node $A \in M_2^n$ satisfying the condition (Equation 1) and so on until the set M_2^n is exhausted. For the next ($n+1$) time step, the algorithm is reproduced without any changes and the computation goes on until step 'k' where all calculation node of the water area are included in the set M_1^k.

For the Pacific Ocean, it has been shown that the travel time charts are accurate to plus or minus one minute, for each hour of travel. There is no reason to expect that the travel time charts will be less precise for the Indian Ocean. This level of error is considered acceptable for tsunami warning purposes. Since, at present one cannot predict precisely the location and time of occurrence of a tsunami-genic earthquake; it is not possible to construct tsunami travel time charts for all possible future tsunamis. In any case, travel time information is required for coastal locations, where disaster mitigation procedures have to be invoked during real tsunami events. However, tsunami travel time charts are reversible, in the sense that the travel times are exactly the same, no matter in which direction the tsunami travels on a given chart, i.e., from an epicenter in the ocean to a coastal site or vice-versa (i.e., from a coastal site to an epicenter in the ocean). Once a reasonable number of tsunami travel time charts are prepared for the Indian Ocean, for selected coastal and island locations, as well as for all historical tsunami events, it is quite probable that for any future tsunami events, the travel time information that is required could be quickly and effortlessly obtained from these charts.

6. APPLICATION OF ARTIFICIAL NEURAL NETWORK IN TSUNAMI TRAVEL TIME PREDICTION

Artificial Neural Networks (ANNs) are inspired by the biological nervous system. Composed of elements operating in parallel, the network function is primarily determined by connections (weights) between elements. A neural network can be trained to perform a particular function by adjusting the values of these weights between elements. Commonly neural networks are adjusted, or trained so that a particular input leads to a specific target output. The network can be trained based on a comparison of the output and the target until the network output matches the target within a specified error level (supervised learning). In this study, different combinations of input were used to train the network. The trained network can be validated with the data that has not been used for training. If the system performs well with the validation data, the system can be deployed real time to perform a specific job.

Neural networks can be trained to solve problems that are difficult for conventional computers. Neural networks have been trained to perform complex functions in diverse fields of application which include nonlinear regression, classification, identification, pattern recognition and control systems (Hinton, 1992; Navone and Ceccatto, 1994; Bishop, 1995; Venkatesan et al., 1997; Silverman and Dracup, 2000; Li et al., 2005). The supervised training methods are commonly used, but other networks can be obtained from unsupervised training techniques which can be used where there are no input/output pairs as such but only input data. This for instance may be used to identify groups of data (Hopfield Networks).

Neural network approach has been also used to solve inverse problems which include a methodology to assess the severity of a tsunami based on real-time water-level data near the source (Yong et al., 2003). This inverse method which uses tsunami signals in water-level data to infer seismic source parameters is extended to predict the tsunami waveforms away from the source. The present study which involves the prediction of *Tsunami Arrival Time* is within the domain of nonlinear regression where MLP is used to tackle the nonlinearity in the data. A brief description of the MLP used to tackle the prediction problem is provided below.

An MLP (Fig. 4) is a feed forward network consisting of three classes of layers: the input layer, the hidden layers (a network can have several hidden layers) and the output layer. The inputs to the network are given at the input layer and the number of input nodes would be equal to the number of input parameters. The input nodes receive the data and pass them onto the hidden layer nodes (neurons). The neuron model and the architecture of a neural network then determine how the input is transformed into an output. This transformation involves computation and can be represented with detailed mathematical algorithms. The perceptron then computes single output from multiple real-valued inputs by forming a linear combination according to its input weights and then possibly putting the output through a nonlinear activation function. This can be expressed mathematically as (Simon, 1998):

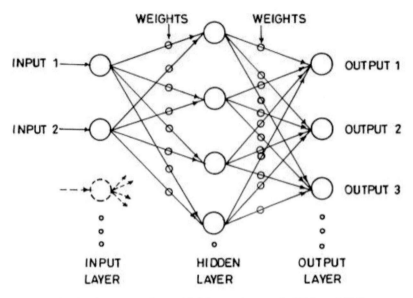

Fig. 4 Structure of an artificial neural network (Bishop, 1995).

$y = \psi \sum_{i=1}^{N}(w_i x_i + b)$, where w denotes the vector of weights, x the vector of inputs, b the bias and ψ the activation function. The supervised learning problem of the MLP can be solved with the back-propagation algorithm. This algorithm consists of two steps. In the forward pass, the predicted outputs corresponding to the given inputs are evaluated and in the backward pass, partial derivatives of the cost function with respect to the different parameters are propagated back through the network (Bishop, 1995). It may be noted that chain rule of differentiation gives very similar computational rules for the backward pass as the one in the forward pass. The networked weights can then be adapted using a gradient-based optimization algorithm. The entire computational process is then iterated until the weights have converged (Simon, 1998).

7. DEVELOPMENT OF MODEL FOR TSUNAMI ARRIVAL TIME PREDICTION

The technique used to compute travel times over the study domain is an application of Huygens principle which states that all points on a wave-front are point sources for secondary spherical waves. From the starting point (actual epicenter location), travel times are computed to all surrounding equidistant grid points. The grid point with minimum time is then taken as the next starting point (new location) and computation is performed thereafter to all surrounding points. The starting point is then continually moved to the point with minimum total travel time until all grid points have been evaluated. For the countries surrounding the Indian Ocean rim, 250 coastal locations were selected for this study. The travel time of tsunami waves from the epicenter to various coastal regions were evaluated for all the sample points identified. This technique was used for multiple tsunami-genic locations in the Indian Ocean rim facilitating the development of a comprehensive ETA database (Prasad et al., 2005), which has been used for training the neural network in the present study. For the benefit of the readers, a detailed methodology of ETA computations and further skill assessment with available observations for the 26 December 2004 event in the Indian Ocean was investigated (Prasad et al., 2006).

7.1 Data for Artificial Neural Network

This ANN-based study computes a new travel time chart which is highly efficient in the prediction of ETA to several coastal destinations in the Indian Ocean for any given epicenter location in the study domain. The input to the ANN model comprises the location of the underwater earthquake source viz., latitude and longitude and the ETA from the comprehensive database. In the 250 coastal destinations of countries surrounding the Indian Ocean rim (Fig. 1), ETA for 47 coastal destinations is within the purview of Indian sub-continent. The ANN model for this study has been trained to perform prediction to these 47 coastal destinations. The coastal locations used for this study is chosen mainly from high density population centers depicted in Fig. 2. Minutiae shown in Table 3 list the country name and number of locations taken from the comprehensive database of ETA for the present study. The epicenters of all past tsunamigenic events in the Indian Ocean were compiled from information obtained from NGDC (National Geophysical Data Center, NOAA, Boulder, USA; http://www.ngdc.noaa.gov/hazard/earthqks.html). Figure 2 illustrates the epicenter locations of the past earthquake events (250 locations) having magnitude greater than 6.0 on the Richter scale used in the comprehensive ETA database.

Table 3. Computed Expected Arrival Time (ETA) to 250 coastal stations for the December 26, 2004 tsunami event in the Indian Ocean

Country	City	Population[#]	Location		Arrival time (hour)
			Latitude (Degree)	Longitude (Degree)	
1. South Africa	Mtunzini	12,050	28.97S	31.77E	11.222
	Durban	2,117,650	29.87S	30.99E	11.002
	Ladysmith	89,087	28.02S	32.66E	10.6756
	Port St Johns	1,46,132	31.62S	29.53E	11.032
	East London	212,323	32.97S	27.87E	11.4353
	Port Elizabeth	1,100,000	33.96S	25.59E	12.131
	Grahamstown	62,640	33.19S	26.31E	12.22
	Mosselbaai	55,100	34.18S	22.13E	12.9163
	Cape Agulhas	26,182	34.83S	20.00E	13.3881
	Pietermaritzburg	229,000	29.36 S	30.23 E	11.1826
	Cape Town	2,350,000	33.93S	18.47E	13.5143
2. Mozambique	Pemba	84,897	12.58S	40.30E	9.34164
	Nacala	158,248	14.31S	40.34E	9.38824
	Angoche	74,624	16.17S	39.97E	9.64704
	Pebane	1,274	17.23S	38.17E	10.2653
	Quelimane	150,116	17.53S	36.58E	11.7402
	Beira	397,368	19.50S	34.52E	13.8161
	Vilanculos	19,371,057	22.02S	35.32E	10.8631
	Inhambane	52,370	23.02S	35.92E	10.3011
	Maputo	966,837	25.58S	32.32E	11.3468
3. Tanzania	Mtwara	1,128,523	10.20S	40.20E	9.07369
	Lindi	791,306	09.56S	39.61E	9.06953
	Kilwa	171,057	08.55S	39.30E	9.31335
	Dar es Salaam	1,292,973	06.50S	39.12E	9.52655
	Tanga	1,642,015	05.05S	39.02E	9.27646

(Contd.)

Table 3. (*Contd.*)

Country	City	Population[#]	Location Latitude (Degree)	Longitude (Degree)	Arrival time (hour)
	Zanzibar	391,002	06.10S	39.20E	9.68088
	Wete	186,013	05.04S	39.43E	9.1367
4. Kenya	Lamu	2,249	02.28S	40.90E	9.14886
	Malindi	64,300	03.12S	40.05E	9.06343
	Mombasa	1,880,000	04.02S	39.43E	9.04207
5. Somalia	Berbera	200,000	10.47N	45.03E	9.17937
	Boosaso	90,100	11.28N	49.18E	8.27345
	Ras Hafun	5,000	10.48N	51.33E	7.79386
	Eyl	682	08.00N	49.82E	7.69955
	Obbia	386	05.33N	48.50E	7.82298
	Mogadishu	1,262,000	02.06S	45.37E	7.8039
	Merca	173,100	01.48N	44.50E	8.18546
	Kisimayo	201,600	00.22S	42.32E	8.51144
6. Sudan	Suakin	10,500	19.08 S	37.17 E	10.3177
	Port Sudan	730,000	19.38 N	37.08 E	13.5179
7. Egypt	Hurghada	182,526	27.15 N	33.50 E	15.492
	Suez	417,610	30.00 N	32.30 E	18.5576
	Al-Ghardaqah	71,800	23.88 N	35.27 E	14.6119
8. Yemen	Aden	519,822	12.45N	45.00E	8.96302
	Al Mukalla	890,246	14.33N	49.02E	8.12144
9. Saudi Arabia	Jeddah	2801481	21.53N	39.17E	14.1548
	Rabigh	31,963	22.50 N	39.05 E	14.0922
	Al Qunfudhah	1,772	19.03N	41.04E	13.8529
10. U.A.E.	Sharjah	320,095	25.20N	55.24E	11.8153
	Abu Dhabi	398,695	24.28N	54.25E	14.8918
	Dubai	669,181	25.271N	55.329E	11.4325
11. Qatar	Doha	285,000	25.15N	51.36E	15.3089
	Dukhan	9,835	25.3N	50.8E	18.6514
12. Bahrain	Manama	143,035	26.236N	50.583E	16.4257
13. Kuwait	Kuwait	28,747	28.59N	47.52E	19.3276
14. Iran	Bandar-e-Bushehr	143,641	28.59N	50.46E	16.2342
	Bandar Abbas	273,578	27.12N	56.15E	10.3892
	Jask	66,128,965	25.642N	57.772E	8.2053
15. Oman	Salalah	156,530	16.56N	53.59E	7.45856
	Sur	66,785	22.34N	59.32E	7.46266
	Muscat	540,000	23.37N	58.36E	7.90009
	Duqm	4,269	19.65N	57.7E	7.60163
	Masirah	841	20.417N	58.833E	7.29651
16. Pakistan	Karachi	9,856,318	24.53N	67.00E	8.43242
	Gwadar	185,498	25.10N	62.18E	7.66909
	Jiwani	25,000	25.117N	61.733E	8.20665
17. Bangladesh	Chittagong	6,545,078	24.05N	91.00E	6.37145
	Cox Bazaar	1,757,321	21.26N	91.59E	4.44494
	Dhulasar	27,046	21.87N	90.23E	5.20506

(*Contd.*)

(*Contd.*)

18.	Myanmar	Sittwe	107,620	20.15N	92.09E	3.76803
		Kyaukpyu	19,456	19.45N	93.55E	4.47832
		Sandoway	4,000	18.47N	94.45E	3.94103
		Kadonkani	47,382,633	15.83N	95.18E	5.6117
		Yangon	2,513,023	16.47N	96.10E	4.37541
		Mawlamyine (Moulmein)	219,961	16.30N	97.37E	4.53773
		Tavoy	139,900	14.12N	98.30E	6.14122
		Mergui	177,961	12.43N	98.56E	4.62463
		Kawthaung	41,994,678	10.02N	98.53E	3.87474
19.	Thailand	Bangkok	7,506,700	13.45N	100.35E	26.6819
		Surat Thani	153,500	09.06N	99.20E	24.5288
		Songkhla	294,200	07.13N	100.37E	22.0887
		Phuket	211,000	07.53N	98.24E	2.1819
		Ranong	163,160	9.962N	98.638E	4.27555
		Satun	270,802	6.617N	100.067E	4.61174
20.	Taiwan	Hsin-chu	384,384	24.80N	120.98E	12.0184
		Kao-hsiung	1,500,000	22.60N	120.28E	10.563
21.	Malaysia	Georgetown	180,573	05.25N	100.20E	4.28901
		Klang(Kelang)	563,173	03.02N	101.26E	6.98943
		Kuala Lumpur	1,297,526	03.09N	101.41E	6.98113
		Melaka	369,222	02.15N	102.15E	8.50881
		Johor Bahru	384,613	01.28N	103.46E	10.8117
		Kuantan	283,041	03.49N	103.20E	16.611
		Kuala Terengganu	250,528	05.20N	103.08E	18.0761
		Kota Bahru	233,673	06.07N	102.14E	20.1728
		Kuching	579,900	01.53N	110.33E	15.4757
		Bintulu	116,600	03.20N	113.02E	12.503
		Kota Kinabalu	24,821,286	05.98N	116.06E	11.1984
		Sandakan	24,385,858	05.86N	118.06E	9.18603
22.	Brunei	Bandar Seri	46,229	04.93N	114.96E	11.8372
23.	Indonesia	Jayapura	145,200	02.28S	140.38E	9.13068
		Namlea	124,084	03.25S	127.12E	6.28019
		Ambon	313,100	03.43S	128.12E	6.35841
		Bula	57,474	03.12S	130.45E	6.94673
		Tobelo	3,860	01.75N	127.98E	7.77472
		Manado	398,900	01.29N	124.51E	7.55753
		Majene	1,268,500	03.55S	118.98E	5.94927
		Ujung Pandang	1,091,800	05.10S	119.20E	5.7257
		Kupang	165,500	10.22S	123.63E	5.12295
		Sumbawa Besar	52,654	08.50S	117.42E	4.87692
		Mataram	306,600	08.35S	116.07E	4.21647
		Denpasar	435,000	08.39S	115.13E	4.10912
		Surabaya	3,092,400	07.17S	112.45E	8.02103
		Semarang	1,366,500	07.00S	110.26E	7.81216
		Yogyakarta	419,500	07.49S	110.22E	3.41068
		Jakarta	8,987,800	06.09S	106.49E	4.7177

(*Contd.*)

Table 3. (*Contd.*)

Country	City	Population#	Latitude (Degree)	Longitude (Degree)	Arrival time (hour)
	Genteng	79,652	07.35S	106.33E	2.6125
	Bandar Lampung	832,400	05.30S	104.30E	2.45704
	Mentok	26,709	02.07S	105.20E	10.3857
	Tanjungbalai	142,506	01.00N	103.32E	11.2742
	Langsa	117,256	04.47N	97.98E	3.17407
	Banda Aceh	291,300	05.35N	95.20E	0.761689
	Meulaboh	4,775	04.17N	96.15E	0.567799
	Sibolga	22,513	01.70N	98.80E	1.97274
	Padang	721,500	01.00S	100.20E	1.80631
	Bengkulu	262,100	03.50S	102.12E	2.24429
	Pontianak	449,100	00.03S	109.15E	12.8599
	Ketapang	1,680	01.83S	109.98E	10.7732
	Bandjarmasin	534,600	03.20S	114.35E	10.1546
	Balikpapan	448,700	01.25S	116.83E	7.54061
	Tarakan	98,800	03.33N	117.63E	7.71036
24. Madagascar	Antsiranana	220,000	12.25S	49.20E	8.3672
	Antalaha	75,000	14.88S	50.27E	7.67597
	Toamasina	230,000	18.10S	49.25E	7.91618
	Manakara	25,689	22.15S	48.00E	8.2238
	Cape St Marie	31,592,805	25.57S	45.17E	9.50308
	Toliary	150,000	21.50S	43.74E	10.3438
	Morondava	33,372	20.32S	44.28E	10.4581
	Tambohorano	406,564	17.50S	43.59E	9.63051
	Mahajanga	200,000	15.40S	46.25E	8.71684
	Hell Ville	23,050	13.40S	48.28E	8.62156
25. Seychelles	Victoria	20,050	04.63S	55.47E	7.37542
26. Comoros	Moroni	629,000	11.67S	43.27E	8.69479
	Dzaoudzi	690,948	12.80S	45.30E	8.42018
27. Maldives	Male	74,069	04.00N	73.00E	4.09193
28. Mauritius	Port Louis	127,855	20.10S	57.30E	6.87434
29. Reunion	Saint-Benoit	101,804	21.03S	55.71E	7.01025
	Saint-Denis	236,599	20.87S	55.46E	7.07017
	Saint-Paul	138,551	21.00S	55.27E	7.3678
	Saint-Pierre	229,346	21.27S	55.53E	7.14395
30. Sri Lanka	Kankesanturai	31,506	9.85N	80.08E	4.09997
	Mullaittivu	7,900	09.25N	80.80E	3.09918
	Trincomalee	91,000	08.38N	81.15E	3.13302
	Batticaloa	515,707	7.72N	77.73E	3.46255
	Okanda	9,594	06.65N	81.77E	2.55246
	Hambantota	11,734	6.12N	81.12E	2.67229
	Matara	643,786	5.95N	80.55E	2.73498
	Galle	97,000	06.05N	80.10E	2.88227
	Moratuwa	177,190	06.45N	79.55E	2.93046

(*Contd.*)

(*Contd.*)

		Dehiwala-Lavinia	209,787	06.51N	79.52E	2.94426
		Colombo	642,163	05.56N	79.58E	2.80593
		Negombo	121,933	07.12N	79.50E	3.08903
		Talalla	78,023	08.13N	79.70E	3.38355
		Mannar	106,235	8.98N	79.92E	4.41793
		Jaffna	177,190	09.45N	80.02E	4.71506
31.	Singapore	SingaporeCity	4,163,700	01.22N	103.55E	10.948
32.	Philippines	Laoag	89,468,677	18.23N	120.60E	10.2071
		Quezon City	2,173,831	14.38N	121.00E	11.1002
		Manila	1,581,082	14.40N	121.03E	11.1002
		Bulan	28,529	12.66N	123.88E	9.406
		Mindaro	81,159,644	13.00N	121.00E	9.89557
		Iloilo	365,820	10.68N	122.55E	9.27341
		Cebu City	718,821	10.18N	123.54E	9.73164
		Siaton	64,258	09.08N	123.08E	8.67981
		Palawan Is	737,000	9.50N	118.50E	8.81906
		Zamboanga City	601,794	06.54N	122.04E	7.76695
		Davao	1,725,355	7.08N	125.63E	8.23129
33.	Australia	Melbourne	3,488,800	37.50S	145.00E	12.5851
		Adelaide	1,110,500	34.52S	138.30E	12.121
		Nhulunbuy	3,202	12.50S	136.93E	14.293
		Crocker	14,375	11.03S	136.63E	11.7589
		Bathurst I	37,001	11.75S	130.68E	9.78129
		Darwin	108,200	12.25S	130.51E	9.47687
		C. St Lambert	20,976	14.28S	127.71E	9.12529
		C. Leveque	12,330	16.41S	122.91E	6.05995
		Broome	13,218	17.97S	122.25E	6.53677
		Port Hedland	14,288	20.40S	118.60E	7.00693
		Dampier Downs	770	18.52S	123.45E	9.1031
		Onslow	700	21.68S	115.20E	5.95066
		Exmouth	2,400	21.90S	114.16E	5.10908
		Carnarvon	7,392	24.85S	113.75E	6.38753
		Kalgoorlie	36,852	27.70S	114.16E	6.50292
		Geraldton	27,258	28.81S	114.60E	6.23176
		Perth	1,397,000	31.57S	115.52E	6.28896
		Bunbury	26,369	33.33S	115.56E	6.77029
		Albany	23,913	34.95S	117.90E	7.10817
34.	Vietnam	Haiphong	1,447,523	20.47N	106.41E	15.1392
		Hon Gai	129,394	20.57N	107.05E	14.7076
		Vinh	175,167	18.45N	105.38E	14.4124
		Dong Hoi	40,290	17.53N	106.58E	13.7196
		Hue	260,489	16.30N	107.35E	13.3015
		Da Nang	369,734	16.04N	108.13E	12.9678
		Qui Nhon	201,972	13.40N	109.13E	12.1071
		Nha Trang	263,093	12.16N	109.10E	12.1088
35.	India	Rapur	5,380	23.05N	68.83E	8.71781
		Kandla	175,000	23.00N	70.10E	10.1083
		Dwarka	33,614	22.25N	69.05E	7.96334

(*Contd.*)

Table 3. (*Contd.*)

Country	City	Population#	Location		Arrival time (hour)
			Latitude (Degree)	Longitude (Degree)	
	Porbandar	133,083	21.44N	69.43E	6.99742
	Veraval	141,207	20.53N	70.27E	7.22848
	Diu	21,576	20.40N	71.02E	7.74814
	Bhavnagar	510,958	21.45N	72.10E	9.10254
	Daman	113,949	20.25N	72.57E	8.5291
	Dadar & Nagar Haveli	220,451	20.05N	73.00E	8.63183
	Mahim	42,798	19.66N	72.76E	8.55693
	Mumbai	11,914,398	18.55N	72.50E	7.47182
	Ratnagiri	70,335	17.13N	73.32E	7.14562
	Malvan	18,675	16.05N	73.50E	6.58475
	Panaji	1,170,000	15.25N	73.50E	5.92902
	Murmagao	189,383	15.25N	73.56E	5.99745
	Karwar	62,960	14.83N	74.15E	6.28768
	Kumta	27,597	14.48N	74.41E	6.26771
	Bhatkal	31,785	13.96N	74.58E	6.04996
	Mangalore	398,745	12.55N	74.47E	5.01031
	Kozhikode	2,613,683	11.15N	75.43E	4.7295
	Cochin(Kochi)	550,000	09.58N	76.20E	4.64342
	Quilon(Kollam)	391,300	08.90N	76.63E	4.64503
	Trivandrum	744,739	08.41N	77.00E	4.17903
	Kanyakumari	208,149	08.07N	77.58E	3.92992
	Thoothukkudi	216,058	08.50N	78.12E	3.73327
	Rameswaram	38,035	09.28N	79.37E	3.90142
	Nagapattinam	94,965	10.77N	79.88E	3.56652
	Karaikal	170,640	10.59N	79.50E	4.57374
	Pondicherry	735,004	11.59N	79.50E	3.36741
	Chennai	4,216,268	13.08N	80.19E	3.57933
	Nellore	378,947	14.27N	79.59E	3.71164
	Chirala	85,455	15.98N	80.08E	4.22424
	Machilipatnam	215,043	16.15N	81.20E	4.08971
	Visakhapatnam	969,608	17.45N	83.20E	3.64451
	Gopalpur	114,189	19.27N	84.95E	4.21065
	Puri	157,610	19.50N	85.58E	3.79441
	Haldia	170,695	22.03N	88.03E	5.62307
	Henhoaha	1,018	06.80N	93.81E	0.961126
	Misha	2,660	08.00N	93.36E	1.44183
	Kakana	4,291	09.11N	92.81E	1.58246
	Nachuge	2,233	10.71N	92.35E	2.06832
	Port Blair	100,186	11.68N	92.77E	1.92749
	Coco Channel	2,233	14.08N	93.30E	2.46952
	Kavaratti Is	10,113	10.53N	72.71E	4.52211
	Androth Is	10,000	11.00N	73.16E	4.54489
	Chetlat Is	51,707	11.76N	76.83E	5.38175
	Minicoy Is	9,957	08.48N	73.02E	4.20262

Population database is based on the latest information available through various sources from internet.

7.2 Network Learning Principles and Algorithms

The input is transformed to an output through the hidden layers. Input vectors and the corresponding output (target) vectors are used to train the network until it can approximate a function, associate input vectors with specific output vectors, or classify input vectors in an appropriate way as defined. It has been shown that networks with biases, sigmoid layers and a linear output layer are capable of approximating any function with a finite number of discontinuities. In the present work, the back-propagation feed forward type network is used for training the system where the objective is to minimize the global error E given as:

$$E = \frac{1}{P}\sum E_p \tag{4}$$

and

$$E_P = \frac{1}{2}\sum (o_k - t_k^2) \tag{5}$$

where P is the total number of training patterns (the number of input/output pairs used for training), E_p is the error for the p-th pattern, o_k is the network output at the k-th output node and t_k is the target output at the k-th output pattern.

In this type of network the error between the target output and the network output are calculated and this is back propagated. The term back-propagation refers to the manner in which the gradient is computed for nonlinear multilayer networks. Standard back-propagation is a gradient descent algorithm. There are a number of variations on the basic algorithm that are based on other standard optimization techniques. A brief description on the working principle of a back-propagation neural network is given below.

7.3 Back-Propagation Learning

Back-propagation is a widely used algorithm for supervised learning with a multilayer feed forward layer network which implements the repeated application of the chain rule to compute the influence of each weight in the network with respect to an arbitrary error function E (Dayhoff, 1990):

$$\frac{\partial E}{\partial w_{ij}} = \frac{\partial E}{\partial s_i} \frac{\partial s_i}{\partial net_i} \frac{\partial net_i}{\partial w_{ij}} \tag{6}$$

In the above equation [Eqn. (6)], w_{ij} is the weight from the neuron j to neuron i, s_i is the output, and net_i is the weighted sum of the inputs of neuron i. Once the partial derivative for each weight is known, the error function is minimized by performing a simple gradient descent (Zurada, 1992):

$$w_{ij}(t+1) = w_{ij}(t) - \varepsilon \frac{\partial E(t)}{\partial w_{ij}} \tag{7}$$

7.4 Resilient Back-Propagation Learning

Multilayer networks typically use sigmoid transfer functions in the hidden layers. Sigmoid functions cause a problem while training a multilayer network using steepest descent since they compress an infinite input range into a finite output range and are characterized by the fact that their slope must approach zero as the input gets large. The gradient can have a very small magnitude; and therefore, cause small changes in the weights and biases, even though the weights and biases are far from their optimal values. The purpose of the resilient back-propagation training algorithm is to eliminate contamination in the magnitudes of the partial derivatives. Only the sign of the derivative is used to determine the direction of the weight update; the magnitude of the derivative has no effect on the weight update. This adaptive update value evolves during the learning process based on its local sight on the error function E, according to the following learning rule (Reidmiller and Braun, 1993):

$$\Delta_{ij}^{(t)} = \begin{cases} \eta^+ \Delta_{ij}^{(t-1)}, & \text{if } \dfrac{\delta E^{(t-1)}}{\delta w_{ij}} \dfrac{\delta E^{(t-1)}}{\delta w_{ij}} > 0 \\ \eta^- \Delta_{ij}^{(t-1)}, & \text{if } \dfrac{\delta E^{(t-1)}}{\delta w_{ij}} \dfrac{\delta E^{(t-1)}}{\delta w_{ij}} < 0 \\ \Delta_{ij}^{(t-1)}, & \text{else} \end{cases} \quad \text{where } 0 < \eta^- < 1 < \eta^+ \qquad (8)$$

$$\Delta w_{ij}^{(t)} = \begin{cases} -\Delta_{ij}^{(t)}, & \text{if } \dfrac{\delta E^{(t)}}{\delta w_{ij}} > 0 \\ +\Delta_{ij}^{(t)}, & \text{if } \dfrac{\delta E^{(t)}}{\delta w_{ij}} < 0 \\ 0, & \text{else} \end{cases} \qquad (9)$$

$$w_{ij}^{(t+1)} = w_{ij}^{(t+1)} + \Delta w_{ij}^{(t)} \qquad (10)$$

The resilient back-propagation algorithm has been used in this study to train the neural network system.

8. NETWORK ARCHITECTURE AND PARAMETERS

The performance of the network depends on the network architecture and network parameters chosen to model the system. Extensive trial and error tests have been performed with the comprehensive ETA database (Barman et al., 2006). Based on this study, configurations mentioned below are near optimum for the current prediction model.

8.1 Network Architecture

The optimum parameters of network architecture for prediction of tsunami travel time were chosen with the configuration comprising two hidden layers with number of neurons ranging from 25 to 30 in each hidden layer (recommended value for this study is 30). The transfer functions comprise the tan-sigmoid functions in the hidden layers and the linear transfer function in the output layer. This is a useful structure for regression problems.

8.2 Network Parameters

Based on an earlier study (Barman et al., 2006), the following parameters are recommended as an optimum configuration of the ANN network:

(i) Mean Squared Error (MSE) used in this study is 0.25.
(ii) Learning Rate and Momentum Factor: It was found that the performance of resilient back-propagation is not very sensitive to the settings of these training parameters.

The resilient back-propagation has been used to train the system as we noticed convergence to the specified MSE is obtained very fast using this algorithm. In fact, not many algorithms are able to converge to the specified error level. Besides Resilient back-propagation, we used another algorithm called Levenberg-Marquardt (More, 1978) which is also able to converge, but it was found that the time of convergence is much greater than the former, typically about 100 times or so.

Prudently trained back-propagation network performs significantly well and tends to deliver reasonable estimates when presented with input parameters. Typically, a new input leads to an output similar to the correct output for input vectors used in training that are similar to the new input being presented. This generalization property makes it possible to train a network on a representative set of input/target pairs and get good results without training the network on all possible input/output pairs.

8.3 Training and Testing of ANN Model

For this study, 240 data points of past earthquake locations in the Indian Ocean taken from the ETA database were used. For each of this earthquake event, we have computed tsunami travel time to 250 coastal destinations in the 35 countries surrounding the Indian Ocean rim which is published in the TTT atlas for Indian Ocean. Using the above defined network parameters, we demonstrated that ANN model is quite robust in tackling the non-linearity in the ETA database (Barman et al., 2006). Numerical experiments were conducted with different combinations of data-points within the ETA database. In this experiment, grouping of data within ETA is under two steps: (i) first set for training and testing, and (ii) second set for validation which is not exposed to ANN during the learning stage. This set is used to check the performance of the network with unseen data, and thus gives a measure of the network skill with real-time data. Since performance of ANN solely depends on the nature of data being trained, the performance of ANN model is questionable for a new earthquake location which is not within close vicinity of TTT model. Hence, a situation which is rare to the training and testing set cannot be predicted well. For the Indian Ocean, as seen in Fig. 2, considering the homogeneity in the distribution of earthquake locations from past tsunami-genic events the highest degree of probability for an earthquake is within the vicinity of Sumatra, Indonesia, etc. Therefore, under this circumstance the nonlinear technique based on ANN can be suitably used for a real-time prediction of ETA for countries surrounding the Indian Ocean rim.

9. SIMULATION RESULTS

The results of ETA simulations using ANN with different network configurations for coastal destinations in India (47 locations) are presented below. To study the effectiveness of the nonlinear technique, the selection of validation points from ETA database was judicially grouped into three categories. Of the total 240 locations, skill assessments were performed with 80, 60 and 48 points, respectively as shown in Tables 4, 5 and 6. High values of correlation (around 95%) between the model outputs and the observations shown in Tables 4, 5 and 6 suggest the feasibility of the model that can predict tsunami arrival time when the underwater earthquake locations is provided within the study domain.

The correlation between the known and predicted ETA using three different combinations of data for an optimally trained network with two hidden layers using 30 neurons each trained to a mean square error of 0.25 using the resilient back-propagation is shown in Figs 5, 6 and 7 corresponding to Tables 4, 5 and 6, respectively. An optimum selection of 25 neurons each was found to be sufficient keeping in mind the skill factor of prediction of ETA using ANN.

Table 4. Simulation results for 80 validation data when trained with 160 data (Combination 1)

	Validation Data: 1 out of every 3 in ETA Database					
Total number of data		240		Number of inputs		2
Training-Testing data		160		Number of outputs		47
Validation data		80				

Neurons	Goal	Training time (minutes)	Training correlation	Validation correlation		
				Minimum	Maximum	Average
[20 20]	0.3	1	0.9799	0.945	0.9603	0.9553
[20 20]	0.25	1	0.9833	0.929	0.9632	0.9525
[20 20]	0.2	2	0.9866	0.9465	0.9658	0.9563
[25 25]	0.3	1	0.9802	0.9406	0.9692	0.9537
[25 25]	0.25	1	0.9834	0.9289	0.9539	0.9429
[25 25]	0.2	1	0.9866	0.9103	0.9475	0.9252
[30 30]	0.3	1	0.9801	0.9322	0.966	0.955
[30 30]	0.25	1	0.9834	0.9436	0.968	0.959
[30 30]	0.2	1	0.9866	0.9486	0.9655	0.9585

Table 5. Simulation results for 60 validation data when trained with 180 data (Combination 2)

	Validation Data: 1 out of every 4 in ETA Database					
Total number of data		240		Number of inputs		2
Training-Testing data		180		Number of outputs		47
Validation data		60				

Neurons	Goal	Training time (minutes)	Training correlation	Validation correlation		
				Minimum	Maximum	Average
[20 20]	0.3	1	0.979	0.9866	0.9886	0.9874
[20 20]	0.25	2	0.9819	0.9711	0.9797	0.9755
[20 20]	0.2	3	0.9855	0.979	0.9844	0.9809
[25 25]	0.3	1	0.9782	0.961	0.9772	0.9686
[25 25]	0.25	1	0.9819	0.9601	0.9718	0.967
[25 25]	0.2	8	0.9855	0.8206	0.8592	0.8431
[30 30]	0.3	1	0.9783	0.9544	0.9688	0.9615
[30 30]	0.25	1	0.9819	0.9652	0.9754	0.9667
[30 30]	0.2	> 10	0.9842	0.9585	0.9699	0.9651

Table 6. Simulation results for 48 validation data when trained with 192 data (Combination 3)

	Validation Data: 1 out of every 5 in ETA Database					
Total number of data	240		Number of inputs		2	
Training-Testing data	192		Number of outputs		47	
Validation data	48					

Neurons	Goal	Training time (minutes)	Training correlation	Validation correlation		
				Minimum	Maximum	Average
[20 20]	0.3	1	0.9784	0.943	0.9518	0.9471
[20 20]	0.25	1	0.9821	0.941	0.9567	0.9513
[20 20]	0.2	1	0.9856	0.951	0.9597	0.9573
[25 25]	0.3	1	0.9785	0.9452	0.9624	0.9545
[25 25]	0.25	1	0.9821	0.9503	0.9572	0.9534
[25 25]	0.2	1	0.9857	0.9464	0.9557	0.9511
[30 30]	0.3	1	0.9786	0.9489	0.9612	0.9543
[30 30]	0.25	1	0.984	0.9497	0.9588	0.9548
[30 30]	0.2	1	0.9856	0.9402	0.9554	0.9485

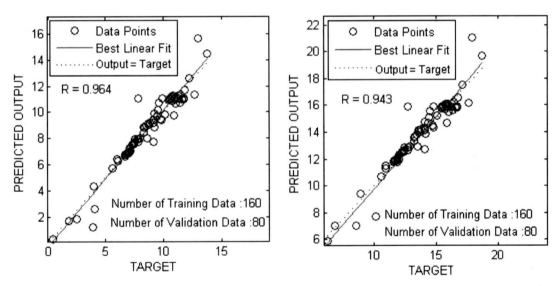

Fig. 5 Actual and predicted outputs (Validation data, Combination 1) for the best correlated output (Left) and the worst correlated output (Right) for 47 coastal locations in Indian sub-continent.

10. CONCLUDING REMARKS

An effective tsunami early warning system is achieved when all persons in vulnerable coastal communities are prepared and respond appropriately, and in a timely manner, upon recognition that a potentially destructive tsunami is approaching. Timely tsunami warnings issued by a recognized tsunami warning center are essential. When these warning messages are received by the designated government agency, tsunami emergency response plans must already be in place so that well-known and practiced actions are

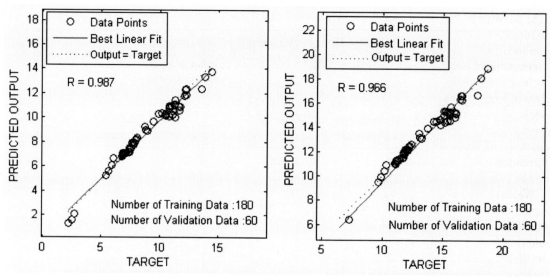

Fig. 6 Actual and predicted outputs (Validation data, Combination 2) for the best correlated output (Left) and the worst correlated output (Right) for 47 coastal locations in Indian sub-continent.

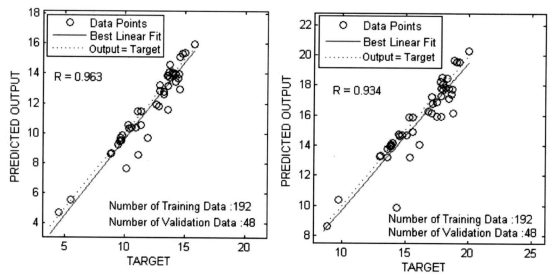

Fig. 7 Actual and predicted outputs (Validation data, Combination 3) for the best correlated output (Left) and the worst correlated output (Right) for 47 coastal locations in Indian sub-continent.

immediately taken to evaluate the scientifically-based warning, and communicate an appropriate course of action to ordinary citizens. Tsunami preparedness programs must be implemented on a national level so that good decisions can be made without delay.

In this chapter, we discussed travel times for the first wave of tsunami approaching the coast which has been subsequently verified by the computation carried out for the recent December 2004 Indian Ocean tsunami from tide gauges as well as signatures of satellite tracks. It does not provide information on the arrival times of subsequent waves, nor does it provide information on how many waves will be in the tsunami event, which wave in succession will be the highest, at what time each wave will arrive at a given location on the coast line, the run-up along littoral belts and resulting inundation, how strong the currents will be in each wave, exactly at what locations should people and domestic animals be evacuated, how long should they be evacuated, at what time it will be safe for them to return, nonlinear dispersive effects of tsunami waves, etc. To obtain detailed information about all these parameters, separate numerical models of tsunami generation and propagation, and of coastal inundation should be developed. The importance of such study in computing TTT is to extend this computational algorithm for multiple tsunami-genic locations in the Indian Ocean rim facilitating development of ETA database. This basic information can be used as extremely important database for early response in the development of tsunami warning systems.

We demonstrated here how soft computing tools like ANN could handle the non-linear system where the prediction of ETA at different coastal destinations of Indian sub-continent is achieved in a real-time mode. The algorithm uses earthquake locations and computed travel time from the ETA database. It could be advocated that the major advantage of using ANN in a real-time tsunami travel time prediction is its high merit in producing ETA at a much faster time and simultaneously preserving the consistency of prediction. The model using ANN performs a rapid computation of ETA (on average of four seconds) compared to the conventional travel time model which takes approximately 60 minutes. The correlation is found very high for the unseen data as noted from different combinations of training and testing the ANN. The importance of this model is highly justified for a tsunami warning system where the time involved in computing ETA and the issue of warning messages to coastal destination is a critical factor. The proposed method is expected to have direct practical applications for a real-time tsunami warning system for the Indian Ocean as well as the global oceans. For the benefit of the readers, an atlas on comprehensive database of Tsunami Travel Time for the Indian Ocean is available at http://www.iitkgp.ac.in/topfiles/tsunami.html. Finally, the validation capability of the model was found to be satisfactory and reliable which suggests its applicability for real-time prediction.

In context of known natural disasters such as earthquakes, floods, cyclones, storm surges, landslides, volcanic eruptions, etc., warning systems exist for most of these disasters except for earthquakes. Developed nations like the USA and Japan are prepared to cope even with earthquakes. Hence, developing nations need to gear up with technology which can be helpful in the event of natural disasters. In case of a tsunami event, the basic and vital information is the time of arrival at various coastal destinations of a nation and hinterland. In this context, we demonstrated the importance of ANN as a tool for an effective tsunami warning system. However, an effective tsunami warning system is a composite feedback mechanism starting from *in situ* bottom pressure sensors, communication linkage from sensors via satellite to land-based stations and final processing of information at a nodal agency for an alert warning to various coastal destinations. The discussion in this chapter pertains to the computation of TTT and how this vital information can be expedited using soft computing tools like ANN. For the operational purpose, database within ETA needs to be periodically updated by TTT computations for earthquake locations which is the future scope not covered in this study. Overall, it can be concluded that modern technology can prevent or help in minimizing the loss of life and property provided we integrate all essential components in the warning system and put it to the best possible use.

REFERENCES

Bapat, A., Kulkarni, R.C. and Guha, S.K. (1983). Catalog of Earthquakes in India and Neighborhood from Historical Period up to 1979. Ind. Soc. Earthq. Tech., Roorkee, 211 pp.

Barman, R., Prasad, K.B., Pandey, P.C. and Dube, S.K. (2006). Tsunami travel time prediction using neural networks. *Geophysical Research Letters,* **33:** L16612, doi:10.1029/2006GL026688.

Bindra, S. (2005). Tsunami: 7 Hours that Shook the World. Harper Collins Publications, New Delhi, India, 291 pp.

Bishop, C.M. (1995). Neural Networks for Pattern Recognition. Oxford University Press, Oxford, U.K., pp. 364-369.

Chapman, C. (2005). The Asian tsunami in Sri Lanka: A personal experience. EOS, *Transactions American Geophysical Union,* **86(1):** 13-14.

Dayhoff, J.E. (1990). Neural Network Architecture: An Introduction. Van Nostrand Reinhold, New York, 259 pp.

Hanson, J.A. and Bowman, J.R. (2005). Dispersive and reflected tsunami signals from the 2004 Indian Ocean tsunami observed on hydrophones and seismic stations. *Geophysical Research Letters,* **32:** L17606, doi:10.1029/2005GL023783.

Hinton, G.E. (1992). How neural networks learn from experience. *Sci. Amer.,* **9:** 144-151.

Holloway, G., Murty, T.S. and Fok, E. (1986). Effects of bathymetric roughness upon tsunami travel time. *Science of Tsunami Hazards,* **4(3):** 165-172.

Li, S., Hsieh, W.W. and Wu, A. (2005). Hybrid coupled modeling of the tropical Pacific using neural networks. *J. Geophys. Res.,* **110:** C09024, doi:10.1029/2004JC002595.

Lietzin, E. (1974). Sea Level Changes. Elsevier Oceanographic Series, No. 8, New York, 273 pp.

More, J.J. (1978). The Levenberg-Marquardt Algorithm: Implementation and Theory. *In:* G.A. Watson (editor), Lecture Notes in Mathematics, 630, Springer-Verlag, Berlin, pp. 105-116.

Murty, T.S. (1984). Storm Surges – Meteorological Ocean Tides. *Canadian Bulletin of Fisheries and Aquatic Sciences,* **212,** Department of Fisheries and Oceans, Ottawa, Canada, 897 pp.

Murty, T.S., Saxena, N.K., Sloss, P.W. and Lockridge, P.A. (1987). Accuracy of tsunami travel time charts. *Marine Geodesy,* **11:** 89-102.

Navone, H.D. and Ceccatto, H.A. (1994) Predicting Indian monsoon rainfall: A neural network approach. *Climate Dynamics,* **10(6):** 305-312.

Okal, E.A., Plafker, G., Synolakis, C.E. and Borrero, J.C. (2002). Near-field survey of the 1946 Aleutian tsunami on Unimak and Sanak Islands. *Bulletin Seismological Society of America,* **93:** 1226-1234.

Orman, J.V., Cochran, J.R., Weissel, J.K. and Jestin, F. (1995). Distribution of shortening between the Indian and Australian plates in the central Indian Ocean. *Earth and Planetary Science Letters,* **133(1-2):** 35-46.

Pendse, C.G. (1945). The Mekran Earthquake of the 28th November 1945. *India Meteor. Department Scientific Notes,* **10(25):** 142-145.

Prasad, K.B., Dube, S.K., Murty, T.S., Gangopadhyay, A., Chaudhuri, A. and Rao, A.D. (2005). Tsunami Travel Time Atlas for the Indian Ocean. CORAL, Indian Institute of Technology Kharagpur, West Bengal, India, 286 pp.

Prasad, K.B., Rajesh, K.R., Dube, S.K., Murty, T.S., Gangopadhyay, A., Chaudhuri, A. and Rao, A.D. (2006). Tsunami travel time computation and skill assessment for the 26 December 2004 event in the Indian Ocean. *Coastal Engineering Journal,* **48(2):** 147-166.

Reidmiller, M. and Braun, H. (1993). A direct adaptive method for faster back-propagation learning: The RPROP algorithm. Proc. IEEE Int. Conf. on Neural Networks, San Francisco, pp. 586-591.

Silverman, D. and Dracup, J. (2000). Artificial neural networks and long range precipitation prediction in California. Journal *of Applied Meteorology,* **31(1):** 57-66.

Simon, H. (1998). Neural Networks: A Comprehensive Foundation. 2nd Edition, Prentice-Hall, Englewood Cliffs, New Jersey, 842 pp.

Synolakis, C.E. (1995). Tsunami prediction. *Science,* **270:** 15-16.

Tandon, A.N. and Srivastava, H.N. (1974). Earthquake Occurrence in India: Earthquake Engineering. Sarita Prakashan, Meerut, 48 pp.

Venkatesan, C., Raskar, S.D., Tambe, S.S., Kulkarni, B.D. and Keshavamurty, R.N. (1997). Prediction of all India summer monsoon rainfall using error-back-propagation neural networks. *Meteorology and Atmospheric Physics*, **62**: 225-240.

Yeh, H., Liu, P., Briggs, M. and Synolakis, C.E. (1994). Tsunami catastrophe in Babi Island. *Nature*, **372**: 6503-6508.

Yong, W., Kwok, F.C., George, D.C. and Charles, S.M. (2003). Inverse algorithm for tsunami forecasts. *Journal of Waterway, Port, Coastal and Ocean Engineering*, **129(2)**: 60-69.

Yuri, I.S. and Leonid, B.C. (1995). Mathematical modeling in mitigating the hazardous effect of tsunami waves in the ocean: A priori analysis and timely on-line forecast. *Science of Tsunami Hazards*, **13(1)**: 27-44.

Zetler, B.D. (1947). Travel time of seismic sea waves to Honolulu. *Pacific Science*, **1(203)**: 185-188.

Zurada, J.M. (1992). Introduction to Artificial Neural Systems. West Publishing Company, St. Paul, Minnesota, 785 pp.

Flood Hazards in India and Management Strategies

Dhrubajyoti Sen

1. INTRODUCTION

India suffers from a large variability of precipitation both spatially and temporally. It is generally known that though the average rainfall for the country is about 1160 mm, the highest anywhere in the world of a comparable size (Kumar et al., 2005), the spatial variability ranges from an average of 2800 mm for most of the north-eastern states, Andaman and Nicobar Islands and northern areas of West Bengal to about 300 mm in the western part of Rajasthan (http://www.rainwaterharvesting.org/urban/Rainfall.htm). Again, except for the States of Assam, Jammu and Kashmir and the southern peninsula, more than 75% of India's annual rainfall is received during the southwest monsoon season, i.e., June through September (Jagannathan and Bhalme, 1973). An effect of this skewed distribution of rainfall is excess water for a region received during a short interval of time, leading to flooding of the surroundings, if not drained off suitably. The region may be small as an urban space, like cities or towns, for which an intense rainfall of even half a day may cause flooding, or larger areas like the over bank and floodplain areas of a river where the state of flooding may extend for several days due to continuous rainfall of a couple of days in the upper catchment. Examples of the former include flooding events in the cities of Kolkata and Mumbai, which occasionally get flooded due to drainage congestion aggravated primarily because of insufficient slopes of drainage channels and tidal influences at the outfalls. Examples of floodplain inundation are common for the rivers of the eastern and north-eastern states of the country, though a few others from the other parts of the country are in the news sometimes. In addition, rivers flowing through the hills often suffer from flash floods due to occasional cloud bursts and the coastal regions, especially in the eastern part of the country, is prone to flooding due to cyclonic storms either by the associated intense rainfall or the impact of storm surge waves, or both. One other aspect that is being investigated lately by many researchers is the probable impact of global warming or a climate change scenario on the risk associated with flooding. There are quite a few reports, though not all agreeing to a common conclusion, wherein it is generally believed that the variability in extreme rainfall is likely to increase in the future which in turn may lead to larger incidents of flash floods (Pal and Al-Tabbaa, 2009).

It may be emphasized that flooding as such would not be a concern unless associated with consequential socio-economic losses. In India, the flooding of the rural areas within the inundated floodplains of rivers result mainly in the losses of agricultural produces with occasional loss of human lives and livestock, whereas those of the cities and towns result in the losses of infrastructure and domestic assets. Similar situation exists for other densely populated countries as well, like those in Europe namely United Kingdom, Germany, Spain, Austria, Italy, etc. and some other places around the world (Christensen and Christensen, 2003; Zbigniew et al., 2005). In fact, there are several institutions and organizations in these countries working on monitoring and providing information on floods, and assessing and managing their risks. Some of these are: Flood Hazard Research Centre (http://www.fhrc.mdx.ac.uk/), Integrated Flood Risk Analysis and Management Methodologies (http://www.floodsite.net/), Water Information System for Europe (http://ec.europa.eu/environment/water/flood_risk/), European Commission-Joint

Research Centre (http://natural-hazards.jrc.ec.europa.eu/activities_flood_riskmapping.html), etc. There has also been an increasing awareness in flood management and monitoring programs in India over the past few decades. Newer tools and policies are being implemented for reducing the risks associated with flooding in the country. A paradigm shift in flood management from the rehabilitation of a post-flood scenario to building up resilience as a pre-flood measure is gradually being introduced across the country (MHA, 2004).

This chapter reviews the hazards associated with different types of floods affecting India, in particular the broad areas of river and coastal flooding as well as the measures to mitigate consequential damages. The specific regions prone to each type of floods, the mechanisms of flood generation and the intensity of flooding for the areas are discussed. Thereafter, the long-term management measures adopted for regulating the developmental activities taking place in and around the flood-prone areas and the possible damages expected for the existing developments are discussed. The current methods of early warning system for flood and the emergency preparedness initiatives adopted for different flooding scenarios have also been described in this chapter.

2. FLOOD HAZARD IN INDIA: AN OVERVIEW

Of the different natural hazards affecting India, the threat due to floods appears to be the most recurring, widespread and disastrous (Kale, 2003, 2004). In recent times, the number of people affected by flooding in India by hydrological disasters overwhelmingly exceeds that by meteorological, climatological and geophysical disasters (Scheuren et al., 2008) as available in the OFDA/CRED International Disaster Database (www.em-dat.net), one of the comprehensive disaster databases currently available. However, it must be remembered that the consequences of flooding is experienced within a much smaller time span, requiring more emergency mitigation measures than the calamity of drought. Moreover, apart from the immediate loss of life and property due to the action of flood water (e.g., drowning and fatal injuries), the post-flood health hazard (by the contamination of drinking water and sanitation) continues to inflict a lingering loss to the affected populace long after the recession of the flood peak (Ahern and Kovats, 2006). As indicated by Ray et al. (1999), the average annual loss of human lives from meteorological disasters in India for the period 1975 to 1996 was maximum due to floods triggered by heavy rains (1441) followed by cyclonic storms (348) and those due to other reasons like cold and heat waves being much lower. The maximum economic loss in India from natural hazards, according to EM-DAT (www.em-dat.net), during the past couple of decades is also due to flooding. According to BMTPC (1998a), an analysis of damage figures since 1953 shows that on an average every year about 7.5 million hectare area is affected by floods, which involve damages of the order of Rs. 9800 million including damages to 11,68,000 houses with a loss of Rs. 1350 million and damage to public utilities costing about Rs. 3750 million. In addition, about 1500 human lives and about 97,000 cattle heads are lost each year due to floods. Table 1 presents the losses incurred in India due to the combined actions of natural hazards, including flooding, at the end of last century (MHA, 2004). Thus, it can be inferred that the hazard due to floods is an important issue for the Indian society and comprehensive and dedicated studies focusing on the causes of floods and the risks involved can enable one to seek the solution for the mitigation of consequential damages.

3. CAUSES AND TYPES OF FLOODING

Although the term 'flooding' is used in common parlance in India to refer to the overtopping of the banks by river water, the broad causes of flooding could be due to either river or coastal submergences. Of

Table 1. Annual damages due to heavy rains, landslides and floods in India from 1999 to 2001 (MHA, 2004)

Year	Districts affected	Villages affected	Population affected (million)	Crop area affected (Mha)	Houses damaged	Loss of lives		Estimated value of losses (million Rupees)	
						Humans	Cattle	Private houses	Public properties
1999	202	33,158	32.81	0.85	884,823	1,375	3,861	7.2	-
2000	200	29,964	41.62	3.48	2,736,355	3,048	102,121	6312.5	3897.2
2001	122	32,363	21.07	187.2	346,878	834	21,269	1955.7	6760.5

course, the coastal regions of India in the vicinity of river estuaries often get submerged due to flooding from heavy local precipitation resulting from cyclonic storms as well as over-bank spill of rivers caused by the high flows being conveyed from the upper catchment and it is often difficult to distinguish specifically between the two. An example is the inundation of the coastal region of the State of Orissa where many significant rivers discharge into the Bay of Bengal and also where the probability of the cyclonic influence is very high. Figure 1 shows an inundated area in the coastal area of Orissa (http://orissafloods.wordpress.com/). Closer to the coastline, the direct impact of the high cyclonic storm surges may additionally have an effect on the habitation and infrastructure in the vicinity. Additionally, the force of the high winds of the cyclones may cause damages or destruction.

Fig. 1 An inundated region of Orissa in eastern India under the impact of 2008 floods (*Source:* http://orissafloods.wordpress.com/).

Flooding of rivers, that is, overflowing of its banks due to sudden bursts of high discharges which exceeds the river's conveying capacity may occur due to the following reasons (Smith, 2004):

(1) Atmospheric: Flood is caused by heavy rainfall, snowmelt or ice jam,
(2) Tectonic: Flood is caused by landslides, and
(3) Technological: Flood is caused due to the incidents like failure of dams or intentional breaching of river embankments.

The first of the above causes is by far the most common cause of occurrences of flooding of rivers in India and is almost invariably due to heavy rainfall within the catchment and the inadequate capacity of rivers to contain within their banks the high flows brought down. Kale (2003), however, differentiates between the *rainfall floods* and *rainstorm floods* as two basic types of meteorological flooding phenomena. The former is stated to be most predominant of the causes of river flooding that result from heavy or intense precipitation in association with active to vigorous monsoon conditions for a number of days. The *rainstorm floods* are defined as those generated by excessively heavy rainfall associated with lows, depressions and cyclonic storms originating over the Bay of Bengal or the adjoining coastal belt. Again, the temporal characteristics of the flood event may lead it to be classified as: *flash floods, single-peak floods, multiple-peak floods* and *synchronized floods* (Dhar and Nandargi, 1998). *Flash floods* are characterized by very fast rise and recession of flow of relatively small volume but very fast flowing discharge, which causes high damages because of its suddenness. Usually, this sort of river flooding is noticed in the hilly and not too hilly regions and sloping lands where heavy rainfall and thunderstorms are common. *Single-peak floods* are characterized by a rising and falling limb of the discharge hydrograph in contrast to the *multiple-peak floods* which have multiple rises and falls. *Synchronization of floods* in more than one tributary of a main river may cause severe impact on the downstream flows by their combined impact. Though meteorological factors are the primary driving force behind river flooding in India, other reasons like riverbed aggradation due to siltation and back-water flooding caused by higher downstream water levels, often due to high sea levels for the estuarine rivers are equally responsible for the increase in the extent of devastation. Construction of flood embankments or levees, as has been practiced for major Indian rivers over the second half of the last century, has also induced rise of riverbeds thus increasing the risk of flooding for rivers carrying large amount of silt from its upper catchments.

The incidence of rivers flooding due to snowmelt and ice jam is rather unheard of in India, except perhaps for some regions of extreme north, in the Himalayas. There have been however, albeit rare, occurrences of flooding caused by landslide as the Alaknanda tragedy (Bhandari, 2003) when a dam was created in the river by the landslide and caused devastation in the valley downstream when it ultimately breached. Other instances of the formation of such natural dams and devastations caused by their failures have been listed by Kale (2003). As for the flooding of rivers by the failure of man-made dams, the extent of damage depends upon the degree of dam collapse, the head of water behind the dam at the time of the disaster and the distance of the point of damage assessment from the location of the dam. There have been quite a few incidences of dam failures in India and brief overviews are presented by Thandaveswara and Mahesh Kumar (1993) and Kale (2003), among others.

On the other hand, coastal flooding may be generated by the effects of the following two prime reasons (Smith, 2004):

(1) Atmospheric: Floods occur due to surges in ocean produced by cyclonic storms, and
(2) Tectonic: Floods occur due to tsunami wave produced by high intensity underwater earthquakes.

Though the losses due to the tsunami wave flooding of the coastal tracts of South India by an undersea earthquake-triggered landslide in the Indian Ocean in December 2004 was enormous (Sheth et al., 2006), such events are extremely rare for the Indian coastline. In fact, flooding of Indian coasts by storm surges is relatively more frequent (Murty and Flather, 1994). This type of flooding is caused by the tropical cyclone, usually originating somewhere in the Bay of Bengal and heading north-westwards, pushing the surface of water on its track. As the eye of the cyclone nears the coast, there is generally a pile-up of water against the coast aided by the shallow ocean bottom nearer coast, which ultimately breaks upon the coastal shoreline at the time of the cyclone's landfall. The resulting wave running up the beach is similar to that caused by the tsunami wave and the extent of run-up depends upon the energy of the wave, represented by the wave height and speed. According to NDMA (2008b), the degree of disaster potential depends on the storm surge amplitude associated with the cyclone at the time of landfall, characteristics of the coast, phases of the tides and vulnerability of the area and community. The world's highest recorded storm tide has been about 12.5 m (about 41 ft) associated with the Backergunj cyclone in 1876 near the Meghna estuary in present-day Bangladesh. A storm tide of closer magnitude (12.1 m) has been estimated to have been observed in West Bengal at the mouth of the River Hooghly in association with a severe cyclone in October 1737. In recent times, the super cyclone of October 1999 generated an estimated wind speed of 252 km/h with an ensuing surge of 7 to 9 m close to Paradip in Orissa (eastern India) which caused unprecedented inland inundation up to 35 km from the coast. It is worth mentioning that, at times, persistent standing water was identified in the satellite images even 11 days after the cyclone landfall, as it happened in the Krishna delta (in the State of Andhra Pradesh, southern India) in May 1990 and in several other instances. The cyclone of 1977 which hit the same delta area, particularly at Divi Seema, also generated winds exceeding 250 km per hour.

It is worth mentioning that since cyclones rotate counter-clock wise in the northern hemisphere of the earth, on approaching the coast the right forward sector of the cyclone experiences wind from ocean to land (*on-shore wind*) which pushes the seawater towards the coast and finally appears as storm surge. The direction of the wind on the left forward sector of the cyclone is from land to ocean (*off-shore wind*) which pushes the water from the coast towards the ocean producing even negative surge.

4. AREAS PRONE TO FLOODING

Maps of flood-prone areas in India have been brought out by various expert committees, like the Building Materials & Technology Promotion Council (BMTPC), Ministry of Housing & Urban Poverty Alleviation, Government of India and the National Disaster Management Authority (NDMA), under the Chairmanship of the Prime Minister of India, and made available via their respective websites, viz., BMTPC: http://www.bmtpc.org/ and NDMA: http://ndma.gov.in/. Figures 2 and 3, adapted from these references, indicate the flood-prone regions of India due to river flooding as well as coastal flooding, respectively.

In comparison to the flooding due to tectonic actions or technological disasters, which are rather unpredictable, those due to meteorological factors follow a definite pattern both spatially and temporally. The geographical extent of the country which suffers from riverine flood hazards may be broadly divided into the following four regions (Mohapatra and Singh, 2003; NDMA, 2008a): (i) Brahmaputra River Region, (ii) Ganga River Region, (iii) North-West Rivers Region, and (iv) Central India and Deccan Region, which are briefly described in the subsequent sections.

4.1 Brahmaputra River Region

Consisting of the rivers Brahmaputra and Barak and their tributaries, this region covers the States of Arunachal Pradesh, Meghalaya, Mizoram, Manipur, Tripura, Nagaland, Sikkim and the northern parts of West Bengal. The catchments of these rivers receive very heavy rainfall ranging from 1100 mm to 6350

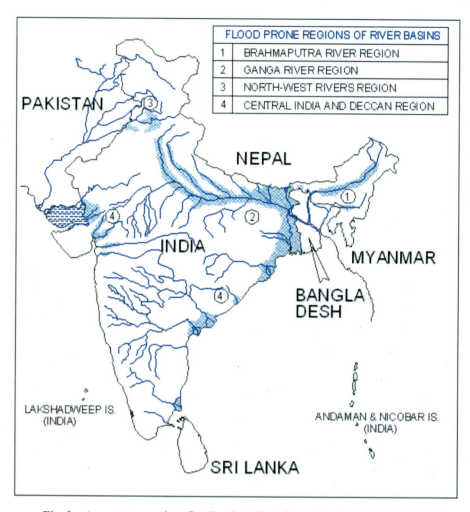

Fig. 2 Areas prone to river flooding in India (adapted from NDMA, 2008a).

mm a year which occurs mostly during the months of May-June to September. As a result, floods in this region are severe and quite frequent. Further, the hills where the rivers originate are fragile and susceptible to erosion and thereby contribute exceptionally high silt discharge to the river flow. The predominant problems in this region are cloud bursts followed by flash floods, soil erosion in the watershed and bank erosion along the rivers, the flooding caused by spilling of rivers over their banks, drainage congestion, and the tendency of some rivers to change their courses. The plain areas of the region suffer from the inundation caused by the spilling of the rivers Brahmaputra and Barak (in the Kachhar region of Assam).

4.2 Ganga River Region

The river Ganga has many tributaries, the important ones being Yamuna, Sone, Ghagra, Rapti, Gandak, Bagmati, Kamla Balan, Adhwara group of rivers, Kosi and the Mahananda. This region covers the States of Uttarakhand, Uttar Pradesh, Jharkhand, Bihar, south and central parts of West Bengal, Punjab and

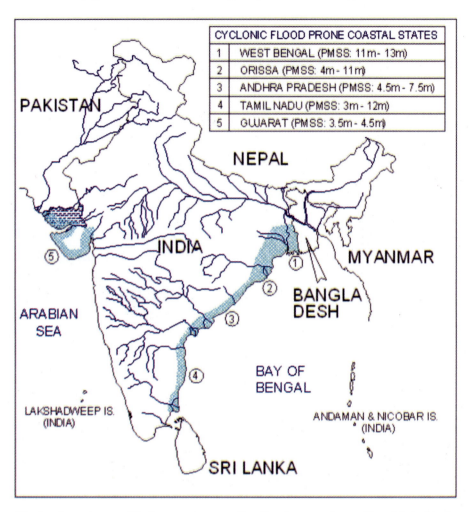

Fig. 3 Coastal areas of India prone to severe flooding due to cyclones. The table inside the figure indicates the States with maximum hazard and corresponding Probable Maximum Storm Surge (PMSS) heights (adapted from NDMA, 2008b).

parts of Haryana, Himachal Pradesh, Rajasthan, Madhya Pradesh and Delhi. The normal annual rainfall in this region varies from about 600 mm to 1900 mm of which more than 80% occurs during the southwest monsoon. The rainfall increases from west to east and from south to north. Most of the damage is caused by the northern tributaries of the Ganga, which spill over banks and frequently change courses. Even though the peak flood of river Ganga may be high, in the order of 57,000 to 85,000 m^3/s, the inundation and erosion problems are confined to relatively few places. In general, the flood problem increases from the west to the east and from south to north. In the north-western parts of the region, the problem of drainage congestion prevails in some places. The drainage problem also exists in the southern parts of West Bengal, where the outflowing discharges from smaller tributaries of the river Hooghly (which itself is a distributary of river Ganga) are sometimes prevented from flowing out by the tidal influences on the Hooghly river.

4.3 North-West Rivers Region

The important rivers in this region, Sutlej, Beas, Ravi, Chenab and Jhelum, are the tributaries of the river Indus and carry quite substantial discharges during the monsoon season and also large volumes of sediment. In the plains, these rivers change their courses frequently and leave behind vast tracts of sandy waste. This region covers the States of Jammu and Kashmir, Punjab and parts of Himachal Pradesh, Haryana and Rajasthan. Compared to the Ganga and the Brahmaputra river regions, the flood problem is relatively less in this region. The major problem is that of inadequate surface drainage which causes inundation and waterlogging over vast areas.

4.4 Central India and Deccan Region

Rivers of importance in this region are the Narmada, Tapi, Mahanadi, Godavari, Krishna and Cauvery, which mostly have well defined and stable courses. In general, these rivers have adequate capacities within the natural banks to carry the flood discharge except in the delta area. The lower reaches of the important rivers on the east coast have been embanked, thus largely eliminating the flood problem. This region covers the States of Andhra Pradesh, Karnataka, Tamil Nadu, Kerala, Orissa, Maharashtra, Gujarat and parts of Madhya Pradesh. The region does not have serious flood problem except for some of the rivers in the State of Orissa namely Mahanadi, Brahmini, Baitarni, and Subarnarekha and in the stretches closer to the sea. The problem of flooding and drainage congestion gets accentuated when the floods synchronize with high tide. Rivers Tapi and Narmada occasionally affect the areas in lower reaches in Gujarat during periods of high floods.

In addition to the above regions of mainland India, the flooding and drainage characteristics of the islands of the Andaman and Nicobar archipelago and Lakshadweep suffer from drainage congestion due to very flat slopes.

The long coastal shoreline of the country, measuring about 5700 km on the mainland (about 7500 km including the islands), is prone to flooding by cyclonic storm surges. According to NDMA (2008b), though the frequency of tropical cyclones in the North Indian Ocean covering the Bay of Bengal and the Arabian Sea is the least in the world (about 7% of the global total), their impact on the east coast of India and the adjoining Bangladesh coast is relatively more devastating. This is evident from the fact that in the last 270 years, 21 of the 23 major cyclones (with a loss of about 10,000 lives or more) worldwide occurred over the area surrounding the Indian subcontinent (India and Bangladesh). This is primarily due to the serious storm-tide effect in the area.

It may be observed that the above mentioned flood-prone regions may be classified under one of the following geographical environments (Smith, 2004):

(1) Low-lying areas of major floodplains
(2) Low-lying coasts and deltas
(3) Small basins and valleys affected by flash floods
(4) Areas below unsafe or inadequate dams
(5) Low-lying inland shorelines
(6) Alluvial fans

However, the vulnerability to river flooding has increased over the years due to anthropological reasons like increase in population resulting in the occupancy of areas previously uninhabited. Plate (2002) reports that earlier human settlements that grew up near rivers were, in general, located at safer places determined by the years of accumulated societal wisdom. Houses near rivers prone to regular flooding were either provided with high plinth or, as in some parts of Assam and Bengal, on raised bamboo platform on stilts (bamboo columns) so as to remain above the dominant flood water levels. This harmonious attitude of "living with floods" has been greatly reduced in the last century with the rapid increase in population of cities through migration from rural areas and construction of flood embankments along rivers, providing a false security for the bank-side environment and encouraging new floodplain development. Smith (2004) calls this the "Levee Effect" and is a global phenomenon.

The phenomenon of riverbank erosion, though not a case of widespread land inundation by water, is often associated with river flooding because of its almost simultaneous occurrence during high floods of rivers (Mohapatra and Singh, 2003). This aspect of river floods also causes enormous losses by damaging the land and property of inhabited areas.

4.5 Urban Flooding

Another aspect highlighted by Mohapatra and Singh (2003) is the rising flooding events of urban spaces in the country, which may once again be attributed to the encroachment of the flood plain areas, presence of several structures of permanent nature within the rightful way of the river floodway and absence of proper regulations for maintenance of the drainage arteries.

4.6 Coastal Flooding

The extensive coastal belt of India is very vulnerable to the deadly storms known as *tropical cyclones*. About 4 to 6 such storms originate in the Bay of Bengal and the Arabian Sea every year. Tropical cyclones, which are characterized by torrential rain, gales and storm surges cause heavy loss of human lives and destruction of property. The States that are most prone to coastal flooding are: West Bengal, Orissa, Andhra Pradesh, Tamil Nadu and Gujarat.

As for the assessment of vulnerability of coastal tracts, an analysis combining storm risk and poverty suggests that the State of Orissa in eastern India is the most vulnerable due to its low coping capacity determined by the per capita income of about Rs. 6,767 and high likelihood cyclonic storms (NDMA, 2008b). Tamil Nadu and Andhra Pradesh in south India as well as West Bengal in eastern India are also vulnerable because they are situated in the high cyclone hazard zones and also have low per capita incomes. Maharashtra and Goa are the two States which are also affected by cyclones, but they are less vulnerable as the cyclones are less frequent and people living there have better coping capacity because of higher incomes (per capita income is more than Rs. 18,365).

5. MANAGEMENT MEASURES FOR RIVER FLOODS

The globally accepted flood management measures for river floods include the following broad strategies (Smith, 2004; NDMA 2008a), which are explained in detail in the subsequent sections:

(1) Prevention or protection from floods by abatement, control or proofing of floods;
(2) Adaptation to existing flood situations like preparedness, forecasting and warning and proper land-use planning; and
(3) Mitigation of flood related losses by disaster aids or insurance coverage.

According to some experts (e.g., Mohapatra and Singh, 2003) the above-mentioned flood management strategies may also be classified as: (a) *structural* (involving structural constructions), and (b) *non-structural*. The former refers to the structural constructions that are installed across the river or in its vicinity with an aim to, primarily, reduce the peak of the flood discharge. On the other hand, the non-structural measures include the strategies adopted other than any structural constructions for reducing flood damage such as forecasting of flood and dissemination of flood warnings, floodplain zoning and regulation of land use within the floodplains, implementation of flood proofing measures, modifying upstream land management practices, etc. The following sections elaborate the above-mentioned flood management strategies in the Indian context.

5.1 Prevention from Floods

The prevention of large-scale flooding may be reduced by truncating the flood peak through construction of large reservoir dams or utilization of low-lying depressions and tanks for accommodating part of the flood discharge. The Hirakud Dam on River Mahanadi in Orissa or the series of dams on the Damodar River system (Maithon, Panchet, Tilayia and Tenughat) in Jharkhand (eastern India) had been primarily constructed for this purpose. In fact, most of the high dams in India are multi-purpose and serve the cause of flood abatement quite efficiently. However, as the capacity of these reservoirs gets reduced with time by siltation and the "live storage" of the reservoir is affected, the utility of these dams in carrying out flood moderation starts reducing.

Other means of surviving from the impact of flooding is by protecting a segment of river floodplains by the construction of flood embankment or levees. According to NDMA (2008a), divergent views have however emerged on the utility of embankments as a means for flood protection. While some NGOs have voiced serious criticisms of existing embankments and advocated their removal, others favor construction of additional lengths of embankments as the only practical medium-/short-term solution for the flood problem. Positive benefits provided by embankments include, apart from flood protection, roads that provide useful communication link in the area. They also provide shelter to the villagers during floods, though as a result they often get damaged by the actions of the villagers themselves. On the negative side, breaches in embankments have often resulted in large-scale flooding in the protected areas. Poor drainage in the protected area also leads to drainage congestion. The worst impact of the embankments is the deposition of silt and rise in riverbed levels, thereby decreasing the carrying capacity of the river and aggravating drainage congestion. This situation exists for many Indian rivers, for example, River Ganga near Patna and Teesta near Jalpaiguri. Also, the prevalent designs of flood protection embankments presently adopted by the different states of India appear to be rather weak to withstand erosion (e.g., Ghosh, 1997). For example, the flood embankments of the Ajoy River in West Bengal, whose bed level has also risen over the past decades due to embanking, regularly breach during floods with consequent high river water levels. As such, the alignments of the embankments are generally arbitrarily decided and often very close to the rivers, disregarding the need to consider an appropriate floodway corresponding to a certain return period of flood discharge as determined by the accepted societal risk.

A recent example of embankment breaching is the devastating failure of the eastern flood embankment of the Kosi River at about 12 km upstream of the Kosi Barrage in Nepal on 18 August 2008, which resulted in widespread inundation of many districts of Bihar, eastern India. After the 18 August breach of the Kosi embankment, the Kosi River has taken a new course almost similar to that of the 1930s and hence, the 18 August 2008 flood caused huge damages to the agricultural land, houses and other properties as well as loss of several lives, besides rendering a large number of people homeless and devoid of resources to sustain their livelihoods. Figure 4, adapted from the images available on the website of the Disaster Management Department, Government of Bihar (http://disastermgmt.bih.nic.in/), shows the extent of flooding due to the breach and the new course of River Kosi. The extent of damage caused by this incident may be gauged from the statistics presented in Table 2 (http://disastermgmt.bih.nic.in/).

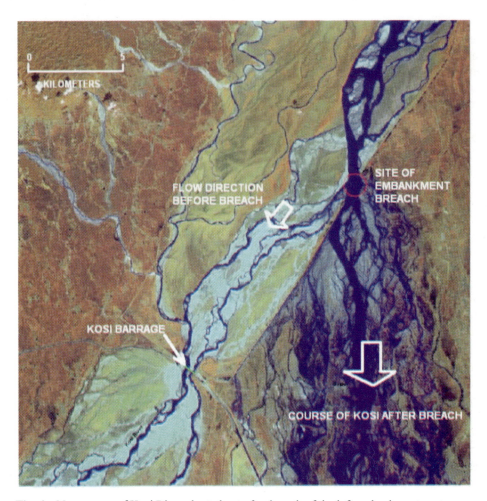

Fig. 4 New course of Kosi River charted out after breach of the left embankment upstream of Kosi Barrage on 18 August, 2008 (http://disastermgmt.bih.nic.in/).

Table 2. Status of 18 August 2008 Kosi flood impact (as on 24 February 2009)

Impact	Districts affected					Total
	Supaul	Madhepura	Araria	Saharsa	Purnea	
Blocks affected	5	11	4	6	9	35
Panchayats affected	65	140	71	59	77	412
Villages affected	178	370	141	169	140	993
Population affected	670709	1419856	626062	448796	164000	3329423
Livestock affected	417704	303640	80000	161000	35000	997344
Area affected (ha)	75000	157000	45000	44000	47000	368000
Human deaths	211	272	2	41	1	527
Livestock deaths	8585	10725	0	13	0	19323

Desilting and dredging of rivers to increase conveyance capacity remains the only viable option when the other (structural) methods for protection against floods are found unsuitable. The difficulty in this case lies in the excavation of the huge quantity of material and its disposal. In fact, this method is rather expensive, but it has been attempted for some rivers as the River Ichhamati in West Bengal-Bangladesh border.

Flood proofing is a means of either designing or retrofitting buildings and their contents to make them more resistant to flood losses (Smith, 2004). The methods normally applied are: (a) Raising of habitable parts of property above flood level; (b) Making uninhabited parts of property resistant to flood damage and allowing water to enter these parts; (c) Sealing the property to prevent flood water from entering; and (d) Relocating properties. According to Mohapatra and Singh (2003), however, the flood proofing measures taken in the past in India consisted of raising of a few flood-prone villages above a pre-determined flood level and connecting them to nearby roads or high lands. Under this program, several villages in West Bengal, Uttar Pradesh and Assam were raised. This method was subsequently discontinued, as it was observed that it could not protect the nearby agricultural areas.

Moreover, for the prevention of riverbank erosion during high floods, the properly designed riverbank protection techniques using synthetic geo-textile, geo-jute fabric or the low-cost alternative of *"darma-mat"* was adopted with equally enduring and sustainable effect. In fact, the latter has been successfully tested in the lower reaches of the River Ganga at Farakka (Sen, 2008). The *"darma-mat"*, made of interwoven bamboo splicings, has to be manually fabricated and is thus labor intensive and employment generative.

Rising river waters normally do not make a forced impact on buildings and structures. However, the inundation may remain for days at some places and if the structures are not strong enough to resist this inundation, for example, the rural mud houses, then the walls may collapse leading to the losses of life and property. Hence, for the flood-prone areas, BMTPC (1998b) recommends certain modifications in material and building construction technology that may help to increase the life of the structures. Apart from such structural modifications of the existing housing technology, recommendations have been provided like raising plinth level above the recurrent flood inundation level of the region. However, a scientific evaluation of the plausible inundation levels for floods of different intensities has to be decided by floodplain zoning studies, as discussed in the following sections.

5.2 Adaptation to Flooding

For societies in human habitations accepting the hazards of flooding, the need to be prepared in the eventuality of any flooding becomes a part of standard adaptive practice. This requires preparedness for food, shelter, rescue and medical aid, and needs to be activated on the basis of an efficient flood forecasting and warning system. The preparedness processes and warning schemes are generally looked after by the local civic administration, though some NGOs also come forward to help according to their capabilities. The key step, however, lies in accurately predicting the arrival time of the flood (i.e., rise, peak and fall) and the expected water level rise. It is on this basis that the social administration offices need to deploy its strength in reducing possible losses in the event of flood hazard. However, since the methods for predicting floods are highly technical, the prediction of flood is mostly entrusted to specialized technical organizations like Central Water Commission or the Water Management organizations of the states (for example, the Department of Irrigation of the Government of Andhra Pradesh, the Department of Irrigation and Waterways of the Government of West Bengal, the Irrigation and Flood Control Department of the Government of Delhi, etc.).

The Central Water Commission under the Ministry of Water Resources, Government of India (http://cwc.nic.in/), has a specialized River Management Wing that is responsible for the collection, compilation, storage and retrieval of hydrological and hydro-meteorological data, formulation and issue of early warnings on flood forecast on all major flood-prone rivers and inflow forecasts for selected important reservoirs. The Central Water Commission is currently maintaining 175 flood-forecasting stations (NDMA, 2008a) of which 147 stations are for river stage forecast and the rest for inflow forecast. These forecast stations, located on various river basins, communicate advises to the respective State Administrations on the expected level of floods. The prevalent technique of issuing flood forecast by the Central Water Commission is the gauge-to-gauge and rainfall-runoff correlation techniques. The gauge to gauge correlation is based upon many years of recorded stage (i.e., water level) data of a river at a number of locations on the upstream of the forecasting station. The forecast is issued based upon this correlation and the current water level at an upstream station is transmitted to the forecasting station through a network of wireless radio system (CWC, 1989). At present, there are over 500 VHF/HF wireless stations of the Central Water Commission spread throughout the country that transmit the rainfall and gauge data on a real-time basis. However, steps are being taken to improve the lead time for forecast by attempting to predict using real-time data of rainfall recorded through automatic rainfall recorders with the data relayed down using satellite communication and internet. This kind of setup has been installed recently in the catchments of Mahanadi and Damodar Rivers, among others. Further improvement in the lead time is possible, if atmospheric models are developed and adapted for the respective regions which can then serve to forecast the amount of rainfall ahead of the actual rainfall data collected by the automatic recorders.

Though a society may be well adapted to the threats of flood hazards, it is prudent to understand and appreciate the risks of development in the flood-prone areas vis-à-vis their proximity to the river itself. This brings about the issue of land use planning for the region taking into account the zoning of floodplains classified according to a given risk. Though this demarcation of river adjacent areas is essential for future development and planning as well as for flood preparedness, the response of the State Governments towards the enactment of floodplain zoning bill has not been encouraging (Mohapatra and Singh, 2003). Though some of the states (e.g., Manipur and Rajasthan) have enacted the floodplain zoning legislation, major flood affected states such as Assam, Goa, Himachal Pradesh, West Bengal and Sikkim have not considered such legislation.

According to NDMA (2008a), one of the main requirements for the implementation of floodplain zoning measure is the availability of high accuracy survey maps. The Central Water Commission had initiated in 1978 a program for surveying areas prone to floods under the central sector through the Survey of India as a pilot scheme. However, of the 106,000 km^2 of area identified in the country as prone to frequent floods, about half (around 55,000 km^2) had been surveyed in the states of Bihar, Assam, Uttar Pradesh, West Bengal, Punjab, Haryana and Jammu and Kashmir. These maps, on the scale of 1:15,000 with a contour interval of 0.3 m or 0.5 m, have been made available to the respective states but no progress has been made so far by any of the states to produce floodplain zoning maps.

It may be noted that though high precision survey maps are required to generate floodplain zoning maps, the validation if these maps may be possible only with accurate mapping of actual flood inundation extent areas for certain given flood events. Also, the manual survey of inundation extents may be extremely difficult and time consuming, if not inaccurate, and satellite pictures using remote sensing technology can provide great help (Bhanumurthy, 1999). Though remotely sensed flood image usually connotes recording of the extent of flooding using satellite imagery, more precise information may be collected using the airborne Synthetic Aperture Radar (SAR) surveys as carried out by the National Remote Sensing Centre, Hyderabad. These surveys, carried out using special airplanes flying at low altitudes, help in gathering finer details which may not be possible to be captured using satellite-borne sensors. In fact, the manual survey of floodplains, as mentioned earlier, may also be substantially improved by employing the airborne Laser Terrain Mapping (LIDAR) survey, which is carried out by the National Remote Sensing Centre.

5.3 Mitigation of Flood Disaster Losses

In the event of the failure of preventing or managing floods from inflicting losses, the society has to rely upon the rescue measures for the minimization of losses. Though the response to such disasters in India have been the duty of the respective State administration, rescue operations during devastating floods, like the recent breaching of the flood embankments of the Kosi River (eastern India), have been carried out with the help of the military and para-military forces. Alleviation of flood losses would be effective if the response system is fast and efficient. This requires proper institutionalization of the system as well as the formulation of a sound evacuation plan. Apart from the official setup, the NDMA (2008a) also stresses the needs of having community-level local search and rescue teams because having such groups would help to act faster, at least as a first aid, than the official team which might have a delayed response time.

Another way of minimizing the losses from flood inundation is by having compensation from flood insurances. However, according to Mohapatra and Singh (2003), though flood risk has been included in the list of items covered by the general insurance companies in India, it has so far been popular in the urban areas only, as seen from the vehicle loss claims after the floods of Mumbai in 2005 and of Surat (Gujarat) in 2006. The insurance companies have also not been able to arrive at different rates of insurance premiums for different areas.

6. MANAGEMENT MEASURES FOR COASTAL FLOODS

Management practices for coastal flooding are similar to those for river floods as mentioned in the previous sections and may be grouped in similar order as discussed below.

6.1 Protection from Floods

For the protection of flood-prone areas of the coastal tracts from the impact of waves and inundation by flooding resulting from tropical cyclonic activities, the recommended measures can be classified as either *structural* or *non-structural*. The principal structural recommendations are those for cyclone shelters and local buildings since the loss of life due to cyclones has been observed to be largely due to the lack of an adequate number of safe shelters which can withstand the fury of cyclones, including wind and storm surge (BMTPC, 1998c). The provisions in the guidelines consider the strength of a building to be adequate in resisting the impact of storm surge waves and also remain stable against high cyclonic winds.

Compared to normal buildings constructed away from the shore, additional requirements for the design of these structures have been recommended. For example, wind, a primary force in destabilizing a building by producing negative lift pressures, is recommended to be taken as 1.3 times the value adopted for a normal structure. The design of the building is also recommended for being able to withstand seismic forces in the regions vulnerable to earthquake hazard. The minimum reinforced concrete grade is recommended to be M30. Though the material specifications are considered important, the planning and design of cyclone shelters have been advised to include special provisions like being located preferably at least 1.5 km away from the coast. The plinth height on stilts is suggested to be of 1.5 m with the height varying from 2.5 m to 4.5 m if the storm surge level is in the range of 1.5 m to 4.5 m. In all the cases, the floor level of a shelter is recommended to be at least 0.5 m above the possible maximum surge level. Additionally, the foundation depths of the shelters are required to be extended to depths equal to that of the surge level to avoid scouring, with a minimum value of 1.5 m.

It has been noticed that during inundation of coastal areas by cyclonic flooding, the communication between neighboring human habitations get snapped. This is because many of the coastal villages do not have all-weather approach roads. Hence, properly designed roads with appropriate base and raised at an elevation higher than the expected inundation level are required for connecting the villages. The road crossing of coastal drainage system should also to be provided with suitably designed bridges or culverts as found appropriate.

Among the non-structural measures, the protection provided by the coastal environment has been highly commended by the experts to play a major role in reducing the impact of storm surge waves. After the tsunami disaster of 26 December 2004 along the Indian coast, it was observed that shorelines endowed with mangroves, forests, sand dunes or coastal cliffs provided the best natural barriers against the tsunami waves. In contrast, heavy damage was reported in the areas where sand dunes had been heavily mined (for example, at Nagapattinam and Kolachal in the State of Tamil Nadu) and where the coastal vegetation was less. Thus, the stretches of mangrove vegetation act as natural bio-shield against the fury of cyclones, coastal storms, tidal waves and tsunamis. The present management principles in mitigating the impact of cyclones, therefore, encourage the regeneration of mangrove vegetation or at least prevent degradation of the already existing stock. Among the coastal states of the country, the State of West Bengal has been found to contain the largest and densest mangrove cover (about 48% of the country's total), especially in and around the Sunderban delta region (MoEF, 2005). Though the other coastal states of the country are not equally gifted with this bio-wealth for cyclone protection, steps for the conservation of mangrove forests have been undertaken to various degrees.

Those coastal belts not suitable for mangrove growth (mangroves normally thrive in sheltered shores enriched with abundant silt brought down by the estuarine rivers) have been encouraging the growth of other non-mangrove bio-shields popularly known as *shelterbelts*. Raising coastal shelterbelts to mitigate the adverse impact of cyclonic winds has been identified as one of the short-term objectives of the National Afforestation Program through Community Participation scheme of the National Afforestation and Eco-Development Board. Based on the Coastal Regulation Zone Notification of 1991 of the Ministry of Environment and Forests, which recognizes the mangrove areas as ecologically sensitive and categorizes them as CRZ-I areas implying the highest order of protection for such areas and from the best land use practices, steps are being taken by the various states of the country to plan for conservation and restoration of mangroves and raise tree shelterbelts extensively in all potential coastal zones.

Coastal shorelines may also be affected by the impact of cyclonic storms and consequential surge waves. Several solutions exist and those similar to the suggestions for riverbank erosion protection have been employed with a varying degree of success. A major difference between the riverbank erosion and coastal shoreline erosion is that the former is subject to toe undercutting by river currents, whereas the latter is predominantly forced with a direct impact of the ocean waves.

6.2 Adaptation to Flooding

Strategies quite similar to those provided for coping with river flooding is adopted for coastal flooding as well. It is natural that habitations may not be completely protected from the vagaries of cyclonic winds and resulting storm surges. In fact, the structural recommendations discussed in the previous section can only be effective in resisting the impact of cyclones up to a given degree. Nature, however, may induce forces that may exceed the design values which are decided on accepted risks and economic constraints. Hence, it is normally thought prudent to be prepared in case of disastrous eventualities. Such preparedness is somewhat similar to those adopted for river flooding. However, the forecasting strategies of an impending coastal disaster differ considerably from river related disasters, as described below.

Tropical cyclonic storms brew up in the Bay of Bengal and the northern part of Indian Ocean, which ultimately approach the east coast of the country. Similarly, those originating in the Arabian Sea move towards the west coast. As tropical cyclones originate over the ocean from a given combination of meteorological factors the origin of which is generally difficult to predict, a constant watch is kept on the Arabian Sea and the Bay of Bengal by the India Meteorological Department (IMD) for its likely genesis with the help of satellite imagery, particularly those from the Indian geostationary satellite, INSAT. The data from ships and ocean buoys are also analyzed.

As a particular cyclonic system develops and gathers strength and moves towards the coast, it becomes necessary to both track it as well as forecast its probable course and the likely place on the coast where it is going to strike, often called the landfall point. The tracking of the path of a cyclone is mostly done from satellite imagery. However, these images are not continuous and are snapshots taken from the satellites (non-geostationary) when they pass above the specific region encompassing the main body of the storm. At present, the prediction of cyclone tracks is done by the IMD using a quasi-Lagrangian dynamic model along with other synoptic, climatological and empirical techniques. The forecast advisories received from different international agencies such as the Joint Typhoon Warning Centre, USA; United Kingdom Meteorological Office; European Centre for Medium Range Weather Forecasting; and the National Centre for Medium-Range Weather Forecasting, Government of India are also considered while finalizing the forecast.

As a cyclonic system closes on to the Indian coastline, its subsequent development and movement is monitored by a chain of Cyclone Detection Radars set up by IMD to cover the entire coastal belt. The latest S-band Doppler weather radars or the somewhat older storm detention X-band radars are used to provide the necessary data. This information is used to prepare and disseminate warning messages from the Area Cyclone Warning Centres at Kolkata, Chennai and Mumbai and Cyclone Warning Centres at Vishakhapatnam, Bhubaneshwar and Ahmedabad.

These early warnings are issued in four stages, viz., (i) Pre-cyclone watch indicating the potential development of a cyclonic system; (ii) Cyclone alert, issued 48 hours prior to the expected time of commencement of adverse weather over a specific coastal area; (iii) Cyclone warning provided 24 hours in advance of a cyclone's landfall; and (iv) Post landfall outlook, issued 12 hours before the landfall indicating the effect the system is likely to cause once it strikes the land. Though the final warnings predicting, normally, the storm intensity and velocity is important, the probable height of storm surge as the cyclone makes a landfall is equally significant in predicting the extent of coastal area that is likely to get submerged. For the prediction of storm surge, IMD uses both nomograms developed from past records and a dynamical storm surge model developed by the Indian Institute of Technology, Delhi. Very recently, IMD has also implemented the dynamical storm surge model of National Institute of Ocean Technology, Chennai for the east coast of India on an experimental mode.

Although the real-time availability of cyclone warnings is helpful for bracing up against an impending disaster, a long-term planning like coastal belt flood inundation zoning, similar to the river floodplain zoning, is helpful for the developments taking place in the coastal region and the likely risks associated thereof. Satellite images showing past inundation extents corresponding to various levels of storm surges combined with high resolution terrain elevation data are being used to demarcate the probable inundation extent areas. The procedure involves mathematical techniques quite similar to those used for demarcating probable flood inundation zones for rivers, except that the possible impact of a dynamic surge wave has to be additionally taken into account.

6.3 Mitigation of Flood Disaster Losses

This is the emergency response part which is activated once a cyclonic disaster has struck. Since the most affected parts of the country by cyclones are the states of Tamil Nadu, Andhra Pradesh, Orissa and West Bengal and the Union Teritorry of Puducherry (previously Pondicherry) on the east coast and the State of Gujarat, the emergency preparedness plans for coastal disasters are generally stronger in these states than in others. Of these, Andhra Pradesh having faced quite a few disastrous cyclones in the recent past has set up a separate Disaster Management Unit to implement the World Bank-funded Andhra Pradesh Hazard Mitigation and Emergency Cyclone Recovery Project, and is currently functioning as the Andhra Pradesh State Disaster Mitigation Society. In Orissa too, the Orissa State Disaster Mitigation Authority, as a Government owned autonomous body, was constituted after the October 1999 super cyclone in order to have a systematic and planned approach for disaster management in the state. This organization coordinates various activities of disaster mitigation in the state, including the capacity building of the community and disaster managers, strengthening of infrastructure, improvement in communication system, etc.

7. DISASTER PREPAREDNESS AND MITIGATION

The Ministry of Home Affairs, Government of India, in its publication on the status of disaster management in the country (MHA, 2004), have very elaborately discussed the strategies adopted by the Central and State Governments of the country in the context of preparedness against a probable disaster and its mitigation. The guidelines prepared by the National Disaster Management Authority for the steps that need to be taken by various states of the country for safeguarding against the hazards of floods due to heavy rainfall in rivers and intense cyclonic precipitation and storm surges are also in place (NDMA, 2008a, b). Although many of the states are yet to formulate a comprehensive disaster management plan, including that of floods, some actions have already been taken in this regard. The following sections discuss some of these measures in the context of disaster preparedness and mitigation of flood hazards in India.

7.1 National Policy on Disaster Management

Disaster management is a multidisciplinary activity involving a number of departments and agencies spanning across all the sectors of development. Where a number of departments and/or agencies are involved, it is essential to have a policy in place, as it serves as a framework for action by all the relevant departments and agencies. According to MHA (2004), a National Policy on Disaster Management has been drafted, and is in the process of finalization. In line with the changed focus, the policy proposes to integrate disaster mitigation into development planning. The National Disaster Management Guidelines for preparing State Disaster Management Plans (NDMA, 2007) has outlined the steps to be followed by the states while preparing the plans. This is in accordance with the Disaster Management (DM) Act 2005 passed by the Government of India on 23 December 2005. The highlights of this Act include a shift from a response and relief-centric approach to a proactive and comprehensive mindset towards disaster management covering all aspects from prevention, mitigation and preparedness to rehabilitation, reconstruction and recovery. It also provides for the following:

- The creation of a policy, legal and institutional framework backed by effective statutory and financial support.
- The mainstreaming of multi-sectoral disaster management concerns into the developmental process and mitigation measures through projects.
- A continuous and integrated process of planning, organizing, coordinating and implementing policies and plans in a holistic, community-based, participatory and sustainable manner.

The DM Act mandates the National Disaster Management Authority (NDMA) to lay down policies and guidelines for the statutory authorities to draw their plans. In essence, the NDMA will concentrate on the prevention, mitigation, preparedness, rehabilitation and reconstruction and will also formulate appropriate policies and guidelines for effective and synergized national disaster response and relief. It will coordinate the enforcement and implementation of policies and plans.

7.2 Mitigation and Preparedness Plans/Measures

In order to respond effectively to floods, the Ministry of Home Affairs (MHA) has initiated National Disaster Risk Management Program in all the flood-prone states. Assistance is being provided to the states to draw up disaster management plans at the State, District, Block/Taluka and Village levels. Awareness generation campaigns are being organized to sensitize the stakeholders on the need for flood preparedness and mitigation measures. Elected representatives and officials are being trained in flood disaster management under this program. Bihar, Orissa, West Bengal, Assam and Uttar Pradesh are among the 17 multi-hazard-prone States where this program is being implemented with assistance from international organizations, viz., United Nations Development Program, United States Agency for International Development and European Commission (MHA, 2004). It is also reported that the Central Government is now in the process of training and equipping eight battalions of Central Para-Military Force (CPMF) as specialist response teams. Each team consists of 45 personnel including doctors, paramedics, structural engineers, etc. and thus there will be 144 Specialist Search and Rescue Teams in the earmarked eight battalions. The process of training and equipping of the 144 specialist search and rescue teams has been initiated and so far 18 teams have been trained. These teams are being trained in collapsed structure search and rescue, medical first response, rescue and evacuation in flood and cyclone situations, underwater rescue, etc. In effect, they will have the capability to operate in all types of terrain in all contingencies/disasters. It is proposed to group together the eight battalions of CPMFs earmarked for specialized emergency response as *"National Emergency Response Force"*. These specialist response teams are being provided modern equipments and also dog squads for search and rescue. They will be provided with special uniforms made of fire retardant materials with enhanced visibility in low light and having equipment carrying capacity.

The use of Information Technology (IT) in disaster management is reflected by the fact that there are plans to set up a Geographical Information System (GIS) database for emergency responders to access information in terms of crucial parameters for the disaster affected areas. The crucial parameters include location of the public facilities, communication links and transportation network at national, state and district levels. The GIS database already available with different agencies of the Government is being upgraded and the gaps are proposed to be bridged. The database will provide multi-layered maps on a district-wise basis. These maps taken in conjunction with the satellite images available for a particular area will enable the district administration as well as State Governments to carry out hazard zonation and vulnerability assessment and coordinate response after a disaster.

8. CONCLUSIONS

According to the estimates of the National Commission on Floods (RBA, 1980), the area prone to floods in India is around 40 million hectares, which is about one-eighth of the country's geographical area. Though the area affected by floods each year is of the order of about 7.4 million hectares (Kale, 2004), most of this falls in the States of Uttar Pradesh, Bihar, Punjab, Himachal Pradesh, Assam, West Bengal, Haryana, Orissa, Andhra Pradesh and Gujarat, in decreasing order of distribution. With rising population and enhanced migration to the cities, there has been a recent spate of urban flooding for many of the Indian cities and towns. Hence, coping with flood disasters is an important priority for the administration of the Central or State Governments and a lot of investment is made each year in this context. Also, a number of initiatives have been taken to institutionalize the actions of flood prevention and loss minimization, flood occurrence prediction, warning dissemination and emergency response. However, given the economic constraints, not all the measures, as would have been ideally desirable, have so far been implemented. For example, a lot remains to be done in terms of implementing satellite data for on-line river flood monitoring or its accurate prediction using numerical simulation models. Further, though the real-time data collection and transmission of hydro-meteorological data using satellite communication has been initiated by the Central Water Commission for a few river basins, the program has faced some operational difficulties and is yet to be made completely functional. Ideally, such monitoring and surveillance of rainfall and occurrence of floods should be provided for every major river basin that can be coupled with real-time flood propagation and inundation spread prediction models. Furthermore for the coastal flooding hazard, establishment of a denser network of S-band radars for better capturing of approaching cyclones and application of numerical models to predict flood inundation due to storm surges and heavy rainfall in the coastal regions need to be implemented. Also, over the past couple of years, there has been a new approach to the disaster management paradigm, as emphasized by MHA (2004), which proceeds from the conviction that development cannot be sustainable unless disaster mitigation is built into the development process. Another corner-stone of the approach is that mitigation has to be multi-disciplinary spanning across all the sectors of development. The new policy also emanates from the belief that the investments in mitigation are much more cost effective than expenditure on relief and rehabilitation. It is hoped that with the initiatives taken by the Building Materials & Technology Promotion Council, the National Disaster Management Authority, and the National Institute of Disaster Management, the security against flood hazards in India can be increased in near future.

REFERENCES

Ahern, M. and Kovats, S. (2006). The Health Impacts of Floods. *In:* R. Few and F. Matthies (editors), Flood Hazards and Health: Responding to Present and Future Risks. Earthscan Publications Ltd., London, U.K., pp. 28-53.

Bhandari, R.K. (2003). Two great landslides of India. *In:* H.K. Gupta (editor), Disaster Management. Orient Longman, New Delhi, pp. 110-127.

Bhanumurthy, V. (1999). Remote sensing for flood management. In: K.M.B. Rahim (editor), Natural Disasters: Some Issues and Concerns. Natural Disasters Management Cell, Visva Bharati, Santiniketan, West Bengal, India, pp. 104-115.

BMTPC (1998a). Major Natural Disasters in India. *In:* Report of Expert Group on Natural Disaster Prevention, Preparedness and Mitigation having bearing on Housing and Related Infrastructure; Part I: Techno-Legal Aspects of Earthquake, Windstorm and Flood Hazards and Land Use Zoning, Building Material and Technology Promotion Council (BMTPC), Ministry of Urban Affairs and Employment, Government of India, New Delhi.

BMTPC (1998b). Improving Flood Resistance of Housing – Guidelines. *In:* Report of Expert Group on Natural Disaster Prevention, Preparedness and Mitigation having bearing on Housing and Related Infrastructure; Part III: Guidelines for Improving Hazard Resistant Construction of Buildings and Land Use Zoning (for Earthquake, Windstorm and Flood Hazards). Building Material and Technology Promotion Council (BMTPC), Ministry of Urban Affairs and Employment, Government of India, New Delhi.

BMTPC (1998c). Improving Wind/Cyclone Resistance of Housing – Guidelines. *In:* Report of Expert Group on Natural Disaster Prevention, Preparedness and Mitigation having bearing on Housing and Related Infrastructure; Part III: Guidelines for Improving Hazard Resistant Construction of Buildings and Land Use Zoning (for Earthquake, Windstorm and Flood Hazards). Building Material and Technology Promotion Council (BMTPC), Ministry of Urban Affairs and Employment, Government of India, New Delhi.

Christensen, J.H. and Christensen, O.B. (2003). Severe summertime flooding in Europe. *Nature,* **421:** 805-806.

CWC (1989). Manual on Flood Forecasting. River Management Wing, Central Water Commission (CWC), Ministry of Water Resources, Government of India, New Delhi.

Das, P.K. (1981). Storm surges in the Bay of Bengal. Proceedings of the Indian Academy of Science (Engineering Science Section), September 1981, Vol. 4, Part 3, pp. 269-276.

Dhar, O.N. and Nandargi, S.S. (1998). Floods in Indian rivers and their meteorological aspects. *In:* V.S. Kale (editor), Flood Studies in India. Memoir 41, Geological Society of India, Bangalore, India, pp. 53-76.

Ghosh, S.N. (1997). Flood Control and Drainage Engineering. Second Edition. Oxford and IBH Publishing, New Delhi.

Jagannathan, P. and Bhalme, H.N. (1973). Changes in the pattern of distribution of southwest monsoon rainfall over India associated with sunspots. *Monthly Weather Review,* **101(9):** 691-700.

Kale, V.S. (2003). The spatio-temporal aspects of monsoon floods in India: Implications for flood hazard management. *In:* H.K. Gupta (editor), Disaster Management. Orient Longman, New Delhi, pp. 22-47.

Kale, V.S. (2004). Floods in India: Their frequency and pattern. *In:* K.S. Valdiya (editor), Coping with Natural Hazards: Indian Context. Orient Longman, New Delhi, pp. 91-103.

MHA (2004). Disaster Management in India — A Status Report. Ministry of Home Affairs, National Disaster Management Division, Government of India, New Delhi.

MoEF (2005). State of Forest Report. Forest Survey of India, Ministry of Environment & Forests, Government of India, Dehradun, Uttarakhand.

Mohapatra, P.K. and Singh, R.D. (2003). Flood management in India. *Natural Hazards,* **28:** 131-143.

Murty, T.S. and Flather, R.A. (1994). Impact of Storm Surges in the Bay of Bengal. *In:* C.W. Finkl Jr. (editor), Coastal Hazards: Perception, Susceptibility and Mitigation. *Journal of Coastal Research,* Special Issue No. **12,** pp. 149-161.

NDMA (2007). National Disaster Management Guidelines: Preparation of State Disaster Management Plans. National Disaster Management Authority, Ministry of Home Affairs, Government of India.

NDMA (2008a). National Disaster Management Guidelines: Management of Floods. National Disaster Management Authority, Ministry of Home Affairs, Government of India, New Delhi.

NDMA (2008b). National Disaster Management Guidelines: Management of Cyclones, National Disaster Management Authority, Ministry of Home Affairs, Government of India, New Delhi.

Pal, I. and Al-Tabbaa, A. (2009). Trends in seasonal precipitation extremes: An indicator of climate change in Kerala, India. *Journal of Hydrology,* **367:** 62-69.

Plate, E.J. (2002). Flood risk and flood management. *Journal of Hydrology,* **267:** 2-11.

Ray, K.C., Mukhopadhyay, R.K. and De, U.S. (1999). Meteorological Disasters during Last Twenty Two Years. *In:* K.M.B. Rahim (editor), Natural Disasters: Some Issues and Concerns. Natural Disasters Management Cell, Visva Bharati, Santiniketan, West Bengal, pp. 10-23.

RBA (1980). Report of the Rashtriya Barh Ayog (National Commission on Floods). Volume 1, Ministry of Energy and Irrigation, Government of India, New Delhi.

Scheuren, J-M., le Polain de Waroux, O., Below, R., Guha-Sapir, D. and Ponserre, S. (2008). Annual Disaster Statistical Review: The Numbers and Trends 2007. Center for Research on the Epidemiology of Disasters, Université Catholique de Louvain, Brussels, Belgium.

Sen, P. (2008). Riverbank erosion and protection in Gangetic Delta. *In:* K.M.B. Rahim, M. Mukhopadhyay and D. Sarkar (editors), Riverbank Erosion. Natural Disasters Management Cell, Visva Bharati, Santiniketan, West Bengal, India, pp. 1-31.

Sheth, A., Sanyal, S., Jaiswal, A. and Prathibha Gandhi, P. (2006). Effects of the December 2004 Indian Ocean Tsunami on the Indian Mainland. Earthquake Spectra, Earthquake Engineering Research Institute, Oakland, California, U.S.A., Volume 22, No. S3, pp. S435-S473.

Smith, K. (2004). Environmental Hazards: Assessing Risk and Reducing Disaster. Fourth Edition, Routledge, USA.

Thandaveswara, B.S. and Mahesh Kumar, D.V. (1993). Dam break analysis using MacCormack Scheme for Kali River. Proceedings of the International Conference on Dam Engineering, Jahore Behru, Malaysia, pp. 12-13.

Zbigniew, W.K., Ulbrich, U., Brucher, T., Graczyk, D., Kruger, A., Leckebusch, G.C., Menzel, L., Pinskwar, I., Radziejewski, M. and Szwedl, M. (2005). Summer Floods in Central Europe – Climate Change Track? *Natural Hazards*, **36:** 165-189.

8

Modeling for Flood Control and Management

D. Nagesh Kumar*, K. Srinivasa Raju and Falguni Baliarsingh

1. INTRODUCTION

Flood hazard has long been recognized as one of the most recurring, wide-spread and disastrous natural hazards. In many parts of the Indian subcontinent, flooding reaches catastrophic proportions during monsoon season. For centuries, floods in the Ganga, the Brahmaputra and the Godavari basins have brought number of disasters to the inhabitants in the flood plains apart from inundating large tracts of fertile land (Kale, 1998).

In the last few decades, there have been many serious attempts to improve the understanding of the causes and effects of floods for better control and management of floods. Recognizing the gravity and the extent of the problem, several countries, including India, have taken up extensive flood management and flood protection works. High priority has been given to flood forecasting and warning systems on major flood prone rivers and development of different strategies for flood control and management. In addition, most of the impacts due to climate change (Nakicenovic et al., 2000) are expected to increase the frequency of extreme events of hydrology, i.e., floods and droughts in the near future. This will further necessitate the development of proper measures for flood control and management.

The purpose of this chapter is to study flood control and management with special reference to modeling aspects to suggest suitable management strategies. Initially, various structural and non-structural approaches are presented for flood control and management. Literature review of mathematical models and their applications to flood aspects are presented next. A case study is presented to demonstrate the development of operating policy of a reservoir with flood control as a major criterion. To conclude, various measures are suggested for flood control.

2. STRUCTURAL AND NON-STRUCTURAL APPROACHES

Structural methods to safely dispose floods include: (a) construction of reservoirs for storage of flood water that can be utilized for other purposes once flood recedes, (b) embankments to retain the flood water far away from the flood prone areas, and (c) construction and improvements of channels to adequately discharge the flood waters. Structural measures can be supplemented with non-structural measures such as floodplain zoning as well as flood forecasting as a cost-effective strategy. Both structural and non-structural methods are to be taken into consideration while planning flood mitigation measures. Salient factors that affect data collection are cost, speed with which the data can be collected and their accuracy.

Data is required about the existing vegetation/soils, level of urbanization, land use, watershed characteristics, rainfall intensity and its duration, runoff, degree of flood protection/severity of flood damage, stage-discharge relationships, climatic records, geology and geomorphology, depth of

---

*Corresponding Author

groundwater, etc. Database of as many flood events as possible for different locations is essential to get a comprehensive overview of flood events over the entire area. This will avoid any ambiguous inferences due to incomplete data. Additional data can also be collected from secondary sources such as by interviews and earlier reports. Data thus collected are to be processed and analyzed in such a manner as to minimize errors which otherwise affect the accuracy of general analyses as well as calibration and validation of hydrologic simulation models. Statistical analysis is very helpful in this regard. The processed data can be utilized for various aspects including development of maximum probable flood, standard project flood, design flood, flood frequency analysis, flood sediment, flood envelope curve, stage-discharge curves and floodplain zoning. The data can also be used for analyzing past flood situations, managing present flood situation or for forecasting future floods of different magnitudes.

In India, many agencies are involved in water resources planning, development and disaster management such as Central Water Commission (CWC), Central Ground Water Board (CGWB), Indian Meteorological Department (IMD), National Disaster Management Authority (NDMA), National Water Development Agency (NWDA), National Institute of Hydrology (NIH), Water and Land Management Institutes (WALMI) of various State Governments, National Remote Sensing Centre (NRSC) and State Remote Sensing Application Centers, Geological Survey of India (GSI), etc. Various mathematical modeling approaches based on soft computing and related fields are available for flood control and management (e.g., ASCE, 2000b; Chau et al., 2005; Jain and Singh, 2006).

3. OVERVIEW OF TOOLS AND TECHNIQUES FOR FLOOD MODELING

Modeling for flood control and management has attracted considerable attention from number of researchers and various tools are developed to combat floods. A brief overview of some of the tools developed for modeling floods is presented below.

3.1 Artificial Neural Network (ANN) and Fuzzy Logic

The Back Propagation Algorithm (Rumelhart et al., 1986) is a procedure to train feed forward ANN models, in which the outputs can be sent only to the immediate next layers. The selection of a suitable architecture for the problem on hand can be done in three steps: fixing the architecture, training the network and testing the network. Main parameters concerned are network architecture, learning rate, type of activation function, definition of error, number of epochs etc. The procedures for training the ANN network are described in ASCE (2000a).

Radial Basis Function network (ASCE, 2000a) is a three-layer network in which the hidden layer performs a fixed non-linear transformation with no adjustable parameters. This layer consists of a number of nodes and a parameter vector called a center which can be considered as its weight vector. For each node, the Euclidean distance between the center and the input vector of the network input is computed and transformed by a non-linear function that determines the output of the nodes in the hidden layer. The output layer then combines these results in a linear fashion. The common activation functions used in Radial Basis Function are Sigmoidal and the Gaussian Kernel functions. More details about Radial Basis Functions are available in ASCE (2000a, b). Applications of ANN with reference to flood modeling are discussed in ASCE (2000b).

Moreover, there are often many situations with components that are intrinsically vague, called uncertain. Different situations that lead to uncertainty are unquantifiable information, incomplete information, non-obtainable information and partial ignorance (Bojadziev and Bojadziev, 1997). In such cases, fuzzy approach is the most suitable to tackle the vagueness of data. The advantage of fuzzy approach is that the quantitative as well as qualitative analysis can be integrated and can be represented using fuzzy membership functions (Deka and Chandramouli, 2005; Venkatesh et al., 2008). Adaptive Neuro-Fuzzy Inference System (ANFIS) is also utilized for flood related studies (e.g., Chau et al., 2005; Akbari et al., 2009).

3.2 Optimization Methods

Traditional optimization methods such as Linear Programming, Nonlinear Programming, Dynamic Programming, etc. have played a major role in flood management studies (Mays, 1996; Lund, 2002; Jain and Singh, 2006; Karamouz et al., 2009). However, some real-world flood management problems present situations such as a complex search space which are difficult to be solved by such exact methods. In such cases, more efficient optimization strategies such as evolutionary algorithms are required (Deb, 2001; Ranjithan, 2005). The evolutionary algorithms have the following characteristics and advantages:

- Applicability to any problem that can be formulated as a function optimization task.
- Flexibility of the procedures, as well as ability to self-adapt the search for optimum solutions.
- Ability to use knowledge and hybridize with other methods.
- Adapt solutions to changing circumstances as compared to traditional methods of optimization which require a complete restart in order to provide a solution.
- Find multiple optimal solutions simultaneously in multimodal function optimization problems.

3.3 Geographical Information System

Geographical Information Systems (GIS) technology provides tools for effective and efficient storage and manipulation of remotely sensed information and other spatial as well as non-spatial information (e.g., Engman and Gurney, 1991; De Vantier and Feldman, 1993; Singh and Fiorentino, 1996; Vieux, 2001; Islam and Sado, 2002; Nagesh Kumar, 2002). The strength of GIS results from its ability to analyze data representing a particular point, line or polygon as all features of any landscape can be reduced to one of these three spatial categories. Remote sensing images, effectively integrated within GIS, can be used to facilitate measurement, mapping, monitoring and modeling activities. Building a GIS database includes import or entry of data from different sources and digitizing the data from different source documents. GIS analyses allow the user to perform a wide variety of investigations such as (DeMers, 1997; Heywood et al. 1998; Clarke, 1999; Verbyala, 2002):

- Proximity analyses, neighborhood operations (for example identifying objects within a certain neighborhood fulfilling specific criteria).
- Determine the relationships between datasets within such a neighborhood.
- Temporal operations and analyses.
- Generation of new information by combining several data layers and attributes.

Clark (1998) suggested three different approaches for models using GIS for flood management: (i) Data process in the pre-modeling phase; (ii) direct GIS support for flood modeling; and (iii) integrating GIS and flood models for post-processing of data such as for flood risk mapping.

3.4 Expert System

Expert System is a branch of Artificial Intelligence which uses the knowledge and inference procedures to solve problems (Ignizio, 1991; Simonovic, 1991; Varas and Von Chrismar, 1995; Nagesh Kumar,

2006). These are designed to carry information in the form of knowledge base (rules) inferred from experts and provides this knowledge to users for further decision making. Expert Systems are getting prominence due to their ability to provide consistent answers for repetitive decisions, processes and tasks; hold and maintain significant levels of information; centralize the decision making process; create efficiencies and reduce time needed to solve problems; combine multiple human expertise, reduce the human errors and review transactions that human experts may have overlooked (Simonovic, 1991).

3.5 Multicriteria Decision Making

Multicriteria Decision Making (MCDM) is an emerging area of decision making where the role of all stake holders (all agencies directly or indirectly involved in the decision making including those affected) can be considered in a structured environment. Levy (2005) discussed about flood risk management for which MCDM was used for selecting a suitable flood control strategy with inputs from models such as GIS, hydrological models and real-time flood information systems. Bana e Costa et al. (2004) evaluated flood control options for the catchment of Livramento creek in the peninsula of Setúbal, in Portugal in Multicriterion context using MACBETH approach. Yu et al. (2004) proposed multi-person multiobjective decision-making model for the problems of flood control operation for Fengman Reservoir located in Songhua River Basin in China. Hajkowicz and Collins (2007) present an excellent review of the applications of multicriteria analysis to water resources planning and management.

4. LITERATURE REVIEW ON FLOOD MODELING AND MANAGEMENT

A brief review of the literature related to the general aspects of floods and different tools used for flood modeling and management are presented in this section.

4.1 General

Seth (1998) discussed about flood estimation, flood routing, flood inundation, flood mapping, floodplain zoning and flood insurance. Garde (1998) suggested non-structural methods of flood control including the installation of temporary openings or permanent closure of existing openings in structures, raising the existing structures, installation of flood warning systems with an appropriate evacuation plan, tax incentive to encourage proper use of flood plains, adoption of appropriate development policies for facilities in or near floodplain land and flood insurance.

Lekuthai and Vongvisessomjai (2001) proposed Anxiety-Productivity and Income interrelationship approach (API) to quantify the intangible damage in monetary terms. Hall et al. (2003) discussed about Integrated Flood Risk Management. Mohapatra and Singh (2003) discussed about flood management in India. They presented some special flood problems in India such as river bank/bed erosion, sediment transport by rivers, dam break flows, urban drainage, flash floods, floods due to snow melt etc. They also discussed current status of flood management in India such as progress of structural measures and nonstructural measures. They recommended legislation for floodplain zoning, flood cushion in reservoirs, flood insurance, flood data center, community participation, international cooperation etc. Carsell et al. (2004) quantified the benefit of a flood warning system. Simonovic and Li (2004) discussed sensitivity of the Red River Basin Flood Protection System, Canada, to climate variability and change. Kumar and Chatterjee (2005) developed regional flood frequency relationships for the estimation of floods of various return periods for gauged and ungauged catchments in North Brahmaputra region of India. Sanders et al.

(2005) studied national flood modeling for insurance purposes. The study focused on the flood risk information needs of insurers and how these can be met. Data requirements of national and regional flood models are also addressed in the context of accuracy of available data on property location. Insurance implications of recent flood events in Europe and the issues surrounding insurance of potential future events are also analyzed. Holz et al. (2006) stressed the potential of Information and Communication Technology (ICT) for flood management and discussed the role of various stake holders. Slutzman and Smith (2006) presented a methodology to estimate flood response in urban watersheds for assessing the performance of systems of small flood control reservoirs for extreme floods.

4.2 Artificial Neural Network and Fuzzy Logic

Solomatine and Xue (2004) compared ANN with M5 model tree machine learning approach and observed that accuracy of M5 trees is similar to that of ANN. Deka and Chandramouli (2005) studied river flow prediction using a fuzzy neural network (FNN) model and validated on the River Brahmaputra. Wu et al. (2005) demonstrated application of ANN for watershed-runoff and streamflow forecasts and concluded that ANN-hydrologic forecasting models can be considered as an alternate and practical tool for streamflow forecast. Ahmad and Simonovic (2006) developed intelligent DSS including ANN and expert systems to assist decision makers during different phases of flood management. Akbari et al. (2009) presented fuzzy rule-based (FRB) control model using a Takagi-Sugeno fuzzy system and a model modification algorithm.

4.3 Optimization Methods

Needham et al. (2000) presented a mixed integer linear programming model for a reservoir system analysis of three projects on the Iowa and Des Moines Rivers. They reviewed operating procedures for the flood of 1993 and illustrated how much damage could have been reduced if inflows could be predicted months in advance or if the existing operating rules were more averse to extreme flood events. Lund (2002) proposed two-stage linear programming formulation which provides an explicit economic basis for developing integrated floodplain management plans. Chau et al. (2005) employed the genetic algorithm-based artificial neural network (ANN-GA) and the adaptive-network-based fuzzy inference system (ANFIS) for flood forecasting.

Travis and Mays (2008) introduced discrete dynamic programming technique to determine the ideal locations and geometry of retention basins within a watershed. Karamouz et al. (2009) presented two optimization models based on GA. The first model is to determine economical combination of permanent and emergency flood control options and the second one is to determine the optimal crop pattern. Wei and Hsu (2008) developed a generalized multipurpose multi-reservoir optimization model based on feed-forward back-propagation neural network for basin scale flood control in which one constraint is linear channel routing.

4.4 Geographical Information System

Rao et al. (1998) discussed role of GIS and remote sensing in flood mapping, flood damage assessment, identification of erosion-prone areas and flood risk zone mapping in India. Noman et al. (2003) presented a process for delineating flood plains accurately from a digital terrain model (DTM) and hydraulic model in an automatic environment, providing the flexibility of incorporating professional judgment in the process. Liu and Smedt (2005) employed Wetspa for calculating flood discharges. Jain et al. (2005)

applied remote sensing techniques for the Koa catchment, Bihar. Jain et al. (2006) used NOAA-AVHRR data for mapping of flood-affected area during the year 2003 for the River Brahmaputra.

Khan et al. (2007) quantified the impact of flood and rain events on spatial scale for a case study of shallow water-table levels and salinity problems in the Murray irrigation area using GIS. Bahremand et al. (2007) employed Wetspa model within GIS to predict flood hydrographs and spatial distribution of hydrologic characteristics in a watershed. Machado and Ahmad (2007) assessed the flood hazard caused by Atrato River in Quibdo, northwest of Colombia using statistical modeling techniques, hydraulic modeling using HEC-RAS and GIS. Three flood hazard maps for return periods of 10, 20 and 50 years are generated. It is concluded that results can be useful for evacuation planning, estimation of damages and post-flood recovery efforts.

Venkatesh et al. (2008) presented key issues associated with uncertainty in flood inundation mapping and proposed an integrated approach in producing probabilistic flood inundation maps. Li and Zhang (2008) compared three GIS-based distributed hydrological models for flood simulation and forecasting.

Based on the above literature review, a step-by-step approach is evolved for choosing the most appropriate flood control strategy, which is presented in Table 1.

Table 1. Tasks and related activities for effective flood control strategy

Task	Activities
1. Define the problem	To select the suitable/best flood control strategy
2. Identification of decision makers	Policy makers who are related with flood control events
3. Identification of flood related criteria that will influence decision making	Incorporation of both quantitative and qualitative criteria
4. Prioritize flood related criteria	Incorporation of their preference structure among identified criteria, i.e., assign suitable priority to each criterion.
5. Identification of modeling approaches	Optimization, simulation, hydrological models, etc. whichever is applicable. For example, ANN, Fuzzy, GIS, MIKE, Evolutionary algorithms, Expert systems, MCDM, etc.
6. Establishment of payoff matrix of feasible and efficient flood control strategies	Flood control strategies can be generated based on modeling approaches or based on the existing information or by combination of both.
7. Classification/clustering of payoff matrix	Sometimes feasible set may be too large to be handled. In that case, these alternatives can be reduced in number using clustering approaches.
8. Analyze each flood control strategy using inputs from decision makers and selection of appropriate MCDM method to solve the problem	MCDM methods can be chosen from four groups i.e., distance, outranking, priority/utility, and mixed category. Selection of MCDM method is mainly based on data availability, acquaintance of the method(s) by the decision maker(s).
9. Selection of the best/suitable flood control strategy	Choosing one or more of the best/suitable flood control strategies for further analysis. If satisfied after further analysis, implement the strategy. Evaluate the outcome of the decision. Document the lessons that were learnt in this process for further improvement of decision making skills.

5. DEVELOPMENT OF LONG-TERM OPERATING POLICY FOR FLOOD CONTROL: A CASE STUDY

In this section, a case study of Hirakud reservoir for the development of long-term optimal operating policy for flood control using Folded Dynamic Programming (FDP) is presented.

5.1 Description of Study Area

The project under the present study, Hirakud single reservoir system is situated in Mahanadi River basin. The Mahanadi basin lies mostly in Madhya Pradesh and Orissa States. It is bounded on the north by Central India Hills, on the south and east by Eastern Ghats and on west by the Maikela range, the south east part of Deccan Plateau. The basin is situated between 80° 30′ and 86° 50′ East longitudes and 19° 20′ and 23° 35′ North latitudes. It is roughly circular in shape with a diameter of about 400 km with an exit passage of 160 km in length and 60 km in breadth. Area of this basin is 1,41,600 km^2 and is broadly divisible into three distinct zones, the upper plateau, the central hill part flanked by Eastern Ghats, and the delta area. Hirakud dam across the Mahanadi River is located in the second zone.

The Mahanadi River originates in Raipur district of Madhya Pradesh at an elevation of 442 m above mean sea level. The total length of this east flowing river from its origin to its outfall into the Bay of Bengal is 851 km of which 357 km is in Madhya Pradesh and remaining 494 km in Orissa state. After a run of 450 km from its starting point, the river reaches Sambalpur district of Orissa, where the Hirakud dam is built on the main river. Below the dam, the river gets water mainly from two sub-basins, Ong and Tel, in addition to free catchment along the river. The river flows down to Naraj, the head of delta and finally joins the Bay of Bengal. The catchment area up to Naraj is 1,32,200 km^2. On the downstream of Naraj, the river divides into several branches, namely, Birupa, Chitrotpala, Devi, Kushabhadra, Bhargabi, Daya etc. and runs 80 km before discharging into the Bay of Bengal.

The multipurpose Hirakud reservoir is utilized mainly for three purposes, flood control, irrigation, and hydropower production. There is expectation from Hirakud reservoir to control flood at coastal delta area by limiting the flow at Naraj within 25,500 m^3/s. There are three head regulators, which can draw 128.8 m^3/s from the reservoir for irrigation purpose. Areas of 0.16 and 0.11 million hectares are irrigable from the reservoir during *Kharif* and *Rabi* seasons, respectively. The total installed hydropower capacity of the project is 307.5 MW, out of which 235.5 MW can be produced from the seven units of Hirakud hydropower station, and 72 MW from three units of Chipilima hydropower station, located further d/s of the Hirakud dam. The water, used for power generation at Hirakud, flows from the Hirakud hydropower station to the Chipilima hydropower station through a power channel of 22.4 km long. After generating power at Chipilima, water flows back into the river.

Flood control is the first preferred objective for this reservoir with hydropower generation and irrigation as the secondary objectives. This reservoir is situated 400 km upstream of confluence of Mahanadi River with the Bay of Bengal. There is no other flood controlling structure downstream of Hirakud reservoir. During monsoon season, the coastal delta part, between Naraj and the Bay of Bengal, is severely affected by floods. This flood-prone area gets water from the Hirakud reservoir and from rainfall in the downstream catchment. Naraj is situated at the head of the delta area, where the flow of the Mahanadi River is measured. The flow of the Mahanadi River at Naraj is used as an indicator of occurrence of flood in the coastal delta area by the Hirakud authority. As the Hirakud reservoir is on the upstream side of delta area in the basin, it plays an important role in alleviating the severity of the flood in this area. This is done by regulating release from the reservoir. The schematic diagram of Hirakud project is shown in Fig. 1.

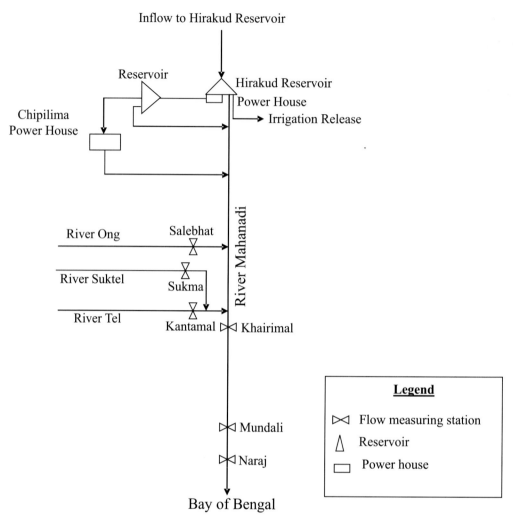

Fig. 1 Schematic diagram of the Hirakud project.

5.2 Methodology

5.2.1 Folded Dynamic Programming

Before discussing the algorithm of FDP, it is necessary to explain the way of finding maximum and minimum possible storages at the beginning of each time period, hereafter called time step, of the whole operating horizon for reservoir(s). The operating horizon, the duration of reservoir operation, is considered as stage and the storage of reservoir is considered as state variable in DP formulation. In physical terms, the storage state variable can be at any point between the dead storage level and full reservoir level. In the FDP algorithm, the entire storage state space at each time period is required to be divided into four equal state increments to form five grid points. The storage can be changed from any grid point of one time step to any grid point of adjacent time step.

Step-wise procedure of the proposed algorithm (FDP) is explained below:

(1) Depending on the natural inflow, release capacity, and boundary conditions of storage, the minimum and maximum possible storage values for each reservoir (in case of multiple reservoir system) at every time step of operating horizon are found out.
(2) Considering the minimum and maximum possible storages as two extreme grid points, three middle grid points are determined adopting uniform state increments. This means that the possible storage space at each time step is divided into four equal state increments to get five grid points. So, there are $5M$ grid points in M dimensional problem (M reservoirs in multiple reservoir system) at each time step. State increment is different for different time steps as well as for different reservoirs. The mesh of these grid points for the whole operating horizon of all the reservoirs forms the corridor.
(3) Conventional DP is run through this corridor to find the trajectory, P, which gives maximum objective function value, F.
(4) For finding the trajectory for next iteration, if this trajectory (P) is either the minimum or maximum storage value, i.e., extreme grid points at any time step, these points are changed to the next interior grid points to form the revised trajectory. This revised trajectory will be within the corridor for the next iteration.
(5) In the next iteration, the state increment is halved at each time step. A fresh corridor is formed by taking two state increments or grid points on each side of the trajectory. Then conventional DP is run through this corridor as in Step 3 to find the best trajectory, P', whose objective function value is F'.
(6) The iterations are continued with half value of state increments of the previous one at each time step. There can be two stopping rules. First, the decrement of state increment at a time step stops, where state increment happens to be less than a predefined value. The iteration stops, when decrement of state increment process stops at each time step. Second, the iteration stops, when $F'' < \xi$ is satisfied, where $F'' = (F' - F)/F$ and ξ is a predefined ratio. In the present case, second stopping rule is applied.

Further details of FDP could be found in Baliarsingh (2000), and Nagesh Kumar and Baliarsingh (2003).

Various phases are formulated to derive the long-term operating policy for the reservoir. These include data collection, computation of downstream catchment contribution, preparation of block period, formulation of basis for rule curves for floods and the application of Folded Dynamic Programming (FDP) to evolve rule curves for reservoir management during floods. These are described below.

5.2.1 Data Collection

Data of sixty-eight floods such as peaks of inflow hydrograph, peak of outflow from reservoir during each flood, and peak of flow hydrograph at Naraj, spread between 1958 and 1995, without downstream catchment contribution, are available in the Hirakud data book (Patri, 1993). The outflow from the reservoir is a combination of release for power and spill from reservoir. The distance between the Hirakud reservoir and Naraj is 320 km. Hence, apart from the measured runoff of Ong and Tel basins (Fig. 1), there is a considerable amount of lateral flow coming from the free catchment between Hirakud and Naraj joining the main river, which was not measured and therefore is arithmetically assessed. The representative inflow into the reservoir and downstream catchment contribution series are used subsequently for evolving the rule curve.

5.2.2 Computation of Downstream Catchment Contribution

The downstream catchment contribution during the sixty-eight historical floods is calculated from the known values of outflow from the reservoir and flow at Naraj. The hydrograph at upstream section of the Hirakud-Naraj River reach is the outflow of the Hirakud reservoir, which is the combination of release for power generation and spill from reservoir. This outflow collects the downstream catchment contribution with it while traveling down to Naraj. As the downstream catchment contribution was not measured, this data is not available for the sixty-eight floods during 1958 and 1995. However, this quantity is calculated theoretically, by subtracting the routed quantity of outflow of Hirakud reservoir from the flow at Naraj. The relationship between the peak outflow from Hirakud reservoir and its routed quantity at Naraj is obtained from the data of nine floods between 1992 and 1995. In this regard, the reach of the river from Hirakud to Naraj is considered for application of Muskingum routing method. Muskingum equation for the flood routing is

$$QH_t = C_0 RPS_{t-1} + C_1 RPS_t + C_2 QH_{t-1} \qquad (1)$$

where RPS_t = ordinate of inflow (outflow from reservoir) hydrograph at the upstream section of Hirakud-Naraj routing reach at beginning of routing interval t; QH_t = ordinate of outflow (routed quantity of outflow from reservoir at Naraj) hydrograph at the downstream section of routing reach at beginning of routing interval t; C_0, C_1, C_2 = coefficients of Muskingum equation; and t = routing interval.

The values of the coefficients of Muskingum equation are arrived at as $C_0 = 0.471$, $C_1 = 0.117$, and $C_2 = 0.412$. Time interval of the ordinates of the hydrographs for routing procedure is chosen as 24 hours (Subramanya, 1994). For all the nine floods considered, the initial flow at Naraj is taken to be same as the initial value of outflow from the reservoir and routing procedure is carried out to get the relationship between peak outflow from reservoir and its routed quantity at Naraj.

The percentage of attenuation varies from one flood to another. The floods are characterized by peak outflow from reservoir. It is observed from this analysis that there is no distinct correlation between the percentage of attenuation and peak outflow from the reservoir. It is also observed that attenuation varies from 26 to 8% among all the floods. The higher the percentage of attenuation considered for routing, the lower would be the routed outflow quantity of Hirakud reservoir. Hence, the calculated downstream catchment contribution will be more. The rule curve, obtained for this higher downstream catchment contribution will also serve for the lower value of downstream catchment contribution. Therefore, 26% of attenuation is considered for the calculation of downstream catchment contribution, which is used for finding the representative flows. All the floods whose peak is more than 5,000 m³/s are taken into consideration for this purpose. Applying 26% attenuation, the peak flow at Naraj from the outflow from Hirakud reservoir alone is calculated which is assumed to occur simultaneously with the total peak flow at Naraj. This value is subtracted from the total peak flow at Naraj to get the peak flow of downstream catchment contribution.

5.2.3 Preparation of Block Period

As the major objective of the Hirakud reservoir is flood control, the reservoir operation is considered only for the monsoon season (floods occur only in this season) in the present study. Hence, for control purpose, the reservoir is made empty at beginning of monsoon and is made full by the end of monsoon to utilize the water for conservative purposes during the following non-monsoon period. The monsoon period for reservoir operation is taken as 1st July to 30th September. The reservoir is operated for filling up from the dead storage level of 179.83 m on 1st July to the full reservoir level of 192.02 m by 1st October every year. There are 92 days in the duration from 1st July to 30th September. Hereafter, 1st July is termed 1st day and 30th September 92nd day. The outflow from reservoir on the 92nd day will take more than a day to reach Naraj. The outflow from reservoir on 92nd day will collect the downstream catchment contribution of 93rd day to form the flow at Naraj on 93rd day. The predicted value of downstream catchment contribution of 93rd day is to be known, to take necessary reservoir operation decision for 92nd day. Therefore, the representative downstream catchment contribution series is generated for 93 days and the representative inflow into the reservoir series is generated for 92 days. Nine blocks of 10 days each are considered for placing one flood in each of the blocks. Though there is no chance of occurrence of flood in each of the 10 days block period simultaneously, it is assumed so, to get the rule curve and to tackle the flood during any duration individually. The peaks of inflow hydrograph to the Hirakud reservoir of the historical sixty-eight floods are arranged in nine different block periods and are shown in Table 2. Similarly, the peaks of calculated downstream catchment contributions are shown in Table 3. The peaks of individual block period with respect to the peak of the whole monsoon season in terms of percentage for both these components are also shown in Tables 2 and 3.

During transit from the Bay of Bengal on to the land surface, the monsoon brings rain first to the d/s catchment of Hirakud reservoir. Then it takes another day to travel to interior area, contributing to the inflow into the reservoir. That is why, the two representative hydrographs are made to lag by a day. The patterns of these two representative hydrographs in each block period are kept same as that for probable maximum flood. In this process, the hydrograph of inflow into the reservoir is made to lag by a day from the hydrograph of downstream catchment contribution. Here, flow on any day means the flow at the beginning of that day. After getting the pattern of the above two series, it is necessary to obtain the flow rate of each day for various possible peaks. When the total flow reaches 9,000 m^3/s of inflow hydrograph at Naraj, it is treated as flood by the Hirakud authority. The peak of maximum probable flood is 80,000

Table 2. Peaks of inflow hydrograph into the reservoir (thousand m^3/s) of historical sixty-eight floods (1958-1995) during various block periods

Sl. No of flood	Block period									Non-monsoon season
	1/7-10/7 1	11/7-20/7 2	21/7-31/7 3	1/8-10/8 4	11/8-20/8 5	21/8-31/8 6	1/9-10/9 7	11/9-20/9 8	21/9-30/9 9	
1	43.22	23.83	14.42	27.49	23.46	15.73	25.79	24.43	16.86	14.4(29.10.73)
2	21.54	23.12	23.83	7.88	18.42	29.81	18.82	19.69		24.37(29.6.86)
3	18.87	25.84	12.47	23.52	24.4	18.08	15.87	11.76		20.63(21.6.94)
4	18.36		8.08	7.91	20.66	13.94	12.64	13.15		
5			15.22	16.18	18.05	17.82	17.23	37.75		
6			13.49	12.64	17.34	22.61		15.3		
7			19.84	17.99	21.37	18.5				
8			15.61	15.3	16.63	13.43				
9				15.73	17	26.92				
10				14.74	15.98	13.04				
11				13.29	19.16	14.74				
12				17.77	13.21	14				
13						16.58				
14						11.82				
Max	43.22	25.84	23.83	27.49	23.46	29.81	25.79	37.75	16.86	
% of season's peak	100	59.8	55.1	63.6	54.3	69.0	59.7	87.3	39.0	
No. of floods	4	3	8	12	12	14	5	6	1	3

Table 3. Peaks of calculated downstream catchment contribution (thousand m³/s) of the historical sixty-eight floods (1958-1995) during various block periods

Sl. No of flood	Block period									Non-monsoon season
	1/7-10/7 1	11/7-20/7 2	21/7-31/7 3	1/8-10/8 4	11/8-20/8 5	21/8-31/8 6	1/9-10/9 7	11/9-20/9 8	21/9-30/9 9	
1	14.02	20.23	19.66	05.38	02.99	13.38	24.56	24.95	07.15	9.69(31.10.73)
2	25.53	13.54	11.08	16.84	12.26	13.62	12.73	23.94	11.27	18.20(29.6.86)
3		20.15	07.08	24.75	16.79	13.45	12.13	10.18		11.31(22.6.94)
4		16.66	32.14	28.22	11.59	14.02	16.73	14.24		
5		15.91	09.63	16.31	13.06	12.72	17.90	11.83		
6			25.20	17.33	20.55	10.79	22.50			
7				08.12	10.10	11.91				
8				16.11	15.62	19.88				
9				13.84	16.81	12.40				
10				21.24	34.13	44.89				
11				12.47	19.89	25.45				
12				19.31		05.55				
13				08.68		13.38				
14						30.23				
15						15.77				
Max	25.53	20.23	32.14	28.22	34.13	44.89	24.56	24.95	11.27	
% of season's peak	56.9	45.1	71.6	62.9	76.0	100	54.7	55.6	25.1	
No. of floods	2	5	6	13	11	15	6	5	2	3

m³/s. Therefore, an operating policy is to be developed for peaks of inflow lying between 9,000 and 80,000 m³/s. The various hypothetical series for inflow into the reservoir are calculated with peaks of 9, 23, 37, 51, 60, 74, and 80 thousand m³/s. Similarly, 3, 9, 14, 20, and 25.5 thousand m³/s of peaks of downstream catchment contribution series are considered. Upper bound is taken as 25,500 m³/s, which is required to keep the flow at Naraj below this value. It can be observed from Fig. 2 that the data of inflow into the reservoir is generated for 92 days and the highest peak (37 thousand m³/s) occurs on the 4th day. Similarly, the data of downstream catchment contribution is generated for 93 days and the highest peak (25.5 thousand m³/s) occurs on the 53rd day (Fig. 3).

5.2.4 Application of Folded Dynamic Programming to Develop Rule Curves for Floods

As the major objective of the Hirakud reservoir is flood control, operating policy is developed only for the monsoon season, i.e., from 1st July to 30th September. The minimum and maximum possible storages at each time step are dependent on the inflow into the reservoir, maximum release capacity during that time step and the maximum permissible release from reservoir for flood control purpose at Naraj, which varies with respect to time. The initial storage state on 1st July is taken as the dead storage and final desired storage state on 1st October is the full capacity of reservoir. Hence, forward pass alone is sufficient to find out the maximum possible storage state, starting from initial time step and backward pass alone is sufficient to find out the minimum possible storage state starting from last time step. In case of Hirakud reservoir system, release for power and spill from the reservoir join back the Mahanadi River downstream

Modeling for Flood Control and Management **159**

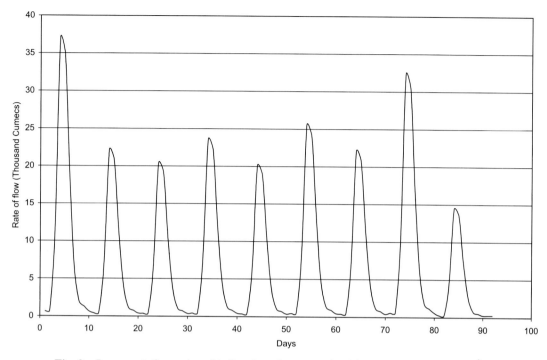

Fig. 2 Representative series of inflow into the reservoir with peak of 37 thousand m^3/s.

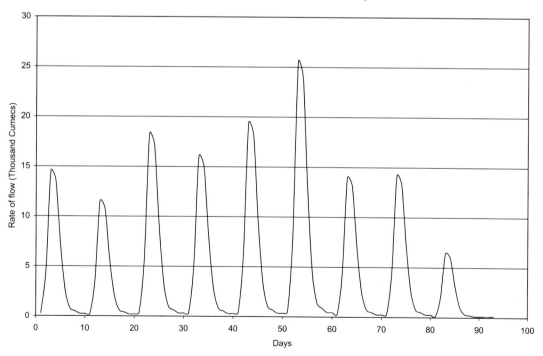

Fig. 3 Representative downstream catchment contribution series for Hirakud reservoir with peak of 25.5 thousand m^3/s.

of the Hirakud dam and ultimately contribute to the flow at Naraj. This quantity is to be regulated at reservoir during flood to restrict the flow at Naraj below non-damaging flow (25.5 thousand m³/s). Therefore, this factor is to be taken care of while finding minimum and maximum possible storages. Accordingly

$$S_{max,t+1} = S_{max,t} + (I_t - IR_t) - RPS_t \qquad \forall\, t \qquad (2)$$

$$S_{min,t+1} = S_{min,t} + (I_t - IR_t) - RPS_t \qquad \forall\, t \qquad (3)$$

$$S_{DSL} \leq S_{max,t} \leq S_{FRL} \qquad \forall\, t \qquad (4)$$

$$S_{DSL} \leq S_{min,t} \leq S_{FRL} \qquad \forall\, t \qquad (5)$$

$$RPSMIN \leq RPS_t \leq RPSMAX \qquad \forall\, t \qquad (6)$$

where $S_{max,t}$ = maximum possible storage at beginning of the time period t; $S_{min,t}$ = minimum possible storage at beginning of the time period t; I_t = inflow into the reservoir during time period t; IR_t = release for irrigation during time period t; RPS_t = release for power and spill, if any, during time period t; S_{DSL} = reservoir storage at dead storage level; S_{FRL} = reservoir storage at full reservoir level; $RPSMIN$ = minimum required release from reservoir through spillway, sluice, and power house; and $RPSMAX$ = maximum release capacity of reservoir (summation of maximum release through spillway, sluice, and power house).

The difference between non-damaging flow at Naraj and downstream catchment contribution is the maximum permissible release, which can be made from reservoir to keep the flow at Naraj within non-damaging flow. For calculating the permissible release from reservoir for t^{th} day, downstream contribution of $(t + 1)^{th}$ day is considered to take into account the lag time between Hirakud reservoir and Naraj. It is observed in the calculations of percentage of attenuation from the historical nine floods that the percentage of attenuation varies from a minimum of 8% to a maximum of 26%. If the percentage of attenuation is more than 8%, then the routed quantity of flow at Naraj corresponding to outflow from reservoir will be less than the flow with 8% attenuation, without affecting the objective of filling up the reservoir. The result obtained by 8% attenuation will hold good for the flood, whose percentage of attenuation is more than this value. Therefore, a minimum value of 8% attenuation is used for this purpose. The factor 1/(1 – 0.08) is used to account for the attenuation of routing the outflow from reservoir to Naraj. RPS_t in any period should be such an amount that the flow at Naraj will be less than non-damaging flow (NDF), i.e., 25.5 thousand m³/s. Accordingly,

$$0 \leq RPS_t \leq \frac{1}{1-0.08}(NDF - DC_{t+1}) \qquad \forall\, t \qquad (7)$$

where NDF is the non-damaging flow at Naraj (i.e., 25.5 thousand m³/s), and DC_t is the downstream catchment contribution during time period t.

The forward pass uses Eqns. (2), (4), (6), and (7) and the backward pass uses Eqns. (3), (5), (6), and (7). Releases for irrigation and downstream catchment contribution in each time duration are the known components. Non-damaging flow at Naraj is 25.5 thousand m³/s, as adopted by Hirakud reservoir authority. Flood control objective is to try and keep the flow at Naraj, comprising release from dam with proper routing and downstream catchment contribution, to be within 25.5 thousand m³/s (Baliarsingh, 2000).

The level of 179.83 m [dead storage level corresponding to a volume of 1.816 Thousand Million Cubic Meter (TMCM)] of the reservoir on 1st July is expected to go to a level of 192.02 m (full reservoir level corresponding to a volume of 7.197 TMCM) by 1st October. So, both minimum and maximum possible storages on 1st July are 1.816 TMCM. Similarly, minimum and maximum possible storages on 1st October are 7.197 TMCM. For finding maximum possible storages of the remaining days of monsoon, Eqns. (2), (4), (6), and (7) are used, starting from 1st July to 1st October. For finding minimum possible storage, Eqns. (3), (5), (6), and (7) are used in backward direction, i.e., from 1st October to 1st July. The limits of release for power and spill (RPS_t) for any time duration are given by Eqns. (6) and (7). In case of Hirakud reservoir, RPSMIN is zero. Hence, RPS_t can vary from a minimum value of 0 to a maximum value of $(NDF - DC_{t+1})/(1 - 0.08)$ or RPSMAX, whichever is less. It is always tried to keep RPS_t as 0 in Eqns. (2) and (3). Two values, minimum and maximum possible storages are different for different combinations of inflow into reservoir and d/s catchment contribution. In the process, if the maximum possible storage is more than full reservoir level in the forward pass [Eqn. (2)], only then the RPS_t is allowed to be positive. Similarly, if the minimum possible storage is less than the dead storage level in backward pass [Eqn. (3)], only then RPS_t will be positive. If it is not possible to keep these two factors within the limits by utilizing the RPS_t to its full extent, then that particular combination of inflow into the reservoir and d/s catchment contribution cannot be handled by Hirakud reservoir to keep the flow at Naraj within non-damaging flow. When this combination of inflow into the reservoir and d/s catchment contribution occurs, the flow at Naraj will exceed non-damaging flow. One sample of minimum and maximum possible storages for the entire monsoon season are shown in Table 4 corresponding to a representative inflow with peak of 37 thousand m³/s and a representative downstream catchment contribution with a peak of 25.5 thousand m³/s.

5.3 Results and Conclusion

In this chapter, the rule curves corresponding to flood conditions are found out by using Folded Dynamic Programming (FDP). It is observed that the combination of 51 thousand m³/s as peak inflow into reservoir and any value up to 25.5 thousand m³/s as peak downstream catchment contribution can be handled by the Hirakud reservoir, but combination of 60 thousand m³/s inflow into the reservoir and any value of downstream catchment contribution cannot be handled. The minimum and maximum possible storages for the combination of 3, 9, 14, 20, 25.5 thousand m³/s as peak downstream catchment contribution and 9, 23, 51 thousand m³/s as peak inflow into reservoir are computed. It is observed from the analysis that there is very little change in minimum and maximum possible storages for the variation in downstream catchment contribution for any specified value of inflow into the reservoir. Then the process was tried with the peak inflow into the reservoir from 51 to 60 thousand m³/s at an increment interval of three thousand m³/s. It is observed that the combination of 51, 54, 57 thousand m³/s as peak inflow into reservoir with 25.5, 20, 14 thousand m³/s, respectively as peak downstream catchment contribution can be handled by the Hirakud reservoir and are considered as critical combinations for flood situation. The minimum and maximum possible storages for these three critical combinations are shown in Fig. 4. Any combination with more flow rate than these three critical combinations will either create flow at Naraj to be more than non-damaging flow and/or overflow of reservoir.

Now, FDP is applied to these critical combinations of inflow into the reservoir and d/s catchment contribution to get rule curves for flood conditions. The process of FDP is applied with the objective function (maximization) and constraints applicable to the Hirakud reservoir which are as follows.
The backward recursive relationship is

Maximize $p_t(S_t) = \max_{PR_t \leq PR_{cap}} [B_t(S_t, PR_t) + p_{t+1}(S_{t+1})]$ $t = 0, 1, \ldots, T-1$ (8)

where $p_T(S_T) = 0$ (9)

subject to the following constraints:

$S_{t+1} = S_t + CF[(I_t + I_{t+1})/2 - (IR_t + IR_{t+1})/2]$
$\quad - CF[(PR_t + PR_{t+1})/2 + (SR_t + SR_{t+1})/2]$ $t = 0, 1, \ldots, T-1$ (10)

Table 4. Minimum and maximum possible storages for the combination of 37 thousand m³/s as peak of inflow into the reservoir and 25.5 thousand m³/s as peak of downstream catchment contribution

Day	Minimum possible storage	Maximum possible storage	Day	Minimum possible storage	Maximum possible storage	Day	Minimum possible storage	Maximum possible storage
	TMCM	TMCM		TMCM	TMCM		TMCM	TMCM
1	1.816	1.816	32	1.816	7.197	63	1.816	7.197
2	1.816	1.874	33	1.816	7.174	64	1.816	7.028
3	1.816	1.931	34	1.816	6.805	65	1.985	7.197
4	1.816	2.890	35	2.208	7.197	66	1.816	7.197
5	3.299	6.087	36	1.991	7.197	67	1.816	7.197
6	4.151	7.197	37	1.816	7.197	68	1.816	7.197
7	3.427	7.197	38	1.816	7.197	69	1.816	7.197
8	1.816	7.197	39	1.816	7.197	70	1.816	7.197
9	1.816	7.197	40	1.816	7.197	71	1.816	7.197
10	1.816	7.197	41	1.816	7.197	72	1.816	6.990
11	1.816	7.197	42	1.816	7.197	73	1.816	5.977
12	1.816	7.197	43	1.816	7.109	74	1.816	5.671
13	1.816	7.197	44	1.816	6.946	75	2.878	6.733
14	1.816	7.140	45	2.067	7.197	76	3.342	7.197
15	1.873	7.197	46	1.816	7.197	77	2.549	7.197
16	1.816	7.197	47	1.816	7.197	78	3.042	7.197
17	1.816	7.197	48	1.816	7.197	79	3.190	7.197
18	1.816	7.197	49	1.816	7.197	80	3.288	7.197
19	1.816	7.197	50	1.816	7.197	81	3.338	7.197
20	1.816	7.197	51	1.816	7.165	82	3.361	7.197
21	1.816	7.197	52	1.816	5.528	83	3.383	7.197
22	1.816	7.197	53	1.856	5.568	84	3.757	7.197
23	1.816	7.197	54	2.374	6.086	85	5.004	7.197
24	1.816	6.973	55	3.382	7.094	86	6.176	7.197
25	2.040	7.197	56	3.485	7.197	87	6.799	7.197
26	1.816	7.197	57	2.320	7.197	88	7.020	7.197
27	1.816	7.197	58	1.816	7.197	89	7.086	7.197
28	1.816	7.197	59	1.816	7.197	90	7.130	7.197
29	1.816	7.197	60	1.816	7.197	91	7.152	7.197
30	1.816	7.197	61	1.816	7.197	92	7.175	7.197
31	1.816	7.197	62	1.816	7.197	93	7.197	7.197

$$B_t(S_t, PR_t) = [H_t(PR_t + PR_{t+1})]/(13800*2) \qquad t = 0, 1, \ldots, T-1 \qquad (11)$$

$$H_t = fs\,[(S_t + S_{t+1})/2] \qquad t = 0, 1, \ldots, T-1 \qquad (12)$$

$$[1/(1-0.08)]\,(PR_t + SR_t) \leq (NDF - DC_{t+1}) \qquad \forall\, t \qquad (13)$$

$$RPSMIN \leq PR_t + SR_t \leq RPSMAX \qquad \forall\, t \qquad (14)$$

$$S_{min,t} \leq S_t \leq S_{max,t} \qquad \forall\, t \qquad (15)$$

$$PR_t \geq 0;\ SR_t \geq 0 \qquad \forall\, t \qquad (16)$$

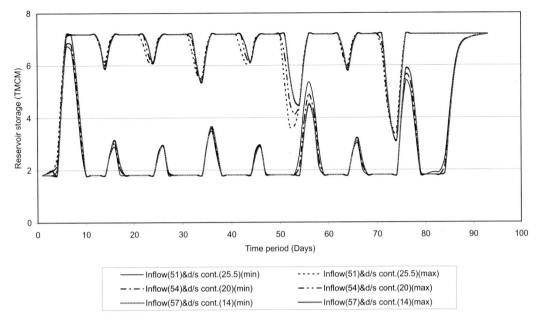

Fig. 4 Minimum and maximum possible storages for critical combinations of inflow into the reservoir and downstream catchment contribution for flood condition.

where $p_t(S_t)$ = total power produced in mega watts for the remaining periods of $t, t+1, ..., T-1$ with S_t as initial storage state at the beginning of time period t; S_t = storage state in TMCM at beginning of time period t; PR_t is the rate of release for hydropower in m³/s [ft³/s only in Eqn. (11)] at beginning of time period t; PR_{cap} = turbine capacity in m³/s; CF = conversion factor to convert from rate to volume of any variable; I_t = rate of inflow into the reservoir in m³/s at beginning of time period t; IR_t = release rate for irrigation in m³/s at beginning of time period t; SR_t = rate of spill from dam in m³/s at beginning of time period t; $B_t(S_t, PR_t)$ = power produced in mega watts during time period t; H_t = average head available for power production in feet during time period t; fs = relationship of storage-elevation curve; NDF = non-damaging flow at Naraj in m³/s; DC_t = downstream catchment contribution at beginning of time period t; $RPSMIN$ = minimum required release from reservoir; $RPSMAX$ = maximum release capacity of reservoir; $S_{min,t}$ = minimum possible storage at the beginning of time period t; and $S_{max,t}$ = maximum possible storage at the beginning of time period t.

There are T time periods in the operation horizon. Here, $T = 92$ days, which is designated as 0, 1, ..., 91. In Eqn. (8), $p_t(S_t)$ is the objective function, which is to be maximized. Equation (9) shows the objective function value at $t = T$. This is taken as zero here, but any value other than zero will not affect the result. Mass balance of the reservoir is answered by Eqn. (10). Equation (11) shows the amount of hydropower produced within time period t, which is a function of average release for power and the head of water available to the turbine. This equation is used by Hirakud authority, in which actual turbine efficiency is incorporated. In this equation alone, PR_t is expressed by the unit of cusecs and H_t by the unit of feet. The average reservoir storage level during time period t can be found out from the reservoir storage by Eqn. (12). This is obtained from the elevation-storage table of Hirakud reservoir (Patri, 1993). The quantity released for power comes back to Mahanadi River and joins with spill and d/s catchment contribution to make up the total flow at Naraj. This combined release from reservoir (release for power and spill) is restricted by Eqns (13) and (14) taking due consideration of lag time and attenuation to keep the flow at Naraj below the non-damaging flow, i.e., 25.5 thousand m³/s. Equation (15) shows the restriction of reservoir storage state, which should be in between minimum and maximum possible storages. Equation (16) shows the non-negative constraints on release for power and spill from the dam.

As discussed earlier, the combinations of 51, 54, 57 thousand m³/s as peak inflow into reservoir and 25.5, 20, 14 thousand m³/s, respectively as peak downstream catchment contribution form the critical combinations for flood situation. The rule curve for critical combination of representative series (inflow into the reservoir and downstream catchment contribution) for flood condition is shown in Fig. 5. The continuous wave like pattern, seen in the rule curve is because of the hypothetical inflow series, where a flood is assumed during every 10 days block period. In actual flood condition, the chance of occurrence of these types of series may not be there, but there is every chance of occurrence of single flood in the corresponding 10 days block period as shown in the historical sixty-eight floods. To make the reservoir ready to tackle these floods, should be brought at least equal to the level corresponding to the trough of each wave. If the reservoir is at a level more than this and that particular flood occurs, either the reservoir will overflow or flow at Naraj will be more than 25.5 thousand m³/s. The final rule curve for flood condition corresponds to the trough of the waves of rule curves, obtained from the critical combination of two flow series. It may be observed from Fig. 5 that the combination of 57 thousand m³/s of inflow into reservoir and 14 thousand m³/s for downstream catchment contribution is the most critical among the above three critical combinations of flow series. It is always necessary to keep the reservoir at the lowest level for tackling these combinations. Hence, the rule curve, corresponding to this combination, is chosen for finding the final rule curve for flood condition. The level starts from 179.83 m at 0th time period. The lowest level among all troughs of the whole season occurs in 73rd time period and corresponds to 184.3 m. Hence, there is no point in increasing the level beyond 184.3 m before 73rd time period. At the end of monsoon season, i.e., at end of 92nd time period, the reservoir level is 192.02 m. The levels for all intermediate time periods are obtained by linear interpolation. This is the final rule curve for flood

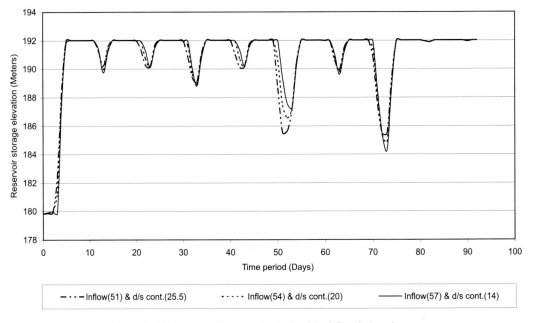

Fig. 5 Rule curve for hypothetical critical flood situation.

condition. Although in this particular case study, only one lowest value of the troughs (corresponding to 73rd time period) is used for arriving at the final rule curve, in a general case, all the troughs may have to be used if there is a gradual increase in the lowest value of troughs with increase in time. As such, this represents the most conservative attitude, which is vital for the present study area because the main objective of reservoir is flood control.

The following conclusions could be drawn from the present case study:

- FDP is a useful tool to develop the optimal operation policy of a reservoir system for flood control.
- The long-term operating policy developed for the Hirakud reservoir system can be actually adopted for better utilization of the system.
- In case of need to change the operating policy due to change in hydrology regime or increased risk aversion, the methodology developed in this study can be used to develop a new rule curve.

6. MEASURES SUGGESTED FOR EVOLVING A FEASIBLE AND IMPLEMENTABLE FLOOD CONTROL STRATEGY

Based on the number of flood control strategies being adopted throughout the world, suitable measures for flood control strategy are listed below which are generic in nature and are not based on the case study presented alone.

- Reasonably accurate forecast, coordination, flood preparedness, feasible and practical evacuation plans, and creation of suitable infrastructure.
- Integrated regional and flexible water resources development/watershed management plan such as dredging of rivers/streams and re-excavation of abandoned channels based on the ground level situation.
- Interlinking of rivers at a regional level may also reduce the flood damage.
- Development of more sophisticated and robust mathematical models with suitable database in close collaboration with academia.
- Research, education and training at all levels such as universities, R&D organizations (e.g., NIH, CWC, CSIR), water resources departments of various state governments, and River Valley Development authorities.
- Stakeholder's survey and involvement of stakeholders at relevant level for improving participatory flood management situation. This also helps to get effective feedback to learn lessons from the past events.
- Establishment of National Flood Control Authority which will coordinate all the relevant agencies.
- Environment impact assessment of each flood.

7. CONCLUDING REMARKS

In this chapter, modeling for the flood control and management is described to suggest suitable management strategies. A case study is presented to demonstrate the development of an optimal operating policy for a reservoir with flood control as a major criterion. In addition, the following issues are addressed: various structural and non-structural approaches for flood control and management, a step-by-step approach for choosing the most appropriate flood control strategy and measures for evolving a feasible and implementable flood control strategy. To conclude, it is essential to develop different modeling approaches for effective flood control and management so as to ensure sustainable development of flood-prone regions.

REFERENCES

Ahmad, S. and Simonovic, S.P. (2006). An Intelligent decision support system for management of floods. *Water Resources Management*, **20(3)**: 391-410.

Akbari, M., Afshar, A. and Sadrabadi, M.R. (2009). Fuzzy rule based models modification by new data: Application to flood flow forecasting. *Water Resources Management*, **23(12)**: 2491-2504.

ASCE Task Committee on Application of Artificial Neural Networks in Hydrology (2000a). Artificial neural networks in hydrology. I: Preliminary concepts. *Journal of Hydrologic Engineering, ASCE*, **5(2)**: 115-123.

ASCE Task Committee on Application of Artificial Neural Networks in Hydrology (2000b). Artificial neural networks in hydrology. II: Hydrologic applications. *Journal of Hydrologic Engineering, ASCE*, **5(2)**: 124-137.

Bahremand, A., Smedt, F.D., Corluy, J., Liu, Y.B., Poorova, J., Velcicka, L. and Kunikova, E. (2007). WetSpa model application for assessing reforestation impacts on floods in Margecany-Hornad Watershed, Slovakia. *Water Resources Management*, **21(8)**: 1373-1391.

Baliarsingh, F. (2000). Long-Term and Short-Term Optimal Reservoir Operation for Flood Control. Ph.D. Thesis, Indian Institute of Technology, Kharagpur, West Bengal, India.

Bana e Costa, C.A., Silva, P.A.D. and Correia, F.N. (2004). Multicriteria evaluation of flood control measures: The case of Ribeira do Livramento. *Water Resources Management*, **18(3)**: 263-283.

Bojadziev, G. and Bojadziev, M. (1997). Fuzzy Logic for Business: Finance and Management. World Scientific, Singapore, 232 pp.

Carsell, K.M., Pingel, N.D. and Ford, D.T. (2004). Quantifying the benefit of a flood warning system. *Natural Hazards Review*, **5(3)**: 131-140.

Chau, K.W., Wu, C.L. and Li, Y.S. (2005). Comparison of several flood forecasting models in Yangtze River. *Journal of Hydrologic Engineering, ASCE*, **10(6)**: 485-491.

Clark, M.J. (1998). Putting water in its place: A perspective on GIS in hydrology and water management. *Hydrological Processes*, **12**: 823-834.

Clarke, K.C. (1999). Getting started with geographic information systems. Prentice Hall, New Jersey, 338 pp.

De Vantier, B.A. and Feldman, A.D. (1993). Review of GIS applications in hydrologic modeling. *Journal of Water Resources Planning and Management, ASCE*, **119(2)**: 246-261.

Deb, K. (2001). Multiobjective Optimization using Evolutionary Algorithms. Wiley, Singapore, 515 pp.

Deka, P. and Chandramouli, V. (2005). Fuzzy neural network model for hydrologic flow routing. *Journal of Hydrologic Engineering, ASCE*, **10(4)**: 302-314.

DeMers, M.N. (1997). Fundamentals of Geographic Information Systems. John Wiley & Sons, New York, 480 pp.

Dhar, O.N. and Nandagiri, S. (1998). Floods in Indian Rivers and their Meteorological Aspects. *In:* V.S. Kale (editor), Flood Studies in India, Memoir 41, Geological Society of India, Bangalore, pp. 1-25.

Engman, E.T. and Gurney, R.J. (1991). Remote Sensing in Hydrology. Chapman and Hall, London, 225 pp.

Garde, R.J. (1998). Floods and flood control: Engineering approach. *In:* V.S. Kale (editor), Flood Studies in India, Memoir 41, Geological Society of India, Bangalore, pp. 173-193.

Hajkowicz, S. and Collins, K. (2007). A review of multiple criteria analysis for water resource planning and management. *Water Resources Management*, **21(9)**: 1553-1566.

Hall, J.W., Meadowcroft, I.C., Sayers, P.B. and Bramley, M.E. (2003). Integrated flood risk management in England and Wales. *Natural Hazards Review*, **4(3)**: 126-135.

Heywood, I., Cornelius, S. and Carver, S. (1998). An Introduction to Geographical Information Systems. Pearson Education, 279 pp.

Holz, K.P., Hildebrandt, G. and Weber, L. (2006). Concept for a web-based information system for flood management. *Natural Hazards*, **38(1-2)**: 121-140.

Ignizio, J P. (1991). An Introduction to Expert Systems: Development and Implementation of Rule-Based Expert Systems. McGraw-Hill Inc, New York, 384 pp.

Islam, M.M. and Sado, K. (2002). Development priority map for flood countermeasures by remote sensing data with geographical information system. *Journal of Hydrologic Engineering, ASCE*, **7(5)**: 346-355.

Jain, S.K. and Singh, V.P. (2006). Water Resources Systems Planning and Management: Developments in Water Science. Elsevier, New Delhi, 858 pp.

Jain, S.K., Saraf, A.K., Goswami, A. and Ahmad, T. (2006). Flood inundation mapping using NOAA AVHRR data. *Water Resources Management*, **20(6)**: 949-959.

Jain, S.K., Singh, R.D., Jain, M.K. and Lohani, A.K. (2005). Delineation of flood-prone areas using remote sensing techniques. *Water Resources Management*, **19(4)**: 333-347.

Kale, V.S. (1998), Flood studies in India. Edited volume, Memoir 41, Geological Society of India, Bangalore, 256 pp.

Karamouz, M., Abesi, O., Moridi, A. and Ahmadi, A. (2009). Development of optimization schemes for floodplain management: A case study. *Water Resources Management*, **23(9)**: 1743-1761.

Khan, S., Ahmad, A. and Wang, B. (2007). Quantifying rainfall and flooding impacts on groundwater levels in irrigation areas: GIS approach. *Journal of Irrigation and Drainage Engineering, ASCE*, **133(4)**: 359-367.

Kumar, R. and Chatterjee, C. (2005). Regional flood frequency analysis using L-moments for North Brahmaputra Region of India. *Journal of Hydrologic Engineering, ASCE*, **10(1)**: 1-7.

Lekuthai, A. and Vongvisessomjai, S. (2001). Intangible flood damage quantification. *Water Resources Management*, **15(5)**: 343-362.

Levy, J.K. (2005). Multiple criteria decision making and decision support systems for flood risk management. *Stochastic Environmental Research and Risk Assessment*, **19(6)**: 438-447.

Li, Z.J. and Zhang, K. (2008). Comparison of three GIS-based hydrological models. *Journal of Hydrologic Engineering, ASCE*, **13(5)**: 364-370.

Liu, Y.B. and Smedt, F.D. (2005). Flood modeling for complex terrain using GIS and remote sensed information. *Water Resources Management*, **19(5)**: 605-624.

Lund, J.R. (2002). Floodplain planning with risk-based optimization. *Journal of Water Resources Planning and Management, ASCE*, **128(3)**: 202-207.

Machado, S.M. and Ahmad, S. (2007). Flood hazard assessment of Atrato River in Colombia. *Water Resources Management*, **21(3)**: 591-609.

Mays, L.W. (1996). Water Resources Hand Book. McGraw-Hill, New York, 1568 pp.

Mohapatra, P.K. and Singh, R.D. (2003). Flood management in India. *Natural Hazards*, **28**: 131-143.

Nagesh Kumar, D. (2002). Remote sensing applications to water resources. *In:* Rama Prasad and S. Vedula (editors), Research Perspectives in Hydromechanics and Water Resources Engineering. World Scientific, Singapore, pp. 287-316.

Nagesh Kumar, D. (2006). Lecture Notes on Soft Computing. Indian Institute of Science, Bangalore, India, 120 pp.

Nagesh Kumar, D. and Baliarsingh, F. (2003). Folded dynamic programming for optimal operation of multireservoir system. *Water Resources Management*, **17**: 337-353.

Nakicenovic, N., Alcamo, J., Davis, G., de Vries, B., Fenhann, J., Gaffin, S., Gregory, K., Grübler, A., Jung, T.Y., Kram, T., La Rovere, E.L., Michaelis, L., Mori, S., Morita, T., Pepper, W., Pitcher, H., Price, L., Raihi, K., Roehrl, A., Rogner, H.H., Sankovski, A., Schlesinger, M., Shukla, P., Smith, S., Swart, R., van Rooijen, S., Victor, N., Dadi, Z. (2000). IPCC Special Report on Emissions Scenarios. Cambridge University Press, Cambridge, United Kingdom and New York, NY, 599 pp.

Needham, J.T., Watkins, D.W., Lund, J.R. and Nanda, S.K. (2000). Linear programming for flood control on the Iowa and Des Moines Rivers. *Journal of Water Resources Planning and Management, ASCE*, **126(3)**: 118-127.

Noman, N.S., Nelson, E.J. and Zundel, A.K. (2003). Improved process for floodplain delineation from digital terrain models. *Journal of Water Resources Planning and Management, ASCE*, **129(5)**: 427-436.

Patri, S. (1993). Data on Flood Control Operation of Hirakud Dam. Irrigation Department, Government of Orissa, Bhubaneswar, India, 200 pp.

Ranjithan, S.R. (2005). Role of evolutionary computation in environmental and water resources systems analysis. *Journal of Water Resources Planning and Management, ASCE*, **131(1)**: 1-2.

Rao, D.P., Bhanumurthy, V., Rao, G.S. and Manjusri, P. (1998). Remote sensing and GIS in flood management in India. *In:* V.S. Kale (editor), Flood Studies in India. Memoir 41, Geological Society of India, Bangalore, pp. 195-218.

Rumelhart, D.E., Hinton, G.E. and Williams, R.J. (1986). Learning internal representation by back-propagating errors. *Nature*, **32(3)**: 533-536.

Sanders, R., Shaw, F., Galy, M.H. and Foote, M. (2005). National flood modeling for insurance purposes using IFSAR for flood risk estimation in Europe. *Hydrology and Earth Systems Sciences*, **9(4)**: 449-456.

Seth, S.M. (1998). Flood hydrology and flood management in India. *In:* V.S. Kale (editor), Flood Studies in India. Memoir 41, Geological Society of India, Bangalore, pp. 155-172.

Simonovic, S.P. (1991). Knowledge-based systems and operational Hydrology. *Canadian Journal of Civil Engineering*, **18(1)**: 1-11.

Simonovic, S.P. and Li, L. (2004). Sensitivity of the Red River Basin flood protection system to climate variability and change. *Water Resources Management*, **18(2)**: 89-110.

Singh, V.P. and Fiorentino, M. (1996). Geographical Information Systems in Hydrology. Kluwer Academic Publishers, Dordrecht, Germany, 468 pp.

Slutzman, J.E. and Smith, J.A. (2006). Effects of flood control structures on flood response for hurricane floyd in the Brandywine Creek Watershed, Pennsylvania. *Journal of Hydrologic Engineering, ASCE*, **11(5)**: 432-441.

Solomatine, D.P. and Xue, Y. (2004). M5 model trees and neural networks: Application to flood forecasting in the upper reach of the Huai River in China. *Journal of Hydrologic Engineering, ASCE*, **9(6)**: 491-501.

Subramanya, K. (1994). Engineering Hydrology. Tata McGraw-Hill Publishing, New Delhi, 392 pp.

Travis, Q.B. and Mays, L.W. (2008). Optimizing retention basin networks. *Journal of Water Resources Planning and Management, ASCE*, **134(5)**: 432-439.

Varas, E.A. and Von Chrismar, M. (1995). Expert system for the selection of methods to calculate design flood flows. *Hydrological Sciences Journal*, **40(6)**: 739-749.

Venkatesh, M., Olivera, F., Arabi, M. and Edleman, S. (2008). Uncertainty in flood inundation mapping: current issues and future directions. *Journal of Hydrologic Engineering, ASCE*, **13(7)**: 608-620.

Verbyala, D.L. (2002). Practical GIS analysis. Taylor and Francis, London, 294 pp.

Vieux, B.E. (2001). Distributed Hydrologic Modeling using GIS. Kluwer Academic Publishers, Dordrecht, Germany, 312 pp.

Wei, C.C. and Hsu, N.S. (2008). Multireservoir flood-control optimization with neural-based linear channel level routing under tidal effects. *Water Resources Management*, **22(11)**: 1625-1647.

Wu, J.S., Han, J., Annambhotla, S. and Bryant, S. (2005). Artificial neural networks for forecasting watershed runoff and stream flows. *Journal of Hydrologic Engineering, ASCE*, **10(3)**: 216-222.

Yu, Y.B., Wang, B.D., Wang, G.L. and Li, W. (2004). Multi-person multiobjective fuzzy decision-making model for reservoir flood control operation. *Water Resources Management*, **18(2)**: 111-124.

9 Real-Time Flood Forecasting by a Hydrometric Data-Based Technique

Muthiah Perumal* and Bhabagrahi Sahoo

1. INTRODUCTION

Flooding, an age-old global problem, has been increasing at a worrisome pace in recent years. However, natural flooding of large areas did not create more dangerous situation in a prehistoric world. With the expansion of anthropogenic activities, there is destruction of natural drainage systems and with increasing urbanization there is reduction in the opportunity for rainwater infiltration, resulting in an increase of surface runoff potential, and thereby increasing the loss of lives and economic damages due to floods. Further, global climate change induced extreme rainfall events trigger flash floods in both urban and rural watersheds making the severity of flood hazards many fold. According to the Report of the Sub-Group on Flood Management for the XI Five Year Plan (2007-2012), Central Water Commission, Ministry of Water Resources, Government of India (CWC, 2006), on an average, 7.55 million hectares of land in India is affected by flood annually, of which 3.54 million hectares is the cropped area. Besides these, 1589 lives and 94,839 heads of cattle are lost in addition to damaging 1.22 million houses annually. The value of damages to crops, houses, and public utilities alone accounts for about Rs. 1805 crore (US $361 million) per annum. Some recent instances of such type of devastating floods could be the 2008 Bihar flood in the Kosi River, 2007 Bangladesh flood in the Padma River as shown in Fig. 1 and 2005 Mumbai

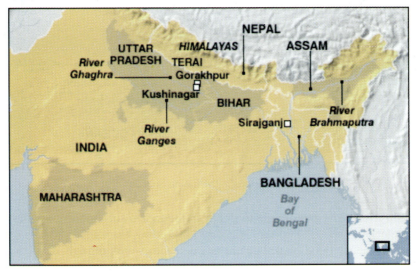

Fig. 1 Map of South Asia showing some recently flood-affected areas. (*Source:* http://news.bbc.co.uk/1/hi/world/south_asia/6927389.stm)

---

*Corresponding Author

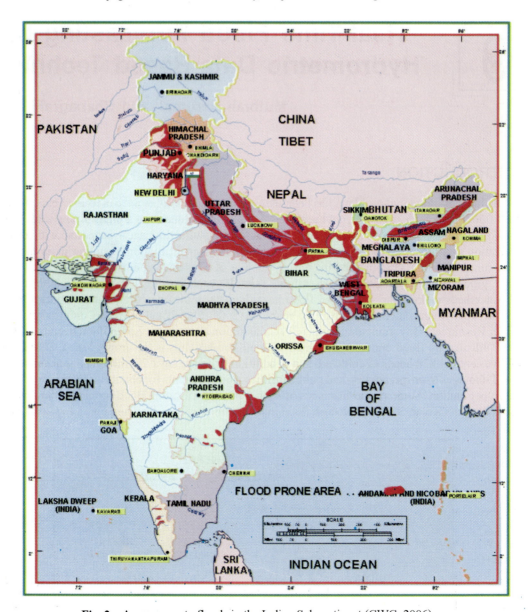

Fig. 2 Areas prone to floods in the Indian Subcontinent (CWC, 2006).

flood in the Mithi River. The areas prone to floods in the Indian subcontinent are illustrated in Fig. 2 (shaded in red color).

Early efforts to reduce flood-related deaths and damages were primarily devoted to flood control measures such as construction of levees, flood walls, dams, storage reservoirs, channelization of rivers, channel improvements, drainage improvements, diversion of flood waters, and watershed management – all of these measures fall under the category of structural measures to control flood problems. Given the impossibility of building larger and larger structures to cope up with the extremely low probability flood events, the structural measures alone cannot completely circumvent risk of flood hazards. Hence, an important role is left to the non-structural measures to be compared, evaluated and actuated in real-time, which implies the need for accurate flood forecasts with a sufficient lead-time to allow for proper response action. Consequently, flood forecasting with sufficient lead-time has become an important non-structural measure for flood hazard mitigation and for minimizing flood related deaths. Therefore, it is essential that reliable flood forecasting methods be employed which is physically based, less data intensive and, over and above, is easily understood by the field engineers.

In light of the above facts, the analysis presented in this chapter focuses on this specific aspect of flood forecasting by studying the use of a variable parameter Muskingum stage-hydrograph (VPMS) routing method as a component model of a hydrometric data-based deterministic forecasting model. It would be shown later in this chapter that the use of a physically based component in a forecasting model enables the use of a simple stochastic error-updating model to estimate the forecast error. The estimation of forecast error is made in the proposed model using a two-parameter linear autoregressive model with its parameters updated at every time interval of 30 minutes at which the stage observations are made. The proposed forecasting model is tested considering several flood events which occurred in a 15 km river reach selected along the Tiber River, in Central Italy, bounded by upstream Pierantonio and downstream Ponte Felcino gauging stations.

This chapter deals with an overview of the flood forecasting and early warning system being adopted worldwide, technical aspects of the proposed hydrometric data-based VPMS routing model and its frameworks for floodplain flow condition as well as real-time application, and a case study demonstrating the application of the proposed model.

2. FLOOD FORECASTING AND EARLY WARNING SYSTEM: AN OVERVIEW

Flood forecasting and warning systems are cost-effective means of reducing the damaging impacts of floods. These real-time flood forecasting methods in practice can broadly be categorized as *'discharge forecasting'* and *'stage forecasting'*. The real-time discharge forecasting obtained by rainfall-runoff modeling is generally less accurate than that obtained by the channel routing of an upstream discharge hydrograph (Srikanthan et al., 1994). However, in headwater catchments with no information on discharge hydrographs, these forecasting methods are very much useful. The various rainfall-runoff modeling approaches used in flood forecasting are: (i) Unit Hydrograph (UH) approach for linear catchment modeling; (ii) Non-linear catchment routing models, viz., RORB model, Watershed Bounded Network Model (WBNM), RAFTS model, and URBS model (Carroll, 1992); (iii) Loss models, viz., Constant loss rate (Φ-index) method (Reed, 1982), Variable loss rate method, Constant proportional loss model (NERC, 1975), Variable proportional loss model (NERC, 1975), and Initial loss-continuing loss model of Bureau of Meteorology, Australia; (iv) Non-linear storage models, viz., Inflow-Storage-Outflow model (Lambert, 1969, 1972), Isolated event model (NERC, 1975), and Generalized non-linear storage model (O'Connell, 1980); (v) Conceptual models, viz., Sacramento model, SAMFIL model (Vermuleulen and Snoeker, 1991), IPH-II model (Bertoni et al., 1992), Australian Water Balance Model (AWBM) (Boughton, 1993), NAM model (Refsgaard et al., 1988), SBV model (Bergstrom, 1976, 1992), Tank model (Sugawara, 1979), Probability distributed model (Moore and Jones, 1991), Alabama rainfall-runoff model (Henry et al., 1988), Xinanjiang model (Zhao et al., 1980), ARNO model (Todini, 1996), and Variable Infiltration Capacity (VIC) model (Liang et al., 1996); (vi) Spatially distributed models, viz., SIMPLE model (Kouwen, 1988), TOPOG model (O'Loughlin et al., 1989), TOPMODEL (Beven and Kirkby, 1979), and SHE model (Abbott et al., 1986a, 1986b); (vii) Transfer function models (Harpin, 1982; Powell,

1985; Cluckie and Ede, 1985; Owens, 1986; Troch et al., 1991); and (viii) Statistical methods, viz., Constrained linear systems (Natale and Todini, 1977), IHACRES model, and Filter separation autoregressive model.

Furthermore, the real-time discharge forecasting in long river systems can be obtained by different flood routing methods. These flood routing methods can be categorized as: (a) Statistical routing methods, (b) Artificial intelligence-based methods (e.g., Artificial Neural Network, Fuzzy logic, and Genetic Algorithm); (c) Hydrologic routing methods, viz., Muskingum method, Muskingum-Cunge method (Cunge, 1969), variable parameter Muskingum-Cunge method (Ponce and Yevjevich, 1978), multilinear Muskingum discharge routing method (Perumal, 1992), multilinear discrete cascade model (Perumal, 1994a), and non-linear reservoir-type channel routing method (Georgakakos and Bras, 1982); and (d) Hydraulic routing methods, viz. kinematic wave routing method, diffusion wave routing method, and variable parameter Muskingum discharge (VPMD) routing method (Perumal, 1994b, 1994c). However, there are a few stage routing methods available in the literature which can be used for real-time flood forecasting. The hydrologic stage routing methods include 'statistical methods', 'artificial intelligence-based methods', 'multilinear Muskingum stage routing method' (Perumal et al., 2009b), and 'multilinear discrete cascade model'. Similarly, the hydraulic stage routing methods include 'de Saint-Venant equations and their simplifications', and 'variable parameter Muskingum stage (VPMS) routing method' (Perumal and Ranga Raju, 1998a, 1998b, Perumal et al., 2007). Among the various models enlisted above, this chapter focuses only on the VPMS routing method amenable for real-time flood forecasting.

The current developments in computer technology, coupled with advances in telemetry system for automatic data acquisition, not only allows the improved forecasting of any magnitude of flow at any point in a watershed, but also enables the automatic operation of the hydro-systems affected by the forecasted flow. The development of hydrological forecasting system involves various sub-systems which deal with historical and real-time data collection, data transmission, database management, forecasting procedure (modeling), forecast dissemination, and forecast evaluation and updating. Technological advances in the field of flood forecasting under these subsystems, which deal with technical aspects of forecasting, can be mainly divided into three groups: (1) data collection, (2) transmission, and (3) analysis for developing a forecasting model. With the recently improved hydrological instrumentation, automatic acquisition system of a wide range of hydrological data that includes automated rain gauges and river stage recorders, radar to detect the likely areas that would receive precipitation and its intensity, satellite based methods, and radio and satellite telemetry for transmission of data, the modern flood forecasting service systems have reached at a high point of development. The relevance of the flood forecasting as a significant flood abatement measure can be recognized from the fact that many agencies are now involved in real-time flood forecasting services in the world as discussed briefly herein.

At the beginning of modernization of flood forecasting services long before, only the government agencies were able to provide this service as they were the prime agencies responsible for the collection and dissemination of hydrologic data in real-time, its archival and processing needed for the forecasting purposes. However, with the rapid development of technology in data acquisition, telemetry, launching of many communication satellites and, above all, rapid development in computer technology resulting in cheaper desktop computers, it has become now that real-time satellite down-link systems for basin-wide hydrologic data are no longer the prerogative of government agencies. Affordable, easy-to-use hardware and software installations are within the reach of even small hydro-systems operators and they can get timely hydrologic forecasts for the efficient and environment friendly operations of the hydro-systems. While this is the scenario of flood forecasting technology in developed countries, the same is not prevailing in many developing countries. Since in most developing countries management of water resources is under the control of government agencies and flood evacuation and relief measures are their responsibility, the implementation of flood forecasting systems is mostly carried out by these government agencies or the agencies designated by them such as the World Meteorological Organization (WMO), aid agencies and, in some countries, by private companies.

The WMO Hydrological Operational Multipurpose Sub-program (HOMS) is intended to promote the transfer of hydrological technology between the member countries of WMO for use in their water resources projects. The technology is made available to users in the form of various kinds of components, for instance, manual and computerized techniques for data collection, processing and analysis; commonly used hydrological models; manuals describing field or office procedures; and instruments specifications. The WMO has been entrusted with technical supervision of the establishment of a large number of Forecasting Operational Real-Time Hydrological Systems (FORTH) in developing countries of Asia, Latin America, and Africa. However, over the past several years, many of these countries which have installed the hydrological forecasting systems with the technical assistance of WMO are modernizing these systems with the improved data acquisition systems, telemetry and advanced forecasting software, which have been well-tested and operational in developed countries. Some of the other factors responsible for the modernization are the assistance of the developed countries in installing improved flood forecasting systems in the form of bilateral aid, emergence of many private companies in developed countries for manufacturing electronic sensors and telemetry systems needed for real-time flood forecasting, and ever reducing costs of desktop computers. The notable countries which are playing major role in advancing the flood forecasting technology in other parts of the world, apart from their own countries, are the United States of America (USA), Canada, Denmark and The Netherlands. While the USA and Canada have developed easy-to-install advanced data collection instruments and telemetry systems, countries such as Canada, Denmark and The Netherlands have developed improved hydrological forecasting tools. A brief description of the activities of these agencies in real-time flood forecasting is presented herein.

The National Weather Service (NWS) of National Oceanic and Atmospheric Administration (NOAA) is the federal agency responsible to issue forecasts and warning of floods in the USA. Although many cities, counties or other local flood management agencies are involved now-a-days in the operation of local flood warning systems, the NWS is still the principal national agency responsible for flood forecasting and warning in the entire USA. The local agencies coordinate with NWS in getting the forecasting technology implemented in their area and continuously get technical guidance. The NWS uses the National Weather Service River Forecast System (NWSRFS) as the foundation of hydrologic forecast system. NWSRFS is a suite of hydrologic and hydraulic models that contains all the programs necessary to produce hydrologic forecasts for a river basin. The NWSRFS provides the basic framework for a national river forecast and flood warning system. The current available operations, which form the components of the NWSRFS are:

(a) Temperature index snow accumulation and ablation model
(b) Sacramento soil-moisture accounting model
(c) Antecedent Precipitation Index (API) rainfall-runoff models used in the Missouri Basin, Ohio basin, Middle Atlantic States, North-Central USA, and South-Western USA
(d) Unit hydrograph with a constant and variable baseflow option
(e) Lag and K, Muskingum, layered coefficient and Tatum routing procedures
(f) Flood Wave Model (FLDWAV)
(g) Reservoir model that allows the user to select and combine thirteen modes of regulation to simulate the operation of a single independently controlled reservoir
(h) Stage/discharge conversion using single-valued rating curves with log or hydraulic extensions and dynamically induced loop ratings

(i) Simple flow adjustment and blend procedure
(j) Simplified channel loss procedure
(k) Computation of mean discharge from instantaneous values
(l) Set timeseries values to zero
(m) Add and subtract timeseries
(n) Weight timeseries
(o) Change the time interval of a timeseries
(p) Plot instantaneous discharge
(q) Operational hydrograph display
(r) General timeseries plot
(s) Daily flow plots (calibration used only), and
(t) Statistical package (calibration)

While most of the precipitation data are collected by the NWS or its designated agencies in real time, the stream gauge data for most of the streams are received from the USGS stream network. Expanded use of telemetry at the USGS streamflow stations and refinement of telemetry equipment continue to improve the timeliness and reliability of data that are transmitted for forecasting purposes. The automated surface observatory systems are now replacing the manual weather observations, and advanced telecommunication systems are improving the integration and distribution of data. The US Army Corps of Engineers has developed the HEC-RAS and HEC-HMS modeling packages for flood forecasting in large river basins.

The Danish Hydraulic Institute (DHI) of Denmark has developed a range of most advanced flood management software, viz., MIKE 11, MIKE 11 GIS, MIKE 21, MIKE FLOOD WATCH, and MIKE SHE, and they have been used in many of the flood forecasting systems implemented for the large river basins in the world. In the majority of flood management projects carried out, DHI-Water and Environment has also been responsible for the overall project management at the implementation stage. Some of the flood forecasting projects implemented by DHI-Water and Environment are: Flood management in Poland; Flood management in Czech Republic; Anglican flow forecasting modeling system, UK; Flood forecasting for barrage operation, Malaysia; Inflow forecasting for hydropower optimization, Wales, UK; Expansion of flood forecasting and warning services, Bangladesh; Flood forecasting for middle river Yangtze, China; Real-time flood forecasting in Italy; Environment Waikato flood forecasting system in New Zealand; Ping River basin flood forecasting project in Thailand; and Pilot flood forecasting system for Lower Colorado River, Texas, USA.

Flood forecasting and warning in Canada has evolved into a network of forecast system across the country. There are five provisional streamflow centers in Canada. In addition, the power generation companies such as Hydro BC and Hydro-Quebec have developed advanced flood forecasting systems for operating a cluster of reservoirs and run-of-the river plants under their control for power generation. These companies are implementing the technology adopted in their forecasting systems for other countries also. Furthermore, some of the private companies such as Riverside Technology, Inc. (Rti), USA, and Water and Monitoring System International (WMSI), Canada are implementing the flood forecasting systems developed in the USA and in Canada, respectively, and also in other parts of the world.

Technologically, every flood forecasting model operates on two modes, viz., *simulation mode*, and *operation mode* (on-line forecasting). A flood forecasting model in the simulation mode attempts to produce the response of the system for the past recorded precipitation or upstream flow input. The response of the model is compared with the recorded response at the point of forecasting interest and, if both do not match, either the model structure is changed or the parameters are modified till the matching is done satisfactorily. Once the structure of the model and its parameters are identified during the calibration phase, the model can be used for forecasting purposes and it is said to be used in operational mode. While the basic structure of the model is not changed in the operational mode, the parameters need to be changed considering the current catchment conditions due to the variation of the input and subsequent change in other components of the rainfall-runoff process.

Moreover, the flood forecasting models are typically made up of two components: (i) *deterministic flow component*, and (ii) *stochastic flow component*. While the former is determined by the hydrologic/hydraulic model, the latter is determined based on the residual (error) series of the difference between the forecasted flow for a specified lead time and the corresponding observed one. The residual series reflects both the model error, due to the inability of the model used for forecasting to correctly reproduce the flow process, and the observational error while measuring the flow. It is imperative, therefore, to use an appropriate approach to reduce the model error.

The following section describes about the physical basis of the VPMS model used as a component model for flood forecasting.

3. VARIABLE PARAMETER MUSKINGUM STAGE-HYDROGRAPH ROUTING METHOD

The variable parameter Muskingum stage-hydrograph routing method, henceforth referred to as the *VPMS method*, was developed by Perumal and Ranga Raju (1998a, 1998b) directly from the Saint Venant equations. The form of the routing equation developed is same as that of the *Muskingum method*, by replacing the discharge variable by the stage variable and hence, the reason for adherence of the term "*Muskingum*". Further, the parameters vary at every routing time interval and they are related to the channel and flow characteristics by the same relationships as established for the physically based Muskingum method (Apollov et al., 1964; Cunge, 1969; Dooge et al., 1982; Perumal 1994b, 1994c). The detail development of this method is presented below (e.g., Perumal and Ranga Raju, 1998a; Perumal et al., 2007).

3.1 Concept

The VPMS method has been developed using the following concept: During steady flow in a river reach having any shape of prismatic cross-section, the stage and, hence, the cross-sectional area of flow at any point of the reach is uniquely related to the discharge at the same location defining the steady flow rating curve. However, this situation is altered during unsteady flow, as conceptualized in the definition sketch (Fig. 3) of the variable parameter Muskingum routing reach of length Δx, in which the same unique relationship is maintained between the stage and the corresponding steady discharge at any given instant of time, recorded not at the same section, but at a downstream section (section 3 in Fig. 3) preceding the corresponding steady stage section (mid-section in Fig. 3).

3.2 Theoretical Background

The routing method is derived from the Saint Venant equations, which govern the one-dimensional unsteady flow in channels and rivers without considering lateral flow and are given by:

$$\frac{\partial Q}{\partial x} + \frac{\partial A}{\partial t} = 0 \quad \text{(continuity equation)} \tag{1}$$

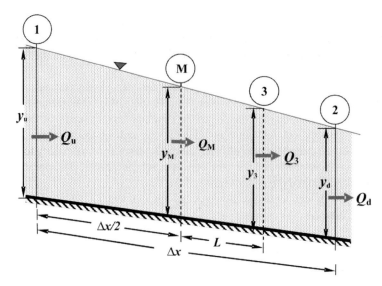

Fig. 3 Definition sketch of the stage–hydrograph routing method.

$$S_f = S_o - \frac{\partial y}{\partial x} - \frac{v}{g}\frac{\partial v}{\partial x} - \frac{1}{g}\frac{\partial v}{\partial t} \quad \text{(momentum equation)} \tag{2}$$

where Q = discharge; A = cross-sectional area of flow; S_o = channel bed slope; S_f = energy slope; g = acceleration due to gravity; v = average velocity over cross-section, and y = depth of flow. The notations x and t denote the space and time variables, respectively.

The derivation of the method involves the assumptions which enable the simplification of the unsteady flow process by assuming the channel reach to be prismatic and the gradients $\partial y/\partial x$, $(v/g)(\partial v/\partial x)$ and $(1/g)(\partial v/\partial t)$, which, respectively, denote the longitudinal water surface gradient, the convective and the local acceleration gradients, remain constant at any instant of time in a given reach. The latter assumption implies that the friction slope S_f is constant over the computational reach length at any instant of time and, hence, the flow depth varies linearly. It has been shown by Perumal and Ranga Raju (1998a, 1998b; 1999) that the use of the assumption of constant S_f and the Manning's friction law governing the unsteady flow enable to arrive at the simplified momentum equation expressed as:

$$\frac{\partial Q}{\partial x} = v\left[\frac{\partial A}{\partial y} + mP\frac{\partial R}{\partial y}\right]\left[\frac{\partial y}{\partial x}\right] \tag{3}$$

where m = an exponent which depends on the friction law used ($m = 2/3$ for the Manning's friction law, $m = 1/2$ for the Chezy's friction law); R = hydraulic radius (A/P); and P = wetted perimeter.

Using Eqn. (3), it may be shown that the celerity of the flood wave can be expressed as:

$$c = \frac{\partial Q}{\partial A} = \left[1 + m\left(\frac{P\partial R/\partial y}{\partial A/\partial y}\right)\right]v \tag{4}$$

Note that the application of Eqn. (4) for unsteady flow in rectangular channels yields the same celerity relationship as given in the report of NERC (1975).

Using Eqns (1), (2), (3) and the expression of discharge using Manning's friction law, S_f can be expressed as follows:

$$S_f = S_o \left[1 - \frac{1}{S_o} \frac{\partial y}{\partial x} \left\{ 1 - \left[mF \left(\frac{P \partial R / \partial y}{\partial A / \partial y} \right) \right]^2 \right\} \right] \qquad (5)$$

where F is the Froude number which is mathematically expressed as:

$$F = \left(\frac{Q^2 \partial A / \partial y}{gA^3} \right)^{1/2} \qquad (6)$$

Use of Eqn. (5) in the expression for discharge Q_M at the middle of the computational channel reach, using the Manning's friction law, and its simplification based on the binomial series expansion leads to the simplified expression for Q_M as:

$$Q_M = Q_3 - \frac{Q_3 \left[1 - m^2 F_M^2 \left(\frac{P \partial R / \partial y}{\partial A / \partial y} \right)_M^2 \right]}{2 S_o \left. \frac{\partial A}{\partial y} \right|_3 \left[1 + m \left(\frac{P \partial R / \partial y}{\partial A / \partial y} \right) \right] v_3} \left. \frac{\partial Q}{\partial x} \right|_3 \qquad (7)$$

Eqn. (7) expresses the discharge Q_M in terms of normal discharge Q_3, corresponding to y_M, the flow depth at the middle of the reach. The section where Q_3 passes corresponding to y_M is located at a distance L downstream of the midsection and is expressed as:

$$L = \frac{Q_3 \left[1 - m^2 F_M^2 \left(\frac{P \partial R / \partial y}{\partial A / \partial y} \right)_M^2 \right]}{2 S_o \left. \frac{\partial A}{\partial y} \right|_3 \left[1 + m \left(\frac{P \partial R / \partial y}{\partial A / \partial y} \right) \right] v_3} \qquad (8)$$

where the subscripts M and 3 attached with different variables denote these variables at midsection and section 3, respectively.

Use of Eqns (1), (3) and (4) leads to the following expression (Perumal and Ranga Raju, 1998a, 1998b):

$$\frac{\partial y}{\partial t} + c \frac{\partial y}{\partial x} = 0 \qquad (9)$$

It was pointed out by Perumal and Ranga Raju (1998a, 1998b) that although the form of Eqn. (9) is same as that of the well-known kinematic-wave equation (Lighthill and Whitham, 1955), it is capable of approximately modeling a flood wave in the transition range between the zero-inertia wave, governed by the convection-diffusion equation (Hayami, 1951) and the kinematic wave, including the latter. The characteristic of this new wave type governed by Eqn. (9), termed as the approximate convection-diffusion (ACD) equation, has been investigated in detail by Perumal and Ranga Raju (1999).

Applying Eqns (4) and (9) at section 3 of Fig. 3 and its simplification leads to (Perumal and Ranga Raju, 1998a, 1998b) the governing differential equation of the Muskingum type routing, using stage as the operating variable in place of discharge, and it is expressed as:

$$y_u - y_d = \frac{\Delta x}{\left[1 + m \left(\frac{P \partial R / \partial y}{\partial A / \partial y} \right)_3 \right] v_3} \times \frac{\partial}{\partial t} \left[y_d + \left(\frac{1}{2} - \frac{L}{\Delta x} \right) (y_u - y_d) \right] \qquad (10)$$

where y_u and y_d denote the stages at the upstream and downstream of the reach, respectively. Using the similarity between the governing differential equation of the Muskingum method in discharge formulation and that of Eqn. (10), it is inferred that the travel time K of the Muskingum type stage routing method can be expressed as:

$$K = \frac{\Delta x}{\left[1 + m\left(\frac{P\partial R/\partial y}{\partial A/\partial y}\right)_3\right]v_3} \tag{11}$$

and the weighting parameter θ, after substituting for L from Eqn. (8), can be obtained as:

$$\theta = \frac{1}{2} - \frac{Q_3\left[1 - m^2 F_M^2\left(\frac{P\partial R/\partial y}{\partial A/\partial y}\right)_M^2\right]}{2S_o \frac{\partial A}{\partial y}\bigg|_3 \left[1 + m\left(\frac{P\partial R/\partial y}{\partial A/\partial y}\right)_3\right]v_3 \Delta x} \tag{12}$$

The product term $(\partial A/\partial y)_3[1 + m(P\partial R/\partial y)/(\partial A/\partial y)]_3 v_3$, present in the denominator of Eqn. (12), can be replaced by the simple expression deduced from Eqn. (4) as:

$$\frac{\partial A}{\partial y}\bigg|_3 \left[1 + m\left(\frac{P\partial R/\partial y}{\partial A/\partial y}\right)_3\right]v_3 = \frac{\partial Q}{\partial y}\bigg|_3 \tag{13a}$$

Expressing in terms of c_3, Eqn. (13a) is modified as:

$$\frac{\partial Q}{\partial y}\bigg|_3 = \frac{\partial A}{\partial y}\bigg|_3 c_3 \tag{13b}$$

Using Eqn. (13a), Eqn. (12) is modified as:

$$\theta = \frac{1}{2} - \frac{Q_3\left[1 - m^2 F_M^2\left(\frac{P\partial R/\partial y}{\partial A/\partial y}\right)_M^2\right]}{2S_o \frac{\partial Q}{\partial y}\bigg|_3 \Delta x} \tag{14a}$$

After neglecting inertial terms of the Saint Venant equations, Eqn. (14a) can be expressed as:

$$\theta = \frac{1}{2} - \frac{Q_3}{2S_o \frac{\partial Q}{\partial y}\bigg|_3 \Delta x} \tag{14b}$$

where the subscript 3 attached with these variables denote the section 3, wherein the discharge passing is the normal discharge corresponding to the stage at the middle of the Muskingum reach.

Use of Eqns (11) and (14b) in Eqn. (10) leads to a form similar to that of the Muskingum routing equation, but using stage as the operating variable instead of discharge, and it is expressed as follows:

$$y_{d,j+1} = C_1 y_{u,j+1} + C_2 y_{u,j} + C_3 y_{d,j} \tag{15}$$

where $y_{u,j+1}$ and $y_{d,j+1}$ denote the upstream and downstream stages at time $(j+1)\Delta t$, respectively; $y_{u,j}$ and $y_{d,j}$ denote the upstream and downstream stages at time $j\Delta t$, respectively, where Δt is the routing time interval; and the coefficients C_1, C_2 and C_3 are expressed as:

$$C_1 = \frac{-K\theta + 0.5\Delta t}{K(1-\theta) + 0.5\Delta t} \qquad (16a)$$

$$C_2 = \frac{K\theta + 0.5\Delta t}{K(1-\theta) + 0.5\Delta t} \qquad (16b)$$

$$C_3 = \frac{K(1-\theta) - 0.5\Delta t}{K(1-\theta) + 0.5\Delta t} \qquad (16c)$$

4. EXTENSION OF THE VPMS METHOD FOR ROUTING IN A TWO-STAGE COMPOUND CROSS-SECTION CHANNEL REACH

4.1 Channel Reach Details

It is assumed that the channel reach is characterized by a two-stage uniform compound cross-section with a trapezoidal main channel flow section and an extended trapezoidal floodplain section as shown in Fig. 4. It is, further, assumed that the entire channel reach is characterized by a uniform or representative Manning's roughness coefficient irrespective of main or floodplain channel. This assumption may not be strictly valid in practice. Since the main aim herein is to develop a simplified hydraulic routing method using stage as the main routing variable, such an assumption helps to reduce complications in the development of the method.

4.2 Development of Celerity-Stage Relationship

Stage-hydrograph routing using the VPMS method, either in a single section (main channel section) or a compound section channel reach, involves the use of Eqns (11), (14b), (15), and (16a–c). Estimation of the parameters K and θ, given by Eqns (11) and (14b) at every routing time interval, involves the variables Q_3, c_3, and $(\partial Q/\partial y)_3$. One of the important parameters in flood routing process is the celerity at which the flood wave travels along the river reach downstream. The average celerity of a flood wave can be estimated as the average travel time of the flood peaks of the hydrographs recorded at either end of a reach (Wong

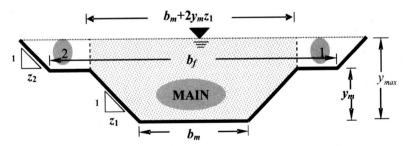

Fig. 4 Compartmentalization of the compound channel section into a main channel (shade) and two floodplains (1 and 2) for celerity computation.

and Laurenson, 1983, 1984). Celerity, corresponding to any discharge, Q can also be estimated using the rating curve at a particular cross-section as:

$$c = \frac{\partial Q}{\partial A} = \frac{\partial Q}{\partial y} \bigg/ \frac{\partial A}{\partial y} \quad (17)$$

Alternatively, Q can be estimated analytically by Eqn. (4).

Note that the wave celerity relationship is not unique during unsteady flow due to differing relationships established when the flood is in the rising and falling stages. Following Eqn. (21), it is expected that the celerity versus stage relationship to have a discontinuity at the intersection of the full-bank flood level (corresponding to the top width of the main channel cross-section) and the bottom width of the floodplain cross-section, due to sudden increase of the wetted perimeter.

4.2.1 Celerity-Stage Relationship for the Main Channel

When the depth of flow in the channel ranges between zero and y_m (see Fig. 4), the unsteady flow corresponds to the main channel flow traversing within the simple trapezoidal section. The celerity of flow at any section of the main channel flow reach can be expressed using Eqn. (4) as:

$$c = \left[\frac{5}{3} - \frac{2}{3} \frac{R_{main}(\partial P_{main}/\partial y)}{(\partial A_{main}/\partial y)} \right] \left(\frac{Q_{main}}{A_{main}} \right) \quad (18)$$

where Q_{main} is the discharge at the section where celerity is computed; and

$$R_{main} = \frac{A_{main}}{P_{main}} \quad (19a)$$

$$A_{main} = b_m y + z_1 y^2 \quad (19b)$$

$$P_{main} = b_m + 2y\sqrt{1+z_1^2} \quad (19c)$$

$$\frac{\partial A_{main}}{\partial y} = b_m + 2yz_1 \quad (19d)$$

$$\frac{\partial P_{main}}{\partial y} = 2\sqrt{1+z_1^2} \quad (19e)$$

The longitudinal gradient of water depth $\partial y/\partial x$ can be estimated using backward difference scheme as:

$$\frac{\partial y}{\partial x} = \frac{y_d - y_u}{\Delta x} \quad (19f)$$

where y_u and y_d, respectively, correspond to flow depths, at any instant of time, at the upstream and the downstream of the conceptual reach length Δx.

4.2.2 Celerity-Stage Relationship for the Floodplain Channel

When the flow depth exceeds y_m, the flow occupies the floodplain. For the estimation of celerity under this condition, the flow is compartmentalized into flow in the main channel and in the two symmetrical compartments above bank-full level as shown in Fig. 4. It may be noted that the main channel flow section above the bank-full level corresponds to a rectangular section having the width corresponding to the top-width of the main trapezoidal channel section. Accordingly, the compound channel discharge encompassing the flow in the floodplain is expressed as follows:

$$Q_{compound} = Q_{main} + Q_1 + Q_2 \tag{20a}$$

where Q_{main} is the flow in the main channel, and Q_1 and Q_2, respectively, are the flow components of the floodplain channel compartments 1 and 2. Equation (20a) may be written in terms of the flow area and velocity of respective compartmentalized sections as:

$$Q_{compound} = A_{main} v_{main} + A_1 v_1 + A_2 v_2 \tag{20b}$$

where v_{main}, v_1, and v_2 are the velocities in the main channel section, and in the floodplain channel compartments 1 and 2, respectively.

Using Eqn. (17), the wave celerity for the compound section may be expressed as:

$$c_{compound} = \frac{\partial Q_{compound}}{\partial A_{compound}} = \left(\frac{\partial Q_{compound}}{\partial y}\right) \bigg/ \left(\frac{\partial A_{compound}}{\partial y}\right) \tag{21}$$

The expression for $(\partial Q_{compound})$ may be estimated from Eqn. (20a) as:

$$\frac{\partial Q_{compound}}{\partial y} = \frac{\partial Q_{main}}{\partial y} + \frac{\partial Q_1}{\partial y} + \frac{\partial Q_2}{\partial y} \tag{22}$$

Each of the derivatives of the right hand side of Eqn. (22) may be expressed using Eqn. (13a) as

$$\frac{\partial Q_{main}}{\partial y} = \left[\frac{5}{3}\frac{\partial A_{main}}{\partial y} - \frac{2}{3}\frac{A_{main}}{P_{main}}\frac{\partial P_{main}}{\partial y}\right] v_{main} \tag{23a}$$

Similarly, the second and the third terms of the Eqn. (22) can be expressed as:

$$\frac{\partial Q_1}{\partial y} = \left[\frac{5}{3}\frac{\partial A_1}{\partial y} - \frac{2}{3}\frac{A_1}{P_1}\frac{\partial P_1}{\partial y}\right] v_1 \tag{23b}$$

$$\frac{\partial Q_2}{\partial y} = \left[\frac{5}{3}\frac{\partial A_2}{\partial y} - \frac{2}{3}\frac{A_2}{P_2}\frac{\partial P_2}{\partial y}\right] v_2 \tag{23c}$$

where the velocities of unsteady flow in different compartments of the compound channel section can be expressed using the Manning's friction law and the momentum Eqn. (2), after neglecting the acceleration terms, as follows:

$$v_{main} = v_{main,0} \sqrt{1 - \frac{1}{S_o}\frac{\partial y}{\partial x}} \; ; \; v_1 = v_{1,0} \sqrt{1 - \frac{1}{S_o}\frac{\partial y}{\partial x}} \; ; \text{ and}$$

$$v_2 = v_{2,0} \sqrt{1 - \frac{1}{S_o}\frac{\partial y}{\partial x}} \tag{23d,e,f}$$

where

$$v_{main,0} = \frac{\sqrt{S_o}}{n}\left(\frac{A_{main}}{P_{main}}\right)^{2/3} \; ; \; v_{1,0} = \frac{\sqrt{S_o}}{n}\left(\frac{A_1}{P_1}\right)^{2/3} \; ; \text{ and } v_{2,0} = \frac{\sqrt{S_o}}{n}\left(\frac{A_2}{P_2}\right)^{2/3} \tag{23g,h,i}$$

Substituting Eqns. (23a–f) in Eqn. (22) leads to

$$\frac{\partial Q_{compound}}{\partial y} = \left[\frac{5}{3}\frac{\partial A_{main}}{\partial y} - \frac{2}{3}\frac{A_{main}}{P_{main}}\frac{\partial P_{main}}{\partial y}\right]v_{main,0}\sqrt{1-\frac{1}{S_o}\frac{\partial y}{\partial x}}$$

$$+ \left[\frac{5}{3}\frac{\partial A_1}{\partial y} - \frac{2}{3}\frac{A_1}{P_1}\frac{\partial P_1}{\partial y}\right]v_{1,0}\sqrt{1-\frac{1}{S_o}\frac{\partial y}{\partial x}}$$

$$+ \left[\frac{5}{3}\frac{\partial A_2}{\partial y} - \frac{2}{3}\frac{A_2}{P_2}\frac{\partial P_2}{\partial y}\right]v_{2,0}\sqrt{1-\frac{1}{S_o}\frac{\partial y}{\partial x}} \qquad (24)$$

Substituting Eqn. (24) in Eqn. (21), the celerity of the flow when it exceeds the main channel section may be modified as (Perumal et al., 2007; Sahoo, 2007):

$$c_{compound} = \left[\left(\frac{5}{3}\frac{\partial A_{main}}{\partial y}\right)v_{main,0}\sqrt{1-\frac{1}{S_o}\frac{\partial y}{\partial x}}\right]\Big/\left[\frac{\partial A_{compound}}{\partial y}\right]$$

$$+ \left[\left(\frac{5}{3}\frac{\partial A_1}{\partial y} - \frac{2}{3}\frac{A_1}{P_1}\frac{\partial P_1}{\partial y}\right)v_{1,0}\sqrt{1-\frac{1}{S_o}\frac{\partial y}{\partial x}}\right]\Big/\left[\frac{\partial A_{compound}}{\partial y}\right]$$

$$+ \left[\left(\frac{5}{3}\frac{\partial A_2}{\partial y} - \frac{2}{3}\frac{A_2}{P_2}\frac{\partial P_2}{\partial y}\right)v_{2,0}\sqrt{1-\frac{1}{S_o}\frac{\partial y}{\partial x}}\right]\Big/\left[\frac{\partial A_{compound}}{\partial y}\right] \qquad (25)$$

where

$$A_{compound} = (b_m + y_m z_1)y_m + \left(b_f + z_2(y-y_m)\right)(y-y_m) \qquad (26a)$$

$$\frac{\partial A_{compound}}{\partial y} = b_f + 2(y-y_m)z_2 \qquad (26b)$$

$$A_{main} = (b_m + y_m z_1)y_m + (b_m + 2y_m z_1)(y-y_m) \qquad (26c)$$

$$P_{main} = b_m + 2y_m\sqrt{1+z_1^2} \qquad (26d)$$

$$\frac{\partial P_{main}}{\partial y} = 0 \qquad (26e)$$

$$A_1 = A_2 = 0.5\left((b_f - b_m - 2y_m z_1) + z_2(y-y_m)\right)(y-y_m) \qquad (26f)$$

$$P_1 = P_2 = 0.5(b_f - b_m - 2y_m z_1) + (y-y_m)\sqrt{1+z_2^2} \qquad (26g)$$

$$\frac{\partial A_1}{\partial y} = \frac{\partial A_2}{\partial y} = 0.5(b_f - b_m - 2y_m z_1) + z_2(y-y_m) \qquad (26h)$$

$$\frac{\partial P_1}{\partial y} = \frac{\partial P_2}{\partial y} = \sqrt{1+z_2^2} \qquad (26i)$$

4.3 Routing Procedure

The procedure described below is adopted while routing using the VPMS method.

The stage at the outlet of the reach is estimated using the recursive Eqn. (15). The parameters K and θ vary at every routing time interval in a two-step process: In the first step, by estimating the unrefined stage estimate $y_{d,j+1}$ for the current routing time interval using the values of K and θ, estimated at the previous time step and, subsequently, using this estimate and $y_{u,j+1}$, the flow depth at the middle of the reach is estimated as

$$y_M = (y_{u,j+1} + y_{d,j+1})/2 \tag{27}$$

The initial values of K and θ are estimated using the initial steady flow in the reach.

Similarly, the depth of flow at section 3 of Fig. 3 is computed as:

$$y_3 = \theta\, y_{u,j+1} + (1-\theta)\, y_{d,j+1} \tag{28}$$

Using y_M, given by Eqn. (27), the discharge at section 3 is computed in the following manner depending on whether y_M is within the main channel section or in the compound channel section:

$$Q_3 = \begin{cases} \dfrac{A_M}{n}\left(\dfrac{A_M}{P_M}\right)^{2/3} S_0^{1/2}, & \text{when } y_M \leq y_m \\ [A_{main}\, v_{main}]_M + [A_1 v_1]_M + [A_2 v_2]_M, & \text{when } y_M > y_m \end{cases} \tag{29a,b}$$

where $A_{main,M}$, $A_{1,M}$ and $A_{2,M}$ are evaluated at the midsection of the reach using Eqns (26c) and (26f), respectively, and

$$v_{main,M} = \frac{\sqrt{S_o}}{n} R_{main,M}^{2/3};\ v_{1,M} = \frac{\sqrt{S_o}}{n} R_{1,M}^{2/3};\text{ and } v_{2,M} = \frac{\sqrt{S_o}}{n} R_{2,M}^{2/3} \tag{30a,b,c}$$

When $y_3 \leq y_m$, the wave celerity c_3 is estimated by using Eqn. (18) corresponding to y_3; and when $y_3 > y_m$, the wave celerity c_3 corresponding to flow depth y_3 of the compound channel section is estimated by using Eqn. (25).

Corresponding to the estimated values of y_M and y_3, obtained using Eqns (27) and (28), respectively, the value of $(\partial Q/\partial y)_3$ is estimated using Eqns (13b), (18) and (19a–f) or (24) and (26a–i) depending on whether y_M is within the main channel section or extends into the floodplain channel section, respectively. In the second step, using these values of Q_3, c_3 and $(\partial Q/\partial y)_3$, the refined values of K and θ are estimated using Eqns (11) and (14b), respectively, for the current routing time interval, which are, then, used to estimate the refined stage-hydrograph estimate using Eqns. (15) and (16a–c).

5. APPLICATION OF VPMS MODEL FOR REAL-TIME FLOOD FORECASTING

The variable parameter Muskingum stage routing model applicable for forecasting purposes is written by modifying Eqn. (15) as (Perumal et al., 2009a):

$$\hat{y}_{d,(j\Delta t + T_L)} = C_1 y_{u,j\Delta t} + C_2 y_{u,j\Delta t} + C_3 \hat{y}_{d,((j-1)\Delta t + T_L)} + e_{est,(j\Delta t + T_L)} \tag{31}$$

where $j\Delta t$ is the time of forecast, $\hat{y}$ denotes the forecasted stage values and T_L is the forecasting lead time. The minimum T_L is Δt, the routing time interval at which the stage measurements are made, and this corresponds to one time interval ahead of forecast. The maximum lead time interval that can be adopted depends on the accuracy of the obtained forecast. The larger the T_L, the poorer would be the accuracy of the forecast.

In order to apply the VPMS routing method in a river reach for real-time flood forecasting purposes, an error updating model also needs to be developed for estimating the forecast error, which when added to the model estimated forecast for a given lead time would yield the final forecasted flow at the site of interest. It is proposed to use a second-order linear autoregressive error updating model of the following form for forecasting the error at time $(j\Delta t+T_L)$ (Perumal et al., 2009a):

$$e_{est,(j\Delta t+T_L)} = a_1 e_{obs,j\Delta t} + a_2 e_{obs,(j-1)\Delta t} + \varepsilon_{(j\Delta t+T_L)} \tag{32}$$

where $e_{obs,j\Delta t}$ and $e_{obs,(j-1)\Delta t}$ are the forecasting errors estimated at time $j\Delta t$ and $(j-1)\Delta t$, respectively, and $\varepsilon_{(j\Delta t+T_L)}$ is the random error (white noise). However, the flow depth forecasting can be made only after the lapse of certain initial period of the forecasting event, known as the warm up period. The difference between the observed stage and the VPMS routed stage in the warm up period is considered as the actual error and its series is assumed to be stochastic in nature. The initial parameters a_1 and a_2 of the error update model are assessed using this error series estimated in the warm up period. The duration of initial warm up period considered for developing the error update model should not be long to render the forecasting exercise to be of no practical use for forecasting the given event, and, at the same time, it should not be too short resulting in numerical problem while estimating the parameters a_1 and a_2 using the least squares approach. The parameters a_1 and a_2 are updated in real-time on the basis of the last available observations.

6. CASE STUDY AND RESULTS

The proposed forecasting model consisting of the VPMS routing method as the basic model, and the second-order linear autoregressive model as the error updating model, is applied for forecasting flow in a 15 km reach along the Tiber River of Central Italy (Fig. 5). The selected reach is bounded by Pierantonio and Ponte Felcino gauging stations and has an average bed slope (S_o) of 0.0016. This average bed slope is estimated from the elevation difference of the two gauging stations considered herein.

Note that the approximation of the VPMS method for routing a given stage-hydrograph in a river reach requires the use of an equivalent prismatic channel reach; this involves the approximation of the actual river reach sections at the two ends to an equivalent prismatic section with a one-to-one relationship established between the flow depth of the actual section of a given flow area with the corresponding flow depth of the prismatic channel section of the same flow area. Based on the surveyed cross-sections at the ends of the actual river reach, it was considered appropriate to approximate the actual reach by a compound trapezoidal section reach. Accordingly, the surveyed cross-sections of the actual reach were overlapped and a two-stage trapezoidal compound section geometry with $b_m = 25$ m, $y_m = 5$ m, $b_f = 59.5$ m and $z_1 = z_2 = 2.5$ (see Fig. 6) as required for the prismatic channel reach conceptualization of the VPMS routing method was finalized by a trial and error approach by fitting the best relationships between the actual flow depths and the equivalent trapezoidal section ones as: $(y_{u-trap} = 0.8887 y_{u-actual} + 0.11)$ for Pierantonio section and $(y_{d-trap} = 1.0582 y_{d-actual} - 0.1308)$ for Ponte Felcino site. y_{u-trap} and y_{d-trap} are the equivalent upstream and downstream flow depths in the trapezoidal channel section corresponding to the flow depths $y_{u-actual}$ and $y_{d-actual}$ in the actual river section. Using the upstream section relationship, the observed stage hydrograph of any event was converted to equivalent trapezoidal section stage hydrograph to enable the routing using the VPMS method and, using the relationship $(y_{d-actual} = 0.945 y_{d-trap} + 0.1236)$, developed on the basis of the downstream relationship, the routed hydrographs of the equivalent trapezoidal section was converted to the actual end section estimated hydrograph.

For studying the applicability of the proposed forecasting model, 12 flood events recorded concurrently at Pierantonio and Ponte Felcino stations were used. The details of these events, each recorded at half-

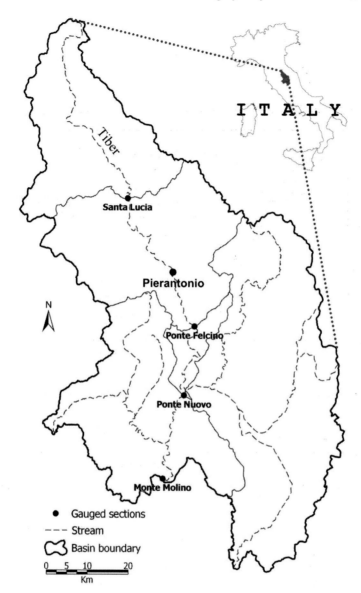

Fig. 5 Index map of the upper Tiber River in Central Italy.

an-hour intervals, are shown in Table 1, where also the details of wave travel time, percentage of lateral flow and actual and equivalent trapezoidal peak flow depths at both stations are reported. The accuracy of the proposed forecasting model was studied using a warm up period of five hours and considering five forecast lead times (1.0, 1.5, 2.0, 2.5 and 3.0 hours). The efficiency of the forecast was evaluated using two criteria: Nash-Sutcliffe (NS) criterion (Nash and Sutcliffe, 1970) and Persistence criterion (PC). As the NS criterion is well known in hydrological literature, only the Persistence criterion is explained herein. The Persistence criterion compares the prediction of the proposed model against that obtained by the no-model, which assumes steady state over the forecasting lead time, and is evaluated as follows:

$$PC = \left(1 - \frac{\sum_i (y_{i\Delta t} - \hat{y}_{i\Delta t})^2}{\sum_i (y_{i\Delta t} - y_{(i\Delta t - T_L)})^2}\right) \times 100 \tag{33}$$

where y and $\hat{y}$ denote the observed and the forecasted flow depth values, respectively.

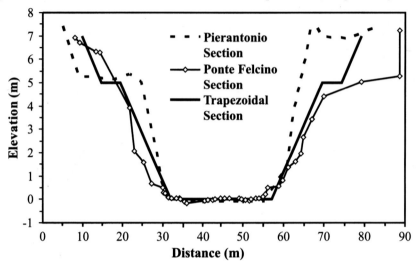

Fig. 6 Cross sections of the upper Tiber River at Pierantonio (upstream) and Ponte Felcino (downstream) gauging stations.

Table 1. Pertinent characteristics of the flood events studied

Flood event	Wave travel time (h)	Lateral inflow (%)	Pierantonio section		Ponte Felcino section	
			Actual peak stage (m)	Equivalent trapezoidal peak stage (m)	Actual peak stage (m)	Equivalent trapezoidal peak stage (m)
December 1996	1.50	1.90	4.74	4.32	4.22	4.33
April 1997	1.50	6.50	5.07	4.62	4.57	4.70
November 1997	1.00	5.40	4.22	3.86	3.81	3.90
February 1999	2.00	4.40	5.06	4.61	4.52	4.65
December 1999	0.00	24.70	2.71	2.52	2.79	2.82
December 2000	2.00	Flooding	5.92	5.37	5.25	5.42
April 2001	2.00	0.20	3.68	3.38	3.23	3.29
November 2005	2.5	Flooding	7.10	6.42	6.92	7.19
3rd December 2005	1.00	3.60	5.10	4.64	4.42	4.55
5th December 2005	1.00	5.70	5.49	4.99	4.76	4.91
30th December 2005	2.00	1.90	4.99	4.54	4.34	4.46
February 2006	1.50	28.40	2.28	2.14	2.64	2.66

Tables 2 to 6 show the forecasting results for peak flow depth forecast at Ponte Felcino station for all the selected flood events and for all the investigated lead times. The results also include the accuracy of peak stage reproduction, error in time to peak stage, Nash-Sutcliffe (NS) efficiency and Persistence criterion (PC) efficiency. Some of the floods studied herein are characterized by the flooding events (December 2000 and November 2005) with flow spilled over the main channel almost in the entire stretch of the reach and also received unaccounted lateral flow.

Figures 7 to 10 show some typical forecasted events for various lead times. The given inflow hydrograph and the corresponding observed outflow hydrograph are also shown in these figures. It is inferred from the results given in Tables 2 to 6 that up to a lead time of 3.0 h, only two flood events (December 1999 and February 2006) could not be successfully forecasted as reflected by their PC estimates (<50%). These two events are characterized by significant lateral flows (>25% of inflow hydrograph volume). As the proposed forecasting model has been developed using the assumption of no lateral flow

Table 2. Results of the forecasting model for a lead time of one hour (err_y_{peak} = percentage error in peak stage; err_t_{peak} = error in time to peak stage)

Event	err_y_{peak} (%)	err_t_{peak} (h)	NS (%)	PC (%)
December 1996	0.08	−1.50	99.82	93.80
April 1997	−0.20	−0.50	99.95	97.80
November 1997	0.97	−3.00	99.87	96.15
February 1999	−0.77	−0.50	99.90	96.59
December 1999	1.95	1.00	99.79	78.68
December 2000	−0.64	0.50	99.80	90.11
April 2001	−0.61	0.50	99.67	95.66
November 2005	0.06	0.00	99.87	90.54
3rd December 2005	−1.29	1.00	99.74	95.26
5th December 2005	−0.17	0.50	99.80	93.66
30th December 2005	0.02	0.00	99.91	92.60
February 2006	1.50	1.00	99.62	81.56
Mean absolute value	0.69	0.83	99.81	91.87

Table 3. As for Table 2, but for a lead time of 1.5 hours

Event	err_y_{peak} (%)	err_t_{peak} (h)	NS (%)	PC (%)
December 1996	0.53	−1.00	99.70	95.33
April 1997	−0.77	0.00	99.85	97.01
November 1997	1.86	−2.50	99.79	97.13
February 1999	−0.27	0.50	99.93	98.94
December 1999	2.53	1.50	99.49	75.67
December 2000	−0.82	−1.00	99.66	92.10
April 2001	1.06	−0.50	99.57	97.44
November 2005	−0.38	0.00	99.66	88.97
3rd December 2005	−0.48	−0.50	98.89	90.59
5th December 2005	0.39	0.00	99.59	94.12
30th December 2005	0.96	0.00	99.87	94.95
February 2006	3.40	0.00	98.86	74.01
Mean absolute value	1.12	0.63	99.57	91.36

Table 4. As for Table 2, but for a lead time of 2.0 hours

Event	err_y_{peak} (%)	err_t_{peak} (h)	NS (%)	PC (%)
December 1996	0.96	−0.50	99.33	94.06
April 1997	−0.31	−2.00	97.38	92.83
November 1997	2.60	−3.50	99.40	95.33
February 1999	0.43	0.50	99.62	96.54
December 1999	2.94	2.00	98.79	66.30
December 2000	−0.12	−8.50	99.30	90.53
April 2001	3.72	0.00	97.79	92.29
November 2005	−0.65	1.00	99.36	88.06
3rd December 2005	1.62	−8.00	95.54	77.98
5th December 2005	1.51	−3.00	98.58	88.34
30th December 2005	1.17	1.50	99.67	92.65
February 2006	5.70	0.50	97.20	62.68
Mean absolute value	1.81	2.58	98.50	86.47

Table 5. As for Table 2, but for a lead time of 2.5 hours

Event	err_y_{peak} (%)	err_t_{peak} (h)	NS (%)	PC (%)
December 1996	3.82	−4.50	97.81	87.22
April 1997	0.77	−1.50	97.93	84.36
November 1997	5.78	−4.00	98.17	90.65
February 1999	3.16	−4.00	98.18	89.21
December 1999	3.63	2.50	97.04	45.34
December 2000	3.95	−8.50	98.17	83.65
April 2001	7.92	−1.00	90.73	78.61
November 2005	−0.94	2.00	98.86	86.05
3rd December 2005	7.74	−7.50	86.25	54.76
5th December 2005	5.87	−4.00	94.66	71.37
30th December 2005	1.75	0.50	98.88	83.90
February 2006	7.81	1.50	94.07	47.23
Mean absolute value	4.43	3.46	95.90	75.20

in the considered reach, it is expected that the efficiency of the model would be poorer in forecasting the flow depth when that event is associated with significant lateral flow. Though the error update model can, to some extent, improve the forecasts in the event of lateral flow, it may not give reliable forecasts when there is significant lateral flow in the reach.

It can be seen from Figs 7 to 10 and from the forecast results of other events (not shown herein) that for almost all the events studied the update error model overestimates the forecast error when the rate of increase of rising limb suddenly decreases resulting in increased forecast error around this time zone. The minimum PC estimated for the forecasted events is greater than 60%, except for three events (December 1999, 3rd December 2005 and February 2006) out of which two events are characterized by significant lateral flow.

Overall, the results presented herein show that the hydrometric data-based VPMS model can be efficiently used as the forecasting model for practical river engineering problems. The other advantages

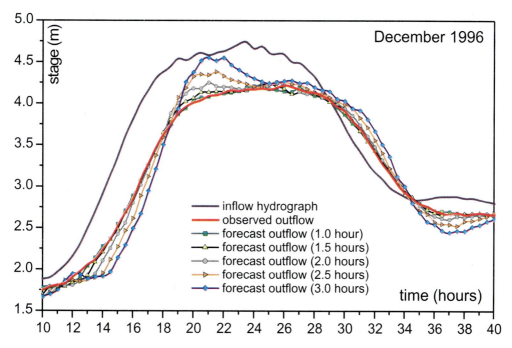

Fig. 7 December 1996 event: comparison between the observed and forecasted stage hydrographs for different lead times at Ponte Felcino section. The input stage hydrograph at Pierantonio site is also shown.

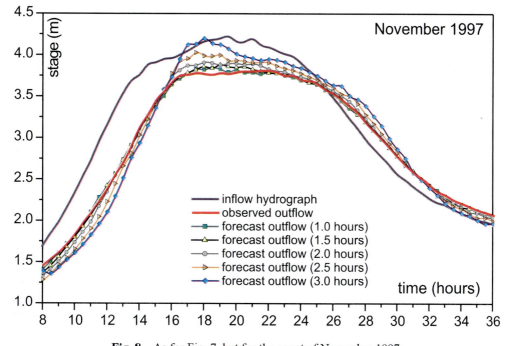

Fig. 8 As for Fig. 7, but for the event of November 1997.

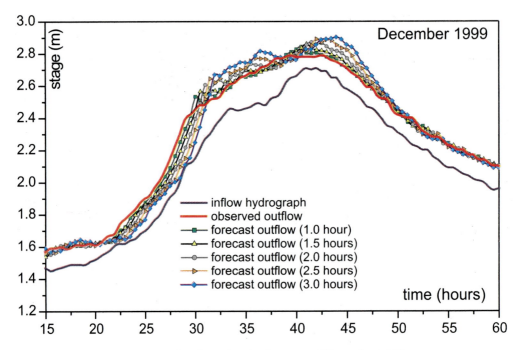

Fig. 9 As for Fig. 7, but for the event of December 1999.

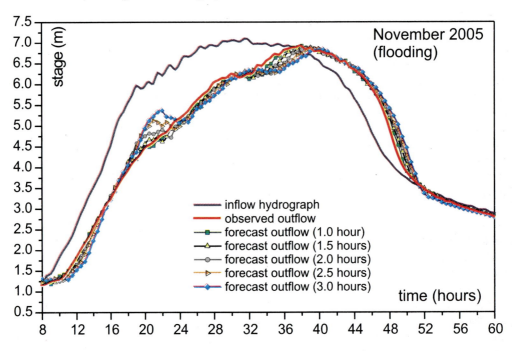

Fig. 10 As for Fig. 7, but for the event of November 2005.

Table 6. As for Table 2, but for a lead time of 3.0 hours

Event	err_y_{peak} (%)	err_t_{peak} (h)	NS (%)	PC (%)
December 1996	7.79	−5.00	94.74	78.10
April 1997	2.65	−7.50	95.48	75.72
November 1997	10.09	−3.50	96.26	86.55
February 1999	11.26	−3.50	95.27	80.04
December 1999	4.02	4.50	95.87	45.41
December 2000	8.79	−8.50	96.23	75.88
April 2001	13.58	−0.50	79.09	65.16
November 2005	−1.22	2.50	98.15	84.07
3rd December 2005	13.06	−6.50	74.75	39.59
5th December 2005	10.42	−3.50	90.50	63.89
30th December 2005	2.46	0.00	97.80	77.68
February 2006	9.85	2.00	90.88	41.20
Mean absolute value	7.93	3.96	92.09	67.77

of this model includes its capability to compute the downstream discharge hydrograph corresponding to the routed or forecasted downstream stage hydrograph, establishment of sectional rating curves, and celerity-stage, celerity-discharge relationships in ungauged and semi-gauged river basins (e.g., Perumal et al., 2007).

7. APPLICABILITY CRITERIA FOR THE VPMS MODEL

Identifying a suitable simplified method for application to a given flood routing problem is a difficult task. Several researchers have attempted to provide criteria for the selection of the appropriate routing methods (Henderson, 1966; Woolhiser and Liggett, 1967; Ponce et al., 1978; Daluz, 1983; Fread, 1985; Ferrick, 1985; Price, 1985; Dooge and Napiorkowski, 1987; Marsalek et al., 1996; Moussa and Bocquillon, 1996; Singh, 1996; Tsai, 2003) for the application to a given routing problem with or without considering any downstream boundary condition. Among these criteria, the one introduced by Ponce et al. (1978) has found its place in standard textbooks (French, 1986; Ponce, 1989; Chaudhry, 1993; Viessman and Lewis, 1996; Singh, 1996). However, the criteria by Ponce et al. (1978) were established on the basis of at least 95% accuracy in the wave amplitude when compared with the dynamic wave after one propagation period. The linear stability analysis used in arriving at these criteria considers the first-order approximation of the shallow water wave propagation which is treated as an infinitesimal disturbance imposed to the initially steady uniform flow. The common features of these criteria include the assumptions of a prismatic channel and a sinusoidal wave of arbitrary amplitude. However, in reality, flood waves found in natural rivers differ significantly from the assumption of sinusoidal shape and also they exhibit nonlinear behavior. Hence, the assumptions behind the development of these criteria are inherently contradictory with the characteristics of real life flood waves.

Ferrick and Goodman (1998) pointed out that large amplitude flow increases of practical interest must be described by the nonlinear equations. Since linear stability theory is valid for small perturbations from the reference flow, and the real world flood waves are frequently very large in amplitude, the linear analysis used in the development of the Ponce et al. (1978) applicability criteria is questionable (Crago and Richards, 2000). Further, Zoppou and O'Neill (1982) tested the criteria of Ponce et al. (1978) for a real life flood routing problem of a 33.2 km reach of the Australian river Yarra between Yarra Grange and Yering, for assessing the applicability of the diffusive and kinematic wave models as approximations to the dynamic wave model. A good agreement was obtained in all cases studied using the kinematic wave model, despite the criteria of Ponce et al. (1978) predicting that it would be unsuitable for routing under these circumstances. On the basis of these considerations, Zoppou and O'Neill (1982) and Perumal and Sahoo (2006) cautioned the river engineers and hydrologists about the limitations of these criteria. In light of subsequent development of improved simplified methods, which have moved from the domain of complete linear models to that of variable parameter models, duly accounting for the non-linear characteristics of a flood wave, this caution seems to have a greater significance. Hence, the applicability criteria advocated by Ponce et al. (1978) for identifying an appropriate flood wave model for a given routing problem may be replaced by alternative criteria with physical significance, given the non-linear mechanism of flood wave propagation in real-world rivers.

Perumal and Sahoo (2007) showed that one of the logical ways of developing these alternative criteria can be by directly incorporating the magnitude of the scaled water profile gradient $(1/S_o)(\partial y/\partial x)$, which is used for the classification of flood waves (Henderson, 1966; NERC, 1975) as kinematic or diffusive. In fact, such an applicability criterion was advocated by Price (1985) for the simplified routing method developed by him, but it is too restrictive with $|(1/S_o)(\partial y/\partial x)| \leq 0.05$. The hydrograph characterized by the presence of $(1/S_o)(\partial y/\partial x)$ signifies a diffusive flood wave and its absence signifies a kinematic flood wave (NERC, 1975). The scaled gradient can be estimated at every routing time level of the given hydrograph at the inlet of the routing reach.

On the basis of an extensive study by Perumal and Sahoo (2007), it is revealed that the applicability of the VPMS model to be assessed at the inlet of the reach for routing a given hydrograph requires satisfying the criteria $(1/S_o)(\partial y/\partial x)_{max} \leq 0.79$ and $(1/S_o)(\partial y/\partial x)_{max} \leq 0.63$ for stage routing and discharge computation, respectively.

8. CONCLUSIONS

River stage forecasting at any downstream ungauged site plays a vital role in a comprehensive and coordinated planning for flood hazard mitigation and evacuation work. The adaptive parameter estimation methods employing the Kalman filtering technique may not be worth the effort for real-time flood forecasting (Ahsan and O'Connor, 1994). In such a scenario, the application of the simplified physically based model like the variable parameter Muskingum stage-hydrograph routing (VPMS) method along with a simple error updating technique, such as the one proposed in this chapter, is much more useful for real-time flood forecasting at a river gauging site. Based on the forecasting performance for different events investigated, it can be inferred that the proposed model has the potential for practical forecasting applications in hydrometric data-based modelling provided that the adopted forecasting lead time is not longer than the mean wave travel time of the selected river reach. Further investigations through different case studies need to be carried out in order to verify the proposed forecasting model accuracy. Furthermore, as a future study, the model formulation presented in this chapter may be extended by accounting for significant lateral flow contribution entering along the selected river reach.

REFERENCES

Abbott, M.B., Bathurst, J.C., Cunge, J., O'Connell, P.E. and Rasmussen, J. (1986a). An introduction to the European Hydrological System: Systeme Hydrologique Europeen, "SHE", 1: History and philosophy of a physically-based, distributed modeling system. *Journal of Hydrology*, **87**: 45-59.

Abbott, M.B., Bathurst, J.C., Cunge, J., O'Connell, P.E. and Rasmussen, J. (1986b). An introduction to the European Hydrological System: Systeme Hydrologique Europeen, "SHE", 2: Structure of a physically-based, distributed modeling system. *Journal of Hydrology*, **87**: 61-77.

Ahsan, M. and O'Connor, K.M. (1994). A reappraisal of the Kalman filtering technique, as applied in river flow forecasting. *Journal of Hydrology*, **161**: 197-226.

Apollov, B.A., Kalinin, G.P. and Komarov, V.D. (1964). Hydrological forecasting (translated from Russian). Israel Program for Scientific Translations, Jerusalem.

Bergstrom, S. (1976). Development and application of a conceptual runoff model for Scandinavian catchment. Swedish Meteorological and Hydrological Institute (SMHI), RHO No. 7, Norrköping, Sweden.

Bergstrom, S. (1992). The HBV Model: Its Structure and Applications. Swedish Meteorological and Hydrological Institute (SMHI), RH No. 4, Norrköping, Sweden, 32 pp.

Bertoni, J.C., Tucci, C.E. and Clarke, R.T. (1992). Rainfall based real-time flood forecasting. *Journal of Hydrology*, **131**: 313-339.

Beven, K.J. and Kirkby, M.J. (1979). A physically based variable contributing area model of basin hydrology. *Hydrological Sciences Bulletin*, **24(1)**: 43-69.

Boughton, W.C. (1993). A hydrograph-based model for estimating the water yield of ungauged catchments. Proceedings of the Hydrology and Water Resources Symposium, June 30-July 2, 1993, Institute of Engineers, Newcastle, Australia, pp. 317-324.

Carroll, D.G. (1992). URBS: The urbanized catchment runoff routing model. Brisbane City Council, Qeensland State, Australia.

Chaudhry, M.H. (1993). Open-Channel Flow. Prentice-Hall, Upper Saddle River, N.J.

Cluckie, I.D. and Ede, P.F. (1985). End-point use of a criterion for model assessment. Proceedings of the 7th IFAC/IFORS Symposium on Identification and System Parameter Estimation, York, U.K.

Crago, R.D. and Richards, S.M. (2000). Non-kinematic effects in storm hydrograph routing. *Journal of Hydrologic Engineering, ASCE*, **5(3)**: 323-326.

Cunge, J.A. (1969). On the subject of a flood propagation method (Muskingum Method). *Journal of Hydraulic Research*, **7(2)**: 205-230.

CWC (2006). Report of the Sub-Group on Flood Management for the XI Five Year Plan (2007-2012). Central Water Commission, Ministry of Water Resources, Government of India, New Delhi, 119 pp.

Daluz, V.J.H. (1983). Conditions governing the use of approximations for the Saint-Venant equations for shallow surface water flow. *Journal of Hydrology*, **60(1)**: 43-58.

Dooge, J.C.I. and Napiorkowski, J.J. (1987). Applicability of diffusion analogy in flood routing. *Acta Geophysica Polonica*, **35(1)**: 66-75.

Dooge, J.C.I., Strupczewski, W.G. and Napiorkowski, J.J. (1982). Hydrodynamic derivation of storage parameters of the Muskingum model. *Journal of Hydrology*, **54(4)**: 371-387.

Ferrick, M.G. (1985). Analysis of river wave types. *Water Resources Research*, **21(2)**: 209-220.

Ferrick, M.G. and Goodman, N.J. (1998). Analysis of linear and monoclinal river wave solutions. *Journal of Hydrologic Engineering, ASCE*, **124(7)**: 728-741.

Fread, D.L. (1985). Applicability Criteria for Kinematic and Diffusion Routing Models: Report. Laboratory of Hydrology, National Weather Service, NOAA, Silver Spring, Md.

French, R.H. (1986). Open-Channel Hydraulics. McGraw-Hill, New York.

Georgakakos, K.P. and Bras, R.L. (1982). Real-time statistically linearized adaptive flood routing. *Water Resources Research*, **20(11)**: 513-524.

Harpin, R. (1982). Real-time Flood Routing with Particular Emphasis on Linear Methods and Recursive Estimation Techniques. Ph.D. Thesis, Department of Civil Engineering, University of Birmingham, Birmingham, UK.

Hayami, S. (1951). On the propagation of flood waves. Bulletin of Disaster Prevention Research Institute, Kyoto University, Japan, Vol. 1, pp. 1-16.

Henderson, F.M. (1966). Open Channel Flow in Flood Routing. Macmillan, New York.

Henry, H.R., Hains, D.K., Burkett, E. and Schoel, W. (1988). Alabama Rainfall-Runoff Model, ARRM. Proceedings of the 3rd Water Resources Operations Management Workshop, Fort Collins, Colorado, pp. 195-209.

Kouwen, N. (1988). WATFLOOD: A micro-computer based flood forecasting system based on real-time weather radar. *Canadian Water Resources Journal*, **13(1)**: 62-77.

Lambert, A.O. (1969). A comprehensive rainfall–runoff model for upland catchment area. *Journal IWES*, **23**: 231-238.

Lambert, A.O. (1972). Catchment models based on ISO functions. *Journal IWES*, **26**: 413-422.

Liang, X., Lettenmaier, D.P. and Wood, E.F. (1996). One-dimensional statistical dynamic representation of subgrid spatial variability of precipitation in the two-layer variable infiltration capacity model. *Journal of Geophysical Research*, **101(D16)**: 21403-21422.

Lighthill, M.J. and Whitham, G.B. (1955). On kinematic waves—I. *Proceedings of Royal Society, Series A*, **229**: 281-316.

Marsalek, J., Maksimovic, C., Zeman, E. and Price, R.K. (editors) (1996). Hydroinformatics Tools for Planning, Design, Operation and Rehabilitation of Sewer Systems. NATO ASI Series, Series 2, Vol. 44. Springer, New York.

Moore, R.J. and Jones, D.A. (1991). A river flow forecasting system for region-wide application. Proceedings of the Ministry of Agriculture, Fisheries and Food (MAFF) Conference of River and Coastal Engineers, Loughboriugh University, Leicestershire, UK, 8-10 July 1991, Lisbon, pp. 1-12.

Moussa, R. and Bocquillon, C. (1996). Criteria for the choice of flood routing methods in natural channels. *Journal of Hydrology*, **186(1)**: 1-30.

Nash, J.E. and Sutcliffe, J.V. (1970). River flow forecasting through conceptual models, Part-1: A discussion of principles. *Journal of Hydrology*, **10(3)**: 282-290.

Natale, L. and Todini, E. (1977). A constrained parameter estimation technique for linear models in hydrology. Mathematical Models for Surface Water Hydrology, Proceedings of the Workshop held at the IBM Scientific Centre, Pisa, Italy. John Wiley and Sons.

NERC (1975). Flood Studies Report. Vol. III. Natural Environment Research Council (NERC), Institute of Hydrology, Wallingford, U.K.

O'Connell, P.E. (editor) (1980). Real-time hydrological forecasting and control. Proceedings of the First International Workshop, 4-29 July 1977. Institute of Hydrology, Wallingford, UK.

O'Loughlin, E.M., Short, D.L. and Dawes, W.R. (1989). Modeling the hydrologic response of catchments to land use change. Proceedings of the Hydrology and Water Resources Symposium, Institution of Engineers Australia, National Conference Publication No. 89/19, pp. 335-340.

Owens, M.D. (1986). Real-time Flood Forecasting using Weather Data. Ph.D. Thesis. Department of Civil Engineering, University of Birmingham, Birmingham, UK.

Perumal, M. (1992). Multilinear Muskingum flood routing method. *Journal of Hydrology*, **133(3-4)**: 259–272.

Perumal, M. (1994a). Multilinear discrete cascade model for channel routing. *Journal of Hydrology*, **158(1-2)**: 135-150.

Perumal, M. (1994b). Hydrodynamic derivation of a variable parameter Muskingum method: 1. Theory and solution procedure. *Hydrological Sciences Journal*, **39(5)**: 431-442.

Perumal, M. (1994c). Hydrodynamic derivation of a variable parameter Muskingum method: 2. Verification. *Hydrological Sciences Journal*, **39(5)**: 443-458.

Perumal, M. and Ranga Raju, K.G. (1998a). Variable-parameter stage-hydrograph routing method. I: Theory. *Journal of Hydrologic Engineering, ASCE*, **3(2)**: 109-114.

Perumal, M. and Ranga Raju, K.G. (1998b). Variable-parameter stage-hydrograph routing method. II: Evaluation. *Journal of Hydrologic Engineering, ASCE*, **3(2)**: 115-121.

Perumal, M. and Ranga Raju, K.G. (1999). Approximate convection diffusion equations. *Journal of Hydrologic Engineering, ASCE*, **4(2)**: 161-164.

Perumal, M. and Sahoo, B. (2006). Evaluation of criteria for the choice of flood routing methods. Proceedings of the 15th Congress of APD-IAHR and the International Symposium on Maritime Hydraulics (ISMH), 7-10 August 2006. Department of Ocean Engineering, Indian Institute of Technology Madras, Chennai, India. Allied Publishers Pvt. Ltd., Chennai, Vol. I, pp. 559-565.

Perumal, M. and Sahoo, B. (2007). Applicability criteria of the variable parameter Muskingum stage and discharge routing methods. *Water Resources Research*, **43(5)**, W05409, doi: 10.1029/2006WR004909, pp. 1-20.

Perumal, M., Moramarco, T., Barbetta, S., Melone, F. and Sahoo, B. (2009a). Real-time flood forecasting using Muskingum stage-hydrograph routing method. Proceedings of the International Conference on Water, Environment, Energy, and Society (WEES-2009), 12-16 January 2009. National Institute of Hydrology, Ministry of Water Resources, Government of India, New Delhi. Allied Publishers Pvt. Ltd., New Delhi, Vol. II (Statistical and System Analysis Techniques), pp. 735-741.

Perumal, M., Moramarco, T., Sahoo, B. and Barbetta, S. (2007). A methodology for discharge estimation and rating curve development at ungaged river sites. *Water Resources Research*, **43(2)**, W02412, doi:10.1029/2005WR004609, pp. 1-22.

Perumal, M., Sahoo, B., Moramarco, T. and Barbetta, S. (2009b). Multilinear Muskingum method for stage-hydrograph routing in compound channels. *Journal of Hydrologic Engineering, ASCE*, **14(7)**: 663-670.

Ponce, V.M. (1989). Engineering Hydrology. Principles and Practices. Prentice-Hall, Upper Saddle River, N.J., USA.

Ponce, V.M. and Yevjevich, V. (1978). Muskingum-Cunge method with variable parameters. *Journal of Hydraulic Division, ASCE*, **104(HY12)**: 1663-1667.

Ponce, V.M., Li, R.N. and Simons, D.B. (1978). Applicability of kinematic and diffusion models. *Journal of Hydraulic Division, ASCE*, **104(3)**: 353-360.

Powell, S.M. (1985). River Basin Models for Operational Forecasting of Flow in Real-Time. Ph.D. Thesis, Department of Civil Engineering, University of Birmingham, Birmingham, UK.

Price, R.K. (1985). Flood routing. In: P. Novak (editor), Developments in Hydraulic Engineering, Chapter 4. Elsevier, New York, pp. 129-173.

Reed, D.W. (1982). Real-time flood forecasting by rainfall/runoff modeling – A case study. Institute of Hydrology, Wallingford, UK.

Refsgaard, J.C., Havno, K., Ammentorp, H.C. and Verwey, A. (1988). Application of hydrological models for flood forecasting and flood control in India and Bangladesh. *Advances in Water Resources*, **11(2)**: 101-105.

Sahoo, B. (2007). Variable Parameter Flood Routing Methods for Hydrological Analyses of Ungauged Basins. Ph.D. Thesis. Department of Hydrology, Indian Institute of Technology Roorkee, Roorkee, India.

Singh, V.P. (1996). Kinematic-Wave Modeling in Water Resources: Surface-Water Hydrology. John Wiley, New York.

Srikanthan, R., Elliott, J.F. and Adams, G.A. (1994). A Review of Real-Time Flood Forecasting Methods. Report No. 94/2, Cooperative Research Centre for Catchment Hydrology, Department of Civil Engineering, Monash University, Australia, 120 pp.

Sugawara, M. (1979). Automatic calibration of the tank model. *Hydrological Sciences Bulletin*, **24(3)**: 375-388.

Todini, E. (1996). The ARNO rainfall-runoff model. *Journal of Hydrology*, **175**: 339-382.

Troch, P.A., de Troch, F.P. and van Erdeghem, D. (1991). Operational flood forecasting on the River Meuse using online identification. *In:* Hydrology for the Water Management of Large River Basins, IAHS Publication No. 201, IAHS, Institute of Hydrology, Wallingford, U.K., pp. 379-389.

Tsai, C.W.S. (2003). Applicability of kinematic, noninertia, and quasi-steady dynamic wave models to unsteady flow routing. *Journal of Hydraulic Engineering, ASCE*, **129(8)**: 613-627.

Vermeulen, C.J.M. and Snoeker, X.C. (1991). SAMFIL version 2.0, a program to simulate the rainfall–runoff process in real-time. SAMFIL Program Description, Delft Hydraulics, Delft, The Netherlands.

Viessman, W. Jr. and Lewis, G.L. (1996). Introduction to Hydrology. Harper Collins, New York.

Wong, T.H.F. and Laurenson, E.M. (1983). Wave speed-discharge relation in natural channels. *Water Resources Research*, **19(3):** 701-706.

Wong, T.H.F. and Laurenson, E.M. (1984). A model of flood wave speed-discharge characteristics of rivers. *Water Resources Research*, **20(12):** 1883-1890.

Woolhiser, D.A. and Liggett, J.A. (1967). Unsteady one-dimensional flow over a plane: The rising hydrograph. *Water Resources Research*, **3(3):** 753-771.

Zhao, R.J., Zuang, Y.L., Fang, L.R., Liu, X.R. and Quan, Z. (1980). The Xinanjiang model. Hydrological Forecasting Symposium-Prévisions Hydrologiques (Proceedings of the Oxford Symposium), 15-18 April 1980. Oxford, IAHS-AISH Publication No. 129, pp. 351-356.

Zoppou, C. and O'Neill, I.C. (1982). Criteria for the choice of flood routing methods in natural channels. Proceedings of the Hydrology and Water Resources Symposium, Melbourne. Institution of Engineers, Barton, A.C.T., pp. 75-81.

Drought Hazards and Mitigation Measures

G. Ravindra Chary*, K.P.R. Vittal, B. Venkateswarlu,
P.K. Mishra, G.G.S.N. Rao, G. Pratibha, K.V. Rao,
K.L. Sharma and G. Rajeshwara Rao

1. INTRODUCTION

Droughts are manifestations of significant shortages in all domains of the water cycle. They have adverse impacts on the environment, water availability and water quality, water supply system, hydropower generation, navigation, groundwater balances, vegetation cover, agricultural production, etc. of the affected region. Drought is a regular part of natural cycles and single-most weather related natural disaster affecting livelihoods, developmental activities, natural resources (water, soil, and biodiversity) and economy of a country (http://en.wikipedia.org/wiki/Drought). Although droughts may persist for several years, even a short, intense drought can cause significant damage and severely affect local economy. This global phenomenon has a widespread impact on agriculture. Indeed drought is one of the most serious problems arising from climate variability for human societies and ecosystems (Yurekli and Kurune, 2006). The occurrence of droughts is not limited to a particular region. It has been observed that their impact has been completely different in developed and developing nations because of several socio-economic and political factors influencing both behavioral and management patterns. Even within the developing countries, the effects of droughts can vary significantly, but the fact remains that the economically weaker countries or groups in a country are most severely affected by the droughts.

According to the Fourth Assessment Report of the Intergovernmental Panel on Climate Change (IPCC, 2007), many parts of the developing world are very likely to warm during this century. South Asia is likely to experience extreme climatic events associated with climate change, such as increasing severity and frequency of floods and droughts, and the length of growing season and yield potential of crops is expected to decrease in areas suitable for agriculture, particularly along the margins of semi-arid and arid areas which are already drought prone. India may experience a rise of temperature of 1°C by the year 2050 (IPCC, 2007; Sivakumar, 2008). By the end of the century, there will be a change in precipitation by 5-25% over India with more reductions in the winter rainfall than the summer leading to droughts during summer months (Lal et al., 2001; Prabhakar and Shaw, 2008).

Drought has been a recurring feature of Indian agriculture. Major parts of India having the probability of three to four-year drought in a ten-year period in which, there is again a probability of getting one or two years moderate and half year to one year severe drought. Due to the variability of climate in recent past, more intense and longer droughts have been observed in wider areas since 1970, particularly in the tropics and subtropics. The rainfed regions encompassing arid, semi-arid and dry sub-humid regions (covering regions having rainfall less than 1150 to 1200 mm) are more prone to climatic variability and drought (Ramakrishna et al., 2007). Rainfed agriculture is practiced in 90 Mha accounting 60% of cultivated area and contributes 44% foodgrain production contributing 91, 91, 80, and 60% of coarse

---
*Corresponding Author

cereals, pulses, oilseeds and cotton, respectively. Besides this, 66% of livestock population is also dependent on rainfed areas. Even with the completion of envisaged river linkage project covering various parts of India, it is estimated that 50% would still remain rainfed (Vittal et al., 2006). On the top of it, the projected changes in temperature and precipitation due to climatic variability will further aggravate the water scarcity problem in already suffering rainfed regions (Lal et al., 2001; Prabhakar and Shaw, 2008). These issues call for concerted efforts in addressing the drought as an opportunity, rather than as a national calamity. A general belief about natural disasters like drought is that they are inevitable, unavoidable and unmanageable. A change in this attitude following a scientific approach would help in mitigating the effects of drought.

This chapter presents, besides the basics of droughts, the impacts of drought and the management of agricultural drought in various agricultural drought regions of India with an emphasis on drought coping practices in season and drought amelioration on a permanent basis. It also addresses issues concerning meteorological and hydrological droughts, for example, efficient rainwater management, conservation of soil moisture, crop and contingency planning, etc. Along with the existing policies and support systems, new approaches or programs to combat droughts in the drought-prone regions are also discussed.

2. DEFINITION OF DROUGHT AND DROUGHT TYPES

Drought is a climatic anomaly, characterized by deficient supply of moisture resulting either from subnormal rainfall, erratic rainfall distribution, higher water need or a combination of all the three factors. In general, drought means different things to different people. To a meteorologist it is the absence of rain, while to the agriculturist it is the deficiency of soil moisture in the crop root zone to support crop growth and productivity. To the hydrologist, it is the lowering of water levels in lakes, reservoirs, etc., while for the city management it may mean the shortage of drinking water availability. Thus, it is unrealistic to expect a universal definition of drought for all the fields (saarc-sdmc.nic.in).

There are several definitions of drought given by various experts (Thronthwaite, 1948; Ramdas, 1960; Van Bavel, 1953; Palmer, 1965). However there is no universally accepted definition of the term. Drought is a multifaceted concept, which defies attempts at precise and objective domain. Most drought definitions are based on meteorological observations, agricultural problems, hydrological conditions and socio-economic considerations (Sharma et al., 2006). Schneider (1996) defined drought as an extended period—a season, a year or several years of deficient rainfall relative to the statistical multiplayer mean for a region. Dracup et al. (1980) considered four variables to define drought appropriately: (i) nature of the water deficit, precipitation, soil moisture or stream flow; (ii) basic time unit of the data, e.g., month, season or year; (iii) truncation which distinguishes low flows from normal/high flows, e.g., mean, median, mode or other derived threshold value; and (iv) regionalization and/or standardization. Thronthwaite (1948) described four types of droughts as *permanent*, *seasonal*, *contingent* and *invisible*. In India, the definition given for meteorological drought by the India Meteorological Department (IMD) is most widely used by Central and State Governments.

It is worth mentioning that drought is different from *famine* and *desertification* in respect of occurrence, duration, types/kinds and extent and impact. According to the Encyclopedia Britannica, "famine is severe and prolonged hunger in a substantial proportion of the population of a region or country, resulting in widespread and acute malnutrition and death by starvation and disease". Famines usually last for a limited time, ranging from a few months to a few years. They cannot continue indefinitely, if for no other reason than that the affected population would eventually be decimated. On the other hand, "desertification is land degradation in the arid, semi-arid and dry sub-humid regions resulting from various factors, including climatic variations and human activities" (UNCCD, 1995). One of the most important reasons for famine could be severe drought. Land degradation on a long-term basis leads to desertification in a region and recurrent droughts in dry regions hasten the process of land degradation resulting in desertification. Although the nature and magnitude of impacts of drought, famine and desertification vary, the reasons of occurrence of these three hazards are intricately linked in a natural ecosystem in a region impacting natural resources and livelihoods of human populations, survival, health and productivity of animal populations, and economic conditions of a nation as a whole.

Based on the criteria, concept of its utilization and different schools of thought, drought is broadly categorized into *meteorological drought, hydrological drought, agricultural drought* and *socio-economic drought*.

Meteorological drought is said to occur when there is prolonged absence or marked deficiency or poor distribution of precipitation over an area (Sarkar et al., 2005). India Meteorological Department (IMD) defines meteorological drought as a situation when the seasonal rainfall received over an area is less than 75% of its long-term average value and further classifies as 'moderate drought' if the rainfall deficit is between 26-50% and 'severe drought' when the deficit exceeds 50% of the normal. Furthermore, India Meteorological Department (IMD) defines *drought week* as a week in which rainfall is less than half of its normal, *drought year* as a year when the annual rainfall is deficient by 25% of normal or more, and *severe drought year* as a year when the annual rainfall is deficient by 50% of normal or more. **Hydrological drought** is said to occur when there is marked depletion of surface water with dried up lakes, rivers, and reservoirs. Often prolonged meteorological drought results in hydrological drought. Linsley et al. (1988) defined hydrological drought as a period during which stream flows are inadequate to supply established uses of water under a given water-management system, while Whipple (1966) defined hydrological drought as one in which the aggregate runoff is less than the long-term average runoff. Low snow accumulation in mountainous areas can lead to hydrological drought in adjacent lowlands.

Agricultural drought refers to an extended dry period in which the lack of rainfall results in insufficient moisture in the crop root zone causing adverse effects on arable crops. In view of the vagaries in monsoon rainfall, agricultural drought can be anticipated to occur at any time within the crop growth period of the rainfed crops. Hence, depending upon the occurrence, five distinct categories of agricultural drought were identified, viz., *early season drought* (due to delay in the commencement of sowing rains), *mid-season drought* (when there are breaks in the southwest monsoon), *late season drought* (due to early cessation of rains, crops sometimes experience moisture stress in the reproductive phase), *permanent drought* (due to inadequacy of soil moisture/rainfall to meet the water requirement of a crop during most of the year) and *apparent drought* (rainfall in a region may be adequate for one crop but may not be adequate for other crops) (Singh and Ramana Rao, 1988). Finally, **socio-economic drought** is said to occur when the social fabric of a society is affected in terms of increase in unemployment, distress migration, discontent, etc., which might have happened due to the disturbed economy of a region caused by extreme meteorological and hydrological droughts (Sarkar et al., 2005).

3. CAUSES OF DROUGHT

The root cause for deficit rainfall that results in drought is the widespread and persistent atmospheric subsidence arising from the general circulation of the atmosphere, ENSO phenomenon, deforestation and degradation. One of the major reasons for droughts has been a strong link with El Nino Southern Oscillation (ENSO) patterns (Gadgil et al., 2003) with the evidence of occurrence of 10 drought years out of 22 during the ENSO period of 1965-1987, while there were only three drought years during 1921-1964 (DAC, 2004). The recent droughts of 2002 and 2004 suggest the inherent vulnerability of the

Indian monsoon system to El Nino phenomenon (Saith and Singh, 2006) and recent research clearly demonstrated the linkage between ENSO and food grain production in India (Selvaraju, 2003; Prabhakar and Shaw, 2008). With a warmer climate, droughts and floods could become more frequent, severe, and longer-lasting. The potential increase in these hazards is a great concern given the stresses being placed on water resources and the high costs resulting from recent hazards (http://www.drought.unl.edu/whatis/cchange.htm). Due to change in climate, the water cycle will change and intensify, leading to changes in water supply as well as flood and drought patterns (http://www.agu.org/sci_soc/policy/climate_change_position.html). The extensive deforestation in vogue since centuries has altered the water cycle, thus, enhancing the incidence of droughts.

Shortage of rainfall coupled with its erratic distribution during rainy season causes severe water deficit conditions resulting in various intensities of droughts. In India, the seasonal rainfall (monsoon rains) over the Indian sub-continent is a global phenomena associated with large-scale hemispherical movement of air masses. Therefore, identification of the major atmospheric phenomenon that influences the monsoon over Indian sub-continent is essential in drought management research. Two such relationships, viz., (i) sea-surface temperature anomaly around the Indian sub-continent in relation to atmospheric circulation, and (ii) large-scale pressure oscillation in atmosphere over southern Pacific Ocean are found to be useful in this context. The El Nino event is one such phenomenon, which has profound influence on the monsoon activity over Indian sub-continent. The Southern Oscillation Index (SOI) is one important parameter in the predictive sixteen parameters model used by IMD for long-range forecasting purposes. The study of the Indian summer monsoon over the country by India Meteorological Department showed that all the drought years are El Nino years whereas all the El Nino years are not drought years indicating thereby that various other factors also equally influence the monsoon over the sub-continent. In this context, the winter circulation over the sub-continent, extended period of occurrence of western disturbances (late in the season), strengthening of heat low over NW India in summer and shifts in zonal cells over India are some of the important parameters that influence monsoon system over the country. Some of the researchers are of the opinion that the sea-surface temperature anomaly in the monsoon path is more important in predicting the monsoon rather than the pressure difference at far off places in the globe. Such studies have been initiated, but definite conclusions are yet to be arrived at. Therefore, successful prediction of monsoon over different parts of the country is still a problem and any progress in this direction will help in forewarning the occurrence of droughts (saarc-sdmc.nic.in).

4. IMPACTS OF DROUGHT

As mentioned earlier, droughts are extreme hydrologic events causing acute water shortages which persist long enough to trigger detrimental effects on human, vegetation, animal and ecosystem over a considerable area. Unlike other natural disasters such as earthquakes, cyclones and floods, droughts don't occur suddenly, but their effects are prolonged and in some situations equally detrimental. In agriculture dominating countries, the effect of prolonged drought is devastating. While the urban populations may withstand drought to some extent, the vast rural populations are badly affected. The severity of the drought depends on its duration, degree of water/moisture deficiency, and the size of the affected area. Droughts have significant consequences in terms of reduction/elimination of agricultural production, effect on energy generation, effect on livestock systems, population migration, and resources required for mitigating the resulting hazards. The impact of drought over a region varies depending on which economic activity is impaired. Since drought affects many economic and social sectors, scores of definitions have been developed by a variety of disciplines and the approaches taken to define it also reflect regional and ideological variations (saarc-sdmc.nic.in).

Drought affects the overall economy of a country at macro (national) and micro (village and household) levels, and its impact may be direct or indirect. The losses due to drought may be physical, social, economic and environmental (Rathore, 2004). The extent and intensity of drought is determined by prevailing economic conditions, the structure of the agricultural sector, management of water resources, food reserves, internal and external conflicts, etc. Entitlements to produce and procure food are largely affected at the micro level. The major impacts would be on agriculture, human and livestock systems. Droughts in India, in the past, have periodically led to major Indian famines, including the Bengal famine of 1770, in which up to one third of the population in affected areas died; the 1876-1877 and 1899 famine over five million and more than 4.5 million people died, respectively. Loss of assets in the form of crop, livestock (mortality, loss in productivity, health and loss in fertility); and productive capital damage as a direct consequence of water shortage or related power cuts; agro-based industries, domestic water availability, health, household activities, etc. are some of the major casualties due to drought. Droughts affect government policies, expenditure on food distribution, food imports, provisions of subsidies, credit to affected productive sectors, law and order services etc., at one side and impact on regional equality, employment, trade deficits, external debt and inflation on the other. The intensification of household food insecurity, water related health risks and loss of livelihoods in the agricultural sector are immediate. Impacts are also regional and on individual families. Further, the value of other crop production loss was 370 million dollars compared to 125 million dollars averaged over 30 years.

Ramakrishna (1997), while discussing about the importance of rainfall variability vis-à-vis successful crop production in the arid regions of India, indicated that the risk involved in successful raising of crops and their choice depends on the nature of drought (chronic and contingent), probable duration of drought, and the periodicity of occurrence within a season. He further added that the impact of drought on food and agriculture is severe because as much as two thirds of the region's agricultural land is rainfed with large portions in arid and semi-arid regions. The effect of drought is more pronounced on fodder availability compared to that of food grains. In the arid regions, where mean annual rainfall is less than 500 mm, drought is almost an inevitable phenomenon during most parts of the year. In semi-arid regions, where mean annual rainfall varies between 500 and 750 mm, drought occurs in 40 to 60% of the years due to deficit seasonal rainfall or inadequate soil moisture availability between two successive rainfall events. Even dry sub-humid regions, where rainfall varies between 750 and 1200 mm, do experience contingent drought situations due to break monsoon conditions. As a result, drought management strategies have to be identified separately for each climatic region and also for each major crop grown in those regions. The effect of droughts on foodgrain production for the last 30 years reflected more in *kharif* season, rice production, while scarcity of green quality fodder affecting animal health, animal draft power, milk production, and further there would be more casualties of animals, particularly small ruminant as the already overgrazed grasslands might have low quality/quantity forage with limited water-points in these grazing areas.

5. DROUGHT FREQUENCY AND AGRICULTURAL DROUGHT REGIONS IN INDIA

Drought is imminent in a vast country like India, which means insufficient soil moisture availability to maintain normal biophysical and chemical phase for supporting plant growth. In India, the historical analysis of drought over past 200 years given by Kulkshetra (1997) indicates that there was an occurrence of five phenomenal droughts that affected more than 47.7% area of the country, of which two occurred during the last quarter of the 20th Century. The occurrence of major droughts since 1801 in India is summarized in Table 1. There were 23 major droughts years (years with rainfall less than one standard deviation below the mean) in India during the period 1871-2005. The frequency of droughts computed based on the rainfall departures over the last 130 years (Fig. 1) indicated that maximum number of droughts were observed in north-western part followed by central parts and the least in the north-east and hilly regions (Ramakrishna et al., 2005).

Analysis of top six severest droughts during 1877 to 2005 in India (Table 2) indicated that the rainfall deficit varied from −19 to −29.1%, whereas the geographical area affected ranged from 49 to 63%. Deficiency in the month of July (crop sowing period) was agronomically more critical for agricultural

Table 1. Reported drought events in India during past 200 years (Kulkshetra, 1997; Samra, 2006)

Period	Drought Years	Period	Drought Years
1801-1825	1801, 1804, 1806, 1812, 1819, 1825	1901-1925	1901*, 1904, 1905*, 1907, 1911, 1918*, 1920
1826-1850	1832, 1833, 1837	1926-1950	1939, 1941*
1851-1875	1853, 1860, 1862, 1866, 1868, 1873	1951-1975	1951, 1965*, 1966, 1971, 1972*, 1974
1876-1900	1877*+, 1891, 1899*+	1975-2005	1979*, 1982, 1985, 1987*+, 1988, 1992, 2002

*Severe drought years = 10 (>39.5% area affected); + Phenomenal drought years = 5 (> 47.7% area affected).

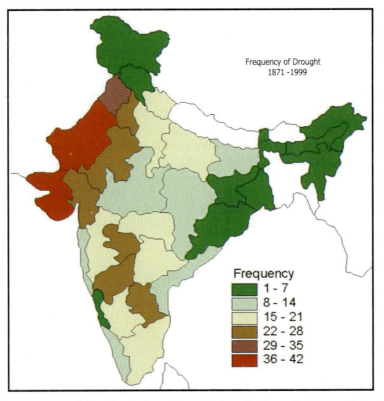

Fig. 1 Frequency of droughts during 1871-1999 in India (Ramakrishna et al., 2005).

Table 2. All India rainfall deficits in the six major drought years in India during 1877-2005 (Samra, 2006)

Features	Drought Years					
	2002	1987	1972	1918	1899	1877
Rainfall deficiency (percentage deviation from normal)	-19.0	-19.0	-23.5	-23.9	-26.8	-29.1
Area under deficient rainfall (%)	56	64	49	68	83	67
Number of meteorological sub-divisions (out of 36 in deficient and scanty categories of rainfall)	21	21	22	25	25	25
Deficiency of rainfall in July (percentage deviation from normal)	-49	-29	-31	-48	-32	-43

production and the deficit was highest in the recent drought of 2002 with severest economic losses (Samra, 2006).

For example, the impact of 2002 drought was such that the water storage in 70 major reservoirs was 33% less than the average of previous 10 years, 22 Mha area was not sown and 47 Mha of the sown area was subsequently damaged and foodgrain production reduced by 29 million tonnes, agricultural GDP was reduced by 3.1% with Rs. 390 (US$9) billion loss of agricultural income, there was a loss of 1250 million person-days employment, and every day about 1.5 billion liters of drinking water was transported by tankers, railways and other means (DAC, 2004). According to the Emergency Database (EM-DAT) of Centre for Research on the Epideomology of Disasters (CRED, 2006), droughts have affected nearly 1,061 million people and killed 4.25 million people in India during 1900-2006.

Agricultural drought is recognized when available soil moisture is inadequate for healthy crop growth causing extreme stress leading to wilting. Thus, drought in simplest terms is insufficiency in precipitation normally required to meet crop productivity and can be a recurring phenomenon across a region (*horizontal drought*) or occurring transient within a season based upon rainfall patterns and dry spells (*vertical drought*). The horizontal and vertical droughts are inseparable. Agricultural drought is a climatic anomaly characterized by deficient supply of moisture in root zone of soil resulting either from sub-normal rainfall, erratic rainfall distribution, higher water need or a combination of all the three factors. This is compounded by poor water holding capacity of soils and land topography. In India, the drought regions can be classified into six regions based on the frequency of drought, climate, rainy season, soil quality and soil orders (Vittal et al., 2003, 2005). They are as follows:

- Region 1: Chronic drought region of arid marginal rainy season - Aridisols.
- Region 2: Chronic drought region of arid sub-marginal rainy season—Vertisols/Alfisols.
- Region 3: Chronic drought region in dry semi-arid delayed rainy season—Vertisols/Alfisols.
- Region 4: Chronic drought region in dry semi-arid post rainy season—Vertic/Vertisols.
- Region 5: Ephemeral drought in wet semi-arid rainy season—Vertisols/Alfisols during early, mid or terminal seasons.
- Region 6: Apparent drought region in dry sub-humid—Alfisols/Oxisols on upland, medium land and low land.

Aridisols, Vertisols/Alfisols, Vertic/Vertisols, Oxisols are the soil orders dominantly occurring in the respective drought regions. The information pertaining to droughts occurring across arid, semi-arid and sub-humid regions, and spatial distribution of agricultural drought regions is shown in Figs 2 and 3, respectively.

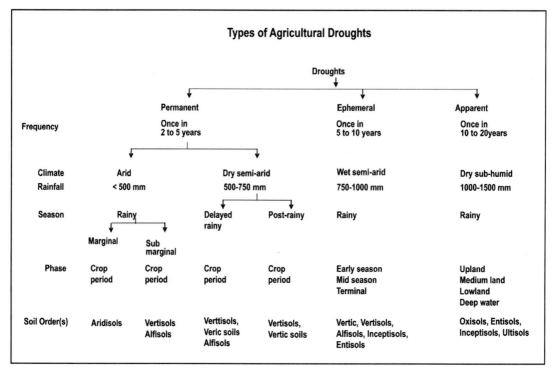

Fig. 2 Types of droughts in rainfed regions in India (Vittal et al., 2003).

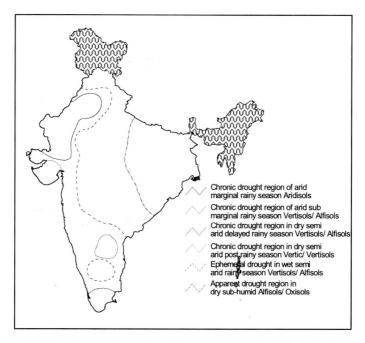

Fig. 3 Spatial distribution of agricultural drought regions in India (Vittal et al., 2003).

The effective length of plant growing period is 12, 19, 20, 22, 31 and 38 weeks with onset of monsoon in 27, 26, 38, 23, 20, and 22 meteorological standard weeks, respectively in the above six regions. Crop production in rainfed areas is generally affected by *chronic* (permanent) *drought* (Southwest monsoon for the *rainy season* region and Northeast monsoon for the *rabi* region with low rainfall), *ephemeral drought* (occurring in early June, July; mid July-August), *terminal drought* (September-early October periods during crop growth in Southwest monsoon season), and *apparent drought* (Southwest monsoon with high rainfall). Horizontal chronic drought usually has a high frequency of occurrence once in less than five years, while the drought in ephemeral regions will be once in 5 to 10 years and once in 10 to 20 years in apparent conditions.

There is a significant difference between chronic and other drought regions. The average annual water balance (computed on monthly basis) of the drought regions comprising precipitation (P), potential evaporation (PE), actual evaporation (AE), water deficit (WD), water surplus (WS), moisture index (MI), mean annual temperature (T) is presented in Table 3.

Identifying periods within a given growing season is important when drought related weather conditions have greatest effect like yield altering impacts. A minimum of 20 mm rainfall per week is a yardstick for meeting the weekly evaporation in rainy and post-rainy seasons and for carrying out agricultural operations in most of the soils. The probability of wet ($\geq$20 mm)/dry ($\leq$ 20 mm) weeks is higher in ephemeral and apparent drought regions. In the chronic region during rainy season, this probability is low but with delayed rainy season (season starting after 31st meteorological standard week during northeast monsoon) the probability increases. In the ephemeral/apparent drought region usually mid season has a higher probability to experience wet and dry spells followed by early season than terminal period. The exception for this case is the chronic drought regions in arid marginal and sub-marginal regions. Thus, the wet and dry weeks affect crop growth in the six drought regions (Table 4).

An operational definition for measuring drought intensity would be one that compares daily precipitation values to evapotranspiration rates to determine the rate of soil moisture diffusion and express these relationships in terms of drought effects on plant behavior at various stages of crop development. Thus, severity of drought is a ratio of actual evapotranspiration (AET) to potential evapotranspiration (PET) during the growing season. The intensity scale of agricultural drought is based on the ratio of AET to PET as given in Table 5.

Table 3. Water balance and mean annual temperature of the drought regions (Vittal et al., 2003)

	Drought region	P (mm)	PE (mm)	AE (mm)	WD (mm)	WS (mm)	MI (%)	T (°C)
1.	Chronic drought in arid marginal rainy season Aridisols (2)	389	1510	389	1121	0	-74	24
2.	Chronic drought in arid sub-marginal rainy season Vertisol/Alfisols (2)	629	2001	642	1359	0	-68	27
3.	Chronic drought in dry semi-arid delayed rainy season Vertisols/Alfisols (1)	659	1817	659	1157	0	-64	-
4.	Chronic drought in dry semi-arid post-rainy season Vertic/Vertisols (3)	594	1729	594	1135	0	-66	28
5.	Ephemeral drought in wet semi-arid Vertisols/Alfisols (6)	813	1490	772	717	39	-45	24
6.	Apparent drought region in dry sub-humid Alfisols/Oxisols (5)	1246	1517	793	723	452	-16	24

Note: Figures in parentheses are the number of regions with real time data (modified from Mandal et al., 1999).

Table 4. Probability of wet or dry weeks (rainfall ≥20 mm) in various drought regions during rainy season (modified from Virmani et al., 1982)

Drought region	Early season	Mid season	Terminal season	Average of rainy season
Chronic drought in arid marginal rainy season Aridisols	0.41	0.37	0.22	0.34
Chronic drought in arid sub-marginal rainy season Vertisol/Alfisols	0.21	0.30	0.34	0.29
Chronic drought in dry semi-arid delayed rainy season Vertisols/Alfisols	0.58	0.55	0.29	0.47
Chronic drought in dry semi-arid post-rainy season Vertic/Vertisols	0.35	0.34	0.36	0.35
Ephemeral drought in wet semi-arid Vertisols/Alfisols	0.61	0.71	0.50	0.60
Apparent drought region in dry sub-humid Alfisols/Oxisols, etc.	0.40	0.59	0.44	0.50

Table 5. Classification of agricultural droughts (modified from Sastry et al., 1981)

Severity of drought	Value of AET/PET
Severe	0.00-0.24
Moderate	0.25-0.49
Mild	0.50-0.74
Low	0.75-0.99
Nil	1.00

6. DROUGHT MANAGEMENT

Drought occurrence is not sudden, hence its mitigation can be planned properly. With the available knowledge and gradually emerging tools and techniques, the effects of droughts can largely be counteracted in many situations, if not eliminated completely. It is rightly said that 'drought is nature's guilt, but the human suffering from drought is man-made'. Effective drought preparedness and management is planning and response process to predict drought and establish timely and appropriate responses to mitigate negative consequences of the drought on lives, livelihoods and environment. The drought risk management may consist of four stages: *preparedness*, *mitigation*, *relief* and *rehabilitation*. A range of administrative and technical measures is needed in dealing with droughts before they occur or when they are in progress. The Indian farmers of diverse agro-climatic regions evolved numerous strategies to live with droughts that include water management, crop management, and contingency crop planning and alternate enterprises. Some of the indigenous rainwater harvesting systems already in vogue in India are *khadins* (Aridisols of Jaisalmer, western Rajasthan), *nadi* system (Inceptisols of Bhilwara, Rajasthan), percolation tanks (Alfisols of Andhra Pradesh and Karnataka), etc.

Scientific management would help in mitigating the effects of drought which are recurrent and mainly affecting the agricultural production and productivity, employment, income and livelihoods of the people in these regions, livestock health, agro-industries, natural resources like water (groundwater, surface water, soil moisture and many water bodies), productive soils and their quality, land degradation and land use/land cover, etc. Thus, droughts influence country's finite natural resources, economy, people's food security, etc. To overcome the adverse effects of droughts, the effective drought management practice is the need of the hour. Magnitude of agricultural drought depends on the type of crops, soil variability, temporal and spatial variability of rainfall, delay in operations in agriculture, and multiplicity in crop management practices. More seasoned and progressive farmers feel that the changes in farming technology may greatly change the nature of future drought impacts, but they do not feel that drought has been overcome.

The earlier studies in India suggest that the productivity of rainfed regions can be enhanced significantly on a sustainable basis, provided the two basic natural resources—*soil* and *rainwater* are well managed. These practices insulate crops against the mild stress and increase crop yields. Above all, drought planning must be viewed as a dynamic process requiring a continued attention and can be tackled in a dual manner by: (i) drought coping practices in a season based on drought intensity, and (ii) drought amelioration on a permanent basis.

6.1 Drought Coping Practices in a Season Based on Drought Intensity

Drought management practices to match the scales of intensity of drought in various drought regions across arid, semi-arid and sub-humid regions are adopted in India (Vittal et al., 2003). Some of the practices are discussed below.

6.1.1 Drought Region 1 (Chronic Drought in Arid Marginal Rainy Season Aridisols)

A drought prone area is defined as one in which the probability of a drought year is greater than 20%. A chronic drought prone area has a possibility of 40%. A drought year occurs when rainfall is less than 75% of the normal rainfall received. Chronic droughts occur in the arid and semi-arid region. In arid region of Western plains, Kutch and part of Kathiawar peninsular hot eco region, the length of growing period (moisture availability period) is mostly less than 60 days. Under assured conditions with deep loamy soils the length of growing period is 60 to 90 days in the hard hyper sandy desert soil. The crops are mostly grown in the deep loamy soils. The low water requirement crops may be cultivated in these regions so that the rainfall is sufficient enough to support the crop. If a high water requirement crop like groundnut or maize is cultivated in this region, it may encounter drought. Some management practices like soil and water conservation may be adopted to achieve higher yields (Fig. 4).

6.1.2 Drought Region 2 (Chronic Drought Region in Arid Sub-marginal Rainy Season Vertisols and Alfisols)

The chronic drought also occurs in the Southwest and Northeast transitional zone of rainfall in Karnataka Plateau (Rayalaseema is included) with Alfisols having low to medium available water capacity and 60-90 days of length of growing period. In dry semi-arid region of Karnataka Plateau, the crops are taken in the *post-rainy season* under receding soil moisture. The crops of the region are groundnut, castor, pearl millet etc. Groundnut is a preferred crop over others due to its commercial value. However, pearl millet matches the length of growing period of the region (Fig. 5).

6.1.3 Drought Region 3 (Chronic Drought Region in Dry Semi-arid Delayed Rainy Season Vertisols and Alfisols)

The *post-rainy season* region occurs mostly in hot semi-arid Deccan Plateau, the available water holding capacity is medium to high with length of growth period varying from 90 to 180 days based on the clay content of soils. In the dry semi-arid region of late *rainy season* and *post-rainy season* Vertisols, the

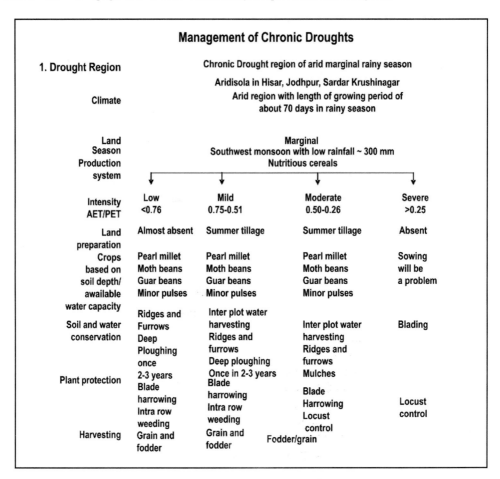

Fig. 4 Management of chronic droughts in the arid marginal region (Vittal et al., 2003).

crop, soil water conservation measures, pest management and harvesting are presented in Fig. 6 based on available water capacity of the profiles.

6.1.4 Drought Region 4 (Chronic Drought in Dry Semi-arid Post Rainy Season Vertic/Vertisols)

In semi-arid Vertisols of down south (Tamil Nadu) uplands and Deccan Plateau, in the late *rainy season*, the available water capacity varies from low to medium. The length of growing period varies from 90 to 150 days depending on the soil depth and clay content. Crops of this region are cotton, pearl millet and sorghum. Minor pulses are gaining importance. Higher yields can be achieved by adopting superior management practices (Fig. 7).

6.1.5 Drought Region 5 (Ephemeral Drought in Wet Semi-arid Rainy Season Vertisols/Alfisols)

Ephemeral drought is usually limited to wet semi-arid regions during southwest monsoon. The regions covered include Northern plains and central high lands including Aravallis, Malwa regions, Gujarat

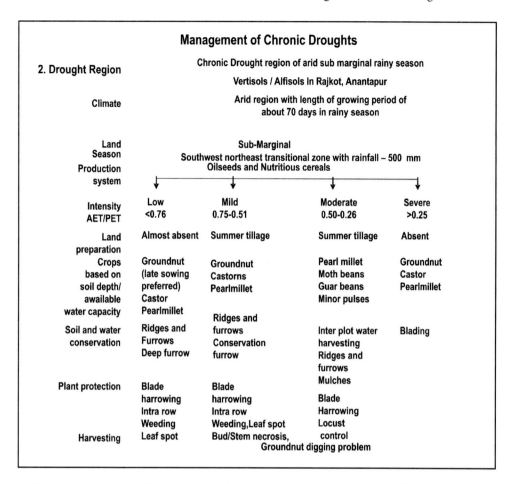

Fig. 5 Management of chronic droughts in the arid sub-marginal region (Vittal et al., 2003).

Plains, Kathiawar Peninsular, Deccan Plateau and Eastern Ghats. The available water capacity varied between 50 and 150 mm in lighter red soils and 100-200 mm in black soils, except in few cases in the lighter soils. The length of growing period is 90-120 days, whereas in medium to deep black soils, the length of growing period is 150-180 days. In the wet semi-arid region with 750-1000 mm annual rainfall, about 85% of rain occurs during June to October months. The drought occurs ephemerally in early season (seeding stage), mid season (grand growth stage) or terminal (flowering to grain filling stage) season. The effect of drought on productivity increases from early to end of season. The contingency crop plans for various weather aberrations have been identified by different centers.

Early season drought: Early season drought generally occurs either due to delayed onset of monsoon or due to prolonged dry spell soon after the onset of rainy season. This may at times result in seedling mortality needing re-sowing or may result in poor crop stand and seedling growth. Further, the duration of water availability for crop growth gets reduced due to the delayed start, and the crops suffer from an acute shortage of water during reproductive stage due to early withdrawal of monsoon. Therefore, for characterization of early season drought, information on optimum sowing period for different crops/ varieties, quantum of initial rainfall spell expected and its ability to wet the soil profile enough to meet

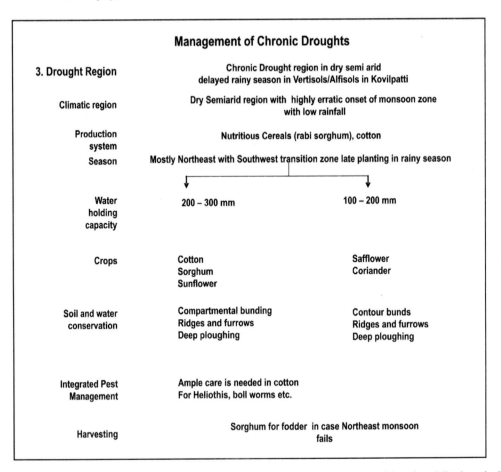

Fig. 6 Management of droughts under receding moisture conditions in dry semi-arid region (Vittal et al., 2003).

the crop water requirements for better germination and establishment is essential. The effect of early season drought is less on the crop, because during this period sowing is carried out. Various operations carried out are primary tillage, sowing, fertilizer application and intercultural operations. On plant establishment, the effect of drought in first 2 to 4 weeks may affect initial vigour but may not have dire consequences on yield. In case of late onset of monsoon by 3 to 4 weeks, contingency crop planning and management are required (Fig. 8).

Mid-season drought: Mid-season drought occurs due to inadequate soil moisture availability between two successive rainfall events during the crop-growing period. Its effect varies with the crop growth stage, duration and intensity of the drought spell. Stunted growth takes place if it occurs at vegetative phase. If it occurs at flowering or early reproductive stage, it will have an adverse effect on the ultimate crop yield. *In situ* soil moisture conservation is a vital component of dryland crop management practices. In cultivated lands with 1-3% slope, runoff is about 25-35% when the rainfall is in the range of 700-800 mm. Though a lot of emphasis has been laid on soil and moisture conservation, the efforts mainly concentrated on construction of various types of bunds across the slope. This helped in controlling erosion and soil loss rather than achieving a uniform moisture distribution. The research results have indicated that bunding increased the crop yields by a mere six percent while a simple inter-terrace management

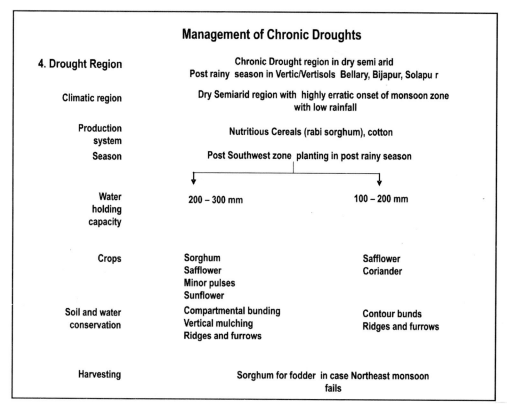

Fig. 7 Management of chronic drought in the dry semi-arid post rainy season region (Vittal et al., 2003).

such as contour cultivation that helped in uniform distribution of moisture, raised crop yields by 15-20%, the impact being more pronounced in years of scanty rainfall. During mid-season, plant protection, top dressing of fertilizer, intercultural and supplemental irrigation are the usual practices, while in case of severe drought for few crops can be tried (Fig. 9). In case of long dry spells, crop based production system (location) related specific contingency plans are needed.

Terminal drought: If there is a terminal drought, crop management strategies like plant protection, soil and water conservation, interculture, supplemental irrigation and harvesting are to be adopted. Late season or terminal droughts occur as a result of early cessation of monsoon rains and it can be anticipated to occur with greater certainty during the years with late commencement or weak monsoon activity. Terminal droughts are more critical as the grain yield is strongly related to water availability during the reproductive stage. Further, these conditions are often associated with an increase in ambient temperatures leading to forced maturity. Fodder needs may be met from contingency plans. Management strategies of ephemeral drought in terminal season are described in Fig. 10.

A **seasonal drought** during rainy season from July to September leads to complete crop failure. Effect of an early season drought in July, mid-season drought in August and the terminal drought in September has a varying effect on the plant growth and production. Experience reveals that the early season drought is of little consequence due to less growth and leaf area index, mid-season drought reduces the potential yield, while the terminal drought is devastating leading to heavy reduction in

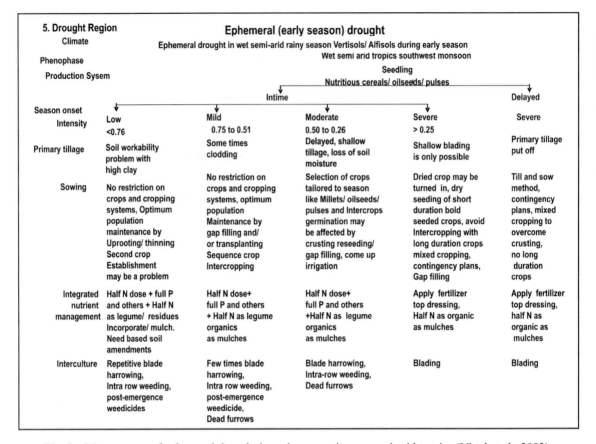

Fig. 8 Management of ephemeral drought in early season in wet semi-arid tropics (Vittal et al., 2003).

productivity and even crop failure. The results of a case study on the performance of sorghum in semi-arid rainy season revealed that the sorghum crop yield was 0.8 t ha^{-1} with 250 mm rainfall received in 16 rainy days (in 1979), while the yield was 2.5 t ha^{-1} with 251 mm rainfall in 28 rainy days (in 1980). While the total rainfall was similar with about 25 cm spread over 100 days, an increase in distribution by 12 rainy days had tripled the grain yield. Thus, the distribution of rainfall is more important for crop productivity. On an average, one rainy day occurred in 6.25 days in 1979, while it was for every 3.50 days in 1980. Thus, a rainless period of five days [(6.25 + 3.50)/2 = 4.95)] caused irreversible damage to the production potential of short duration (100 days) sorghum.

6.1.6 Drought Region 6 (Management of Apparent Drought in Dry Sub-humid Alfisols/Oxisols Regions)

Apparent drought occurs in dry sub-humid area with high rainfall under rainfed rice based production system, in Northern plains, central high lands of Malwa and Bundelkhand, Eastern Plateau, Plains and Ghats region. The length of growing period is 150-180 days. However, in deep loamy clays, the length of growing period is 210 days, while in shallow soils the length of growing period is 120-150 days. Soils have 100-200 mm of medium water holding capacity. Rainfed rice in dry sub-humid regions is often subjected to such terminal droughts due to failure of September rains, which are crucial at the reproductive

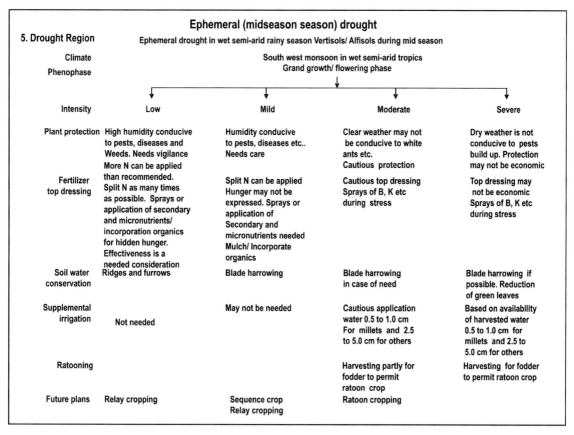

Fig. 9 Management of ephemeral drought in mid-season in wet semi-arid regions (Vittal et al., 2003).

stage. The probabilities of intermittent dry spells of greater than five days duration in the areas are about 40 to 50% during September. Contingency plans are described based on topo-sequence. Management of apparent drought in dry to moist sub-humid region is described in Fig. 11. Livestock also forms an essential part of the production system. Either moderate drought intensity in medium and low lands or mild drought intensity in case of deep-water rice leads to non-planting of rice. Change in land use towards rice with less water requirement may be adopted. This needs advance weather prediction to advise the farmers about the future weather condition.

6.2 Permanent Drought Amelioration

In drought prone regions, rainwater management is the foremost activity to be undertaken. The core strategies of rainwater conservation are based on improving the water availability to the crops and increasing the groundwater recharge. The approaches are on building *in situ* moisture reserves to tide over the recurring drought spells, disallowing subsequent loss of soil-profile stored moisture, permitting safe runoff disposal, its collection above or below ground, and tactical recycling of harvested runoff. Applicability of techniques so developed varied with the soil hydraulic characteristics. Salient findings and/or interventions in the theme area of *rainwater management*, which emerged as research outcomes

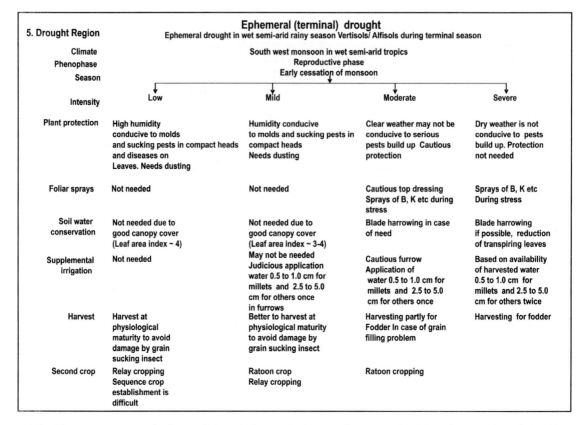

Fig. 10 Management of ephemeral drought in terminal season in wet semi-arid tropics (Vittal et al., 2003).

from the network centers of All India Coordinated Research Project for Dryland Agriculture (AICRPDA) could be effective for drought management (AICRPDA, 2003; Vittal et al., 2003); they are discussed below.

6.2.1 Integrated Watershed Management (IWM)

Integrated watershed management (IWM) is the key to conservation and efficient utilization of vital natural resources viz., 'soil' and 'water', particularly in rainfed agriculture where water is the foremost limiting factor for agricultural production. The prioritized steps involved in resource conservation are the use of practices based on the existing traditional systems. Adoption of contour bunding will be effective in new areas being brought under cultivation and with large farm plots. Strengthening of the existing bunds across the slope and providing of weirs is important. In case of drainage, treatment starts from the ridgeline to the bottom for tackling drainage problem. The traditional loose boulder structures as well as grassing need to be adopted. Later, gully plugging and construction of small check dams on the drainage line are carried out. After bunding and provision of waterway, cultivation of dual-purpose plant species on the bunds should be taken up and provide small cross section bunds with a small furrow on the upper side, preferably with a heavy country plough or mould board plough. The watercourse is covered with vegetation, preferably before other treatments are superimposed (Fig. 12).

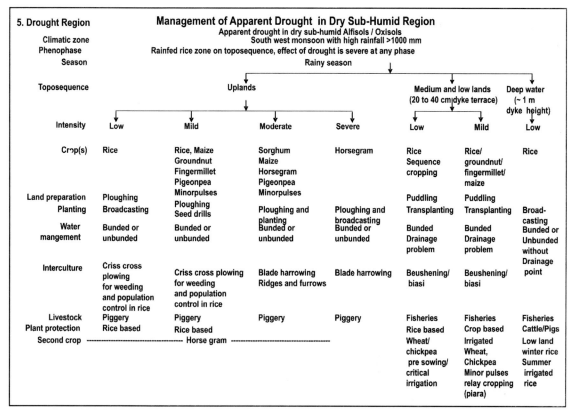

Fig. 11 Management of apparent drought in dry sub-humid regions (Vittal et al., 2003).

Silted tanks could be converted into percolation tanks, particularly in light soils. Structures such as percolation tanks meant for groundwater recharge are expensive and should be constructed only when there are enough funds left after the normal treatment of the watershed. All the resource conservation measures and other water harvesting systems enhance groundwater recharge. When these structures are built, water availability will be more in the region. This will trigger sinking of wells including bore wells. The wells existing in the area should be accounted for and calculate the potential water supply as well as the increased supply through the various field and drainage line treatments. Strict water budgeting for sustainable use of the harvested rainwater needs to be followed allowing only low duty crops. Sugarcane, rice, and wheat should be avoided at the same time encouraging pulses and oilseeds. Other rainwater harvesting systems meant for recession cropping should also receive equal attention, for example, *Khadin* in Western Rajasthan, *Nadi* system in Southern Rajasthan and *Bandh* in Baghelkhand region of Madhya Pradesh. Rainwater harvesting for recession cropping is unique and has some location specificities including socio economic considerations that must be considered first before going in for their imposition in new areas? The *bandh* system (otherwise known as *Haveli* system) can be transformed wherever possible to two cropping system instead of *rabi* (post rainy season) cropping only. Farm ponds are essential primarily for horticulture and multipurpose trees in Class V and above lands. Farm ponds or embankments should be constructed. A pond of 250 m^3/ha catchments is recommended. Even small ponds can be dug and each plastered as a cistern (50 m^3). Water can be harvested or transported into them

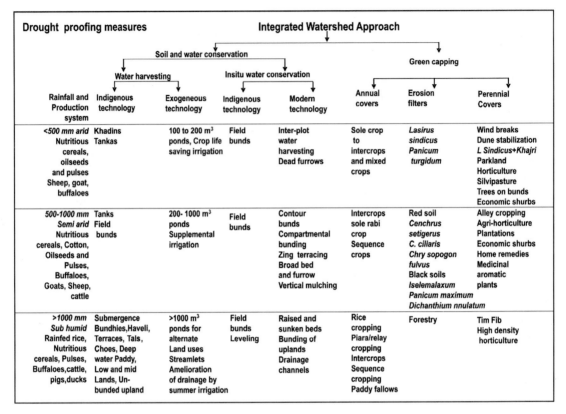

Fig. 12 Drought proofing measures in arid, semi-arid and sub-humid regions (Vittal et al., 2003).

in the arid eco-system. Such water should be treated as of immense value for sustaining tree vegetation during the post rainy periods. Irrigation cover for the first two years is essential for trees and fruit trees should not be irrigated without providing proper basins, as they would lead to wastage of water.

6.2.2 Building in situ Moisture Reserves to Tide over the Recurring Drought Spells

Off-season tillage (summer tillage) coinciding with pre-monsoon showers helped in increasing rainwater infiltration, moisture conservation and efficient weed control. This resulted in an increase of crop yields in Alfisols more with textural discontinuity. Deep tap rooted crops sown a few weeks after the onset of rainy season were not significantly benefited by this practice. Land leveling in deep alluviums helped in increasing yields especially in sub-normal seasons. In chronic drought prone areas, deep tillage (20-30 cm) was found specifically applicable to soils having textural profiles or hard pans. Under uni-modal (<500 mm) rainfall situation in semi-arid regions with shallow Alfisols, sowing across the slope and ridging later was useful. It assisted in greater rainwater absorption and control of pernicious weeds. In black soils, deep tillage alone was of no use. Tillage combined with compartmental bunding was found to be a most effective soil management practice for Vertic Inceptisols. With rainy season cropped Vertisols (uni-modal rainfall regions), water surplussing is an integral part of *in situ* moisture conservation. In bi-modal medium (500-750 mm) rainfall representing semi-arid Alfisol, graded border strips were found advantageous. Extensive tillage of sandy soils (common to desert areas) made them more vulnerable to wind erosion. Superficial scraping to eliminate weeds was found adequate.

6.2.3 Permitting Safe Runoff Disposal, its Collection Above or Below the Ground

Well-established vegetative barriers, similar to mechanical bunds helped in the conservation of soil and rainwater. Vegetative barriers can be established over *Cenchrus ciliaris, Cymbopogan slaxuosus, Pennisetum hohenackeri* and *Leucaena leucocephala* across the centers. Vertisols, medium rainfall regions, broad bed and furrow system (popularly known as ICRISAT technology), which combines benefits of good aeration and safe disposal of excess rainwater, was found useful for post-rainy season crop. A technology for water harvesting in small farm ponds (100-250 m^3) was developed. Among the various sealants tested to prevent percolation losses, particularly in Alfisol regions, brick and mortar lining was found to be most efficient and economical in the long run. Water resource thus developed, could be used for life saving irrigation. Inter-plot water harvesting (1:1 cropped) increased yields in Aridisols by 50% and Alfisols in dry sub-humid region by more than 17%.

If there is a mid-season and terminal drought, the protective irrigations with the harvested water increased the yields considerably. An irrigation to groundnut during severe drought period increased the haulm in yields by 62 kg/ha whereas irrigation to rainfed upland rice increased the yields by 698 kg/ha. At Arjia and Agra, protective irrigations, at terminal drought, to maize and pearl millet increased the grain yields by 38% and 109% respectively. Similarly, in rice at Varanasi, 26% increase was noticed. Higher response to supplemental irrigation was observed after withdrawal of monsoon.

6.2.4 Tillage

Tillage combined with seeding across the slope reduced runoff losses by 10 to 20%. Runoff thus averted could sustain crop growth for another two weeks following intermittent failure of rains. Effectiveness of cultivation across the slope significantly increased if post-planting intercultivation was directed to develop a set of ridges and furrows between the crop rows. The yield increases jointly attributable to sowing across the slope and ridging later varied between 30 and 40%. Ridges and furrows are required to be laid on a minor gradient of 0.3 to 0.4% to prevent temporary water logging. The ridge-and-furrow system is not practicable with closely spaced row-sown crops (row spacing <30 cm). In such instances, the alternative technology involved creation of deep furrows (popular as 'conservation furrow') after 8 to 10 crop rows (~ 3 m interval) for rainwater capture and disposal, if necessary.

In rainy season-cropped, unimodal low rainfall regions (mean annual rainfall 600-750 mm), ridges and furrows with water surplus were found useful. Sowing on flat bed without ridging later on, as a part of cultural operations is usually a traditional practice. The serration of the soil surface is an efficient way, which provides permitting more time for water entry into the soil profile. The ridges and furrows always increase the yield but its effectiveness is more so in moderate drought. In other situations like severe drought and less prone to drought, the effectiveness is constrained by the lack of water or excess of water. In rainy season-cropped, unimodal medium to high rainfall regions (mean annual rainfall <800 mm), raised and sunken bed system was found most appropriate to provide drainage and storage of runoff. The *in situ* conservation practices for post rainy season crop depend upon the soil type, and rainfall. Bimodal low-rainfall regions (mean annual rainfall <750 mm), scooping, compartment bunding and tied ridging were found effective for *in situ* water conservation. In medium deep Vertisols, dividing the fields into sectors of 3 m × 3 m by compartment bunds led to yield increases up to 50%. Where heavy black soils (mean annual rainfall 550-600 mm), vertical mulch increased water intake and reduced runoff. Effect of vertical mulch in increasing yield was more visible during sub-normal rainfall seasons.

6.2.5 Crops and Cropping Systems

Location specific crop/contingency plans for mitigating the effects of drought have been recommended by various research stations of state agricultural universities and also by the network centers of All India

Coordinated Research Project for Dryland Agriculture (Ravindra Chary et al., 2008a; Subba Reddy et al., 2008), applicable to their respective domains and an example each for chronic, apparent and ephemeral drought situations (Vittal at el., 2003) is given in Table 6. Venkateswarlu and Bhaskara Rao (1999) indicated that the delay in sowing situation also might arise due to a possible resowing of an already sown crop. Minimizing evaporation losses by creating soil mulch and minimizing the weed infestation could extend the life of standing crop that is affected by drought. If the rain is received after a prolonged drought, at least 20 kg nitrogen/ha should be applied to the standing cereal crops before soil working. Extended rains would be useful for already established perennial crops like cotton and also be utilized by growing short duration pulse crops. Even if there were no rains, these pulse crops would provide much needed fodder and also enrich the soil.

Organic or chemical mulches may be used to conserve soil moisture and reduce evaporation. Mulching with *Leucaena* loppings, glyricidia and other green leaf manures helped in improving the soil properties. Use of organic mulches not only reduces evaporation losses of soil-stored water but supplies nutrients to crop which help in increasing stability and sustainability of crop yields particularly of post-rainy season crops.

In the drought recurrent regions, length of growing period (LGP), i.e., moisture availability period of a given soil type provides better index than total-rainfall-based crop planning. The LGP is defined as the period when the moisture and temperature regimes are suitable for crop growth and the period is determined by the FAO method (FAO, 1983). LGP is computed as the sum of the periods when moisture is more than 0.5 PET plus time taken to utilize stored soil moisture (assumed 100 mm) after P falls short of PET. For example, Nagpur in Maharashtra receives a rainfall of 1120 mm and LGP determination indicates a LGP of 210 days in deep black soils, suitable for growing a single long duration crop or a short duration crop (soybean) in *kharif* followed a short duration sequence crop (chickpea) (Velayutham, 1999).

6.2.6 Soil Fertility Management

Soils in drought prone regions are universally deficient in nitrogen. Most of them are also deficient in available phosphorus, sulphur and some micro-nutrients (mainly Zn and Fe). The soils are highly degraded with a large number of physical and chemical constraints. Shallow depth is a familiar feature of rainfed soils. Research in this area has primarily focused on overcoming the textural problems posed by dryland soils, tillage and water intake properties, soil fertility management through use of chemical fertilizers, organics and biofertilizers. Contrary to the past belief on excessive risks associated with the use of chemical fertilizers, research findings conclusively established that there is vast potential for increasing crop yields through fertilization, typically N fertilizers, across contrasting rainfed environments. The cost-benefit ratio is highly favourable when crops are fertilized. A fertilized crop is able to withstand drought better than a non-fertilized one. A stressed crop can also recover faster if it is fertilized following relief from stress. Pest attack was low. Split application ensures against loss in nutrients and deep root system for reaching subsurface stored moisture in the aberrant weather more so against loss in nutrients with occurrence of above normal rainfall during crop growth period.

Integrated nutrient management studies have established the value of a number of naturally occurring nutrients containing (organic manures) and generating (biofertilizers) sources to augment overall nutrient turnovers for soil fertility management. Green manure was found to be a dependable source of several plant nutrients. Typically, it could meet half the N requirements of a crop. Schemes to generate green manure in a non-competitive way during the no cropping season and bund farming have been worked out. This has opened up a new vista to make green manuring a viable option. Inclusion of legumes in a rotation benefited the succeeding crop equivalent to 10-30 kg N ha^{-1}. Short-duration legumes such as

Table 6. Crop and contingency plans for different drought types based on climate and soil orders (Vittal et al., 2003)

Recommendation domain	Crop plan							
	June second fortnight to July first fortnight	Second fortnight of July	First fortnight of August	Second fortnight of August	September first fortnight	September second fortnight to October first fortnight	October second fortnight to November first fortnight

1. Chronic Drought

1.1 Chronic drought in arid region in marginal soils with rainfall < 500 mm

Hisar, Bhiwani, Sirsa, Mahendragarh, Gurgaon and part of Rohtak district of Haryana in marginal hot typic arid aridisols	Pearl millet Cluster bean Greengram	Short duration crops Pearl millet (HHB-67) Greengram (Asha) Blackgram (T-9) Cowpea (Charodi) Clusterbeans (HG 365)	Nursery of Pearl millet (HHB 67) nursery may be kept ready for transplanting. Crops Fodder	Toria with supplemental irrigation Rains received after mid of August may be conserved for *rabi* sowing.	Clusterbean Cowpea	Chickpea Taramira Rapeseed	Mustard Taramira Use water in farm ponds, tanks and *tankas (Kund)* for drip and sprinkler irrigation for the horticulture, vegetables and high value crops.

1.2 Chronic drought in arid region in sub-marginal soils with rainfall 500-750 mm

Rajkot, Surendranagar, Jammagar, parts of Junagarh, Bhavanagar and Amreli districts of Gujarat in hot arid Vertisols	Sorghum, Pearl millet, Groundnut, Castor, Cotton	Erect groundnut (GG-2, GG-5, GG-7) Sesame (G. Til-1, G. Til-2); Pearl millet (GHB-235 GHB-316, GHB-558) Greengram (K-851, GM-4); Blackgram (T-9;) Pigeonpea (ICPL-87, GT-101)	Blackgram (T-9), Forage maize/ sorghum (Gundri, GFS-5), Castor (GAUCH-1) Sesame (Purva-1)	Forage maize Sorghum (Gundri, GFS-5) Sesame (Purva-1)	Usually no rainfed crop is sown		

1.3 Chronic drought in dry semi-arid region with rainfall 500-750 mm (late rainy season to post-rainy season)

Madurai, Ramanthapuram and Tirunalveli districts of Tamil Nadu in hot moist	Onset of monsoon is from September				Cotton Sorghum Pearl millet Setaria Cotton	First fortnight of Grain sorghum Third week of Chillies Fourth week	First week *Karunganni* Cotton Sunflower and Sesame

(*Contd.*)

Table 6. (Contd.)

Recommendation domain	Crop plan							
	June second fortnight to July first fortnight	Second fortnight of July	First fortnight of August	Second fortnight of August	September first fortnight	September second fortnight to October first fortnight	October second fortnight to November first fortnight	

Recommendation domain	June second fortnight to July first fortnight	Second fortnight of July	First fortnight of August	Second fortnight of August	September first fortnight	September second fortnight to October first fortnight	October second fortnight to November first fortnight
semi arid alfisols/Vertisols					Chillies Medium-duration pulses, Flowers, vegetables	*Panicum miliaceum* Chickpea Short-duration pulses Safflower Setaria Coriander Pearl millet	sowings can be done during first week of November Senna No rainfed crop sowing beyond third week of November

1.4 Chronic drought in dry semi-arid region with rainfall 500-750 mm (post-rainy season)

Recommendation domain	June second fortnight to July first fortnight	Second fortnight of July	First fortnight of August	Second fortnight of August	September first fortnight	September second fortnight to October first fortnight	October second fortnight to November first fortnight
Bijapur and Gulbarga districts, parts of Belgaum, Linguagur of Raichur district of Karnataka and southern parts of Maharashtra in hot semi-arid Vertisols	Greengram Blackgram Pigeonpea Pearl millet	Groundnut (spreading) Hybrid pearl millet Sunflower and Setaria in *kharif* areas pure pigeonpea/cowpea/horsegram in light soils	Cotton in middle of August. Early sowing of cotton is advantageous Grow *herbaceum* cottons in place of *hirsutums* Sunflower Pigeonpea, castor and setaria in light soils. Pigeonpea in medium to deep black soils	In medium to deep black soils, on contour bunds Castor Relay cotton in groundnut in medium black soils.	Early sowing is more beneficial. Cotton (Bhagya or Laxmi before first fortnight of September). If rains are not received seeding of sunflower, *rabi* sorghum, and chickpea with 1½ times the normal seed rate. Sorghum, Sunflower (after first fortnight of September)	Sorghum, chickpea safflower (before September) *Rabi* sorghum up to first fortnight of October *Rabi* sorghum + chickpea (2:1), *rabi* sorghum and Chickpea as mixed crops Chickpea and safflower (4:2 or 3:1). Wider row spacing or paired rows helps in moisture conservation especially during drought	Fodder crops

2.0. Ephemeral Drought
2.1 Ephemeral drought in wet semi-arid region with rainfall 500-750 mm (rainy season)

Recommendation domain	June second fortnight to July first fortnight	Second fortnight of July	First fortnight of August	Second fortnight of August	September first fortnight	September second fortnight to October first fortnight	October second fortnight to November first fortnight
Bhilwara, Tonk,	Pigeonpea	Sesame (RT-46)	Sesame (RT-125)	Sorghum (Fodder)	Greengram	Greengram	Chickpea

(Contd.)

Location	Crops	Practices	Notes	Crops 2	Crops 3	Crops 4
Dungarpur, Ajmer districts and parts of Bundi, Chittaurgarh, Rajasamand in hot dry semi-arid inceptisols/aridisols	Cowpea	Greengram (K-851 and RMG-62); Sorghum (Fodder); Cowpea (Fodder) (Raj Chari-1 and 2, C-152)	Greengram (RMG-62) Sorghum (fodder) (Raj Chari-1) (single cut) Soil water conservation measures for *in situ* management & runoff harvest for recycling during later part of crop season. No sowing of cereals Only short duration pulses and oil seeds or fodder crops should be sown. Soil mulching and interculture to conserve soil moisture is beneficial.	(Raj Chari –1) Toria (TL-15) Tarmaira (T-27) Rain received after first fortnight of August should be conserved for early *rabi* seeding of toria/taramira during first week of September. Any heavy downpour occurs, harvest the water for pre-sowing supplemental irrigation to *rabi* crops.	Blackgram Cowpea	Blackgram Cowpea	Safflower Mustard Barley

3.0. Apparent Drought
3.1 Apparent drought in dry sub-humid region with rainfall 1000-1500 mm (rainy season)

Location	Crops	Practices	Notes	Crops 2	Crops 3	Crops 4
Uplands and medium lands of Balasore, Cuttack, Puri and Ganjam districts of Orissa in hot moist sub-humid inceptisols	Pigeonpea, Mesta, Maize, Groundnut, Fingermillet, Rice, Sorghum, Cowpea, Blackgram, Greengram	Upland Blackgram Setaria (Pant -30); Greengram (PDM54/K 851); Sesame (Uma or local); Early Pigeonpea (UPAS 120/ICPL-87) Planting of short duration vegetables as radish (Pusa Chetki), okra, cowpea (SEB-2/SEB 1) and clusterbeans.	Upland Sowing of niger, blackgram, sesame, greengram, Planting of vegetables as radish, beans, cowpea, Early Pigeonpea (ICPL-87/UPAS-120)/W Medium and shallow submerged low land Direct line sowing of extra early rice (Heera, Vandana, Kalinga-III, Z HU 11-26, Rudra, Sankar and Jaldi-5).	Upland Horsegram; Sesame Niger; Cowpea	Medium land Making land preparation for sowing pre-*rabi* crops as Mustard/ Greengram/early pigeonpea, which can be sown in the month of September Pigeonpea (short duration), Horsegram	Medium and low land conditions: Direct sown rice, transplanted rice Horsegram

cowpea benefited much more. An integration of FYM (10 t/ha) + recommended NPK at Bangalore not only improved productivity and profitability but also improved sustainability. The increase in enzymatic activities like urease, phosphotase and dehydrogenase activity was more in integrated treatments. pH was reduced in continuously fertilized plots, the calcium in FYM helped in buffering capacity in soils. The phosphorus showed built up when fertilizer, manure was applied while it was not where no fertilizer, manure applied. The nutrient status was above critical limit in FYM applied plots. The tendency of improvement in various parameters was also noticed in application of FYM. Integrated nutrient management systems, besides nutrient supplementation, enhanced soils' ability to hold additional water and produced results in favorable soil biological interactions. Green manuring increased the yield by improving the physical and chemical properties of the soil.

In most of the situations, the yield sustainability was higher when the recommended dose of fertilizer was applied. In case of cereals, higher sustainability was obtained when the recommended dose of nutrients was applied through chemical sources. In case of oilseeds, however, the recommended dose applied half through chemical fertilizer and the other half through organic source led to higher sustainability values. Available nitrogen, organic carbon and phosphorus content in soil were increased with organic fertilizer application. Application of crop residues in combination with chemical fertilizer resulted in higher sustainable yield and maintained higher levels of nitrogen, phosphorus and organic carbon. Green leaf manure proved promising in increasing the sustainability in yield and improving the organic carbon, infiltration rate and hydraulic conductivity of the soil.

Thus, a recommendation milieu of integrated nutrient management should produce a low-cost complex fertilizer with suitable pulse crops. The crops should be supplied essential nutrients considering fertilizer use efficiency, off-season losses in nutrients and nutrient requirement. Limiting nutrients also need to be supplemented so that crop completes its life cycle without serious limitation. Balanced nutrition of rainfed crops could be achieved through application of right proportion of inorganics, organics, biofertilizers etc. by correcting major, secondary and micronutrient deficiencies across varying soil types.

6.2.7 Productive Farming Systems: A 3 × 3 Matrix Approach

Productive farming systems are identified for drought prone regions based on rainfall; land capability and soil order (Vittal et al., 2003). Land use based farmstead plan state-of-the-art based agroforestry models linked to livestock and watershed management for soil and water conservation including water harvesting. Some of the subjects are hedge fencing, multipurpose tree species, bush farming, cereals/millets. Pulse/oilseeds/cotton, parkland horticulture, olericulture, floriculture cum IPM, home remedies, water harvesting, livestock, poultry, fisheries, apiary, etc. are some of the models suggested (Fig. 13). Diversification into higher value agricultural crops (medicinal, aromatic, dye yielding crops, etc.), and non-farm activities like value addition to agricultural products offer good scope for sustained increase in per capita income. Part of the farmstead could also be used for generating seed spices. For the development of commons, these may be divided into small plots of 5-10 ha and can put on long lease of about 19 years to the user groups. The combination of systems such as fruit trees, silvipastures, multipurpose trees, or even pastures may be adopted on commons. Maximum number of trees per hectare may be limited by quantum of annual rainfall (product of rainfall in m and area in m^2) divided by volume of water one full grown tree transpires annually (a product of canopy area, surface area in m^2, and potential evapotranspiration in m per annum). Improved variety or new plant species suitable for the ecosystem and rejuvenation of social fencing of improved plant species may be attempted. In tree farming, the general cleanliness of the area is lost, thus, encouraging new diseases and pests. Therefore, it is important to carry out weeding and form basins for the trees and furrows for *in situ* rainwater harvesting in the case of shrubs, grasses and fodder legumes. Diversification strategies should be based on low external input strategies (Vittal et al., 2006).

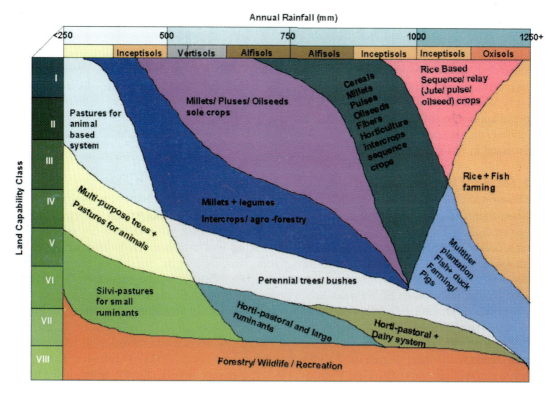

Fig. 13 Productive farming systems matrix in rainfed agriculture (Vittal et al., 2003).

6.2.8 Selective Dryland Mechanization

Traditional field operations practiced with indigenous equipment by farmers in the rainfed regions are time consuming, cost ineffective, and do not respond to poor management. In dryland regions, due to limited soil moisture availability and a very short sowing window, improved management of resources through mechanization to achieve timeliness of operations with required precision is very much important. Selective mechanization in dryland regions, as a critical input, resulted in economic advantages, viz., increased productivity by 12-34%, saving in seeds and fertilizers by 20% and 15-20%, respectively, and net returns up to 29-49% (Srivastava, 1993; Singh, 1995; Mayande, 2006). Timeliness and precision of these operations with selective mechanization help to mitigate early season drought, mid-season drought situations and mid-season corrections. Selective mechanization is needed for tillage, seeding and fertilizer application, interculture and weeding, plant protection, harvesting, post-harvest processing and value addition. Cluster/village level custom hiring centers would not only mitigate early drought situations, but also create employment and income to rural youth. Presently, many institutional and policy back up is in place through focused central and state programs and also by National Bank for Agriculture and Rural Development (NABARD) and other nationalized banks.

6.3 Drought Assessment and Monitoring Techniques

Drought assessment and monitoring at different scales and timely dissemination of information constitute vital part of drought management system. Appropriate mechanism, reliable data, standard procedures for indices of drought prevalence and intensity are prerequisite for developing efficient drought management strategies. Thus, it is essential to have a sound, operationally feasible, economically viable system for drought assessment and monitoring. Conventionally, agricultural drought conditions are characterized by ground observations on meteorological parameters such as rainfall, aridity and agricultural parameters such as crop area sown, crop condition, crop yield etc. The Drought Monitor of USA, Global Information and Early Warning System (GIEWS) of FAO and NADAMS of India are the proven examples for the successful application of geo-informatics for drought assessment (Roy et al., 2006). The methodologies, tools and techniques used in India and other emerging new concepts for drought assessment and monitoring are discussed in this section.

6.3.1 Drought Indices

Identification and development of drought threshold indices for various activities like agriculture, water resources, energy, etc. are important to formulate and initiate drought management strategy approaches for a given region. For this purpose, monitoring soil moisture status, water availability indices such as Palmer's drought index (Palmer, 1965), current crop status and irrigation needs to complete the crop growth cycle, current water status in reservoirs, lakes, irrigation wells, etc. are of great importance. Besides these, monitoring present synoptic conditions, viz., break monsoon condition; the position of monsoon trough, occurrence of low-pressure systems, their intensification and subsequent movement are also essential. Monitoring of these indices need to be carried out at least weekly during the cropping season. This information when coupled with the climate of a region, i.e., information of the region regarding its drought proneness and the pattern of drought incidence and spread, will enable drought prognostication with higher degree of confidence (Ramakrishna et al., 2005). Different drought indices developed are: drought anomaly index (Vanrooy, 1965) (based on rainfall), climatic aridity index (defined during UNESCO Conference on desertification in 1977) and moisture availability index (Hargreaves, 1975) (based on rainfall and potential evapotranspiration). Other drought indices based on water balance parameters are: soil moisture as an index of drought (Van Bavel, 1953), aridity anomaly as an index of drought (Subrahmanyam, 1964), moisture adequacy index developed by Central Arid Zone Research Institute (CAZRI, India) and Palmer's drought index.

6.3.2 Remote Sensing and GIS Techniques

An information system that provides updated information about drought situation is of immense use to the decision makers. Satellite derived Vegetation Index (VI), which is sensitive to vegetation stress is used to monitor drought conditions on near real-time basis. This would help to initiate strategies for mid-season corrections and other crop contingency planning. This is being done by National Remote Sensing Centre (NRSC), Hyderabad now by using high-resolution data and through concerted programs (Rao et al., 1999). Schemes like space-based National Agricultural Drought Assessment and Monitoring System (N-ADAMS) operational at district/sub-district level and Forecasting agricultural output using space agro-meteorology and land-based observation (FASAL) made some impact, but need to be strengthened and made more practicable and implementable. The drought assessment and monitoring system needs to be strengthened with robust operational procedures for early drought detection and assessing the quantitative impact of drought on agricultural production through the use of global satellite and assimilation of data from ground segments, routinely collected by various agricultural departments of the country. In order to achieve these objectives, a Decision Support Centre (DSC) has been set up at the Department of Space at National Remote Sensing Centre, Hyderabad to address various issues related to the disaster management.

6.3.3 Water Balance Method

The India Meteorological Department (IMD), Government of India, New Delhi, is monitoring agricultural drought using Thronthwaite water balance technique (Thronthwaite, 1948), during both *kharif* and *rabi* seasons using information generated on real-time basis every fortnight during cropping season through its wider network stations (Sarkar et al., 2005). This information can be used by IMD itself and research organizations of Indian Council of Agricultural Research (ICAR) and research stations of state agricultural universities engaged in rainfed agriculture research, particularly in developing crop and contingency planning based on the index of moisture adequacy matching the crops and varieties for a particular aberrant weather. These plans can be translated into agro-advisories and disseminated to state line departments and farming community using information and communication technologies.

6.3.4 Scientific Rainfed Land Use Planning

Scientific land use planning in drought prone regions is one of the rational approaches for drought mitigation. The learning experiences from ICAR—National Agricultural Technology Project (NATP)—Mission Mode Project on Land Use Planning for Rainfed Agro-ecosystem has distinctly brought out that in a microwatershed, cadastral level (1:10,000/25,000) both biophysical evaluation (soil resource inventory and socioeconomic evaluation (Participatory Rural Appraisal (PRA), etc.) followed by allocating land parcels with appropriate land utilization types viz., cotton, greengram, groundnut, pearl millet, grasses etc., on a toposequence, could enhance the land productivity from 20 to 50% compared to traditional land use.

Ravindra Chary et al. (2005) suggested that the delineation of Soil Conservation Units (SCUs), Soil Quality Units (SQUs) and Land Management Units (LMUs) from the detailed soil survey maps at cadastral level in a micro watershed would help in land resource management information since these units are homogeneous and has a wider application. A resilient, less risk prone farming system based on the land requirements and farmers' capacities is developed to mitigate the drought and to address the unabated land degradation and imminent climate change. Further, SCUs are basically for soil and water conservation prioritized activities to mitigate drought and could be linked to programs like National Rural Employment Guarantee Scheme (NREGS) in a watershed/village to create physical assets like farm ponds etc. SQUs are to address soil resilience and improve soil organic carbon, problem soils amelioration and wastelands treatment and linked to various schemes and programs in operation like National Horticultural Mission (NHM), Rashtriya Krishi Vikas Yojana (RKVY), etc. SCUs and SQUs are merged in GIS environment to delineate land parcels into homogenous Land Management Units with farm boundaries. LMUs would be operationalized at farm level for taking decisions on arable, non-arable and common lands for cropping, agroforstry, agrohorticulture, etc., and further, for leaving the most fragile land parcels for ecorestoration. Rainfed land use planning modules should be conceptualized and developed considering SCUs, SQUs and LMUs for risk minimization, enhanced land productivity and income, finally for drought proofing. An example of delineation of these units for the Kaulagi watershed, Bijapur district, Karnataka is shown in Figs 14, 15 and 16.

Participatory land use planning is a buzzword for achieving the different goals of the various stakeholders. In stressed ecosystems like rainfed wherein the major crop based production systems are established as best land use planning over a period of time, no single land use or single criteria has sustained the land productivities, incomes, ecosystem and finally the livelihoods—reasons being highly complex situations of risk, diverse socioeconomic settings and subsistence agriculture. Thus, land use planning in drought prone areas mostly occurring in rainfed regions should aim at increased land productivity in totality through various means from annuals to perennials by coping with aberrant weather causing drought and inherent unabated land degradation. The final aim is to build a bio-diverse mixed

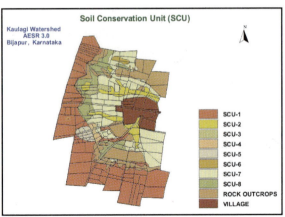

SCU	Details	%
2	Deep + low erosion	50.23
3	Shallow + very low erosion	7.68
4	Shallow + moderate erosion	1.61
5	Deep + moderate erosion	1.17
6	Deep + high & very high erosion	1.99
7	Very shallow + High erosion	0.83
8	Very shallow + moderate erosion	27.31
9	Very shallow + high & very high erosion	9.18

Fig. 14 Soil Conservation Units (SCUs) in the Kaulagi watershed, Bijapur district, Karnataka.

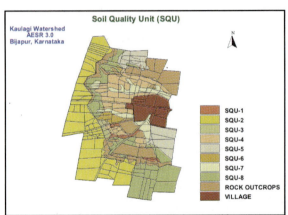

SQU	Details	(%)
1	Non gravel + neutral(pH) + low(OC) + normal(EC) + calcareous(CaCo3)	1.67
2	Non gravel + neutral (Ph) + low (OC) + high(EC) + calcareous(CaCo 3)	36.8
3	Non gravel + alkaline (pH) + low (OC) + normal(EC) +calcareous(CaCo3)	1.17
4	Gravel + alkaline(pH) + high(OC) + normal(C) + calcareous(CaCo3)	15.7
5	Gravel + alkaline(pH) + medium (OC)+ normal(EC) + calcareous(CaCo3)	10.1
6	Gravel + alkaline(pH) + low(OC) + normal(EC) + calcareous(CaCo3)	2.12
7	Gravel + alkaline(pH) + low(OC) + critical (EC)+ calcareous(CaCo3)	7.51
8	Extremely gravel+alkaline(pH)+low(OC) + normal(EC)+calcareous(CaCo3)	25

Fig. 15 Soil Quality Units (SQUs) in the Kaulagi watershed, Bijapur district, Karnataka.

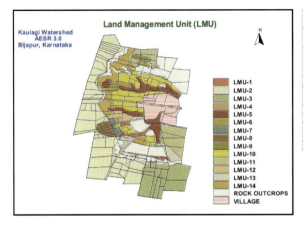

LMU	SCU	SQU	Soil Series
LMU-1	SCU-1	SQU-1	15F
LMU-2	SCU-1	SQU-2	1A/2A
LMU-3	SCU-1	SQU-8	5B
LMU-4	SCU-2	SQU-6	9D
LMU-5	SCU-2	SQU-7	8D/9D/10D/11D
LMU-6	SCU-3	SQU-4	12E
LMU-7	SCU-4	SQU-3	4C
LMU-8	SCU-5	SQU-2	3A
LMU-9	SCU-6	SQU-5	17G
LMU-10	SCU-7	SQU-4	12E/132E/14E
LMU-11	SCU-7	SQU-5	16G
LMU-12	SCU-7	SQU-6	13E
LMU-13	SCU-7	SQU-8	18H
LMU-14	SCU-8	SQU-8	18H

Fig. 16 Land Management Units (LMUs) in the Kaulagi watershed of Bijapur district, Karnataka.

farming system model for individual farmer to sustain the system and achieving the goals of food, nutritional economic and ecological securities with complimentary benefits of drought irrigation or drought proofing and land management and a buffer to impact of land use change. Hence, a framework of rainfed land use plan for participatory self-sustaining viable bio-diverse mixed farming system with activities, relevant partners, strategies and outcomes at different scales is suggested (Ravindra Chary et al., 2004) (Fig. 17).

6.3.5 Weather-based Information System

Activities like agro-meteorological data monitoring, archival weather forecasts for agriculture, and dissemination of advisories to the farmers in different agro-climatic zones, need-based information on soil-crop-weather relationships is being developed/updated by the All India Coordinated Research Project on Agro-meteorology (AICRPAM), Central Research Institute for Dryland Agriculture (CRIDA), Hyderabad, Indian Council of Agriculture Research (ICAR), New Delhi, India Meteorological Department (IMD), Pune and the National Centre for Medium Range Weather Forecasting (NCMRWF), New Delhi (Singh et al., 1999). Both weather-sensitive operations (those that are directly affected by weather) and weather-information sensitive operational decisions (that are indirectly affected by weather) have greater potential value in rainfed agriculture. Detailed analysis of every cropping system that is sensitive to fluctuations in weather is required for identification of the nature of the weather/climate events that have

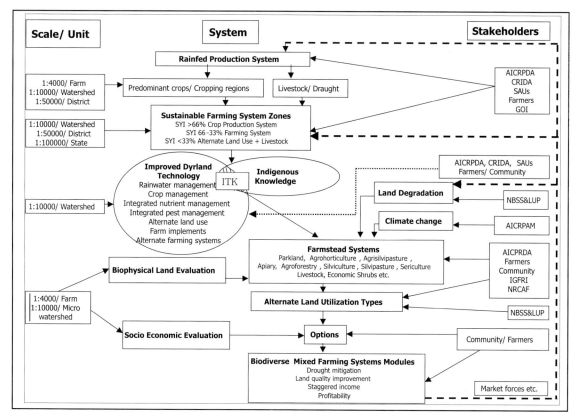

Fig. 17 A framework of rainfed land use plan for participatory self-sustaining viable bio-diverse mixed farming system beneficiaries.

a large impact on productivity. IMD has been issuing short-range and long-range weather forecast for many years (Biswas, 1990). Farm weather services to the farming community through Farm Weather Bulletins have been broadcasted since 1945. Agromet Advisory Services (AASs) on a weekly basis are now available through press, All India Radio and Doordarshan (TV). CRIDA through its website (www.cropweatheroutlook.org) is flashing crop and contingency plans for the rainfed regions on a weekly basis based on the crop and weather information received from its network centers under All India Coordinated Research Project on Dryland Agriculture and All India Coordinated Research Project on Agro-meteorology. With the advancement of information and communication technologies (ICTs), many private institutions are engaged in weather information services that have to be supported with an appropriate policy by the government for the benefit of multiple beneficiaries, particularly the farming community.

6.3.6 Drought Forecasting

Drought prediction in advance would help taking water management, food grains transport from surplus areas to scarce areas, production and distribution of inputs (seed and fertilizers), food grains imports, decisions, etc. at different levels by various stakeholders from governments to farmer levels. Countries in marginal climatic types affected by recurrent droughts adopt short-term and long-term responses and management (Singh et al., 1999). Since drought is the manifestation of rainfall deficiency on a large scale, any forecasting technique should indicate rainfall amounts, particularly on a long-range basis, which is adequate to give an idea of impending drought. For the long-range forecast of monsoon rainfall (June to September) for India as a whole, the India Meteorological Department (IMD), Pune, uses multiple regression model which was first introduced by Walker (1910) and later on improved by many researchers; parametric model (Thapliyal, 1986; Gowarikar et al., 1991; Rajeevan et al., 2004); and power regression model (Rajeevan et al., 2004). Countries in marginal climatic types affected by recurrent drought types adopt short-term and long-term responses and management (Sarkar et al., 2005).

7. GOVERNMENT POLICY AND SUPPORT SYSTEMS

Droughts are recurrent phenomena in India. Both the Central and State governments have made several efforts from time to time to mitigate the impact of these droughts. Early efforts include remittance of land revenue, creation of irrigation facilities, streamlined communications, provision of employment through relief works, and constitution of Famine Commissions (1898 and 1901). During the seventies and eighties, attempts were made to improve disaster management. In order to alleviate drought affected people, Government and non-governmental agencies worked together by initiating programs like groundwater exploitation, minor irrigation works, and construction of soil conservation programmes, afforestation programmes, etc. Subsequently the other programs included were livestock development, assistance to small and marginal farmers as well as landless laborers. Drought management policies are classified into short-term, medium-term and long-term. Short-term policies are aimed at reducing the impact of drought and mitigating the suffering of people from drought, e.g., relief programs, strengthening public distribution system, employment generation programs, contingency planning, etc. Medium-term policies are aimed at tackling some of the causes of drought and aims at corrective measures, e.g., soil and water conservation measures, affforestation programs, crop insurance scheme, National Watershed Development Project for Rainfed Areas (NWDPRA), Drought Prone Area Program (DPAP), Desert Development Programme (DDP), etc. Long-term policies are aimed at preventing the drought, e.g., bringing more area under irrigation, groundwater augmentation, linking of rivers, etc. (Subrahmanyam and Nagasree, 2005). Following are the few policy and support systems/interventions that are needed or strengthened to efficiently manage droughts.

7.1 Weather Code

For drought proofing, as a long-term plan, it is very important to have a good weather code. The approach for drought management should include short-, medium- and long-term prescriptions, with inputs assimilated from various sources, towards preparedness for drought; these approaches, together with a uniform code of assessment, aid in the selection of appropriate drought management measures at the local as well as at macro level (Krishna Rao, 2005).

7.2 Crop/Weather Insurance

Managing risk is one of the key elements in overcoming the drought effects. Crop insurance appears to have a favorable impact but it entails high cost to the government. A pioneering rain insurance scheme, which was an area approach, for old Mysore and India was proposed by Chakravorthi (1920) to protect farmers against drought. Subsequently in 1970, an individual approach was introduced which was not practicable due to huge financial loss. Again in 1979, an area based approach was formulated and was taken up as pilot scheme in three states and extended to 12 states by 1984 which was discontinued, though there was a good financial performance, due to introduction of Comprehensive Crop Insurance Scheme (CCIS) and operated by GIC with the main objective of providing a measure of financial support to the farmers in the event of crop failure as a result of drought, flood, etc., and other two objectives. This scheme is an area approach and linked to institutional credit (Venkateswarlu, 2005). Mishra (1999) pointed out that though this scheme encouraged more inputs use by insured farmers and more net returns and more employment, the country could not benefit much out of the scheme with the enhanced production in subsequent years; hence he suggested rainfall insurance scheme on a pilot scale. Kerr et al. (1999) made similar proposal. Rao (1999) suggested that crop insurance should be a private insurance with full backstopping in many sectors like roads, marketing etc., and fine-tuning and upgradation technologies in use on external inputs besides improving the water use efficiency.

The National Agricultural Insurance Scheme, introduced from *rabi* 1999-2000, has some good features like covering all farmers, crops including annual commercial crops, farmers having option to go in for a high sum insured equal to 150 percent of average yield of notified area etc. However, few recent learning examples from some states in India where weather insurance scheme was implemented by the corporate sector indicates that the effective implementation and benefits to the real beneficiaries would happen when weather data are site-specific and interpreted scientifically. Scientists should look into almost real-time assessment and communication technologies to facilitate better servicing of the insurance policy by instant estimation of economic damages and scientists should also endeavor to diverse sensors for remote sensing of soil moisture, vegetation cover, interpretation methodologies and real-time communication systems, which could be acceptable to the insurance companies as well as policy holders (Samra, 2006). Swaminathan (2007) in the report on 'National Policy for Farmers', submitted to the Government of India, proposed an *Agriculture Risk Fund* to be set up to insulate farmers from risks arising from droughts and other weather aberrations.

7.3 Rainfed Agro-Economic Zones: A Land Use Policy for Recurrent Drought Regions

The earlier developmental and relief measures focused through Drought Prone Area Program (DPAP) and Desert Development Program (DDP) in drought prone areas, which are primarily located in core rainfed agriculture regions, by the central and state governments could address some of the drought impacting issues. Presently, the focused issues in these areas are resource conservation and management, increased productivity and profitability, making rainfed agriculture a viable profession and improving the livelihood of the farmers/people in these areas. Hence, these issues call for an entirely new target domain approach by delineating core *'Rainfed Agro-Economic Zones'* in a district or part of a region in a state (Ravindra Chary et al., 2008b). The important criteria for delineation of these RAEZs could be a predominantly rainfed region with predominant rainfed production system, source and percentage net irrigated area and livelihoods mostly dependent on rainfed agriculture. In this approach, addressing the core issues of improving livelihood of the rainfed farmers while sustaining the land resources is focal, where care is taken for issues related to production through processing, profitability, improved livelihoods in harmony with conservation and maintenance of land resources. Instead of individual and piecemeal interventions the entire production system will be targeted to develop as Rainfed Agro-Economic Zones (RAEZs) which would act as hubs of rainfed agriculture development. Tools like crop or weather insurance, where the insurance product may have an inbuilt condition that a given product is applicable only if a particular crop or commodity is grown in RAEZs wherein the govt. can subsidize part of the premium for farmers who adopt rational water use. New opportunities are also arising in the area of CDM and carbon credits, which can be implemented in RAEZs where farmers can be compensated for adopting conservation practices, which contributes to drought amelioration and sustainable productivity on a long-term basis, but relatively lower returns on short term.

The functional mechanism has to be developed at RAEZ level in a district/region in a consortium mode, involving all the concerned stakeholders and converging all the relevant resources, programs, schemes and pooling resources (human, financial, etc.). Then redesign the strategic interventions, now available and also develop new, which will cover crop production, horticulture, livestock, NRM, post-harvest technology and value addition and market related issues. A sound *Land Use Policy*, separately for *rainfed regions*, at national level, equipped with adequate legislative/judicial powers combined with strong political will, shall promote creation of RAEZs. These RAEZs are contemplated to achieve much desired sustained development in drought prone area and, further, pave the way towards achieving "Second Green Revolution" from rainfed areas.

7.4 Recent Government Policies

A recent collaborative initiative between Government of India and United Nations Development Programme (UNDP) on disaster risk mitigation (DRM) envisages preparing disaster management plans for effective preparedness against all sudden disasters at community, village, block, district and provincial levels (National Institute of Disaster Management, 2006). Subsequently, the Parliament of India passed the "Disaster Management Act 2005" enabling local authorities such as panchayat raj institutions, municipalities, town and district planning authorities to prepare community based disaster management plans for effective disaster mitigation (MoLJ, 2005). More recent policies and initiatives undertaken by the Central Government, which either directly or indirectly help drought mitigation and management, are National Rural Employment Guarantee Programme (nrega.nic.in) in most drought affected areas, National Horticultural Mission (www.nhm.nic.in), expansion of institutional credit to farmers, establishment of National Rainfed Area Authority (http://www.nraa.in), watershed development and micro-irrigation programs, agribusiness development through venture capital participation by the Small Farmer Agribusiness Consortium, Reform and Support for Agriculture Extension Services, National Food Security Mission (www.indg.in/agriculture/rural-employment-schemes/national-food-security-mission), Rashtriya Krishi Vikas Yojana (www.agricoop.nic.in), etc. These policies or programs should be converged and coordinated at a district level for sustainable development of the region, livelihood improvement of the local people and for a better environment, particularly in the drought-prone regions of India. Recently set up National Rainfed Area Authority (NRRA) developed common guidelines for watershed development projects in consultation with various ministries in the Government of India, viz., Ministry of Agriculture (MoA), Ministry of Rural Development (MoRD), Ministry of Environment and Forestry (MoEF), and Ministry of Water Resources (MoWR) as well as the Panchayat Raj, Planning Commission, and National Bank for Agriculture and Rural Development (NABARD). These guidelines have become operational since April 2008 for the implementation of new watershed development projects in the country, which also broadly address drought mitigation at different scales (Samra and Sharma, 2009).

7.5 Capacity Building

Local governments and communities often lack capacity to deal with the catastrophic disasters, and hence there is a need to enhance their capacity (Ivery et al., 2004). This signifies to improve the capacities of local communities and of local governments including the states, which are chronically drought prone. By involving communities in disaster management planning at the local level preparedness, vulnerability mapping while preparing the community level drought management plans, local governments and disaster management can be enabled to gain better understanding about the vulnerability of the communities. Drought preparedness planning will increase the society's capacity to cope more effectively with the extremes of climate (Wilhite, 2002; Pearce, 2003; Hayes et al., 2004). Drought preparedness planning is presently lacking in India and it can be addressed through recent programs on drought risk mitigation (DRM). Further, a combination of mitigation programs guided by the rules for watershed/natural resource management programs of the Ministry of Rural Development, Government of India and preparedness planning by various community drought management plans is essential. There is also need for a long-term strategy to build the capacities of local governments, which can lead to less dependency on the central government. It is also necessary for the local governments to collaborate with community based organizations/NGOs in identified areas within the broad spectrum of drought risk mitigation. Such a clear identification of roles and responsibilities would certainly enhance to lead the local capacities to deal with the future risks (Prabhakar and Shaw, 2008).

Moreover, capacity building is needed for research, development and upscaling of drought management for various stakeholders. Some of the themes are: Farming Systems Research (FSR), Participatory Technology Development (PTD), crop modeling, relevant theme areas in agro-meteorological studies like resource characterization, weather forecasting and so on, application of information and communication technologies (ICTs), etc. Many NGOs, community-based organizations (CBOs), and common interest groups (CIGs) are involved in many activities related to drought management at various levels. Therefore, Public-Private-Partnership (PPP) even for the capacity building in drought management should be considered.

8. CONCLUDING REMARKS

Drought, as a climatic anomaly, is likely to be recurrent in the rainfed regions of India. Their occurrence is not as sudden as some other natural hazards, but in some situations their effects can be equally devastating. Drought is a complex phenomenon demanding multidisciplinary management approach. There are technologies, policies and programs available to moderate the effects of drought. Appropriate management of water, land, crop and natural resources can alleviate the effects of drought to a considerable extent. To meet the future challenges, the issues for drought monitoring and management that need proper attention are: drought prediction, detection and monitoring, impact assessment, and adoption and response. The issues in drought prediction are prediction of break-monsoon conditions and revival, rainfall forecasts for next 15 days, rainfall forecasts at a district level, quantification of current soil moisture status, assessment of crop status, assessment of the period up to which the currently available moisture can support the crop in case of no subsequent rainfall, drought prediction at a district level, development of location-specific crop contingency measures and dissemination to the farmers. A good weather code is the need of the hour. High-resolution satellite images, for example, from CARTOSAT could help in drought monitoring and assessment in the near future.

Many policies and programs, both at central and state levels, are focusing on the development of drought-prone regions. Convergence of the policies and programs at a national level with common guidelines while providing scope for state governments to slightly modify the guidelines to suit local settings, could address the drought-prone regions effectively with available resources. Since many institutions and/or agencies, both governmental and non-governmental, are generating the data on drought and its related information, concerted R&D efforts should be made by involving concerned institutions for database harmonization for the classification of droughts, delineating country into core drought-prone zones and developing guidelines at a national level for priority development. Agro advisories and weather insurance backed up with location-specific data would help agricultural development at large and the real beneficiaries in particular. Application of information and communication technologies (remote sensing, GIS, GPS, modeling, etc.) is to be strengthened for the quicker delivery of relief, insurance and adaptation objectively. Capacity building of local governments and communities involved, with roles defined, is one key area for drought management.

In the drought-prone regions, implementations of natural resource management (NRM) technologies are more important which need proper institutional and policy support, and hence the solution for such problems does not lie in mere research. Since many of the NRM technologies like community-based soil and water conservation measures are labor and energy intensive, institutions such as custom hiring centers and policies like linking rural employment guarantee programs with watershed development activities would help in faster upscaling of technologies and effective implementation of government programs. A national land use policy for the rainfed regions is needed wherein development of drought-prone regions is integral. Adequate policy and funding with appropriate institutional framework, coordination and cooperation between the agencies and stakeholders involved are crucial in addressing the '*drought as an opportunity not as a hazard*' at different scales in the country. These efforts are likely to address the drought disaster in a holistic manner, which in turn can pave the way for achieving the nation's aspiration of bringing "Second Green Revolution" from rainfed areas.

REFERENCES

AICRPDA (2003). Annual Reports of 22 Centers (1971-2001). All India Coordinated Research Project for Dryland Agriculture (AICRPDA), Central Research Institute for Dryland Agriculture (CRIDA), Hyderabad, India, 76,357 pp.

Biswas, B.C. (1990). Forecasting for agricultural application. *Mausam*, **41**: 3230-3234.

Chakravorthi, J.S. (1920). Agricultural Insurance: A Practical Scheme Suited to Indian Conditions. Government Press, Bangalore, India.

CRED (2006). India country profile of natural disasters. Centre for Research on Epidemiology of Disasters (CRED), the International Disaster Database (EM-DAT). http://www.em-da.net/disasters/Visualition/profiles/countryprofile.php (accessed on 6 March 2009).

DAC (2004). Drought 2002. Department of Agriculture and Cooperation (DAC), Ministry of Agriculture, Government of India, New Delhi, 190 pp.

Dracup, J.A., Lee, K.S. and Paulson Jr., E.G. (1980). On the definition of droughts. *Water Resources Research*, **16(2):** 297-302.

FAO (1983). Guidelines: Land Evaluation of Rainfed Agriculture. *Soil Bulletin*, **52,** Food and Agriculture Organization (FAO), Rome, Italy, 237 pp.

Gadgil, S., Vinaychandran, P.N. and Francis, P.A. (2003). Droughts of the Indian summer monsoon: Role of clouds over the Indian Ocean. *Current Science*, **85(12):** 1712-1719.

Gowarikar, V., Thapliyal, V., Kulkshetra, S.M., Manda, G.S., Sen Roy, N. and Sikka, D.R. (1991). A power regression model in long range forecasting of monsoon rainfall in India. *Mausam*, **42:** 125-130.

Hargreaves, G.H. (1975). Precipitation dependability and potential for Agricultural Production in North-east Brazil. EMBRAPA and United State University Publication No. 74-D 159.

Hayes, M.J., Wilhelmi, O.V. and Knutson, C.L. (2004). Reducing drought risk: Bridging theory and practice. *National Hazards Review*, **5(2):** 106-113.

IPCC (2007). Climate Change 2007: Impacts, Adaptations and Vulnerability. Inter-governmental Panel on Climate Change (IPCC), Working Group II, www.ipcc.ch.

Ivery, J.L., Smithers, J., De Loe, R.C., Kreutzwise, R.D. (2004). Community capacity adaptation to climate induced water shortages: Linking institutional complexity and local actors. *Environmental Management*, **33(1):** 36-47.

Kerr, J., Hazell, P. and Jha, D. (1999). Sustainable development of rainfed agriculture in India. EPTD Discussion Paper No. 20 (Module IV of ICAR/World Bank), 71 pp.

Krishna Rao, M.V. (2005). Agricultural drought assessment and monitoring using satellite data. *In:* K.D. Sharma and K.S Ramasastri (editors), Drought Management. Allied Publishers, New Delhi, pp. 132-138.

Kulkshetra, S.M. (1997). Drought management in India and potential contribution of climate prediction. Joint COLA/CARE Technical Report No. 1. Centers of Institute of Global Environment and Society, 105 pp.

Lal, M., Nozawa, T., Emori, S., Harasawa, H., Takahashi, K., Kimoto, M., Abe-Ouchi, A., Nkajima, T., Takemura, T. and Numaguti, A. (2001). Future climate change: Implications for Indian summer monsoon and its variability. *Current Science*, **81:** 1196-1207.

Linsley, R.K., Kohler, M.A. and Paulhus, J.L.H. (1988). Hydrology for Engineers. McGraw-Hill Book Company Ltd., London, U.K.

Mandal, C., Mandal, D.K., Srinivas, C.V., Sehgal, J. and Velayutham, M. (1999). Soil-Climatic Database for Crop Planning in India. National Bureau of Soil Survey and Land Use Planning (NBSS& LUP), Indian Council of Agricultural Research (ICAR), Nagpur, India, 994 pp.

Mayande, V. (2006). Improved farm implements for rainfed regions. *In:* K.D. Sharma and B. Soni, (editors), Land Use Diversification for Sustainable Rainfed Agriculture. Atlantic Publishers, New Delhi, pp. 149-162.

Mishra, P.K. (1999). India's Comprehensive Crop Insurance Scheme: Lessons and Insights. Development and Operation of Agricultural Insurance Schemes in Asia. Asian Productivity Organization, Tokyo, pp. 41-60.

MoLJ (2005). National Disaster Management Act. Ministry of Law and Justice (MoLJ), Government of India, New Delhi.

National Institute of Disaster Management (2006). National Disaster Risk Management Program. Mid-term Review Report, National Institute of Disaster Management (NIDM), Ministry of Home Affairs, Government of India, New Delhi.

Palmer, W.C. (1965). Meteorological Drought. Research Paper 45, U.S. Weather Bureau (now National Weather Service), Silver Spring, Maryland, USA, 58 pp.

Pearce, L. (2003). Disaster management and community planning, and public participation: How to achieve sustainable hazard mitigation? *Natural Hazards*, **28(2-3):** 211-228.

Prabhakar, S.V.R.K. and Shaw, R. (2008). Climate change adaptation implications for drought risk mitigation: A perspective for India. *Climate Change*, **88:** 113-130.

Rajeevan, M., Pai, D.S., Dikshit, S.K. and Kelkar, R.R. (2004). IMD's new operational models for long range forecast of southwest monsoon rainfall over India; their verification for 2003. *Current Science*, **86(3):** 422-431.

Ramakrishna, Y.S. (1997). Climate features of Indian Arid Zone. *In:* S. Singh and A. Kar (editors), Desertification Control in the Arid Ecosystem of India for Sustainable Development. Agro-Botanical Publishers, Bikaner, Rajasthan, India, pp. 27-35.

Ramakrishna, Y.S., Rao, G.G.S.N., Rao, V.U.M., Rao, A.V.S.M. and Rao, K.V. (2007). Water management: Water use in rainfed regions of India. *In:* M.V.K. Sivakumar and R.P. Motha (editors), Managing Weather and Climate Risks in Agriculture. Springer, Berlin, pp. 245-264.

Ramakrishna, Y.S., Rao, G.G.S.N., Vijayakumar, P. and Kesava Rao, A.V.R. (2005). Droughts and their impact. *In:* K.D. Sharma and K.S. Ramasastri (editors), Drought Management. Allied Publishers, New Delhi, pp. 147-156.

Ramdas, L.A. (1960). Crops and Weather in India. Indian Council of Agricultural Research (ICAR), New Delhi, 127 pp.

Rao, D.P., Venkataratnam, L., Krishna Rao, M.V., Ravi Sankar, T. and Rao, S.V.C.K. (1999). Role of remote sensing for resource characterization in dryland areas. *In:* H.P. Singh, Y.S. Ramakrishna, K.L. Sharma and B. Venkateswarlu (editors), Fifty Years of Dryland Agricultural Research in India. Central Research Institute for Dryland Agriculture (CRIDA), Indian Council of Agricultural Research (ICAR), Hyderabad, India, pp. 81-92.

Rao, H. (1999). Declining per capita demand for food grains in India: Why and what does this imply? Presidential Address, 13th National Congress on Agricultural Marketing, Hyderabad, India, 15 pp.

Rathore, M.S. (2004). State Level Analysis of Drought Policies and Impacts in Rajasthan, India. IWMI Working Paper 93, Drought Series Paper No. 6, IWMI, Colombo, Sri Lanka, 40 pp.

Ravindra Chary, G., Maruthi Sankar, G.R., Subba Reddy, G., Ramakrishna, Y.S., Singh, A.K., Gogoi, A.K. and Rao, K.V. (editors) (2008a). District-wise Promising Technologies for Rainfed Cereals. Based Production Systems in India. All India Coordinated Research Project for Dryland Agriculture, Central Research Institute for Dryland Agriculture (CRIDA), Indian Council of Agricultural Research (ICAR), Hyderabad, India, 204 pp.

Ravindra Chary, G., Venkateswarlu, B., Maruthi Sankar, G.R., Dixit, S., Rao, K.V., Pratibha, G., Osman, M. and Kareemulla, K. (2008b). Rainfed Agro-Economic Zones (RAEZs): A step towards sustainable land resource management and improved livelihoods. Proceedings of the National Seminar on Land Resource Management and Livelihood Security. Indian Society of Soil Survey and Land Use Planning, 10-12 September 2008, Nagpur, India, p. 70.

Ravindra Chary, G., Vittal, K.P.R. and Maruthi Sankar, G.R. (2004). A framework of rainfed land use plan for participatory self-sustaining viable bio-diverse mixed farming system. Lead Paper, Proceedings of the National Seminar on Soil Survey for Land Use Planning, 24-25 January 2004, Indian Society of Soil Survey and Land Use Planning, National Bureau of Soil Survey and Land Use Planning (NBSS&LUP), Nagpur, India, p. 2.

Ravindra Chary, G., Vittal, K.P.R., Ramakrishna, Y.S., Sankar, G.R.M., Arunachalam, M., Srijaya, T. and Bhanu, U. (2005). Delineation of soil conservation units (SCUs), soil quality units (SQUs) and land management units (LMUs) for land resource appraisal and management in rainfed agro-ecosystem of India: A conceptual approach. Lead Paper, Proceedings of the National Seminar on Land Resources Appraisal and Management for Food Security. Indian Society of Soil Survey and Land Use Planning, National Bureau of Soil Survey and Land Use Planning (NBSS&LUP), Nagpur, India, 10-11 April 2005, p. 212.

Roy, P.S., Joshi, P.C., Murthy, C.S. and Kishtawal, C.M. (2006). Geoinformatics for drought assessment. *In:* J.S. Samra, G. Singh and J.C. Dagar (editors), Drought Management Strategies in India. Indian Council of Agricultural Research (ICAR), New Delhi, pp. 23-60.

Saith, N. and Singh, J. (2006). The role of the Midde-Julian Oscillation in the El Nino and Indian drought of 2002. *International Journal of Climatology*, **26**: 1361-1378.

Samra, J.S. (2006). Droughts, risks, insurance and management assessment in India. *In:* J.S. Samra, G. Singh and J.C. Dagar (editors), Drought Management Strategies in India. Indian Council of Agricultural Research (ICAR), New Delhi, India, pp. 1-22.

Samra, J.S. and Sharma, K.D. (2009). Watershed development: How to make 'invisible' impact 'visible'? *Current Science*, **96(2):** 203-205.

Sarkar, J., Shewale, M.P., Dikshit, S.K. and Shyamala, B. (2005). Drought concepts, its monitoring and management. *In:* K.D. Sharma and K.S. Ramasastri (editors), Drought Management. Allied Publishers, New Delhi, pp. 107-131.

Sastry, A.S.R.A.S., Ramakrishna, Y.S. and Ramana Rao, B.V. (1981). A new method of classification of agricultural droughts. *Arch. Fur. Met. Geoph. Und. Biokl. Ser.*, **29**: 293-297.

Schneider, S.H. (1996). Encyclopedia of Climate and Weather. Oxford University Press, New York, pp. 256-257.

Selvaraju, R. (2003). Impact of El Nino Southern Oscillation on Indian food grain production. *International Journal of Climatology*, **23**: 187-206.

Sharma, K.D., Pandey, R.P. and Mishra, P.K. (2006). Hydrological drought: Assessment methods for stream flow and groundwater. *In:* J.S. Samra, G. Singh and J.C. Dagar (editors), Drought Management Strategies in India. Indian Council of Agricultural Research (ICAR), New Delhi, pp. 80-102.

Singh, R.P. and Raman Rao, B.V. (1988). Agricultural Drought Management in India: Principles and Practices. Central Research Institute for Dryland Agriculture (CRIDA), Indian Council of Agricultural Research (ICAR), Hyderabad, India.

Singh, S. (1995). Annual Progress Report. Indo-US Project on Research on Mechanization of Dryland Agriculture. Institute of Agricultural Sciences, Varanasi, India.

Singh, S.V., Datta, R.K., Rathore L.S. and Sanjeeva Rao, P. (1999). Role of weather forecasts in dryland agriculture. *In:* H.P. Singh, Y.S. Ramakrishna, K.L. Sharma and B. Venkateswarlu (editors), Fifty Years of Dryland Agricultural Research in India. Central Research Institute for Dryland Agriculture (CRIDA), Indian Council of Agricultural Research (ICAR), Hyderabad, India, pp. 227-234.

Sivakumar, M.V.K. (2008). Climate change and food security. Proceedings of the International Symposium on Agrometeorolgy and Food Security. Association of Agrometeorologists and Central Research Institute for Dryland Agriculture (CRIDA), Hyderabad, India, 18-21 February 2008, CRIDA, Hyderabad, India.

Srivastava, N.S.I. (1993). Projected demand of agricultural machinery for the year 2000 A.D. Proceedings of the 28th Convention of Indian Society of Agricultural Engineers (ISAE). Central Institute for Agricultural Engineering (CIAE), Bhopal, India.

Subba Reddy, G., Ramakrishna, Y.S., Ravindra Chary, G. and Maruthi Sankar, G.R. (editors) (2008). Crop and Contingency for Rainfed Regions of India: A Compendium by AICRPDA. All India Coordinated Research Project for Dryland Agriculture, Central Research Institute for Dryland Agriculture (CRIDA), Indian Council of Agricultural Research, Hyderabad, India, 174 pp.

Subrahmanyam, K.V. and Nagasree, K. (2005). Drought management: Present policies and future requirements. *In:* K.D. Sharma and K.S Ramasastri (editors), Drought Management. Allied Publishers, New Delhi, pp. 12-24.

Subrahmanyam, V.P. (1964). Climatic water balance of the Indian arid zone. Proceedings of the National Symposium on Problems of Arid Zone, Jodhpur, Rajasthan, India, pp. 405-411.

Swaminathan, M.S. (2007). National Policy for Farmers. Report Submitted to the Ministry of Agriculture, Government of India, New Delhi.

Thapliyal, V. (1986). Long Range Forecasting of Monsoon Rainfall in India. World Meteorological Organization (WMO), Geneva, Volume 87, pp. 723-732.

Thronthwaite, C.W. (1948). An approach towards a rational classification of climate. *Geogr. Rev.*, **38**: 85-94.

UNCCD (1995). United Nations Convention to Combat Desertification. Interim Secretariat for the Convention to Combat Desertification, Geneva, 71 pp.

Van Bavel, C.H.M. (1953). A drought criterion and its applications in evaluating drought incidence of hazard. *Agronomy Journal*, **5**: 167-172.

Vanrooy, M.P. (1965). A rainfall anomaly index independent of time and space. NATO Weather Bureau, South Africa, **14**: 43-48.

Velayutham, M. (1999). Crop and land use planning in dryland agriculture. *In:* H.P. Singh, Y.S. Ramakrishna, K.L. Sharma and B. Venkateswarlu (editors), Fifty Years of Dryland Agricultural Research in India. Central Research Institute for Dryland Agriculture (CRIDA), Indian Council of Agricultural Research, Hyderabad, India, pp. 73-80.

Venkateswarlu, J. (2005). Moisture stress in rainfed farming. *In:* J. Venkateswarlu (editor), Rainfed Agriculture in India: Research and Development Scenario. Indian Council of Agricultural Research (ICAR), New Delhi, pp. 178-219.

Venkateswarlu, J. and Bhaskara Rao, U.M. (1999). Drought management. *In:* H.P. Singh, Y.S. Ramakrishna, K.L. Sharma and B. Venkateswarlu (editors), Fifty Years of Dryland Agricultural Research in India. Central Research Institute for Dryland Agriculture (CRIDA), Indian Council of Agricultural Research, Hyderabad, India, pp. 235-248.

Virmani, S.M., Siva Kumar, M.V.K. and Reddy, S.J. (1982). Rainfall Probability Estimates for Selected Locations of Semi-Arid India. Research Bulletin No. 1, 2nd Edition, International Crops Research Institute for Semi Arid Tropics (ICRISAT), Hyderabad, India.

Vittal, K.P.R., Rao, K.V., Sharma, K.L., Victor, U.S., Ravindra Chary, G. and Sankar, G.R.M. (2005). Guidelines for rainfed production systems selection and management. *In:* K.D. Sharma and K.S Ramasastri (editors), Drought Management. Allied Publishers, New Delhi, pp. 76-106.

Vittal, K.P.R., Ravindra Chary, G., Rama Rao, C.A. and Sankar, G.R.M. (2006). Crop diversification in rainfed regions of India. *In:* K.D. Sharma and B. Soni (editors), Land Use Diversification for Sustainable Agriculture. Atlantic Publishers and Distributors, New Delhi, pp. 24-70.

Vittal, K.P.R., Singh, H.P., Rao, K.V., Sharma, K.L., Victor, U.S., Ravindra Chary, G., Sankar, G.R.M., Samra, J.S. and Singh, G. (editors) (2003). Guidelines on Drought Coping Plans for Rainfed Production Systems. All India Coordinated Research Project for Dryland Agriculture, Central Research Institute for Dryland Agriculture (CRIDA), Indian Council of Agricultural Research, Hyderabad, India, 39 pp.

Walker, G.T. (1910). Correlation in Seasonal Variation of Weather. Memoirs XXII, India Meteorological Department, New Delhi.

Whipple, W. (1966). Regional drought frequency analysis. Proceedings of the American Society of Civil Engineering, **92(IR2):** 11-31.

Wilhite, D.A. (2002). Combating drought through preparedness. *Natural Resource Forum,* **26:** 275-285.

Yurekli, K. and Kurune, A. (2006). Simulating drought periods based on daily rainfall and crop water consumption. *Journal of Arid Environment,* **67:** 629-640.

Indicators for Assessing Drought Hazard in Arid Regions of India

K.P.R. Vittal, Amal Kar* and A.S. Rao

1. INTRODUCTION

Drought is a normal phenomenon of earth's climate, and is a common feature in the drylands of India. Indian agriculture, which is highly dependent on the monsoon rainfall, is a major victim of drought. Since nearly 70% of the net sown area in the country is rainfed, aberrant behavior of the monsoon such as low and poor rainfall distribution, its delayed onset, or prolonged dry spells during cropping season, often result in low crop yields. Arid western part of Rajasthan state is a frequent victim of moderate to severe droughts, resulting in huge economic loss and natural resources. Severe droughts have reduced food grain production in western Rajasthan by 70% in 1987-1988, and by 50% in 2002-2003 (Anon., 2004a; Narain and Kar, 2005).

Severity and recurrence of drought in the hot arid zones of India (32 million ha; 62% in Rajasthan, 20% in Gujarat) compels the government to spend a huge sum on drought relief and rehabilitation measures, but there is often some confusion or delay in reaching the affected sections of the population, or parts of the region, especially due to poor infrastructural facilities, but also due to delay in proper assessment and warning, which reflect the inadequacies in drought assessment and monitoring tools. Thus a proper assessment of the severity of drought not only depends on the duration, frequency, intensity and geographical distribution of rainfall, but also on the effects on human, animal, crops and vegetation cover of a region. Seasonal temperature, wind velocity, sunshine, density of vegetation and moisture retaining capacity of soil and soil moisture balance in surface and sub-soil, and groundwater influence water demand during drought years.

In this chapter, an overview of various internationally accepted drought indicators and their application to the drought-prone arid Rajasthan, India, has been provided.

2. TYPES OF DROUGHT

Droughts are classified into four main categories as described below.

2.1 Meteorological Drought

Meteorological drought is a situation when the seasonal rainfall over an area is less than 75% of its long-term average. It is further classified as "*moderate drought*", when the rainfall deficit is between 26% and 50% and "*severe drought*" when it exceeds 50%. Meteorological drought can be at local, regional or extensive scale, varying in extent from a few clusters of *tehsils*/districts to several meteorological sub-divisions. In temporal scale, a drought can last for a few weeks or longer.

---
*Corresponding Author

2.2 Hydrological Drought

Prolonged meteorological drought can result in hydrological drought with marked depletion in surface water input, and consequent drying up of reservoirs, lakes, decline in stream flow and also fall in groundwater table.

2.3 Agricultural Drought

An agricultural drought occurs when soil moisture and rainfall are inadequate during the growing season to support a healthy crop growth till maturity, causing extreme crop stress and drastic reduction in yield.

2.4 Socio-economic Drought

It is a situation, where water shortage ultimately adversely affects the economy of the region. It combines the impact of meteorological, hydrological and agricultural droughts on society, especially in terms of supply and demand of commodities and purchasing power of the people. Severe societal drought may even lead to mass migration in search of food, fodder, water and work, leading to famine, death and social unrest. The worst hit sections of the society during drought are the people below the poverty line and the landless people. In an agriculture-dependent country like India, once the agricultural production declines due to drought, it sets in a chain reaction, leading to lower availability of commodities, lower purchasing power and lower economic growth down the spiral of poverty, hunger and survival of poor people.

3. KINDS OF DROUGHT INDICATORS

Monitoring of drought is usually carried out on the basis of interpretation of a set of indicators on a regular basis during critical periods. For quantitative measurement of the severity of drought, specific limits or threshold values below which the economy is not affected are put to the measured parameters. Various indicators used for drought monitoring can be classified as: *physical, biological* or *socio-economic indicators* depending on the aspect covered by the indicator. The *physical indicators* include rainfall or effective soil moisture, surface water availability, depth to groundwater, etc. The *biological or agricultural indicators* are usually comprised of vegetation cover and composition, crop and fodder yield, condition of domestic animals, pest incidence, etc. The *social indicators* are essentially the impact indicators and include food and feed availability, land use conditions, livelihood shifts, migration of human and livestock population, etc. However, the commonly used indicators measure only the levels of precipitation to meet: (a) agricultural need, (b) drinking water supply for both human and livestock, and (c) storage of water reservoirs.

4. METEOROLOGICAL DROUGHT INDICATORS

4.1 Deciles of Precipitation

In this approach, monthly precipitation totals from a long-term record are first ranked from the highest to the lowest to construct a cumulative frequency distribution. The distribution is then split into ten parts (tenths of distribution or deciles). The first decile is the precipitation value not exceeded by the lowest 10% of all precipitation values in a record, the second is between the lowest 10 and 20 per cent, etc. Any precipitation value can be compared with and interpreted in terms of these deciles. A reasonably long precipitation record (30-50 years) is required for this approach.

Decile Indices (DI) are grouped into five classes, two deciles per class. If the precipitation falls into: (i) Deciles 1 and 2 (the lowest 20%) is classified as "much below normal" precipitation; (ii) Deciles 3 and 4 (20 to 40%) as "below normal" precipitation; (iii) Deciles 5 and 6 (40 to 60%) as "near normal" precipitation; (iv) Deciles 7 and 8 (60 to 80%) as "above normal" precipitation; and (v) Deciles 9 and 10 (80 to 100%) as "much above normal".

Merits and demerits: DI is relatively simple to calculate, requires only precipitation data and fewer assumptions than more comprehensive indices like PDSI or SWSI. Deciles are widely used in Australia to trigger drought relief programs (Gibbs and Mather, 1967).

4.2 Precipitation Departures from Normal

India Meteorological Department (IMD) describes meteorological drought from rainfall departure from its long-term averages and declares meteorological drought on a weekly/monthly basis. The percent rainfall anomalies from normal to be qualified as drought of a severity class are shown in Table 1 (Kulshreshtha and Sikka, 1989).

Table 1. Meteorological drought classification

Departure of annual rainfall from normal (%)	*Meteorological drought condition*
> 0	No drought
–1 to –25	Mild drought
–26 to –50	Moderate drought
< –50	Severe drought

Merits and demerits: This is the most accepted measure of drought in India because of its simplicity. When more than 50% area of the country is under moderate or severe drought, the country is described as severely affected; when the affected area is 26-50% of the country, it is categorized as moderate drought. One of its disadvantages is that the average precipitation is often not the same as the median precipitation, which is the value exceeded by 50% of the precipitation occurrences in a long-term climate record. Another drawback is that the distribution or time-scale of rainfall is not taken into account.

4.3 Palmer Drought Severity Index (PDSI)

Palmer (1965) developed a soil moisture algorithm (a model), which uses data on precipitation, temperature and local available water content (AWC) of the soil. It uses the computed values of CAFEC (Climatically Appropriate for Existing Conditions) rainfall, which is the normal value for the established human activities at a place. This parameter can be obtained by water balance technique. The anomaly (PDSI), which is a difference between the actual and the CAFEC precipitation, is used as a drought indicator. PDSI generally varies between –4.0 (extreme drought) and +4.0 (adequate moisture condition). The index values for categories of drought are given in Table 2.

Merits and demerits: The PDSI is a popular index and is widely used for a variety of applications in the United States, including agriculture. It has also been used for drought monitoring to initiate actions associated with drought contingency plans. The index provides decision makers with a measurement of the abnormality of recent weather for a region. It also provides an opportunity to place current conditions

Table 2. PDSI values for drought categories

Index value	Drought class
−1.00 to −1.99	Mild drought
−2.00 to −2.99	Moderate drought
−3.00 to −3.99	Severe drought
< −4.00	Extreme drought

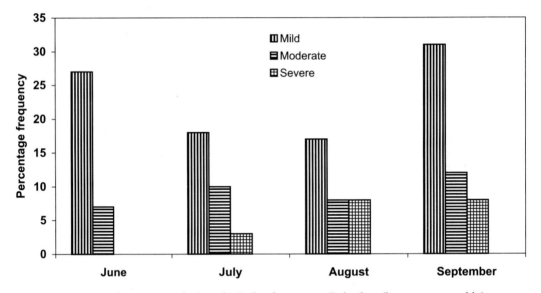

Fig. 1 Palmer's Drought Severity Index for western Rajasthan (long-term, monthly).

in historical perspective, and also spatial and temporal representations of historical droughts. The PDSI is a meteorological drought index, and responds to weather conditions that have been abnormally dry or abnormally wet. However, the Palmer values may lag emerging droughts by several months or weeks and is not suited for mountainous areas or areas that experience frequent climatic extremes. Figure 1 illustrates the pattern of long-term PDSI for the summer monsoon months in western Rajasthan.

4.4 Standardized Precipitation Index (SPI)

The SPI is an index based on the probability of precipitation for any time scale (McKee et al., 1993). Table 3 shows the drought classes based on SPI values. The SPI is calculated as follows:

Table 3. Drought classes based on SPI values

SPI	Drought class
Less than −2.0	Extreme drought
−1.50 to −1.99	Severe drought
−1.0 to −1.49	Moderate drought
−0.99 to −0.0	Mild drought

$$SPI = \frac{X - X_{mean}}{\sigma} \tag{1}$$

where X = precipitation for the station, X_{mean} = mean precipitation, and σ = standard deviation.

Merits and demerits: The SPI can be computed for different time scales, can provide early warning of drought, help assess the drought severity, and is less complex than the PDSI. Soil moisture conditions respond to precipitation anomalies on a relatively short time scale. Groundwater, stream flow, and reservoir storage reflect the longer-term precipitation anomalies. For these reasons, the SPI is calculated for 3-, 6-, 12-, 24-, and 48-month time scales. The measured drought condition in western Rajasthan during 2002 (one of the most severe droughts in the region during recorded period) through SPI values, and that during 2003 (overall a normal year), is depicted in Fig. 2. The calculated relationship between SPI and pearl millet (a summer monsoon crop) yield in Jodhpur district between 1971 and 2006 is shown in Fig. 3.

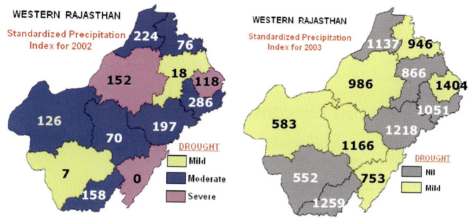

Fig. 2 Pearl millet yield (kg/ha) during severe drought of 2002 and that during a normal year 2003 in western Rajasthan.

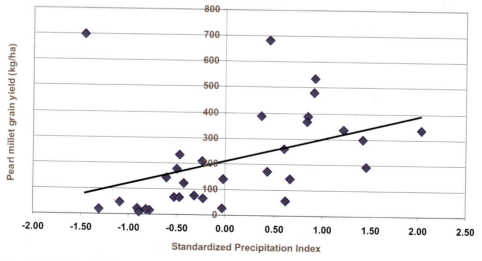

Fig. 3 Relationship between Standardized Precipitation Index and pearl millet yield in Jodhpur district (1971-2006).

5. HYDROLOGICAL DROUGHT INDICATORS

Measurement of both groundwater and surface water levels provide important clues to drought occurrence. Monitoring reservoir water levels and groundwater table through a closed well observation network is important for drought assessment.

5.1 Standardized Water Level Index (SWI)

The SWI is an index based on the probability of water level in a reservoir for any time scale. It is calculated as:

$$SWI = \frac{W_{ij} - W_{im}}{\sigma} \tag{2}$$

where W_{ij} = seasonal water level for the ith and jth observations, W_{im} = seasonal mean water level, and σ = standard deviation of water levels.

Merits: The SWI can be computed for different time scales, and it can provide early warning of water shortage and hydrological drought severity.

5.2 Surface Water Supply Index (SWSI)

This index integrates reservoir storage, stream flow and two precipitation types (snow and rain) into a single index number (Shafer and Dezman, 1982). SWSI is expressed as:

$$SWSI = \frac{aP_{snow} + bP_{prec} + cP_{strm} + dP_{resv} - 50}{12} \tag{3}$$

where a, b, c and d = weights for snow, rain, stream flow and reservoir storage, respectively ($a + b + c + d = 1$), and P_i = probability (%) of non-exceedance for the ith component of water balance. Calculations are performed with a monthly time step. The SWSI is designed for river basins where mountain snow pack is a key element of water supply.

Merits and Demerits: SWSI calculation is unique to a basin or a region, and thus has better focus on the characteristics of the area for a time-series analysis. However, it is difficult to compare SWSI values between different basins or regions. Extreme events also cause a problem if the events are beyond the historical time series, and the index will need to be re-evaluated to include these events within the frequency distribution of a basin component.

5.3 Reclamation Drought Index (RDI)

The RDI has been developed for defining drought severity and duration, and for predicting the onset and end of drought (Table 4). Like the SWSI, the RDI is calculated at the river basin level, incorporating temperature and precipitation, snow pack, streamflow and reservoir levels as inputs.

Merits and demerits: The RDI differs from the SWSI in that it builds a temperature-based demand component and duration into the index. The RDI is adaptable to each particular region and its main strength is its ability to account for both climate and water supply factors. Because the index is unique to each river basin, inter-basin comparisons are limited. The RDI values and severity designations are similar to the SPI, PDSI and SWSI.

Table 4. RDI classification of drought

Values of RDI	Drought classification
4.0 or more	Extremely wet
1.5 to 4.0	Moderately wet
1 to 1.5	Normal to mild wetness
0 to –1.5	Normal to mild drought
–1.5 to –4.0	Moderate drought
–4.0 or less	Extreme drought

6. AGRICULTURAL DROUGHT INDICATORS

In a country like India, where agriculture provides livelihood to maximum population, the impact of drought on agriculture, including livestock, assumes the highest priority. Several indices are available to measure agricultural droughts (Hounam, 1975) of which some commonly used indicators are described in subsequent sections.

6.1 Aridity Index

Aridity index (Thornthwaite and Mather, 1955) is the percentage ratio of annual water deficit to annual water need or annual potential evapotranspiration (PE). Aridity anomaly (I_a) index is the departure of aridity index value from normal, expressed as a percentage (Table 5). India Meteorological Department (IMD) monitors agricultural drought during *kharif* season for the country as a whole and during *rabi* season for those areas which receive rainfall during northeast monsoon. The values of aridity anomaly (I_a) index are then plotted on a map and analyzed for identification of drought intensity.

Table 5. Drought categories based on aridity index

Drought category	*Aridity anomaly value*
Mild drought	Up to 25%
Moderate drought	26-50%
Severe drought	More than 50%

Merits and demerits: The aridity index indicates the water deficit conditions in a region. Crop-water requirements, which vary not only for each crop, but also with geographical location, are not considered in this index. The water balance calculations have also limitations for not properly accounting rainfall-runoff before stored moisture is estimated.

6.2 Moisture Adequacy Index (MAI)

Central Arid Zone Research Institute (CAZRI) developed a technique for the quantification of agricultural drought by taking MAI (AE/PE in percentage) during different phenological stages of a crop (Ramana Rao et al., 1981; Sastri et al., 1981). The MAI is obtained from weekly water balance studies. The drought impact is more important at certain stages than others, and is sought to be identified for a crop through this method. Subsequently, the drought conditions at different phenophases are integrated into a single drought code like mild, moderate and severe. Depending on the values of AE/PE during different phenophases, the drought code varies as S_0, V_1, R_2, S_1, V_2, R_3, etc. When the crop factor is introduced, the drought code in three syllables is unified into a single drought code applicable to one particular crop for a specific region. The scheme for determining drought intensity code at different phenophases is elaborated in Table 6.

Table 6. Phenophase-wise drought intensity codes from MAI

AE/PE (%)	Drought intensity	Phenophase-wise code for stages		
		Seedling (S)	Vegetative (V)	Reproductive (R)
76 to 100	No drought	S_0	V_0	R_0
51 to 75	Mild drought	S_1	V_1	R_1
26 to 50	Moderate drought	S_2	V_2	R_2
25 or less	Severe drought	S_3	V_3	R_3

The scheme may be illustrated with an example of pearl millet crop in western Rajasthan. The average growing season of pearl millet crop is about 14 weeks. The duration of different phenophases of the crop in this region are: seedling (S) three weeks, vegetative (V) four weeks, reproductive (R) four weeks, and maturity (M) three weeks. As the water stress during maturity stage does not have much influence compared to the water stress during other three stages, the maturity stage is eliminated and the three syllable agricultural drought code can be unified into a single code. The possible combinations that could normally be derived for the arid region of Rajasthan are shown in Table 7.

Interpreting Table 7, the agricultural drought for pearl millet in western Rajasthan can be defined in the following terms (Ramana Rao et al., 1981; Sastri et al., 1981):

- Agricultural drought for pearl millet crop is severe (A_3) when both vegetative (V) and reproductive (R) stages experience severe drought with any combination of S_0, S_1, S_2 or S_3, as the water requirement during seedling stage is usually less. In these circumstances even the natural grasses too suffer from drought conditions.
- Agricultural drought for the crop is moderate (A_2) when vegetative (V) and reproductive (R) stages experience one moderate and one severe drought each, with any combination of S_0, S_1, S_2 or S_3. During this situation short-duration crops also suffer from drought.
- Pearl millet crop escapes drought situation (A_0) even when mild drought prevails in one or two growth stages, with no drought condition in the third stage.

Table 7. Possible combinations of phenophase-wise drought for pearl millet crop in western Rajasthan

Intensity of agricultural drought			
No drought	Mild drought	Moderate drought	Severe drought
$S_0 V_0 R_0$	$S_0 V_0 R_2$	$S_0 V_0 R_3$	$S_0 V_2 R_3$
$S_0 V_0 R_1$	$S_0 V_1 R_2$	$S_0 V_1 R_3$	$S_0 V_3 R_2$
$S_0 V_1 R_0$	$S_0 V_2 R_0$	$S_0 V_2 R_2$	$S_0 V_3 R_3$
$S_1 V_0 R_0$	$S_0 V_2 R_1$	$S_0 V_3 R_0$	$S_1 V_2 R_3$
$S_1 V_0 R_1$	$S_1 V_0 R_2$	$S_0 V_3 R_1$	$S_1 V_3 R_2$
$S_1 V_1 R_0$	$S_1 V_1 R_1$	$S_1 V_0 R_3$	$S_1 V_3 R_3$
$S_0 V_1 R_1$	$S_1 V_1 R_2$	$S_1 V_1 R_3$	$S_2 V_0 R_3$

Table 8. Influence of commencement of sowing rains on the occurrence of agricultural drought in western Rajasthan (1901-1995)

Station	Commencement of sowing rains	Frequency of occurrence of agricultural drought with intensity				
		Nil	Mild	Moderate	Severe	Total
1. Sikar	Early	8	11	8	3	30
	Normal	16	12	7	8	43
	Late	7	5	3	7	22
2. Jodhpur	Early	6	1	8	4	19
	Normal	13	9	8	13	43
	Late	3	10	4	16	33
3. Barmer	Early	2	3	6	4	15
	Normal	5	11	6	7	29
	Late	0	13	4	34	51
4. Jaisalmer	Early	1	0	2	6	9
	Normal	1	2	6	10	19
	Late	2	2	0	63	67

The rest of the situations result in mild drought (A_1) for pearl millet. Short-duration crops like pulses escape drought under these circumstances. Using the above index, Rao (1997) assessed the influence of sowing rains on agricultural drought for some stations across western Rajasthan (Table 8).

Merits and demerits: The water balance calculation takes into account soil characteristics, crop growing period and water requirement of major crops. The drought is specified crop-wise and on a real-time basis. In other words, the calculation is highly data-intensive, and may require field-level information from a large network of monitoring sites. With the gradual advancements in satellite remote sensing and remote monitoring of weather, soil moisture status and crop growth conditions, however, this method has a better prospect.

6.3 Crop Moisture Index (CMI)

The Crop Moisture Index (CMI) uses a meteorological approach to monitor weekly crop conditions. It was developed by Palmer (1968) following the procedures within the calculation of the PDSI. While the PDSI monitors the long-term meteorological wet and dry spells, the CMI has been designed to evaluate short-term moisture conditions across the major crop-producing regions. It is based on the mean temperature and total precipitation for each week within a climate division, as well as the CMI value from the previous week. The CMI responds rapidly to changing conditions, and it is weighted by location and time so that maps, which commonly display the weekly CMI, can be used to compare moisture conditions at different locations.

Merits and demerits: Since CMI is designed to monitor short-term moisture conditions affecting a growing crop, it is not a good long-term drought-monitoring tool. Another limiting characteristic of CMI as a long-term drought-monitoring tool is that it typically begins and ends each growing season near zero. This limitation prevents CMI from being used to monitor moisture conditions outside the general growing season, especially in droughts that extend over several years.

6.4 Crop Water Stress Index (CWSI)

The CWSI value is a daily integration of plant-available soil water, evaporative demand and plant phenological stage susceptibility. It is defined for the growing season as follows (Saxton, 1989):

$$CWSI = \sum_{Planting}^{Harvest} \left(1 - \frac{T}{T_p}\right) SUS \qquad (4)$$

where T = computed actual transpiration (mm/day), T_p = potential transpiration (mm/day), and SUS = seasonally dependent weighting factor for grain-yield susceptibility.

Merits and demerits: This index needs the running of a dynamic simulation model, Soil-Plant-Atmosphere-Water Model (SPAW), for simulation of soil water and calculation of effective rainfall used for plant transpiration. The SPAW model needs calibration for each crop and region, and hence has a limitation for its use. However, the estimates obtained from this model are reasonably good (Rao and Saxton, 1995).

6.5 Performance of Different Agricultural Drought Indicators

The comparative performance of different agricultural drought indicators in Jodhpur district of Rajasthan is presented in Table 9. Standardized Precipitation Index, which is a meteorological drought indicator, performed the least (with correlation coefficient = 0.216) and Crop Water Stress Index (SPAW model) performed the best (with correlation coefficient 0.890). Unfortunately, its execution at field-level over a large area demands a huge database on weather, soil and crop parameters that are still difficult to gather reliably in near-real time. Lack of good field-scale details on soils, crop and weather precludes the testing of its actual potential at a zonal to regional scale. Moisture Adequacy Index (MAI), which is simple to calculate on the basis of rainfall and temperature data, has shown more promises in drought monitoring.

Table 9. Comparative performance of agricultural drought indicators in Jodhpur district

Indicator	Correlation with pearl millet yield (1971-2005) R^2 value	Merits and demerits
1. Standardized Precipitation Index	0.216	Can be computed for different time scales, and is less complex than the PDSI.
2. Aridity Index	0.380	Water balance calculations have limitations for not accounting rainfall-runoff.
3. Palmer's Drought Severity Index	0.473	Meteorological index responds better to abnormally dry or wet weathers.
4. Moisture Adequacy Index	0.673	Can specify drought crop-wise and on a real-time basis.
5. Crop Water Stress Index (SPAW model)	0.890	Simulates soil-water and calculates effective rainfall for plant transpiration, with reasonable results. Model needs calibration for each crop and region.

7. DROUGHT-RELATED INDICES FROM REMOTE SENSING

Several indices for drought monitoring have been developed over the past few decades using remote sensing data. These are calculated from the reflectance of vegetation and other land covers in different wavelength bands of satellite sensors, and may be obtained for each pixel (the size of a pixel depends on sensor resolution). These indices have some advantages over conventional indices, as they cover large areas, and may show, through vegetation condition signatures, how a drought progresses over the area. However, the indices may not always reflect the actual meteorological conditions on the ground, and need some field verification. Some of the indices may have lagged vegetation response to drought (see e.g., Kogan 1990, 1995; Moran et al., 1994; Jupp et al., 1998; Mcvicar and Jupp, 1998; Peters et al., 2002; Sandholt et al., 2002; Wan et al., 2004; Tadesse, et al., 2005; Jiang et al., 2006; Ghulam et al., 2007 for development of indices and limitations). A summary of some useful techniques in the context of Indian Subcontinent is provided in Singh et al. (2003) and Thenkabail et al. (2004).

7.1 Normalized Difference Vegetation Index (NDVI)

Since 1989, National Agricultural Drought Assessment and Monitoring System (NADAMS) is providing bi-weekly drought bulletins for *kharif* season covering 246 districts in India. These bulletins describe prevalence, relative severity level, and persistence through the season at the district level. Drought assessment is based on a comparative evaluation of the satellite-observed green vegetation cover (both area and greenness) in a district during any specific time period to cover in similar periods in the previous year, or a long-term mean of the specified period. The drought interpretation takes into account rainfall and aridity anomaly trends. This nationwide early warning service has been found to be useful for providing early assessment of drought conditions. NDVI is calculated as (Kidwell, 1990):

$$NDVI = \frac{\lambda_{NIR} - \lambda_{RED}}{\lambda_{NIR} + \lambda_{RED}} \tag{5}$$

where λ_{NIR} and λ_{RED} are the reflectances in the near infra-red and red band, respectively. NDVI ranges from –1 to 1. Drought severity may be evaluated as the difference between the NDVI for the current month (e.g., September 2007) and a long-term mean NDVI for this month (e.g., a 30-year long mean NDVI for September).

$$\text{Drought Severity} = NDVI_i - NDVI_{mean,m} \tag{6}$$

where $NDVI_i$ is the current NDVI for the ith month and $NDVI_{mean,m}$ is the long-term mean NDVI for the month m ($m = 1, 2, \ldots, 12$). Positive departure from the mean NDVI indicates that the vegetation condition is better than the normal in this month (i.e., wetter than usual). Negative departure from the mean NDVI points to a dryer condition for the current month than usual. The more negative the departure is, the drier the month is. Bayarjargal et al. (2006) provided a recent assessment of drought severity index and other related indices.

7.2 Enhanced Vegetation Index (EVI)

This index has been developed for use with MODIS data (Liu and Huete, 1995). Unlike NDVI, it takes the advantage of multiple bands. The EVI is calculated by using following formula:

$$EVI = G \times \frac{\rho_{NIR} - \rho_{RED}}{\rho_{NIR} + C_1 \times \rho_{RED} - C_2 \times \rho_{Blue} + L} \tag{7}$$

where ρ_{NIR} = NIR reflectance, ρ_{RED} = Red reflectance, ρ_{Blue} = Blue reflectance, C_1 = atmosphere resistance red correction coefficient, C_2 = atmosphere resistance blue correction factor, L = canopy background brightness correction factor, and G = gain factor.

The coefficients adopted in the EVI algorithm were $L = 1$, $C1 = 6$, $C2 = 7.5$, and G (gain factor) = 2.5. EVI is more sensitive in high biomass regions and ensures the improved monitoring through a reduction in atmosphere influences. At the same time, it is computationally intensive and is not widely used at present. EVI may be used in Drought Severity formula above.

7.3 Vegetation Condition Index (VCI)

Vegetation Condition Index (VCI) shows how close is the current month's NDVI to the minimum NDVI calculated from the long-term record of satellite images. It is mathematically expressed as (Liu and Kogan, 1996):

$$VCI_j = \frac{(NDVI_j - NDVI_{min})}{(NDVI_{max} - NDVI_{min})} \times 100 \qquad (8)$$

where $NDVI_{max}$ and $NDVI_{min}$ are calculated from the long-term record (e.g., 20 years) for that month or week, and j is the index of the current month or week. The condition (health) of vegetation presented by VCI is reported in percent and may serve as an approximate measure of how dry the current month is. In the case of extremely dry month, the vegetation condition is poor and the VCI is close or equal to zero. The VCI of 50% reflects a fair vegetation condition. At optimal condition of vegetation, the VCI is close to 100%. At this condition, NDVI for the current time step (month or week) is equal to $NDVI_{max}$. Singh et al. (2003) have demonstrated the application of VCI and TCI (Temperature Condition Index) from AVHRR for drought monitoring in India.

In order to exploit the signatures of lesser green biomass in the dryland vegetation cover on satellite images and to find out deterioration in conditions from changes in such signatures, Pickup and Chewing (1988) developed a model called PD54, which is a 'perpendicular vegetation index' (PVI), rather than a non-dimensional vegetation index (NDVI). It was calculated using graphically plotted perpendicular distances of soil and vegetation signatures in bands 5 and 4 (i.e., red and green spectral bands, respectively of Landsat MSS). A recent study by CAZRI and CSIRO (Australia) to find out the performance of NDVI and PD54 in a rangeland near Jodhpur showed better results from the latter (Anon., 2004b). The PD54 was later improved as SAVI (Soil Adjusted Vegetation Index; Huete, 1988) and then further modified as MSAVI (Qi et al., 1994).

7.4 Temperature Condition Index (TCI)

Temperature Condition Index (TCI) is calculated similarly to VCI. However, in contrast to the VCI, TCI includes the deviation of the current month's (week's) value from the recorded maximum:

$$TCI_j = \frac{(BT_{max} - BT_j)}{(BT_{max} - BT_{min})} \times 100 \qquad (9)$$

where BT is the brightness temperature (e.g., AVHRR Band 4. Under the atmospheric conditions, objects emit heat in this thermal band). The maximum and minimum values of BT are calculated from the long-term (e.g., 20 years) record of satellite images for each calendar week or month j (Kogan, 1995, 2002).

The low TCI value (close to 0%) indicates the very high temperature in that month or week. Consistently low TCI values over several consecutive time intervals indicate drought occurrence.

7.5 Merits and Demerits of Available Remote Sensing-Based Indicators

Vegetation conditions, including crops, being an indicator of rainfall, the maps on NDVI and other such indices (EVI, VCI, etc.) can be integrated with meteorological indices as these have larger coverage of area and provide data for areas where ground data is not accessible (Huete et al., 2002). However, during monsoon season the remote sensing data collection for vegetation cover becomes a constraint under cloudy conditions. As a result, analysis of satellite data for NDVI becomes limited by the availability of cloud-free data.

Moreover, in the sparsely vegetated arid regions, NDVI becomes a poor representative of vegetation cover as the background soils provide signatures almost similarly to the vegetation signatures. Also, the signatures of open rangelands are not very dissimilar to that of sparse distribution of *kharif* crops in the arid areas, which add to the problem of estimating actual crop growth and natural vegetation growth.

As we have described above, a number of wavelength bands and processing techniques have been tested and sensors have been experimented to negate the soil effects. Based on the encouraging results, regular monitoring of vegetation conditions are now made using satellite sensors like SPOT Vegetation (France) and MODIS (USA), and analytical products on drought, etc. are produced (e.g., Hayes et al., 2005; Wilhite, 2005). The assessment, mapping and monitoring in the USA involves a coordinated program of activities between the National Drought Mitigation Center (NDMC) at the University of Nebraska, Lincoln, US Department of Agriculture (USDA), and NOAA's Climate Prediction Center at the US National Weather Service (Wilhite, 1990).

A recent development in linking observations from different platforms is the assimilation of indices from ground-based atmospheric monitoring products, and the land information products (especially on land use and land cover, irrigation status, soil-available water capacity and eco-regions) with the satellite-derived indices on greenness conditions and start of season anomaly, to produce a Vegetation Drought Response Index (VegDRI) at 1 km resolution for application in USA (Tadesse et al., 2005; Brown et al., 2008). The products are now regularly displayed in the websites of NDMC and the US Geological Survey for monitoring the conditions of croplands and rangelands at county level, with a statistical database of the phenomenon. Concurrently, based on the meteorological indices (SPI, PDSI and percent long-term average precipitation) a US Drought Monitor (USDM) is prepared by NOAA, with data from many organizations, including USDA, and is regularly published in NDMC and NOAA websites. A variant of this product, the North American Drought Monitor (NADM), is a joint venture of the North American nations. The USDA uses the USDM and VegDRI data along with the predictions of medium range weather forecasting to analyze stresses on crops, rangelands and livestock, as well as the likely impacts on yield.

A different decile-based drought index, the Floating Month Drought Index (FMDI), which is based on calculation of precipitation percentile for current month, length and begin month of the current dry spell, and precipitation percentile for the current dry spell, has been developed for the Australian continent and is also being tested in the USA for inter-comparison of results from USDM. FMDI has the capability to show both wet and dry spell conditions.

One of the major inputs for monitoring crop performance is soil moisture condition, which is generally calculated from the rainfall, temperature and soil texture data, and is an approximation. Microwave satellite sensors are being tested to provide reliable global soil moisture data, and a new satellite, carrying L-band sensors to measure microwave radiation emitted from earth surface at L-band (1.4 GHz) to eliminate the effects of cloud, haze, other weather and atmospheric conditions, as well as vegetation effects, is expected to be launched in the near future by the European Space Agency.

In India, regular satellite-based assessment of crop growth conditions and agricultural drought condition during summer monsoon period are made by the National Remote Sensing Agency (NRSA) for the Ministry of Agriculture, Government of India, which also gathers data from other sources, including the States and Indian Council of Agricultural Research to review the situation on a weekly to fortnightly basis for addressing the emerging issues. NDVI from AWiFS is the major output that is currently analyzed by NRSA, and the results are assessed with soil moisture assessment data and agro-meteorological yield models (Murthy et al., 2006). Samra (2004) provides a summary of the systems of monitoring followed in India.

8. SOCIETAL DROUGHT

Societal drought relates to the impact of meteorological, hydrological and agricultural droughts on the society. Although the indices for measuring meteorological, hydrological and agricultural droughts are now reasonably well developed, those for measuring the societal aspects of drought are not. This is despite the fact that drought hits the agriculture-based economy the most, which in turn impacts the nation's economy through lower production that creates an imbalance between the supply and demand of commodities, lowers the purchasing power of the people, and slows down the nation's economy, as has been shown in a recent study on the impact of the drought of 2002 in western Rajasthan (Narain and Kar, 2005). Severe societal drought can lead to mass migration in search of feed, fodder and water. However, there are not many studies on the development of indices. Vulnerability of the society's different segments to drought is a key issue that ideally determines the kind of counter-measures required.

Considering the above facts, it is necessary to develop a set of social indicators that identify the drought-vulnerable segments of the society and that could be used to link with the indicators on physical aspects in a Decision Support System (DSS) developed in the GIS environment. This will help in not only identifying the vulnerable areas of drought and the vulnerable segments of a society, but also in working out the quantum of relief and rehabilitation measures, as well as identifying the needs of infrastructure development as a long-term strategy for drought proofing.

9. ROLE OF DECISION SUPPORT SYSTEM IN DROUGHT MANAGEMENT

A knowledge-based spatial decision support system (SDSS) is an effective computer tool for drought management by farmers, experts and general end-users. This type of DSS is crucial for a National Drought Management Institute to effectively communicate the location-specific drought-related problems in advance to the stakeholders, risk assessment and options available to mitigate the problems. The DSS can utilize information database (information layer) and develop domain-specific knowledge-based algorithms. Knowledge layer tools even allow experts to interpret the complex problem of drought-risk management. Therefore, there is an urgent need for better data management and computing facilities to analyze climatic and natural resources on spatial and temporal scales and to prepare vulnerability maps, as well as forecast economic and environmental impacts. Some such systems are already in use globally for early warning of droughts.

As has been shown earlier, an elaborate system of assessment and monitoring is followed by the NDMC of the USA that has the potentials for a global coverage. An automated drought analyses system for water-resources analysis called "SPATSIM" (**Spa**tial and **T**ime **S**eries **I**nformation **M**odeling) software package, has been developed jointly by the Institute for Water Research (IWR), Grahamstown, South

Africa, and the International Water Management Institute (IWMI), Colombo, Sri Lanka (Smakhtin and Hughes, 2004). In India, different assessment tools are being used by different affected states for declaring drought. In Rajasthan, more emphasis is given on yield loss estimates at a village level by the village revenue officials (*Patwaris*), which are collated at *tehsildar* level and upwards for a state-level assessment. In Karnataka, the assessment is based on the village/*taluka*-level collection of rainfall amounts and dry weeks, for which both satellite-based information and information from key field functionaries are used. In a novel departure from the IMD-station-based information system, the state is now getting their ground-based rainfall data on a near-real time basis from the automatic rain gauge stations installed in the villages through cell phones. In Andhra Pradesh, drought condition is analyzed from *Mandal*-level data on rainfall and crop sown area, as well as an estimation of yield and length and time of dry spells in relation to sowing period. Both satellite and ground observations are taken into account.

Although simple and less data-demanding for the government-level technocrats and bureaucrats to initiate remedial action in near-real time and monitor future needs, these lack the scientific depth, and do not consider the socio-economic variables (except in Rajasthan; see e.g., Anon., 2007) for a holistic assessment of the situation for targeting the most affected segments of the society. To overcome the drawbacks, and to systematize the integration of different data layers, efforts have been made in Chhattisgarh state to use a number of 'ecological', 'production system' and 'socio-economic' indicators, provide weight and ranking to them, and to find out statistically from their relationships the vulnerable Development Blocks (Gupta, 2002).

Biophysical and socio-economic parameters used by some of the drought early warning and food security systems working mainly for the African continent have been reviewed by CeSIA, Italy (Anon., 1999), and is summarized in Table 10. Although the indicators provide useful information for the organizations carrying out emergency relief and rehabilitation work in the affected African nations, the above review (Anon., 1999) suggested that the indicators could also be used for early warning of desertification. Kar and Takeuchi (2003) made a critical analysis of the indicators of drought early warning and those needed for a desertification early warning to show that the indicators of drought early warning system are inadequate for desertification early warning. Samra (2004) provides a review of drought monitoring and declaration policy followed in India, while Rathore (2005) discusses the system followed in Rajasthan.

10. CONCLUSIONS

There are many drought-monitoring indices, which have different merits and demerits. Our experience suggests that for the Indian arid region, where rural livelihood depends mostly on agriculture and related activities, a simple and reliable set of indicators, including bio-physical and socio-economic, that is amenable to quantitative measurements and modeling, and for which data could be gathered at least at the *Tehsil* level, needs to be developed for forecasting. The availability of real-time remote sensing data products, especially on soil moisture conditions, percent crop and natural vegetation cover (or, biomass yield from croplands and rangelands) at moderate to high resolution will help much in modeling the system that will also take as input ground-based data on meteorological parameters and socio-economic conditions. The information on rangeland production estimates is crucial due to the greater roles played by the livestock component in the rural economy of the region, especially during the droughts when crops fail. The models should be able to estimate the biological production from the land at different time steps, availability of water in the reservoirs, vulnerable areas and segments of population, infrastructure bottlenecks, if any, possible migration routes and areas for animals, and other related aspects that can help in proper agro-advisory, and in relief and rehabilitation. Models should also be developed to provide

Table 10. Indicators of early warning systems for food security used by major systems

Indicator	AP3A	FIVIMS	GIEWS	SADC	FEWS	VAM
1. Food crop performance	✓		✓	✓	✓	✓
2. Crop conditions	✓		✓	✓	✓	✓
3. Crop production forecast	✓		✓	✓		
4. Marketing and price information		✓	✓	✓	✓	✓
5. Food supply/demand		✓	✓	✓	✓	✓
6. Health conditions					✓	✓
7. Food crops and shortages			✓	✓	✓	✓
8. Food supply			✓	✓		
9. Food consumption			✓	✓		
10. Crop areas	✓	✓	✓	✓	✓	✓
11. Pests			✓	✓		
12. Food balance		✓	✓	✓	✓	✓
13. Vegetation front	✓					
14. CCD	✓		✓	✓	✓	✓
15. NDVI	✓		✓	✓	✓	✓
16. Biomass	✓					
17. Seeding risk areas	✓					
18. Expected season length	✓				✓	✓
19. Estimated seeded areas	✓			✓		
20. Estimated seeding date	✓					
21. Vegetation cover						
22. Agro-ecological zones						✓
23. Crop use intensity						✓
24. Variation coefficient of agricultural production						✓
25. Cash crop production area	✓		✓	✓		✓
26. Coping strategies						✓
27. Average cost to travel to nearest market						✓
28. Livestock production	✓					
29. Population density	✓		✓	✓	✓	✓
30. Access to water						✓
31. Children education						✓
32. Rainfall	✓		✓	✓	✓	✓

Notes: AP3A = Alerte Precoce et Prevision des Production Agricoles (by AGRHYMET); FIVIMS = Food Insecurity and Vulnerability Information and Mapping Systems (by FAO); GIEWS = Global Information and Early Warning System (by FAO); SADC = Southern Africa Development Community (from Zimbabwe); FEWS = Famine Early Warning System (by USAID); and VAM = Vulnerability Analysis and Mapping (by World Food Programme).

information on the performance of the infrastructures developed as a long-term measure for drought mitigation. The ultimate goal of researchers is to produce an International Drought Early Warning System that would help to plan the response system to drought, including planning, mitigation and recovery, and to create a database and map that could be accessed globally at near-real time as an effective warning tool (Wilhite et al., 2000; Wilhite, 2005).

REFERENCES

Anon. (1999). Early Warning Systems and Desertification. CeSIA: Accademia dei Georgofili, Florence, Italy.

Anon. (2004a). Drought 2002: A Report. Department of Agriculture and Cooperation, Ministry of Agriculture, Government of India, New Delhi.

Anon. (2004b). Assessing the extent and causes of degradation in India's arid rangelands. ACIAR Research Note 27. Australian Centre for International Agricultural Research, Canberra, pp. 1-4.

Anon. (2007). Drought Management Manual. Disaster Management and Relief Department, Government of Rajasthan, Jaipur, India.

Bayarjargal, Y., Karnieli, A., Bayasgalan, M., Khudulmur, S., Gndush, C. and Tucker, C.J. (2006). A comparative study of NOAA-AVHRR derived drought indices using change vector analysis. *Remote Sensing of Environment*, **105**: 9-22.

Brown, J.F., Wardlow, B.D., Tadesse, T., Hayes, M.J. and Reed, B.C. (2008). The Vegetation Drought Response Index (VegDRI): A new integrated approach for monitoring drought stress in vegetation. *GIScience and Remote Sensing*, **45**: 16-46.

Ghulam, A., Qin, Q., Teyip, T. and Li, Z. (2007). Modified perpendicular drought index (MPDI): A real-time drought monitoring method. *ISPRS Journal of Photogrammetry and Remote Sensing*, **62**: 150-164.

Gibbs, W.J. and Mather, J.V. (1967). Rainfall Deciles as Drought Indicators. Bureau of Meteorology. Bulletin No. 48, Commonwealth of Australia, Melborne.

Gupta, S. (2002). Water Policy for Drought Proofing Chhattisgarh: Report. Institute for Human Development, New Delhi.

Hayes, M., Svoboda, M., Le Comte, D., Redmond, K. and Pasteris, P. (2005). Drought monitoring: New tools for the 21st century. *In:* D.A. Wilhite (editor), Drought and Water Crises: Science, Technology and Management Issues. CRC Press, Boca Raton, pp. 53-69.

Hounam, C.E. (editor) (1975). Drought and Agriculture. WMO Technical Bulletin, 392 pp.

Huete, A., Didan, K., Miura, T., Rodreguez, E., Gao, X. and Ferreira, L. (2002). Overview of the radiometric and biophysical performance of the MODIS vegetation indices. *Remote Sensing of Environment*, **83**: 195-213.

Huete, A.R. (1988). A soil adjusted vegetation index (SAVI). *Remote Sensing of Environment*, **25**: 295-309.

Jiang, Z., Huete, A., Chen, J., Chen, Y., Li, J., Yan, G. and Zhang, X. (2006). Analysis of NDVI and scaled difference vegetation index retrievals of vegetation fraction. *Remote Sensing of Environment*, **101**: 366-378.

Jupp, D.L.B., Tian, G., McVicar, T.R., Qin, Y. and Fuqin, L. (1998). Soil Moisture and Drought Monitoring Using Remote Sensing. I: Theoretical Background and Methods. CSIRO Earth Observation Centre, Canberra, Australia.

Kar, A. and Takeuchi, K. (2003). Towards an early warning system for desertification. *In:* Early Warning Systems, UNCCD Ad Hoc Panel, Committee on Science and Technology, UN Convention to Combat Desertification, Bonn, pp. 37-72.

Kidwell, K.B. (1990). Global Vegetation Index User's Guide. NOAA/National Climatic Data Center/Satellite Data Services Division.

Kogan, F.N. (1990). Remote sensing of weather impacts on vegetation in non-homogeneous areas. *International Journal of Remote Sensing*, **11**: 1405-1419.

Kogan, F.N. (1995). Application of vegetation index and brightness temperature for drought detection. *Advances in Space Research*, **15(11)**: 91-100.

Kogan, F.N. (2002). World droughts in the new millennium from AVHRR-based vegetation health indices. *EOS Transactions of the American Geophysical Union*, **83**: 562-563.

Kulshreshtra, S.M. and Sikka, D.R. (1989). Monsoons and droughts in India: Long-term trend and policy choices. Proceedings of the National Workshop on Drought Management, New Delhi.

Liu, H.Q. and Huete, A. (1995). A feedback based modification of NDVI to minimize canopy background and atmospheric noise. *IEEE Transactions on Geoscience and Remote Sensing*, **38**: 457-463.

Liu, W.T. and Kogan, F.N. (1996). Monitoring regional drought using the Vegetation Condition Index. *International Journal of Remote Sensing*, **17**: 2761-2782.

McKee, T.B., Docsken, N.J. and Kleist, J. (1993). The relationship of drought frequency and duration to time scale. Pre-prints, Eighth Conference on Applied Climatology, Anchel, C.A., pp. 179-184.

Mcvicar, T.R. and Jupp, D.L.B. (1998). The current and potential operational uses of remote sensing to aid decisions on drought exceptional circumstances in Australia: A review. *Agriculture System*, **57**: 399-468.

Moran, M.S., Clarke, T.R., Inoue, Y. and Vidal, A. (1994). Estimating crop water deficit using the relation between surface-air temperature and spectral vegetation index. *Remote Sensing of Environment*, **49**: 246-263.

Murthy, C.S., Sesha Sai, M.V.R., Bhanuja Kumari, V. and Roy, P.S. (2006). Agricultural drought assessment at disaggregated level using AWiFS/WiFS data of Indian Remote Sensing satellites. *Geocarta International*, **22**: 127-140.

Narain, P. and Kar, A. (editors) (2005). Drought in Western Rajasthan: Impact, Coping Mechanism and Management Strategies. Central Arid Zone Research Institute, Jodhpur, Rajasthan, 104 pp.

Palmer, W.C. (1965). Meteorological Drought. Research Paper No. 45, U.S. Department of Commerce, Weather Bureau, Washington D.C.

Palmer, W.C. (1968). Keeping track of crop moisture conditions, nationwide: The new crop moisture index. *Weatherwise*, **21**: 156-161.

Peters, A.J., Walter-Shea, E.A., Lei, J., Vina, A., Hayes, M. and Svoboda, M.R. (2002). Drought monitoring with NDVI-based standardized vegetation index. *Photogrammetric Engineering and Remote Sensing*, **65**: 71-75.

Pickup, G. and Chewings, V.H. (1988). Forecasting patterns of soil erosion in arid lands from Landsat MSS data. *International Journal of Remote Sensing*, **9**: 69-84.

Qi, J., Chehbouni, A., Huete, A.R., Kerr, Y.H. and Sorooshian. (1994). A modified soil adjusted vegetation index. *Remote Sensing Environment*, **48**: 119-126.

Ramana Rao, B.V., Sastri, A.S.R.A.S. and Ramakrishna, Y.S. (1981). An integrated scheme of drought classification as applicable to Indian arid region. *Idojaras*, **85**: 317-322.

Rao, A.S. (1997). Impact of droughts on Indian arid ecosystem. *In:* S. Singh and A. Kar (editors), Desertification and Its Control in the Arid Ecosystem of India for Sustainable Development. Agro-Botanical Publishers, Jodhpur, Rajasthan, India, pp. 120-130.

Rao, A.S. and Saxton, K.E. (1995). Analysis of soil water and water stress for pearl millet in an Indian arid region using the SPAW Model. *Journal of Arid Environments*, **29**: 155-167.

Rathore, M.S. (2005). State Level Analysis of Drought Policies and Impacts in Rajasthan, India. Working Paper 93, International Water Management Institute (IWMI), Colombo.

Samra, J.S. (2004). Review and Analysis of Drought Monitoring, Declaration and Management in India. Working Paper 84, International Water Management Institute (IWMI), Colombo.

Sandholt, I., Rasmussen, K. and Andersen, J. (2002). A simple interpretation of the surface temperature/vegetation index space for assessment of surface moisture status. *Remote Sensing of Environment*, **79**: 213-224.

Sastri, A.S.R.A.S, Ramana Rao, B.V., Ramakrishna, Y.S. and Rao, G.G.S.N. (1981). Agricultural droughts and crop production in the Indian arid zone. *Arch. Met. Geoph. and Bioklim.*, **31**: 127-132.

Saxton, K.E. (1989). Users Manual for SPAW: A Soil-Plant-Atmosphere-Water Model. Pullman, Washington, USDA-ARS, 89 pp.

Shafer, B.A. and Dezman, L.E. (1982). Development of a Surface Water Supply Index (SWSI) to assess the severity of drought conditions in snowpack runoff areas. Proceedings of the Western Snow Conference, Colorado State University, Fort Collins, Colorado, pp. 164-175.

Singh, R.P., Roy, S. and Kogan, F. (2003). Vegetation and temperature condition indices from NOAA AVHRR data for drought monitoring over India. *International Journal of Remote Sensing*, **24**: 4393-4402.

Smakhtin, V.U. and Hughes, D.A. (2004). Review of Automated Estimation and Analyses of Drought Indices in South Asia. Working Paper 83, International Water Management Institute (IWMI), Colombo.

Tadesse, T., Brown, J. and Hayes, M. (2005). A new approach for predicting drought-related vegetation stress: Integrating satellite, climate, and biophysical data over the U.S. central plains. *ISPRS Journal of Photogrammetry and Remote Sensing*, **59**: 244-253.

Thenkabail, P.S., Gamage, M.S.D.N. and Smakhtin, V.U. (2004). The Use of Remote Sensing Data for Drought Assessment and Monitoring in Southwest Asia. Report 85, International Water Management Institute, Colombo.

Thornthwaite, C.W. and Mather, J.R. (1955). The water balance. *Publications in Climatology*, **VIII(1)**, Drexel Institute of Climatology, Centerton, New Jersey, 104 pp.

Wan, Z., Wang, P. and Li, X. (2004). Using MODIS land surface temperature and normalized difference vegetation index products for monitoring drought in the southern Great Plains, USA. *International Journal of Remote Sensing*, **25**: 61-72.

Wilhite, D.A. (1990). Planning for Drought: A Process for State Government. IDIC Technical Report Series 90-1, International Drought Information Center, University of Nebraska, Lincoln.

Wilhite, D.A. (editor) (2005). Drought and Water Crises: Science, Technology, and Management Issues. CRC Press, Boca Raton.

Wilhite, D.A., Sivakumar, M.V.K. and Wood, D.A. (editors) (2000). Early warning systems for drought preparedness and drought management. Proceedings of an Expert Group Meeting. Lisbon, Portugal, September 5-7, World Meteorological Organization, Geneva.

12

Tropical Cyclones: Trends, Forecasting and Mitigation

O.P. Singh

1. INTRODUCTION

The Tropical Cyclones (TCs) are among the deadliest natural disasters of the world. These intense weather systems develop over the warm tropical ocean waters. On an average about 80 tropical cyclones form every year in the tropical oceans. Highest annual frequency occurs over the northwestern Pacific Ocean which is about 30. The hazardous impacts associated with the tropical cyclones are primarily due to three factors, namely strong winds, phenomenal storm surge and exceptionally heavy rainfall. The storm surge caused due to severe cyclones is generally responsible for inundation of coastal areas causing tremendous loss of life and property.

The North Indian Ocean (NIO), which accounts for about 5% of total global tropical cyclones, produces about four TCs per year out of which three TCs form in the Bay of Bengal (BOB) and one TC forms in the Arabian Sea (AS). Due to various socio-economic factors these cyclones inflict heavy loss of life and property in the NIO rim countries. The east coast of India and the coasts of Bangladesh, Myanmar and Sri Lanka are vulnerable to the incidence of tropical cyclones of the Bay of Bengal. The frequency of tropical cyclones in the Arabian Sea is significantly low compared to the Bay of Bengal and except Gujarat and north Maharashtra coasts the west coast of India is generally unaffected by the tropical cyclones. The debate on the impacts of global climate change on the frequency and intensity of tropical cyclones in different ocean basins is still on among the tropical cyclone experts of the world. Global climate change resulting from anthropogenic activity is likely to manifest itself in the weather and climate of the Indian subcontinent and adjoining seas also. The trends and variabilities in the frequency and intensity of tropical cyclones during intense cyclonic months of May, October and November is one such problem which has been addressed in the present chapter.

Most of the severe cyclones of the Bay of Bengal form during the post-monsoon season in the months of October and November. A few severe cyclones form during May also, but the post-monsoon cyclones are the severest, and hence this season is also known as *storm season* in South Asia. The frequency of tropical cyclones in the Bay of Bengal during the period 1877-2005 is presented in Table 1.

In recent years, a few studies on long-term trends and oscillations in the tropical cyclone (TC) frequency and intensity in the Bay of Bengal have been conducted. Ali (1995) and Joseph (1995) have examined the trends in TC frequency in the Bay of Bengal. However, these studies could not bring out

Table 1. Frequency of tropical cyclones in the Bay of Bengal (1877-2005)

Type of Tropical Disturbance	Month				
	May	*June*	*September*	*October*	*November*
Cyclonic Storm (CS)	59	35	41	92	116
Severe Cyclonic Storm (SCS)	44	5	16	40	65

clear-cut trends in the frequency of severe cyclones of the Bay of Bengal during the intense cyclonic period of the year. Mooley (1980, 1981) and Sikka (2006) have studied the trends in the annual frequency of cyclonic storms. Singh (2001) has reported a decreasing trend in the frequency of cyclones formed during the monsoon season (June to September) during past decades. Singh and Khan (1999) have examined the trends in the cyclogenesis over the north Indian Ocean during past decades comprehensively and have reported that there is indeed a tendency for the enhanced cyclogenesis during the intense cyclonic months on a long-term basis, though the annual frequency has not changed much. Cyclogenesis refers to the genesis of cyclonic disturbances. Here, the trends with a special reference to severe cyclonic storms, i.e., the cyclonic storms having maximum sustained wind speed of 48 knots or more have been presented. Similarly, the intensification rate of cyclonic disturbances to the severe cyclonic storm stage alone has been considered. Mooley (1980, 1981), Srivastav et al. (2000), Singh et al. (2000, 2001), and Singh (2007) have demonstrated the variabilities in the frequency of tropical cyclones of the north Indian Ocean well. There have been some ambiguities in the reliability of TC data before the satellite detection of TCs which commenced in 1970s. For the North Indian Ocean (NIO), the satellite detection of TCs started in 1972 by the Joint Typhoon Warning Centre (JTWC), Guam (now shifted to Pearl Harbor, Hawaii), USA. The India Meteorological Department (IMD) started the satellite detection of TCs from early 1980s onwards when the Indian geostationary satellite, INSAT was launched. Thus, the JTWC satellite era data set is little longer and it provides information on various categories of TCs, namely Categories 1 to 5 which can be used to examine the trends in the number of TCs of different categories. Keeping this objective in view, the JTWC data on TCs of NIO for 1972-2006 have been used along with the IMD data. Utilizing IMD's existing data sets, a few earlier studies by Singh et al. (2000, 2001) and Singh (2007) have shown that there is an uptrend in the frequency of intense TCs in the NIO basin. In these studies an attempt has been made to derive firm conclusions on recent trends in the annual number of stronger TCs in NIO removing the ambiguity of data of pre-satellite era.

The Indian Ocean Dipole Mode (IODM) is a coupled ocean-atmosphere phenomenon observed in the Indian Ocean in the form of an east-west dipole in the sea-surface temperature (SST) anomalies (Webster et al., 1999). IODM index (IODMI) is defined as the difference in SST anomaly between the tropical western Indian Ocean; 50° E-70° E, 10° S-10° N and the tropical southeastern Indian Ocean; 90° E-110° E, 10° S-equator (Saji et al., 1999). Positive IODMI is associated with warm SST anomaly over the western tropical Indian Ocean and cold SST anomaly over the southeastern tropical Indian Ocean. Sign of index reverses when the SST anomalies swing to the opposite phase. The IODM phenomenon seems to play a key role in the occurrence of droughts over the Indonesian region (Behera et al., 1999; Iizuka et al., 2000; Behera and Yamagata, 2003). When IODMI is negative, it leads to drought over Indonesia and floods over East Africa and vice versa. Positive IODMI seems to correspond to more monsoon rains over India.

In the present chapter, the relationships between the IODM and the cyclone frequency in the Bay of Bengal during the post-monsoon season (October to December), which is also known as *storm season* in South Asia, are discussed. The probable impact of IODM on the frequency of monsoon depressions and storms, which are significant rainfall-producing systems during the monsoon season from June to September, has also been examined. The monthly time-series of IODMI from January 1958 to December 2002 has been used to determine the correlation with cyclone frequencies. Saji and Yamagata (2003a,b) used the same data in their studies.

2. STUDY AREA

The study area for the present work is the North Indian Ocean (NIO) region bounded by 5°-30° N and 50°-100° E comprising the Arabian Sea and the Bay of Bengal.

3. DATA ACQUISITION AND ANALYSES

The tropical cyclone data for the period 1877-1979 have been obtained from the Storm Atlas published by the India Meteorological Department (IMD, 1979). The data for recent years (1980-2005) have been obtained from the IMD records. The Joint Typhoon Warning Centre (JTWC) data have been downloaded from the JTWC website (http://metocph.nmci.navy.mil/jtwc.php). The time-series of monthly Indian Ocean Dipole Mode index used by Saji and Yamagata (2003a,b) has been utilized. The satellite-derived sea-surface temperature data have been obtained from the NASA Physical Oceanography Distributed Active Archive Center at the Jet Propulsion Laboratory, California, USA (Casey and Cornillon, 1998).

Using the above-mentioned data, the trend analyses on the cyclone frequencies and the intensification rates have been performed using the method of least squares. The significance of correlations has been tested using two-tailed t-test. The intensification rate has been computed using the ratio of number of severe cyclones and the total number of cyclonic disturbances, i.e., *depression* (maximum sustained wind speed of 17-33 knots) plus *cyclonic storm* (maximum sustained wind speed of 34-47 knots) plus *severe cyclonic storm* (maximum sustained wind speed more than or equal to 48 knots).

Average Sea Surface Temperatures (SSTs) over the south (5° N-13° N) and central (13° N- 18.5° N) Bay of Bengal have been obtained by averaging out all grid point SST values lying in the respective areas. SSTs have been analyzed for the south and central Bay of Bengal only because of the fact that pre- and post-monsoon cyclones form over these areas. Furthermore, simulation experiments have been conducted using the regional climate model HadRM2 of the Hadley Centre for Climate Prediction and Research, U.K. (Singh et al., 2006).

4. RESULTS AND DISCUSSION

4.1 Trends in the Frequency of Severe Cyclonic Storms in the Bay of Bengal

Post-monsoon cyclones of October and November in the Bay of Bengal are most disastrous. The entire east coast of India and the coasts of Sri Lanka, Bangladesh and Myanmar are vulnerable to the incidence of severe cyclones of the post-monsoon season. The implications of the changes in cyclone frequency are enormous due to high vulnerability of the Bay of Bengal rim countries where the incidence of only one severe cyclone is capable of setting back the economic advancement of small developing nations by many years (Obasi, 1977). It is due to this reason that any increasing trend in the severe cyclone frequency in the Bay of Bengal assumes greater significance.

The frequencies of severe cyclonic storms (SCS) formed in the Bay of Bengal during intense cyclone months of May, October and November in each pentad (five years' period) from 1881-2005 have been presented in Table 2. As the purpose of the analysis is to examine the long-term trends in the frequency of SCS only and that too during the period of the year when their normal frequency is maximum, i.e., May, October and November, the intensification rate (IR) concept was introduced to determine the trends in the intensity patterns. When the annual frequencies of SCS are considered, the trends get smoothened as the cyclogenesis patterns in the Bay of Bengal are different in different seasons. Even during a particular season, the characteristics may vary from month to month. For instance, cyclogenesis in November is entirely different from December though both are post-monsoon months. Thus, when the combined cyclonic frequency during post-monsoon is considered, it dilutes the trends during peak activity month (i.e., November). These features are clearly shown in Table 2.

Table 2. Pentad frequency of Severe Cyclone Storms (SCS) over the Bay of Bengal during intense cyclonic months May, October and November

Pentad	May	October	November
1881-1885	1	0	0
1886-1890	1	1	2
1891-1895	1	1	3
1896-1900	1	1	1
1901-1905	1	0	1
1906-1910	1	4	1
1911-1915	1	2	0
1916-1920	2	1	2
1921-1925	2	1	5
1926-1930	2	1	3
1931-1935	1	0	2
1936-1940	2	1	3
1941-1945	2	2	2
1946-1950	0	2	0
1951-1955	1	1	2
1956-1960	1	4	1
1961-1965	5	3	0
1966-1970	3	4	7
1971-1975	1	1	5
1976-1980	3	1	6
1981-1985	3	5	4
1986-1990	2	1	4
1991-1995	0	1	4
1996-2000	2	2	5
2001-2005	2	0	1

Table 2 reveals several salient features of SCS frequency trends in the Bay of Bengal during intense cyclonic period of the year, i.e. May, October and November. It is evident that the last four decades of 20th century did witness a spurt in the SCS activity in the Bay of Bengal during these months, especially during November. A total of 35 SCS formed in the Bay of Bengal during November in the 40 years period (1961 to 2000) implies an average of about one SCS every year against 18 during 1921-1960 and 10 during 1881-1920 which shows a monotonic increase in the SCS frequency on a four-decade scale. It may be mentioned that superimposed on the linear trends are the decadal-scale fluctuations and the increasing trend need not imply a monotonic (continuous) increase decade after decade. Statistically, the trend is significantly positive as the last decade's (1991-2000) SCS frequency of 9 is significantly higher than the decadal average of 5.25 for the 12 decades period under consideration (1881-2000). Also, it is more than the decadal average of 8.75 for the last four decades (1961-2000). The long-term trends determined by the statistical analyses are discussed below.

The pentad running total frequencies of SCS and corresponding trends in the Bay of Bengal during November are shown in Fig. 1. The uptrend in the frequency of severe cyclones during November as revealed by Fig. 1 is highly significant. The trend correlation coefficient for November is more than 0.5, which is significant at 99.5% level. The increasing trend in the cyclone frequency is +0.67 per hundred years which implies that every five years about three more cyclones are now forming in the Bay of Bengal during the month of November which is known for the severest cyclones in South Asia. Keeping

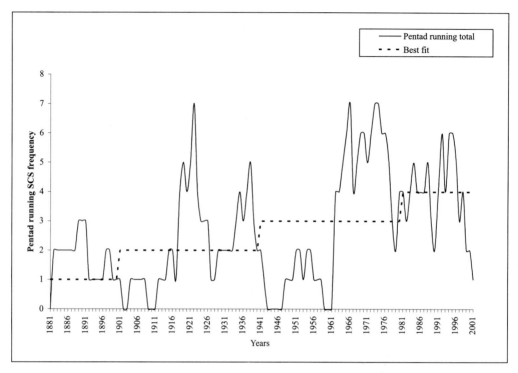

Fig. 1 Long-term trend in the frequency of severe cyclonic storms in the Bay of Bengal during November.

in view the highest average of cyclone frequency in November (Table 1), the significant increasing trend in the cyclone frequency during this month is important.

In Fig. 1, the pentad running totals along with the best polynomial fit (second degree) are also shown. The polynomial equation obtained is $Y = 0.753 + 0.156X - 0.001X^2$ which shows that the SCS frequency during November has increased almost linearly as the coefficient of X^2 is very small (i.e., 0.001). Here, X is the pentad number (i.e., 1, 2, and 25 starting from the pentad 1881-1885) and Y is the number of SCS in that pentad. Thus, for $X = 1$, $Y = 1$ and for $X = 25$, $Y = 4$, which shows that the pentad SCS frequency has increased from 1 to 4 during last 125 years. Statistically, this trend is highly significant (99.5% level of significance). It should be noted that the best fit depicted in Fig. 1 is obtained from the above-mentioned polynomial equation after rounding off the SCS frequencies to whole numbers.

October accounts for the second highest monthly cyclone frequency in the Bay of Bengal. The cyclones formed during later half of October have a tendency to become more severe as compared to those forming during the beginning of the month. The trend correlation coefficient is significant during October also, but one remarkable difference between cyclone frequency trends during November and October is that in recent four decades, the frequency jump during November has been highly significant. The SCS of pre-monsoon month (i.e., May) generally form in the southeast Bay of Bengal and move northwestwards initially. They have a tendency to recurve northward and then northeastward to strike Orissa/West Bengal coasts of India or Bangladesh/Myanmar coasts. May cyclones are quite severe and have very high probability of reaching to very intense stage (Table 1) due to long sea travel. The frequency of cyclones formed in the Bay of Bengal during May has also registered a significant increasing trend on the century scale (+ 0.27 per hundred years).

4.2 Trends in the Frequency of Severe Cyclonic Storms of the North Indian Ocean

The severe cyclone frequency in the north Indian Ocean (i.e., Bay of Bengal and Arabian Sea) during intense cyclone months May, October and November has registered about three-fold increase during past decades. As compared to the previous decades when about one severe cyclone was expected to form in the north Indian Ocean, every year during the intense cyclonic period (i.e., May, October and November) the number has now gone up to about three per year.

4.3 Trends in the Intensification Rate

In the north Indian Ocean, maximum probability of a disturbance reaching to SCS stage is during the month of November followed by May and October (Table 1). Therefore, it is interesting to look into the intensification rates in addition to the absolute numbers. As mentioned earlier, the "Intensification Rate (IR)" is defined as the ratio between SCS frequency and the frequency of total disturbances (i.e., Depressions, CS and SCS). The average intensification rates during each month and pentad were computed for the period 1881-2005. Maximum increasing trend in the intensification rate has been observed during November followed by October and May. The results for November are shown in Fig. 2. The trend in IR during November has been almost linear as the coefficient of X^2 in the second degree polynomial fit is only 0.002. The increasing trend in IR during November is highly significant. In Fig. 1, the pentad running average curve not touching X-axis signifies this aspect. In total contrast, earlier decades have

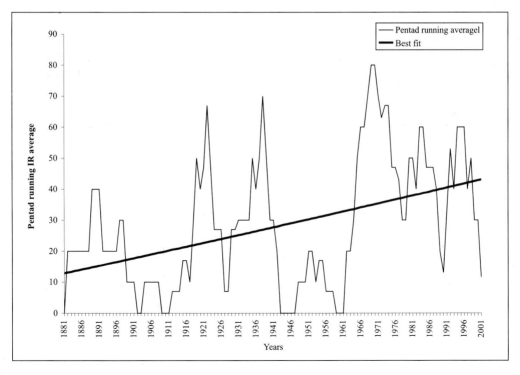

Fig. 2 Long-term trend in the intensification rate of tropical disturbances to SCS stage in the Bay of Bengal during November.

been characterized by the pentads not having a single SCS. There is about three-fold increase in the probability of intensification of a cyclonic disturbance to SCS stage in the Bay of Bengal during November. In October, the trend in IR shows that the probability to reach SCS stage is doubled. In May, the IR has increased by about 50% during past 125 years. Therefore, the analysis of intensification rates has revealed that the probability of a tropical depression formation in the Bay of Bengal during intense cyclone months, especially November, to reach the SCS stage has gone up substantially during the past century.

4.4 IMD and JTWC Classifications of Tropical Cyclones

IMD classifies a tropical disturbance in NIO as *cyclone* if the maximum sustained wind (MSW) in the disturbance is 34 knots and above. The system is termed *severe cyclone, very severe cyclone* and *super cyclone* if the MSWs are 48 knots and above, 64 knots and above and in excess of 120 knots, respectively. In the JTWC's classification (Table 3), there are five categories of tropical cyclones (TCs) starting from MSW 64 knots. Thus, there is a difference between IMD and JTWC classifications.

The classification given in Table 3 is based on the Saffir-Simpson scale. The JTWC uses same classification for all the ocean basins including the NIO. As per the JTWC's classification, the disturbances having MSW ≥34 knots are termed *tropical storm* (TS) and the disturbances having MSW <34 knots are termed *tropical depression* (TD).

As mentioned earlier, IMD's dataset on cyclone numbers and intensity in NIO prior to 1980s belong to the pre-satellite era and the works on cyclone trends using this dataset are often questioned. There is a feeling among many experts that the observed uptrend in the frequency of *severe cyclonic storms* (SCS) may be due to better detection of cyclones in the satellite era, rather than the climate change impacts. As mentioned earlier, using IMD's dataset of pre- and post-satellite era, some studies have established an uptrend in the frequency of SCS in the Bay of Bengal during intense cyclone period of the year. The JTWC's dataset allows us to examine the trends in the frequency of stronger tropical cyclones (Category 3 and above) in the satellite era separately.

4.5 Trends in the Frequency of Stronger TCs (MSW 96 knots and above)

The JTWC started the classification (Table 3) from 1972 onwards. Before that no information on MSW is available in their dataset. Therefore, the JTWC dataset from 1972 onwards for NIO is not only authentic, but it also removes the ambiguity of pre-satellite era. As more than three decades data of satellite era is available, it is possible to assess the trends in cyclone numbers and intensity in the NIO. In Fig. 3, the frequency of stronger TCs in NIO for the three decades viz., 1972-1981, 1982-1991 and 1992-2001 have

Table 3. JTWC's classification of TCs

	Category	MSW (knots)	Storm Surge (ft)
1.	Tropical Depression (TD)	<34	*Not Applicable*
2.	Tropical Storm (TS)	34-63	*Not Applicable*
3.	TC 1	64-82	4-5
4.	TC 2	83-95	6-8
5.	TC 3	96-112	9-12
6.	TC 4	113-135	13-18
7.	TC 5	>135	>18

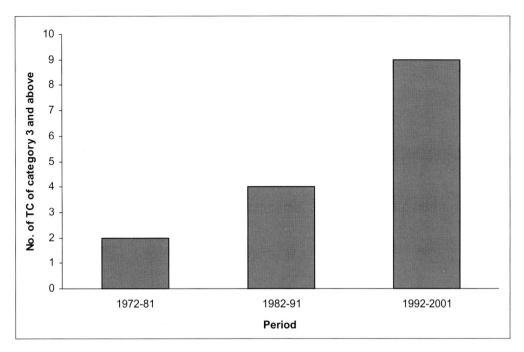

Fig. 3 Decadal frequency of stronger TCs (Category 3 and above) in the NIO basin (Bay of Bengal and Arabian Sea) during the satellite era.

been shown. Only two stronger TCs formed in the NIO (Bay of Bengal and Arabian Sea) during 1972-1981. The frequency increased to four and nine stronger TCs in next two decades, respectively. The implications of this increase in stronger TC number in the NIO may be enormous for the NIO rim countries like Bangladesh, India, Myanmar, Sri Lanka and even Pakistan and Oman.

The observed trend in the frequency of stronger TCs in NIO during past decades shows that on an average the rim countries have to face about one TC of Category 3 and above every year. The situation three decades ago was one TC of Category 3 and above every five years. Thus, there is almost a five-fold increase in the occurrence of a stronger TC in NIO.

Table 4 summarizes all the 15 cases of stronger TCs that occurred in the NIO during the three decades period, i.e., from 1972 to 2001. It shows many salient features of stronger TC occurrence in the NIO, the most important one being the fact that the months of November and May account for 80% of the total annual number of TCs of Category 3 and above in the NIO. The probability of occurrence of stronger TCs in the Bay of Bengal (BOB) is three times higher than the Arabian Sea (AS). In the BOB, stronger TCs could be expected in both the cyclone seasons, namely April-May and October-November, whereas in AS stronger TCs could be expected only in the months of May and June. During 1972-2001, all TCs of Categories 4 and 5 have formed in the BOB only. Category 5 TCs have formed in the 1990s which is another indication of enhanced frequency of stronger TCs in recent years. The first decade of 21st century is not yet over and one Category 5 TC has already formed in the Arabian Sea during June 2007. This statistics is useful for the managers dealing with cyclones preparedness/mitigation programs in NIO rim countries.

Table 4. Summary of all TCs of Category 3 and above in the NIO during 1972-2001

S. No.	Period	TC category	MSW (knots)	Sea area
1	9-23 November, 1977	3	110	BOB
2	14-20 November, 1977	3	111	BOB
3	30 April to 5 May, 1982	4	120	BOB
4	21-30 November, 1988	3	110	BOB
5	3-11 May, 1990	4	125	BOB
6	22-30 April, 1991	5	140	BOB
7	26 April to 3 May, 1994	4	125	BOB
8	18-25 November, 1995	3	105	BOB
9	1-7 November, 1996	4	115	BOB
10	13-20 May, 1997	4	115	BOB
11	1-9 June, 1998	3	105	AS
12	15-21 May, 1999	3	110	AS
13	15-18 October, 1999	4	120	BOB
14	25 October to 3 November, 1999	5	140	BOB
15	21-29 May, 2001	3	110	AS

4.6 Trends in the Frequency of All Cyclones with MSW 64 knots and above

In order to assess the probable impacts of global climate change on the intensity of TCs, it is necessary to look into the trends in stronger and weaker cyclones separately. When all tropical disturbances from tropical depression onwards are clubbed together and the trends are determined then it becomes misleading. To bring out this aspect, the annual numbers of all TCs of Category 1 and above in the NIO basin during the 35 years period (1972-2006) are presented in Fig. 4. It can be seen from Fig. 4 that there is only marginal uptrend in the annual frequency of all cyclones (weaker + stronger) which is statistically not significant. The trend line is: $Y = 1.29 + 0.007X$, where Y is the annual frequency of all TCs of Category 1 and above and X is the period (i.e., 1972 = 1, 1973 = 2, etc.). It shows that the annual frequency of all TCs (with MSW 64 knots and above) in the NIO has increased at a rate of 0.7 cyclones/hundred years. It is interesting to note that when all disturbances with MSW 34 knots and above are considered, the trend in annual numbers becomes negative (–0.8 cyclones/hundred years). This is mainly due to decreasing trends during the monsoon. Thus, undoubtedly, the frequency of stronger TCs has increased in the NIO in recent decades and the frequency of weaker disturbances like tropical depressions has decreased. It is the large average annual number of tropical depressions/storms like monsoon depressions/storms in the NIO, especially BOB which obscures the enhanced occurrences of stronger TCs.

4.7 Recent Trends in the Sea-surface Temperature over South and Central Bay of Bengal

As discussed earlier, south and adjoining central Bay of Bengal is the seat of intense cyclogenesis during pre- and post-monsoon seasons. The observed trends in the frequency of SCS would tempt an investigator to examine the probable causes of such trends. It is well known that SST is one of the parameters, which determines the cyclogenesis at sea (Gray, 1968). However, due to the scarcity of SST data over the Bay of Bengal, it becomes very difficult to construct long time-series of SST for smaller spatial resolutions. An attempt was made to examine the recent SST trends over the Bay of Bengal during 1985-1998 for

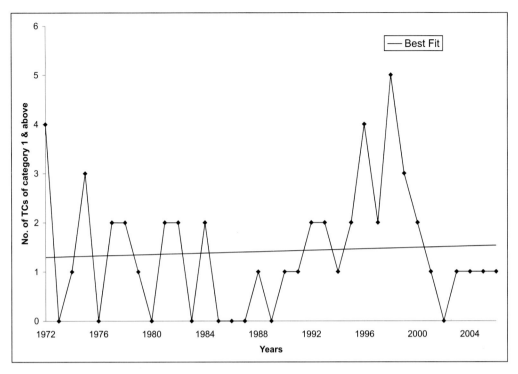

Fig. 4 Trend in the annual number of TCs of Category 1 and above in the NIO basin during the satellite era.

which reliable satellite-derived datasets were available. The SSTs during post-monsoon have registered rising trends during 1990s and it is found that 1990s did witness an uptrend in the post-monsoons cyclogenesis over the Bay of Bengal. However, due to shorter length of the satellite-derived SST time-series, it is not possible to reveal the SST trends in the Bay of Bengal similar to cyclone frequency trends.

4.8 Simulation of Global Climate Change Impacts on Cyclone Frequency

In order to simulate the impacts of global climate change (due to increased anthropogenic emissions) on the cyclogenesis in the Bay of Bengal, two experiments, first with the fixed amount of greenhouse gas concentration corresponding to 1990 levels called the *'control'* (CTL) and second with annual compound increase of 1% in the greenhouse gas concentration for 2041-2060 from 1990 onwards called the *'greenhouse gas'* (GHG) were conducted. The annual compound increment of 1% in the greenhouse gas concentration has been adopted from the projections of Intergovernmental Panel for Climate Change (IPCC). The model used was HadRM2 of Hadley Centre for Climate Prediction and Research, U.K. (Singh et al., 2006). The horizontal resolution of the model is 0.44° × 0.44°, i.e., minimum resolution of 50 km × 50 km at the equator (Singh et al., 2006). The criteria adopted for the identification of storms, in addition to a local minimum in sea level pressure, were: (i) Sea level pressure departure <–5 hPa, (ii) Maximum wind speed >15 m/s, and (iii) Duration of the storm at least two days. It is worth mentioning that all storms (vortices) could easily be identified in simulation results.

4.8.1 Simulation of Frequency

The simulation experiments showed that the frequency of post-monsoon tropical disturbances in the Bay of Bengal increased from 21 in CTL to 31 in GHG implying an increase of about 50% (Table 5). During pre-monsoon the increase is about 25%, i.e., from 11 in CTL to 15 in GHG. Thus, under warmer conditions due to increased emissions, the model simulated enhancement in the frequency of pre- and post-monsoon tropical storms in the Bay of Bengal. As these frequencies pertain to the period 1941-2060, the results suggest an increase of about 50% in the post-monsoon storm frequency and 25% in the pre-monsoon storm frequency in the Bay of Bengal during next 50 years. Even the annual frequency is likely to increase slightly during next 50 years, though the observed trends in the annual storm frequency till now have been slightly negative mainly due to decreasing trends during the monsoon.

Table 5. Simulated frequency of tropical disturbances in the Bay of Bengal for 2041-2060

Experiment	Pre-monsoon (March-May)	Post-monsoon (October-November)	Annual
CLT	11	21	113
GHG	15	31	121

4.8.2 Simulation of Intensity

The results on intensity simulations for May, October and November months are illustrated in Fig. 5. During all the three months, the model has simulated an enhancement in the average maximum wind

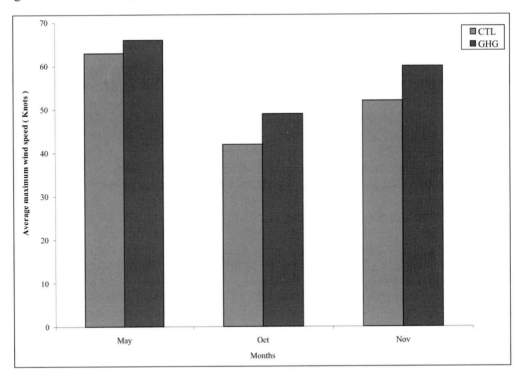

Fig. 5 Simulated intensities (in terms of maximum winds) during May, October and November. CTL and GHG refer to 'control' and 'Greenhouse Gas' experiments, respectively.

speed of the storms. In October, the average wind speed has gone up from 42 knots in CTL to 48 knots in GHG and in November it has gone up from 52 knots in CTL to 60 knots in GHG. That is, during both intense cyclone months of the post-monsoon season, the intensity has increased and the average cyclone during these months will be a severe cyclone (maximum wind speed equal to or greater than 48 knots), which is not the case at present during October. Similarly, during May also, the average intensity has increased slightly in GHG as compared to CTL. Thus, the model has simulated an increase in the average maximum wind speed of cyclones to be formed during May, October and November months.

4.9 Relationship between IODMI and Post-Monsoon Cyclone Frequency in the Bay of Bengal

As mentioned earlier, the post-monsoon season from October to December is known for maximum number of intense cyclones in the north Indian Ocean (Singh et al., 2001). Monthly frequency of severe cyclones (with a maximum wind speed exceeding 48 nautical miles per hour) is highest during November. As about 80% of the north Indian Ocean tropical cyclones form in the Bay of Bengal, the focus of present work was on the relationship between the IODM and the tropical cyclone frequency in the Bay of Bengal.

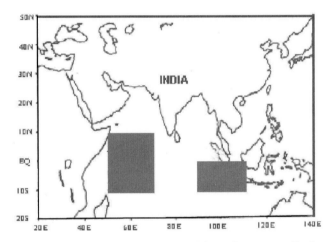

Fig. 6 Western and eastern poles of the Indian Ocean dipole.

The locations of western and eastern poles of IODM have been depicted in Fig. 6. The time-series of IODMI during September-October and the cyclone frequency in the Bay of Bengal during November are illustrated in Fig. 7. The aim is to demonstrate the lag relationships which have forecasting applications. It is apparent from Fig. 7 that the IODM indices of preceding two months have inverse relation with the cyclone frequency in November. In other words, the negative IODMI during September-October corresponds to the enhanced cyclone frequency during November. Therefore, colder SST anomalies over the western tropical Indian Ocean and warmer SST anomalies over the southeastern tropical Indian Ocean during September-October will correspond to the enhanced cyclone activity in the Bay of Bengal during November.

Figure 8 depicts the correlation coefficients (CCs) between IODMI (September-October) and the cyclone frequency (November). The CCs between September-October IODMI and November cyclone frequency are about –0.4, which are statistically significant at the 99% level of significance. Thus, the IODMI can provide useful indications of November cyclone frequency two months in advance, which could be a potential tool in seasonal tropical cyclone forecasting. It is interesting to note that the simultaneous correlation between November IODMI and November cyclone frequency is less (–0.34) than the lag correlations, which implies that SST anomalies during preceding two months play a more important role in the cyclogenesis in the Bay of Bengal during November than the anomalies during November itself.

When the seasonal frequency of tropical cyclones during all the three months of post-monsoon period (October to December) is considered, the correlations get diluted. For instance, the lag correlations

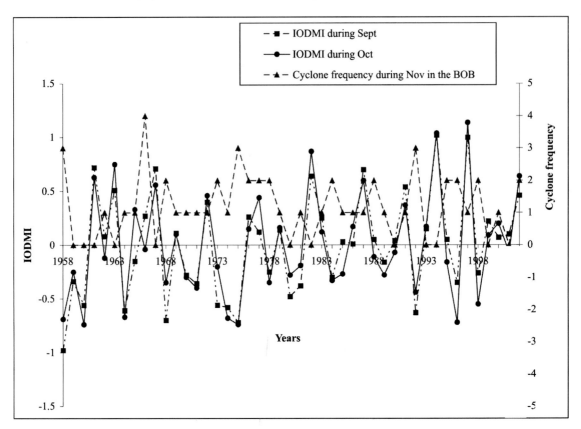

Fig. 7 Time-series of IODMI during September-October and the tropical cyclone frequency during November.

between August-September IODMI and the post-monsoon cyclone frequency are approximately –0.2. Although these correlations are not very significant (level of significance 90% only), they can provide good indications of the seasonal tropical cyclone frequency in the Bay of Bengal during the post-monsoon season.

4.10 Relationship between IODMI and the Frequency of Monsoon Depressions and Cyclones

Monsoon depressions and cyclones are important rainfall producing systems in India during the monsoon season (June to September). Substantial percentage of monsoon rainfall over the central parts of India is associated with these monsoon systems. These systems generally form over the North and adjoining Central Bay of Bengal and move in a northwesterly direction along the Monsoon Trough. It was discovered that the IODMI of May has a correlation of –0.22 with the seasonal frequency of monsoon depressions and cyclones in the Bay of Bengal during the monsoon season. The correlation coefficient is almost significant at the 95% level of significance. Therefore, with the lead time of one month, IODMI could be a predictor for the seasonal frequency of monsoon depressions in the Bay of Bengal.

When the lead time was increased and the correlation between April IODMI and the frequency of monsoon depressions and cyclones was computed, it was found that IODMI of April and the seasonal

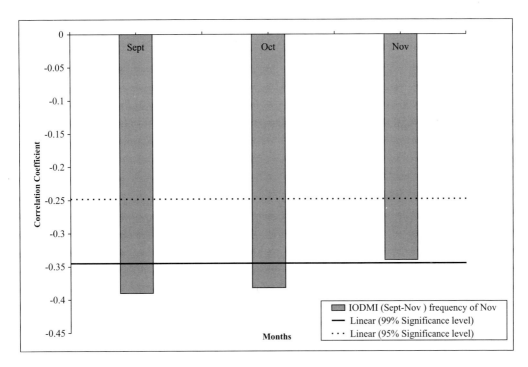

Fig. 8 Lag and simultaneous correlations between IODMI (September-November) and cyclone frequency during November.

frequency of monsoon depressions and cyclones were not related. Thus, in case of monsoonal cyclogenesis, the IODMI of only preceding month (i.e., May) has same prognostic utility, whereas in case of November (post-monsoon) cyclogenesis, the IODMI of preceding two months are significantly correlated with the cyclone frequency.

4.11 Lag Correlations between IODMI and the Pre-Monsoon Tropical Cyclone Frequency in the Bay of Bengal

During the pre-monsoon season (March to May), the cyclone frequency is maximum in May. The lag correlations between IODMI and the cyclone frequency during pre-monsoon are insignificant, which imply that IODMI cannot provide any predictive indications of the cyclone frequency during the pre-monsoon season.

4.12 Cyclone Monitoring, Early Warning System and Mitigation

As mentioned earlier, the tropical cyclones are among the deadliest natural disasters of the world. Their impacts are more pronounced in the developing countries like India, Bangladesh, Pakistan, Sri Lanka and Myanmar. The India Meteorological Department (IMD) has been designated by the World Meteorological Organization (WMO) to issue warnings and advisories in respect of tropical cyclones that form in the north Indian Ocean and affect the rim countries (WMO, 1986). IMD has necessary infrastructure and facilities to detect the tropical cyclones in the formative stages and predict their future movement. At present, IMD's cyclone warning work is carried out by the Regional Specialized Meteorological Centre (RSMC), New Delhi, through three Area Cyclone Warning Centers (ACWCs) and three Cyclone Warning Centers (CWCs) located at Kolkata, Chennai, Mumbai, Bhubaneswar, Visakhapatnam and Ahmedabad. The warnings are disseminated through 352 satellite-based Cyclone Warning Dissemination Systems (CWDS) installed along eastern and western coasts of India. The early warnings issued by IMD since 1877 have saved millions of lives not only in India, but in the neighboring countries as well. In the IMD's modernization plan, the observational network will be strengthened for better prediction skills of cyclone genesis and tracks. The CWDS network will be upgraded and made at par with the best systems in the world.

The launch of geostationary satellite INSAT in early 1980s heralded a quantum jump in the field of tropical cyclone forecasting (Dvorak, 1984; Koba et al., 1991). The satellite images have enabled the detection of cyclone genesis out at sea in sufficient advance. An accurate prediction of track and intensity of tropical cyclones is particularly required for disaster management programs.

4.12.1 Diagnosis

First important step in the prediction process of cyclonic disturbances is the accurate determination of the position and intensity of the disturbance. Presently, the forecasters rely heavily on the satellite information for this purpose. The recent developments in this field have shown that the satellite data can be used to predict the formation/genesis of the tropical cyclones with a lead time of 2-3 days (Leinder et al., 2003; Wang et al., 2007). This requires the monitoring of satellite-derived relative vorticity fields over the areas of cyclogenesis. If the relative vorticity is significantly higher than the normal, the formation may occur. The signal is observed 2-3 days in advance. Comprehensive work is required to arrive at the threshold values in different cyclone seasons over the Bay of Bengal and the Arabian Sea. The main genesis parameters are: (i) location of the center of the disturbance (latitude/longitude), and (ii) intensity (maximum sustained winds).

The following are the important satellite inputs for diagnosing the genesis of tropical cyclones:

(a) Satellite imagery
(b) Scatterometer-based surface winds and relative vorticity
(c) Relative vorticity in the lower troposphere (850 hPa)
(d) Lower level convergence
(e) Cloud top temperature (CTT)
(f) Outgoing Long wave Radiation (OLR)

4.12.2 Prediction

The main satellite-derived products which find applications in the intensity and track prediction are: (a) wind shear and wind shear tendency, (b) upper level divergence (300-150 hPa), (c) SST, and (d) water vapor winds. At present, not all of these parameters are available from the Indian satellites. However, after the proposed launch of the Indian remote sensing satellites INSAT-3D, OCEANSAT-2 and MEGHATROPIQUES in 2009, a majority of satellite products like temperature and humidity profiles in the atmosphere and sea surface winds and relative vorticity, which are required for diagnosing and predicting tropical cyclones, will become available from the Indian satellites. These satellite-derived parameters will be assimilated into the Numerical Weather Prediction (NWP) models to minimize the errors in cyclone track prediction for effective mitigation of TC hazards in India and neighboring countries. The major challenge ahead is the validation of these products before putting them into operational use. A few examples of satellite products used in tropical cyclone forecasting are depicted in Figs 9 to 11.

Fig. 9 Satellite images of tropical cyclones *Gonu* and *Sidr*. These images are used for the determination of cyclone center and intensity (*Source:* Cooperative Institute for Meteorological Satellite Studies, USA).

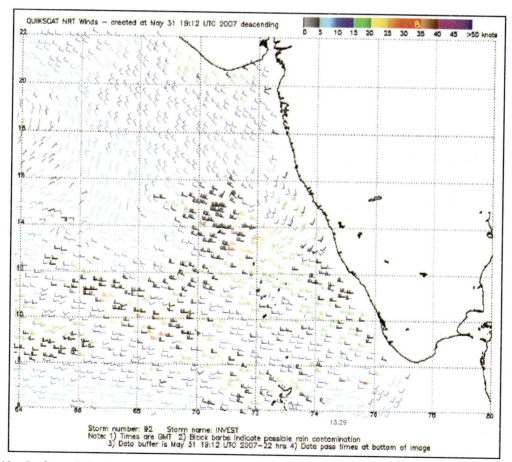

Fig. 10 Surface winds derived from QUICKSCAT satellite. Arrows indicate wind direction and bars indicate wind speed in nautical miles per hour (1 bar = 10 nautical miles/h). *Source:* Naval Research Laboratory (NRL), USA.

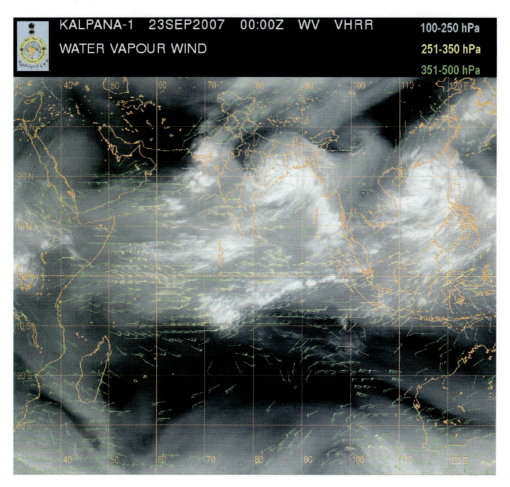

Fig. 11 Water vapor winds derived from the KALPANA satellite. These winds are used for predicting cyclone movement (*Source:* India Meteorological Department).

The satellite imagery shown in Fig. 9 is useful for the determination of cyclone center and intensity. The scatterometer-based sea surface winds shown in Fig. 10 provide vital inputs for the forecasting of cyclone genesis and accurate determination of cyclone center. The upper atmospheric winds shown in Fig. 11 are used for predicting the movement of cyclone.

5. CONCLUSIONS

Based on the results of various analyses and the results of simulation modeling presented in this chapter, the following conclusions could be drawn:

- On a long-term basis, the frequency of severe cyclonic storms in the Bay of Bengal has registered a significant increasing trend in November during the past century. The increasing trends get smoothened when the annual or seasonal frequencies of all storms are considered and, therefore, do not reflect the trends in the frequency of severe cyclones during intense cyclone months.

- The intensification rate of tropical disturbances to severe cyclonic storm stage has risen in the Bay of Bengal during intense cyclone months (May, October and November), with November witnessing the maximum uptrend of about 26% per hundred years.
- Undoubtedly, the frequency of stronger tropical cyclones (TCs) has increased in the North Indian Ocean (NIO) in recent decades and the frequency of weaker disturbances like tropical depressions has decreased. It is the large average annual number of tropical depressions/storms like monsoon depressions/storms in the NIO, especially the Bay of Bengal (BOB) which obscures the enhanced occurrences of stronger TCs.
- The results of simulation experiments using a regional climate model indicated an enhancement in the frequency and intensity of tropical disturbances in the Bay of Bengal during intense cyclone months, May, October and November, due to the climate change arising from increased greenhouse gas concentrations in the atmosphere. The experiments indicated that the frequency of post-monsoon tropical disturbances in the Bay of Bengal will increase by 50% by the year 2050, which is in good agreement with the observed trends.
- The Indian Ocean Dipole can provide a good indication of the tropical cyclone frequency in the Bay of Bengal during November with a lead time of two months, and it could be a potential tool in the forecasting of November tropical cyclone frequency. It was also found that the IODMI of May is correlated to the monsoon depressions/cyclones frequency during the monsoon season, and it can provide predictive indications of the seasonal monsoon depressions/cyclones frequency. However, the IODMI of preceding months cannot provide indications of the tropical cyclone frequency during the pre-monsoon season.
- Keeping in view the increasing trends in the frequency of stronger tropical cyclones in the North Indian Ocean in recent decades, there is an urgent need for better preparedness to mitigate their disastrous impacts in the coastal regions of the affected countries. In India, the National Disaster Management Authority (NDMA) can play a lead role in coordinating the efforts of concerned central and state agencies for mitigating tropical cyclone hazards. The scientific community needs to develop better forecasting skills, particularly in the field of seasonal forecasting of tropical cyclone frequency in the North Indian Ocean basin.

REFERENCES

Ali, A. (1995). Impacts of global climate on tropical cyclones in the Bay of Bengal and Bangladesh. Proceedings of the Workshop on Global Climate Change and Tropical Cyclones, 18-21 December 1995, Dhaka, Bangladesh.

Behera, S.K. and Yamagata, T. (2003). Influence of the Indian Ocean dipole on the southern oscillation. *J. Met. Soc. Japan*, **81(1):** 169-177.

Behera, S.K., Krishnan, R. and Yamagata, T. (1999). Unusual ocean-atmosphere conditions in the tropical Indian Ocean during 1994. *Geophys. Res. Lett.*, **26:** 3001-3004.

Casey, K.S. and Cornillon, P. (1998). A comparison of satellite and in-situ based sea surface climatologies. *Journal of Climate*, **12:** 1848-1863.

Dvorak, V.F. (1984). Tropical cyclone intensity analysis using satellite data. NOAA Tech. Rep. NESDIS 11.

Gray, W.M. (1968). Global view of the origin of tropical disturbances and storms. *Monthly Weather Review*, **96(10):** 669-700.

Iizuka, S., Matsuura, T. and Yamagata, T. (2000). The Indian Ocean SST dipole simulated in a coupled general circulation model. *Geophys. Res. Lett.*, **27(20):** 3369-3372.

IMD (1979). Tracks of Storms and Depressions in the Bay of Bengal and Arabian Sea. India Meteorological Department (IMD), New Delhi.

Joseph, P.V. (1985). Change in the frequency and tracks of tropical cyclones in the Indian Ocean Seas. Proceedings of the Workshop on Global Change and Tropical Cyclone, 18-21 December 1995, Dhaka, Bangladesh.

Koba, H., Hagiwara, T., Osana, S. and Akashi, S. (1991). Relationship between CI number and minimum sea level pressure/maximum wind speed of tropical cyclones. *Geophysical Magazine*, **44**: 15-24.

Leinder, S.M., Isaksen, L. and Hoffman, R.N. (2003). Impact of NSCAT winds on tropical cyclones in the ECMWF 4D VAR assimilation system. *Monthly Weather Review*, **131**: 4-26.

Mooley, D.A. (1980). Severe cyclonic storms in the Bay of Bengal, 1877-1977. *Monthly Weather Review*, **108**: 1647-1655.

Mooley, D.A. (1981). Increase in annual frequency of the severe cyclonic storms of the Bay after 1964: Possible causes. *Mausam*, **32(1)**: 35-40.

Obasi, G.O.P. (1997). WMO's programme on tropical cyclone. *Mausam*, **48(2)**: 103-112.

Saji, N.H. and Yamagata, T. (2003a). Possible roles of Indian Ocean Dipole Mode Events in global climate. *Climate Research*, **25**: 151-169.

Saji, N.H. and Yamagata, T. (2003b). Structure of SST and surface wind variability during Indian Ocean Dipole Mode Events: COADS observation. *Journal of Climate*, **16**: 2735-2751.

Saji, N.H., Goswami, B.N., Vinayachandran, P.N. and Yamagata, T. (1999). A dipole mode in the tropical Indian Ocean. *Nature*, **401**: 360-363.

Sikka, D.R. (2006). Major advances in understanding and prediction of tropical cyclones over the north Indian Ocean: A perspective. *Mausam*, **57(1)**: 165-196.

Singh, O.P. (2001). Long-term trends in the frequency of monsoonal cyclonic disturbances over the North Indian Ocean. *Mausam*, **52(4)**: 655-658.

Singh, O.P. (2007). Long-term trends in the frequency of severe cyclones of Bay of Bengal: Observations and simulations. *Mausam*, **58**: 59-66.

Singh, O.P. and Khan, T.M.A. (1999). Changes in the Frequencies of Cyclonic Storms and Depressions over the Bay of Bengal and the Arabian Sea. SAARC Meteorological Research Centre (SMRC), Report No. 2, Dhaka, Bangladesh.

Singh, O.P., Khan, T.M.A. and Rahman, M.S. (2001). Has the frequency of intense tropical cyclones increased in the North Indian Ocean? *Current Science*, **80**: 575-580.

Singh, O.P., Khan, T.M.A. and Rahman, S. (2000). Changes in the frequency of tropical cyclones over the north Indian Ocean. *Meteorology and Atmospheric Physics*, **75**: 11-20.

Singh, O.P., Rupakumar, K., Mishra, P.K., Krishnakumar, K. and Patwardhan, S.K. (2006). Simulation of characteristic features of Asian summer monsoon using a regional climate model. *Mausam*, **57(2)**: 221-230.

Srivastav, A.K., Sinha Ray, K.C. and De, U.S. (2000). Trends in the frequency of cyclonic disturbances and their intensification over Indian Seas. *Mausam*, **51(2)**: 113-118.

Wang, L., Lau, K.H., Fung, C.H. and Gan, J.P. (2007). The relative vorticity of ocean surface winds from the QUICKSCAT satellite and its effects on the genesis of tropical cyclones in the South China Sea. *Tellus-A*, **59**: 562-569.

Webster, P.J., Moore, A.M., Loschnigg, J.P. and Leben, R.R. (1999). Coupled ocean-atmosphere dynamics in the Indian Ocean during 1997-98. *Nature*, **401**: 356-360.

WMO (1986). Tropical Cyclone Operational Plan for the Bay of Bengal and the Arabian Sea. WMO TD No. 84, TCP-21, WMO, Geneva.

Temperature Extremes over India and their Relationship with El Niño-Southern Oscillation

J.V. Revadekar*, H.P. Borgaonkar and D.R. Kothawale

1. INTRODUCTION

The climate of a place represents the average weather in that place over more than thirty years' time period. While the weather can change in just a few hours, climate changes over longer time frames. Climate change is the variation in the Earth's global climate or in regional climates over time. It involves the variability or average state of the atmosphere over durations ranging from decades to millions of years. These changes can be caused by dynamic processes on the earth, external forces and human activities. Detection of change in climate against its variability is a key issue in climate research. Climate change is often expressed simply in terms of changes in the mean climate. However, regional climatic change in terms of extremes could have more significant socio-economic consequences than the changes in mean climate conditions as they may not show an appreciable change but may be characterized by a variety of extreme situations. There is a general agreement within the climate community that changes in the frequency as well as the intensity of extreme climate events would have profound impacts on natural and social systems. They can have serious and detrimental effects on human society and infrastructure as well as on ecosystems and wildlife. In general, the relationship between climate, extreme events and the extent of damage is extremely complex. The extreme events such as heat waves, floods, storms and dangerous avalanches have caused repeated concern in recent years. Our present knowledge of meteorological processes suggests that frequency and intensity of certain extreme events— for example, number of hot days and nights, heat waves, heavy precipitation and flood situations—will increase with the change in climate. However it remains uncertain whether or not to expect changes in some other extremes. The information based on current meteorological processes and model simulations must be taken into account in risk assessment, planning of precautionary measures and in development processes.

With the growing concerns of the regional manifestations of global warming worldwide, the most important aspects of temperature and precipitation variability are the occurrence of extremes. There have been some significant increases and decreases in extremes over time in some regions. Clear trend to fewer extremely low minimum temperatures in several areas is observed in recent decades, e.g., Australia, United States of America, Russia and China. Widespread extended periods of extremely high temperatures are also expected to become more frequent. Higher temperatures lead to higher rates of evaporation and capacity of atmosphere to hold the moisture also increases with increase in temperature. Therefore, we expect more precipitation as the earth warms and it is also expected to fall over shorter intervals of time leading to increase in frequency of heavy precipitation. Due to more rapid evaporations from soil, lakes, etc., some areas are expected to have frequent and prolonged drought conditions also. Therefore, extreme climate and weather events are increasingly being recognized as key aspects of climate change and they

---

*Corresponding Author

adversely affect the biosphere and have serious socio-economic implications (Bijlsma, 1996; Turner et al., 1998; Wheeler et al., 2000; Bugmann and Pfister, 2000; Tol, 2002). For example, decline in the rate of national economic growth leads to increase of import and high prices of food. Other commodities may lead to civil strife, loss of traditional export markets, reduced earnings from tourism, loss of biodiversity, and shortage of energy sources. Many of the assessments also emphasized effects of climate change acting on coastal systems that already are under stress. Temperature extremes constitute an integral component of climate variability and change on the regional scale (Revadekar et al., 2009). Extremes in the temperatures are characterized by daily temperature level exceeding tolerable limits, and their frequency and spell duration are of great interest in terms of human impacts. Extremes of heat and cold have a broad and far-reaching set of impacts on the nation. These include significant loss of life and illness, economic costs in transportation, agriculture, production, energy and infrastructure (Kilbourne, 1997; Hassi, 2005).

The Fourth Assessment Report of the Intergovernmental Panel on Climate Change (IPCC) concluded that global mean surface temperatures have risen by 0.74°C ± 0.18°C as estimated by a linear trend over the recent 100-year period 1906-2005. The rate of warming over the last 50 years in this period is almost double that over the entire 100 years, i.e., 0.13°C ± 0.03°C versus 0.07°C ± 0.02°C per decade (IPCC, 2007).

Most of Europe has experienced an increase in surface air temperature during the 20th century that averaged across the continent amounts to 0.8 °C in annual temperature (Beniston et al., 1998). This warming has been stronger in winter than in summer (Maugeri and Nanni, 1998; Domonkos and Tar, 2003; Feidas et al., 2004). Over Mediterranean region, the warming rate appears to be slightly lower, i.e., between 0.2 and 0.4 °C per 100 years (e.g., Kutiel and Maheras, 1998; Hasanean, 2001). In several regions of the world, indications of changes in various types of extreme climate events have also been noticed. The IPCC report on climate change (IPCC, 2007) indicated significant changes in various types of climate extreme events in several regions across the globe. The number of drought affected regions has been found to be increased as precipitation over land has marginally decreased while evaporation has increased due to warmer conditions. This report summarizes the various studies on extreme climate from different parts of the globe that there has been a significant decrease in the annual occurrence of cold nights and a significant increase in the annual occurrence of warm nights. Decreases in the occurrence of cold days and increases in hot days are generally less marked. The report also pointed out that the cold extremes have warmed more than the warm extremes over the last 50 years. Many regional studies worldwide have attempted the analysis of the changes in frequencies of cold and hot events and have generally indicated that cold extremes are decreasing and warm extremes are increasing (Karl et al., 1996; Plummer et al., 1999; Easterling et al., 2000; Hyun et al., 2002).

IPCC (2007) concluded that changes in extremes of temperature are consistent with the observed warming of the climate. Over India, Kothawale and Rupa Kumar (2005) reported that although the all-India mean annual temperature has shown a significant warming trend of 0.05°C/10 years during the period 1901-2003, the recent period 1971-2003 has seen a relatively accelerated warming of 0.22°C/10 years, which they noted to be largely due to unprecedented warming during the last decade. They also observed that the recent accelerated warming over India was manifest equally in day and night time temperatures. Therefore, information on the changing pattern of climate on global as well as regional scale and the changes in extreme climatic events is very important in the context of socio-economic effects. The vulnerability and adaptation aspects in climate change should be given higher priority to understand the impact of climate change and changing climate extremes on various socio-economic parameters like health, natural ecosystems, agriculture, water sector, forest, coastal zone, etc.

El Niño and La Niña are well known to be associated with significant monthly/seasonal climate anomalies at many places around the globe. It has also been shown that El Niño/La Niña events have a significant relationship with the relative frequency of climatic extremes (e.g., Swolter et al., 1999). Kiladis and Diaz (1989) have shown the relationship between temperature across east Asia and west Pacific with El Niño/Southern Oscillation (ENSO). Regional climate anomalies associated with the major El Niño episode during the 1877-1878 boreal winter were destructive, particularly in the Northern Hemisphere, where starvation due to intense droughts in Asia, Southeast Asia and Africa took the lives of more than 20 million people. At the same time anomalous intense precipitation were reported in other parts of world e.g. coastal areas of southern Ecuador and northern Peru, etc. It may be noted that during the period human influence was negligible to cause greenhouse effect (Aceituno et al., 2009). While there have been several studies on temperature changes over the Indian region (Kothawale and Rupa Kumar, 2005; Klein-Tank et al., 2006; Alexander et al., 2006; Rupa Kumar et al., 2006), and also the impact of ENSO on monsoon rainfall seasonal extremes (Pant and Rupa Kumar, 1997; Krishna Kumar et al., 2006), very little information is available on the role of El Niño/La Niña in temperature extremes over the Indian region.

In view of the above facts, the present study examines the changes in extremes over Indian region using objectively defined indices of observed temperature extremes. An attempt has also been made to examine the relationship between El Niño-Southern Oscillation (ENSO) and the temperature extremes over India in terms of their monthly/seasonal frequencies as well as intensities by computing anomalies in indices of temperature extremes during El Niño and La Niña and also by computing correlation coefficients between temperature extremes over India with NINO3.4 sea surface temperatures.

2. OVERVIEW OF EL NIÑO-SOUTHERN OSCILLATION

2.1 What is ENSO?

The El Niño-Southern Oscillation (ENSO) is a global coupled ocean-atmosphere phenomenon, i.e., an interaction between the ocean and the atmosphere and their combined effect on climate. It is an irregular phenomenon that alternates between its two phases, El Niño and La Niña approximately every 2-7 years. The Pacific Ocean signatures, El Niño (warming) and La Niña (cooling) are important temperature fluctuations in surface of tropical Eastern Pacific Ocean. It is a disruption of the ocean-atmosphere system in the tropical Pacific having important consequences for weather around the globe. Along the equator, western Pacific has some of the world's warmest ocean water, while in the eastern Pacific cool water wells up. After every two to seven years, strong westward flowing trade winds subside. This interrupts the upwelling of cool water at eastern Pacific. Peruvians named this phenomenon as El Niño. Under normal conditions, the atmosphere of the eastern south Pacific is dominated by an eastern center of high pressure. A zone of lower pressure prevails to the west and the resulting pressure difference causes the trade winds to blow east to west; however, trade winds may relax or sometimes even reverse. Since ocean currents are greatly influenced by the winds blowing above them, this ease of the trade winds also affects surface ocean currents. The Southern Oscillation refers to the pressure difference between southeastern tropical Pacific and Australian-Indonesian regions. When the waters of the eastern Pacific warm abnormally, the sea level pressure drops in the eastern Pacific and rises over the west. The combined effect of El Niño and Southern Oscillation (ENSO) affects the local climate and also global impact as well (Bjerknes, 1966,1969; Allan et al., 1996; Glantz, 2001). In contrast, La Niña events cause cooling in tropical Pacific and Indian Oceans that enhances rainfall in eastern Pacific Region (Allan et al., 1996; Joelle and Fowler, 2009).

2.2 Impacts of ENSO

The ENSO is known to be a major force of earth's year-to-year climate variability. ENSO is associated with floods, droughts and other disturbances in a range of locations around the world. These effects of ENSO and the irregularity of the ENSO phenomena make its prediction of high interest. This prominent source of inter-annual variability in climate has 2 to 7 years periodicity. Though not all, many of the countries are severely affected by current as well as Lag-1 ENSO events. Over certain regions, particularly parts of the Tropics, evidence for such relationship is overwhelming (Hastenrath, 1991).

India is largely dependent on the summer monsoon rainfall for agricultural and fishery sectors for food, water, etc. Indian subcontinent is one of the prime locations which is greatly affected by ENSO. Several studies in the past have established the relationship between SSTs over the ENSO region and the climate over India, particularly the ENSO-monsoon teleconnections (Pant and Rupa Kumar, 1997). Association between ENSO and summer monsoon rainfall over India has been rigorously studied (Sikka, 1980; Rasmusson and Carpener, 1983; Parthasarathy and Pant, 1985; Mooley, 1997). Suppression of convection in the El Niño phase and enhancement of convection in the La Niña phase have been observed over the Indian Ocean and land regions (Gadgil et al., 2004). It has been observed that the relationship has been weakening in recent two decades (Shukla, 1995; Kripalani and Kulkarni, 1997; Krishna Kumar et al., 1999; Kinter et al., 2002). Such weakening is possibly a short-lived feature, considering the conflicting results reported by global warming simulations (Ashrit et al., 2001) and some subsequent instances of monsoon failures in association with moderate El Niño situations such as 2002 and 2004.

Composite analysis in mean temperature over India during ENSO events shows negative anomalies in winter and pre-monsoon. However, the winter anomalies are stronger than the pre-monsoon anomalies. In monsoon and post-monsoon anomalies changes are towards the positive side (Kothawale, 2005). Kiladis and Diaz (1989) have shown the relationship of temperature across the east Asia and west Pacific with El Niño/Southern Oscillation (ENSO). Trewin (2001) also observed similar relationship for many parts of the Australian region and suggested that the relationship was strong enough to allow prediction. An effect of the 1997-1998 El Niño on ocean salinity variability has been reported by Maes (2000). The effect of ENSO on Total Ozone Column (TOC) deduced by ground observation and satellite sensors has been studied on a regional scale (Chandra et al., 1998; Langford et al., 1998). It has also been shown that El Niño/La Niña events have a significant relationship with the relative frequency of climatic extremes (Swolter et al., 1999). Han et al. (2001) discussed the ENSO signal in the inter-annual variation of Tibet total ozone based on the analysis of satellite ozone observation and atmospheric circulation data. They have shown that the Tibet ozone increases in El Niño events and decreases in La Niña events. The effect of El Niño in 1997-1998 on TOC has also been discussed by Singh et al. (2002) in light of the prevailing SSTs anomaly over the Indian Ocean and the Arabian Ocean, and concluded that the anomalous TOC is likely due to an El Niño effect which shows a close relation with the observed SSTs over the Indian Ocean and the Arabian Ocean. Sarkar and Singh (2000) also studied the inter-annual variability of total ozone deduced from Global Ozone Monitoring Experiment (GOME) and its relation to the observed El Niño of 1997-1998.

It was also noticed that the El Niño and La Niña events have a significant relationship with the relative frequency of climatic extremes (Swolter et al., 1999). They also have examined the relationship between short-term climate extremes over the continental United States and ENSO. Their results show that the greatest geographical coverage of statistically significant relationship between ENSO and seasonal temperature extremes occur in winter and spring, especially with the SOI leading by one season. For the Indian region, ENSO has been shown a strong relationship with variations in frequencies of extreme precipitation in winter season (Revadekar and Kulkarni, 2008). Jayanthi and Govindachari (1999) observed that Tamil Nadu had received record rainfall in 1997 that happened to be one of strongest El Niño years in which sea surface temperature (SST) anomalies were abnormally high throughout the year over equatorial Pacific. Goswami and Xavier (2005) have shown that ENSO controls the south Asian monsoon through the length of the rainy season with strong negative correlations between the length of rainy season and the ENSO related SSTs. According to Nicholls et al. (2005), the onset of El Niño leads to increased numbers of warm extremes (and fewer cool extremes) in east Asia and western Pacific. While analyzing the annual extremes, they pointed out that further research is necessary to examine seasonal variations in the relationships between climate extremes and ENSO. Deviations in temperature due to warm and cold phases of ENSO are geographically variable (Sittel, 1994).

3. DATA AND METHODOLOGY

3.1 Data Used in the Study

In this study, a good even coverage was sought to ensure that results could be representative of the whole country (India). With the network of fixed number of stations for the entire period, it became possible to perform an in-depth quality control process on data for each site whilst producing results giving a good coverage of India. Daily maximum and minimum temperature data were collected from Indian Daily Weather Report (IDWR), India Meteorological Department (IMD), New Delhi. Both daily maximum and minimum temperatures data of 121 well distributed meteorological stations for the period 1971-2003 were used in the study.

3.2 Quality Control

The data pre-processing is an important part of research. In order to ensure that extremes are not discarded with an over-enthusiastic or a purely automated data cleansing, a well considered pre-processing is crucial to ensure a meaningful analysis (Burn, 2004). An automated statistical system could not have managed the checks performed on the data in this study. Therefore, the following basic manual quality checks were applied to daily temperature data at each station:

- Maximum temperature is less than minimum temperature of the day.
- Same data value is repeated for several consecutive days.
- Same extreme is repeated.
- Mean $- n \times$ SD $> x >$ mean $+ n \times$ SD, where 'x' is data value to be inspected, 'mean' is average for that day of the year, 'SD' is standard deviation of the mean and 'n' is an integer set by the user. All the data values not lying in the range are treated as outliers and are examined relative to the surrounding days and documentary evidences and switched or removed as necessary.

3.3 Calculation of Temperature Extreme Indices

The joint World Meteorological Organization (WMO) Commission for Climatology (CCl)/World Climate Research Program (WCRP) project on Climate Variability and Predictability (CLIVAR), and Expert Team on Climate Change Detection, Monitoring and Indices (ETCCDMI) coordinated the development of a suite of climate change indices which primarily focus on extremes (Peterson et al., 2001). These indices are derived from daily temperature and precipitation data. The definitions of the indices including a user-friendly software package are freely available to the international research community (http://ccma/seos.uvic.ca/ETCCDMI). These indices have been widely used for global and regional analyses of climate extremes (Alexandar et al., 2006; Klein Tank et al., 2006). In the present study, temperature

Table 1. List of temperature extremes used in the present study

Index with unit	Description	Definition
(i) Frequency with Statistical Thresholds		
TN10 (% days)	Cold nights	Number of days in a month or season with minimum temperature below 10th percentile of the daily minimum temperature distribution in the 1971-2000 baseline period.
TX10 (% days)	Cold days	Number of days in a month/season with maximum temperature below 10th percentile of the daily maximum temperature distribution in the 1971-2000 baseline period.
TN90 (% days)	Hot nights	Number of days in a month/season with minimum temperature above 90th percentile of the daily minimum temperature distribution in the 1971-2000 baseline period.
TN90 (% days)	Hot days	Number of days in a month/season with maximum temperature above 90th percentile of the daily maximum temperature distribution in the 1971-2000 baseline period.
(ii) Intensities		
TNn (°C)	Coldest night	Monthly or seasonal lowest minimum temperature
TXx (°C)	Hottest day	Monthly or seasonal highest maximum temperature
TNx (°C)	Hottest night	Monthly or seasonal highest minimum temperature
TXn (°C)	Coldest day	Monthly or seasonal lowest maximum temperature
(iii) Range		
DTR (°C)	Diurnal temperature range	Mean monthly difference between maximum and minimum temperature

extreme indices shown in Table 1 were used to examine the role of ENSO on temperature extremes over India. These indices are computed for each of 121 stations over Indian landmass for the period 1971-2003 and are used further for comprehensive analysis.

3.4 Development of ENSO-Extreme Relationships

During the data period, the El Niño, La Niña and normal years were grouped into separate categories to analyze the temperature extremes. The El Niño years categorized for the analysis are: 1972, 1976, 1982, 1987, 1991 and 1997, whereas the La Niña years are: 1970, 1973, 1975, 1988 and 1999 (Halpert and Ropelewski, 1992; Kothawale et al., 2007). For each season, composite anomalies of extreme temperature indices were calculated for El Niño and La Niña with respect to neutral phase to examine their spatial distributions over the Indian region.

Moreover, seasonal composite anomalies of extreme temperature indices for El Niño and La Niña were also calculated for each station and their spatial distributions over the Indian region have been examined. For this purpose, extreme indices computed for each of the 121 stations were gridded onto a 1° × 1° grid. Each seasonal all-India mean index of extreme was correlated with monthly NINO3.4 HadSST values starting from January to December of the previous year (year 1).

4. RESULTS AND DISCUSSION

Until recently, most of the analyses of long-term global climate change using temperature have focused on changes in mean values. For example, Kothawale and Rupa Kumar (2005) have shown that the mean, maximum as well as minimum temperatures have increased by about 0.2 °C per decade during the period 1971-2003, for the country as a whole.

Rupa Kumar et al. (2003) studied the future scenarios of changes in rainfall and temperature in 21st century over the Indian region based on the simulations of various General Circulation Models (GCMs). Their analysis on extreme temperature based on the HadRM2 simulation projected an increase in extreme maximum temperature over the country in the 21st century due to an increase in Green House Gases (GHGs) concentrations. Over the region south of 25° N, this increase will be of order of 2-4°C. In the northern region, the increase in maximum temperature may exceed 4 °C. They also indicated that there will be a general increase in minimum temperature up to 4 °C over the southern peninsula and some parts of north India.

Rupa Kumar et al. (2006) analysed high resolution climate change scenarios of a state-of-art regional climate modeling system, known as PRECIS (Providing Regional Climates for Impacts Studies) developed by the Hadley Centre for Climate Prediction and Research for India. They have shown that PRECIS simulations under scenarios of increasing greenhouse gas concentrations and sulphate aerosols suggest marked increase in temperature towards the end of the 21st century (2071-2100). This study also indicated that extremes in maximum and minimum temperatures are also expected to increase into the future, but the night temperatures are increasing faster than the day temperatures.

However, extreme temperature has important impacts on vital aspects of society including crop yield, power production and consumption and human health. Therefore, detailed analysis on observed temperature extremes using daily station data for well-distributed network of 121 stations for minimum and maximum temperature have been carried out. Results are presented in following sections.

4.1 Changes in Temperature Extremes over India

4.1.1 Summary Statistics for Observed Trends in Temperature Extremes over India

Summary statistics of the stations indicating number of stations with positive/negative and significant trends in seasonal temperature extreme indices over Indian region have been prepared as shown in Table 2. It is clear from this table that all intensity indices of both cold and hot events (TNn, TXn, TNx and TXx) show a widespread increasing trend in temperature. However, more number of stations have an increasing trend for the winter season for all intensity indices. For hot intensities, minimum number of stations have shown an increasing trend during pre-monsoon season, which is the hottest season of India. Whereas for the coldest day temperature, maximum number of stations show an increasing trend during winter season which is the cold season of India.

All frequency indices of hot events (TX90p and TN90p) show a widespread increasing trend and all frequency indices of cold events (TX10p and TN10p) show a widespread decreasing trend (Table 3). For frequency indices also, more number of stations show an increasing trend in hot frequency and a decreasing trend in cold frequency for the winter season.

Considering the changes in both intensity and frequency indices, it is clear that in general higher changes are in post-monsoon season. Out of 121 stations, 94 stations show increasing trend in number of hot days and 97 stations show increasing trend in number of hot nights. Similarly for cold events also, more number of stations (113 for number of cold days and 105 for number of cold nights) show decreasing trends. This indicates the tendency of Indian winter towards the warming. Minimum changes in frequency indices are seen in pre-monsoon season. Analysis in general indicate reduction in seasonality, as hot season is slowly warming while winter show rapid changes.

Slightly more number of stations show decreasing trend in seasonal as well annual Diurnal Temperature Range (DTR) in Table 4. For the winter season, JF number of stations having decreasing

Table 2. Summary statistics of the stations for positive and negative trends in seasonal intensity indices of temperature extremes

	Hot events									
Intensity	Hottest day temperature					Hottest night temperature				
Number of Stations with	JF	MAM	JJAS	OND	Ann	JF	MAM	JJAS	OND	Ann
+ve trends	97	82	91	103	95	83	75	87	82	70
Significant +ve trends	32	29	27	40	36	27	23	32	23	20
-ve trends	24	39	30	18	26	38	46	34	39	51
Significant -ve trends	2	2	4	0	3	2	12	9	4	3
	Cold Events									
Intensity	Coldest day temperature					Coldest night temperature				
Number of Stations with	JF	MAM	JJAS	OND	Ann	JF	MAM	JJAS	OND	Ann
+ve trends	100	67	84	94	87	94	73	86	77	81
Significant +ve trends	36	30	23	38	37	36	30	37	19	39
-ve trends	21	54	37	27	34	27	48	35	44	40
Significant -ve trends	2	4	7	1	7	2	9	8	8	14

Table 3. Summary statistics of the stations for positive and negative trends in seasonal frequency indices of temperature extremes

	Hot events									
Frequency	Hot days					Hot nights				
Number of Stations with	JF	MAM	JJAS	OND	Ann	JF	MAM	JJAS	OND	Ann
+ve trends	72	81	85	94	69	88	90	84	97	86
Significant +ve trends	22	21	25	30	21	37	32	32	46	35
-ve trends	49	40	36	27	52	32	31	37	24	35
Significant -ve trends	5	0	6	1	8	4	5	6	6	8
	Cold events									
Frequency	Cold days					Cold nights				
Number of stations with	JF	MAM	JJAS	OND	Ann	JF	MAM	JJAS	OND	Ann
+ve trends	34	39	18	8	25	43	35	23	16	18
Significant +ve trends	2	5	4	1	3	6	5	3	6	7
-ve trends	85	82	103	113	96	78	86	98	105	103
Significant -ve trends	25	15	34	38	52	29	32	59	52	67

Table 4. Summary statistics of the stations for positive and negative trends in seasonal Diurnal Temperature Range

Number of stations with	JF	MAM	JJAS	OND	Ann
+ve trends	46	59	52	53	56
Significant +ve trends	16	22	19	16	24
-ve trends	75	62	69	68	65
Significant -ve trends	25	23	15	14	26

trend is maximum (~62% stations), while it is minimum (~51% stations) for MAM. Also, maximum number of stations shows a significant negative trend for the winter season. Similar to seasonality, daily variation in temperature also has decreasing tendencies.

4.1.2 Trends in All-India Time Series of Temperature Extremes

Simple arithmetic means of extreme indices at all the stations under study are considered to characterize all-India mean temperature extremes. Therefore for all seasons (JF, MAM, JJAS and OND), yearly time series of extreme temperature indices are constructed by taking arithmetic mean of corresponding seasonal index for each of 121 stations. These time series of spatially aggregated extreme indices are tested for the existence of any trends. All the intensity (TXx, TXn, TNx, and TNn) indices show an increasing trend in the all-India time series indicating warmer event towards the recent period. For frequency indices i.e. TX10p, TN10p, TX90p and TN90p, the frequency of hot events (TX90p and TN90p) shows an increasing trend and the frequency of cold events (TX10p an TN10p) shows a decreasing trend indicating that warm events are frequent during the recent period. All these seasonal trend values and their significance are summarized in Table 5.

A rise of 0.02 °C per year in the highest maximum temperature during JF and OND season and in the lowest minimum temperature during JF, MAM and OND are statistically significant and all have the same rate. However, the decrease in cold days/nights in all the seasons is at a higher rate than the increase in hot days/nights. Also, it can be clearly seen from Table 5 that the decrease in frequency of cold nights is at a higher rate than any other seasonal index of temperature extremes of frequency. Thus, it is obvious from the analysis that warming through night temperature (i.e., minimum temperature) is widespread and at a higher rate than that of day time temperature (i.e., maximum temperature). Diurnal Temperature Range (DTR), which is a combination of both these temperatures, shows a decreasing trend in all the seasons except MAM though no trend in DTR is statistically significant.

Table 5. Trends in all-India seasonal time series for indices of temperature extremes

Index	JF	MAM	JJAS	OND
TXx	0.02*	0.01	0.01	0.02*
TNn	0.02*	0.02*	0.01	0.02*
TN10p	−0.19*	−0.13*	−0.20*	−0.33*
TN90p	0.13*	0.047	0.13	0.07
TX10p	−0.15*	−0.10	−0.12*	−0.19*
TX90p	0.19*	0.06	0.08	0.13
DTR	−0.004	0.001	−0.003	−0.005

* Significant at 5% level.

4.2 Monthly Composite Anomalies in Temperature Extremes during El Niño and La Niña

It is known that all-India mean temperature start to increase from February and attain a peak value in month of May. Thereafter, the temperatures start decreasing and attain a secondary maximum in the month of October (Kothawale, 2004). In case of annual cycle of temperature extremes also, the all-India mean highest maximum temperature and lowest minimum temperature, both start to increase from January and attain their respective peak values in the month of May. Thereafter, the highest maximum temperature decreases sharply with the onset of the summer monsoon and attains secondary maximum in October and then decreases rapidly. However, lowest minimum temperature remains elevated at constant level throughout summer monsoon season (JJAS) and gradually decreases thereafter (Revadekar, 2009). So it is quite interesting to know whether variation in monthly temperature extremes are modulated by variation NINO3.4 SSTs or not. Therefore, this section presents monthly composite anomalies in various temperature extremes for India as a whole during El Niño and La Niña years with respect to neutral phase. The composite all-India monthly anomalies for the two categories of El Niño and La Niña with respect to normal years, for the frequencies (TX90p, TN90p, TX10p and TN10p) and intensities (TXx, TNx, TXn and TNn) of temperature extremes are as shown in Tables 6 and 7, respectively. These tables clearly indicate an association between monthly temperature extremes over the Indian region with El Niño/La Niña events in terms of frequencies as well as for intensities. The frequencies of hot days and hot nights (TX90p and TN90p respectively) presented in Table 6 show negative anomalies in winter (DJF) and pre-monsoon months (MAM), and positive anomalies during monsoon (JJAS) and post-monsoon (ON). Opposite features can be seen in La Niña events which show positive anomalies in winter and pre-monsoon months and negative anomalies during monsoon and post-monsoon months. Similarly, frequencies of cold days and cold nights (TX10p and TN10p respectively) also show a conspicuous association with El Niño/La Niña events with opposite patterns to that of hot events. All the intensity indices of hottest and coldest temperatures show negative anomalies during winter and pre-monsoon months and thereafter they reverse its sign in El Niño years (Table 7). However, opposite features can be seen during La Niña years.

Table 6. Composite anomalies in frequencies of hot day/night and cold day/night during El Niño and La Niña years

Month	TX90p		TN90p		TX10p		TN10p	
	El Niño	La Niña	El Niño	La Niña	El Niño	La Niña	El Niño	La Niña
January	−3.8	0.4	−1.8	−1.1	0.0	−1.8	−0.2	0.6
February	−2.1	2.5	−3.0	2.3	1.3	−5.3	2.6	−5.6
March	−4.1	−0.8	−2.7	1.6	−2.0	−3.6	−2.0	−0.9
April	−6.0	8.0	−4.5	5.9	4.2	−5.6	2.7	−3.8
May	−2.9	3.3	−4.3	4.2	0.2	−1.4	1.9	−3.3
June	4.2	−4.8	0.9	−2.7	−2.2	0.9	−0.0	0.1
July	7.9	−4.1	3.8	−3.6	−3.3	−0.5	−2.1	0.1
August	2.4	−4.0	−0.2	−2.9	−1.4	0.7	−0.4	−1.1
September	2.9	−4.9	−0.3	−0.9	−3.7	4.6	0.3	−2.7
October	4.1	−5.4	−2.6	−0.9	−1.9	1.3	−0.2	−2.8
November	−1.0	−4.0	3.8	−6.1	1.9	0.0	−4.3	7.1
December	1.2	−3.6	5.1	−6.3	2.8	−0.9	−3.9	6.7

Table 7. Composite anomalies in hottest day/night temperature and coldest day/night temperature during El Niño and La Niña years

Month	TXx		TNx		TXn		TNn	
	El Niño	La Niña	El Niño	La Niña	El Niño	La Niña	El Niño	La Niña
January	−0.07	0.03	−0.07	−0.01	0.09	0.68	−0.07	0.03
February	−0.01	0.42	−0.30	0.63	−0.07	0.77	−0.09	0.58
March	−0.10	0.39	0.08	0.03	0.15	0.29	−0.06	0.23
April	−0.54	0.41	−0.49	0.41	−0.58	1.37	−0.29	0.35
May	−0.22	0.44	−0.24	0.45	−0.16	0.15	−0.35	0.42
June	0.30	−0.50	0.11	−0.18	0.68	−0.17	−0.13	−0.10
July	0.98	−0.26	0.39	−0.13	0.44	−0.07	0.03	0.00
August	0.33	−0.16	0.09	−0.02	0.12	−0.23	−0.11	−0.05
September	0.34	−0.57	0.06	−0.04	0.40	−0.52	−0.43	0.28
October	0.39	−0.54	−0.35	0.08	−0.04	−0.08	0.09	0.47
November	0.14	0.22	0.39	0.01	−0.62	0.51	0.64	−0.72
December	0.06	−0.23	0.65	−1.06	−0.56	0.41	0.29	−0.27

4.3 Seasonal Spatial Patterns of Temperature Extremes during El Niño and La Niña

As mentioned in the earlier section the composite all-India mean monthly anomalies in temperature extremes have an association with El Niño/La Niña events. It should also be noted that all the months of each season show similar pattern of relationship with El Niño/La Niña, with characteristic change in the ENSO-associated extremes with the onset of the summer monsoon in June. Pre-monsoon season (MAM) is the hottest part of the year during which country faces extreme temperatures and heat wave conditions (Kothawale, 2004). It is well known that the Indian monsoon is one of the most dominant circulations of the atmosphere through its transport of heat and moisture from the tropics (Rajendran and Kotoh, 2008). Therefore, the indices of temperature extremes were analyzed to explore spatial coherence during pre-monsoon (MAM) and monsoon (JJAS). Figures 1(a, b) show MAM and JJAS seasonal spatial distribution of composite anomalies of frequencies of hot days and hot nights during El Niño and La Niña years.

The occurrence of hot days has a strong association with El Niño and La Niña events and a marked spatial coherence. Frequencies of hot days show a strong association as compared to hot nights. Also, for MAM, strong association has been observed over central and northern parts of India. Inconsistent with known association of ENSO with the Indian summer monsoon activity (Pant and Rupa Kumar, 1997), overall increase in hot frequencies has been noticed during El Niño events. Figures 2(a, b) show a weak but opposite association of frequencies in cold days and cold nights with El Niño or La Niña events. For the frequency of cold events also, day time frequencies have a strong association with ENSO; however, night time frequencies have a weak association. The spatial distribution of composite anomalies can also be seen for intensity indices, viz., highest maximum temperature and lowest minimum temperature [Figs. 3(a, b)]. El Niño leads to an increase in extreme highest temperature during monsoon. It is worth mentioning that the cloudy and humid conditions during the summer monsoon season are primarily responsible for the elevated temperature. Both the highest maximum temperature and the lowest minimum temperature show a decrease in temperature during pre-monsoon season. However, the lowest minimum temperatures do not show negative anomalies during monsoon, especially over northwest India.

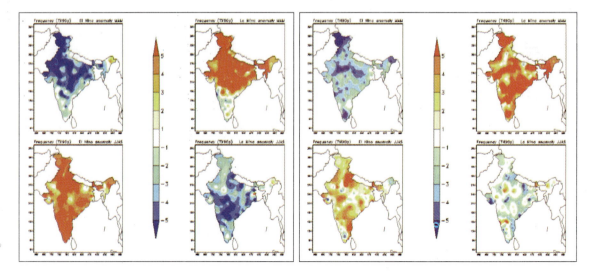

Fig. 1(a) Composite anomalies of frequencies of hot days (TX90p) during El Niño (left) and La Niña events (right). Top panel is for the pre-monsoon season (MAM) and bottom panel is for the monsoon season (JJAS).

Fig. 1(b) Composite anomalies of frequencies of hot nights (TN90p) during El Niño (left) and La Niña events (right). Top panel is for the pre-monsoon season (MAM) and bottom panel is for the monsoon season (JJAS).

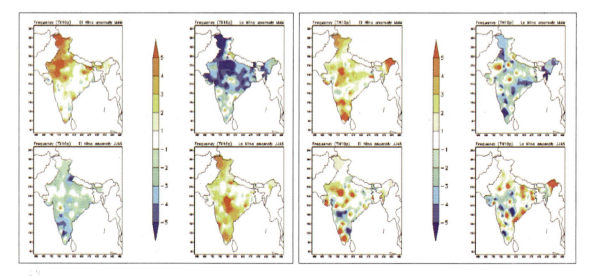

Fig. 2(a) Composite anomalies of frequencies of cold days (TX10p) during El Niño (left) and La Niña events (right). Top panel is for the pre-monsoon season (MAM) and bottom panel is for the monsoon season (JJAS).

Fig. 2(b) Composite anomalies of frequencies of cold nights (TN10p) during El Niño (left) and La Niña events (right). Top panel is for the pre-monsoon season (MAM) and bottom panel is for the monsoon season (JJAS).

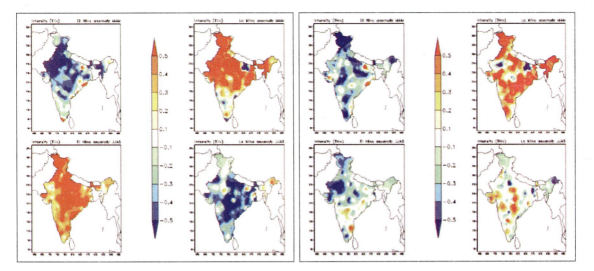

Fig. 3(a) Composite anomalies of hottest day temperature (TXx) during El Niño (left) and La Niña events (right). Top panel is for the pre-monsoon season (MAM) and bottom panel is for the monsoon season (JJAS).

Fig. 3(b) Composite anomalies of coldest day temperature (TNn) during El Niño (left) and La Niña events (right). Top panel is for the pre-monsoon season (MAM) and bottom panel is for the monsoon season (JJAS).

4.4 Lag-Correlation of Extreme Temperature Indices with NINO3.4 SSTs

For the Indian region, pre-monsoon (MAM) is the hottest and winter (JF) is the coldest season of a year. Therefore, the existence of lag-correlation is examined for the indices of temperature extremes occurring during JF and MAM with NINO3.4 sea surface temperature, which would be helpful to have a prior indication of cold or hot events. Table 8 presents lag (−1) correlations of hot events during MAM/cold events during JF with NINO3.4 SSTs. Frequencies of both hot days and nights during MAM have positive correlations with NINO3.4 SSTs for all months from January to December of preceding year. Correlations are weak during the beginning of the year, then start increasing and become strong in May onwards and again decrease towards the end of the year. Similar relationship is seen for the highest maximum temperature also, with positive correlations in all lag-1 months and a significant relationship in mid of the year. However, the relationship is weak compared to the frequencies of hot day and night. For the frequencies of cold day and night occurring during JF, opposite relationship can be seen to that of frequencies of hot day and night occurring during MAM. For these indices, positive but weak relationship was observed at the beginning of previous year. It changes its sign to negative in the month of March and becomes stronger from March onwards. For the cold nights, correlations are strong negative with statistically significant at 1% level towards the end of the previous year. The lowest minimum temperature during JF shows strong positive correlations towards the end of previous year. On the whole, extremes in minimum temperature have strong correlations. The frequencies of hot (cold) nights have a strong correlation than the frequencies of hot (cold) days. Magnitudes of the correlations of hot nights are higher than those of cold nights. Correlations of frequencies in hot and cold events are opposite to each other in sign.

Table 8. Correlations of seasonal temperature extreme indices over India with monthly NINO3.4 SSTs

Month	JF			MAM		
	TN10p	TX10p	TNn	TN90p	TX90p	TXx
January	0.06	0.05	−0.09	0.08	0.24	0.25
February	0.01	0.01	−0.04	0.13	0.28	0.26
March	−0.06	−0.07	0.02	0.19	0.34*	0.30
April	−0.17	−0.18	0.13	0.31	0.41*	0.34*
May	−0.26	−0.13	0.24	0.40*	0.44**	0.34*
June	−0.37*	−0.19	0.33	0.45**	0.42*	0.32
July	−0.45**	−0.29	0.43*	0.55**	0.47**	0.39*
August	−0.47**	−0.30	0.45**	0.59**	0.49**	0.38*
September	−0.49**	−0.34*	0.44**	0.53**	0.42*	0.31
October	−0.46**	−0.28	0.43*	0.40*	0.28	0.20
November	−0.47**	−0.27	0.45**	0.38*	0.27	0.19
December	−0.44**	−0.23	0.42*	0.35*	0.26	0.17

* Significant at 5% level; ** Significant at 1% level.

5. CONCLUSIONS

Analysis based on the indices of temperature extremes derived from daily station data for maximum and minimum temperature over the period 1971-2003 for well distributed network of 121 stations suggest tendency of Indian regions toward the warmer climate and also the warm events are frequent during the recent period. An analysis also suggests that the temperature extremes over Indian region have conspicuous association with El Niño and La Niña events. The analysis of this study indicated that the onset of summer monsoon in June shows a characteristic change in ENSO-associated temperature extremes. This is consistent with the known associations of ENSO with summer monsoon rainfall activity. El Niño leads to an increase in the frequencies of hot events and a decrease in the frequencies of cold events. In addition, El Niño is associated with an increase in the temperature of intensity indices, i.e., highest maximum temperature and lowest minimum temperature. Spatial distribution of composite anomalies of various indices of temperature extremes shows a spatial coherence during El Niño and La Niña. Correlation analysis suggests that the frequencies and intensities are modulated by NINO3.4 SSTs. Strong correlations were found in extremes in minimum temperature and in frequencies of hot events. An ENSO is well known to be associated with weakening of Indian summer monsoon and it is also known to be associated with the strengthening of northeast monsoon rainfall activities. Therefore, it causes increase in temperature over the country during summer monsoon and decrease in temperature over northeast monsoon.

Strong relationship of seasonal temperature extremes over Indian region with the ENSO is consistent with other regional studies made by Muller et al. (2000), Rusticucci and Varagas (2002) and Nicholls (2005) for Pampa Humeda region, Argentina and East Asia/west Pacific respectively.

ACKNOWLEDGEMENTS

The authors are grateful to Prof. B.N. Goswami, Director, IITM, Pune, India for providing facilities and continuous encouragement during the course of this study. The authors are also thankful to Dr. Nityanand Singh, Head, Climatology and Hydrometeorology Division, for his keen interest in the work. The basic daily maximum and minimum temperature data are provided by India Meteorological Department.

REFERENCES

Aceituno, P., Prieto, M. del R., Solari, M.E., Martínez, A., Poveda, G. and Falvey, M. (2009). The 1877-1878 El Niño episode: Associated impacts in South America. Climatic Change, 92: 389-416.

Alexander, L.V., Zhang, X., Peterson, T.C., Caesar, J., Gleason, B., Tank, A.K., Haylock, M., Collins, D., Trewin, B., Rahimzadeh, F., Tagipour, A., Ambenje, P., Rupa Kumar, K., Revadekar, J.V., Griffiths, G., Vincent, L., Stephenson, D., Burn, J., Aguilar, E., Brunet, M., Taylor, M., New, M., Zhai, P., Rusticucci, M. and Vazquez-Aguirre, J.L. (2006). Global observed changes in daily climate extremes of temperature and precipitation. *Journal of Geophysical Research*, **111**: D05109, doi: 10.1029/2005JD006290.

Allan, R., Lindesay, J. and Parker, D. (1996). El Niño Southern Oscillation and Climate Variability. CSIRO Publishing, Melbourne, Australia.

Ashrit, R.G., Rupa Kumar, K. and Krishna Kumar, K. (2001). ENSO-monsoon relationships in a greenhouse warming scenario. *Geophysical Research Letters*, **28**: 1727-1730.

Beniston, M., Tol, R.S.J., Delécolle, R., Hoermann, G., Iglesias, A., Innes, J., McmIchael, A.J., Martens, W.J.M., Nemesova, I., Nicholls, R., Toth, F.L., Kovats, S., Leemans, R. and Stojic, Z. (1998). Regional Impacts of Climatic Change on Europe. Special Report of the Intergovernmental Panel on Climate Change (IPCC), Chapter 5, Cambridge University Press, pp. 149-185.

Bijlsma, L. (1996). Coastal zone and small islands. *In:* R.T. Watson, M.C. Zinyowera, and R.H. Moss (editors), Climate Change 1995: Impacts, Adaptations, and Mitigation of Climate Change—Scientific-Technical Analyses. Contribution of Working Group II to the Second Assessment Report of the Intergovernmental Panel on Climate Change (IPCC), Cambridge University Press, pp. 289-324.

Bjerknes, J. (1966). A possible response of the atmospheric Hadley circulation to equatorial anomalies of ocean temperature. *Tellus*, **XVIII**: 820-829.

Bjerknes, J. (1969). Atmospheric teleconnections from the equatorial Pacific. *Monthly Weather Review*, **97**: 163-172

Bugmann, H. and Pfister, C. (2000). Impacts of interannual climate variability on past and future forest composition. *Regional Environmental Change*, **1(3-4)**: 112-125.

Burn, J. (2004). Climatic trends in daily weather extremes over India. M.Sc. Thesis, the University of Reading, U.K.

Chandra, S., Ziemke, J.R., Min, W. and Read, W.G. (1998). Effects of 1997-1998 El Niño on tropospheric ozone and water vapor. *Geophysical Research Letters*, **25**: 3867-3870.

Domonkos, P. and Tar, K. (2003). Long-term changes in observed temperature and precipitation series 1901-1998 from Hungary and their relations to larger scale changes. *Theoretical and Applied Climatology*, **75**: 131-147.

Easterling, D.R., Evans, J.L., Gorssmam, P.Y., Karl, T.R., Kunkel, K.E. and Ambenje, P. (2000). Observed variability and trends in extreme climate events: Brief review. *Bulletin of the American Meteorological Society*, **81**: 417-425.

Feidas, H., Makrogiannis, T. and Bora-Senta, E. (2004). Trend analysis of air temperature time series in Greece and their relationship with circulation using surface and satellite data: 1955-2001. *Theoretical and Applied Climatology*, **79**: 185-208.

Gadgil, S., Vinayachandran, P.N., Francis, P.A. and Gadgil, S. (2004). Extremes of the Indian summer monsoon rainfall, ENSO and equatorial Indian Ocean oscillation. *Geophysical Research Letters*, **31**, L12213, doi:10.1029/2004GL019733.

Glantz, M.H. (2001). Currents of Change: El Niño and La Niña Impacts on Climate and Society. 2nd edition. Cambridge University Press, Cambridge, U.K., 252 pp.

Goswami, B.N. and Xavier, P.K. (2005). ENSO control on the south Asian monsoon through the length of the rainy season. *Geophysical Research Letters*, **32(L18717)**: 1-4.

Halpert, M.S. and Ropelewski, C.F. (1992). Surface temperature patterns associated with the Southern Oscillation. *Journal of Climate*, **5**: 557-593.

Han, Z., Chongping, J., Libo, Z., Wei, W. and Yongxiao, J. (2001). ENSO signal in total ozone over Tibet. *Advances in Atmospheric Sciences*, **18**: 231-238.

Hasanean, H.M. (2001). Fluctuations of surface temperature in the Eastern Mediterranean. *Theoretical and Applied Climatology*, **68**: 75-87.

Hassi, J. (2005). Cold extremes and impacts on health. Extreme Weather Events and Public Health Responses, Editors Wilhelm Kirch, Roberto Bertollini and Bettina Menne. Springer, Berlin Heidelberg, 10.1007/3-540-28862-76, 59-67.

Hastenrath, S. (1991). Climate Dynamics of the Tropics. Kluwer Academic Publishers, Dordrecht, the Netherlands, 488 pp.

Hyun, S.J., Choi, Y. and Gyu-Holim, J. (2002). Recent trends in temperature and precipitation over south Korea. *International Journal of Climatology*, **22**: 1327-1337.

IPCC (2007). Climate Change, 2007: The Scientific Basis. Contribution of Working Group I to the Fourth Assessment Report of Intergovernmental Panel on Climate Change (IPCC). Cambridge University Press, Cambridge, 746 pp.

Jayanthi, N. and Govindachari, S. (1999). El Niño and northeast monsoon rainfall over Tamil Nadu. *Mausam*, **50(2)**: 217-218.

Joelle, L.G. and Fowler, A.M. (2009). A history of ENSO events since A.D. 1525: Implications for future climate change. *Climatic Change*, **92**: 343-387.

Karl, T.R., Knight, R.W., Easterling, D.R. and Quayle, R.G. (1996). Indices of climate change for United States. *Bulletin of the American Meteorological Society*, **77**: 279-292.

Kiladis, G.N. and Diaz, H.F. (1989). Global climatic anomalies associated with extremes in the Southern Oscillation. *Journal of Climate*, **2**: 1069-1090.

Kilbourne, E.M. (1997). Heat Waves and Hot Environments in Noji, Eric K., editor, The Public Health Consequences of Disasters. Oxford University Press, New York, 245-269.

Kinter, J.L., Miyakoda, K. and Yang, S. (2002). Recent changes in the connection from the Asian monsoon ENSO. *Journal of Climate*, **15**: 1203-1214.

Klein Tank, Peterson, T.C., Quadir, D.A., Doorji, S., Xukai, Z., Hongyu, T., Santhosh, K., Joshi, U.R., Jaswal, A.K., Rupa Kumar, K., Sikder, A., Deshpande, N.R., Revadekar, J.V., Yeleuova, K., Vandasheva, S., Faleveva, M., Gomboluudev, P., Budhathoki, K.P., Hussain, A., Afzaal, M., Chandrapala, L., Anvar, H., Amanmurad, D., Asanova, V.S., Jones, P.D., New, M.G. and Spektorman, T. (2006). Changes in daily temperature and precipitation extremes in central and south Asia. *Journal of Geophysical Research*, **111**: D16105, doi. 10.1029/2005JD006316.

Kothawale, D.R. (2004). Surface and upper air temperature variability over India: Its influence on the summer monsoon rainfall. Ph.D. Thesis, University of Pune, Pune, India.

Kothawale, D.R. and Rupa Kumar, K. (2005). On the recent changes in surface temperature trends over India. *Geophysical Research Letters*, **2**: L18714, doi:10.1029/2005GL023528.

Kothawale, D.R., Munot, A.A. and Borgaonkar, H.P. (2007). Temperature variability over Indian Ocean and its relationship with Indian summer monsoon rainfall. *Theoretical and Applied Climatology*, doi 10.1007/s00704-006-0291-z.

Kripalani, R.H. and Kulkarni, A. (1997). Climatic impact of El Niño/La Niña on the Indian monsoon: A new perspective. *Weather*, **52**: 39-46.

Krishna Kumar, K., Rajagopalan, B. and Cane, M.A. (1999). On the weakening relationship between the monsoon and ENSO. *Science*, **284**: 2156-2159.

Krishna Kumar, K., Rajagopalan, B., Hoerting, M., Bates, G. and Cane, M.A. (2006). Unraveling the mystery of Indian monsoon failure during El Niño. *Science*, **314**: 115-119.

Kutiel, H. and Maheras, P. (1998). Variations in the temperature regime across the mediterranean during the last century and their relationship with circulation indices. *Theoretical and Applied Climatology*, **61**: 39-53.

Langford, A.O., O'leary, T.J., Masters, C.D., Aikin, K.C. and Proffitt, M.H. (1998). Modulation of middle and upper tropospheric ozone at Northern midlatitudes by the El Niño/Southern Oscillation. *Geophysical Research Letters*, **25**: 2667-2670.

Maes, C. (2000). Salinity variability in the equatorial Pacific Ocean during the 1993-98 period. *Geophysical Research Letters*, **27**: 1659-1662.

Maugeri, M. and Nanni, T. (1998). Surface air temperature variations in Italy: Recent trends and an update to 1993. *Theoretical and Applied Climatology*, **61**: 191-196.

Mooley, D.A. (1997). Variation of summer monsoon rainfall over India in El Niño. *Mausam*, **48**: 413-420.

Muller, G., Nunez, M. and Seluchi, M. (2000). Relationship between ENSO cycles and frost vents within the Pampa Humeda region. *International Journal of Climatology*, **20**: 1619-1637.

Nicholls, N., Back, H.J., Gosai, A., Chambers, L.E., Choi, Y., Collins, D., Della-Marta, P.M., Griffths, G.M., Haylock, M.R., Iga, N., Lata, R., Maitrepierre, L., Manton, M.J., Nakamigawa, H., Ouprasitwong, N., Solofa, D., Tahani, L., Thuy, D.T., Tibig, L., Trewin, B., Vediapan, K. and Zai, P. (2005). The El Niño: Southern Oscillation and daily temperature extremes in east Asia and the west Pacific. *Geophysical Research Letters*, **32**, L16714, doi: 10.1028/2005GL022621.

Pant, G.B. and Rupa Kumar, K. (1997). Climates of South Asia. John Wiley & Sons, Chichester, U.K., 320 pp.

Parthasarathy, B. and Pant, G.B. (1985). Seasonal relationships between Indian summer rainfall and the southern oscillation. *Journal of Climate*, **5**: 369-378.

Plummer, N., Salinger, M.J., Nicholls, N., Suppich, R., Hennessy, K.J., Leigthan, R.M., Treulin, B.C., Page, C.M. and Lough, J.M. (1999). Changes in climate extreme over the Australian region and New Zealand during the twentieth century. *Climate Change*, **42(1)**: 183-202.

Rajendran, K. and Kitoh, A. (2008). Indian summer monsoon in future climate projection by a super high-resolution global model. *Current Science*, **95(11)**: 1560-1569.

Rasmusson, E.M. and Carpenter, T.H. (1983). The relationship between eastern equatorial Pacific sea surface temperatures and rainfall over India and Sri Lanka. *Monthly Weather Review*, **111**: 517-528.

Revadekar, J.V. and Kulkarni, A. (2008). The El Niño-Southern Oscillation and winter precipitation extremes over India. *International Journal of Climatology*, **28**: 1445-1452.

Revadekar, J.V., Kothawale, D.R. and Rupa Kumar, K. (2009). Role of El Niño/La Niña in temperature extremes over India. *International Journal of Climatology*, doi:10.1002.joc. 1851, 1-9.

Rupa Kumar, K., Krishna Kumar, K., Prasanna, V., Deshpande, N.R., Patwardhan, S. and Pant, G.B. (2003). Future climate scenarios. *In:* P.R. Shukla, S.K. Sharma, N.H. Ravindranath, A. Garg and S. Bhattacharya (editors), Climate Change and India: Vulnerability Assessment and Adaptation. University Press (India), Hyderabad, pp. 69-127.

Rupa Kumar, K., Sahai, A.K., Krishna Kumar, K., Patwardhan, S.K., Mishra, P.K., Revadekar, J.V., Kamala, K. and Pant, G.B. (2006). High-resolution climate change scenarios for India for the 21st century. *Current Science*, **90**: 334-345.

Rusticucci, M. and Vargas, W. (2002). Cold and warm events over Argentina and their relationship with the ENSO phases: Risk evaluation analysis. *International Journal of Climatology*, **22**: 467-483.

Sarkar, S. and Singh, R.P. (2000). Inter-annual variability of total ozone deduced from GOME and its relation to observed El Niño of 1997-1998. *Current Science*, **79**: 79-82.

Shukla, J. (1995). Predictability of the tropical atmosphere, the Tropical Ocean and Global Atmosphere. Proceedings of the International Conference on the Tropical Ocean and Global Atmosphere (TOGA) Program, WCRP-91. World Climate Research Program, Geneva, Switzerland.

Sikka, D.R. (1980). Some aspects of large scale fluctuations of summer monsoon rainfall over India in relation to fluctuations in the planetary and regional scale circulation parameters. *Proceedings of the Indian Academy of Sciences* (Earth Planet Science), **89**: 179-195.

Singh, R.P., Sarkar, S. and Singh, A. (2002). Effect of El Niño on inter-annual variability of ozone during the period 1978-2000 over the Indian sub-continent and China. *International Journal of Remote Sensing*, **23**: 2449-2456.

Sittel, M. (1994). Differences in the Means of ENSO Extremes for Temperature and Precipitation in the United States. COAPS Technical Report Series. Centre for Ocean Atmospheric Prediction Studies, the Florida State University, Tallahassee.

Swolter, K., Dole, R.M. and Smith, C.A. (1999). Short-term climate extremes over the continental United States and ENSO, Part I: Seasonal temperatures. *Journal of Climate*, **12:** 3255-3277.

Tol, R.S.J. (2002). Estimates of the damage costs of climate change. Part I: Benchmark estimates. *Environmental and Resource Economics*, **21:** 47-73.

Trewin, B.C. (2001). Extreme Temperature Events in Australia. Ph.D. Thesis. University of Melbourne, Melbourne, Australia.

Turner, R.K., Lorenzoni, I., Beaumont, N., Bateman, I.J., Langford, I.H. and McDonald, A.L. (1998). Coastal management for sustainable development: Analyzing environmental and socio-economic changes on the UK coast. *The Geographical Journal*, **164(3):** 269-281.

Wheeler, T.R., Craufurd, P.Q., Ellis, R.H., Porter, J.R. and Vara, P.V. (2000). Temperature variability and the yield of annual crops. *Agriculture, Ecosystems and Environment*, **82(1-3):** 159-167.

14 Monitoring Physiographic Rainfall Variation for Sustainable Management of Water Bodies in India

N.A. Sontakke*, H.N. Singh and Nityanand Singh

1. INTRODUCTION

About 90% of the natural disasters are of meteorological or hydrometeorological origin. Disasters such as earthquakes and tsunamis are occasional, but rainfall related hazards are frequent and needs constant monitoring. The annual average rainfall over India is 1166 mm out of which 78% is received in the hydrological wet season (mostly summer monsoon rainfall) of 135 days from 30 May to 11 October (Ranade et al., 2007) and is widespread over the country with large spatial variability due to large variation in intensity and frequency of rain-inducing disturbances such as monsoon depressions/lows, monsoon trough position, and physiography. There is a large intra-seasonal variability in this hydrological wet season. Rainfall during the remaining period though restricted to limited areas (i.e. during winter over northernmost and east peninsular India, during summer over northeast and southwestern India and during post-monsoon season over south peninsula, east coast and northeastern part of India) is crucial.

The large temporal variability in the Indian rainfall creates frequent flood/drought and epochs of above/below normal rainfall period for last several centuries. Areas of high rainfall and runoff are flooded in the summer monsoon season whereas in rest of the period severe water stress is faced. The rainfall extreme events are also increasing. Most Indian communities have a history of traditional mechanism of coping with such calamities. Irrigation, bunds, canals and several other techniques are adopted. However during recent decades with increasing population, large urbanization and industrialization, rapid economic and technological growth, water demand is increasing in all sectors such as agriculture, industry, drinking water etc. whereas climate change/increase in global temperature and related global atmospheric changes are seen in the form of higher incidences and intensity of natural disasters. In view of this we need improved policies of adaptation and mitigation to overcome such disasters. Ghosh-Bobba et al. (1997) and Niemczyniwicz et al. (1998) have analyzed changes facing integrated management of water systems over India. The need for interdisciplinary research and management of water systems over South Asia has been discussed by Bandyopadhyay (2007). Prabhakar and Shaw (2008) have considered the available evidences for the climate change over Indian subcontinent and assessed the existing preparedness and mitigation mechanism for drought risk reduction in the country. Revi (2008) has considered the likely changes that climate change will bring in temperature, precipitation and extreme rainfall, drought, river and inland flooding, storms/storm surges/coastal flooding, sea-level rise, environmental health risks and needed adaptation and mitigation agenda for cities in India where urban population is rapidly growing.

Collection, compilation, analysis of rainfall data and display in the form of maps and graphs on smaller spatial scales showing rainfall variability are required to decide adaptation and mitigation plans

---
*Corresponding Author

and policies. We are fortunate to have very long period instrumental rainfall data. These data are analyzed for each physiographical division and subdivision/province and latest tendency in annual, seasonal and monsoon monthly rainfall are reported in this chapter. Bandyopadhyay (2007) has emphasized that developing a framework for water systems research, the physiographic diversity of the region deserves serious attention, since the mountains and upland watersheds, the foothills and the floodplains and the delta areas differ widely both in terms of the natural environment and the social relations with water systems. Studies have reported rainfall fluctuation over administrative (states), meteorological subdivisions and hydrologic units (basins/catchments) of the country, but perhaps none for physiographic units. Thus, the objectives of the present study are:

(i) To understand chief features of annual, seasonal and monsoon monthly rainfall fluctuations over different physiographic divisions and subdivisions/provinces of India using longest instrumental data from sufficiently dense network of raingauge stations;
(ii) To demarcate area of the country showing different recent tendencies — increase, decrease and stationary, in annual, seasonal and monsoon monthly rainfalls; and
(iii) To investigate association between intensity of monsoon circulation and large-scale rainfall over the country on monthly and seasonal scales.

2. PHYSIOGRAPHY AND NATURAL ECOSYSTEMS OF INDIA

The physiography relates to the study of all the natural features on the earth's surface (landforms, climate, soil, vegetation, hydrology and distribution of flora and fauna) and delineation and description of regions from the viewpoint of landforms, including their origin and evolution, and the processes that shape them (Strahler, 1969). Surrounded between the parallels of 8° 4′ 28″ N and 37° 17′ 53″ N, and between the meridians of 68° 7′ 53″ E and 97° 24′ 47″ E a beehive-shaped India occupies geographical area of 3,287,263 sq. km. (including territorial sea) on the southern plank of the Asian landmass. The contiguous land area of the country is 3,279,501 sq. km; its north-south length is about 3,214 km and east-west breadth about 2,933 km. The land frontier is 15,200 km and the coastline 7,516.5 km. Andaman and Nicobar Islands in the Bay of Bengal and Lakshadweep in the Arabian Sea are parts of India. On the west of the country are Pakistan and Afghanistan, on the east Bangladesh and Burma, on the north Sinkiang province of China, Tibet, Nepal and Bhutan, and on the south Sri Lanka separated by a narrow channel of sea formed by the Palk Strait and the Gulf of Mannar.

Physiography of the country shows large regional variations. There are seven mountain ranges the Himalayas, the Patkai, the Vindhyas, the Satpura, the Aravalli, the Sahyadri and the Eastern Ghats. According to plate tectonics the Himalayas appear to have risen as a result of a collision between the northward drifting Indo-Australian plate at the rate of 6 mm/year, and the relatively stable Eurasian plate about 50 million years ago. Its east-west extension is 2,500 km and covers an area of about 500,000 sq. km. World's highest peak Everest in the Himalayas is 8,848 m above sea level, and there are 10 more peaks rising above 7,500 km. Patkai and the mountain ranges along the Indo-Bangladesh-Burma Border have come into existence along with the Himalayas. The present Aravalli range is only a remnant of gigantic mountain system that existed in pre-historic time with several of its summits rising above the snow line and nourishing glaciers of stupendous magnitude which in turn fed many great rivers. Vindhyan range in central India is about 1,050 km long and its average elevation is some 300 meters. The range has been formed by the wastes created by weathering of the ancient Aravalli ranges. Satpura range with many of its peaks rising above 1000 meters near Ratnagiri extends for a distance of 900 km between the Narmada and Tapi rivers. Sahyadri (or Western Ghats) with an average height of 1,200 meters and length 1,600 km runs along the west coast of the Deccan Plateau between the river Tapi and Kanyakumari, the southern-most point in India. The mountain range along the East Coast (or Eastern Ghats) is cut up by the powerful rivers into discontinuous blocks. Between the Godavari and Mahanadi rivers some of the peaks rise above 1,000 meters.

The main groups of the Indian soil types and their state-wise distribution are as follows (Raychaudhury et al., 1963): *Alluvial soil* (Andhra Pradesh, Assam, West Bengal, Bihar, Gujarat, Jammu and Kashmir, Kerala, Madhya Pradesh, Tamilnadu, Punjab, Rajasthan, Uttar Pradesh, Delhi and Tripura); *Black soil* (Andhra Pradesh, Gujarat, Madhya Pradesh, Tamilnadu, Maharashtra, Karnataka, Orissa, Rajasthan and Uttar Pradesh); *Red Soil* (Andhra Pradesh, Assam, West Bengal, Orissa, Rajasthan, Uttar Pradesh and Manipur); *Laterite and Lateritic soil* (Assam, West Bengal, Kerala, Tamilnadu and Karnataka); *Mixed Red and Black soil* (Tamilnadu, Orissa, Rajasthan and Uttar Pradesh); *Coastal soil* (West Bengal, Maharashtra, Karnataka, Orissa and Pondicherry); *Terai soil* (West Bengal and Uttar Pradesh); *Gray and Brown soils* (Jammu and Kashmir); *Tea soil* (West Bengal); *Brown soil* (Jammu & Kashmir); *Podzolized soil* (Jammu & Kashmir); *Skeletal soil* (Madhya Pradesh); *Forest and Hill soils* (Tamilnadu, Tripura and Kerala); *Peaty soil* (Kerala); *Red and Grey soil* (Maharashtra); *Mixed Red, Yellow and Grey soils* (Karnataka); *Saline and Deltaic soils* (Orissa); *Red and Yellow soil* (Rajasthan and Tripura); *Saline and Alkali soils* (Uttar Pradesh); *Desert soil* (Rajasthan); *Mountain soil* (Jammu & Kashmir and Himachal Pradesh); and *Hill soil* (Jammu & Kashmir and Himachal Pradesh).

Natural ecosystems of the country display great diversity. The panorama of different forests spans over a wide range from evergreen tropical rain forests in the Andaman and Nicobar Islands to dry alpine scrub high in the Himalaya. One of the earliest forests classifications of Greater India is due to Champion (1936), which was later modified by Champion and Seth (1968) for present-day India. In their four-tier classification of the Indian forests types Champion and Seth (1968) suggested six major types, 16 groups, 20 subgroups and 221 micro-groups. The six major types (and 16 groups) and their distribution in the country are as: *Moist Tropical Forests* (Tropical Wet Evergreen Forests, Tropical Semi Evergreen Forests, Tropical Moist Deciduous Forests and Littoral and Swamp Forests) – extreme northeastern India, Brahmaputra Valley, northern Indo-Gangetic plains, Jharkhand, eastern Orissa, coastal Andhra Pradesh, Sahyadri Range, Rann of Kutch and coastal Tamilnadu; *Dry Tropical Forests* (Tropical Dry Deciduous Forests, Tropical Thorn Forests and Tropical Dry Evergreen Forests)– most part of the country excluding western Himalaya, northeast India, eastern plateau and Sahyadri range; *Montane Subtropical Forests* (Subtropical Broad-Leaved Hill Forests, Subtropical Pine Forests and Subtropical Dry Evergreen Forests)– Jammu and Kashmir, Himachal Pradesh, Uttaranchal, Arunachal Pradesh, Nagaland, Manipur and Meghalaya; *Montane Temperate Forests* (Montane Wet Temperate Forests, Himalayan Moist Temperate Forests and Himalayan Dry Temperate Forests)– Arunachal Pradesh, Nagaland, Manipur, Mizoram, Uttaranchal, Himachal Pradesh and Jammu and Kashmir; *Subalpine Forests* and *Alpine Scrubs* (Sub-Alpine Forests, Moist Alpine Scrubs and Dry Alpine Scrubs)– Jammu and Kashmir, Himachal Pradesh and Uttaranchal. But the forest resource is fast depleting due to its overuse and misuse by modern civilized growing human population, about 25 crore in 1880 to more than 100 crore in 2001. Forest and other land use/land cover changes (LUCC) in the country as percentage of total geographical area from 1880 to 1980 are as: forest 20.35% to 12.27%; net cultivated 31.53% to 44.56%; built-up 1.17% to 3.15%; intermittent woods 11.9% to 7.61%; grasses 19.92% to 18.93%; barren land 11.31% to 9.49%; wetlands 1.98% to 1.27% and surface water 2.71% to 2.72%.

Physiography of India has large diversity constituting plains, valleys, plateaus, scarplands, mountain ranges, basins, hills and peninsula. The Himalayan Mountains have a major role in maintaining the climate of the whole country by obstructing the monsoon winds and providing rain throughout the country as well as protecting from the cold winds which blow from the North. Orographic rainfall on windward side of mountain and hill ranges such as Aravalli, Sahyadri, Khasi and Jayanti hills play important role in isohyets over India giving wetter climate over windward side and drier/less wetter over the leeward side. While occurrences of water bodies (rivers and canals; reservoirs, tanks and ponds; beels, oxbow lakes and derelict water; and brackish water) across the country depend upon physiographic settings and rainfall conditions, the recharging of the water bodies depends mostly on rainfall which is a highly variable parameter.

The National Atlas & Thematic Mapping Organization (NATMO, 1986) has published a map of India 'Physiographic Regions' on the projection system *'Conical Equal Area Projection with two Standard Parallels'* and scale 1:6 M showing boundary of physiographic divisions and subdivisions/provinces. Based on homogeneity in physiographic features the main land of the country is divided into 14 major divisions and 50 subdivisions/provinces. The divisions (and subdivisions/provinces) are as: The Western Himalaya (North Kashmir Himalaya, South Kashmir Himalaya, Punjab Himalaya and Kumaun Himalaya), The Eastern Himalaya, The Northern Plains (Punjab Plains, Ganga-Yamuna Doab, Rohilkhand Plains and Avadh Plains), The Eastern Plains (North Bihar Plains, South Bihar Plains, Bengal Plains and Bengal Basins), The North Eastern Range (Assam Valley, Meghalaya and Purvanchal), The Western Plains (Marusthali and Rajasthan Bagar), The North Central Highlands (Aravalli Range, East Rajasthan Uplands, Madhya Bharat Pathar and Bundelkhand), The South Central Highlands (Malwa Plateau, Vindhyan Scarplands, Vindhya Range and Narmada Valley), The North Deccan (Satpura Range and Maharashtra Plateau), The South Deccan (Karnataka Plateau and Telangana Plateau), The Eastern Plateaus (Baghelkhand Plateau, Chhotanagpur Plateau, Mahanadi Basin, Garhjat Hills and Dandakaranya), The Western Hills (North Sahyadri, Central Sahyadri, South Sahyadri and Nilgiri Hills), The Eastern Hills (Eastern Ghats (north), Eastern Ghats (south) and Tamilnadu Uplands), The West Coastal Plains (Kutch Peninsula, Kathiawar Peninsula, Gujarat Plains, Konkan, Karnataka Coast and Kerala Plains) and The East Coastal Plains (Utkal Plains, Andhra Plains and Tamilnadu Plains). The combined area of the Northern Plains and the Eastern Plains is known as the Indo-Gangetic Plains (IGPs) and is also considered additionally in this study as major physiographic division. Paralleling the Southern Himalayan Province, the IGPs are very important for food security of the South Asia. Due to large physiographic diversity coupled with large rainfall diversity a wide spectrum of natural and man-made fresh-water bodies (94 wetlands, 4291 dams and reservoirs, 500,000 minor irrigation tanks and ponds, and numerous canals, beels, oxbow lakes, abandoned quarries and derelict water) occur across the country. Development, management and conservation of these water bodies are highly dependent upon rainfall variability. Long period area-average rainfall series of physiographic regions (units) provides vital input to the study on influence of rainfall variability on land formation, flora and fauna, as well as environmental consequences of dynamics of wetland, dryland, desert, forest, grassland, drainage, etc.

3. RAINFALL DATA USED

Instrumental monthly rainfall data from a well spread network of 316 raingauge stations (Sontakke et al., 2008b) from earliest available year up to 2006 have been used in the present study. For the period 1901-2006 all the 316 stations data available have been used. Prior to 1901, the number of available stations from this network progressively decreases back in time—decreases to 314 raingauges up to 1900, to 312 up to 1871, to 190 up to 1870, to 100 up to 1861, to 80 up to 1860, to 70 up to 1851, to 60 up to 1846, to 56 up to 1844, to 12 up to 1843, to 6 up to 1829, to 4 (Chennai, Mumbai, Pune and Nagpur) up to 1826, to 2 (Chennai and Mumbai) up to 1817 and to lone station Chennai up to 1813 (Sontakke and Singh, 1996). Missing observation in the continuous data sequence has been filled by the ratio method (Rainbird, 1967) using nearest available observation as reference value. Number of filled values is less than 2% of the total number of monthly rainfall records. Data up to 1900 is obtained from the India Meteorological

Department (IMD) publication (Eliot, 1902), and for the 1901-2006 period from the National Data Center and Hydrology Section of the IMD, Pune. An account of this dataset is described by Mooley and Parthasarathy (1984).

Blanford (1886) checked the reliability of then available data by applying two types of tests. First, he compared the average and variation of the earlier years at any given station with those of the last 10-12 years, during which much more care and attention had been given to rainfall registration. Second, he compared the variation in corresponding years and months between neighboring stations situated under approximately similar conditions. Records which failed to satisfy either of the above tests were rejected. He concluded that selected data were 'free from any serious error'. In his attempt to compile and publish rainfall data for all the gauges over British India up to 1900 AD, Eliot (1902) also checked them thoroughly, but he did not mention the method employed. Regarding reliability of rainfall data of the Indian region, Walker (1910) had stated that 'long established observatories like Madras (Chennai), Bombay (Mumbai) and Calcutta (Kolkata) which have rainfall records available for earlier periods in the nineteenth century are trustworthy'.

4. DATA GENERATION USING GIS

The National Atlas & Thematic Mapping Organization (NATMO, 1986) has prepared a map of India 'Physiographic Regions' on *Conical Equal Area Projection with two Standard Parallels*' projection system and 1:6M scale showing boundary of physiographic divisions and provinces. The country has been divided into 14 physiographic divisions and 49 provinces. We have digitized the map of India 'Physiographic Regions' in four layers, (i) all-India boundary; (ii) boundaries of physiographic divisions; (iii) boundaries of physiographic provinces; and (iv) location of 316 raingauge stations. The physiographic divisions and provinces have been labeled by appropriate names (Fig. 1) and area of each physiographic division and province has been measured by creating attributes. Stations in different physiographic divisions and provinces out of total 316 available raingauge stations over the country have been identified with this map for development of longest rainfall series. The map is prepared using GeoMedia Professional 5.1 Geographic Information System (GIS) software package.

5. DEVELOPMENT OF LONGEST RAINFALL SEQUENCE

Longest instrumental area-averaged annual, seasonal and summer monsoon monthly rainfall series for different physiographic regions (divisions and subdivisions/provinces) have been developed in two parts: (i) simple arithmetic mean for the period with all available observations from the selected network and (ii) construction by applying established objective method for the period with lesser available observations. The complete process is described step by step for the Western Himalaya. In this region rainfall observation started in 1844 at Dehradun, in 1849 at Nainital, in 1853 at Dharamsala, in 1862 at Simla, in 1864 at Pithoragarh, in 1868 at Kulu, in 1889 at Joshimath, in 1891 at Srinagar and in 1901 at Jammu and Udhampur. Thus, the monthly rainfall data for 10 stations is available from 1901 onwards. The computational steps are as follows (Singh, 1994; Wigley et al., 1984; Sontakke et al., 2008b):

1. Prepare the area-averaged annual rainfall series for the period 1901-2006 of the Western Himalaya through simple arithmetic mean of all the 10 raingauges in the region.
2. Prepare the mean annual rainfall series (1901-2000) of the eight raingauges for which the annual rainfall observations extend back to 1891.
   (As indicated annual rainfall data of only eight raingauges from the 10-gauge network extends back to 1891.)

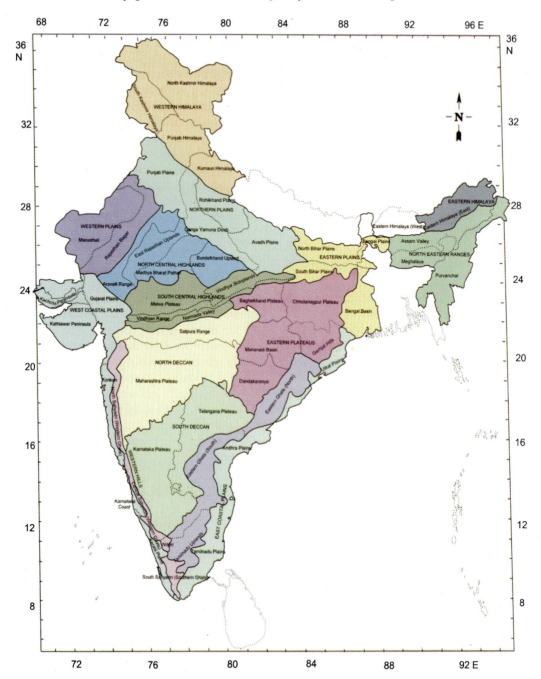

Fig. 1 Map showing physiographic divisions and subdivisions/provinces of the country.

3. Estimate the linear regression ($Y = a + bX$) of the representative 10-gauge mean series $\{Y_i\}$ on the 8-gauge mean series $\{X_i\}$ based on the period 1901-2000.

(All constructions in this report have been done with respect to the *'reference period'* 1901-2000.). Theoretically derived mathematical expression for the correlation ($R_{m,M}$) between *M*-gauge mean rainfall series and *m*-gauge mean series (*m* is a subset of *M*) is given by (Wigley et al., 1984):

$$R_{m,M} = \frac{1}{m \, s(m)} \sum_{i=1}^{m} s_i r_{i,M} \qquad (1)$$

In the present example, *M* is 10 and *m* is 8; $s(m)$ is the standard deviation of the 8-gauge (here $m = 8$) mean series; s_i is the standard deviation of each of the eight series; $r_{i,M}$ the correlation coefficient between each of the 8-gauge series and the *M*-gauge mean series. The correlation coefficient directly calculated between 10-gauge mean series (Y_i) and the 8-gauge mean series (X_i) and that calculated with the above equation were found equal.

4. Substitute the mean annual rainfall of the eight gauges available during 1900 in the regression and estimate the representative mean annual rainfall for the Western Himalaya for the year 1900.
5. Inflate the variance of the estimated rainfall amount of the year 1900 by dividing its departure from long term mean by the correlation coefficient between 10-gauge mean series (1901-2000) and the corresponding 8-gauge mean series ($R_{8,10}$), and get the constructed mean annual rainfall of the Western Himalaya for the year 1900 (Klein et al., 1959).
6. Repeat the above process to estimate the rainfall of each of the four seasons (winter JF, summer MAM, monsoon JJAS and post-monsoon OND) for the year 1900.
7. Check if total of the estimated four seasonal rainfalls is equal to the estimated annual rainfall amount.
8. For any discrepancy between the two rainfall figures (generally of a few millimeters) proportionately increase/decrease the seasonal rainfall amounts to get their finally constructed amounts.
9. Estimate the monthly rainfall for June, July, August and September in a similar way, compare them with corresponding constructed monsoon (JJAS) rainfall amount, and get the constructed monsoon monthly rainfalls after suitable correction for the year 1900.
10. Take up the year 1899, 1898, … …1891 and repeat the above process sequentially.
11. Repeat the entire procedure depending on available number of gauges for each year of the period 1889-1890 ($R_{7,10}$), 1868-1888 ($R_{6,10}$), 1864-1867 ($R_{5,10}$), 1862-1863 ($R_{4,10}$), 1853-1861 ($R_{3,10}$), 1849-1852 ($R_{2,10}$) and 1844-1848 ($R_{1,10}$).

The correlation coefficient $R_{m,M}$ is an indicator of reliability of the constructed rainfall amount. The constructed amount is retained if $R_{m,M}$ exceeded the 5% statistical significance level otherwise rejected. In general, the $R_{m,M}$ ranged from 0.71 to 0.99 for the constructed rainfall amount of the period 1813-1900. Especially in early years for very few cases the $R_{m,M}$ is weak (statistically significant but less than 0.7) when very few stations are available. In the absence of any quantitative data, these constructed values based on instrumental observations could provide vital information.

6. PHYSICAL FEATURES AND LONGEST RAINFALL SEQUENCE OF DIFFERENT PHYSIOGRAPHIC DIVISIONS AND SUBDIVISIONS/PROVINCES

For the 15 physiographic divisions, 49 sub-divisions/provinces and whole India the longest rainfall sequences have been prepared up to 2006. Construction details for divisions have been described and for subdivsions/provinces tabulated in Table 1. Area measured by the GeoMedia GIS Package, the mean annual PE, the mean annual rainfall, the seasonal rainfall as percent of annual rainfall, the annual number of rainy days, starting date, ending date and duration of the climatological rainy season for the physiographic divisions as well as the subdivisions/provinces are given in Table 2. Number of rainy days are the days with rainfall greater than 2.5 mm (IMD, 1961) and rainy season is the 'continuous period with monthly rainfall greater than 50 mm' (Singh, 1986). Abstract statistics (mean, median, standard deviation (SD), coefficient of variation and the quintiles (Q_1, Q_2, Q_3, Q_4)) for the rainfall series of physiographic divisions and all-India based on the period 1901-2000 are tabulated in Table 3 for the benefit of the users. Mean is measure of central tendency, standard deviation and coefficient of variation gives the nature of variability while quintiles are measures of the frequency distribution which divides the rainfall sequences in five equal parts. The table also lists the Index of Areal Representativeness (IAR), the mean correlation coefficient (CC) between the area-averaged series and the individual stations ($\bar{R}$) and the mean of CC between all possible combinations of raingauges that are averaged ($\bar{r}$). The IAR is defined as the ratio of "variance of area-averaged rainfall series" to the "mean variance of the individual rainfall series averaged" expressed in percentage and it provides a measure of spatial representation of the area-averaged rainfall series (Singh, 1994). The mathematical expression of IAR is:

$$\text{IAR} = \frac{S_R^2}{\frac{1}{M}\sum_{i=1}^{M} s_i^2} \times 100 \tag{2a}$$

That is, $\quad \text{IAR} = \frac{\text{Variance of the area-averaged rainfall series}}{\text{Mean variance of the rainfall series at all stations}} \times 100 \tag{2b}$

The $\bar{R}$ is also an indicator of areal representation of area-averaged rainfall series and is very useful in theoretical derivation dealing with the problem of rainfall spatial variability. The $\bar{r}$ is a measure of spatial coherency in the rainfall field and the inverse of $\bar{r}$ provides a broad idea of number of observations optimally required to prepare that particular representative area-averaged rainfall series.

Fluctuation characteristics, i.e., tendencies (increasing, decreasing or no trend) of annual, seasonal (winter, summer, summer monsoon, post-monsoon) and summer monsoon monthly (June, July, August and September) rainfall for 15 divisions and 49 subdivsions/provinces and whole India have been derived by examining the actual and the 9-point Gaussian low-pass filtered value plots (Sontakke et al., 2008a) as well as Cramer's t_k statistic applied on 31-term running means. Tendencies for annual, seasonal and monsoon monthly rainfall for the 15 physiographic divisions and whole India are described below for the total constructed period. The recent period tendencies for the 49 physiographic subdivisions/provinces are given in Table 1. Tendencies for total reconstructed period are available in Sontakke et al. (2008a). Each of the series is showing large fluctuations of increasing/decreasing epochs.

The Western Himalaya: The physiographic division is composed of the states of the Uttaranchal, the Himachal Pradesh and the Jammu & Kashmir. Earliest monthly rainfall record for Dehradun is available from 1844; and the continuous data for 10 stations is available since 1901. Longest area-averaged annual, seasonal and monsoon monthly rainfall series of the region could be developed for the period 1844-2006. The actual series along with 9-point Gaussian low-pass filtered values are shown in Fig. 2. Important statistics of the rainfall series are given in Table 3. Chief features of the rainfall fluctuations are: annual—1845-1894 I (increase), 1895-1902 D (decrease), 1903-1960 I, 1961-2006 D; winter—1845-1893 I, 1894-1963 D, 1964-2006 I; summer—1845-1938 D, 1939-2006 I (1983-2006 above normal but decreasing tendency); summer monsoon—1844-1960 I, 1961-2006 D; post-monsoon—1844-2006 I; June—1844-2006 D; July—1844-1959 I, 1960-2006 D; August—1844-1885 I, 1886-2006 D; and

Table 1. Monthly rainfall data availability position, longest rainfall sequence developed and recent tendency in rainfall fluctuations for different physiographic subdivisions/provinces of the country (I denotes 'increasing', D 'decreasing', and N 'no trend')

Sl. No.	Subdivisions/ Provinces	No. of available raingauges- Starting year	Longest sequence (ending yr 2006)	Recent tendencies in rainfall fluctuations									
				Annual	Winter	Summer	Monsoon	Post-monsoon	June	July	August	Sept
1	South Kashmir Himalaya	1-1891 & 3-1901	1891	1997 D	1964 I	1984 D	1960 D	1983 I	1961 I	1960 D	1956 D	1955 D
2	Punjab Himalaya	1-1853 & 3-1868	1853	1968 D	1862 D	1922 I	1968 D	1861 N	1861 D	1968 D	1861 D	1961 D
3	Kumaun Himalaya	1-1844 & 4-1889	1844	1955 D	1899 D	1938 I	1937 D	1957 D	1844 D	1935 D	1886 D	1976 D
4	Punjab Plains	7-1844 & 22-1871	1844	1975 D	1964 I	1959 I	1975 D	1844 N	1948 I	1981 D	1977 D	1946 D
5	Ganga-Yamuna Doab	15-1844 & 16-1860	1844	1962 D	1955 D	1961 I	1962 D	1957 D	1966 I	1978 D	1962 D	1844 I
6	Rohilkhand Plains	5-1844, 6-1864 & 9-1867	1844	1961 D	1860 N	1955 I	1961 D	1961 D	1860 I	1860 D	1962 D	1860 N
7	Avadh Plains	7-1844 & 19-1871	1844	1981 D	1943 D	1925 I	1981 D	1844 D	1960 I	1981 D	1939 D	1983 D
8	North Bihar Plains	1-1842 & 10-1871	1842	1988 D	1943 D	1958 I	1988 D	1978 D	1973 I	1982 D	1915 D	1921 D
9	South Bihar Plains	1-1842 & 6-1870	1842	1988 D	1958 D	1925 I	1988 D	1962 D	1961 D	1959 I	1936 D	1842 N
10	Bengal Plains	1-1837 & 5-1871	1837	1998 D	1968 I	1958 I	1998 D	1998 I	1910 D	1986 D	1860 D	1995 D
11	Bengal Basin	1-1829 & 12-1871	1829	1963 I	1829 N	1958 I	1962 I	1982 I	1914 D	1829 I	1997 D	1923 I
12	Assam Valley	4-1848 & 6-1849	1848	1979 I	1979 I	1980 I	1919 D	1989 I	1918 D	1977 D	1905 D	1988 D
13	Meghalaya	1-1866 & 2-1870	1866	1949 D	1964 I	1958 I	1866 D	1866 I	1957 D	1866 D	1866 D	1866 D
14	Purvanchal	1-1848 & 6-1871	1848	1948 D	1863 D	1966 I	1966 D	1988 D	1967 D	1968 D	1863 D	1970 I
15	Marusthali	1-1861 & 4-1871	1861	1994 D	1975 I	1950 I	1994 D	1943 I	1951 I	1978 D	1945 D	1994 D
16	Rajasthan Bagar	1-1867 & 9-1871	1867	1976 D	1867 N	1983 D	1976 D	1998 D	1969 I	1979 D	1973 D	1976 D
17	Aravalli Range	1-1856 & 5-1871	1856	1957 D	1948 D	1949 I	1951 D	1856 N	1856 I	1937 D	1999 I	1962 D
18	East Rajasthan Uplands	1-1866 & 6-1871	1866	1943 D	1908 D	1948 I	1943 D	1866 I	1962 I	1942 D	1947 D	1947 D
19	Madhya Bharat Pathar	6-1871	1871	1962 D	1871 N	1972 I	1962 D	1871 I	1932 N	1926 D	1961 D	1962 D
20	Bundelkhand Upland	2-1844, 3-1860, 4-1861, 5-1868 & 7-1871	1844	1962 D	1943 D	1973 I	1962 D	1962 D	1900 N	1860 D	1962 D	1990 D
21	Malwa Plateau	1-1844 & 15-1871	1844	1945 D	1860 N	1973 I	1945 D	1997 I	1860 D	1945 D	2000 I	1962 D
22	Vindhyan Scarplands	1-1844 & 4-1871	1844	1965 I	1978 D	1984 I	1965 I	1864 N	1998 I	1993 I	1997 D	1974 I
23	Vindhya Range	1-1868 & 4-1871	1868	1945 D	1982 D	1966 I	1945 D	1928 D	1868 D	1943 D	2000 I	1962 D
24	Narmada Valley	3-1844 & 4-1871	1844	1945 D	1844 I	1972 I	1945 D	1844 D	1971 D	1980 I	1844 N	1844 I
25	Satpura Range	3-1844, 4-1863, 5-1864, 6-1867 & 8-1871	1844	1945 D	1863 I	1985 I	1945 D	1928 D	1863 D	1945 D	1985 D	1961 D
26	Maharashtra Plateau	2-1828 & 22-1871	1828	1934 N	1973 I	1984 I	1997 I	1908 I	1967 I	1945 D	1922 I	1959 D

(Contd.)

Table 1. (*Contd.*)

Sl. No.	Subdivisions/ Provinces	No. of available raingauges- Starting year	Longest sequence (ending yr 2006)	Recent tendencies in rainfall fluctuations							
				Annual	Winter	Summer	Monsoon	Post-monsoon	June	July	August	Sept
27	Karnataka Plateau	1-1835 & 17-1871	1835	1957 D	1929 D	1963 D	1960 D	1993 D	1929 I	1961 D	1968 I	1982 D
28	Telangana Plateau	1-1843 & 11-1871	1843	1989 D	1955 I	1835 N	1989 D	1909 I	1978 D	1989 D	1984 D	1964 D
29	Baghelkhand Plateau	3-1871	1871	1926 D	1979 D	1871 N	1926 D	1928 D	1871 D	1930 D	1924 D	1966 I
30	Chhotanagpur Plateau	4-1848, 5-1869, 7-1870 & 9-1871	1848	1967 I	1928 D	1957 I	1957 I	1968 I	1927 I	1967 N	1943 D	1860 I
31	Mahanadi Basin	1-1861, 3-1863 & 6-1871	1861	1937 D	1992 I	1988 I	1937 D	1981 I	1861 D	1941 D	1947 D	1962 D
32	Garhjat Hills	3-1871	1871	1947 D	1870 D	1973 I	1944 D	1982 I	1977 I	1941 D	1908 N	1957 D
33	Dandakaranya	2-1871	1871	2002 I	1956 N	1974 I	1946 D	1982 I	1871 D	1945 D	1999 I	1945 D
34	North Sahyadri	1-1871	1871	1871 I	1871 N	1871 N	1871 I	1871 I	1871 I	1871 I	1871 I	1908 I
35	Central Sahyadri	1-1863	1863	1863 N	1907 D	1987 I	1863 N	1996 D	1967 I	1961 D	1863 I	1950 D
36	South Sahyadri	1-1871	1871	2000 D	1992 I	1937 D	1925 D	1961 I	1871 D	1925 D	1871 N	1871 N
37	Nilgiri	1-1829	1829	1912 D	1910 D	1956 D	1898 D	1939 D	1897 D	1961 D	1933 D	1958 I
38	Eastern Ghats (North)	2-1871	1871	1995 D	1945 N	1990 D	1995 D	1982 I	1936 D	1940 D	1981 I	1954 D
39	Eastern Ghats (South)	1-1852, 2-1863 & 3-1871	1852	1852 I	1952 I	1965 I	1988 D	1951 I	1852 I	1989 D	1931 I	1852 N
40	Tamilnadu Uplands	1-1852, 2-1853 & 3-1871	1852	1863 I	1933 D	2002 I	1997 D	1863 I	1996 D	1990 D	1999 D	1989 D
41	Kachchh Peninsula	1-1861	1861	1861 N	1942 D	1861 N	1861 N	1943 I	1951 I	1988 I	1999 D	1861 N
42	Kathiawar Peninsula	1-1861 & 6-1871	1861	1919 I	1905 D	1861 N	1919 I	1982 D	1951 I	1960 D	2000 I	2000 I
43	Gujarat Plains	1-1843 & 10-1871	1843	1960 D	1920 D	1993 D	1999 I	1856 N	1949 I	1972 I	1994 I	1955 D
44	Konkan	1-1817, 3-1844, 4-1860 & 5-1871	1817	1847 I	1847 D	1847 I	1919 I	1932 D	1926 N	1955 D	2000 I	1817 N
45	Karnataka Coast	1-1853 & 2-1861	1853	1921 I	1853 D	1988 I	1997 D	1990 I	1962 I	1994 D	1983 D	1962 D
46	Kerala Plains	1-1838 & 10-1871	1838	1925 D	1929 D	1997 I	1925 D	1983 I	1838 D	1926 D	1932 D	1838 I
47	Utkal Plains	1-1848, 2-1859, 3-1867 & 5-1871	1848	1980 I	1924 D	1958 I	1950 I	1983 I	1965 I	1956 N	1848 I	1996 I
48	Andhra Plains	4-1852, 5-1861, 6-1866, 7-1870 & 8-1871	1852	1852 I	1976 I	1852 I	1852 I	1966 I	1945 I	1989 D	1868 I	1969 I
49	Tamilnadu Plains	1-1813 & 11-1871	1813	1953 I	1924 D	1986 I	1962 D	1949 I	1997 D	1985 D	1967 D	1935 I

Table 2. Geographical area, mean annual evapotranspiration (PE), mean annual rainfall, seasonal rainfall (% of annual rainfall) and the parameters of wet season for different physiographic divisions and subdivisions/provinces of the country

Sl. No.	Physiographic divisions Subdivisions/Provinces	Area (km²)	Annual PE (mm)	Mean annual rainfall (mm)	Contribution to annual rainfall (%)				Annual rainy days	Rainy season (date)		Duration (days)
					Win	Sum	Mon	Post-mon		Start	End	
1	*The Western Himalaya*	308,486	1005.6	1562.1	10.1	12.5	71.6	5.8	76.1	28 Jun	22 Sep	238
i.	South Kashmir Himalaya	54,538	874.0	1084.4	15.7	19.8	56.3	8.2	57.9	18 Jan	7 Apr	80
ii.	Punjab Himalaya	55,020	870.8	1820.1	10.5	12.5	71.5	5.5	86.3	18 Jan	23 Sep	249
iii.	Kumaun Himalaya	47,517	1072.2	1726.8	7.3	9.1	78.7	4.9	82.1	27 Jan	6 Mar	39
										27 May	30 Sep	127
2	*The Northern Plains*	291,998	1484.2	846.8	4.8	4.5	85.3	5.4	41.4	19 Jun	28 Sep	94
i.	Punjab Plains	85,222	1482.7	637.8	7.6	7.3	80.1	5.0	31.8	30 Jun	15 Sep	78
ii.	Ganga-Yamuna Doab	69,791	1514.4	802.8	4.4	3.7	86.6	5.3	40.1	23 Jun	28 Sep	90
iii.	Rohilkhand Plains	49,486	1422.1	1028.3	4.3	4.2	85.8	5.7	45.6	14 Jun	22 Sep	102
iv.	Avadh Plains	87,499	1473.7	1048.5	3.1	3.1	88.2	5.6	51.0	14 Jun	23 Sep	103
3	*The Eastern Plains*	162,434	1460.4	1460.4	2.4	10.4	79.1	8.1	71.2	17 May	15 Oct	152
i.	North Bihar Plains	48,929	1393.5	1213.8	2.3	6.7	84.6	6.4	57.2	30 May	7 Oct	131
ii.	South Bihar Plains	35,080	1518.4	1141.1	3.1	5.3	84.8	6.8	58.0	11 Jun	7 Oct	119
iii.	Bengal Plains	16,523	1131.3	2521.9	1.1	13.6	79.3	6.0	96.7	19 Apr	19 Oct	184
vi.	Bengal Basin	61,902	1534.4	1534.4	2.5	12.7	75.0	9.8	78.9	14 May	18 Oct	158
4	*The Indo-Gangetic Plains*	454,432	1431.7	1025.7	3.8	7.0	82.7	6.5	51.2	13 Jun	30 Sep	110
5	*The North Eastern Range*	194,803	1093.0	2287.0	2.1	24.3	65.1	8.5	114.4	22 Mar	21 Oct	214
i.	Assam Valley	56,756	1143.4	2196.1	2.3	25.0	65.8	6.9	109.6	23 Mar	18 Oct	210
ii.	Meghalaya	37,344	1023.8	2720.2	1.2	21.2	68.3	9.3	122.8	30 Mar	24 Oct	209
iii.	Purvanchal	100,709	1078.2	2233.5	2.1	24.9	63.3	9.7	116.5	19 Mar	21 Oct	216
6	*The Western Plains*	197,411	1753.9	363.2	3.0	5.7	87.7	3.6	20.1	14 Jul	31 Aug	49
i.	Marusthali	117,005	1838.9	243.8	3.9	8.1	84.3	3.7	14.9	22 Jul	11 Aug	21
ii.	Rajasthan Bagar	80,406	1667.5	398.4	3.0	5.3	88.0	3.7	21.5	13 Jul	3 Sep	53
7	*The North Central Highlands*	209,648	1525.5	785.1	2.4	2.5	90.7	4.4	39.0	23 Jun	18 Sep	88
i.	Aravalli Range	52,167	1473.8	695.4	1.6	2.9	92.0	3.5	35.9	21 Jun	17 Sep	89
ii.	East Rajasthan Uplands	53,698	1745.2	705.6	2.0	3.0	91.3	3.7	35.0	25 Jun	15 Sep	83
iii.	Madhya Bharat Pathar	55,426	1520.8	814.2	2.1	2.0	91.7	4.3	40.1	23 Jun	19 Sep	89
iv.	Bundelkhand Upland	48,357	1501.1	935.4	3.0	2.1	89.7	5.2	43.6	28 Jun	21 Sep	94
8	*The South Central Highlands*	179,593	1503.7	1051.4	2.1	1.9	91.1	4.9	51.0	13 Jun	22 Sep	102
i.	Malwa Plateau	81,421	1590.3	1001.9	1.6	1.7	91.8	4.9	47.5	14 Jun	21 Sep	100
ii.	Vindhyan Scarplands	39,333	1452.7	1174.5	3.4	2.4	89.5	4.7	57.5	14 Jun	22 Sep	101
iii.	Vindhya Range	27,662	1813.2	1094.0	2.1	1.9	90.9	5.1	52.9	12 Jun	23 Sep	104
iv.	Narmada Valley	31,177	1347.3	1122.1	2.4	2.2	90.1	5.3	55.8	12 Jun	22 Sep	103

(Contd.)

Table 2. (Contd.)

Sl. No.	Physiographic divisions Subdivisions/Provinces	Area (km²)	Annual PE (mm)	Mean annual rainfall (mm)	Contribution to annual rainfall (%)					Annual rainy days	Rainy season (date)		Duration (days)
					Win	Sum	Mon	Post-mon			Start	End	
9	*The North Deccan*	351,838	1606.6	956.6	2.1	4.3	84.2	9.4	53.8	11 Jun	5 Oct	117
i.	Satpura Range	99,241	1432.2	1169.5	3.1	3.3	87.2	6.4	61.4	10 Jun	22 Sep	105
ii.	Maharashtra Plateau	252,597	1661.7	887.4	1.6	4.6	83.4	10.4	51.1	11 Jun	6 Oct	118
10	*The South Deccan*	313,452	1653.6	919.9	1.2	11.1	70.2	17.5	55.1	26 May	17 Oct	145
i.	Karnataka Plateau	175,120	1611.3	956.3	0.9	14.0	64.6	20.5	58.0	19 May	31 Oct	166
ii.	Telangana Plateau	138,332	1719.4	862.5	1.8	7.1	77.8	13.3	50.5	13 Jun	12 Oct	122
11	*The Eastern Plateaus*	337,487	1438.6	1361.7	3.0	6.0	84.1	6.9	71.4	9 Jun	10 Oct	124
i.	Baghelkhand Plateau	48,114	1407.5	1335.5	3.8	3.1	88.1	5.0	66.6	10 Jun	23 Sep	106
ii.	Chhotanagpur Plateau	102,078	1405.5	1318.9	3.2	7.6	81.3	7.9	72.8	28 May	13 Oct	139
iii.	Mahanadi Basin	79,134	1487.1	1379.4	2.5	4.2	88.1	5.2	68.5	8 Jun	3 Oct	118
iv.	Garhjat Hills	47,769	1539.9	1434.3	3.4	8.4	79.5	8.7	76.3	24 May	15 Oct	145
v.	Dandakaranya	60,392	1422.3	1378.1	1.7	6.0	84.7	7.6	74.4	8 Jun	12 Oct	127
12	*The Western Hills*	75,223	1222.9	2141.7	0.7	7.9	79.6	11.8	114.3	20 May	10 Nov	175
i.	North Sahyadri	34,555	No data	1934.3	0.2	1.1	94.2	4.5	81.5	7 Jun	7 Oct	123
ii.	Central Sahyadri	20,369	1264.6	3249.8	0.3	7.1	83.4	9.2	132.4	22 Apr	10 Nov	203
iii.	South Sahyadri	16,283	1413.3	2923.8	1.9	20.0	53.7	24.4	138.4	17 Mar	4 Dec	263
iv.	Nilgiri	4,016	1036.9	1002.5	2.7	20.3	46.8	30.2	104.7	21 Apr	30 Nov	224
13	*The Eastern Hills*	195,098	1731.1	1015.3	2.1	11.1	63.4	23.4	59.2	25 May	11 Nov	171
i.	Eastern Ghats (North)	70,672	No data	1510.4	1.7	7.9	81.0	9.4	82.8	27 May	16 Oct	143
ii.	Eastern Ghats (South)	78,901	1696.7	873.7	2.4	8.2	63.6	25.8	49.9	16 Jun	11 Nov	149
iii.	Tamilnadu Uplands	45,568	1765.5	826.7	2.2	18.0	42.0	37.8	52.8	18 May	16 Nov	183
14	*The West Coastal Plains*	223,923	1642.3	1765.2	0.7	7.8	80.3	11.2	72.9	18 May	5 Nov	172
i.	Kachchh Peninsula	15,564	1897.1	347.1	1.5	3.5	89.3	5.7	16.2	10 Jul	7 Sep	60
ii.	Kathiawar Peninsula	58,467	1826.7	543.4	0.6	1.6	92.6	5.2	24.9	18 Jun	11 Sep	87
iii.	Gujarat Plains	66,904	1678.0	925.8	0.4	1.1	95.1	3.4	40.5	12 Jun	28 Sep	101
iv.	Konkan	23,876	1624.0	2445.9	0.1	1.7	93.4	4.8	85.1	24 Jun	14 Oct	114
v.	Karnataka Coast	8,426	1400.4	3341.2	0.1	5.4	87.1	7.4	111.6	11 May	6 Nov	180
vi.	Kerala Plains	24,144	1590.1	2824.0	1.1	14.0	66.8	18.1	121.6	15 Apr	21 Nov	221
15	*The East Coastal Plains*	168,481	1699.0	1111.0	3.4	9.6	49.7	37.3	56.3	27 May	6 Dec	194
i.	Utkal Plains	28,690	1586.6	1478.0	2.8	9.8	71.0	16.4	70.6	28 May	3 Nov	160
ii.	Andhra Plains	61,323	1655.3	1005.2	2.2	7.3	54.5	36.0	51.4	16 Jun	18 Nov	156
iii.	Tamilnadu Plains	78,467	1762.6	1021.2	4.7	11.1	32.2	52.0	53.4	28 May	17 Dec	204
16	*The Whole India*	3,188,111	1522.2	1165.9	2.7	8.7	77.8	10.8	57.4	30 May	11 Oct	135

Table 3. Important statistics and the parameters of the spatial coherency and representation of annual, seasonal and monsoon monthly area-averaged rainfall series (1901-2000) of the physiographic divisions and the whole country

	Ann	Win	Sum	Sum Mon	Post Mon	Jun	Jul	Aug	Sep
The Western Himalaya									
Mean (mm)	1562.1	158.4	195.0	1118.8	89.9	139.5	397.6	400.1	181.5
Median (mm)	1540.0	152.7	185.7	1116.0	81.8	127.8	406.1	385.8	174.4
SD (mm)	225.5	63.6	75.7	220.8	59.4	66.2	107.5	103.5	89.4
CV (%)	14.4	40.2	38.8	19.7	66.1	47.5	27.0	25.9	49.3
Q_1 (mm)	1336.1	97.2	129.2	948.5	40.7	84.1	306.8	316.9	107.9
Q_2 (mm)	1488.4	137.1	165.3	1035.1	68.5	115.4	372.4	361.4	143.6
Q_3 (mm)	1592.7	168.0	201.0	1160.4	88.9	144.0	421.3	413.1	203.6
Q_4 (mm)	1768.5	217.1	260.9	1306.8	118.6	187.5	476.5	478.9	253.3
$\bar{r}$	0.14	0.45	0.49	0.20	0.50	0.29	0.17	0.17	0.32
$\bar{R}$	0.47	0.70	0.72	0.49	0.74	0.59	0.47	0.48	0.61
IAR (%)	20.1	58.0	59.4	22.2	48.2	27.0	21.5	21.6	31.6
The Northern Plains									
Mean (mm)	846.8	40.6	38.5	722.0	45.8	81.5	245.5	241.8	153.2
Median (mm)	848.3	35.7	31.8	729.1	35.2	76.5	244.8	236.7	151.7
SD (mm)	148.2	24.0	24.0	139.4	42.0	44.7	75.8	66.4	63.7
CV (%)	17.5	59.1	62.3	19.3	91.9	54.9	30.9	27.4	41.6
Q_1 (mm)	700.1	19.4	19.9	599.3	11.2	41.2	182.8	188.3	100.1
Q_2 (mm)	822.0	31.2	26.5	684.0	26.0	63.3	228.0	217.5	137.8
Q_3 (mm)	889.0	42.5	36.3	768.7	41.6	83.4	258.9	256.4	167.4
Q_4 (mm)	961.2	61.2	55.3	836.1	67.6	114.4	310.5	297.4	201.5
$\bar{r}$	0.30	0.52	0.46	0.30	0.45	0.36	0.32	0.25	0.27
$\bar{R}$	0.56	0.72	0.68	0.56	0.67	0.60	0.57	0.51	0.53
IAR (%)	30.7	49.8	43.6	30.1	43.5	34.3	31.5	26.0	26.5
The Eastern Plains									
Mean (mm)	1460.4	34.4	152.5	1155.1	118.4	232.8	345.9	325.0	251.4
Median (mm)	1466.7	30.3	147.7	1152.8	107.1	215.9	344.6	318.1	245.6
SD (mm)	184.7	23.6	55.2	146.9	64.7	79.7	74.5	66.5	68.1
CV (%)	12.6	68.6	36.2	12.7	54.6	34.2	21.5	20.5	27.1
Q_1 (mm)	1279.7	14.7	99.4	1012.0	63.4	168.2	277.4	271.8	192.9
Q_2 (mm)	1404.4	25.4	137.0	1115.1	90.1	198.8	327.3	300.9	235.6
Q_3 (mm)	1509.7	36.6	166.3	1182.1	117.0	232.5	357.5	334.6	258.1
Q_4 (mm)	1603.3	53.7	192.4	1266.1	167.0	285.3	402.8	373.6	301.5
$\bar{r}$	0.30	0.43	0.43	0.25	0.45	0.40	0.24	0.20	0.25
$\bar{R}$	0.57	0.67	0.66	0.52	0.68	0.65	0.52	0.47	0.53
IAR (%)	30.4	36.9	37.4	25.5	43.0	37.9	23.7	20.5	25.8
The Indo-gangetic Plains									
Mean (mm)	1025.7	38.7	71.4	848.6	66.9	125.5	274.7	266.2	182.2
Median (mm)	1012.6	36.4	68.3	851.7	55.3	115.2	272.8	264.9	182.6
SD (mm)	128.3	21.3	28.0	115.3	41.3	48.2	65.6	51.6	53.3

(Contd.)

Table 3. (*Contd.*)

	Ann	Win	Sum	Sum Mon	Post Mon	Jun	Jul	Aug	Sep
CV (%)	12.5	55.1	39.2	13.6	61.6	38.4	23.9	19.4	29.2
Q_1 (mm)	924.1	18.7	45.4	738.7	32.6	79.3	217.1	223.2	134.1
Q_2 (mm)	986.2	29.5	59.9	818.6	49.2	104.5	254.5	252.0	169.2
Q_3 (mm)	1051.8	39.5	78.0	876.3	61.9	131.3	290.8	275.4	191.0
Q_4 (mm)	1136.9	57.5	95.8	940.1	101.8	164.5	331.1	308.9	226.0
$\bar{r}$	0.21	0.41	0.36	0.19	0.34	0.30	0.23	0.15	0.18
$\bar{R}$	0.46	0.64	0.58	0.45	0.58	0.54	0.48	0.39	0.44
IAR (%)	19.7	36.2	23.8	18.9	29.7	25.7	21.8	6	17.6
				The North Eastern Ranges					
Mean (mm)	2287.0	47.2	556.0	1489.1	194.7	426.9	419.3	364.1	278.8
Median (mm)	2286.9	43.1	549.1	1483.2	195.2	424.4	406.9	359.5	275.5
SD (mm)	215.1	26.2	115.9	146.9	76.5	81.6	83.1	70.5	63.9
CV (%)	9.4	55.5	20.8	9.9	39.3	19.1	19.8	19.4	22.9
Q_1 (mm)	2087.4	28.1	456.4	1380.8	126.1	365.8	350.9	300.2	219.3
Q_2 (mm)	2212.7	37.4	514.1	1440.2	172.3	395.8	389.5	343.5	261.7
Q_3 (mm)	2334.2	52.2	572.8	1526.3	212.3	444.9	436.1	381.9	288.1
Q_4 (mm)	2467.2	64.7	650.5	1591.1	249.3	489.1	488.4	414.3	333.6
$\bar{r}$	0.23	0.48	0.33	0.15	0.42	0.22	0.19	0.19	0.23
$\bar{R}$	0.53	0.72	0.61	0.46	0.67	0.52	0.50	0.49	0.53
IAR (%)	26.4	48.4	34.4	21.5	41.6	26.3	26.3	24.0	27.1
				The Western Plains					
Mean (mm)	363.2	11.0	20.7	318.2	13.2	36.3	115.9	115.6	50.5
Median (mm)	346.0	7.5	14.6	300.4	7.0	29.9	107.5	101.3	36.6
SD (mm)	126.5	11.0	20.0	116.1	17.9	28.7	62.4	74.9	46.1
CV (%)	34.8	99.7	96.6	36.5	135.7	79.3	53.9	64.8	91.2
Q_1 (mm)	262.4	2.1	7.2	215.4	1.3	11.5	62.0	51.0	9.1
Q_2 (mm)	308.5	5.5	10.7	276.5	3.9	25.0	83.9	83.0	20.6
Q_3 (mm)	368.9	9.0	18.7	323.9	9.8	35.0	129.5	114.0	46.7
Q_4 (mm)	453.7	16.3	30.8	408.0	18.0	54.1	161.5	155.5	91.5
$\bar{r}$	0.47	0.47	0.51	0.44	0.48	0.38	0.42	0.46	0.40
$\bar{R}$	0.72	0.71	0.74	0.69	0.72	0.65	0.68	0.70	0.67
IAR (%)	50.1	49.1	53.0	46.7	47.8	41.9	45.2	48.7	42.1
				The North Central Highlands					
Mean (mm)	785.1	18.5	19.7	712.8	34.2	70.2	256.4	257.7	128.5
Median (mm)	795.3	13.0	14.8	721.6	22.1	62.9	254.8	266.4	129.2
SD (mm)	168.4	15.2	17.5	156.2	36.7	45.8	87.1	91.8	75.8
CV (%)	21.5	82.0	89.2	21.9	107.2	65.3	34.0	35.6	59.0
Q_1 (mm)	637.0	6.2	6.9	580.2	7.0	29.2	186.9	165.5	54.6
Q_2 (mm)	737.9	10.9	11.7	668.7	16.0	49.3	235.5	219.1	96.6
Q_3 (mm)	816.3	15.6	16.9	744.9	26.8	70.4	274.3	280.6	142.9
Q_4 (mm)	927.4	31.1	24.9	834.6	56.6	106.1	329.2	323.6	178.8
$\bar{r}$	0.38	0.45	0.46	0.37	0.50	0.43	0.35	0.39	0.43
$\bar{R}$	0.64	0.68	0.69	0.63	0.72	0.67	0.61	0.64	0.67
IAR (%)	38.2	44.7	37.7	37.0	48.5	42.8	37.6	40.8	42.1

(*Contd.*)

(Contd.)

The South Central Highlands

Mean (mm)	1051.4	21.7	20.0	958.0	51.7	120.9	333.0	325.1	179.0
Median (mm)	1035.6	17.6	16.8	947.4	45.9	108.3	325.8	318.6	163.3
SD (mm)	188.9	17.2	15.3	173.9	41.2	64.8	97.3	103.6	100.1
CV (%)	18.0	79.0	76.4	18.2	79.7	53.6	29.2	31.9	55.9
Q_1 (mm)	893.6	7.3	7.3	808.6	14.6	64.8	248.6	229.5	86.8
Q_2 (mm)	989.8	12.8	12.7	888.3	31.0	92.9	302.5	275.3	135.2
Q_3 (mm)	1097.0	20.1	19.6	1000.3	57.4	129.3	358.6	349.4	187.4
Q_4 (mm)	1193.4	36.3	30.7	1116.7	82.1	176.5	410.3	413.1	266.0
$\bar{r}$	0.40	0.41	0.36	0.37	0.47	0.49	0.34	0.40	0.49
$\bar{R}$	0.65	0.64	0.62	0.63	0.70	0.72	0.61	0.65	0.72
IAR (%)	41.4	40.2	36.4	39.0	48.9	50.0	36.9	41.2	51.1

The North Deccan

Mean (mm)	956.6	20.1	41.1	805.2	90.2	148.9	264.2	222.5	169.6
Median (mm)	950.2	13.5	36.4	802.9	90.7	149.4	258.3	220.3	171.9
SD (mm)	150.1	16.6	23.5	127.6	50.1	53.0	58.7	60.3	66.4
CV (%)	15.7	82.6	57.1	15.8	55.6	35.6	22.2	27.1	39.1
Q_1 (mm)	826.2	6.7	20.5	683.5	39.6	101.6	215.3	164.1	106.3
Q_2 (mm)	911.3	11.1	30.0	760.1	69.5	134.3	246.5	207.7	154.8
Q_3 (mm)	990.0	17.0	41.7	847.0	100.9	159.9	276.4	235.9	179.9
Q_4 (mm)	1086.3	35.7	59.8	927.2	131.8	194.9	320.6	279.3	226.9
$\bar{r}$	0.36	0.35	0.35	0.32	0.43	0.31	0.24	0.29	0.33
$\bar{R}$	0.61	0.59	0.61	0.58	0.67	0.58	0.51	0.55	0.59
IAR(%)	35.7	33.8	34.1	32.1	43.5	33.3	25.4	28.2	35.5

The South Deccan

Mean (mm)	919.9	11.3	102.2	645.3	161.0	123.6	197.9	166.0	157.8
Median (mm)	919.7	6.9	95.2	636.3	160.7	124.5	197.1	163.3	151.5
SD (mm)	124.2	11.7	30.6	103.8	61.8	32.7	52.2	50.5	50.2
CV (%)	13.5	103.5	29.9	16.1	38.4	26.4	26.4	30.4	31.8
Q_1 (mm)	811.8	1.4	77.0	560.2	109.3	93.4	158.2	124.7	107.5
Q_2 (mm)	879.2	5.1	87.0	605.3	138.0	111.4	181.7	152.0	139.1
Q_3 (mm)	943.4	10.4	104.4	648.8	172.7	134.1	207.8	177.6	160.7
Q_4 (mm)	1000.4	19.2	125.0	745.5	206.4	153.4	236.5	209.1	200.8
$\bar{r}$	0.28	0.30	0.24	0.30	0.37	0.17	0.28	0.30	0.26
$\bar{R}$	0.55	0.56	0.51	0.56	0.63	0.43	0.53	0.56	0.53
IAR (%)	26.6	27.4	23.5	26.1	38.3	17.0	20.5	25.2	28.1

The Eastern Plateaus

Mean (mm)	1361.7	40.8	81.7	1145.4	93.8	199.5	360.6	358.8	226.6
Median (mm)	1355.8	36.2	78.1	1145.9	85.8	177.1	358.1	350.1	225.3
SD (mm)	172.8	29.9	36.0	140.2	54.0	85.8	76.5	71.1	61.1
CV (%)	12.7	73.3	44.1	12.2	57.6	43.0	21.2	19.8	27.0
Q_1 (mm)	1205.8	15.4	49.3	1032.5	43.5	123.1	289.6	293.2	167.8
Q_2 (mm)	1330.4	32.1	61.5	1113.3	71.9	160.9	344.2	329.8	210.1
Q_3 (mm)	1391.0	41.5	86.8	1180.8	99.7	211.5	372.4	375.5	241.4
Q_4 (mm)	1485.3	55.8	113.2	1250.8	139.5	275.1	414.2	425.4	272.8

(Contd.)

Table 3. (*Contd.*)

	Ann	Win	Sum	Sum Mon	Post Mon	Jun	Jul	Aug	Sep
$\bar{r}$	0.32	0.49	0.40	0.25	0.44	0.46	0.24	0.24	0.25
$\bar{R}$	0.59	0.71	0.65	0.53	0.68	0.70	0.52	0.53	0.53
IAR (%)	34.9	51.1	39.9	28.4	44.9	48.7	28.2	27.8	27.9
The Western Hills									
Mean (mm)	2141.7	15.7	168.1	1704.6	253.3	331.6	658.7	468.7	245.6
Median (mm)	2147.3	10.0	156.5	1701.1	241.6	320.6	640.4	467.2	232.9
SD (mm)	294.0	19.4	65.3	284.2	81.7	118.7	190.3	136.8	105.0
CV (%)	13.7	123.9	38.8	16.7	32.3	35.8	28.9	29.2	42.7
Q_1 (mm)	1898.2	1.9	112.9	1491.4	178.0	211.9	539.5	347.0	151.5
Q_2 (mm)	2060.9	6.0	138.9	1594.6	209.0	292.7	591.8	403.5	198.8
Q_3 (mm)	2207.8	12.3	175.3	1763.3	270.8	347.0	670.6	483.1	261.8
Q_4 (mm)	2354.7	23.5	214.9	1913.7	341.2	415.9	800.8	573.6	349.5
$\bar{r}$	0.00	0.23	0.26	0.06	0.20	0.17	0.13	0.06	0.08
$\bar{R}$	0.55	0.68	0.68	0.57	0.68	0.65	0.62	0.56	0.58
IAR (%)	33.4	47.8	49.9	36.0	47.8	45.9	38.4	40.8	42.9
The Eastern Hills									
Mean (mm)	1015.3	21.7	112.6	643.8	237.2	113.5	179.2	186.7	164.4
Median (mm)	1008.3	16.0	109.4	634.1	230.0	109.4	177.1	185.8	160.7
SD (mm)	123.6	19.6	36.2	88.1	74.9	38.8	45.2	45.8	35.8
CV (%)	12.2	90.3	32.1	13.7	31.6	34.2	25.3	24.5	21.8
Q_1 (mm)	901.9	4.6	80.9	569.3	172.9	80.5	143.7	147.7	137.3
Q_2 (mm)	965.5	11.8	99.8	614.7	208.1	102.3	167.9	173.5	152.2
Q_3 (mm)	1049.0	19.1	115.0	648.3	250.7	119.2	187.5	196.2	167.5
Q_4 (mm)	1126.1	36.3	137.0	724.9	308.0	134.1	209.6	223.8	193.6
$\bar{r}$	0.16	0.29	0.21	0.12	0.24	0.14	0.13	0.10	0.07
$\bar{R}$	0.50	0.61	0.55	0.44	0.58	0.45	0.43	0.42	0.42
IAR (%)	23.6	31.9	30.9	18.7	34.3	20.7	19.0	16.8	16.9
The West Coastal Plains									
Mean (mm)	1765.2	11.5	138.2	1418.4	197.2	402.6	513.2	314.9	187.6
Median (mm)	1770.1	9.2	123.2	1428.9	199.1	416.0	509.1	308.3	186.1
SD (mm)	223.8	8.2	63.2	213.9	58.4	92.5	122.6	111.1	91.6
CV (%)	12.7	71.3	45.8	15.1	29.6	23.0	23.9	35.3	48.8
Q_1 (mm)	1597.8	4.2	86.5	1254.3	145.1	314.8	426.8	205.8	103.9
Q_2 (mm)	1712.5	7.1	115.1	1360.1	188.2	376.0	471.6	286.6	148.8
Q_3 (mm)	1815.2	11.9	132.6	1487.9	207.2	432.9	544.1	338.5	206.5
Q_4 (mm)	1934.6	18.6	185.8	1566.1	239.4	482.9	616.9	396.3	271.1
$\bar{r}$	0.26	0.16	0.25	0.27	0.27	0.25	0.26	0.35	0.37
$\bar{R}$	0.52	0.31	0.44	0.54	0.48	0.49	0.52	0.60	0.62
IAR (%)	24.4	11.4	25.1	26.8	21.7	22.7	25.9	34.2	37.5
The East Coastal Plains									
Mean (mm)	1111.0	37.9	106.6	551.9	414.6	98.9	140.0	157.4	155.6
Median (mm)	1105.7	29.7	97.6	545.5	413.1	94.4	141.1	158.9	151.1

(*Contd.*)

(*Contd.*)

SD (mm)	127.8	31.0	48.1	73.9	101.2	30.4	29.7	30.5	34.7
CV (%)	11.5	81.8	45.1	13.4	24.4	30.7	21.2	19.4	22.3
Q_1 (mm)	1008.0	10.8	72.7	489.0	325.3	77.0	112.2	131.1	127.1
Q_2 (mm)	1073.7	23.7	89.1	526.2	391.9	89.8	129.6	146.1	142.4
Q_3 (mm)	1137.6	34.9	103.2	566.4	443.7	102.7	151.9	165.7	159.3
Q_4 (mm)	1229.5	52.5	130.8	615.5	504.6	113.5	165.0	180.8	184.4
$\bar{r}$	0.17	0.29	0.28	0.18	0.22	0.16	0.13	0.11	0.13
$\bar{R}$	0.45	0.56	0.55	0.45	0.50	0.41	0.38	0.37	0.40
IAR (%)	20.7	31.9	30.2	18.2	24.6	15.7	13.2	11.4	15.5
The Whole India									
Mean (mm)	1165.9	31.5	101.6	906.5	126.3	169.1	297.4	263.4	176.6
Median (mm)	1177.2	29.9	100.5	919.4	126.3	166.0	300.4	264.6	177.0
SD (mm)	105.8	13.4	21.3	88.2	35.2	37.1	40.5	37.6	42.2
CV (%)	9.1	42.5	21.0	9.7	27.8	21.9	13.6	14.3	23.9
Q_1 (mm)	1060.7	18.2	81.9	831.9	92.8	138.0	270.5	232.8	137.0
Q_2 (mm)	1143.2	26.5	91.7	893.3	113.8	161.4	291.1	250.1	161.8
Q_3 (mm)	1190.6	33.0	106.2	944.1	130.7	176.1	310.4	279.4	185.5
Q_4 (mm)	1244.0	42.7	119.3	978.7	150.6	201.7	328.3	296.7	213.5
$\bar{r}$	0.13	0.17	0.15	0.11	0.19	0.14	0.08	0.07	0.11
$\bar{R}$	0.37	0.41	0.36	0.34	0.42	0.37	0.28	0.28	0.34
IAR (%)	11.1	13.5	6.9	9.8	12.2	10.8	7.0	6.9	11.7

September—1844-1961 I, 1962-2006 D. There are four provinces viz. North Kashmir Himalaya (No rainfall data available), South Kashmir Himalaya, Punjab Himalaya and Kumaun Himalaya in this division.

The Northern Plains: The combined area of the Punjab, the Haryana and the Uttar Pradesh states is called the northern plains. The monthly rainfall data for 32 stations in the region is available since 1844. The data of additional 31 stations became available from 1871. Thus the total of 63 stations is available from 1871. Table 3 gives the important statistics. Chief features of the rainfall fluctuations are: annual—1844-1998 I, 1999-2006 D; winter—1844-1941 N (no trend), 1942-1967 D, 1968-2006 I; summer—1844-1972 D, 1973-2006 I; summer monsoon—1844-1975 I, 1976-2006 D; post-monsoon—1844-1956 D, 1957-2006 N; June—1844-1965 D, 1966-2006 I; July—1844-1911 D, 1912-1980 I, 1981-2006 D; August—1844-1961 I, 1962-2006 D; and September—1844-2006 I. There are four provinces viz. Punjab Plains, Ganga-Yamuna Doab, Rohilkhand Plains and Avadh Plains in this division.

The Eastern Plains: The combined area of the Bihar and the West Bengal states is called the eastern plains. Earliest rainfall record of Kolkata is available from 1829. The data of all the 26 stations from the selected network is available from 1871. Longest rainfall sequence for the division could be developed for the period 1829-2006. The important statistics of the rainfall sequences are given in Table 3. Chief features of the rainfall fluctuations are as: annual—1829-1913 I, 1914-1966 D, 1967-2006 I; winter—1829-1937 I, 1938-2006 D; summer—1829-1957 D, 1958-2006 I; summer monsoon—1829-1922 I, 1923-1966 D, 1967-2006 I; post-monsoon—1829-1907 D, 1908-2006 I; June—1829-1913 I, 1914-1972 D, 1973-2006 I; July—1829-1926 I, 1927-1961 D, 1962-2006 I; August—1829-1941 I, 1942-2006 D; and September—1829-2006 I. There are four provinces viz. North Bihar Plains, South Bihar Plains, Bengal Plains and Bengal Basin, in the division.

The Indo-Gangetic Plains: The combined area of the northern plains and the eastern plains is called Indo-Gangetic plains. The area is very important for Indian summer monsoon studies as well as for the food security of south Asia. Earliest rainfall record is available for Kolkata from 1829; Bankura

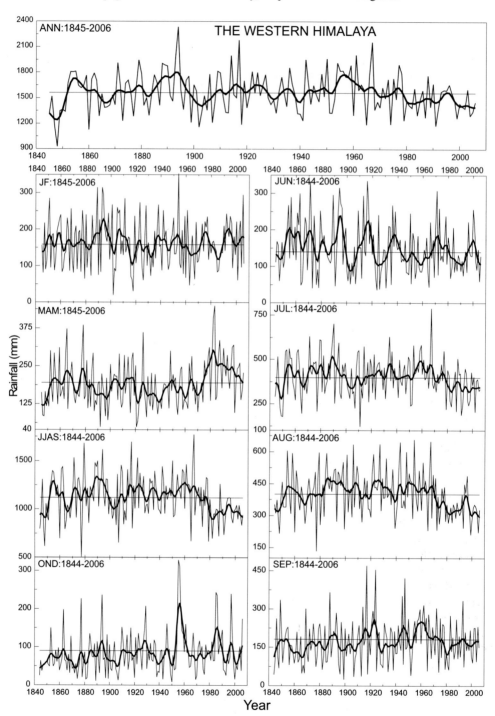

Fig. 2 Instrumental period time series plots of annual, seasonal and monsoon monthly rainfall over the Western Himalaya.

started in 1831, Patna in 1842 and 32 stations (Mainpuri, Gorakhpur, Jaunpur, Mirzapur, Varanasi, Ghazipur, Azamgarh, Farrukhabad, Etawah, Kanpur, Hamirpur, Fatehpur, Allahabad, Bijnor, Moradabad, Bareilly, Budaun, Shahjahanpur, Gurgaon, Rohtak, Saharanpur, Muzaffarnagar, Meerut, Delhi, Bulandshahar, Aligarh, Mathura, Agra, Firozpur, Karnal, Sirsa and Hissar) in 1844. The data for 55 more stations became available as follows: for five stations (Muzaffarpur, Bhagalpur, Malda, Suri and Berhampore) from 1848, for four stations (Motihari, Chapra, Arrah, Gaya) from 1849, for two stations (Ludhiana, Ambala) from 1850, for Jalandhar from 1852, for two stations (Hoogly, Midnapore) from 1854, for two stations (Gurudaspur, Hoshiarpur) from 1857, for three stations (Amritsar, Etah, Burdwan) from 1860, for Krishnanagar from 1861, for Pilibhit from 1864, for *Basti* from 1866, for 12 stations (Hardoi, Faizabad, Sitapur, Pratapgarh, Kheri, Sultanpur, Lucknow, Bahraich, Nawabgunj, Unnao, Rai-Bareilly and Sagar Island) from 1867, for two stations (Howrah and Gonda) from 1868, for two stations (Jalpaiguri and Chaibassa) from 1869, for three stations (Madhipura, Jamui and Purnea) from 1870, and for 14 stations (Moga, Rupar, Patiala, Ranike, Kaithal, Bhiwani, Jind, Jatusana, Sonepat, Deoria, Ballia, Darbhanga, Itahar and Baripada) from 1871. In all data for 90 stations from selected network is available from 1871. Longest rainfall sequence could be developed for the period 1829-2006. Chief features of the rainfall fluctuations are: annual—1829-1998 I, 1999-2006 D; winter—1829-1942 I, 1943-1964 D, 1965-1995 I, 1996-2006 D; summer—1829-1922 D, 1923-2006 I; summer monsoon—1829-1890 I, 1891-1907 D, 1908-1998 I, 1999-2006 D; post-monsoon—1829-2006 N; June—1829-1958 D, 1959-2006 I; July—1829-1861 I, 1862-1911 D, 1912-1980 I, 1981-2006 D; August—1829-1855 D, 1856-1995 I, 1996-2006 sudden D; and September—1829-2006 N.

The North Eastern Range: The North Eastern India excluding the Arunachal Pradesh State is called as the North Eastern Range. Rainfall observations for five stations (Silchar, Dibrugarh, Sibsagar, Nowgong and Guwahati) are available from 1848; Goalpara and Tezpur stations were included in the network in 1849, Shillong in 1866, Tura in 1870, and five more stations (Kohima, Imphal, Haflong, Agartala and Demagiri) in 1871. Longest rainfall sequence could be developed for the period 1848-2006. The actual and filtered rainfall series are displayed in Fig. 3. The important statistics of the rainfall sequences are given in Table 3. Chief features of the rainfall fluctuations are as: annual—1848-1947 N, 1948-1982 D, 1983-2006 I; winter—1848-1978 D, 1979-1993 I, 1994-2006 D; summer—1848-1947 N, 1948-1960 D, 1961-2006 I; summer monsoon—1848-2006 D; post-monsoon—1848-1887 D, 1888-2006 I; June—1848-1896 D, 1897-1934 I, 1935-2006 D; July—1848-1859 I, 1860-1887 D, 1888-1974 I, 1975-2006 D; August—1848-1905 I, 1906-2006 D; and September—1848-1878 I, 1879-1962 D, 1963-1987 I, 1988-2006 D. There are three provinces, viz., Assam Valley, Meghalaya and Purvanchal in the division.

The Western Plains: Desert and semi-desert regions of the western Rajasthan, east of Aravallis, form the Western Plains. Earliest record for Jaisalmer is available from 1861; for the Pali station the records became available from 1867. The data for all 12 stations of the selected network in the region is available from 1871. Longest rainfall sequence could be developed for the period 1861-2006. The actual and filtered rainfall series are displayed in Fig. 4. The important statistics of the rainfall sequences are given in Table 3. Chief features of the rainfall fluctuations are as: annual—1861-1975 I, 1976-2006 D; winter—1861-2006 D; summer—1861-1916 N, 1917-1954 D, 1955-2006 I; summer monsoon—1861-1975 I, 1976-2006 D; post-monsoon—1861-1971 N, 1972-2006 I; June—1861-1968 D, 1969-2006 I; July—1861-1978 I, 1979-2006 D; August—1861-1944 I, 1945-2006 D; and September—1861-1958 N, 1959-2006 D. There are two provinces viz. Marusthali and Rajasthan Bagar in the division.

The North Central Highlands: The region is spread over East Rajasthan, northwest Madhya Pradesh and southern Uttar Pradesh. Earliest rainfall record for Hamirpur is available from 1844; the data for all 22 stations of the selected network is available from 1871. Longest rainfall sequence could be developed

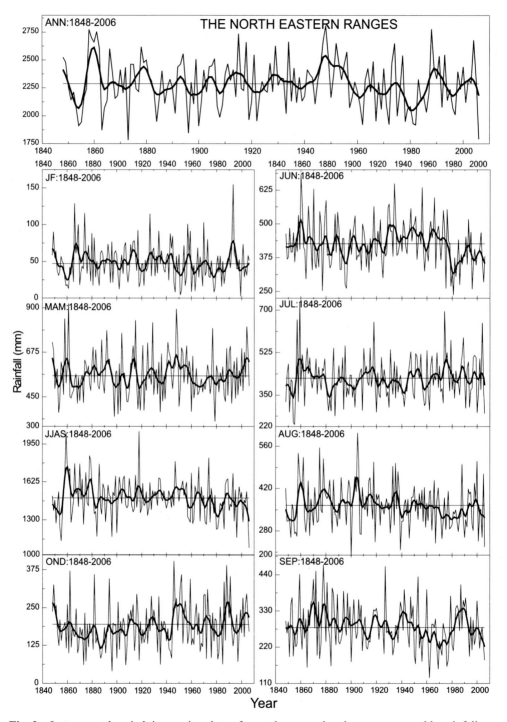

Fig. 3 Instrumental period time series plots of annual, seasonal and monsoon monthly rainfall over the North Eastern Ranges.

Monitoring Physiographic Rainfall Variation for Sustainable Management of Water Bodies in India   313

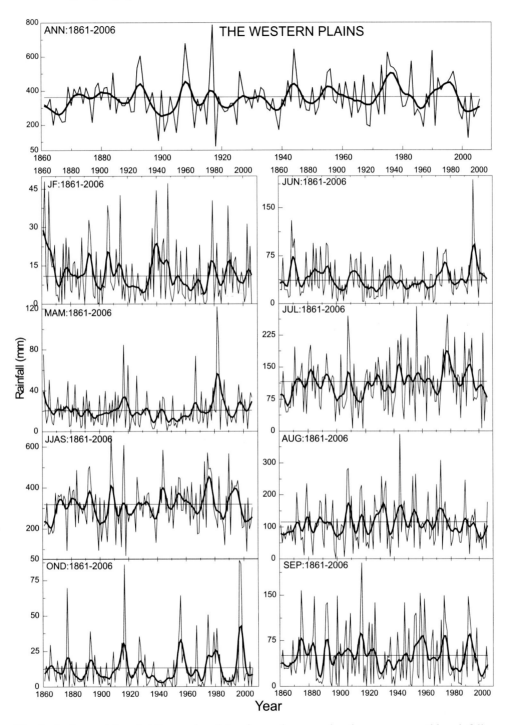

Fig. 4 Instrumental period time series plots of annual, seasonal and monsoon monthly rainfall over the Western Plains.

for the period 1844-2006. Chief features of the rainfall fluctuations areas: annual—1857-1905 D, 1906-1961 I, 1962-2006 D; winter—1857-1978 N, 1979-2006 D; summer—1857-1895 N, 1896-1917 I, 1918-1955 D, 1956-2006 I; summer monsoon—1844-1904 N, 1905-1961 I, 1962-2006 D; post-monsoon—1857-1985 I, 1986-2006 D; June—1844-1992 D, 1993-2006 I; July—1844-1911 D, 1912-1976 N, 1977-2006 D; August—1844-1960 I, 1961-2006 D; and September—1844-1927 N, 1928-1961 I, 1962-2006 D. There are four provinces, viz., Aravalli Range, East Rajasthan Uplands, Madhya Bharat Pathar and Bundelkhand Upland in the division.

The South Central Highlands: The region is in the central parts of the Madhya Pradesh state. Earliest rainfall records for five stations are available from 1844; the data for all 25 stations of the selected network is available from 1871. Longest rainfall sequence could be developed for the period 1844-2006. The important statistics of the rainfall sequences are given in Table 3. Chief features of the rainfall fluctuations are as: annual—1844-1893 I, 1894-1899 D, 1900-1944 I, 1945-2006 slight D; winter—1844-1942 I, 1943-1967 D, 1968-2003 I and mostly above normal; summer—1844-1972 N, 1973-2006 I; summer monsoon—1844-1874 I, 1875-1899 D, 1900-1944 I, 1945-2006 slight D; post-monsoon—1844-2006 I; June—1844-1893 I, 1894-1902 D, 1903-2006 N; July—1844-1876 I, 1877-1918 D, 1919-1944 I, 1945-2006 D; August—1844-2006 I; and September—1844-1960 N, 1961-1979 slight D, 1980-2006 slight I. There are four provinces, viz., Malwa Plateau, Vindhyan Scarplands, Vindhya Range and Narmada Valley in the division.

The North Deccan: The region is spread over Maharashtra and Madhya Pradesh states. The earliest record for Pune and Nagpur is available from 1826; the data for all 30 stations of the selected network is available since 1871. Longest rainfall sequence could be developed for the period 1826-2006. The actual and filtered rainfall series are displayed in Fig. 5. The important statistics of the rainfall sequences are given in Table 3. Chief features of the rainfall fluctuations are as: annual—1853-1892 I, 1893-1899 D, 1900-1961 I, 1962-2006 slight D; winter—1851-1947 I, 1948-1976 D, 1977-2003 I; summer—1851-1911 D, 1912-1957 I, 1958-2006 D; summer monsoon—1853-1892 I, 1893-1899 D, 1900-1961 I, 1962-2003 slight D, 2004-2006 I; post-monsoon—1851-1886 I, 1887-1909 D, 1910-2006 I; June—1851-1962 highly fluctuating, 1963-2006 I; July—1853-1887 I, 1888-1918 D, 1919-1944 I, 1945-1999 D, 2000-2006 I; August—1851-2006 I; and September—1851-1891 I, 1892-1958 N, 1959-2006 D. There are two provinces, viz., Satpura Range and Maharashtra Plateau in the division.

The South Deccan: The region is spread over Interior Karnataka and Eastern Andhra Pradesh. Earliest record for Bangalore is available from 1835; the data for all 27 stations of the selected network is available from 1871. Longest rainfall sequence could be developed for the period 1835-2006. The important statistics of the rainfall sequences are given in Table 3. Chief features of the rainfall fluctuations are as: annual—1835-1956 I, 1957-2006 D; winter—1835-1886 D, 1887-1936 I, 1937-1955 D, 1956-2006 I; summer—1835-1913 N, 1914-1962 I, 1963-2002 D, 2003-2006 I; summer monsoon—1835-1988 I, 1989-2006 D; post-monsoon—1835-2006 slightly I; June—1835-1991 slight I, 1992-2006 D; July—1835-1961 I, 1962-2006 D; August—1835-1919 highly oscillating, 1920-2006 I; and September—1835-1857 D, 1858-1981 I, 1982-2006 D. There are two provinces namely Karnataka Plateau and Telangana Plateau in the division.

The Eastern Plateaus: It is spread over Chhattisgarh, Jharkhand, Orissa and West Bengal states. Earliest rainfall data for three stations is available from 1848; the data for all 21 stations of the selected network is available from 1871. Longest rainfall sequence could be developed for the period 1848-2006. The important statistics of the rainfall sequences are given in Table 3. Chief features of the rainfall fluctuations are as: annual—1860-1943 I, 1944-2006 D; winter—1860-1944 I, 1945-2006 D and mostly below normal; summer—1860-1954 D, 1955-2006 I; summer monsoon—1860-1943 I, 1944-2006 D; post-monsoon—1860-1905 D, 1906-1959 I, 1960-1981 D, 1982-2006 I; June—1860-1910 N and above

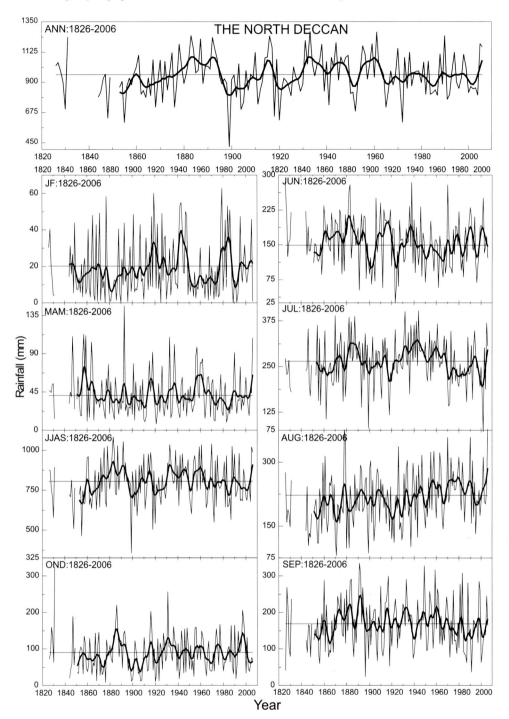

Fig. 5 Instrumental period time series plots of annual, seasonal and monsoon monthly rainfall over the North Deccan.

normal, 1911-1974 D, 1975-2006 I; July—1860-1892 I, 1893-1918 D, 1919-1929 I, 1930-2006 D; August—1860-1944 I, 1945-2006 D; and September—1860-1961 I, 1962-2006 D. There are five provinces namely Baghelkhand Plateau, Chhotanagpur Plateau, Mahanadi Basin, Garhjat Hills and Dandakaranya in the division.

The Western Hills: The upper portion of Sahyadri Range is referred to as the Western Hills. Earliest record from 1829 is available for Ootacamund; Mercara became available from 1863 and Ahwa and Punalur from 1871. Longest rainfall sequence could be developed for the period 1829-2006. The important statistics of the rainfall sequences are given in Table 3. Chief features of the rainfall fluctuations are as: annual—1863-1961 I, 1962-1965 D, 1966-2006 I; winter—1863-1909 I, 1910-2006 D; summer—1863-1926 N, 1927-1955 I, 1956-1985 D, 1986-2006 I; summer monsoon—1863-2006 I sharp from 1964 onwards; post-monsoon—1863-2006 N; June—1863-1896 I, 1897-1966 D, 1967-2006 I; July—1863-1960 N, 1961-2006 D; August—1863-2006 steep I; and September—1863-1954 I, 1955-1982 D, 1983-2006 steep I. There are four provinces namely North Sahyadri, Central Sahyadri, South Sahyadri and Nilgiri Hills in the division.

The Eastern Hills: The elongated strips spread over Tamilnadu, Coastal Andhra Pradesh and Orissa states, west of east coastal plains, is referred to as Eastern Hills. Earliest record from 1852 is available for Cuddappah and Salem; the data for all eight stations of the selected network is available from 1871. Longest rainfall sequence could be developed for the period 1852-2006. Chief features of the rainfall fluctuations are as: annual—1852-1956 I, 1957-1980 D, 1981-2006 I; winter—1852-1926 I, 1927-1974 D, 1975-2006 I; summer—1852-2006 N; summer monsoon—1852-1956 I, 1957-1979 D, 1980-1989 I, 1990-2006 D; post-monsoon—1852-2006 I; June—1852-2006 N; July—1852-1964 I, 1965-2006 D; August—1852-1967 N, 1968-1991 I, 1992-2006 D; and September—1852-1897 I, 1898-1976 D, 1977-1988 I, 1989-2006 D. There are three provinces namely Eastern Ghats (North), Eastern Ghats (South) and Tamilnadu Uplands in the division.

The West Coastal Plains: The plain area parallel to the west coast from Gujarat to Kerala forms the West Coastal Plains. Earliest record of Mumbai is available from 1817. The data for all the 34 stations of the selected network is available from 1871. Longest rainfall sequence could be developed for the period 1817-2006. The actual and filtered rainfall series are displayed in Fig. 6. The important statistics of the rainfall sequences are given in Table 3. Chief features of the rainfall fluctuations are as: annual—1838-1961 I, 1962-2006 D; winter- 1838-1928 I, 1929-2006 D; summer—1838-1866 D, 1867-1960 I, 1961-1983 D, 1984-2006 I; summer monsoon—1817-1855 D, 1856-1961 I, 1962-2006 D; post-monsoon—1838-1932 I, 1933-1971 D, 1972-2006 I; June—1817-1962 D, 1963-2006 I; July—1817-1959 I, 1960-2006 D; August—1817-1845 D, 1846-1983 I, 1984-1990 D, 1991-2006 I; and September—1817-1839 D, 1840-1954 I, 1955-1999 D, 2000-2006 I. There are six provinces namely Kachchh Peninsula, Kathiawar Peninsula, Gujarat Plains, Konkan, Karnataka Coast and Kerala Plains in the division.

The East Coastal Plains: Relatively plain area along the east coast stretching from Tamilnadu, Andhra Pradesh and Orissa states is referred to as the East Coastal Plains. Earliest record from 1813 is available for Chennai. The data of other 23 stations from the selected network is available from 1871. Longest rainfall sequence could be developed for the period 1813-2006. The actual and filtered rainfall series are displayed in Fig. 7. The important statistics of the rainfall sequences are given in Table 3. Chief features of the rainfall fluctuations are as: annual—1813-1865 D, 1866-2006 I; winter—1813-1887 D, 1888-1923 I, 1924-1973 D, 1974-2006 I; summer—1813-1921 D, 1922-1943 I, 1944-1985 D, 1986-2006 I; summer monsoon—1813-1931 N, 1932-2006 I; post-monsoon—1813-2006 slight I; June—1813-1995 N, 1996-2006 D; July—1813-1865 D, 1866-2006 I; August—1813-1922 N, 1923-2006 I; and September—1813-1902 N, 1903-1934 D, 1935-2006 I. There are three provinces, viz., Utkal Plains, Andhra Plains and Tamilnadu Plains in the division.

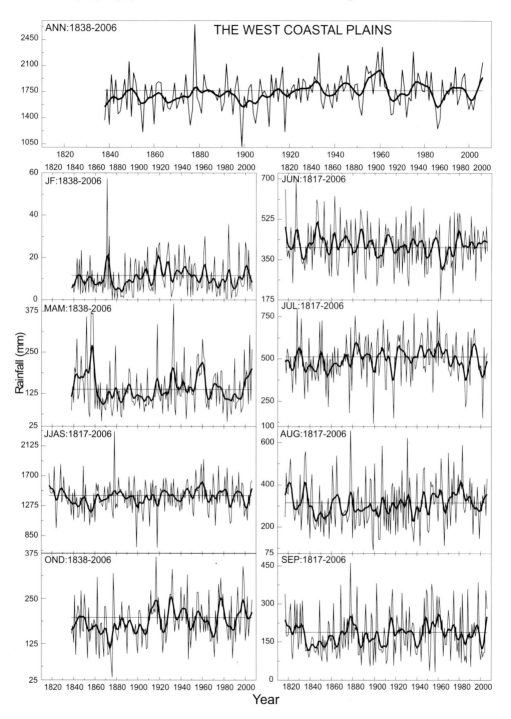

Fig. 6 Instrumental period time series plots of annual, seasonal and monsoon monthly rainfall over the West Coastal Plains.

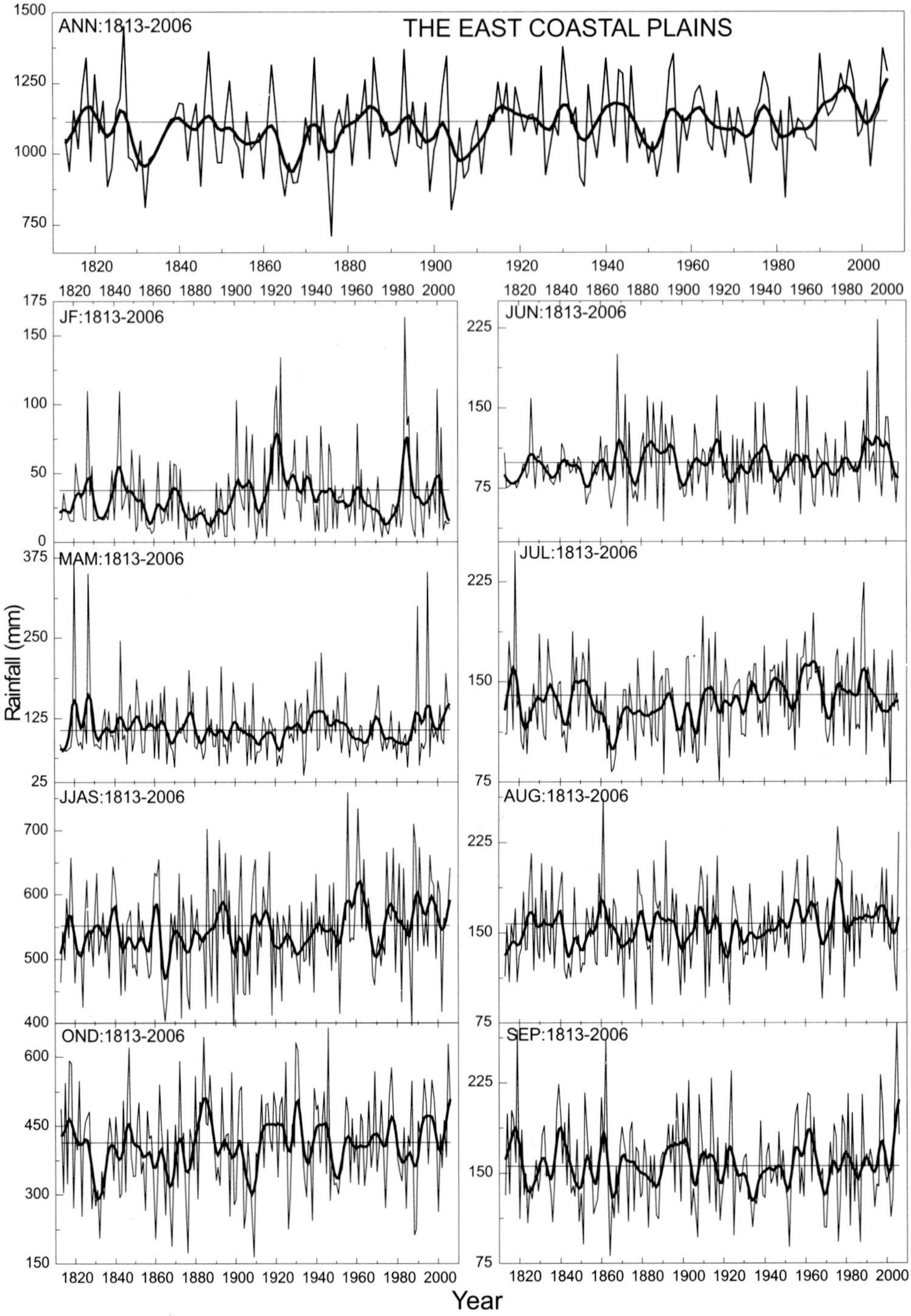

Fig. 7 Instrumental period time series plots of annual, seasonal and monsoon monthly rainfall over the East Coastal Plains.

The Whole of India: Longest monthly, seasonal and annual rainfall series have also been prepared for the whole country using all the 316 stations available. Longest annual, seasonal and monsoon monthly rainfall sequence of the whole country could be developed for the period 1813-2006. The actual and filtered rainfall series are shown in Fig. 8. The important statistics of the rainfall sequences are given in Table 3. Chief features of the rainfall fluctuations are as: annual—1813-1838 D, 1839-1893 I,1894-1899 D, 1900-1961 I, 1962-2006 slight D; winter—1813-1944 I, 1945-1964 D, 1965-1995 I, 1996-2006 D; summer—1813-1922 D, 1923-1943 I, 1944-1966 D, 1967-2006 I; summer monsoon—1813-1848 D, 1849-1892 I, 1893-1899 D, 1900-1961 I, 1962-2006 D; post-monsoon—1813-1908 D, 1909-2006 I; June—1813-1893 I, 1894-1962 D, 1963-2006 I; July—1813-1882 I, 1883-1918 D, 1919-1942 I, 1943-2006 D; August—1813-1855 D, 1856-1963 I, 1964-2006 D; and September—1813-1845 D, 1846-1961 I, 1962-2006 D.

7. SUMMARY OF RECENT TENDENCY IN RAINFALL FLUCTUATIONS OVER THE COUNTRY

The physiographic subdivisions/provinces are grouped in three categories based on recent tendency (increase, decrease or no trend) in each of annual, seasonal and monsoon monthly rainfall fluctuations. The percentage area of the country under different tendency has been worked out from total area of the subdivisions/provinces and is depicted in the pie diagram (Fig. 9). The details are as follows:

Annual rainfall: 22.4% area of the country shows *increasing trend* (Bengal Basin, Assam Valley, Vindhyan Scarplands, Chhotanagpur Plateau, North Sahyadri, Eastern Ghats (South), Tamilnadu Uplands, Kathiawar Peninsula, Konkan, Karnataka Coast, Utkal Plains, Andhra Plains, Tamilnadu Plains), 68.1% *decreasing trend* (Punjab Plains, Ganga-Yamuna Doab, Rohilkhand Plains, Avadh Plains, North Bihar Plains, South Bihar Plains, Bengal Plains, Meghalaya, Purvanchal, Marusthali, Rajasthan Bagar, Aravalli Range, East Rajasthan Uplands, Madhya Bharat Pathar, Bundelkhand Upland, Malwa Plateau, Vindhya Range, Narmada Valley, Satpura Range, Karnataka Plateau, Telangana Plateau, Baghelkhand Plateau, Mahanadi Basin, Garhjat Hills, Dandakaranya, South Sahyadri, Nilgiri, Eastern Ghats (North), Gujarat Plains, Kerala Plains) and remaining 9.5% *no trend* (Maharashtra Plateau, Central Sahyadri, Kachchh Peninsula). Figure 10 depicts geographical distribution of recent tendencies in annual rainfall and pie diagram shows area under different tendencies. Annual rainfall over India from 1944-1969 to 1970-2006 has decreased by 4.11%.

Winter rainfall: 37.1% area of the country shows *increasing trend* (South Kashmir Himalaya, Punjab Plains, Bengal Plains, Assam Valley, Meghalaya, Marusthali, Narmada Valley, Satpura Range, Maharashtra Plateau, Telangana Plateau, Mahanadi Basin, South Sahyadri, Eastern Ghats (South), Andhra Plains), 46.6% *decreasing trend* (Punjab Himalaya, Kumaun Himalaya, Ganga-Yamuna Doab, Avadh Plains, North Bihar Plains, South Bihar Plains, Purvanchal, Aravalli Range, East Rajasthan Uplands, Bundelkhand Upland, Vindhyan Scarplands, Vindhya Range, Karnataka Plateau, Baghelkhand Plateau, Chhotanagpur Plateau, Garhjat Hills, Central Sahyadri, Nilgiri, Tamilnadu Uplands, Kachchh Peninsula, Kathiawar Peninsula, Gujarat Plains, Konkan, Karnataka Coast, Kerala Plains, Utkal Plains, Tamilnadu Plains) and remaining 16.3% *no trend* (Rohilkhand Plains, Bengal Basin, Rajasthan Bagar, Madhya Bharat Pathar, Malwa Plateau, Dandakaranya, North Sahyadri, Eastern Ghats (North)) (Fig. 11). The winter rainfall over the country from 1892-1948 to 1949-2006 has decreased by 17.18%.

Summer rainfall: 73.3% area of the country shows *increasing trend* (Punjab Himalaya, Kumaun Himalaya, Punjab Plains, Ganga-Yamuna Doab, Rohilkhand Plains, Avadh Plains, North Bihar Plains, South Bihar Plains, Bengal Plains, Bengal Basin, Assam Valley, Meghalaya, Purvanchal, Marusthali, Aravalli Range, East Rajasthan Uplands, Madhya Bharat Pathar, Bundelkhand Upland, Malwa Plateau,

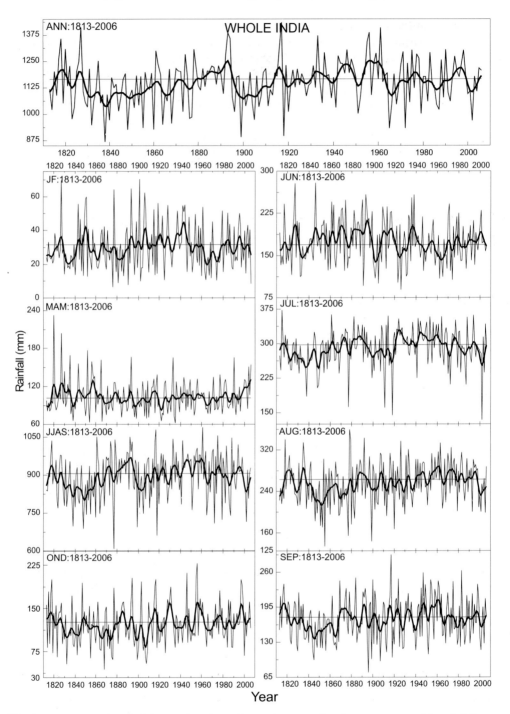

Fig. 8 Instrumental period time series plots of annual, seasonal and monsoon monthly rainfall over the whole country.

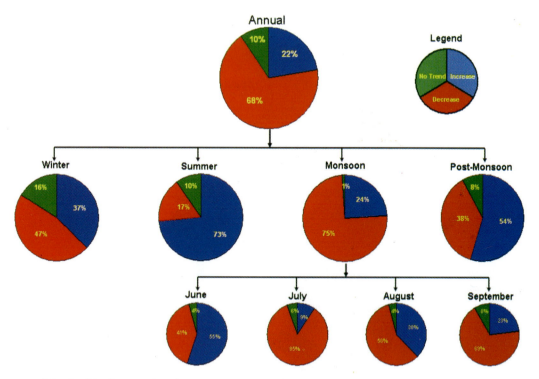

Fig. 9 Pie diagram showing percentage area of the country under different recent rainfall tendencies for annual, seasonal and four monsoon months.

Vindhyan Scarplands, Vindhya Range, Narmada Valley, Satpura Range, Maharashtra Plateau, Chhotanagpur Plateau, Mahanadi Basin, Garhjat Hills, Dandakaranya, Central Sahyadri, Eastern Ghat (South), Konkan, Karnataka Coast, Kerala Plains, Utkal Plains, Andhra Plains, Tamilnadu Plains), 16.9% *decreasing trend* (South Kashmir Himalaya, Rajasthan Bagar, Karnataka Plateau, South Sahyadri, Nilgiri, Eastern Ghats (North), Tamilnadu Uplands, Gujarat Plains) and remaining 9.7% *no trend* (Telangana Plateau, Baghelkhand Plateau, North Sahyadri, Kachchh Peninsula, Kathiawar Peninsula) (Fig. 11). The summer rainfall over the country from 1962-1996 to 1997-2006 has increased by 18.10%.

Summer monsoon rainfall: 24.1% area of the country shows *increasing trend* (Bengal Basin, Vindhyan Scarplands, Maharashtra Plateau, Chhotanagpur Plateau, North Sahyadri, Kathiawar Peninsula, Gujarat Plains, Konkan, Utkal Plains, Andhra Plains), 74.7% *decreasing trend* (South Kashmir Himalaya, Punjab Himalaya, Kumaun Himalaya, Punjab Plains, Ganga-Yamuna Doab, Rohilkhand Plains, Avadh Plains, North Bihar Plains, South Bihar Plains, Bengal Plains, Assam Valley, Meghalaya, Purvanchal, Marusthali, Rajasthan Bagar, Aravalli Range, East Rajasthan Uplands, Madhya Bharat Pathar, Bundelkhand Upland, Malwa Plateau, Vindhya Range, Narmada Valley, Satpura Range, Karnataka Plateau, Telangana Plateau, Baghelkhand Plateau, Mahanadi Basin, Garhjat Hills, Dandakaranya, South Sahyadri, Nilgiri, Eastern Ghats (North), Eastern Ghats (South), Tamilnadu Uplands, Karnataka Coast, Kerala Plains) and remaining 1.2% *no trend* (Central Sahyadri, Kachchh Peninsula, Tamilnadu Plains) (Fig. 11). The summer monsoon rainfall over the country from 1931-1964 to 1965-2006 has decreased by 4.72%.

Post-monsoon rainfall: 54.4% area of the country shows *increasing trend* (South Kashmir Himalaya, Bengal Plains, Bengal Basin, Assam Valley, Meghalaya, Marusthali, Aravalli Range, East Rajasthan

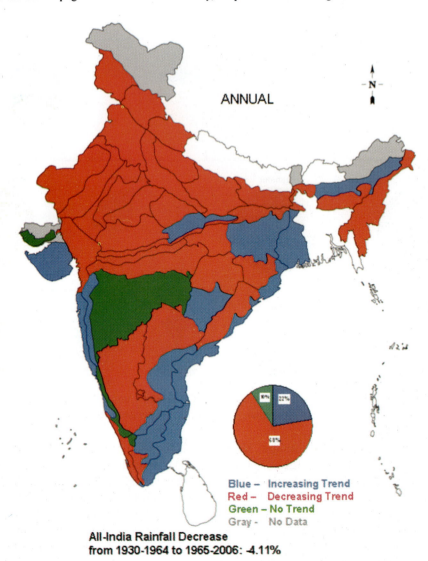

Fig. 10 Geographical distribution of recent tendency in annual rainfall and pie diagram showing percentage area of the country under different tendencies.

Uplands, Madhya Bharat Pathar, Maharashtra Plateau, Telangana Plateau, Chhotanagpur Plateau, Mahanadi Basin, Garhjat Hills, Dandakaranya, North Sahyadri, South Sahyadri, Eastern Ghats (North), Eastern Ghats (South), Tamilnadu Uplands, Kachchh Peninsula, Karnataka Coast, Kerala Plains, Utkal Plains, Andhra Plains, Tamilnadu Plains), 37.5% *decreasing trend* (Kumaun Himalaya, Ganga-Yamuna Doab, Rohilkhand Plains, Avadh Plains, North Bihar Plains, South Bihar Plains, Purvanchal, Rajasthan Bagar, Bundelkhand Upland, Malwa Plateau, Vindhya Range, Narmada Valley, Satpura Range, Karnataka Plateau, Baghelkhand Plateau, Central Sahyadri, Nilgiri, Kathiawar Peninsula, Konkan) and remaining 8.1% *no trend* (Punjab Himalaya, Punjab Plains, Vindhyan Scarplands, Gujarat Plains) (Fig. 11). The post-monsoon rainfall over the country from 1827-1914 to 1915-2006 has increased by 13.32%.

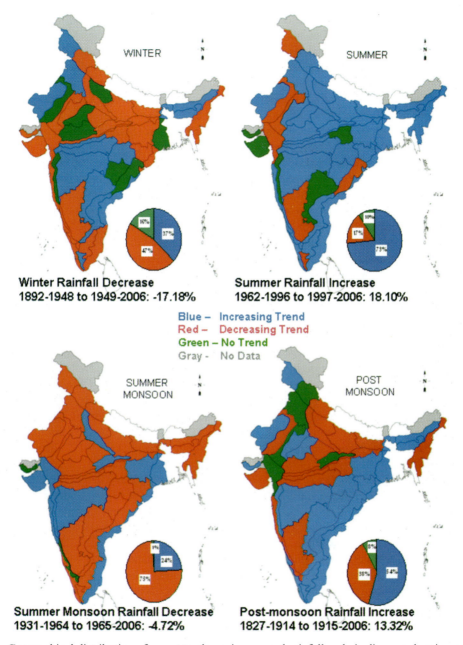

Fig. 11 Geographical distribution of recent tendency in seasonal rainfall and pie diagram showing percentage area of the country under different tendencies.

June rainfall: 55.1% area of the country shows *increasing trend* (South Kashmir Himalaya, Punjab Plains, Ganga-Yamuna Doab, Rohilkhand Plains, Avadh Plains, North Bihar Plains, South Bihar Plains, Marusthali, Rajasthan Bagar, East Rajasthan Uplands, Vindhyan Scarplands, Maharashtra Plateau, Karnataka Plateau, Chhotanagpur Plateau, Garhjat Hills, North Sahyadri, Central Sahyadri, Eastern Ghats

(South), Kachchh Peninsula, Kathiawar Peninsula, Gujarat Plains, Karnataka Coast, Utkal Plains, Andhra Plains), 40.7% *decreasing trend* (Punjab Himalaya, Kumaun Himalaya, Bengal Plains, Bengal Basin, Assam Valley, Meghalaya, Purvanchal, Aravalli Range, Malwa Plateau, Vindhya Range, Narmada Valley, Satpura Range, Telangana Plateau, Baghelkhand Plateau, Mahanadi Basin, Dandakaranya, South Sahyadri, Nilgiri, Eastern Ghats (North), Tamilnadu Uplands, Kerala Plains, Tamilnadu Plains) and remaining 4.2% *no trend* (Madhya Bharat Pathar, Bundelkhand Upland, Konkan) (Fig. 12). The June rainfall over the country from 1944-1969 to 1970-2006 has increased by 11.33%.

July rainfall: 9.4% area of the country shows *increasing trend* (South Bihar Plains, Bengal Basin, Vindhyan Scarplands, Narmada Valley, North Sahyadri, Kachchh Peninsula, Gujarat Plains), 85.1% *decreasing trend* (South Kashmir Himalaya, Punjab Himalaya, Kumaun Himalaya, Punjab Plains, Ganga-Yamuna Doab, Rohilkhand Plains, Avadh Plains, North Bihar Plains, Bengal Plains, Assam Valley, Purvanchal, Marusthali, Rajasthan Bagar, Aravalli Range, East Rajasthan Uplands, Madhya Bharat Pathar, Bundelkhand Upland, Malwa Plateau, Vindhya Range, Satpura Range, Maharashtra Plateau, Karnataka Plateau, Telangana Plateau, Baghelkhand Plateau, Mahanadi Basin, Garjhat Hills, Dandakaranya, Central Sahyadri, South Sahyadri, Nilgiri, Eastern Ghats (North), Eastern Ghats (South), Tamilnadu Uplands, Kathiawar Peninsula, Konkan, Karnataka Coast, Kerala Plains, Andhra Plains, Tamilnadu Plains) and remaining 5.5% *no trend* (Meghalaya, Chhotanagpur Plateau, Utkal Plains) (Fig. 12). The July rainfall over the country from 1920-1969 to 1970-2006 has decreased by 7.99%.

August rainfall: 37.6% area of the country shows *increasing trend* (Aravalli Range, Malwa Plateau, Vindhya Range, Narmada Valley, Maharashtra Plateau, Karnataka Plateau, Dandakaranya, North Sahyadri, Central Sahyadri, Eastern Ghats (North), Eastern Ghats (South), Kachchh Peninsula, Kathiawar Peninsula, Gujarat Plains, Konkan, Utkal Plains, Andhra Plains), 58.3% *decreasing trend* (South Kashmir Himalaya, Punjab Himalaya, Kumaun Himalaya, Punjab Plains, Ganga-Yamuna Doab, Rohilkhand Plains, Avadh Plains, North Bihar Plains, South Bihar Plains, Bengal Plains, Assam Valley, Meghalaya, Purvanchal, Marusthali, Rajasthan Bagar, East Rajasthan Uplands, Madhya Bharat Pathar, Bundelkhand Upland, Vindhyan Scarplands, Satpura Range, Telangana Plateau, Baghelkhand Plateau, Chhotanagpur Plateau, Mahanadi Basin, Nilgiri, Tamilnadu Uplands, Karnataka Coast, Kerala Plains, Tamilnadu Plains) and remaining 4.2% *no trend* (Bengal Basin, Garjhat Hills, South Sahyadri) (Fig. 12). The August rainfall over the country from 1939-1997 to 1998-2006 has decreased by 10.90%.

September rainfall: 23.5% area of the country shows *increasing trend* (Ganga-Yamuna Doab, Bengal Basin, Purvanchal, Vindhyan Scarplands, Baghelkhand Plateau, Chhotanagpur Plateau, North Sahyadri, Nilgiri, Kathiawar Peninsula, Kerala Plains, Utkal Plains, Andhra Plains, Tamilnadu Plains), 68.3% *decreasing trend* (South Kashmir Himalaya, Punjab Himalaya, Kumaun Himalaya, Punjab Plains, Avadh Plains, North Bihar Plains, Bengal Plains, Assam Valley, Meghalaya, Marusthali, Rajasthan Bagar, East Rajasthan Uplands, Aravalli Range, Madhya Bharat Pathar, Bundelkhand Upland, Malwa Plateau, Vindhya Range, Satpura Range, Maharashtra Plateau, Karnataka Plateau, Telangana Plateau, Mahanadi Basin, Garjhat Hills, Dandakaranya, Central Sahyadri, Eastern Ghats (North), Tamilnadu Uplands, Gujarat Plains, Karnataka Coast) and remaining 8.3% *no trend* (Rohilkhand Plains, South Bihar Plains, Narmada Valley, South Sahyadri, Eastern Ghats (South), Kachchh Peninsula, Konkan) (Fig. 12). The September rainfall over the country from 1942-1964 to 1965-2006 has decreased by 12.23%.

8. POSSIBLE REASON OF RECENT CHANGES IN MONSOON RAINFALL OVER THE COUNTRY

Gradient in the geopotential height (GPH) of the upper tropospheric isobaric levels from Tibetan Anticyclone (TA) region to Southern Tropical Indian Ocean High (STIOH) region is useful as index of

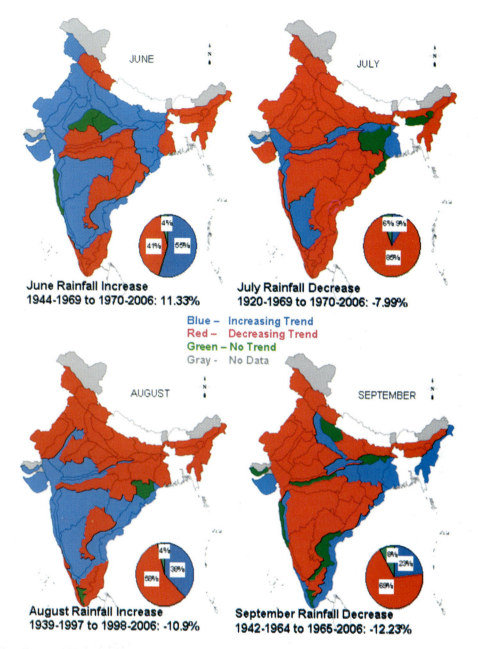

Fig. 12 Geographical distribution of recent tendency in monsoon monthly rainfall and pie diagram showing percentage area of the country under different tendencies.

intensity of Indian monsoon circulation (Singh et al., 2006; Sontakke et al., 2008b). Besides STIOH, the North Pacific High (NPH), the Azores High (AH) and the North Sub Polar Low (NSPL) have been identified and gradient between TA and NPH, TA and AH and TA and NSPL have also been found useful as index of intensity of the Indian monsoon circulation. The TA, STIOH, NPH, AH and NSPL regions are shown in Fig. 13. The combined area of the Iranian High (32.5°N-40°N; 50.0E-72.5°E) and the West Tibet (27.5°N-37.5°N; 72.5E-85°E) is considered as TA region, the area between parallels 40.0°E to 110°E and meridians 5°S to 20°S as STIOH region, the area between parallels 135°E to 120°W and meridians 20.0°N to 45°N as NPH region, the area between parallels 60°W to 10°W and meridians 20.0°N to 50°N as AH region, and the area between parallels 40°E to 110°E and meridians 55.0°N to 75°N as NSPL region.

For June, the interannual variation (1949-2006) of the geopotential height gradient (GPHG) at 300 hPa (NCEP/NCAR reanalysis data, Kalnay et al., 1996) from TA to STIOH region is shown in Fig. 14a on standardized scale along with rainfall over northern India (north of 18°N). The correlation coefficient (CC) of 0.77 between the two parameters is significant at 0.1% level and above. For July, August and September the GPHG of the 200 hPa level is used, the time series of the GPHG and the corresponding period rainfall on standardized scale are shown in Figs 14(b), 14(c) and 14(d), respectively. Figure 14(e) depicts the time series of the two parameters for the whole season. For July CC is 0.70, for August 0.52, for September 0.60 and whole season 0.73 and in every case the CC is significant at 0.1% level and above. Interesting to notice is that the two parameters are in excellent agreement during extreme years and in the latest period from 2001 which suggest possibility of exploring cause and effect relationship between them.

The two parameters GPH and rainfall for June show rising tendency from late 1950s, but for other three monsoon months and the season they show declining tendency. The June GPH of the 300 hPa is rising over the TA region at the rate of 5.96 m/decade (significant at 1% level) and over the STIOH region at the rate of 4.93 m/decade (significant at 0.1% level). Therefore the GPHG shows mild increasing tendency which is producing slight increase in June rainfall over the country. During July, August, September (200 hPa level) and the whole season the GPH over the STIOH region is rising at the rate of 5.03 m, 5.47 m, 4.59 m and 5.0 m per decade respectively, all significant at 0.1% level, but over the TA region the rising rate is 3.84 m/decade (significant at 10% level), 5.03 m/decade (significant at 1% level), 1.27 m/decade (not significant) and 4.03 m/decade (significant at 5% level), respectively. The rising tendency in the 200 hPa GPH is at a faster rate over the STIOH region compared to the TA region; therefore, the GPHG from the TA region to STIOH region shows slight decreasing tendency (statistically not significant). The lower GPH in the upper tropospheric level contributes partially to reduction in rainfall over the country.

Relationship between GPHG at NPH, AH and NSPL with TA and rainfall over the country have been examined for June, July, August, September and whole summer monsoon season and are shown in Fig. 14 along with the correlation coefficient (CC) obtained. In brief, the GPHG from the TA region to the STIOH region showed highest correlation with the corresponding rainfall over Indian region, that from TA to NPH region showed almost of the same order, but from TA to AH region and from TA to NSPL region showed weaker correlation in the same order.

9. MECHANISM

During heightened summer heating of the Afro-Eurasian dry province (Arabia to Mongolia) and the land area over and around Tibet-Himalayan highlands the Tibetan Anticyclone intensifies. A major portion of the outflows from the Tibetan Anticyclone sinks over the North Pacific and intensifies the high over there. The intense North Pacific High produces strong easterlies over central and western North Pacific. Part of these easterlies turns towards China, Korea and Japan, and part enters into Bay of Bengal after crossing Vietnam-Laos-Thailand-Combodia-Mynamar sector (Southeast Asia) where it merges with the main southwest monsoon current and the combined wind blows over the Indo-Gangetic Plains of northern

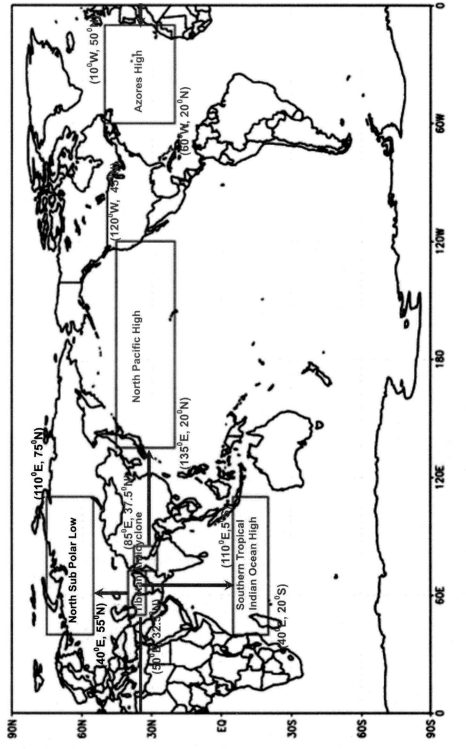

Fig. 13 Map showing STIOH, NPH, AH and NSPL regions identified for computing upper tropospheric geopotential height gradient from TA region as index of intensity of Indian summer monsoon circulation.

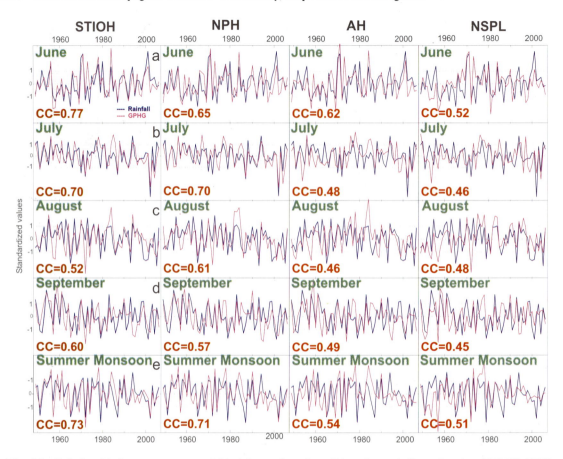

Fig. 14 Relationship between geopotential height gradient from TA region to indicated region (STIOH, NPH, AH and NSPL) and corresponding rainfall over the Core Monsoon Area of India for the period 1949-2006.

India as easterly wind parallel to southern flank of the Himalayas. Part of this easterly wind converges over central India and the rising air flows towards southern hemisphere via the upper troposphere. Also, the large outflows from the Tibet-Pacific Anticyclone in the upper troposphere after crossing the equator above the western Indian Ocean sink over southern Indian Ocean High. Part of this upper tropospheric cross-equatorial flow from northern to southern tropical/subtropical Indian Ocean and while circulating anticyclonically in the lower troposphere it crosses the equator along the Somali coast and becomes southwest monsoon flow, and part merges with the southern temperate westerlies.

The upper tropospheric GPHG from the Tibetan Anticyclone region to the NSPL region provides a majority of intensity of mid-latitude upper tropospheric westerlies. While steep gradient suggests strong poleward shifted westerlies and well developed and intense Tibetan Anticyclone, the reverse is true during shallow gradient. In the upper troposphere the Azores High appears as westward extension of the Tibetan Anticyclone.

Examination of geopotential height and temperature fluctuation of standard isobaric levels over the five centers of circulation revealed the following notable features:

(i) During June, over TA region the GPH at each standard isobaric level from 925 to 200 hPa is rising (non-linearly) at a faster rate compared to temperature, while over other centers of circulation positive trend in GPH and temperature fluctuations are going closely parallel;

(ii) During other three monsoon months, over TA region the temperature showed a mild rising tendency in the lower troposphere and mild decreasing trend in the upper troposphere but the GPH of the different levels showed rising tendency; and

(iii) Over the surrounding four centers of circulation rising trend in the GPH and temperature closely paralleled to each other from surface to tropopause.

From time series plots of low-frequency mode of fluctuations (9-point Gaussian low-pass filtered values) of the GPH and temperature (not shown) it is clearly noticed that within deep highs over oceanic region, e.g., the STIOH, the NPH and the AH both GPH and temperature showed rising tendency at different levels from 925 to 200 hPa. In the TA region, however, the temperature showed a rising trend in the lower troposphere and a falling trend in the upper troposphere. Thus a rising trend in the GPH of different levels in the TA region is essentially due to rising trend of temperature in the lower tropospheric layers.

Possible cause of cooling of upper troposphere over the Tibet-Iran-Afghanistan sector (TA region) is not presently understood. Attempts are being made to investigate these crucial issues by examining composition of the atmosphere such as dust load, greenhouse gases, moisture content, etc. and results will be published in the future.

10. CONCLUSIONS

In this chapter, fluctuation characteristics of annual, seasonal and monsoon monthly rainfall have been reported for 15 physiographic divisions and 49 subdivisions/provinces using longest available instrumental rainfall records from well-spread 316 raingauge stations. For the period 1901-2006, with data available for all the selected raingauges, the area-averaged rainfall series has been prepared from simple arithmetic mean of the gauges in the particular region. Prior to 1901 (sometimes going back to 1813) the series has been constructed by applying established objective technique on the limited available raingauges. For 15 physiographic divisions longest rainfall series could be developed from (ending year is always 2006) 1844 for the Western Himalaya and the Northern Plains, 1829 for the Eastern Plains and the Indo-Gangetic Plains, 1848 for the North Eastern Range, 1861 for the Western Plains, 1844 for the North Central Highlands and the South Central Highlands, 1826 for the North Deccan, 1835 for the South Deccan, 1848 for the Eastern Plateaus, 1829 for the Western Hills, 1852 for the Eastern Hills, 1817 for the West Coastal Plains and 1813 for the East Coastal Plains. The recent trend in annual rainfall fluctuation over physiographic divisions is as: the Western Himalaya decrease since 1961, the Northern Plains decrease since 1999, the Eastern Plains increase since 1967, the Indo-Gangetic Plains decrease since 1999, the North Eastern Range increase since 1983, the Western Plains decrease since 1976, the North Central Highlands decrease since 1962, the South Central Highlands slight decrease since 1945, the North Deccan slight decrease since 1962, the South Deccan decrease since 1957, the Eastern Plateaus decrease since 1944, the Western Hills increase since 1966, the Eastern Hills increase since 1981, the West Coastal Plains decrease since 1962 and the East Coastal Plains increase since 1866. For the country as a whole slight decrease in annual rainfall fluctuation can be seen since 1962.

The recent rainfall trends for annual, seasonal and monsoon monthly period over the 49 physiographic subdivisions/provinces have been identified and geographical distributions of different tendencies have been plotted on map of India. The maps show that annual rainfall is decreasing over 68.1%, winter decreasing over 46.6%, summer increasing over 73.3%, summer monsoon decreasing over 74.7%, post-monsoon increasing over 54.4%, June increasing over 55.1%, July decreasing over 85.1%, August decreasing over 58.3% and September decreasing over 68.3% area of the country.

Epochal analysis of the recent period rainfall fluctuations shows that amount of rainfall is increasing or decreasing as follows:

- Annual rainfall: decrease from 1930-1964 to 1965-2006 is –4.11%;
- Winter rainfall: decrease from 1892-1948 to 1949-2006 is –17.18%;
- Summer rainfall: increase from 1962-1996 to 1997-2006 is +18.1%;
- Summer Monsoon rainfall: decrease from 1931-1964 to 1965-2006 is –4.72%;
- Post-monsoon rainfall: increase from 1827-1914 to 1915-2006 is +13.32%;
- June rainfall: increase from 1944-1969 to 1970-2006 is +11.33%;
- July rainfall: decrease from 1920-1969 to 1970-2006 is –7.99%;
- August rainfall: decrease from 1939-1997 to 1998-2006 is –10.9%; and
- September rainfall: decrease from 1942-1964 to 1965-2006 is –12.23%.

Decrease in the intensity of the Asian Summer Monsoon Circulation (ASMC) appears to be the main cause of decline in monsoon rainfall over India. Gradient in the GPH of upper tropospheric isobaric level from Tibetan Anticyclone (TA) region to southern tropical Indian Ocean High (STIOH) region; to North pacific High (NPH) region; to Azores High (AH); and to North Sub Polar Low (NSPL) region can be used as index of intensity of the ASMC.

Geopotential height of different upper tropospheric isobaric levels over different centers is increasing, but over TA region the increase is slower compared to other regions. Over TA region, the lower troposphere is warming but upper troposphere is cooling while over other centers the ambient temperature is rising throughout, surface to tropopause. Possible cause of decline in the upper tropospheric temperature over TA region is being investigated.

In spite of the fact that there has been a large-scale decline in rainfall in the recent period, the survival of water systems is reasonably well suggests better management practices for the water bodies adopted by the people to combat frequent droughts. However, for long-term survival of the water systems under changing environmental conditions calls for better management practices for the water bodies so as to combat twin problems of flood and drought. Real-time monitoring of the physiographic rainfall variations can provide vital information in this endeavor.

ACKNOWLEDGEMENTS

The authors are extremely grateful to Prof. B.N. Goswami, Director, Indian Institute of Tropical Meteorology, Pune for necessary facilities to pursue this study. The rainfall data used in this study were provided by the India Meteorological Department, Pune, which is thankfully acknowledged. The part of the research in this chapter was funded by Department of Science and Technology, Government of India (DST Project No. ES/48/ICRP/2000).

REFERENCES

Bandyopadhyay, J. (2007). Water systems management in South Asia: Need for a research framework. *Economic and Political Weekly*, March 10, pp. 863-873.

Blanford, H.F. (1886). Rainfall of India. Indian Meteorological Memoirs, **81(III),** 658 pp.

Champion, H.G. (1936). A Preliminary Survey of the Forest Types of India and Burma. Indian Forest Record (New Series) 1, 286 pp.

Champion, H.G. and Seth, S.K. (1968). A Revised Survey of the Forest Types of India. Manager of Publications, Government of India, Delhi, 404 pp.

Eliot, J. (1902). Occasional Discussions and Compilations of Meteorological Data: India and the Neighboring Countries. Indian Meteorological Memoirs, Vol. XIV. Published by Order of His Excellency the Viceroy and Governor General of India in Council, Calcutta, India, 709 pp.

Ghosh-Bobba, A., Singh, V.P. and Bengtsson, L. (1997). Sustainable development of water resources in India. *Environmental Management*, **21(3)**: 367-393.

IMD (1961). Monthly and Annual Normals of Rainfall and of Rainy Days (based on records from 1901-1950). India Meteorological Department (IMD), New Delhi, 206 pp.

Kalnay, E., Kanamitsu, M., Kistler, R., Collins, W., Deaven, D., Gandin, L., Iredell, M., Saha, S., White, G., Woollen, J., Zhu, Y., Chelliah, M., Eblsuzaki, W., Higgins, W., Janowiak, J., Mo, K.C., Ropelewski, C., Wang, J., Leetmaa, A., Reynolds, R., Jenne, R. and Joseph, D. (1996). The NCEP/NCAR 40-Year Reanalysis Project. *Bulletin of the American Meteorological Society*, **77**: 437-471.

Klein, W.H., Lewis, B.M. and Enger, I. (1959). Objective prediction of five-day mean temperatures during winter. *Journal of Meteorology*, **16**: 672-682.

Mooley, D.A. and Parthasarathy, B. (1984). Fluctuations in all-India summer monsoon rainfall during 1871-1978. *Climatic Change*, **6**: 287-301.

NATMO (1986). National Atlas of India Physiographic Regions of India. Third edition, Plate 41, Prepared under the Direction of G.K. Dutt, Director, NATMO (the National Atlas & Thematic Mapping Organization), Kolkata, India.

Niemczyniwicz, J., Tyagi, A. and Dwivedi, V.K. (1998). Water and environment in India: Related problems and possible solutions. *Water Policy*, **1(2)**: 209-222.

Prabhakar, S.V.R.K. and Shaw, R. (2008). Climate change adaptation implications for drought risk mitigation: A perspective for India. *Climatic Change*, **88**: 113-130.

Rainbird, A.F. (1967). Methods of Estimating Areal Average Precipitation. WMO/IHD Report No. 3, WMO, Geneva, 42 pp.

Ranade, A.A., Singh, N., Singh, H.N. and Sontakke, N.A. (2007). Characteristics of Hydrological Wet Season over Different River Basins of India. IITM Research Report No. 119, 155 pp.

Raychoudhury, S.P., Agarwal, N.R., Datta, N.R., Gupta, S.P. and Thomas, P.K. (1963). Soils of India. Indian Council of Agricultural Research (ICAR), New Delhi, 496 pp.

Revi, A. (2008). Climate change risk: An adaptation and mitigation agenda for Indian cities. *Environment and Urbanization*, **20(1)**: 207-229.

Singh, N. (1986). On the duration of the rainy season over different parts of India. *Theoretical and Applied Climatology*, **37**: 51-62.

Singh, N. (1994). Optimizing a network of raingauges over India to monitor summer monsoon rainfall variations. *International Journal of Climatology*, **14**: 61-70.

Singh, N., Sontakke, N.A. and Singh, H.N. (2006). Global Warming, Subtropical Anticyclones and the Indian Summer Monsoon. Proceedings of the National Symposium Tropmet-2006 on Role of Meteorology in National Development, Vol. I, Indian Institute of Tropical Meteorology, Pune, pp. d29-d31.

Sontakke, N.A. and Singh, N. (1996). Longest instrumental regional and all-India summer monsoon rainfall series using optimum observations: Reconstruction and update. *Holocene*, **6(3)**: 315-331.

Sontakke, N.A., Singh, H.N. and Singh, N. (2008a). Chief Features of Physiographic Rainfall Variations across India during Instrumental Period (1813-2006). IITM Research Report No. 121, 132 pp.

Sontakke, N.A., Singh, N. and Singh, H.N. (2008b). Instrumental period rainfall series of the Indian region (1813-2005): Revised reconstruction, update and analysis. *Holocene*, **18(7)**: 1055-1066.

Strahler, A.N. (1969). Physical Geography. John Wiley & Sons, New York, 732 pp.

Walker, G.T. (1910). On the meteorological evidence for supposed changes of climate in India. Indian Meteorological Memoirs, 21, Part I, pp.1-21.

Wigley, T.M.L., Briffa, K.R. and Jones, P.D. (1984). On the average value of correlated time series, with application in dendroclimatology and hydrometeorology. *Journal of Climate and Applied Meteorology*, **23**: 201-213.

15

Emerging Tools and Techniques for Mine Safety and Disaster Management

Dheeraj Kumar

1. INTRODUCTION

Managing disasters in mines is both proactive and reactive. The proactive management deals with various steps taken to prevent and eliminate any disaster to occur by prompt action by operations and emergency staff. Reactive management, on the other hand, deals with the actions taken to reduce the damage caused by a disaster, mitigate sufferings, take up recovery measures, organize rehabilitation, bring normalcy of different operations in the mine, disseminate prompt information to relatives of the victims, civil authorities, print and electronic media and people living nearby. There are detailed and comprehensive legislations, standing orders, circulars, regulations, various codes of practice and plans to deal with various situations which may lead to a disaster in mines (Ghatak, 2008). Now, with advances in technology and forecasting techniques, preventive measures can be planned like drilling advance boreholes while approaching water-logged barriers, installing stone-dust barriers in gassy mines to arrest local methane explosion so as not to spread into devastating coal-dust explosions. The mining industries always strive hard to comply with all the preventive measures as per the laid guidelines to prevent the possible disaster.

Mine safety has improved significantly over the years, yet the recent disasters (Sethi, 2006; Terradaily, 2007; Reuters, 2007) at several underground coalmines worldwide have made the legislators, mine owners, researchers and manufacturers to debate on the inherent dangers in this industry, and their shared responsibilities to ensure the safety and health of mineworkers. These dangers raise serious concerns about coal mine safety among the mining industry worldwide.

There has been a tremendous growth in the use of Geographic Information System (GIS), Global Positioning System (GPS), Remote Sensing (RS), Satellite Communication (SC), and Modeling & Simulation techniques in the recent past. These tools and techniques can help significantly in characterizing infrastructure, risk area and disaster zones, planning and implementation of hazards reduction measures, etc. (Bahuguna and Kumar, 2006; Verma et al., 2008; Venkateswar, 2008; Kumar et al., 2008a,b; Semwal et al., 2008). GIS can improve the quality and power of analysis of hazards assessments, guide development activities as well as can assist planners in the selection of mitigation measures and in the implementation of emergency preparedness and response action. Remote Sensing as a tool can very effectively contribute towards identification of hazardous areas, can monitor the planet for its changes on a real-time basis and can provide early warning to many impending disasters. Communication satellites have become vital for providing emergency communication and timely relief measures. Indeed, the integration of space technology inputs into natural disaster monitoring and mitigation mechanisms is critical for hazard reduction. There is a need for assimilating the advances in information and communication technologies into mining operations. The mining industry could be further modernized by using software packages for an integrated data management, analysis and 3D geological modeling, 3D plant design and advanced real-time control and monitoring systems.

Over the course of history, the success of the mining industry has always remained in the evolution of mining technology. The implementation of novel systems and adoption of improvised equipment in mines help mining companies in two important ways: enhanced mine productivity and improved worker safety. Safety is one of the key factors driving the trend to automation. Efficiency is imperative if a mine is to survive and automation can play a large role in this. By developing and commercializing automated mining technologies for continuous mining equipment, the productivity of each mining machine improves and the operators can run the machine from a safe distance, which reduces associated costs for worker exposure, health benefits, and liability. With the advent of new autonomous equipment used in the mine, the inefficiencies are reduced by limiting human inconsistencies and error. The desired increase in productivity at a mine can sometimes be achieved by changing only a few simple variables.

This chapter focuses on the issues pertaining to mine safety and disaster management, and the recent developments in tools and technologies and their applications in mining industries for safe mining operations with increased productivity. Major emphasis has been given to the recent developments in tools and technologies and their applications in mining industries for safe mining operations and improved productivity.

2. MINE DISASTER AND ITS MANAGEMENT

2.1 Mine Disaster: An Overview

Mining of minerals is considered to be one of the most hazardous occupations as it involves working against nature. It entails constant struggle of human being with reasons and resources against the changing forces of nature. The mining systems, especially underground operations, are constrained by the absence of natural light, fresh air and open space, and the undesirable presence of high temperature, humidity, dust, fumes, noise and rock stresses. Due to these constraints, the hazards and hazard potentials inherent in a mine may trigger disasters unless sound and strong measures are taken to prevent them (Katiyar and Sinha, 2008).

Directorate General of Mines Safety (DGMS), the mining regulatory authority in India, defines *mine disaster* as "an accident where ten or more fatalities have taken place" (Bose, 2008), whereas in US, five or more fatalities in mine at any point of time is termed Mine Disaster. In India, 55 disasters have occurred in mines from 1901 to 2007, killing more than 2200 persons, the maximum being 375 in Chasnala inundation in 1975. In recent years, frequency of disasters has increased, which is a cause of concern to the concerned nations. This is further aggravated by the fact that most of the disasters were caused by human errors and were by and large preventable. Salient examples of mine disasters are presented below.

Fifty miners died at the Nagda mine, BCCL, in Jharia coalfield in eastern India on September 17, 2006, as a consequence of a methane explosion almost half a kilometer underground. The lethal combination of coal dust and carbon monoxide is believed to have killed those who survived the initial shock of the explosion. Only four workers on that shift escaped. They were haulage operators working at Level Zero, 120 m below the surface (Sethi, 2006). Figure 1 shows the rescue operation in progress at the Nagda mine.

China's mines are by far the world's most dangerous. About 6,300 people were killed in 2004 in floods, explosions and fires in China's mines. At least 203 miners had died in northeast China after an earthquake apparently triggered a gas explosion underground at the Sunjiawan mine in Liaoning province, China, on February 14, 2005. The accident happened 242 m below the ground at the Sunjiawan colliery of the Fuxin coal industry group. A blast in the northern province of Shaanxi in November 2005 killed

166 miners. Another explosion in October 2005 killed 148 miners. Prior to these mining accidents, the deadliest reported mining accident in recent years was a fire in 2000 in southern China that killed 162 miners (*Times Online*, 2005). On November 13, 2006, there was a large and fatal gas explosion at the Nanshan Colliery in Lingshi County, Jinzhong, Shanxi Province, China. Twenty-four people were killed. The mine was operating without any safety license as its original had expired. While no official cause has emerged, the news agency Xinhua claims the explosion was triggered by incorrect usage of explosives. Floods, blasts and other accidents in China's coal mines killed 345 in October 2006, 44 per cent higher than the previous month, according to official figures. Despite a 22 per cent decline in fatalities from a year earlier (2005), 3726 Chinese coal miners died in over 2300 accidents in the first 10 months of 2006 (Reuters, 2006). Furthermore, 172 miners were trapped in a flooded mine in eastern China due to heavy rains which caused a nearby river to burst its banks on August 18, 2007, flooding the Zhang Zhuang coal mine near the city of Xintai. An extensive rescue operation was focused on pumping water from the pit, but none of the missing was found (Terradaily, 2007).

Fig. 1 Rescue team members enter a seam of the Nagda mine in Dhanbad, Eastern India, one day after a massive explosion trapped 54 miners in a coal mine (*The Hindu*, 2006).

A gas explosion at a coal mine in northern China on December 7, 2007 killed at least 105 people in one of the country's worst mining accidents in the year 2007. Coal mining accidents are a common occurrence in China, where on an average more than 5,000 miners lose their lives every year, the worst accident rate in the world (Barboza, 2007). Nearly every week, there is news of mining blasts, deaths, rescue efforts and teams of miners trapped in horrific circumstances. By comparison, there are typically about 30 mining deaths a year in the United States. In 2006, China reported that about 4,700 miners were killed in accidents. It was the first time in 30 years that there were fewer than 4,800 mining deaths.

Moreover, twenty-nine coal miners went underground at the International Coal Group's Sago Mine near Buckhannon in Upshur County, West Virginia, on the morning of January 2, 2006. At 6:26 a.m., a methane ignition in a recently sealed area of the mine triggered an explosion that blew out the seals and propelled smoke, dust, debris and lethal carbon monoxide into the working sections of the mine. One miner was killed by the blast. Sixteen escaped. Twelve were unable to escape and retreated to await rescue behind a curtain at the face of the Two Left section. Mine rescuers found the trapped miners approximately 41 hours later. By that time all but one had succumbed due to carbon monoxide asphyxiation (Davitt, 2006). It was the worst mining disaster in the U.S. since a 2001 disaster in Alabama killed 13 and the worst in West Virginia since a 1968 disaster that took 78 lives. The deadliest coal mine disaster in the U.S. history was an explosion in 1907 in Monongah, West Virginia, that killed 362 people.

Total deaths in all types of U.S. mining, which had averaged 1,500 or more during earlier decades, decreased on an average during the 1990s to below 100 and reached a record low of 55 in 2004. There were 65 mining fatalities in 2007. The average annual injuries to miners have also decreased steadily (MSHA, 2009).

Major coal mine disasters in the U.S. in recent years are as follows (Foxnews, 2006):

- 2001: Explosions at a Jim Walter Resources Inc. mine in Brookwood, Ala., killed 13 people.
- 1992: A blast at a South Mountain Coal Co. mine in Norton, Virginia, killed eight people.
- 1989: An explosion at a Pyro Mining Co. mine in Wheatcroft, Ky., killed 10 people.
- 1986: A coal pile collapsed at Consolidation Coal Co.'s mine in Fairview, West Virginia, killing five persons.
- 1984: A fire at the Emery Mining Corp.'s mine in Orangeville, Utah, killed 27 people.

Major coal mine disasters in Russia during 1997-2007 are as follows (Reuters, 2007):

- December 1997: A methane gas explosion rips through the Zyryanovskaya mine in the city of Novokuznetsk in western Siberia killing more than 30 miners.
- January 1998: A powerful explosion at the Tsentralnaya mine in the Arctic town of Vorkuta killed 27 miners.
- January 13, 2002: At least five people died in a methane gas blast at the Vorkutinskaya mine in the northern Komi Republic.
- October 29, 2003: Five miners were killed in a methane blast at a mine in the Far Eastern town of Partizansk.
- April 10, 2004: A gas explosion at the Taizhina colliery in Siberia's Kemerovo region killed 45 people.
- October 28, 2004: A methane blast at the Listvyazhnaya mine in Siberia killed at least 13 people.
- February 9, 2005: An explosion in the Yesaulskaya mine in Kemerovo killed 21 people.
- September 8, 2005: A methane blast at Anzhero-Sudzhensk in Kemerovo region killed three people.
- September 7, 2006: Twenty-five miners died in a fire at a remote gold mine in Eastern Siberia near the border with China.
- February 25, 2007: One miner is killed after a portion of a shaft collapsed at a coal mine in Kemerovo region.
- March 1, 2007: One coal miner was killed and nine were injured after a portion of a shaft collapsed in Kemerovo region.
- March 19, 2007: A gas explosion at the Ulyanovskaya mine in Kemerovo region killed 106 people. It should be noted that the mine in Siberia's Kuznetsk Basin holds some of the biggest coal reserves in the world.

Mine disasters are mostly due to roof-fall, inundation, fire, explosions, etc. Underground methane emission mainly maximizes in the late monsoon, because percolating water drives out the occluded gas from pore-spaces. When the accumulated methane gas gets minimum required oxygen to constitute an explosive mixture, it leads to an explosion trigged by any igniting source apparently as it happened in the Bhatdih-Nagda mine, BCCL, India explosion on 5th September 2005. Many mining inundation disasters have taken place in the monsoon season (i.e., rainy season) by overflowing streams running into the mine (Samanta, 2008).

An analysis revealed that while the fatalities due to explosions were rather high in the first three quarters varying from 65 to 76%, it came down to 11% in the last quarter of the century (Bose, 2008). During the same period, however, the accidents due to inundation were as high as 75% which in the first three quarters varied from 6 to 20%. The scenario in non-coal mines is radically different. Very few accidents in metalliferons mines could be classified as mine disasters mainly because underground mines in this sector are very few compared to the coal sector. Out of 2400 working mines in the mineral sector, only around 80 are worked by underground methods. The accidents in open cast mines could take place due to inadequate stability of the pit slopes and dumps, drilling and blasting, loading and transportation.

However, dumpers/trucks constitute a major cause of fatal accidents (about 70%) both in coal and non-coal mines (Bose, 2008).

Mining in India is primarily labour intensive. Therefore, accident, death and injury rates are generally expressed in terms of units per 1000 persons employed on a yearly average basis. It was observed further that while in the case of coal mines roof falls, rope haulage and dumpers/trucks accounted for a major share of fatal accidents, side falls and transport equipment were responsible for such fatalities in non-coal mines (Bose, 2008).

Figures 2 and 3 illustrate cause-wise fatal accidents in the coal and metal mines of India during the year 2006. From Fig. 2 it can be seen that roof and side fall accidents accounted for about 21% of all fatal accidents during the year 2006. Figure 3, on the other hand, reveals that during 2006, there were 14 fatal accidents due to fall of persons and three fatal accidents due to explosives accounting 23% and 5% of all fatal accidents, respectively.

The data of Canadian mines for number of workdays missed in one year per 100 workers is less than 2.5 over the last few years. The number of fatalities in the USA coal mines in 2007 as reported by MSHA is 33. The major contributors are slip or fall of people and power haulage. The underground coal mines explosion fatalities and injuries in the US during 1980-2006 are shown in Fig. 4. The trend of accidents in the Indian mines is shown in Tables 1 and 2. For coal mines, a consistent decline is observed in the 10-yearly average number of accidents per year since the 1950s and in the 10-year average number of

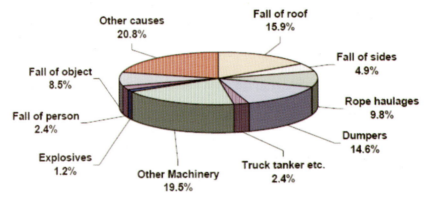

Fig. 2 Cause-wise distribution of fatal accidents in the coal mines of India in 2006 (DGMS, 2007).

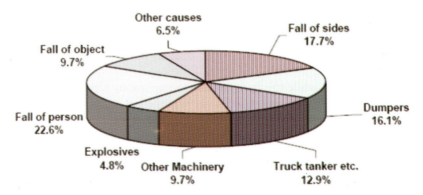

Fig. 3 Cause-wise distribution of fatal accidents in the metal mines of India in 2006 (DGMS, 2007).

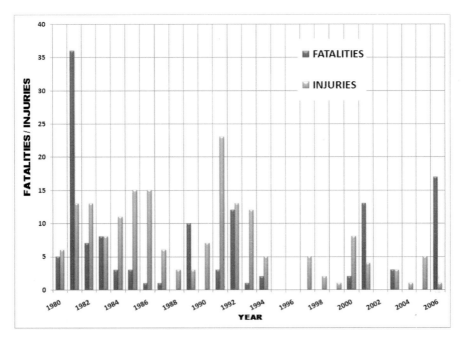

Fig. 4 Underground coal mines explosion fatalities and injuries in the US during 1980-2006 (Gurtunca and Breslin, 2007).

fatalities since the 1970s. The same trend continued for the 6-year period (2001-2006). For the non-coal mines, the average numbers of accidents and fatalities have remained more or less at the same level during the last three decades ending in 2000, while the 6-year average has fallen during the period 2001-2006 (Mukherjee, 2008).

Table 1. Trend in fatal accidents and fatality rates per 1000 persons employed in the Indian mines (10-year average) (DGMS, 2007)

Year	Coal mines				Non-coal mines			
	Average number of accidents	Accident rate	Average number of fatalities	Fatality rate	Average number of accidents	Accident rate	Average number of fatalities	Fatality rate
1901-10	74	0.76	92	0.93	16	0.47	23	0.67
1911-20	139	0.94	176	1.29	29	0.57	37	0.73
1921-30	174	0.99	219	1.24	43	0.54	50	0.66
1931-40	172	0.98	228	1.33	35	0.41	43	0.51
1941-50	226	0.87	273	1.01	26	0.24	31	0.29
1951-60	223	0.61	295	0.82	64	0.27	81	0.34
1961-70	202	0.49	259	0.62	72	0.28	85	0.33
1971-80	187	0.40	264	0.55	66	0.27	74	0.30
1981-90	162	0.30	185	0.34	65	0.27	73	0.31
1991-00	140	0.27	170	0.33	65	0.31	77	0.36
2001-06	90	0.21	118	0.29	140	0.27	170	0.33

Table 2. Trend in fatal and serious accidents, and death and serious injury rates in mines (place-wise) (Mukherjee, 2008)

Year	Number of fatal accidents				Number of serious accidents			
	Below ground	Open cast	Above ground	Overall	Below ground	Open cast	Above ground	Overall
1997	94	27	22	143	440	79	158	677
1998	80	24	24	128	346	72	105	523
1999	74	30	23	127	408	77	110	595
2000	62	38	17	117	444	108	109	661
2001	67	26	12	105	464	73	130	667
2002	48	22	11	81	434	92	103	629
2003	46	23	14	83	380	82	101	563
2004	49	32	6	87	757	82	123	962
2005	51	28	20	99	833	98	162	1093
2006	42	28	12	82	546	62	111	719

Table 3. Chronology of disasters in the Indian mines due to inundation (Nath, 2008)

	Name of mine	Date of occurrence	Number of persons killed
1.	Phularitand Colliery	11.07.1912	21
2.	Jotejanaki Colliery	28.06.1913	13
3.	Loyabad Colliery	16.01.1935	11
4.	Makerwal Colliery	06.07.1942	14
5.	Majri Colliery	05.08.1953	11
6.	Newton Chikli Colliery	10.12.1954	63
7.	Burra Dhemo Colliery	26.09.1956	28
8.	Central Bhowrah Colliery	20.02.1958	23
9.	Damua Colliery	05.01.1960	16
10.	Silewara Colliery	18.11.1975	10
11.	Chasnalla Colliery	27.12.1975	375
12.	Central Saunda Colliery	16.09.1976	10
13.	Hurriladih Colliery	14.09.1983	19
14.	Mahabir Colliery	13.11.1989	6
15.	Rajpura Dariba Lead-Zinc Mine	28.08.1994	13
16.	Gaslitand Colliery	27.09.1995	64
17.	Bagdigi Colliery	02.02.2001	29

Inundation in mines has been the most common type of mine disasters which has continued to happen at regular intervals involving loss of valuable lives in spite of extensive legislations, technical circulars and guidelines issued in this regard. Table 3 presents major disasters that have taken place in the Indian mines due to inundation.

2.2 Disaster Management

Mine disaster management is the process of designing and maintaining the mine environment with zero accident and zero disaster potential in which individuals working together in groups efficiently accomplish the objective of the organization while carrying out the managerial functions of forecasting, planning, organizing, staffing, leading and controlling (Katiyar and Sinha, 2008).

Disaster management in a long-term perspective should focus on (Katiyar and Sinha, 2008):

- Education and training with a view to build a human base with total involvement.
- Mechanization and automation in mining operation.
- Lesser exposure of human beings to hazard.
- Greater use of safety equipment and appliances, preferably in-built safety features.
- Effective supervision; better quality and depth of inspections.
- R&D plans for safety.

For ensuring safety in normal operation of the mine and dealing with post-disaster emergencies, a comprehensive action plan should be laid down incorporating planning, prediction, prevention, preparedness, emergency-response system, mitigation and recovery.

- **Planning:** Long-term conceptual plan for the whole life of new mines and the remaining life of existing mines should be prepared with an eye to identify the sources of danger and measures taken to prevent any source of danger to cause accident or disaster. A three-dimensional scanning of the entire mine area together with effective peripheral area outside the mine boundary is necessary for obviating any danger in the mine during its entire life.
- **Prediction of Mine Disaster Potential:** The technological advancements and rich pool of knowledge in mining, together with digital convergence by IT revolution can go a long way in predicting the *potential and proneness* of mine disasters. The triggering events having hazard potential have to be identified suitably by auditing the system design, layout and operation and the use of latest technologies such as ground penetration radar (GPR) system for mapping and locating water-logged workings; design and development of advanced signaling system to sense fire, gases, etc.; and continuous tele-monitoring system.
- **Comprehensive Risk Assessment and Analysis** of all the four factors causing the accidents as described earlier should be carried out as follows:
  - The unsafe acts and unsafe conditions must be analyzed separately and corrective actions taken.
  - The individual and work factors also need to be examined to obviate any risk of human error of omission or commission.
  - The risk assessment should be mine-specific and hazard-specific.
  - The risk identified should be categorized in accordance with the degree of severity—catastrophic, critical, marginal or negligible.
  - The operation which may lead to catastrophic and critical hazard must be eliminated, substituted or corrective action taken for reducing the risk.
- **Prevention of Mine Disaster:** The prevention of disaster is the essence of disaster management and must follow the corrective action after prediction. For the elimination of condition that could lead to disaster, managerial imperatives lies under the following two components:
  - Human factors or software component involves:
    - Human resource backup system by education and training, quality, system and standards of inspections.
    - Human error audit.
    - Total employee involvement.
    - Morale and motivation of personnel with a view to create more awareness and positive and pro-active attitude.

- Technological factors or hardware component involves:
  - Identification and re-assessment of layouts and installation which pose threat to safety.
  - Hazard and operability studies.
  - Fault tree analysis and disaster consequence analysis.

The challenge lies in recognizing the dangers precisely and taking pre-emptive measures well before it triggers a mine disaster. The identification of hazard and hazard potential, analysis, and control and management are illustrated in Figs 5, 6 and 7, respectively.

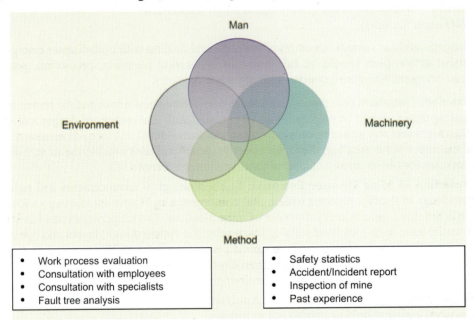

Fig. 5 A typical hazard identification process in a mine system (Katiyar and Sinha, 2008).

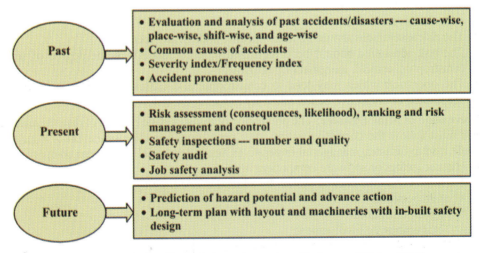

Fig. 6 Analysis of hazards in a mine system (Katiyar and Sinha, 2008).

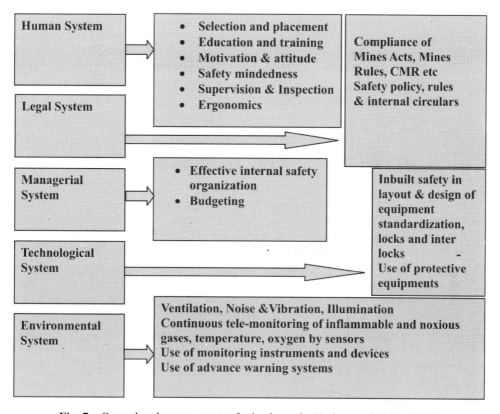

Fig. 7   Control and management of mine hazards (Katiyar and Sinha, 2008).

2.3 Prevention of Mine Disasters

The 9th conference on safety in mines which was held during 2-3 February 2000 at New Delhi recommended the following measures to be taken for preventing mine disasters from inundation and fire along with the effective emergency responses (Samanta, 2008):

2.3.1 Preventing Mine Disasters from Inundation (DGMS, 2000)

(1) Each mine shall be critically examined for its proneness to inundation and deliberated in the Safety Committee of the mine and information disseminated as widely as possible.
(2) Suitable infrastructure at area level may be provided for drilling advance boreholes to detect presence of waterlogged workings in advance.
(3) Embankments provided against river and jore to guard against inundation should be constructed and properly shown in the underground plan and water danger plan.
(4) Surface excavation near major sources of water, which are connected or are likely to be connected to below ground workings should be filled up completely.
(5) Detailed precautions against inundation by framing and implementing standing order for the safe withdrawal of persons, with effective communication system.
(6) Mechanism may be developed for warning mines about impending heavy rains; opening of dams in the rivers on the upstream side should be examined.

(7) Winding system serving as sole means of ingress and egress may be made constantly available even in adverse weather conditions, failure of steam or electricity or any other reasons.
(8) Effective communication may be established for the safe withdrawal of persons. Necessary standing orders in this regard need to be framed and enforced.
(9) Providing infrastructure including computerized facilities at an area level to oversee survey work of the mine may be implemented.
(10) R&D efforts should be continued to develop a system for construction of water-tight chamber as last refuge (shelter) below ground in case of inundation.

2.3.2 Preventing Mine Disasters from Fire and Effective Emergency Response (DGMS, 2000)

(1) Considering the risk of fire, all coal mine companies shall rank its coalmines on a uniform scale according to the scientific basis.
(2) Making the emergency plan responsive, speedy and effective, each mine shall review the existing emergency plan, at a higher level keeping in view the risk from fire.
(3) Establishing rescue rooms in coal mines having high risk of fire and employing more than 350 persons ordinarily employed in a shift below ground may be considered.
(4) Storing oxygen type self rescuer at strategic places below ground in coal mines with a risk of fire in such scale so as to cater to the needs of persons.
(5) Each mining company shall formulate and implement structured training program for the development of awareness and increasing effectiveness of emergency response.
(6) Early detection of heating effect due to reversal of fan, control of fire and other associated aspects may be studied.

2.3.3 Risk Management as a Tool for Developing Appropriate Health and Safety Management Systems (DGMS, 2000)

- Every mining company should identify one or more mines and should undertake a formal risk assessment process aimed at reducing the likelihood and impact of mishaps of all kinds in mines. Subsequently risk assessment process should be extended to other mines.
- Risk assessment process should aim at effective management of risks, by identifying
  - which risks are most in need of reduction, and the options for achieving that risk reduction, and
  - which risks need careful on-going management, and the nature of the on-going attention.
- The risk assessment exercise should follow an appropriate process.
- Risk management plans shall be prepared on the basis of risk assessment and implemented in the identified mines.

2.3.4 Quality Control for Improving Safety (DGMS, 2000)

(1) Each mining company and the manufacturer/supplier shall satisfy themselves that the product has a valid approval wherever applicable and conform to the relevant standards at the time of supply.
(2) Each large mining company may setup a quality control cell or strengthen it where the same exists, identify critical items which require testing for quality assurance at the time of procurement and during use, and arrange testing of the same. Testing facilities may be setup wherever feasible.
(3) Any defect or failure of approved items or those having BIS certification may be promptly brought to the notice of an appropriate authority for further action.

2.3.5 Communication System between DGMS Offices and Mine Management at Site (DGMS, 2000)

(1) Recognizing existence of a large variety of communication systems in mines, a comprehensive review of the existing communication systems at all mines in all mining companies should be undertaken and a consolidated status report prepared for working out a realistic and effective system of communication.

(2) In an organized mining sector, an effective internal and external system of communication besides P&T means shall be established both ways between the mines and Rescue Rooms/stations, hospitals and DGMS offices.

(3) Each mining company in the organized sector shall formulate and implement a comprehensive communication protocol clearly assigning duties and responsibilities of persons at various levels.

(4) In an un-organized mining sector, an effective communication system shall be established in the following manner: in large mines, effective wireless communication within the mine including attendance rooms and managers' office and residence; and, P&T telephone at managers' office and residence.

2.3.6 Other Measures

- Sufficient number of self-contained rescue devices should be made available at mine site. These rescuers must be capable of enhanced performance.
- Improved battery capacity and common connection specifications to enable emergency communication devices to be run from the same portable power source as a headlamp, continuous dust monitor, or other device a miner carries.
- Improved technology for assisting mine rescue teams, including devices to enhance vision during rescue or recovery operations.
- Improved technology and improved protocols for using existing technologies to enable underground conditions to be assessed promptly and continuously in emergencies to help officials determine instructions to give both to miners trapped underground and to mine rescue teams and others engaged in rescue.
- Improvements to underground mine ventilation controls separating mine entries to be more resistant to mine fires and explosions, particularly in entries used for miners' escape ways.
- Mine-wide monitoring systems and strategies that can monitor mine gases, oxygen, air flows, and air quantities at strategic locations throughout the mine that would be functional during normal mining operations and following mine fires, explosions, and roof falls, including systems utilizing monitoring sensors that transfer data to the mine surface and the installation of tubing to draw mine gas samples that are distributed throughout the mine and can quickly deliver samples to the mine surface.
- Protective strategies for the placement of equipment, cables, and devices that are to be utilized during mine emergencies, for example, communication systems, oxygen supplies, and mine atmosphere monitoring systems to protect them from mine fires, roof falls, explosions, and other dangers.

3. APPLICATION OF INFORMATION TECHNOLOGY IN MINE SAFETY AND DISASTER MANAGEMENT

Use of information technology (IT) is becoming more and more common in all imbedded application and mining industry is not lagging behind, and it is not over, rather it is just a noble beginning. Information technology entered the mining industry half a century ago. It was first introduced in 1960s in staff function, especially in clerical activities like preparation of pay rolls, listing of store items, manpower control, etc. In 1970s, it encompassed designing of civil engineering constructions, laying of tracks, roads, etc. In 1980s, it was widely used in the management information system (MIS), mine planning system (MPS) and in truck dispatch system (TDS). Use of global positioning system (GPS) and geographical information system (GIS) in mine was started in 1990 (Kumar, 2007). GIS is a visualizing technology that captures, stores, checks, integrates, manipulates and displays data using digital mapping. It further spreads in areas like preventive maintenance, quality control and other areas of technology and business related to mining industry directly or indirectly.

Development and use of automated equipments, use of robotics in a hazardous environment in mines, better communication and tracking technologies, self rescue devices, and refuge alternatives are some of the areas which have drawn attentions worldwide (Roy et al., 1988; Anon, 1996; Loney and Ross, 2001; Kumar, 2007). New and more effective training programs, emergency procedures, and mine safety practices are being designed using innovative risk analysis and management systems. Any one of these alone would help to improve mine safety, but the effect is expected to be large in combination.

Some recent advancement in IT enabled tools and techniques which have been proved to be very effective in mine safety and disaster management are discussed in subsequent sections.

3.1 Autonomous Mining System

As a result of extensive research, miners today are starting to use externally controlled machines designed to be used by a single operator, the autonomous miner. Because of this technology, miners can be positioned in a non-hazardous location while they steer the mining equipment. A totally autonomous mining system would prevent any health or safety concerns related to noise, vibration or dust because the equipment would be able to work remotely, away from any human that may have a part of its operation.

Robotic sensors can give the operator the "eyes" needed to accurately and smoothly perform a sequence of operations with less risk, and programmed algorithms can keep the production of a mineral online even under most difficult conditions. Ultimately, the switch to autonomous mining equipment is the future of the mining industry.

Mining and allied industries are now moving towards the automation of their sub-systems with a major emphasis on the following:

- Drill and shovel monitoring for ground characterization
- Global Positioning Satellite applications
- Machine health and maintenance monitoring
- Expert systems in maintenance
- Automated mine design
- Geo-sensing and artificial intelligence
- Image analysis
- Communications
- Radio Frequency Identification technology
- Drilling intelligence and control
- Machine guidance and tele-operation
- Systems safety and human factors

3.2 Mine Safety Technology

One of the most important factors leading to mine disasters is ground failure—mainly roof collapse in underground mines. The prevention of fatalities and injuries from failures of the roof, pillars or floor has been a priority area of research, development, and demonstration, to practice activities worldwide for many years. Significant safety improvements have been achieved in these areas. Coal bumps, bounces, and outbursts have been a longstanding safety hazard in some mines worldwide. A coal bump is the sudden and violent failure of highly stressed coal or surrounding strata.

National Institute for Occupational Safety and Health (NIOSH), Pittsburgh, USA, has developed several computer programs to help mine planners design coal pillars. For longwall mining, there is Analysis of Longwall Pillar Stability (ALPS). For room-and-pillar and retreat mines, there is Analysis of Retreat Mining Pillar Stability (ARMPS). Both the programs are widely used throughout the United States. These programs, along with others developed by industry or academia, provide an excellent methodology for the proper design of coal mine pillars for a wide range of mining conditions (Kohler, 2007).

Making longwall coal mining safer and more productive has been the subject of a long-running CSIRO (Commonwealth Scientific and Industrial Research Organization) project funded by the Australian Coal Association Research Program, which has also come up with a new technology designed to locate and guide coal-cutting equipment in longwall mines. CSIRO has produced commercial-standard automatic face-alignment systems, which ensure the cutting drums follow the coal seam accurately. Pre-commercial prototypes are now being used at the Xstrata's Beltana and BMA's Broadmeadow longwall mines. CSIRO has also created prototypes for automatic horizon control and longwall information management systems, which are ready for commercial production (Hill-Douglas, 2007).

An automation technology has been designed to locate and guide coal-cutting equipment in longwall mines (Fig. 8). The Xstrata's Beltana longwall mine uses this technology. Automation has made this form of mining more productive, and safety has been greatly improved.

Letting machines do the repetitive work, and allowing humans to monitor and correct the automated process, has made this form of mining much more productive with improved safety of miners by reducing their exposure to dust and fly rock from the shearer.

The application of seismic monitoring has emerged as a potential technology for predicting ground movements and coal bumps. Today, seismic monitoring is used more in hard rock mining, as part of a risk management program. However, it is very infrequently used in coal mining. Despite advances in technologies, such as geophones, signal processing equipment and computers, many of the barriers that existed 30 years ago remain today (Kohler, 2007). Notwithstanding, there is an advantage of applying seismic monitoring at mines with a history of bumps, as part of a larger risk management program, as is done in Australian and many European coal mines.

Japan applies its mine safety technology to model coal mines in coal-producing developing countries, thereby reducing mine disasters, improving mine safety and promoting stable coal production (Clean Coal Technology, 2006). Important examples are:

- In China, Japanese gas explosion disaster prevention techniques have been introduced to the Zhang-ji mine of the Huainan Mining (Group) Co. Ltd. in Anhui Province through the Central Coal Mining Research Institute. Specifically, the following four improvements were made: installation of a gas monitoring system, enhancement of gas drainage efficiency with directionally controlled gas drainage boring technology, introduction of underground ventilation analysis software and the installation of an underground radio system.
- In Indonesia, Japanese spontaneous combustion disaster prevention techniques have been introduced at the PTBA Ombilin mine through the Mineral and Coal Technology Research and

Fig. 8 New automation technology designed to locate and guide coal-cutting equipment in longwall mines (Hill-Douglas, 2007).

Development Center (TekMIRA), including CO and temperature monitoring technology, gas analysis, underground ventilation network analysis and wall grouting.

- In Vietnam, Japanese mine water inflow disaster prevention technologies have been introduced in Mao Khe coal mine and other sites through the National Coal and Minerals Group (VINACOMIN), including hydrological data collection and analysis, underground water flow analysis using hydrogeologic models, advanced boring technology for water exploration and drainage, and a flow rate measuring/pumping system.

Techniques to prevent roof fall accidents in roadways or at working faces including roof fall risk prediction and roof fall prevention systems have been developed worldwide. These techniques are subjected to evaluation and applied to mine sites (Clean Coal Technology, 2006). An advanced monitoring and communications system for underground mine safety and stable production has been developed through joint research conducted with the Commonwealth Scientific and Industrial Research Organization (CSIRO), Australia. This joint research also aims to build a basic system for risk information management that allows real-time evaluation of different disaster risk factors (relating to work, environment, devices and machinery). As an early detection technique, an odor sensor based on worldwide standards is now being studied in a joint research project with the Safety in Mines, Testing and Research Station (SIMTARS). At the Kushiro mine in Hokkaido, Japan, a "man location system" and an underground communications system were tested during an on-site demonstration.

3.3 Mine Robots

Mine Robot is defined as a self-propelled mining machine with a flexible control for multifunctional use of the working head during mining. Robots are being developed for hazardous duties and for performing 3D mapping and remote sensing in environments such as coal mines at greater depths. Mine Robots will be doing jobs like laying explosives, going underground after blasting to stabilize a mine roof or mining in areas where it is impossible for humans to work or even survive.

Robotic mining technology offers unique benefits as listed below:

- On-line information like geology (geophysical, geotechnical, and geochemical), production rates, quality help in production planning, etc.
- Robotic mining will allow operation of a mine with faster removal rates of ore and with less risk.
- Robotic mining also allows narrower openings and deeper mining operations to remain profitable.
- Increased safety: With no humans present, there will be no need for sophisticated air circulatory systems.
- Increased productivity: Faster removal rates of ore as compared to miners.
- Increased equipment utilization.
- Time-efficient operations.
- Reduced need for maintenance of mine systems.
- Faster reaction to engineering and maintenance issues.
- Improved mine cash-flows.

Rescue robots are proved to be one of the most important tools in dealing with mine disasters. An autonomous, four-wheeled robot with heavy-duty tires, called *Groundhog* (Fig. 9), was sent into an abandoned coal mine near Pittsburgh, USA, in May 2003, and was able to create accurate three-dimensional maps of its surroundings. It proved the ability of a robot to map the rooms, pillars, and corridors left by generations of mining. Rescue robots in the future will certainly enter mines under the unknown conditions of dust, gas, inundation and roof fall, and will be crucial for exploring and characterizing conditions and reporting back to command centers. They can carry gas sensors that characterize the atmosphere of a mine. Typically, they would deploy two of those sensors on each machine to make sure that there is no mistake in the instruments. Once robots have the capability to get in and get around, they could also provide communications and visuals and carry out map sensing; deliver objects to aid trapped people, like oxygen tanks, and detect vital life signs. Robots can carry sensors that would alert people to the presence of methane gas or poisonous air. Mine rescue response is one application where robots would inevitably make significant contribution.

Some of the examples of robotics-based mine equipments are mentioned below:

1. Tele-operated and automated load-haul-dump trucks that self-navigate through tunnels, clearing the walls by centimeters.
2. The world's largest "robot", a 3500 tonne coal dragline featuring automated loading and unloading.
3. A robot device for drilling and bolting mine roofs to stabilize them after blasting.
4. A pilotless burrowing machine for mining in flooded gravels and sands underground, where human operators cannot go.
5. A robotic drilling and blasting device for inducing controlled caving.

Unlike their counterparts commonly found in the manufacturing industry, mining robots have to be smart. They need to sense their world just like humans. Mining robots need sensors to measure the three-dimensional structure of everything around them along with suitable vision power so that they can know their 3D-position geographically within the mine site in real-time and online. CSIRO, Australia, is developing vision systems for robots using cameras and laser devices to make maps of everything around the machine quickly and accurately, as it moves and works in its ever-changing environment.

Salient case studies related to robotics-based mining are presented in subsequent sections.

3.3.1 World's Largest Industrial Robot

The system is being tested on a dragline (weight = 3500 tons, and boom length = 100 m) located at Tarong Coal's Meandu Mine, near Kingaroy in Queensland, Australia. Robot is designed to install a computer "brain" in a dragline, add a radar sensor to the boom to help the computer locate its targets. This dragline is used to scoop up blasted rock in open-cut coal mines. It picks up from 100 to 300 tons of fragmented rock with every scoop, then swings it round and delivers it to the spoil pile, then swings back again. It also swings back and forth once a minute, so the smooth movement makes the operation more efficient, results in less wear and tear on the machine, more rock is moved and there is less strain on the dragline's human operator (CSIRO, 1998). Use of this robotic-based dragline has resulted into increasing the productivity of a dragline by around 4% along with a saving of approx $3 million a year for an Australian coal mine and approx $280 million for Australia as a whole.

3.3.2 Groundhog

Towards achieving the mine mapping goal, a 700 kg custom-built ATV-type robotic platform known as *Groundhog* (Fig. 9) has been constructed that is physically tailored for operation in the harsh conditions of abandoned mines (Morris et al., 2005). Groundhog has been used extensively in both test and abandoned mine environments, accruing hundreds of hours of mine navigation with eight successful portal entry experiments in the abandoned Mathies mine outside of Pittsburgh, United States. From these experiments, log data has generated globally consistent large-scale maps using offline techniques. Over the course of its lifetime, Groundhog has evolved into a system that is highly proficient at autonomously traversing and mapping isolated mine corridors, successfully navigating over 2 km of abandoned mine.

3.3.3 Ferret

The existence of subterranean void spaces, such as the cavities created by mining, is a hazard to active mining operations and a constant threat to surface developments. When abandoned, these underground

Fig. 9 An autonomous, four-wheeled robot with heavy-duty tires (Morris et al., 2005).

spaces can accumulate tremendous quantities of water and threaten to flood encroaching active mines. To address the existence of subterranean void space belowground, a robotic tool has been developed that is capable of reaching a domeout via borehole access, acquiring the measurements necessary for void analysis, and relaying this information to the surface. To reach the mine cavity, deployment and sensing schemes were required to descend the borehole, identify the mine breach, and maintain a sense of orientation throughout the process. The robotic tool for remote subterranean void analysis has several operational advantages over past techniques. Nicknamed Ferret (Fig. 10), it establishes a physical presence in a mine cavity enabling a "first hand" perspective of the void (unlike non-intrusive methods that require information to be inferred). This physical presence is attained without human presence in or around the mine, which removes the risks faced by surveyors in subterranean inspections. In addition, Ferret provides quantitative information on cavity extent that is difficult or impossible to obtain using borehole camera systems. In the following paragraphs, the mechanical, electrical, and software systems that compose the Ferret and empower it to retrieve cavity data in a rugged and unforgiving environment (Morris et al., 2003) are described.

Moving to robots will not eliminate human miners, but it will change their job description. Instead of placing themselves in hazardous areas to do repetitive and arduous tasks, people will manage the operation of the robots. Mines will also need programmers, technicians and repair people. Mining can be a hazardous job. Getting robots to do the job will make mining safer and ensure the long-term viability of the industry. Robots will be doing jobs like laying explosives, going underground after blasting to stabilize a mine roof or mining in areas where it is impossible for humans to work or even survive.

These are following thrust areas where lots of research needs to be carried out by mining fraternities in order to deploy robotics in mining and allied industries:

- Use of remote-controlled manipulators for the mechanization of miner's actions.
- Multifunctional technological robots.

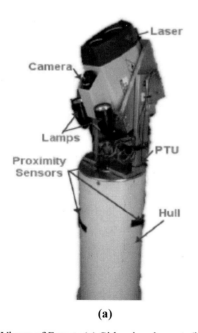

(a)   (b)

Fig. 10 Views of Ferret: (a) Side-view layout; (b) Ferret after deployment in underground (Morris et al., 2003).

- Introduction of information robots or mine rescue robots in case of mine disasters.
- A manipulator equipped with a TV-camera and force sensors.

3.4 Communication and Tracking Technology

In a crisis, any information, particularly accurate information, is fleeting and difficult to capture and to confirm. Meanwhile, the demand for information by families, the public and the media incrementally increases with each passing minute and approaching news deadline. There is an imperative need to not only manage communications but to fiercely guard the quality of the information communicated.

There is a wide variety of applications for systems that can track the position of agents, like persons, vehicles or goods in a reliable and predictable manner. Most indoor applications focus on improving the interaction of mobile robots with people, therefore, requiring good knowledge of a person's location. Outdoor applications of position tracking are more common in the areas of asset tracking, safety or surveillance. Mines generally use reliable and effective communication systems for routine operations, but these systems require hard-wired networks, power supplies, and other infrastructure that are likely to be damaged or destroyed by a catastrophic event, such as a fire, explosion, or water inundation. Technologies are needed that will function in post-disaster environments. Knowing the locations of mine workers would obviously facilitate rescue operations, but reliable signal transmission to the outside of a mine after a catastrophic event is very difficult. An electronic tracking system requires a reliable communication signal to relay location information to the surface. Technical barriers to in-mine or through-the-earth signal propagation must be overcome before reliable emergency communications and miner tracking reach an acceptable functionality.

The most basic requirement of a post-tragedy communication system is a communication link between the underground miner and the surface. A two-way system would be more useful than a one-way system, since escaping or trapped miners could relay valuable information outside. Moreover, a voice rather than text system can be very much helpful in emergency conditions. However, the difficulty in using a voice-based system while wearing a self-contained rescuer with a mouthpiece must be addressed. The emergency communication system should be part of a mine's routine system, not a separate one, to better ensure that it will properly function when an emergency occurs (NIOSH, 2006).

Some of the recent applications of communication and tracking technology in dealing with safety issues in mines are described in the subsequent sections.

3.4.1 Mine Multimedia Rescue Communication

At present, the communication instrument equipped to rescue workers is mine rescue telephone. This equipment cannot exactly reflect the actual status of the mine disaster situation because of its limited capability of information transformation. Besides the underground communication condition especially after disaster is very bad. Traditional wire-communications will be hard to work. Therefore, the command of rescue will be delayed or even blocked. The voice communication system, which lacks capabilities such as data saving, replaying, etc., could not provide effective first-hand data for analyzing disaster, summing up the experience and improving rescue strategies. Hence, development of a mine multimedia rescue communication system which has high-speed data transmitting ability is necessary.

Mine multimedia rescue communication system adopts an integration of MPEG4, VOIP, TCP/IP and self-organizing technology, etc. It can be carried by the rescue workers. The system adopts a cooperation method of wireless communication and wire communication to transmit information. The distance of transmission is 10 to 20 km. The facilities have the following function: Display, save and replay the audio-visual data of the entire rescue process; Achieve multi-user talk; and Collect surroundings parameter of the mine including temperature, oxygen, CO and CH_4. All of the information could be provided for analyzing the cause of the disaster. In addition, the facilities can give an alarm when the parameters go beyond the prescribed limits (Wang et al., 2006).

3.4.2 Internet and Information Superhighway

Quick and accurate information is vital to any decision-making process where numerous technical parameters need to be synthesized and evaluated. A number of important internet websites are now receiving global attention from the mining industries. These websites offer update and relevant information on coal and mineral exploration and exploitation. Information superhighway is a remote PC-based instrumentation system for continuous monitoring of a mining system/sub-system. It is like a highway where all the expertise is readily available for solving all kinds of mining related problems. This could be a forum through which many countries can exchange their views on the aspects of mine accidents. One could gather the data for past 25 years and use the very best of data mining tools to take us to a world of accident-free mining (Roy and Acharya, 2004).

3.4.3 Global Environmental Disaster Information System (GEDIS)

GEDIS, the Global Environmental Disaster Information System (Fig. 11), is a new integrated system for confidential, reliable emergency communication and coherent disaster management information support (Konrad, 1998). This system will aim to overcome existing information access and communication deficiencies by using advanced user-friendly, internet-based, multilingual and multimedia technologies (including tele-/videoconferencing and remote interpretation).

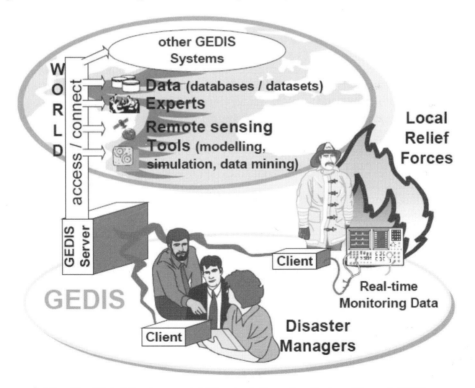

Fig. 11 Global Environmental Disaster Information System (Konrad, 1998).

GEDIS is a distributed client-server system. The internal and external communication flows are confidential and ensured through the use of the best-available and/or most appropriate broad band telecommunication technique (fixed, mobile, satellite based etc.). The basic system is equipped with security, communication (tele- and videoconferencing, data exchange sharing and notification) and data-access utilities as well as multilingual support and real-time GIS tools.

GEDIS accepts input from fixed and mobile sensors for real-time data collection, including acquisition on demand of (near) real-time remote sensing data. It allows to process, integrate, add value and visualize (using the advanced, real-time, internal GIS capabilities), system-internal and (predominantly) system-external information resources. These include information inventories and on-line accessible forecasting, risk assessment and decision support tools. Critical-phase disaster management support is provided by real-time command and control tools. The system employs standard, state-of-the-art (fixed, mobile and satellite-based) telecommunication networks to ensure a broad band reliable information and communication flow. GEDIS lays the foundation for an internet-based early warning system for the citizens in a disaster-prone area (Konrad, 1998).

3.4.4 Close Proximity System (CPS)

The purpose of a CPS is to give drivers a warning/indication that a person or a light vehicle is near the truck, enabling him to react to the threat present. The design of such a CPS can be based on a single sensor or even a multiple-sensor system. At present, different proximity systems are under development (Kloos et al., 2006).

3.4.5 GPS-based Systems

GPS-based systems require all mobile equipments to have a GPS receiver, thus allowing having absolute real-time position information of the equipment. This position information is then exchanged using wireless radio to evaluate relative position, e.g., between a truck and a light vehicle and issue an alarm, if necessary. Full GPS coverage is absolutely necessary for such systems to operate successfully (Naik and Murmu, 2006; Raju and Das, 2006; Kumar, 2007; Kumar and Bahuguna, 2008; Kumar et al., 2008b). Nevertheless, multi-path issues and satellite shadowing may occur, e.g., when a person is close to a truck. Fatalities among equipment operators in open pit mines can be reduced if GPS technology is incorporated in their machines.

With differential GPS equipment, one can quickly determine exact coordinates of a given truck with an accuracy of less than a meter and evaluate whether a given truck is dangerously close to the dumping edge of a waste dump (Fig. 12).

DynaMine—Online Truck Dispatch System (OITDS): a Case Study

A global positioning system (GPS)-based, operator-independent truck dispatch system (OITDS) suitable for open cast mines has been installed and is in operation at The Jayant Opencast Mine, Northern Coalfields Ltd. (NCL), Singrauli, India. The Mine handles 30 million cubic meters of mine overburden (the waste product generated during mining operations) and around 10 million tonnes of coal in a year. It has a fleet of 15 excavators with a capacity ranging from eight to 14 cubic meters, 50 trucks of 85-tonne capacity and 30 trucks of 120-tonne capacity. The OITDS system covers the entire fleet of excavators and trucks. This system was conceptualized in 1999 and was implemented in September 2002 (CMC Ltd., 2007). Figure 13 shows the basic layout of OITDS.

Features of DynaMine

- Global positioning system (GPS)-based on-board equipment with voice and data communication facilities and vital signs monitoring devices (VSMDs) is mounted on the excavators and trucks.

Emerging Tools and Techniques for Mine Safety and Disaster Management 353

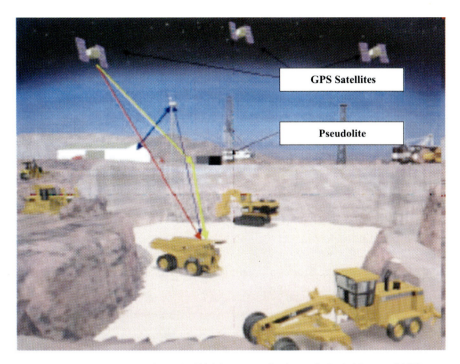

Fig. 12 Improving safety of off-highway trucks through GPS (Kumar, 2007).

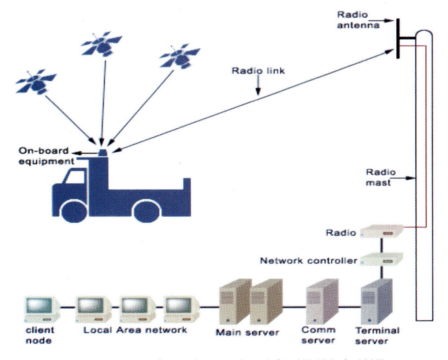

Fig. 13 Layout of OITDS system DynaMine (CMC Ltd., 2007).

- Three communication masts with repeaters are installed at strategic locations to ensure reliable radio communications over an area of around 20 km², covering the entire mine.
- Fibre-optic cables are laid over a length of 18 km for LAN connectivity between the main control room and various user locations.
- Around 25 clients are connected to the host application and database server for online monitoring.
- The system is integrated with an attendance recording system for the operating staff based on card swiping, and manages an automatic crew-mining equipment-allocation facility. These allocations are automatically announced through a public address system as well as displayed on a large screen in scrolling mode, reducing considerable time-loss at the beginning of the shift.
- Important information on production and availability of mining equipment is available on the interactive voice response system (IVRS).
- The system directs each truck to move on optimized route, increasing productivity through a dynamic programming technique.
- The system tracks the movement of mining equipment and maintains a register of warnings against each identified operator/equipment in case of violation of instructions or moving in non-safe zones during the specified time.

3.4.6 Radar-/Laser-based Systems
These systems are based on range and bearing sensors and actively detect objects being close to the truck. They usually don't provide an identification of the object detected. Radar systems also usually don't detect persons being underneath a truck or fail to detect them if they walk under the radar beam (Kloos et al., 2006).

3.4.7 Vision-based Systems
Computer image recognition usually builds the core of such a system. A simpler version of a vision-based system requires the driver himself to identify intrusive objects. A disadvantage is the unreliability due to possibly varying light conditions (Kloos et al., 2006).

3.4.8 RFID-based Systems
These systems usually equip mobile equipment to be protected with a Radio Frequency (RF) tag, which will be detected by a RF tag reader attached to the source of danger (e.g., truck) when in close proximity. Out of the technologies presented, RFID is the most feasible technology from both a technical and financial perspective that can provide complete sensor-coverage of the dangerous areas around and even under a truck. All the other technologies presented here fail to provide complete coverage or can provide it only at dramatically increased cost and installation efforts (Kloos et al., 2006).

Supervisory Control And Data Acquisition (SCADA), mobile vehicle and production tracking, personnel identification, security, vehicle location and ventilation system monitoring as well as control with integrated ventilation system simulation are all part and parcel of the modern mine's everyday needs. However, these activities and tools are dependent on communication technology that needs to operate in an environment that has been described as a close approximation of hell and where the 'shop floor' is forever changing its topology. Underground mobile platforms such as locomotives, LHDs, continuous miners, etc. need reliable connectivity with the central management system which may or may not be situated on the surface. Wherever its location, the unavailability of reliable communication precludes effective management of these costly resources.

In a number of fully mechanized mines worldwide, centralized fire and hazard detection systems are installed that have the unique feature that all the sensors connected to the cable are powered from the surface so that the sensors will continue operating even if the underground power is discontinued in the case of an emergency. Some of these stand-alone applications include (Buisson, 2004):

- Centralized blasting systems: They have found general acceptance and are once more based on a wired "change of state" system.
- Winder control systems: They are mostly stand-alone affairs with historical data on the winder's performance, which, if archived, is usually only available to the selected audience.
- The VLF (Very Low Frequency) wireless tight rope or slack rope system has emerged as a successful tool in preventing severe accidents occurring as a result of common hoisting malfunctions.
- For the mine rescue work, the well-known and universally acclaimed VLF SSB (Single Side Band) radio has found a market all over the world.
- Ventilation fans and water pumps are usually stand-alone and seldom controlled from a single centralized SCADA system.

The melding of carrier media such as radio and fiber-optics can go a long way towards eliminating the lack of communication bandwidth that exists today in underground mines. The acceptance of standard protocols such as Ethernet can eliminate the need to re-invent a wheel while making the mining wheel compatible with those of the other manufacturing and production industries (Buisson, 2004). Although the communication problems in underground mines are severe, technologies are emerging to overcome these difficulties. A bigger problem will possibly arise in the coordination of these technologies at the management level in order to orchestrate synchronized workable and measurable processes. Provision of timely and correct information to the right people can significantly improve plant availability, which in turn will improve efficiency and productivity.

3.5 GIS and Remote Sensing Technologies

Geographic information and imaging systems visually portray layers of information in new ways to reveal relationships, patterns, and trends. GIS and image processing software packages from vendors such as ESRI and ERDAS provide the functions and tools needed to store, analyze, and display information about geo-space.

A Geographic Information System (GIS) is a computer-based tool for mapping and analyzing things that exist and events that happen on the earth. GIS technology integrates common database operations such as query and statistical analysis with unique visualization and geographic analysis benefits that maps offer. These abilities distinguish GIS from other information systems and make it valuable to a wide range of public and private enterprises for explaining events, predicting outcomes, and planning strategies (Paul et al., 2005).

Remote sensing, on the other hand, is the art and science of making measurements of the earth using remote sensors placed on airplanes or satellites. These sensors collect data in the form of images and provide specialized capabilities for manipulating, analyzing, and visualizing those images. Satellite imagery is integrated within a GIS for information extraction, processing and analyses.

The use of GIS as a powerful tool to analyze and display data is gaining momentum in the mining industry. The use of latest mapping technology like GIS and Remote Sensing (RS) is growing in mines worldwide (Fig. 14). The RS technology has been extensively used in mapping the regions affected by underground fires in mine and its surrounding areas. This technology integrated with a GIS has become

356 Natural and Anthropogenic Disasters: Vulnerability, Preparedness and Mitigation

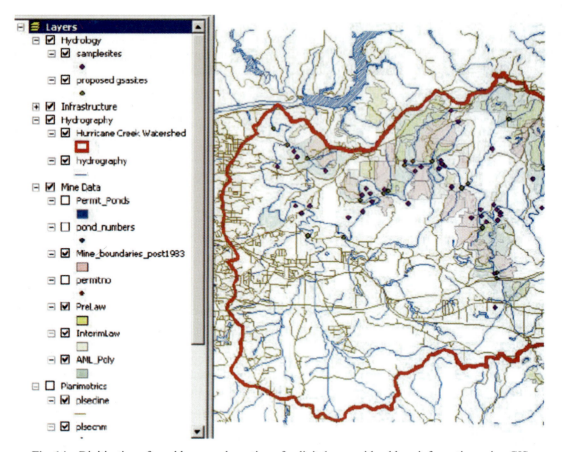

Fig. 14 Digitization of an old map and creation of a digital map with add-on information using GIS.

an effective tool for developing and implementing a rehabilitation plan for a region. GIS technologies create efficiency and productivity opportunities in all aspects of mineral exploration and mining. GIS enables a mineral exploration geologist and mine operator to mine intelligently, efficiently, competitively, safely, and in an environment friendly manner.

3.5.1 GIS for Mining Management

Majority of mining information has some sort of spatial component that can be represented in a map form, including financial and asset information. Management and mineral economists are using GIS for the evaluation of their corporate and competitor assets (Bahuguna and Kumar, 2006). All of the tabular data used to assess a mining prospect or existing operation can be spatially referenced. GIS software allows direct access to data available in most common spreadsheets and databases. Reserve estimates, annual planned production, or cost per ton statistics can be linked to prospect or mine locations and used to control the map symbols. Regional maps can place the mines or prospects in a regional geologic or political setting. Detailed maps of exploration prospects or active mines can be accessed by simple mouse click at a point on the regional map.

3.5.2 GIS for Mining Exploration

GIS is considered to be a versatile tool for data consolidation, archiving, and quick access of information in the process of evaluating various mineral properties. The exploration manager can take advantage of GIS's link between spatial and tabular data to access specific information such as assay, geotechnical, cadastral, metallurgical, environmental, and permitting data to prepare contour maps, derive surfaces, grade distribution, soil types, vegetation, cultural, wildlife, slope and aspect surfaces, neighborhood and zonal statistics, and other thematic layers of interest. Likewise, the exploration manager can get benefitted from GIS's powerful 3D capabilities in performing visibility analysis of surface installations and display underground features in 3D.

3.5.3 GIS for Mining Operations

Pipelines, electric lines, roads, ramps, and other mining facilities change frequently. Engineers and operations staff use GIS for facility planning applications (Fig. 14). Keeping track of existing infrastructure and integrating with the mine plan, block models can be achieved by using GIS. Also, GIS can be used to integrate recent survey data with block models or mine design data from other mining software packages such as TECHBASE, Vulcan, Mine Sight, and SURPAC.

3.5.4 GIS for Environmental Management

Environmentalists involved in reclamation use GIS to analyze and map soils, vegetation, and surface hydrology. For an example, the results of surface geochemistry can be used to calculate required topsoil. Reviewing the location of new operations relative to reclamation areas may reveal an opportunity to move topsoil directly to the reclamation site, therefore eliminating costly stockpiling.

3.5.5 Other Mining Applications

Transport routes: Mine planners utilize GIS in combination with remote sensing to plan the best alternative for transportation of goods and supplies to and from the nearest community to the mine site. Integrating thematic layers such as topography, land ownership, land use, population, geotechnical, and climate can facilitate the selection of an economically and environmentally preferable alternative. The siting of refuge chambers (a safe place in underground mines which is used for the purpose of shelter in the event of any unforeseen circumstance) within a safe distance from production stopes (working panels in underground mines) can easily be achieved by the GIS's proximity analysis. A network of refuge chambers with adequate capacity and first-aid kits proportionate to the miners' concentration can be implemented (Fig. 15). The same applies in mapping and finding the nearest exit and identifying the shortest route from the various working centers by creating a distance grid (Anthony, 2001).

Natural hazards: GIS and remote sensing assist the planners in identifying natural hazards such as potential landslides, floods, earthquakes, and volcanic eruptions prior to the construction of production and housing installations.

Population distribution: Planners of a new mine may need information on population density, socio-economic distribution, labor resources, housing, and recreational infrastructure in the preparation of the environmental impact assessment.

Selection of sites for housing and dumps: GIS is used for the selection of a housing site that meets safety, scenic, and recreational requirement within reasonable proximity to the mining operation. Topographic, vegetation, drainage, and soils coverage together with concentration of toxic substances are incorporated in the analysis (Anthony, 2001).

Study of land use and land resource pattern in the mining areas: New types of remotely sensed data can be easily acquisitioned at present. Older technologies are still available as well. Sifting through the options to determine the appropriate mix of data sources and processing techniques to provide valuable information without excessive cost is a task for which a few organizations have time, money or inclination.

Mine safety is of paramount concern in the mining industry. In the effort to achieve the objective of safety and productivity, GIS can contribute in providing a safe working environment in underground mining by performing network analysis and determine the appropriate sites for refuge chambers and facilitate the prompt evacuation of mine personnel.

The remote sensing technology integrated with GIS can become an effective tool for mapping the regions affected by underground fires in Jharia and its surrounding areas and developing and implementing a rehabilitation plan for the same. Differential satellite interferograms (InSAR) produced from satellite radar images (Single Look Complex) have been used by researchers for the development of procedures leading to the determination of ground deformations and slope stability in the vicinity of open pit mining operations. Validation of the deformation model obtained using InSAR technology is made using the deformation model developed from existing GPS and Prism Monitoring Systems (Jarosz and Wanke, 2003).

3.6 Computer-Aided Mine Planning and Design

Recent advances in computer technology and software development has made it possible to develop an advanced simulation model for the strategic mine planning and risk evaluation (Fig. 16).

Some of the applications of Computer-Aided Mine Plan and Design (CAMPAD) are:

- Use of CAD methods for mine design, geological databases and resource estimates, block modeling techniques, blast pattern designs, 3D ore body modeling and 3D open-pit and underground mine layout.
- Development of software for strategic open pit planning based on object-oriented stochastic simulation.
- Development of simulation/animation models for open pit mine planning.
- Computer modeling for performance estimation and optimization of mechanical excavators.
- Computer simulation of gravity flow of ore in Ore Passes by the Discrete Element Method.
- Project scheduling and costing for major mining projects using CPM and PERT software packages, viz., Microsoft Project, Microsoft Visio, etc.

The occurrence of deposits, grade control and operational planning, all being complex in nature, call for sophisticated computer models for predicting/deciding different alternative courses of action. The important modules in the Mine Planning & Design software covers a comprehensive range of applications as listed below:

- Gridded Seam Model (GSM) module
- Underground Planning module
- Industrial Minerals module
- Operations/Planning module – Surveying
- Operations/Planning module – Ore control system
- Mining engineering module – Pit optimization
- Mining engineering module – Pit design
- Mining engineering module – Scheduling
- Exploration/Geology module – Exploration
- Exploration/Geology module – Modeling
- Exploration/Geology module – Geostatistics

Other software available for the stability analysis, mine planning and design purposes are:

- Numerical modeling software (ITASCA): UDEC, 3DEC, FLAC, FLAC 3D, PFC 2D, PFC 3D. They are essentially for stability analysis and mine design.
- MICROMINE: This software is meant for exploration and mining.
- SURPAC: This software is applied in geological modeling and mine planning.
- TECHBASE: The features available in this package are database generation, statistical modeling, graphics, etc.
- GEMCOM: This is for the geological exploration and evaluation apart from surface underground mine design, environmental engineering and hydrological engineering.
- MINEX 3D: This software is meant for geological modeling and mine planning systems.
- MINEX UG: It is for interactive underground mine planning.
- O.P.MINE: It facilitates open pit mining simulation for designing mining sequences, mining methods and equipment.

Recent developments in computer-aided mine planning and design tools and numerical modeling techniques available in software packages have made the life of a mine planner and designer very easy. These tools and techniques can be adopted for the state-of-the-art mine planning and designs for achieving higher productivity with improved safety.

3.7 Virtual Reality in Mineral Industry

A design starts as a concept and must be given a shape and size. Once these elements are determined, rather than producing two-dimensional drawings, it is more practical now to go directly to a computer-generated three-dimensional model of the design. The models are produced using the modeling tools available in the CAD package or any 3D graphic package based on the Virtual Reality Modeling Language (VRML). This model can then be used to produce a wide range of visuals ranging from rendered images to animations.

In September 2001, the Laurentian University, Sudbury in Ontario, Canada, officially opened a state-of-the-art virtual reality laboratory as part of its center for integrated monitoring technology (CIMTEC). This facility designed to meet the needs of the mineral exploration and mining industries, offers a team interpretation environment for earth modeling application (Fig. 17). The Virtual Reality (VR) technology can greatly assist in mine planning and design through its strong capability to visualize overall impact of various factors in a complex mining environment. Some of these applications are described below with an explanation of the type of visuals that enhance the outcomes.

3.7.1 Mining Equipment Concept Development

Mining and equipment companies are presently developing prototype equipment that will revolutionize underground mining. To make this task as quick and easy as possible, 3D modeling of the equipment and the working environment is a key tool. The use of the model to review the equipment prior to fabrication allows modifications to be made without the costly investment of multiple versions of prototype equipment. The Automated Underground Diamond Core Drill Project Team recently utilized a similar modeling approach for Inco-Mines Technology Department Ltd. [Fig.18(a,b)]. The objective of this project was to minimize the cost of gathering and utilizing geological data (Loney and Ross, 2001). The design review of the project can be completed utilizing a parametric 3D solid model using Pro/ENGINEER, or by any VRML-enabled compiler.

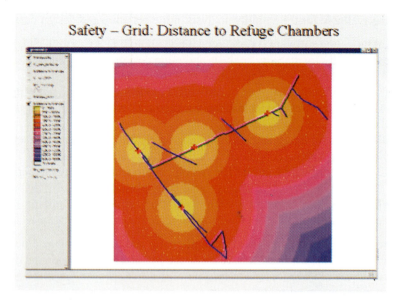

Fig. 15 Mapping and finding the nearest exit and identifying the shortest route from the various working centers by creating a distance grid using GIS (Anthony, 2001).

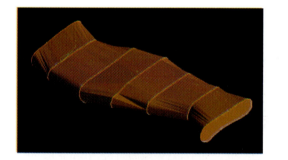 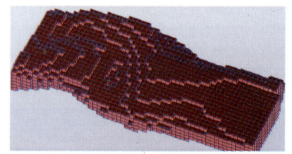

Fig. 16 Solid modeling and block modeling of an iron ore deposit using SURPAC.

Fig. 17 State-of-the-art virtual reality laboratory at the Laurentian University, Sudbury, Ontario (MIRACO, 2008).

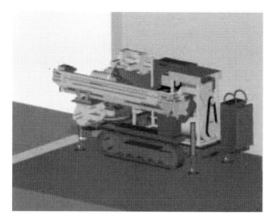 

Fig. 18(a) Diamond core cutting model. **Fig. 18(b)** Diamond drill and underground drill station.

3.7.2 Design Review, Operator Input, Maintenance Procedure Review and Simulations

The design review using VRML allows the client to review the project so that they can get ensured that the required design is being properly developed as per laid specifications. The operator input allows personnel (who will eventually be working with the equipment or in the environment) to get familiar with the project in advance of construction.

Mobile equipment such as Load Haul Dump (LHD) vehicles and haulage trucks are extensively used in the mining industry. However, the designs of both LHDs and haulage trucks, in combination with the constraints of the underground operating environment, have resulted in a number of serious accidents, including fatalities and numerous incidents which had the potential for serious injury. A fully interactive VR underground model combining CAD LHD models or haulage trucks in different mine design layouts can be constructed with an aim of improving operator visibility. MIRARCO and Laurentian University, Canada, are undertaking an investigation to improve operator visibility when driving mobile equipment in underground environments (Delabbio et al., 2007).

3.8 Armchair Mining

Armchair mining is a technology to drive miners out of the tunnels and into the control room. Remote-controlled mining, known as tele-operation, could help drive down injury further by removing miners from hazardous areas and, at the same time, could significantly increase productivity, according to some experts. As companies and automation experts bring tele-operated machinery into working mines, they are facing questions of how this evolution will affect day-to-day operations, worker responsibilities, and mine designs (DeGaspari, 2003). Figuring out the best ways to handle those issues could help pave the way for wider implementation.

Two developments in recent years have dovetailed to make tele-operation in mines possible. One is a robust communication backbone in the mine, capable of handling data, voice, and video signals. The other is 'smart' mining equipment, outfitted with on-board computers and a host of sensors.

With an ore body 4 km long, 80 m thick and reaching a depth of 2 km, LKAB's Kiruna iron ore mine in Sweden is the world's largest, most modern underground iron ore mine (www.mining-technology.com). Since mining began here over 100 years ago, LKAB has produced over 950 Mt of ore. Very few people

Fig. 19 A load-haul-dump vehicle in LKAB's iron ore mine in Sweden scoops rock with the help of a teleoperator working off-site; the vehicle does hauling and dumping automatically (DeGaspari, 2003).

work underground. The seams are drilled by remote operated drills (Fig. 19). Huge Finnish-Built Driverless Wheel Loaders follow computer-controlled routes and only stop at piles of broken rock to collect the ore. At this point, an operator sitting in front of a TV screen on the surface loads the ore and carries it to the shaft where it is dropped to the 1,045 m level. The ore is crushed here and then hoisted to the surface to be processed. Electric-powered, remote-controlled drilling and ore handling equipment supplied by Atlas Copco and Tamrock is widely used here. After blasting, load-haul-dump machines (some of which are fully automated) carry the run-of-mine ore to the nearest ore pass, from which it is loaded automatically on one of the trains operating on the 1,045 m level (www.mining-technology.com).

The backbones of armchair mining are the robust communication system in the mine, capable of handling data, voice, and video signals and 'Smart' mining equipment outfitted with on-board computers and a host of sensors.

4. CONCLUSIONS

There is a substantial need of adoption of state-of-the-art automation technologies in the mines to ensure safety and to protect the health of mineworkers. Issues pertaining to mine safety and disaster management have been discussed in this chapter. In addition, emerging tools and techniques coupled with some of the recent innovations in mine automation that could be deployed in mines for safe mining operations and for avoiding any unforeseen mine disasters have also been highlighted in this chapter. Significant developments have been made in the areas of surface and underground communication, robotics, smart sensors, tracking systems, mine gas monitoring systems and ground movements, etc. Advancement in information technology (IT) in the form of Internet, GIS, remote sensing, GPS, satellite communication, etc. have been proved to be very important tools for hazard reduction and disaster management.

The evaluation and analysis of the past disasters point towards human error, poor maintenance, poor communication, sense of complacency and laxity, the lack of concern and above all absence of safety culture. Apparently, the mine disasters were by and large predictable and preventable. Human factors, if identified and addressed in the right perspective, can go long way in fostering a safety culture in the organization and reducing the risk potential of mines. Over the next few years, semi-automated mining technologies backboned with information technologies will become much more prevalent at large-scale mining operations because these technologies satisfy the two most important goals of any mining operation: 'improved productivity' and 'safer working condition'. Eventually, these imaginative ideas may become a reality for us in near future. The autonomous mining equipment will continue to encourage further improvements in productivity and miners' safety. Inarguably, the near and distant future of mining lies in the development and enhanced application of autonomous mining technologies, especially in developing nations.

REFERENCES

Anon. (1996). American Crane Uses Engineering Productivity as a Competitive Edge. *In:* Pro/NEWS International, Volume 01 Number 01, June 1996.

Anthony, D.H. (2001). An Application of GIS in Underground Mining. http://www.hammond.swayne.com/GIS_mining.htm (accessed on July 30, 2008).

Bahuguna, P.P. and Kumar, D. (2006). Application of Geographical Information System (GIS) for opencast coalmines. Proceedings of the Indian Conference on Mine Surveying (ICMS-2006), 8-9 September 2006, Indian School of Mines (ISM), Dhanbad, Jharkhand, India, pp. 125-132.

Barboza, D. (2007). China Daily, http://www.nytimes.com/2007/12/07/world/asia/07mine.html (accessed on January 29, 2009).

Bose, A.N. (2008). Professor S.K. Bose Memorial Lecture. Proceedings of the 19th National Convention of Mining Engineering and National Seminar on Disaster in Mines, 10-11 March 2008, Indian School of Mines (ISM), Dhanbad, Jharkhand, India, pp. 1-4.

Buisson, D. (2004). www.mmsmag.co.za/script/article_72.htm (accessed on October 12, 2004).

Clean Coal Technology (2006). CCT Overview. Mine Safety Technology, Clean Coal Technologies in Japan. Guide Prepared by the New Energy and Industrial Technology Development Organization (NEDO) and the Japan Coal Energy Center (JCOAL), www.brain-c-jcoal.info/cctinjapan-files/english/2_1A3.pdf (accessed on July 12, 2008).

CMC Ltd. (2007). Case studies on DynaMine, http://www.cmcltd.com/case_studies/mining/dynamine.htm (accessed on July 30, 2008).

CSIRO (1998). Australian Scientists Develop World's Largest Robot. *Science Daily*, May 6, 1998, http://www.sciencedaily.com /releases/1998/05/980506075506.htm (accessed on July 30, 2008).

Davitt, J. (2006). The Sago Mine Disaster: A Preliminary Report to Governor Joe Manchin III. July 2006, Buckhannon, West Virginia.

DeGaspari, J. (2003). Armchair Mining. Feature Article Memagazine, *Mechanical Engineering*, May 2003, www.memagazine.org/contents/current/features/armchair/armchair.html (accessed on July 30, 2008).

Delabbio, F.C., Dunn, P.G., Lynn, I. and Hitchcock, S. (2007). The Application of 3D CAD Visualization and Virtual Reality in the Mining and Mineral Processing Industry. Articles on Mining and Mineral Processing, www.hatch.ca/Mining_Mineral_Processing/Articles/3D_CAD_Visualization_Virtual_reality_in_MMP.pdf (accessed on July 16, 2008).

DGMS (2000). Recommendations of the Ninth Conference on Safety in Mines, 2-3 February 2000, New Delhi, India. DGMS (Tech) Circular No. 3, Dhanbad, Jharkhand, India.

DGMS (2007). Standard Note. Directorate General of Mines Safety (DGMS), Dhanbad, Jharkhand, India, http://www.dgms.gov.in/STANDARDNOTE.pdf (accessed on May 15, 2008).

Foxnews (2006). http://www.foxnews.com/printer_friendly_story/0,3566,180369,00.html (accessed on December 18, 2006).

Ghatak, G.P. (2008). Application of Information and Communication Technology in Management of Mine Disasters: A Systems Concept. Proceedings of the 19th National Convention of Mining Engineering and National Seminar on Disaster in Mines, 10-11 March 2008, Indian School of Mines (ISM), Dhanbad, Jharkhand, India, pp. 67-72.

Gurtunca, R.G. and Breslin, J.A. (2007). Recent developments in coal mining safety in the United States. Proceedings of the 20th International Mining Congress of Turkey (IMCET- 2007), 6-8 June 2007, Ankara, Turkey, http://www.cdc.gov/NIOSH/Mining/pubs/pubreference/outputid2605.htm (accessed on December 15, 2008).

Hill-Douglas, O. (2007). Mine Automation, CSIRO Solve. A CSIRO Review of Scientific Innovation for Australian Industry, Issue 11, May 07, www.solve.csiro.au/0507/article3.htm (accessed on May 20, 2008).

Jarosz, A. and Wanke, D. (2003). Use of InSAR for monitoring of mining deformations. Proceedings SP-550, Fringe 2003, ESA International Workshop on ERS SAR Interferometry, Frascati, Italy, 1-5 December 2003, http://earth.esa.int/fringe03/proceedings/ (accessed on December 15, 2008).

Katiyar, S.N. and Sinha, V.K. (2008). Mine Disaster Prevention: A Human Approach. Proceedings of the 19th National Convention of Mining Engineering & National Seminar on Disaster in Mines, 10-11 March 2008, Indian School of Mines (ISM), Dhanbad, Jharkhand, India, pp. 113-119.

Kloos, G., Guivant, J.E., Nebot, E.M. and Masson, F. (2006). Range based localization using RF and the application to mining safety. Proceedings of the 2006 IEEE/RSJ International Conference on Intelligent Robots and Systems, 9-15 October 2006, Beijing, China, pp.1304-1311.

Kohler, J.L. (2007). Current Mine Safety Disasters: Issues and Challenges. Statement before Committee on Health, Education, Labor and Pensions, United States Senate, www.hhs.gov/asl/testify/2007/10/t20071002d.html (accessed on June 15, 2008).

Konrad, Z. (1998). The concept of GEDIS, the Global Environmental Disaster Information System. The Second GDIN Conference, 11-14 May 1998, Mexico City.

Kumar, D. (2007). Emerging trends in Information Technology enabled Tools for Mining industries. Proceedings of the National Seminar on Emerging Trends in Mining & Allied Industries, National Institute of Technology (NIT), Rourkela, November 30-December 2, 2007, pp.44-56.

Kumar, D. and Bahuguna, P.P. (2008). Application of Information Technology Enabled Tools for Mine Safety & Disaster Management. Proceedings of the 19th National Convention of Mining Engineers & National Seminar on Disaster in Mines, March 10-11, 2008, Indian School of Mines (ISM), Dhanbad, Jharkhand, India, pp. 103-112.

Kumar, L., Viswakarma, J.K. and Singh, V. (2008a). Use of GIS in Underground Mining for Optimization. http://www.gisdevelopment.net/application/geology/mineral/mwf_7abs.htm (accessed on December 28, 2008).

Kumar, S., Bahuguna, P.P., Kumar, D. and Sarkar, B.C. (2008b). Recent applications of GIS & GPS in mining and allied fields. Proceedings of the National Seminar on Environmental Issue on Geotechnics & Mineral Industry, April 4-5, 2008, BIT Sindri, Jharkhand, India.

Loney, E. and Ross, R.E. (2001). Automation of Underground Diamond Core Drills. Presented at Prospectors and Developers Association of Canada (PDAC) as part of Canadian Diamond Drillers Association Meeting, 11-14 March 2001, Toronto, Canada.

MIRACO (2008). Virtual Reality at MIRARCO: Mining Innovation. www.rocscience.com/library/rocnews/fall2002/Miraco.pdf (accessed in May 2008).

Morris, A.C., Kurth, D., Huber, D., Whittaker, C. and Thayer, S. (2003). Case Studies of a Borehole Deployable Robot for Limestone Mine Profiling and Mapping. Proceedings of the International Conference on Field and Service Robotics, July 2003, Lake Yamanaka, Japan.

Morris, A.C., Silver, D., Ferguson, D. and Thayer, S. (2005). Towards topological exploration of abandoned mines. Proceedings of the IEEE International Conference on Robotics and Automation, April 2005, Barcelona, Spain.

MSHA (2009). FactSheets. http://www.msha.gov/MSHAINFO/FactSheets/MSHAFCT2.HTM (accessed on January 29, 2009).

Mukherjee, S.N. (2008). Accidents in Indian Coal Mines: A Review. Proceedings of the 19th National Convention of Mining Engineering and National Seminar on Disaster in Mines, 10-11 March 2008, Indian School of Mines (ISM), Dhanbad, Jharkhand, India, pp. 5-8.

Naik, H.K. and Murmu, B.N. (2006). Application of Global Positioning System in mining. Proceedings of the Indian Conference on Mine Surveying (ICMS-2006), September 8-9, 2006, Indian school of Mines (ISM), Dhanbad, Jharkhand, India, pp. 115-124.

Nath, R. (2008). Inundation in Mines – A Case Study. Proceedings of the 19th National Convention of Mining Engineering and National Seminar on Disaster in Mines, 10-11 March 2008, Indian School of Mines (ISM), Dhanbad, Jharkhand, India, pp. 37-40.

NIOSH (2006). Communication and Tracking, www.cdc.gov/niosh/mining/mineract/communicationsandtracking.htm (accessed on November 19, 2008).

Paul Longley, P., Goodchild, M., Maguire, D. and Rhind, D. (2005). Geographic Information Systems and Science. Second Edition. John Wiley & Sons and ESRI Press, 2005.

Raju, E.V.R. and Das, P. (2006). Global Positioning System (GPS): A potential tool for land degradation monitoring in Jharia coalfield. Proceedings of the Indian Conference on Mine Surveying (ICMS-2006), September 8-9, 2006. Indian School of Mines (ISM), Dhanbad, Jharkhand, India, pp. 149-153.

Reuters (2006). http://www.theaustralian.news.com.au/story/0,20867,20749177-1702,00.html (accessed on January 29, 2009).

Reuters (2007). Chronology: Coal Mine Disasters in Russia. March 20, 2007, http://uk.reuters.com/article/worldNews/idUKL1952464620070320 (accessed on July 10, 2008).

Roy, A.K. and Acharya, T. (2004). Information Technology in Mining and Electrical Load Forecasting. Information Technology Principles and Applications. Prentice Hall of India Pvt. Ltd., New Delhi, pp. 505-506.

Roy, S., Nutter, J.R. and Aldrige, D.M. (1988). Status of Mine Monitoring and Communications. *IEEE Transactions on Industry Applications*, **24(5)**, pp. 820-825.

Samanta, B.K. (2008). Mine Disaster Management Leading to a World Record. Proceedings of the 19th National Convention of Mining Engineering and National Seminar on Disaster in Mines, 10-11 March 2008. Indian School of Mines (ISM), Dhanbad, Jharkhand, India, pp. 53-59.

Semwal, D.P., Naithani, V., Pant, D.N. and Roy, P.S. (2008). Impact Assessment of Coal Mining on Land Use/Land Cover Using IRS-IA Satellite Data, http://www.gisdevelopment.net/application/geology/mineral/geom0006.htm (accessed on December 28, 2008).

Sethi, A. (2006). Disaster Death Trap. *Frontline*, New Delhi, **23(19),** September 23-October 06, 2006, http://www.hinduonnet.com/fline/fl2319/stories/20061006003102800.htm (accessed on May 18, 2008).

Terradaily (2007). China Pronounces 172 Miners Dead in Mining Disaster. Report dated September 6, 2007, www.terradaily.com/reports/China_pronounces_172_miners_dead_in_mining_disaster_999.html (accessed on July 30, 2008).

The Hindu (2006). Online Edition of India's Newspaper, September 08, 2006, http://www.thehindu.com/2006/09/08/stories/2006090809861300.htm (accessed on July 30, 2008).

Timesonline (2005). Times Online. February 15, 2005, http://www.timesonline.co.uk/tol/news/world/article514441.ece (accessed on July 30, 2008).

Venkateshwar, A. (2008). Scope for Application of GPS in Indian Coal Industry, http://www.gisdevelopment.net/application/geology/mineral/geom0011.htm (accessed on December 28, 2008).

Verma, K., Montu, A.K. and Anand, V.K. (2008). GPS Based Truck Dispatch System at West Bokaro Collieries. http://www.gisdevelopment.net/application/utility/transport/utilitytr0021.htm (accessed on December 28, 2008).

Wang, Z., Wenfeng, L. and Xuezhao, Z. (2006). Multimedia communication system for small coal mines. *Coal Science and Technology*, **33(6):** 2865-2868.

Management of Forest Fire Disaster: Perspectives from Swaziland

Wisdom M.D. Dlamini

1. INTRODUCTION

Forest fires (also known as 'wildfires') are extremely powerful and destructive phenomena which occur with significant frequency and intensity on many parts of the Earth. These fires are part of the Earth system and are an important ecosystem disturbance with varying return frequencies, resulting in landscape alteration and change as well as atmospheric changes on multiple time scales (Chuvieco and Kasischke, 2007). Fire is believed to be an ecological imperative in many of the indigenous, fire prone ecosystems of the world. These indigenous vegetation species are regarded as being fire adapted and dependant upon fire for their survival. Under normal circumstances, fire is therefore regarded as vital in maintaining the delicate balance of hundreds of millions of hectares of tropical and subtropical savannas and open forests, as well as coniferous forests of the temperate and the northern boreal zones that are quite well adapted to natural and even human-influenced fire regimes (Bond and Keeley, 2005). These ecosystems can, however, become more prone to forest fires due to invasive plant infestations, and human activities thus leading to substantial losses of bio-diversity (FAO, 2007). Such fires become a risk when their frequency or intensity destroys forests or vegetation beyond what is naturally admitted and threaten humans and their activities. Goldammer (2007) identifies several global issues and trends that are impacting the occurrence and consequences of forest fires on the environment and societies, and these include demographic changes, widespread poverty associated with unemployment, exurban migrations and land tenure conflicts, land-use change, expansion of the wildland-urban interface, climate change, and threats to human health, security and peace.

The socio-economic changes of our society have increased forest fire risk as experienced during the latter part of 20th century and first decade of the 21st century. Over this period, many regions of the world have and continue to experience a growing trend of excessive fire application in the forestry-agriculture interface, land-use systems and land-use change, and an increasing occurrence of extremely severe wildfires or 'megafires' (Goldammer, 2007). Due to the lack of consistent and coherent data on the statistics on the global occurrence and impact of forest fires, it is not possible to precisely determine the trend in the global number of such fires or the area burnt over long periods of time. However, evidence exists which indicates that there is an increase in the number of larger and more destructive fires (FAO, 2007). The high socio-economic and demographic changes experienced by many countries in many regions over the past few decades have been based in part upon the exploitation of natural resources and development of agro-based industries, both of which involve conversion of forests and intensification of land-uses (Murdiyarso et al., 2004). These wildfires cause widespread destruction and damage affecting economic sustainability and productivity for entire regions or countries. Recently, such fires have had an immense impact in areas such as Southern (Mediterranean) Europe, Southern and Eastern Africa, Australia, Southeast Asia, and Central and Northern Latin America. Using daily global observations from the Advanced Very

High Resolution Radiometers (AVHRR) on the series of National Oceanic and Atmospheric Administration (NOAA) meteorological satellites between 1982 and 1999, Carmona-Moreno et al. (2005) found that these regions always have a maximum of fire occurrence probability in a given trimester of the year and considered these as fire-prone areas-ecosystems. Similar patterns have also been confirmed by Chuvieco et al. (2008). The global distribution of fires detected by the MODIS sensor between January 2000 and December 2008 is shown in Fig. 1.

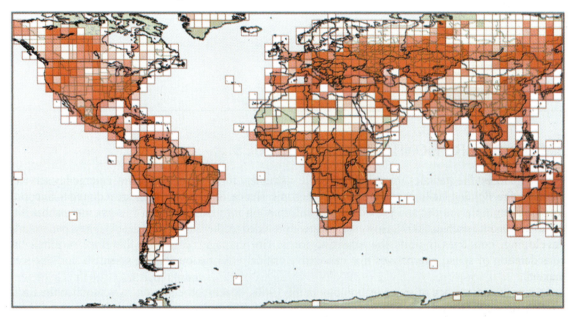

Fig. 1 Global active fire detections from MODIS for January 2000 to December 2008; dark red color in the figure indicates the highest number of fires detected (NASA/University of Maryland, 2002).

Historically, there have been some major forest fire disasters, which had considerable impacts on people and properties. Such include the October 1994 and October/November 1997 forest fires in Indonesia which affected around three million people and killed two hundred people, respectively (EM-DAT, 2008). The 1997 fires resulted in a haze that spread as far as the Philippines to the north, Sri Lanka to the west, and northern Australia to the south and an estimated 750,000 ha was affected. In 2007, forest fire killed more than 50 people in Greece. The forest fire records since the beginning of the twentieth century show many catastrophic forest fires affecting a large number of human populations all over the world. Table 1 shows the world's greatest forest fire disasters in terms of people killed. Sub-Saharan Africa is of exceptional note in this regard. Estimates show that about 17% of sub-equatorial Africa burns annually, accounting for 37% of the dry matter burned globally (Scholes et al., 1996); hence the African continent is often referred to as the "*Fire Continent*" (Pyne et al., 2004) followed at some distance by Australasia. The 2007 and 2008 fire seasons produced ravaging fires which uncharacteristically occurred in South Africa, Mozambique, Botswana and Swaziland.

This chapter highlights forest fire disasters and discusses forest fire management issues with a focus on the experiences from the kingdom of Swaziland. The role of remote sensing and geospatial technologies in the forest fire monitoring and management, possible early warning and mitigation of forest fires are also discussed.

Table 1. The ten most catastrophic forest fires in terms of people killed during 1980-2008 (EM-DAT, 2008)

Name of country	Year of fire occurrence	No. of people killed
Indonesia	1997	240
China	1987	191
Australia	1983	75
Greece	2007	67
Indonesia	1991	57
Nepal	1992	56
Mexico	1998	50
Mozambique	2008	49
Sudan	1998	47
Poland	1992	35

2. FOREST FIRE MONITORING TECHNIQUES

Long-term forest fire statistics are an indispensable instrument for forest management, emergency services planning, fire-fighting facilities and preventive measures. Hence, data and methods are required to integrate data from multiple sources and provide timely information for assessments of fire risk and probability (Chuvieco and Kasischke, 2007). However, as already alluded to, the lack of such data in many, particularly developing, countries inhibits the effective forest fire management. This therefore requires the consideration of sound alternatives that can derive critical information for fire scientists and decision makers.

Since the emergence of space technology in the 1960s, space-borne sensors now monitor the Earth continuously. Due to the large spatial and temporal variability in fire activity, satellite data provide the most useful means and the only practical way to monitor fire activity from a local to a global scale. There exist polar-orbiting and geostationary systems with full operational status and experimental systems providing systematic observations that have been used for the creation of long-term data fire mapping. In addition to providing fire data for use in long-term studies, these sensors provide synoptic snapshots of fire activity in near-real time to support operational fire management (Qu et al., 2008). Two general approaches are commonly used to monitor forest fires or biomass burning: post-fire burned areas mapping and active fires detection (Eva and Lambin, 1998). Post-fire burn detection is mostly done through measuring changes in surface reflectance before and after the fire mainly from red (0.65–0.70 µm) and near infrared (0.7–3.0 µm) information. This is based on the observation that burnt areas have a lower reflectance in the red and near infrared portions of the electromagnetic spectrum than healthy vegetation (Flannigan and Vonder Haar, 1986). In contrast, active fire detection by remote sensing systems is usually based on the detection of hot temperatures above normal environmental temperatures (threshold methods) and with respect to the background (contextual methods) and the detection of smoke plumes produced by fire emissions (San-Miguel-Ayanz et al., 2005). This is because the high temperature of fire enables the emission of thermal radiation with a peak in the middle infrared region, in accordance with Planck's theory of blackbody radiation. Therefore, active fire sensing is often done using middle infrared and also thermal infrared (usually around 3.7–11 µm) information from satellites (Dozier, 1981).

To date, a number of large-scale, multi-year fire datasets have been produced using observations acquired by various satellite-based sensors. Major long-term global records of active fires that have been generated include the Along Track Scanning Radiometer (ATSR) World Fire Atlas by the European

Space Agency (ESA) and the Advanced Very High Resolution Radiometer (AVHRR), Tropical Rainfall Measuring Mission's Visible and Infrared Scanner (TRMM VIRS) and Moderate Resolution Imaging Spectrometer (MODIS) by NASA. Geostationary fire monitoring has been undertaken using the GOES (WF-ABBA) and MSG SEVIRI (EUMETSAT Active Fire Monitoring) instruments. A number of these and other fire products are accessible through Web-based distribution systems, ranging from simple file distribution systems to complex visualization and search utilities using web GIS. Future systems are at advanced design stages and these include the NPP/NPOESS Visible Infrared Imagery Radiometer Suite (VIIRS) and sensors onboard the Global Monitoring for Environment and Security (GMES) Sentinel satellites and the provision of baseline high resolution fire observations for product validation should ensure the continuity of fire mapping and detection capabilities. The use of remote sensing for monitoring the distribution, frequency and impacts of wildfires is, therefore, a maturing scientific field which is gaining widespread applications in various countries of the world (Chuvieco and Kasischke, 2007).

3. CAUSES OF FOREST FIRE

It is basic scientific knowledge that for fire to occur, three elements are necessary for combustion, namely heat, oxygen and fuel and, in addition, the triangle of primary conditions that can affect extreme wildfire behavior consists of topography, fuels, and weather (Trollope et al., 2004). Hot, dry, and windy conditions are generally ideal for the rapid growth and spread of wildfires. Steeper slopes tend to increase the rate of fire spread. However, the determination of the causes of fires is an important undertaking that is necessary for the formulation of appropriate forest fire policies and management strategies.

It has been generally accepted and repeatedly quoted by bureaucrats that fires were caused by the El Nino Southern Oscillation (ENSO), and therefore, a natural disaster, which could not be prevented. However, a majority of fires are caused by people and most are lit deliberately to take advantage of the dry conditions to prepare land (or, as a tool in conflicts). Cyclical climate changes such as ENSO, similar to heat waves and droughts may increase the risk of these fire by providing the climatic conditions suitable for large forest fire episodes, such as the July 2007 Swaziland fire disaster (Dlamini, 2007), but it is hardly the underlying driver or cause. Recently, some researchers suggest the influence of solar activity as a possible cause of large forest fires (e.g., Gomesa and Radovanovic, 2008). In order to understand the topic fully, an appreciation of the difference in the characteristics of a *"natural fire"* and *"anthropogenic fire"* (i.e., fire originating from human intervention) is necessary. *Natural fires* typically occur seasonally and from a species diversity perspective and the more random the fire regime (vegetation age, time of year, weather, etc.), the more likely the fire is to actually maintain the biodiversity or indigenous flora and fauna of that area. These are commonly ignited by lightning in many African and other regions, but earthquakes, volcanoes and landslides are also known to start forest fires (Trollope et al., 2004).

Anthropogenic fires, on the other hand, result from the intervention of humans and purposeful or accidental ignitions are becoming increasingly frequent, leading to land degradation and loss of biodiversity. It is these types of wildfire which are the most damaging and which sow widespread destruction throughout southern Africa. These anthropogenic or unnatural fires can be split into unwanted, unmanaged wildfires and managed, or prescribed fires, which can be beneficial, provided these are implemented by experienced burners under carefully controlled conditions. The unmanaged fires are ignited through arson and accidents from sources such as cigarettes, matches and powerline sparks. The rains in the summer rainfall region, whilst late, are concentrated, leading to extraordinary growth of vegetation. In turn, this will lead to increased biomass or higher fire fuel loads. If these fuel loads are not reduced and there is a protracted dry season, extreme fire danger conditions will prevail. Over the winter rainfall area prevailing dry conditions coupled with high temperatures seem set to continue. Fuel loads can be very high, depending upon age of forest and seasonal rainfall. It is estimated that more approximately 40 to 95% of forest fires are caused by human intervention exceptions being the boreal regions of North America and Russia where lightning is the major cause (Tishkov, 2004; FAO, 2007). Based on domain knowledge, the key characteristic differences between natural and anthropogenic fires can be summarized as shown in Table 2.

Table 2. Key characteristics of natural and anthropogenic fires

Natural fires	*Anthropogenic fires*
The primary source of ignition is lightning.	The primary source of ignition is usually human intervention.
Due to the source of ignition, occurrence is most frequent during the rainy seasons.	Due to the source of ignition, frequency is usually higher and mostly occurring in the dry season (often protracted by late rains).
Typically high moisture content of the vegetation and accompanying rain contains fire to smaller patches.	High fire load due to the low moisture content of the vegetation results in large patches being burnt.
Indigenous vegetation is fire adapted and dependant on fire for biodiversity.	Radical fire spread through invasive alien plants, often with high resin or oil content.
A lower fire intensity and spread occurs in normal circumstances.	A high fire intensity (hot fire) and spread rate.

4. AN OVERVIEW OF SWAZILAND

The Kingdom of Swaziland, located in southern Africa, is 17,364 km^2 in extent straddling latitudes 25°40′ and 27°20′ South and longitudes 30°40′ and 32°10′ East (Fig. 2). The country is sandwiched between South Africa to the north, west and south, and Mozambique to the east. It is also endowed with tremendous natural diversity and complex terrain with elevation that decreases from an average of 1400 m above mean sea level on the west to below 100 m on the eastern part of the country resulting in four major eco-climatic regions (see Fig. 2), namely the Highveld, Middleveld, Lowveld and the Lubombo Plateau (Goudie and Price-Williams, 1983). Climatic variations within the country are largely controlled by the topography and there are four seasons within a year with December being mid-summer and June mid-winter. Mean annual rainfall also varies extensively from above 2000 mm per annum in the Highveld to below 500 mm in the Lowveld. Temperature variations also follow the altitudinal gradients, the Highveld being temperate and seldom hot, while the semi-arid Lowveld can record temperatures of up to 40°C during summer.

5. FOREST FIRES IN SWAZILAND

5.1 Policies and Practices

Together with thunderstorms and drought, forest fires are one of the most frequently occurring and pervasive natural hazards in Swaziland, causing severe damage to the country's economy as well as to the natural and human environment. Dlamini (2005, 2007) has recently investigated the fire management practices and achievements of Swaziland. In Swaziland, fire is recognized as a management tool both in the forestry and agricultural sectors and is often used to facilitate pasture regeneration and in clearing vegetation for farming and settlements (Dlamini, 2005). Most of the burning that takes place in the wooded areas and grasslands is aimed at improving grazing conditions (Dlamini, 2005). However, there is evidence that these fires are used recklessly and proper fire regimes for Swaziland are not yet fully known. The wildland-urban interface problem is also more commonplace now than in the recent past. Urban development towards wilderness areas in such major cities as Piggs Peak, Mbabane and Manzini and other towns adjacent to forested or vegetated areas has exposed people and property to forest fires and increased the risk of damage. The wildland-urban interface fires are more complex to fight, due to the mixture of different fuels (forests and other vegetation) and structures (buildings/settlements) in the interface. This requires concerted fuel load management efforts and the formulation of policies and relevant legislation in order to minimize the negative impacts of such fires. Several attempts have been made and continue to be implemented in both the public and private sectors.

Forest fires have traditionally constrained the development of forestry in Swaziland by hindering forest companies and the society in general from obtaining significant benefits from the investment made. To minimize losses due to fires, plantation forest companies have been designed with networks of fire breaks, which are burnt annually (especially between June and July) to provide a clean belt around the forest compartments (Dlamini, 2005). Some protected areas or national parks also implement early dry season burning policies in the form of fire belts and block burns to counter stray fires and to reduce fuel load, whilst at the same time facilitating the removal of moribund vegetation. This is usually done between the months of June and September. Even with these precautions, such protected areas are not spared from fires. One such protected area, Malolotja Nature Reserve in the north-eastern part of the

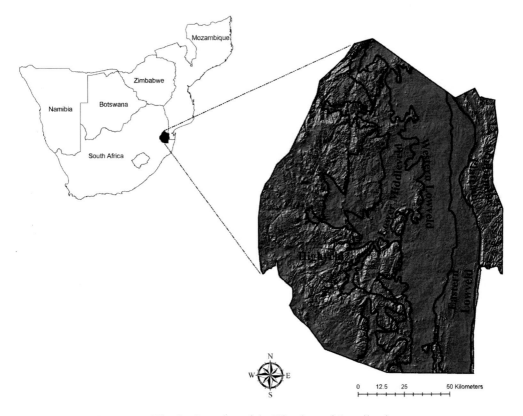

Fig. 2 Location of the Kingdom of Swaziland.

country, experiences regular dry season fires, most of which are a result of arson from poachers and neighboring communities. Lighting also occasionally triggers some of these fires within this protected area during the late winter/early summer season, but these are usually smaller in size and affect only smaller areas (personal observation).

The fire management policy framework of Swaziland—which in essence is not different from the other colonial and fire suppression policies generally adopted in Sub-Saharan Africa— has centered its attention on solutions to the fire problem that can only be effective in the short-term, no matter how sound and systematic their implementation is. For instance, an outdated Grassland Fires Act of 1955 stipulates that the burning of grass or other vegetation on land not cultivated or needed for cultivation requires the issue of a permit from the Director of Agriculture. The Act does not, however, apply to cultivated land such as plantation forests or sugarcane plantations and does not take into account the current land use systems, and the existing climatic and ecological conditions. National Fire Services Order No. 14 of 1975 establishes the Swaziland Fire and Emergency Services, which is responsible for fire-fighting and for attending to emergencies. The National Forest Policy of 2002, in turn, proposed that local Fire Prevention Units should be established in all chiefdoms and on all private farms. Such bodies are to be tasked with developing fire prevention and fire fighting strategies, in close co-operation with neighboring land users and the traditional and national authorities. Unfortunately, this was not implemented or converted into law for enforcement and, as such, these structures have not been established. The Swaziland National Trust Commission Act of 1972 also prohibits the willful or negligent cause of forest fire in a park or reserve. Similarly, the Game (Amendment) Act of 1991 prohibits willful or negligent cause of forest fires in protected areas. The Private Forests Act of 1951 contains few sections relating to forest fires and forest management. For instance, it establishes the legal need for fire belts around private forests. It is evident that the various pieces of legislation related to wild fire management are fragmented and outdated resulting in the continued negligent use of fire and numerous uncontrolled forest fires. The lack of effective policy and legislative and institutional frameworks for forest fire management is of serious concern.

5.2 Spatio-Temporal Patterns: Evidence from Remote Sensing Data

The MODIS instrument on-board Terra and Aqua satellites has 36 spectral bands spanning from the visible to the thermal infrared regions. The thermal bands in the 3 to 4 μm (e.g., bands 21 and 22) and 10 to 13 μm (e.g., band 31) wavelength regions of the electromagnetic spectrum are particularly useful, coupled with the higher sensor saturation level and better geolocation accuracy which enable more reliable detection of active fires compared to earlier and other sensors (Justice et al., 2002; Giglio et al., 2003). The standard fire hotspots detection algorithms developed for MODIS are also based on the conventional two-band thresholding method which uses both fixed and adaptive thresholds for fire detection (Kaufman et al., 1998; Justice et al., 2002; Giglio et al., 2003). The MODIS instrument is also used to map burned areas using an algorithm that takes advantage of the spectral, temporal, and structural changes using a change detection approach (Roy et al., 2005). It detects the approximate date of burning at 500 m by locating the occurrence of rapid changes in daily surface reflectance time series data and requires the consistently calibrated and processed MODIS data provided by the NASA MODIS land production system (Roy et al., 2008). The MODIS data are used to ascertain the spatial and temporal distribution of forest fires in Swaziland. The data were obtained from the Fire Information for Resource Management System (FIRMS), an integrated data system developed under NASA's Applied Sciences Program for easy access to MODIS active fire data to natural resources managers around the world (NASA/University of Maryland, 2002).

Figures 3(a) and (b) illustrate the spatial occurrence of fires and burnt areas in the country as determined by the MODIS sensor. The most evident clusters of fires are on the western part of the country largely from grassland fires and plantation forest fires where the Peak Timbers and Sappi Usuthu companies reportedly lost tens of millions of US dollars worth of property. The Peak Timber Company to the northeast lost an estimated 80% of the total forest area during the July 2007 disaster. It is in the same zone where several homesteads were destroyed by the same fires. The country's plantation forests have a very high fire hazard during the winter months, especially from July through to October. These fires result mainly from uncontrolled honey collection and arson fires due to strained social relations between the forest companies and the neighboring communities (Dlamini, 2005).

This geographic distribution of the fires illustrates the spatial pattern of the current burning practices, land use and the landscape in the country. Other clusters of fire activity are very evident to the east of the country mainly from sugarcane plantations. The steep and rugged topography of the country, more especially the Highveld and Middleveld (Remmelzwaal, 1993), can often generate the hot, dry, and windy environment needed for extreme fire behavior and accelerated fire spread rates. During the time of both 2007 and 2008 fires [Figs 3(c) and (d)], gusty and dry conditions were prevalent in the country creating perfect conditions for fires (Dlamini, 2007, 2009). The observations, therefore, indicate that these eco-climatic zones are riskiest in terms of forest fire disasters under such conditions and as such fire disaster management should pay particular attention to these areas. There is also an observed evidence of transboundary fires between Swaziland and South Africa particularly on the south-western part of the boundaries [Figs 3(c) and (d)]. It is important to observe from Figs 3(c) and (d) that both the 2007 and 2008 fires in Swaziland occurred during the same day as in neighboring South Africa where there were also reports of extensive damage and loss of human lives. This illustrates the transboundary nature and impacts of forest fires irrespective of the source of ignition.

A temporal analysis of both the MODIS active fire and burnt area data shows that the country had anomalously high fire incidents in 2007 and 2008 as compared to the previous years (Fig. 4). These are confirmations of the two extreme and disastrous fire events that the country experienced consecutively in the years 2007 and 2008. The 2007 fire was considered the worst fire experienced on record in the country resulting in it being declared a national disaster. More than 80% of the active fires in 2007 were concentrated in the plantation forests and sugarcane fields with widespread but scattered incidents in bushveld and grasslands (Dlamini, 2007). Although the 2008 reports from the Swaziland National Fire and Emergency Services have not been analyzed, the 2006 and 2007 reports indicated that a substantial amount (approximately 80%) of all fires reported were in the "Grass and Woodlands" category of fires and rest within the dwellings and industrial buildings category (Swaziland National Fire and Emergency Services, 2007).

The greatest chance of large and economically destructive fires occurs when the fuels have reached maximum quantity, are continuous over a large region, are in an extremely dry state, and are subject to a strong dry wind in an unstable atmosphere (Gill, 2005). Meteorological indications had earlier suggested that anomalous drought conditions would occur during the years 2007 and 2008. This proved true when the national government declared the drought a national disaster and since then, a lot of focus from the government was centered on providing food and water for drought victims unaware that other big and destructive disasters were looming (Dlamini, 2007). Research and results from several climate models suggest that El Nino-like conditions are likely to become more frequent in the future which subsequently means that fires in the near future will become the forerunners of larger disasters (Flannigan et al., 2005). Long drought periods contribute to the desiccation of living vegetation, dryness of large-diameter dead fuels on the ground and the elimination of the soil's effect on dead surface-fuel moisture thereby increasing the regional continuity of fuels and the chances that forest fires will spread more widely and be harder to

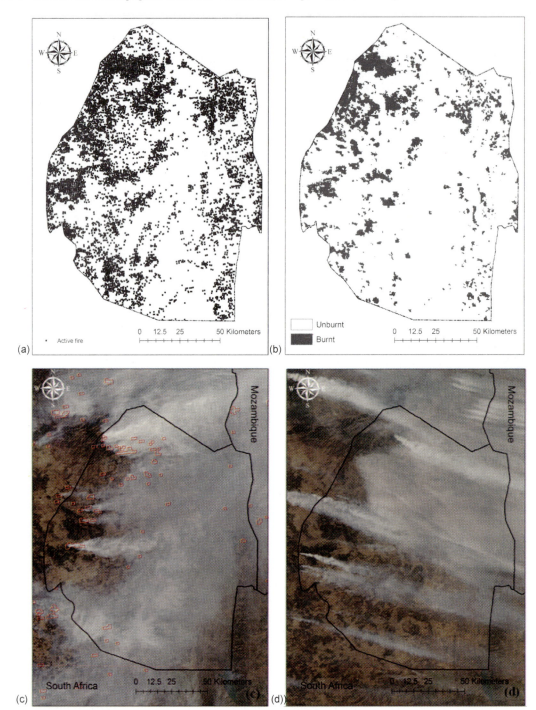

Fig. 3 MODIS active fires (**a**) and burnt area (**b**) for the period January 2003-November 2008, and the MODIS (Aqua) satellite images 28 July 2007 (**c**) and 31 August 2008 (**d**).

control (Gill, 2005). The risk of drought and fires in southern Africa has also been observed to increase dramatically in El Nino years (Anyamba et al., 2003; Riano et al., 2007).

Both the burnt area and active fire data from MODIS (Fig. 4) reveal that the fire season typically runs from May to November with a peak in August, which is an indicator of the country's temporal forest fire risk profile. Using the normalized difference vegetation index (NDVI) as an estimate of vegetation water stress, and hence fire risk, Dlamini (2007) reported a negative NDVI deviation trend from the 2001-2007 mean due to the persistence of the El Nino-like conditions which results in increasing forest fire risk. The persistent low values of NDVI obtained in the periods of 2006-2007 could, therefore, have been indicative of the looming disasters as manifested by the biggest fires ever recorded in the country. This illustrates the effects of the El Nino, especially considering the high risk associated with the phytophysiognomy of grasslands and plantation forests in the areas affected.

5.3 Socio-Economic and Environmental Impacts

Although the study of the use of fire is increasing in importance, research on the evaluation of impacts of fire is not common. This is due to the fact that fire generates a great variety of costs that not only have private effects, but also affect society and are difficult to quantify (de Mendonça et al., 2004). Fires are also an important ecological factor, having a number of effects on the terrestrial and atmospheric environments. Scholes et al. (1996) observed that in savannas, fire suppression can cause increase in woody plants with respect to grass plants (bush encroachment), while repeated late dry-season fires can lead to a decrease in woody vegetation and an increase in grasslands. Altering burning frequency, together with climatic and edaphic factors, has been observed to modify accumulation rates of carbon in biomass and soil and influence species composition and spatial distribution of forest ecosystems (Bond et al., 2003). Impoverished soils, in turn, produce less biomass and render natural regeneration less successful exposing the soil surface to excessive runoff and the erosion of upper layers. This may lower infiltration

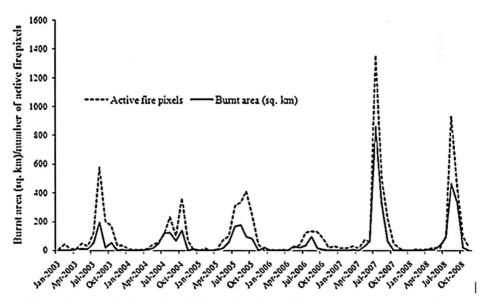

Fig. 4 Temporal variation in MODIS active fires and burnt area between January 2003 and November 2008 in Swaziland.

and keep the water table deeper, making the growth period shorter and eliminating drought susceptible vegetation and degrading wildlife habitat.

Apart from their contributory effects on the global earth system, the 2007 and 2008 forest fires and haze had regional, national and local impacts through multiple and widespread effects on health, transportation and tourism industries in Swaziland and neighboring countries, South Africa and Mozambique. During these fires (as shown in Figs 3(c) and (d)), visibility was adversely affected, bringing traffic to a standstill in some areas. The local press reported some tourists turning back at one of the country's border posts with South Africa after seeing the smoke from the forest fires. The net economic impacts were therefore significant. As a result of the observed health impacts of forest fires, the World Health Organization (WHO), in conjunction with the United Nations Environment Program (UNEP), the World Meteorological Organization (WMO) and Institute of Environmental Epidemiology (Singapore) published the first guidelines for forest fire events (Schwela et al., 1999). The extent and frequency of forest fires discussed in the preceding section point to the possible pattern of emissions from the Swaziland forest fires and the ecological and socio-economic impacts that these might have. Although not quantified in this analysis, based on the observed spatial and temporal patterns, it is important to discuss the possible impacts of such fires.

The social and economic effects of forest fires characteristically manifest through property loss (e.g., buildings, personal property, and timber), fire-fighting costs, injuries, and loss of life. Millions of dollars worth of forest plantations have been lost over the years as a result of forest fires in Swaziland. The July 2007 and August 2008 wildfires put at stake the jobs of hundreds of workers and the profitability of the plantation companies, Peak Timbers and Sappi Usuthu and other smaller companies. Reports from the Swaziland government indicated that some homesteads lost entire homes (169), property, food and cattle, goats and chickens in the fires including the loss of two lives, more than a dozen injuries and a total of 938 people affected (Dlamini, 2007; MORDYA, 2007). The plantation forest industry lost an estimated R465 million (US$45 million) of forests, 20,280 ha of plantation forest, mainly pine and eucalyptus (for sawlogs and pulpwood) comprising 19,000 ha and 1,240 ha of forests planted with pine and eucalyptus, respectively. In addition to the direct losses, the livelihoods of 21,100 people were put at risk resulting from the 728 direct and 4,368 indirect job losses incurred (Godsmark, 2007). In addition to factors that influence fire behaviors such as fuels, weather, and topography, the location and design of infrastructure and the construction materials used are critical in determining the vulnerability of communities to forest fires. These factors make the houses built in rural Swaziland particularly vulnerable to forest fires.

Another large wildfire event in late August/early September 2008 resulted in the loss of more plantation forest and more than 100 homesteads were affected including the loss of one life and several injuries. The widespread smoke from these fires is evidence of the amount of emissions released into the atmosphere and it covered a large portion of the country resulting in intense air pollution and reducing visibility in many areas. The 2008 season in most parts of southern Africa was also devastating with other countries such as Botswana experiencing fires covering three million hectares; South Africa and Swaziland losing thousands of hectares of valuable timber plantations and hundreds of thousands of hectares of grazing grounds, bushveld and savannas. These fires also ravaged settlements and other property damages estimated in the hundreds of millions of Emalangeni (Rands). Two human fatalities and countless injuries were also reported in the 2007 inferno which was eventually declared a national disaster on August 1, 2007 by the country's Prime Minister. The 2008 event resulted in one fatality and more than 100 homesteads were destroyed in Swaziland. All the impacts are not known yet, but it is evident from Figs 3(c) and (d) that there were significant emissions from the smoke.

The use of fire in Swaziland also generates negative externalities for society, among which are damages identified with CO_2 emissions into the atmosphere, genetic resource losses, and human health damage provoked by smoke. Other negative effects that were experienced included traffic/transportation delays, power failures, and recreational losses. The health impacts of biomass smoke episodes have been extensively reviewed (Arbex et al., 2004) and show the adverse health effects associated with particles from combustion of biomass including wood, other vegetation, animal wastes, sugar cane, etc. Since particulate matter produced by incomplete combustion of biomass is mainly less than 1 μm in aerodynamic diameter, both PM10 (particulate matter with an aerodynamic diameter less than 10 μm) and PM2.5 (aerodynamic diameter less than 2.5 μm) concentrations increase during air pollution episodes caused by vegetation fires (Arbex et al., 2004), which can also be transported to great distances. Nearly 200 distinct organic compounds have been identified in the wood smoke aerosol from biomass burning, including polycyclic aromatic hydrocarbons which are known to be linked to increased mortality and morbidity in susceptible persons, and increased risk of hospital and emergency admissions (Schwela et al., 1999).

In Swaziland, as in much of the world, patterns of development are superimposed on patterns of forests and other vegetation in ways that may amplify the consequences of forest fires. In the Lowveld, for example, agricultural growth and sugarcane expansion are often much greater at low elevations and so is the case in the higher, grassland and plantation-forested elevations of the Highveld. Given that Swaziland's population density is increasing in high-risk areas, additional infrastructure investment may fail to offset the increased danger. For example, fire-fighting resources are already diverted to protecting structures in high population-density zones at the expense of the capacity to control the growth in fire perimeter, resulting in larger fires. If present development trends continue, the economic impact of additional forest fires could very well be substantial. While the acceleration in settlements into areas with fire-prone vegetation is an important driver of this upward trend, there are suggestions of the observed increases resulting from a combination of climate change, land use practices and demographic factors (Dlamini, 2007, 2009). Even with no augmentation of fire-fighting mechanisms, fire-fighting costs will also increase as climate change accelerates the development of fires to higher dispatch levels where more expensive resources (e.g., water tankers and bulldozers) will be routinely utilized. Added to this are the possible extensive losses of urban infrastructure (e.g., telecommunication, water, electricity, and transportation systems), the costs of which are likely to be borne largely by local and national government and finally translating to individual costs.

6. MONITORING OF FOREST FIRE AND EARLY WARNING SYSTEM

The efficient management of disasters currently postulates a more generalized as well as integrated approach. The social intervention in such cases requires the relevant application of acquired scientific knowledge and technological achievements based on rational design and adequately organized operational actions. Public and operational agencies involved in disaster management have long recognized that many critical problems faced on an operational level are of endogenously spatial nature. For this reason, spatial information and data constitute an important framework of analysis when dealing with disaster management problems. In the case of forest fires, the need for early fire detection varies according to numerous factors.

Similar to other disasters, forest fires constitute a threat when they occur on either fairly populated regions or areas of high ecological or economic value. Therefore, the urgency for forest fire detection and fire-fighting changes according to the nature of the fire. For instance, natural fires need not be extinguished but can only be extinguished when they are considered a threat to human assets. Human-caused fires, on the contrary, are often extinguished as soon as possible implying that early detection of these fires is clearly aimed at fire-fighting. Therefore, it is assumed that early detection is only needed when fire-fighting or fire control resources are available. An analysis of the requirements of the fire suppression community for early fire detection in Europe resulted in a maximum detection time of 15 min from the start of the fire (INSA, 2000), thereby indicating that the value of the information on fire detection decreases according to a negative exponential curve (San-Miguel-Ayanz et al., 2005). This can be easily explained by the fact that fires are usually easy to extinguish at an early stage; once a fire has reached a fairly large size, operations for fire-fighting become very complicated and the control of the fire depends largely on the meteorological conditions that determine fire spread. In sparsely populated areas, where fires are not extinguished, fire detection is only needed for monitoring environmental impacts. Early detection is therefore often not necessary in such areas.

The continuous monitoring of forest fires involves the observation of active fires and forest fire-causing processes (fuel types and fuel conditions), forest fire risk assessment, mapping of burnt area and fire effects and countries with significant forest fire activity have developed ground- and air-based monitoring networks (San-Miguel-Ayanz et al., 2005; Chuvieco and Kasischke, 2007). However, due to economic and technological difficulties, developing countries such as Swaziland are still lagging behind in taking advantage of such technologies. The emergence of satellite remote sensing provides opportunities for the continuous and large-scale monitoring of forest fires, which may overcome the logistical and financial constraints of ground-based and air-borne observations. Geographic information systems (GIS) and remote sensing technology have been developed in such a way that spatial information is stored and efficiently retrieved and modeling techniques are appropriately embedded to support decision making and operational needs. The added value of GIS technology usage in managing emergencies is directly connected to the benefits expected from the exploitation of such technologies designed for supporting decision-making related to the geographical space, especially in the case of the operational field that intensely needs to make important decisions of spatial nature.

Remote sensing instruments on polar-orbiting and geostationary satellites allow forest fire observations at a broad range of spatial and temporal scales. These satellite-based fire and thermal anomaly detection systems are indispensable for both research and operational use. Satellite instruments that can simultaneously utilize 3.9 µm and 11 µm channels can be used for fire detection due to the 3.9 µm channel's strong thermal sensitivity even if only a small portion of the pixel is covered by fire (Matson and Dozier, 1981). For a long time, including the present, operational systems have been using data from the NOAA Advanced Very High Resolution Radiometer (AVHRR) and GOES (Geostationary Operational Environmental Satellite), including a few others such as the DMSP-OLS (San-Miguel-Ayanz et al., 2005). Remote sensors (e.g., MODIS and MSG-SEVIRI) have been implemented and are operationally used to complement the human visual detection available in most of the ground detection systems such as watch towers, as operational fire detection still relies on human surveillance. However, there is still a considerable gap between research and operation.

Few examples exist in which satellite systems are integrated into operational fire detection and alert systems. Two of these examples are the FIRMS and the Advanced Fire Information Service (AFIS) systems, which are also useful for forest fire management in Swaziland. Both are satellite-based real-time observation and alert systems for forest fire management. The FIRMS system is a result of a collaboration effort between the University of Maryland, NASA, FAO and Conservation International. Within the FIRMS system (Fig. 5), global MODIS data are processed in near-real-time by the MODIS Rapid Response System, active fire locations are generated using a global thermal anomalies algorithm (MOD14) and these fire coordinates are ingested into FIRMS from where they are made available in a number of formats including a web mapping interface called *Web Fire Mapper* (a new open source

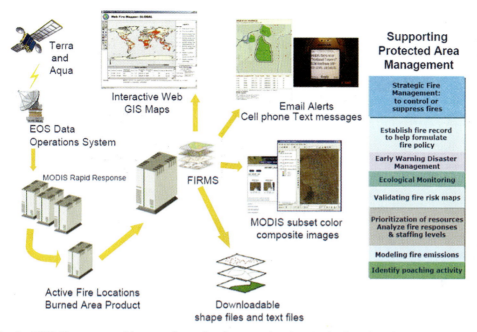

Fig. 5 FIRMS system architecture for active fire mapping (NASA/University of Maryland, 2002).

software-based system called *Firefly* is being tested), as email alerts, text files, ESRI shape files and through files that can readily be viewed in virtual globes such as Google Earth, ArcGlobe and NASA World Wind. FIRMS contributes to the Global Terrestrial Observing System's (GTOS) Global Observation of Forest and Land Cover Dynamics (GOFC/GOLD) program, which seeks to promote improved access to, and use of, satellite data products.

The AFIS system is a joint initiative of South Africa's Council on Industrial Research (CSIR) and the electricity company Eskom. The *AFIS* system, which is based on the combined use of the MODIS and MSG SEVIRI sensors onboard of the NASA and ESA satellites respectively, works operationally in South Africa and covers the whole of Swaziland. Both the *FIRMS* and *AFIS* are stable, reliable, and send email and mobile phone SMS alerts when fire detection occurs (Fig. 6). These systems permit the coverage of a very large area at a relative low-cost, with combined detection rates of up to 70%, and can enable airborne surveys to be directed to critical areas (Frost and Vosloo, 2006). AFIS and FIRMS are an example of how new technology can be incorporated into operational systems for improved forest fire monitoring as currently being utilized (although at a limited scale) in Swaziland.

As sensors for forest fire detection and monitoring improve, and information and communication technologies advance, the capabilities to ingest, process and transmit large amounts of data in (near) real-time are becoming a reality. The development of algorithms for active fire detection and burnt-area mapping and monitoring is also fairly advanced, which brings automatic forest fire surveillance from satellites closer to being realized. It is only required that future satellite missions such as the NPP/NPOESS Visible Infrared Imagery Radiometer Suite (VIIRS), NPP/NPOESS Visible Infrared Imagery Radiometer Suite (VIIRS) and sensors onboard the Global Monitoring for Environment and Security (GMES) provide data with enough frequency to minimize the re-visit time to forest fire risk areas. San-Miguel-Ayanz et al. (2005), Chuvieco and Kasischke (2007), and Qu et al. (2008) provide good overviews of the capabilities and limitations of remotely-sensed data applications for fire emergency management.

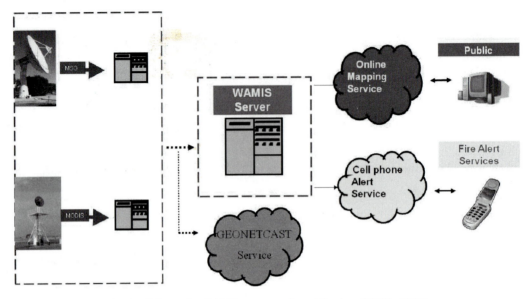

Fig. 6 Schematic of AFIS sensor web architecture (CSIR, 2008).

One of the common shortcomings of space-borne instruments is the lower limits for fire sizes that they can detect and the incapability of detecting fires through clouds or thick smoke. Polar-orbiting satellites, on one hand, cannot see short-duration fires that take place between the satellite overpasses; on the other hand, geostationary satellites have better temporal resolution, but worse spatial resolution and have difficulties when scanning at high viewing angles (Dlamini, 2009). Moreover, caution needs to be taken and considered with remotely-sensed fire observations because the daytime radiance signal at 3.7 mm may be a combination of both emitted thermal, hot background surfaces and a reflected solar radiation which may produce false fire alarms and/or saturate the 3.7 mm channel (Di Bisceglie et al., 2005; Gao et al., 2006). Active fires from polar-orbiting satellites also represent a limited temporal sample due to the satellites' restricted overpass frequency, coupled with the diurnal fire cycle (Hyer et al., 2007).

Despite the advances in forest fire detection and monitoring technology, there is a need for continued coordination to continue improving access to near real-time and archived fire data and its use for resource managers, policy-makers and the scientific community, and to secure long-term fire observing systems from which developing countries such as Swaziland can immensely benefit. The fire theme of Global Observation of Forest Cover/Land Cover Dynamics (GOFC-GOLD) is actively working on the harmonization of remote sensing and in situ observations of forest fires including the development of capabilities for periodic global assessments of fire, capacity building for space-based fire observations, development of a global early fire warning system and a global fire monitoring network from geostationary satellites. In accordance with the United Nations International Strategy for Disaster Reduction (UN-ISDR), the Germany-based Global Fire Monitoring Center (GFMC), the Global Wildland Fire Network and the UN-ISDR Wildland Fire Advisory Group (both of which are coordinated by the GFMC), are also working on improving fire management capacity around the world. Other initiatives include Fire Management Actions Alliance established in May 2007 during the 4th International Wildland Fire Conference in Seville, Spain and based in Rome, Italy under the auspices of the Food and Agriculture Organization.

7. STRATEGIES FOR MITIGATING FOREST FIRE HAZARDS

It seems evident that Swaziland will most likely face the same catastrophe with certain periodicity in the future, unless even better mitigation strategies are established. However, the main problem in applying efficient mitigation measures is the size of the affected area and the cost involved. On the other hand, an important part of the budget has still to be destined to suppression of fires while new policy results are not yet visible, and the real causes are eradicated. It is clear that there are many threats to Swaziland's people and assets as well as many challenges and potential opportunities for the future. The establishment of the Swaziland Fire and Emergency Services and improved planning procedures in the twentieth century, as well as a continuous increase in fire pre-suppression and suppression expenditures did stabilize the burned area in 'acceptable' levels until recent times. The number of forest fires has however remained high, hence denoting failure in preventing fire occurrence because no fire management program can be economically effective when the number of fires is so unregulated. It adds that in the 2007-2008 periods, most of the disastrous fires in Swaziland were of unknown origin, which constitutes an obvious obstacle to defining sound mitigation strategies.

Swaziland's national fire policies have mainly focused on fire suppression and paid very little attention to fuel management thus resulting in an increase of accumulated fuel load and forest vulnerability to fire, setting the scene for larger and more damaging fires even in the future. It is a fact that fire control technology can tackle just a small fraction of the potential intensity of a fire (Gill, 2005). In the Swaziland fires of 2007 and 2008, the limitations of fire fighting were further aggravated by the extent of the wildland-urban interface and capability to halt the spread of the forest fires was seriously diminished when the rough topography and dry, abundant fuel concurred with extreme weather. Fuel management programs, if properly designed and implemented, can reduce the severity of forest fires under any weather and increase the weather thresholds for effective fire fighting (Fernandes and Botelho, 2003). This is the strategy that has been used in the protected areas of the country. Fire suppression and fuel management can, therefore, be thought of as two sides of the same coin. Allocation of the budget between these complementary activities in the framework of a balanced and successful fire management policy is the major challenge. Incoll (1994) indicated that the total costs of fire management and damage caused by wildfire is minimized when the fuel management expense exceeds the cost of fire suppression by a factor of three. However, the above basic question has not received much attention in Swaziland in the past and requires additional research.

Within this context, the interest of having better fire prevention and assessment tools should be emphasized. Fire risk assessment is a critical component of fire prevention, since pre-fire planning resources require objective tools to monitor when and where a fire is more prone to occur, or when it will have more negative effects (Chuvieco et al., 2009). As happens in connection with other kinds of natural risks, forest fire hazard assessment can take place starting within three different conceptual frameworks: static risk assessment, dynamic risk assessment and real time risk assessment (Fiorucci et al., 2005). This involves the evaluation and the risk distribution over the geographical space is carried out on the basis of static or dynamic information (e.g., topography, vegetational cover) and taking into account the main variables involved in the considered process (i.e., fire occurrence, weather conditions and land use). The purpose of forest fire risk assessment could be that of planning the sizing and location of the different kinds of resources and infrastructure necessary to manage forest fire risk over a wide area. This is particularly important in this era of global change, where climate change continues to exacerbate the severity and impacts of forest fires. Unfortunately, forest fire risk assessment is another aspect of forest fire management that has not been addressed in Swaziland. Krawchuk et al. (2009) propose that the expected changes in forest fire activity due to weather conditions need to be integrated with forest landscape models to further understand the complex outcome of the various interacting factors.

One of the most effective approaches to safeguard against the inherent risk of a forest fire disaster is to decrease the community's vulnerability through appropriate mitigation measures. As stated earlier, the current mitigation measures used include systems of fire breaks within plantations and protected areas and controlled block burns. However, an effective disaster mitigation program must include aspects of public awareness, capacity building, and multi-disciplinary collaboration. Public education and safety promotion programs are key components of a successful disaster response which must be developed. In order for the public to be prepared even to a minimal standard, the forest fire disasters that the community may face must be common knowledge. Additionally, the community must be made aware of simple actions that can reduce an individual's personal vulnerability. Unfortunately such measures are least implemented in Swaziland except for ad hoc radio broadcasts during the forest fire season.

Forest fire prevention and mitigation requires knowledge about the weather, ecology, and terrain of an area, the infrastructure for monitoring, road networks, the ability to mobilize and train human resources, and appropriate information, communication and telecommunications infrastructure, all of which are scarce in sub-Saharan Africa. A few African countries, such as Ethiopia and South Africa, have fire danger warning systems. Although the country has relatively good infrastructure, Swaziland does not currently have a fire danger system in place and this needs to be developed as a matter of urgency given the increase in the number and severity of forest fires. The use of satellite data to monitor burnt areas for purposes of estimating biomass-related greenhouse gases has been introduced in some African countries. In support of this, the Southern Africa Fire Network (SAFNet), of which Swaziland is a member, provides a framework for exchanging fire management information and for capacity building, with the emphasis on the use of geo-spatial information technologies. The Global Fire Monitoring Centre (GFMC) also covers forest fires in Africa. The FIRMS and AFIS databases can and will then be used to undertake seasonal assessment of fire risks for different parts of the country, which could be used to develop mitigation strategies in future.

While unprecedented amounts of fuel have accumulated, the population of Swaziland has also shifted temporally and spatially. More and more people are living in or near areas prone to forest fires. In recent decades, the population has also become more dispersed thus increasing the number of people living in heavily vegetated areas where wildlands meet urban development (i.e., wildland/urban interface). The result is increased risk to more homes and other structures, and together with the accumulation of fuel and development in hazardous areas, this poses particular challenges for the insured and insurers as well as the government agencies responsible for fire prevention, mitigation and suppression.

8. CONCLUSIONS AND RECOMMENDATIONS

Fire is one of the disasters causing threats to the forests and the ecosystem from a local level to a global level. Forest fires have adverse effects on the natural environment and humans such as experienced in Swaziland in the disastrous forest fires in the years 2007 and 2008. As revealed from the remote sensing data, both the 2007 and 2008 extreme forest fires occurred at the same time with the South African fires, pointing to the need for a coordinated international approach in the management of forest fires.

In a broad perspective, the challenges and their proposed solutions to forest fire management in Swaziland generally fit into two categories. The first category consists of socio-ecological challenges associated with the accumulation of fuel load and population growth in the areas prone to forest fire. The solutions to those challenges involve mitigating potential losses through increased understanding of fire behavior, public education, improved national and interagency collaboration, capacity building, fire-safe building codes, land use planning, and investment in modern early warning and monitoring technologies including geospatial technologies. The second category consists of the risk decision challenges insurers face in underwriting properties exposed to the forest fire hazards. The solutions to these challenges include: developing and implementing appropriate underwriting guidelines; measuring and managing the aggregate amount of forest fire exposure in an insurer's book of business; managing the geographic distribution of exposures to prevent excessive concentration in any single area or contiguous areas prone to forest fires; and educating agents and the insured about loss mitigation. In both the categories of challenges, Swaziland's newly established Disaster Management Agency should play a key role in ensuring that the proposed solutions are implemented.

Forest fire risk assessment in Swaziland has not been undertaken. This needs to be done as a matter of urgency using remotely sensed data and other spatial data to better quantify values and risks as well as suppression capability. The obtained forest fire risk maps can then be used by the Disaster Management Agency and other decision-makers in making tactical decisions on prioritizing prevention measures. These maps can also give an indication of forest fires on a long-term basis and they can be utilized by decision/policy-makers to protect areas prone to forest fires in line with the Hyogo Framework for Action 2005-2015 (HFA).

REFERENCES

Anyamba, A., Justice, C.O., Tucker, C.J. and Mahoney, R. (2003). Seasonal to interannual variability of vegetation and fires at SAFARI 2000 sites inferred from AVHRR time series data. *Journal of Geophysical Research*, **108(D13)**, 8507, doi: 10.1029/2002JD002464.

Arbex, M.A., Cancãdo, J.E.D., Pereira, L.A.A., Braga, A.L.F. and Saldiva, P.H.N. (2004). Biomass burning and health effects. *Jornal Brasileiro de Pneumologia*, **30:** 158-175.

Bond, W.J. and Keeley, J.E. (2005). Fire as a global 'herbivore': the ecology and evolution of flammable ecosystems. *TRENDS in Ecology and Evolution*, **20(7):** 387-394.

Bond, W.J., Midgley, G.F. and Woodward, G.I. (2003). What controls South African vegetation—climate or fire? *South African Journal of Botany*, **69:** 79-91.

Carmona-Moreno, C., Belward, A., Caperan, P., Hartley, A., Malingreau, J.P., Antonovskiy, M., Buchshtaber, V. and Pivovarov, V. (2005). Global fire calendar probability maps from the analysis of global burned surfaces time series (1982-1999). *In:* J. De la Riva, F. Pérez-Cabello and E. Chuvieco (editors), Proceedings of the 5th International Workshop on Remote Sensing and GIS Applications to Forest Fire Management: Fire Effects Assessment, Universidad de Zaragoza, Zaragoza, Spain, pp. 217-221.

Chuvieco, E, Giglio, L. and Justice, C. (2008). Global characterization of fire activity: Toward defining fire regimes from Earth observation data. *Global Change Biology*, **14:** 1-15.

Chuvieco, E. and Kasischke, E.S. (2007). Remote sensing information for fire management and fire effects assessment. *Journal of Geophysical Research*, **112**, G01S90, doi: 10.1029/2006JG000230.

Chuvieco, E., Aguado, I., Yebra, M., Nietoa, H., Salas, J., Martín, M.P., Vilar, L., Martínez, J., Martín, S., Ibarra, P., de la Riva, J., Baeza, J., Rodríguez, F., Molina, J.R., Herrera, M.A. and Zamora, R. (2009). Development of a framework for fire risk assessment using remote sensing and geographic information system technologies. *Ecological Modeling*, doi:10.1016/j.ecolmodel.2008.11.017.

CSIR (2008). AFIS User Guide. http://divenos.meraka.csir.co.za/afis/AFISUserGuide.pdf (accessed on 10 January 2009).

de Mendonça, M.J.C., Vera Diaz, M.C., Nepstad, D., da Motta, R.S., Alencar, A., Gomesa, J.C. and Ortiza, R.A. (2004). The economic cost of the use of fire in the Amazon. *Ecological Economics*, **49:** 89-105.

Di Bisceglie, M., Episcopo, R., Galdi, C. and Ullo, S.L. (2005). Constant false alarm rate in fire detection for MODIS data. Proceedings of the IEEE International Geoscience and Remote Sensing Symposium (IGARSS), 25-29 July 2005, Seoul, Korea, pp. 5717-5720.

Dlamini, W.M. (2005). Fire situation in Swaziland. Proceedings of the 5th SAFNet Workshop on Towards Meeting Fire Management Challenges in Southern Africa, 9-13 August 2004, Mangochi, Malawi, pp. 24-26.

Dlamini, W.M. (2007). A review of the July 2007 Swaziland fire disaster using GIS and MODIS data. PositionIT, September/October 2007; pp. 61-65.

Dlamini, W.M. (2009). Characterization of the July 2007 Swaziland fire disaster using satellite remote sensing and GIS. *Applied Geography*, doi:10.1016/j.apgeog.2008.10.007.

Dozier, J. (1981). A method for satellite identification of surface temperature fields of sub pixel resolution. *Remote Sensing of Environment*, **11**: 221-229.

EM-DAT (2008). Wild Fire - Data and statistics. http://www.preventionweb.net/english/Wild%20Fire%20-%20Hazards%20-%20PreventionWeb_files/Wild%20Fire%20-%20Hazards%20-%20PreventionWeb.net (accessed on 16 January 2009).

Eva, H.D. and Lambin, E.F. (1998). Remote sensing of biomass burning in tropical regions: sampling issues and multisensor approach. *Remote Sensing of Environment*, **64**: 292-315.

FAO (2007). Fire Management — Global Assessment 2006: A Thematic Study Prepared in the Framework of the Global Forest Resources Assessment 2005. FAO Forestry Paper 151, Food and Agriculture Organization of the United Nations, Rome, Italy.

Fernandes, P.M. and Botelho, H.S. (2003). A review of prescribed burning effectiveness in fire hazard reduction. *International Journal of Wildland Fire*, **12**: 117-128.

Fiorucci, P., Gaetani, F., Minciardi, R. and Trasforini, E. (2005). Natural risk assessment and decision planning for disaster mitigation. *Advances in Geosciences*, **2**: 161-165.

Flannigan, M.D., Amiro, B.D., Logan, K.A., Stocks, B.J. and Wotton, B.M. (2005). Forest fires and climate change in the 21st century. *Mitigation and Adaptation Strategies for Global Change*, **11**: 847-859.

Flannigan, M.D. and Vonder Haar, T.H. (1986). Forest fire monitoring using NOAA satellite AVHRR. *Canadian Journal of Forest Research*, **16**: 975-982.

Frost, P. and Vosloo, H. (2006). Satellite-based early warnings of fires: Reducing fire flashovers on transmission lines. *ESI Africa*, **2**: 48-51.

Gao, B.C., Xiong, X. and Li, R.R. (2006). A study of MODIS fire detecting channel centered at 3.95 µm. Proceedings of the IEEE International Geoscience and Remote Sensing Symposium (IGARSS), 31 July-4 August 2004, Denver, Colorado, pp. 1102-1104.

Giglio, L., Descloitres, J., Justice, C.O. and Kaufman, Y. (2003). An enhanced contextual fire detection algorithm for MODIS. *Remote Sensing of Environment*, **87**: 273-282.

Gill, A.M. (2005). Landscape fires as social disasters: An overview of 'the bushfire problem'. *Environmental Hazards*, **6**: 65-80.

Godsmark, R. (2007). The Impact of the 2007 Plantation Fires on the SA Forestry & Forest Products Industry. http://www.forestry.co.za/uploads/File/forest/forest_protection/Fire%20Damage%202007.ppt (accessed on 12 November 2008).

Goldammer, J.G. (2007). Vegetation fires and the Earth system: Trends and needs for action. *iLEAPS Newsletter*, **3**: 21-22.

Gomesa, J.F.P. and Radovanovicc, M. (2008). Solar activity as a possible cause of large forest fires — a case study: Analysis of the Portuguese forest fires. *Science of the Total Environment*, **394**: 197-205.

Goudie, A.S. and Price-Williams, D. (1983). The Atlas of Swaziland. The Swaziland National Trust Commission Occasional Paper No. 4, Swaziland National Trust Commission, Lobamba, 90 pp.

Hyer, E.J., Kasischke, E.S. and Allen, D.J. (2007). Effects of source temporal resolution on transport simulations of boreal fire emissions. *Journal of Geophysical Research*, **112**, D01302, doi:10.1029/2006JD007234.

Incoll, R. (1994). Asset protection in a fire prone environment. *In:* Fire and Biodiversity: The Effects and Effectiveness of Fire Management. Footscray, Melbourne, Australia, 8-9 October 1994, 9 pp.

INSA (2000). FUEGO Instrument Design, Prototype, Construction and Validation. FUEGO 2 Final Report, INSA Publications, Madrid, Spain.

Justice, C.O., Giglio, L., Korontzi, S., Owens, J., Morisette, J.T., Roy, D., Descloitres, J., Alleaume, S., Petitcolin, F. and Kaufman, Y. (2002). The MODIS fire products. *Remote Sensing of the Environment*, **83**: 244-262.

Kaufman, Y.J., Justice, C.O., Flynn, L.P., Kendall, J.D., Prins, E.M., Giglio, L., Ward, D.E., Menzel, W.P. and Setzer, A.W. (1998). Potential global fire monitoring from EOS-MODIS. *Journal of Geophysical Research*, **103**: 32215-32238.

Krawchuk, M.A., Cumming, S.G. and Flannigan, M.D. (2009). Predicted changes in fire weather suggest increases in lightning fire initiation and future area burned in the mixedwood boreal forest. *Climatic Change*, **92**: 83-97.

Matson, M. and Dozier, J. (1981). Identification of subresolution high temperature sources using a thermal IR sensor. *Photogrammetric Engineering and Remote Sensing*, **47**: 1311-1318.

MORDYA (2007). Fire Disaster: Report. Ministry of Regional Development and Youth Affairs. *The Times of Swaziland*, 25th August 2007, p. 7.

Murdiyarso, D., Lebel, L., Gintings, A.N., Tampubolon, S.M.H., Heil, A. and Wasson, M. (2004). Policy responses to complex environmental problems: Insights from a science–policy activity on transboundary haze from vegetation fires in Southeast Asia. *Agriculture, Ecosystems and Environment*, **104**: 47-56.

NASA/University of Maryland (2002). MODIS Hotspot/Active Fire Detections Data set. MODIS Rapid Response Project, NASA/GSFC, University of Maryland, Fire Information for Resource Management System, http://maps.geog.umd.edu (accessed on 13 January 2008).

Pyne, S.J. (2001). The fires this time, and next. *Science*, **294**: 1005-1006.

Pyne, S.J., Goldammer, J.G., de Ronde, C., Geldenhuys C.J., Bond, W.J. and Trollope, S.W. (2004). Introduction. *In:* J.G. Goldammer and C. de Ronde (editors), Wildland Fire Management Handbook for Sub-Sahara Africa. Global Fire Monitoring Centre (GFMC), Freiburg, Germany, pp. 1-10.

Qu, J.J., Wand, W., Dasgupta, S. and Hao, X. (2008). Active fire monitoring and fire danger potential detection from space: A review. *Frontiers of Earth Science in China*, **2(4)**: 479-486.

Remmelzwaal, A. (1993). Physiographic map of Swaziland, Scale 1:250,000. FAO/UNDP/GOS Land Use Planning for Rational Utilization of Land and Water Resources Project, SWA/89/001, Field Document 41, Mbabane, Swaziland.

Riaño, D., Moreno Ruiz, J.A., Barón Martínez, J. and Ustin, S.L. (2007). Burned area forecasting using past burned area records and Southern Oscillation Index for tropical Africa (1981-1999). *Remote Sensing of Environment*, **107**: 571-581.

Roy, D.P., Boschetti, L., Justice, C.O. and Ju, J. (2008). The collection 5 MODIS burned area product: Global evaluation by comparison with the MODIS active fire product. *Remote Sensing of Environment*, **112**: 3690-3707.

Roy, D.P., Jin, Y., Lewis, P.E. and Justice, C.O. (2005). Prototyping a global algorithm for systematic fire affected area mapping using MODIS time series data. *Remote Sensing of Environment*, **97**: 137-162.

San-Miguel-Ayanz, J., Ravail, N., Kelha, V. and Ollero, A. (2005). Active fire detection for fire emergency management: Potential and limitations for the operational use of remote sensing. *Natural Hazards*, **35**: 361-376.

Scholes, R.J., Kendall, J. and Justice, C.O. (1996). The quantity of biomass burned in Southern Africa. *Journal of Geophysical Research*, **101(D19)**: 23667-23676.

Schwela, D., Glodammer, J.G., Morawska, L. and Simpson, O. (1999). Health Guidelines for Vegetation Fire Events: Guideline Document. World Health Organization (WHO), Geneva.

Swaziland National Fire and Emergency Services (2007). 2006 Annual Report. Swaziland National Fire and Emergency Services, Mbabane, Swaziland.

Tishkov, A.A. (2004). Forest fires and dynamics of forest cover. *In:* V.M. Kotlyakov (editor), Natural Disasters, Encyclopedia of Life Support Systems (EOLSS), UNESCO. Eolss Publishers, Oxford, United Kingdom.

Trollope, S.W., de Ronde, C. and Geldenhuys, C.J. (2004). Fire behavior. *In:* J.G. Goldammer and C. de Ronde (editors), Wildland Fire Handbook of Sub-Saharan Africa. Global Fire Monitoring Centre, Freiburg, Germany, pp. 27-59.

Climate Change and Water Resources in India: Impact Assessment and Adaptation Strategies

Adlul Islam* and Alok K. Sikka

1. INTRODUCTION

Climate change and climate variability has received considerable attention from the scientific community and has been the focus of a multitude of scientific investigation over past two decades (e.g., Gleick, 1986; Lettenmaier and Gan, 1990; Arnell, 1992; Xu, 1999; Nijssen et al., 2001; Rosenzweig et al., 2004; Brumbelow and Georgakakos, 2007; Akhtar et al., 2008). The Third and Fourth Assessment Reports of the Intergovernmental Panel on Climate Change (IPCC) have clearly shown that the earth's climate system has demonstrably changed on both global and regional scales since the pre-industrial era. Thus, the earth's climate system is experiencing a warmer phase. Changes in temperature and precipitation under climate variation have serious impacts on hydrologic processes and water resources availability. Improved understanding of climatic causes of hydrological variability is of paramount importance for developing strategies for the sustainable development and management of vital water resources.

Climate is the long-term average of a region's weather events lumped together. Climate change refers to "a change in the state of the climate that can be identified (e.g. using statistical tests) by changes in mean and/or variability of its properties, and that persists for an extended period, typically decades or longer" (IPCC 2007b). It is the change in climate over time whether due to natural variability or as a result of human activity (IPCC, 2007b). Changes in land use pattern and concentration of greenhouse gases (GHGs) in atmosphere are thought to be the two major anthropogenic causes of climate change and variation. Emissions of GHGs (carbon dioxide, methane, chlorofluorocarbons (CFCs) and nitrous oxide) are causing substantial increase in their concentrations in the atmosphere. This increase would enhance the greenhouse effect, resulting in additional warming of the earth's surface. The *"greenhouse effect"* is the rise in temperature that the earth experiences because certain gases (e.g., water vapor, carbon dioxide, nitrous oxide, and methane) in the atmosphere trap energy from the sun. According to the Intergovernmental Panel on Climate Change, increasing amount of carbon dioxide (CO_2) and other greenhouse gases will rise global temperatures causing what is known as *"global warming"*. The increase in CO_2 concentration started in the industrial revolution with coal burning and continuing today with increasing fossil fuel consumption. The global atmospheric concentration of carbon dioxide has increased from a pre-industrial value of about 280 ppm to 379 ppm in 2005 (IPCC, 2007a). Similarly, anthropogenic nitrogen fixation has doubled as a result of increased fertilizer use, fossil fuel consumption and fixation by leguminous crops (Walker and Steffen, 1999). The atmospheric concentration of methane is reported to have increased at a rate of 1% per year (Glantz and Krenz, 1992) mainly due to wet-paddy rice

---

*Corresponding Author

farming and livestock farming. The global atmospheric concentration of methane has increased from a pre-industrial value of about 715 ppb to 1732 ppb in the early 1990s, and was 1774 ppb in 2005. Similarly, the global atmospheric nitrous oxide concentration increased from a pre-industrial value of about 270 ppb to 319 ppb in 2005 (IPCC, 2007a).

Global climate change induced by increases in greenhouse gas concentrations is likely to increase temperatures, change precipitation patterns and probably raise the frequency of extreme events. According to the Fourth Assessment Report of IPCC, the average global surface temperature is projected to increase by 1.1-2.9 °C for low emission scenarios and 2.4-6.4 °C at high emission scenarios during 2090-2099 relative to 1980-1999 (IPCC, 2007a). Over the same period, the global mean sea level is projected to rise by 18 to 38 cm and 26 to 59 cm for low and high emission scenarios, respectively. The projected sea level rise is extremely alarming because of the effects it will have on low-lying regions of the world. Under the worst case of IPCC modeling scenario, major coastal cities around the world could be substantially inundated and other low-lying areas of the world would also be threatened (IPCC, 2007a). Thus, continuous global warming can be expected to increase the frequency and intensity of weather-related disasters.

One of the most important impacts of future climate changes on society will be changes in regional water availability. Such hydrologic changes will affect nearly every aspect of human well being, from agricultural water productivity and energy use to flood control, municipal and industrial water supply, and fish and wildlife management (Scheraga and Grambsch, 1998; Ragab and Prudhomme, 2002; Gibson et al., 2005; Minville et al., 2008). The potential climatic impacts may have significant ramifications for decision making about water allocation and management that are likely to be made in the coming decades. The tremendous importance of water in both society and nature underscores the necessity of understanding how a change in global climate could affect regional water supplies.

In this chapter, an overview of climate change and climate variability and its impact on water resources in India has been presented, together with the role of simulation modeling in evaluating the impacts of climate change on hydrology and water resources. Finally, a case study dealing with climate change impact assessment on water resources availability in the Brahmani river basin of eastern India has been presented.

2. CLIMATE CHANGE AND CLIMATIC VARIABILITY IN INDIA

India is a land with diverse geographical and climatic endowments. The long-term average annual rainfall for the country as a whole is 116 cm and is highly variable both in space and time. 21% of the area of the country receives less than 750 mm rainfall annually while 15% area of the country receives rainfall in excess of 1500 mm (Kumar et al., 2005). Though the average rainfall is quite adequate, about 75% (88 cm with a standard deviation of ±10) of the long term average rainfall occurs during four months (June to September) of southwest monsoon season. There is considerable intra-seasonal and inter-seasonal variation as well. The mean annual temperature across the country varies from less than 10 °C in the extreme north to more than 28 °C in southern parts of the country. The temperature starts to increase over the country from March onwards and reaches a peak in May/June.

Possibility of increase of variability of Asian summer monsoon precipitation has been projected with increase in *Kharif* rainfall and decrease in *Rabi* rainfall in some areas. IPCC (2001) reported that GCMs show high uncertainty in the future projections of winter and summer precipitation over south Asia. The water and agriculture sectors are likely to be most sensitive to climate change-induced impacts in Asia (IPCC, 2001). Available general circulation models (GCMs) suggest that the area-averaged annual mean warming would be about 3 °C in the decade of the 2050s and about 5 °C in the decade of the 2080s over the land regions of Asia as a result of future increases in atmospheric concentration of greenhouse gases. Under the combined influence of greenhouse gas and sulfate aerosols, surface warming would be restricted to about 2.5 °C in the 2050s and about 4 °C in the 2080s (IPCC, 2001).

It has been observed in several studies that there is a continuous increasing trend in mean surface air temperature over decades in most parts of the Indian subcontinent (Hingane et al., 1985; Rupa Kumar and Hingane, 1988; Govind Rao, 1993; Rupa Kumar et al., 1994). Jäger and Ferguson (1991) observed that temperature, precipitation, and humidity might increase by 1-2 °C, 10% and 5-10%, respectively in Ganga-Brahmaputra Basin. In the above climate change scenarios, high risk of floods in the lower reaches of Ganges and Brahmaputra may further increase (Siedel et al., 2000). Based on the time series analysis of historical temperature data for different seasons over different regions of India, Chattopadhyay and Hulme (1997) reported increasing trend in temperature over most parts of south and central India in all the seasons, and entire India has shown the same trend during the post-monsoon season. The maximum extent of warming in terms of mean temperature is reported in the post-monsoon season in whole of India, except for Gujarat and some parts of west coast. The study also showed that both pan-evaporation and potential evapotranspiration have decreased in all seasons during recent years (1961-1992) in India, and the decrease in evaporation and potential evapotranspiration is strongly associated with increasing relative humidity. An analysis of seasonal and annual surface air temperatures (Pant and Rupa Kumar, 1997) has shown a significant warming trend of 0.57 °C per hundred years. The monsoon temperatures do not show a significant trend in any major part of the country except for a significant negative trend over northwest India. Also, data analyzed in terms of day-time and night-time temperatures indicated that the warming was predominantly due to an increase in the maximum temperatures, while the minimum temperatures remained practically constant during the past century. Spatially, a significant warming trend has been observed along the west coast, in central India, the interior peninsula and over north-east India, while cooling trend has been observed in north-west India and a pocket in southern India.

Singh and Sontakke (2002) reported that in the Indo-Gangetic Plains (IGP) of India, the annual surface air temperature is having rising trend (0.53 °C/100 years, significant at 1% level) during 1875–1958 and decreasing trend (–0.93 °C/100 years, significant at 5% level) during 1958-1997. The post 1958 period cooling of the IGP seems to be due to expansion and intensification of agricultural activities and spreading of irrigation network in the region. Further, summer monsoon rainfall showed increasing trend (170 mm/100 years) from 1900 over western IGP, decreasing trend (5 mm/100 years) from 1939 over central IGP, and decreasing trend (50 mm/100 years) during 1900-1984 and increasing trend (480 mm/100 years) during 1984-99 over eastern IGP. Lal et al. (1995) presented a climate change scenario for the Indian subcontinent, taking projected emissions of greenhouse gases and sulphate aerosols into account. It predicts an increase in annual mean maximum and minimum surface air temperatures of 0.7 °C and 1.0 °C over land in the 2040s with respect to the 1980s. In another study analysis of the time series data of 125 stations distributed over whole of India, showed increase in annual mean temperature, mean maximum temperature, and mean minimum temperature at the rate of 0.42, 0.92 and 0.09 °C/100 years. The stations located in the southern and western regions showed a rising trend of 1.06 to 0.36 °C/100 years, respectively, whereas the stations located at the northern plains showed a falling trend of 0.38 °C/100 years. The seasonal mean temperature has increased by 0.94 and 1.1 °C/100 years during post-monsoon and winter season, respectively (Arora et al., 2005). Significant warming of northwestern Himalayas with rise in temperature by about 1.6 °C in the last century has been reported (Bhutiyani et al., 2007). Dash et al. (2007) reported an increase in the annual mean and maximum temperatures by about 0.7 and 0.8 °C, respectively over India. Maximum increase in the maximum temperature has been predicted in the west coast (1.2 °C) followed by north east (1.0 °C), western Himalayas (0.9 °C), north central (0.8 °C), north west (0.6 °C), east coast (0.6 °C), and interior peninsula (0.5 °C). Mean atmospheric surface temperature in India has increased by 1.0 and 1.1 °C during winter and post-monsoon months, respectively. This study also indicated small increase in rainfall during winter months of January and February, pre-monsoon months of March-May, and post-monsoon months of October-December. Pal and Al-Tabbaa (2009) reported increasing trend in winter and autumn extreme rainfall (indicating more winter and autumn floods), and decreasing trend in spring seasonal extreme rainfall with increasing frequency of dry days (indicating water scarcity in pre-monsoon period and a delaying monsoon onset) in Kerala. Thus, the studies discussed above confirm an increase in temperature and changes in rainfall pattern over India during the last century.

3. IMPACT OF CLIMATE CHANGE ON WATER RESOURCES IN INDIA: AN OVERVIEW

The UN Comprehensive Assessment of the Freshwater Resources of the World estimated that approximately one third of the world's population was living in countries deemed to be suffering from water stress in 1997 and two-thirds of the world's population would be living in water stressed countries by 2025 (WMO, 1997). Based on the SRES (IPCC **S**pecial **R**eport on **E**mission **S**cenarios) socio-economic scenarios and climate projections made using six climate models driven by SRES emission scenarios, Arnell (2004) observed that climate change increases water resources stress in some parts of the world where runoff decreases, including around the Mediterranean, in parts of Europe, central and southern America, and southern Africa. In other water stressed parts of the world, particularly in southern and eastern Asia, climate change increases runoff. However, this increase tends to appear during wet season and extra water may not be available during dry season.

The per capita surface water availability in India came down from 4944 m^3 in 1955 to 2309 m^3 in 1991 and 1902 m^3 in 2001. It is projected to reduce to 1465 m^3 and 1235 m^3 by the year 2025, and 2050, respectively under high population growth scenarios (Kumar et al., 2005). By 2050, it is projected that all the basins except Brahmaputra will be below water stress zone and most of the basins will become water scarce. The water scarcity situation for various uses such as agriculture, drinking water, domestic and industrial needs may still become worse, if anticipated impact of climate change on hydrology and water resources are taken into account. Since the major source of water in India is through rainfall, any temporal and spatial variations in rainfall have reflective effect on water availability in both irrigated and rainfed areas. There are preliminary reports that the recent trend of decline in yields of rice and wheat in Indo-Gangetic Plains could have been partly due to weather changes (Aggarwal et al., 2004). Changes in climatic conditions will affect demand, supply and water quality. In the regions that are currently sensitive to water stress (arid and semi-arid regions of India), any shortfall in water supply will enhance competition for water use for a wide range of economic, social and environmental applications. In the future, larger population will lead to heightened demand for irrigation and perhaps industrialization at the expense of drinking water. A major concern that has recently emerged globally is the impact of climate change on hydrology and water resources (e.g., Nijssen et al., 2001; Ragab and Prudhomme, 2002; Arnell, 2004) and the adaptation and preparedness strategies to meet these challenges, in case of their occurrences (e.g., Smithers and Smit, 1997; Aggarwal et al., 2004).

Hydrologic modeling of different river basins of India using SWAT (Soil and Water Assessment Tool) in combination with the outputs of the HadRM2 regional climate model for the control (1981-2000) and future/GHG (2041-2060) climate data indicated an increase in the severity of drought and intensity of floods in different parts of the country (MOEF, 2004). River basin of Luni is expected to experience acute water scarce condition; Mahi, Pennar, Sabarmati, Krishna and Tapi are likely to experience constant water scarcity; Ganga, Narmada, and Cauvery are likely to experience seasonal or regular water stressed conditions; and Godavari, Brahmani and Mahanadi are expected to experience water shortage only in few locations. The study revealed that the increase in rainfall due to climate change does not result in an increase in the surface run-off as may be generally predicted. For example, in the case of the Cauvery river basin, an increase of 2.7% has been projected in the rainfall, but the runoff is projected to reduce by about 2% and the evapotranspiration to increase by about 2% (Fig. 1). This may be either due to increase in temperature and/or change in rainfall distribution in time. Similarly, a reduction in rainfall in the Narmada is likely to result in an increase in the runoff and a reduction in the evapotranspiration, which is again contrary to the usual myth (MOEF, 2004). Using scenarios data generated from regional climatic model HadRM2, Roy et al. (2004) predicted reduced water availability due to global warming for the 2050 and 2060 scenario in five subcatchments (Talaiya, Konar, Maithon, Panchet, and Durgapur) of Damodar basin. However, the scenarios for 2041, 2045, and 2055 predicted less severe impacts on water availability due to climate change.

According to Seidel et al. (2000), there is an increased risk of summer flood in the low reaches of the rivers Ganges and Brahmaputra under climate change scenario of temperature increase by +1.5 °C, precipitation increase in summer by 10%, and increase in humidity by 5-10%. In another study on the nine sub-basins of the river Ganga, Mirza (1997) found that the changes in mean annual runoff in the range of 27-116% occurred in the sub-basins at doubled CO_2 and that the runoff was more sensitive to climate change in the drier subbasins than in the wetter subbasins. Sharma et al. (2000) reported a decrease in runoff by 2 to 8% in the Kosi basin depending upon the areas considered and models used under the scenario of contemporary precipitation and a rise in temperature of 4 °C. Mehrotra (1999, 2000) observed that basins belonging to relatively dry climatic region are more sensitive to climate change scenarios. The Kolar (moist sub-humid) and the Sher (dry sub-humid) are comparatively more sensitive to climate change, whereas Damanganga (humid) is least sensitive (Table 1). Sikka and Dhruva Narayana (1988) reported that a temperature increase of 1, 2 and 3 °C alone resulted in the reduction of runoff by 7, 12 and 19%, respectively in the Nilgiris. The simulations made through a water balance study for Bikaner indicated that the moderate scenario of 10% reduction in rainfall coupled with 8%

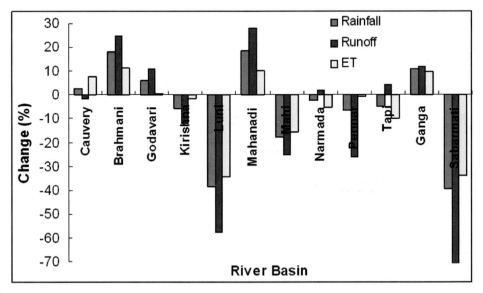

Fig. 1 Changes in water balance under control and GHG climate scenarios in different river basins of India (MOEF, 2004).

increase in PE (i.e., +2 °C temperature) resulted in the 28% reduction of runoff. The percent changes in runoff were relatively higher in arid regions compared to humid Nilgiris, which could be due to the effect of aridity. For the arid zone of Rajasthan, Goyal (2004) reported that even as small as 1% increase in temperature from the base data could result in an increase in evapotranspiration by 15 mm, which means an additional water requirement of 34.275 mcm for Jodhpur district alone and 313.12 mcm for the whole arid zone of Rajasthan. Table 2 summarizes the results of selective studies on climate change impacts on water resources over India during next century.

Table 1. Effect of climate change on different climatic and hydrological variables (Mehrotra, 1999, 2000)

Basin	Temperature	Monsoon rainfall	Runoff	Evapotranspiration
Damanganga (humid)	+1.0 °C (monsoon); +3.0 °C (pre-monsoon)	+5%	—	+13% (monsoon); +3% (annual)
Sher (dry sub-humid)	+1.5 °C (monsoon); +3.5 °C (pre-monsoon)	More than 5%	+15%	+30% (monsoon); +50% (annual)
Kolar (moist sub-humid)	+2.0 °C (monsoon); +4.0 °C (pre-monsoon)	+30%	+20%	+40% (monsoon); +55% (annual)
Hemavati	More than 2 °C in all seasons	−15%	−10%	+10% (monsoon); +20% (annual)

Table 2. Climate change impacts on water resources over India during the next century (Mall et al., 2006)

Region/Location	Impacts
Indian Subcontinent	Increase in monsoon and annual runoff in the central plains; no substantial change in winter runoff; and increase in evaporation and soil wetness during monsoon and on annual basis.
Orissa and West Bengal	One meter sea-level rise would inundate 1700 km^2 of the prime agricultural land.
Indian Coastline	One meter sea-level rise is likely to affect a total area of 5763 km^2 and put 7.1 million people at risk.
All India	Increase in potential evaporation across India.
Central India	Basin located in a comparatively drier region is more sensitive to climate changes.
Kosi Basin	Decrease in runoff by 2-8%.
Southern and Central India	Soil moisture increases marginally by 15-20% during monsoon months.
Chenab River	Increase in river discharge.
River Basins of India	General reduction in the quantity of available runoff, but increase in Mahanadi and Brahmani basins.
Damodar Basin	Decrease in river flow.
Rajasthan	Increase in evapotranspiration.

4. CLIMATE CHANGE IMPACT ASSESSMENT: ROLE OF HYDROLOGIC MODELING

Hydrologic models provide a framework to conceptualize and investigate the relationship between climate, human activities and water resources. The scientific literature of the past two decades contains a large number of studies dealing with the application of hydrologic models to the assessment of the potential effects of climate change on a variety of water resource issues. The use of hydrological models in climate change studies can range from the evaluation of annual and seasonal streamflow variation using simple water balance models (Arnell, 1992) to the evaluation of variations in surface and groundwater quantity, quality and timing using complex distributed parameter models that simulate a wide range of water, energy and biogeochemical processes (Running and Nemani, 1991). The two commonly used approaches to study the effects of climate change on hydrology and water resources are: 'online approach' and 'offline approach'. The

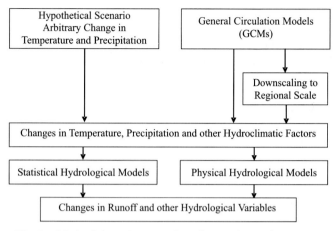

Fig. 2 Methodology for assessing climate change impact on hydrology and water resources (Vicuna and Dracup, 2007).

online approach directly uses and interprets the hydrologic outputs from the general circulation models (GCM). The *offline approach* involves employing hydrological models at the basin or watershed scale driven by climatic data obtained either directly from GCM outputs or from hypothetical or GCM-based scenarios of climate change (Fig. 2). This approach, thus, involves (i) development of climate change scenarios, (ii) modeling hydrological processes, and (iii) sensitivity analysis under different climate change scenarios. Scenario-based studies have been commonly undertaken by hydrologists.

4.1 Climate Change Scenario Generation

Climate change scenarios help to identify the sensitivity or vulnerability of the systems to climate change. A scenario is a coherent, internally consistent and plausible description of a possible future state of the world (IPCC, 1994). A climate change scenario is not a prediction of future climate but refers to a representation of the difference between some plausible future climate and the current or control climate. This is an interim step toward constructing a climate scenario. A climate scenario is the combination of the climate change scenario and the description of the current climate as represented by climate observations (IPCC, 2001).

Results from climate change studies depend critically on the climatic change scenarios used in the study. The simple and direct approach of scenario generation is to develop hypothetical scenarios of changes in temperature and precipitation. The second approach is to obtain the changes in temperature and precipitation from the general circulation models. The general circulation models (GCM) are the primary source of data for use in the assessment of climate change impact. GCMs are mathematical representation of many atmosphere, ocean, and land surface processes based on the law of physics. Such models consider a wide range of physical processes that characterize the climate systems and have been used to examine impact of increased greenhouse gas concentrations in global climate (Gates et al., 1990). Most GCMs simulate reasonable average annual and seasonal features of present climate over large geographic areas but are less reliable in simulating smaller spatial and temporal scale features that are relevant to impact assessment studies (Grotch and MacCarcken, 1991). Further, GCMs accuracy decreases from climate related variables (i.e., temperature, humidity, air pressure and wind) to precipitation, evapotranspiration, runoff and soil moisture, which are of key importance to hydrologic regimes (Xu, 1999).

There are number of GCMs available for use in impact assessment. The output from different GCMs can vary significantly for some regions, posing the problem of which GCM to consider. At least two to three GCMs should be used to create regional climate change scenarios (Benioff et al., 1996). Because, it is time consuming and expensive to run and analyze many climate change scenarios, GCMs be selected on how well they represent the current climate in the region of concern. In order to ensure that climate change scenarios are of most use for impact researchers and policy makers, guidelines prepared by the Task Group on Data and Scenario Support for Impact and Climate Assessment (TGICA) (IPCC-TGCIA, 2007) listed following five criteria to aid scenario selection:

(1) *Consistency with global projections:* The scenarios should be consistent with widely accepted global warming projections based on increased concentrations of greenhouse gases. The IPCC Third Assessment Report (IPCC, 2001) estimated 1.4 °C to 5.8 °C rise in temperature by 2100 compared with 1990.

(2) *Physical plausibility:* Scenarios should not violate the basic laws of physics, which means that not only should the changes in one region be physically consistent with those in another region and globally, but that changes in the different climate variables should also be physically consistent. For example, increase in precipitation should be correlated with increase in cloudiness.

(3) *Applicability in impact assessments:* Scenarios should describe changes in a sufficient number of climate variables on a spatial and temporal scale for impact assessment.

(4) *Representativeness:* Scenarios should be representative of the potential range of future regional climate change in order for a realistic range of possible impacts to be estimated.

(5) *Accessibility:* Scenarios should be straightforward to obtain, interpret and apply in impacts assessments.

The raw outputs from GCM simulations are inadequate for assessing hydrologic impacts of climate change at regional scale. This is because the spatial resolution of GCM grids is too coarse to resolve important catchment scale processes, and the hydrometeorological output produced by the GCMs is unreliable for individual grid points (Hostetler, 1994). A number of different methods exist to construct climate change scenarios, including techniques utilising arbitrary, analogue and global climate model (GCM) information. Synthetic scenarios (also known as 'arbitrary' or 'incremental' scenarios) are the simplest climate change scenarios to construct and apply. Their main use is in sensitivity analyses, i.e., in the determination of the response of a particular 'exposure unit' (e.g., crop yield, stream flow) to a range of climatic variations. A synthetic scenario is constructed by simply perturbing a historical record for a particular climate variable by an arbitrary amount (e.g., by increasing precipitation by 10%). In contrast to synthetic scenarios, analogue scenarios make use of existing climate information either at the site in question (temporal analogues, from paleoclimatic or instrumental information), or from another location which currently experiences a climate anticipated to resemble the future climate of the site under study (spatial analogues).

The GCM largely defines changes in mean monthly climates. Constructing scenarios from GCMs output which are at the same spatial resolution as the GCM is a relatively simple task, but construction of scenarios at finer resolutions which involves spatial and temporal downscaling is more problematic. A common method used in impact studies, known as delta change method, is to estimate average annual changes in precipitation and temperature for a region using one or more GCMs output and then apply these estimates to adjust historic time series of precipitation and temperature. In this procedure adjustment is made by adding temperature change (ΔT) to historic temperature series and by multiplying by percentage change in precipitation ($1+\Delta P/100$) to precipitation series. Hypothetical scenarios using personal estimates or historical measurements of changes, instead of GCMs output, can also be generated using this method. Applying the delta change method assumes that GCMs simulate relative changes more reliably rather than absolute values. This approach for generation of climate change scenario account for changes in the mean historic time series, but do not account for changes in variance. By varying the magnitude of precipitation and temperature changes on monthly basis, changes in variance have also been implemented in hydrologic assessments (Gleick, 1987; Arnell, 1992).

To circumvent the problems stated above, tools for generating the high-resolution meteorological inputs required for modeling ecohydrological processes are needed (Bass, 1996). Downscaling is a set of techniques that relate local and regional scale climate variables to large scale atmospheric forcing (Hewitson and Crane, 1996). The simplest approach to downscale from global climate model scale to finer scale involves interpolation of GCM outputs (changes in temperature and precipitation) from nearest grid boxes using some form of interpolation procedure such as inverse distance interpolation (Brumbelow and Georgakakos, 2007; Notter et al. 2007). Temporal downscaling follows spatial downscaling to convert monthly parameter values to the daily values needed by most of the hydrologic models using different stochastic weather generates such as WGEN (Rechardson, 1981), and LARS-WG (Semenov and Barrow 1997). Using stochastic weather generator, the changes in both climatic means and variability estimated by the GCM can be applied to parameters derived by the weather generator to produce multiple-year change scenarios at the daily time scale. The advantage of using a stochastic weather generator is that a number of different daily time series representing the scenario can be generated by using a different random number to control the stochastic component of the model. Hence, these time series all have the same statistical characteristics, but they vary on a day-to-day basis.

Other downscaling approaches include 'dynamic downscaling', and 'statistical downscaling'. Dynamic downscaling involves nesting of a higher resolution regional climate model (RCM) within a coarse resolution GCM (Giorgi and Mearns, 1999). Dynamic downscaling have been attempted with three approaches (Rummukainen, 1997): (a) running a regional limited area model with coarse GCM data as geographical or spectral boundary conditions (also known as 'one-way nesting'); (b) performing global scale experiments with high resolution atmospheric general circulation models (AGCM) with coarse GCM data as initial boundary conditions; and (c) use of variable resolution global model with highest resolution over the area of interest. The main advantage of RCMs is that they can resolve smaller scale atmospheric features such as orographic precipitation better than host GCMs. The limitation of the dynamic downscaling is that RCMs are computationally as demanding as GCMs. These models still cannot meet the needs of spatially explicit models of ecosystems or hydrological systems, and need to downscale the results from such models to individual sites or localities for impact assessment (Wilby and Wigley, 1997).

The *statistical downscaling* is based on developing empirical relationship between two scales using observed climate data and applies these relationships to simulated coarse scale climate data (Arnell et al., 2003). The assumption of these methods is that the statistical relationship between the large-scale and the local-scale features remains the same even under a changing climate. Empirical methods can generate a large number of realizations; it is therefore possible to assess the uncertainty of the prediction. Further local details, which cannot be examined by the dynamical models, can be considered in these models. Statistical downscaling methods can be classified according to the techniques used (Wilby and Wigley, 1997) or according to the chosen predictor variables (Rummukainen, 1997). Wilby and Wigley (1997) described three categories of statistical downscaling techniques, namely: regression methods; weather pattern-based approaches; and stochastic weather generator. Rummukainen (1997) has classified the statistical downscaling methods as follows:

(1) *Downscaling with surface variables:* This involves the establishment of empirical statistical relationships between large-scale averages of surface variables and local-scale surface variables (e.g., Kim et al., 1984; Wilks, 1989). To develop the relationships, large-scale averages

constructed from local time series are used. In application, the same local-scale surface variables will be the predictands.
(2) *The perfect prog(nosis) (PP) method:*. This involves the development of statistical relationships between large-scale free tropospheric variables and local surface variables. In this method, both the free atmospheric data and the surface data are from observations.
(3) *The model output statistics (MOS) method:* This is similar to the PP method, except that the free atmospheric variables, which are used to develop the statistical relationships, are taken from GCM output.

Wilby et al. (1998) compared six downscaling techniques, namely, weather generator technique (WGEN and SPEL models), two variants of artificial neural network (ANN1 and ANN2) and models based on airflow indices (B-Circ and C-Circ). Comparisons were made using standard set of observed and GCM-derived (i.e., HadCM2) predictor variables and by using standard suite of diagnostic statistics. The weather generator technique yielded the smallest difference between the observed and simulated precipitation, while ANN models performed poorly. In an average sense, changes in diagnostics derived directly from the GCM are generally larger in magnitude than those obtained from area-average statistical downscaling models. The study demonstrated that there are significant differences in the level of skill among the statistical downscaling methods. Wood et al. (2004) evaluated six approaches for downscaling climatic model outputs for use in hydrologic simulations. The six approaches considered for evaluation were: linear interpolation (LI), spatial disaggregation (SD), and bias correction and spatial disaggregation (BCSD) — each applied to both Parallel Climate Model (PCM) output directly, and after dynamical downscaling via a Regional Climatic Model (RCM at 0.5° spatial resolution), for downscaling the climate models output to 0.125° spatial resolution of the hydrologic model. BCSD method was found to reproduce main features of the hydrometeorology from the retrospective climatic simulation, when applied to both PCM and RCM outputs. LI produced better results using RCM output than PCM output, but both methods (i.e, PCM-LI and RCM-LI) lead to an unacceptably biased hydrologic simulation. SD of PCM output produced results similar to those achieved with RCM interpolated results. However, neither PCM nor RCM output was useful for hydrologic simulation purpose without a bias correction step. For the future climate scenarios, only BCSD method (using either RCM or PCM) was able to produce hydrologically plausible results. Further, with BCSD method, the RCM-derived hydrology was found to be more sensitive to climatic changes as compared to PCM-derived hydrology.

4.2 Hydrologic Model Selection

Hydrologic simulation models are often used together with climate scenarios generated from general circulation models to evaluate the impact of climate change on hydrology and water resources. The watershed or regional hydrologic models for assessing impacts of climate change have several attractive characteristics. The different features of regional hydrological models can be summarized as (Xu, 1999):

- A number of hydrologic models are available that are applicable to different climatic/ physiographic conditions at various spatial scales and are capable of representing dominant processes.
- These models can be tailored to fit the characteristics of the available data.
- Regional hydrologic models are considerably easier to manipulate than GCMs.
- They can be used to evaluate the sensitivity of a watershed to both hypothetical changes in climate and the changes predicted by GCMs.

The choice of a particular hydrological model depends on many factors. In general, model should be sufficiently detailed to capture the dominant processes and natural variability, but not unnecessarily refined that computation time is wasted or data availability is limited. For assessing water resources management on regional scale, monthly rainfall-runoff/water balance models have been found useful for identifying consequences in changes in precipitation, temperature and other climatic variables (Gleick, 1986; Mimikou et al., 1991; Arnell, 1992; Xu and Singh, 1998). The conceptual lumped parameter models have been widely used for detailed assessment of surface flows. The Sacramento Soil Moisture Accounting Model, a lumped parameter model, has been widely used by many researchers (Nemec and Schaake, 1982; Gleick, 1987; Lettenmaier and Gan, 1990; Nash and Gleick, 1991) for assessing the impact of climate change. For simulating the spatial patterns of hydrological response within a basin, process based distributed parameter models have been found useful (Beven, 1989; Bathurst and O'Connell, 1992). According to Larson et al. (1982), averaging a certain parameter 'averages' (implicitly) the process being represented. Because of the distributed nature of the climatic inputs and watershed parameters, distributed parameter hydrologic models are more suitable for studying the effect of changes in climate, land use and vegetation on the watershed hydrology (Sikka, 1993). The partitioning of a watershed/basin into smaller homogeneous units in terms of a certain combination of soils, land use, elevation, slope, and aspect is one of the prerequisites for the application of distributed hydrologic models. Spatial variability in the basin characteristics in most distributed models is often captured by partitioning a watershed/basin into smaller homogeneous units in terms of a certain combination of soils, land use, elevation, slope, and aspect. These small sub-basin elements are called *Hydrologic Response Units* (HRUs) (Leavesley and Stannard, 1990) or *Representative Elemental Areas* (REAs) (Wood et al., 1988), or *Grouped Response Unit* (GRU) in a regular grid system (Kouwen et al., 1993). Hydrological modeling at the basin scale using distributed hydrological models requires large input data to describe the spatial variability of watershed characteristics. Manual collection of input data for such models is often difficult and tedious due to level of aggregation and the nature of spatial distribution. GIS has been proven to be an excellent tool to aggregate and organize input data for distributed parameter hydrologic models (Srinivasan and Arnold, 1994; Rosenthal et al., 1995). Passcheir (1996) compared 5 "event" (single runoff event) models and 10 continuous hydrological models for rainfall-runoff modeling of the Rhine and Meuse basin for land use impact modeling, climate change impact modeling, real-time flood forecasting and physically based flood frequency analysis. Four continuous models, namely, Precipitation Runoff Modeling System (PRMS), SACRAMENTO, HBV and SWMM and one event model (HEC-1) were evaluated as the best ones. The HEC-1 and HBV models were found to be the most appropriate for flood frequency analysis, the HBV and SLURP models for climate change impacts on peak discharges, and the PRMS and SACRAMENTO models for assessment of climate change impact on discharge regimes.

Table 3 summarizes the hydrologic models, climate change scenarios and downscaling method used in the selected studies for assessing the impact of climate change on hydrology and water resources. It is apparent from this table that hypothetical scenarios were used in the earliest studies and even in some recent studies because of its simplicity in representing a wide range of alternative scenarios. One major limitation in using GCMs output is its spatial and temporal resolution which does not match with the resolution needed for the hydrologic models. Though several downscaling methods have been developed to downscale GCMs output, the delta change (perturbation) method of modifying the historical time is the preferred method due to its simplicity and limited studies have attempted to use more complex statistical and dynamical downscaling methods. As there is a considerable uncertainty in the GCM simulation of future climate, many studies attempted to use various GCMs output to bracket the plausible changes. Though the application of watershed models ranges from simple regression models to distributed hydrological models, conceptual rainfall-runoff (semi-distributed/distributed) models have been used in most of the studies.

Table 3. Summary of selected studies on climate change impact assessment on hydrology and water resources

Authors	Climate change scenario	Downscaling technique	Hydrologic models	Application area
Akhtar et al. (2008)	PRECIS RCM	No downscaling	HBV model	Hunza, Gilgit, and Astore river basins of Hindukush-Karakoram-Himalaya region
Minville et al. (2008)	HadCM3, ECHAM4, CSIRO, CCSR/NIES and CGCM3	No downscaling (changed factor approach)	Lumped parameter conceptual rainfall-runoff model HSAMI	Chute-du-Diable watershed, Quebec, Canada
Jiang et al. (2007)	Hypothetical	NA	Monthly water balance models (Thornthwaite-Mather, Vrije University Brussel, Xinanjiang, Guo, Schaake, and WatBal)	Dongjiang basin, south China
Lenderink et al. (2007)	HadRM3H	No downscaling	Spatially distributed water balance model RhineFlow	River Rhine
Notter et al. (2007)	ECHAM4	GCM outputs are interpolated using inverse distance interpolation	Semi-distributed grid-based water balance model (NRM3 Streamflow model)	Upper Ewaso Ng'iro basin, Mt. Kenya
Andersson et al. (2006)	HadCM3, CCSR/NIES, CCCma-CGCM2 and GFDL-R30	No downscaling	Pitman hydrological model	Okavango River basin, Africa
Charlton et al. (2006)	HadCM3	Statistical downscaling	HYSIM	Ireland
Toth et al. (2006)	CGCM, CSIRO, ECHAM, GFDL, HadCM2, NCAR and CCSR	No downscaling	Distributed hydrological model (WATFLOOD)	Peace and Athabasca catchment
Booij (2005)	Three GCMs (CGCM1, HadCM3, CSIRO9) and two RCMs (HadRM2, HIRHAM4)	Stochastic precipitation model generated synthetic series	HBV model (HBV96)	Meuse basin
Dibike and Coulibaly (2005)	CGCM1	Statistical downscaling using SDSM and stochastic weather generator LARSWG	HBV-96 (Semi-distributed conceptual model) and CEQUEAU (distributed hydrological model)	Chute-du-Diable sub-basin of Saguenay watershed in northern Quebec, Canada

Contd...

(Contd.)

Authors	Climate change scenario	Downscaling technique	Hydrologic models	Application area
Wurbs et al. (2005)	CCCMA	No downscaling	Soil and Water Assessment Tool (SWAT)	Brazos River basin, USA
Brekke et al. (2004)	HadCM2 and PCM	Statistical downscaling	Sacramento Soil Moisture Accounting Model (SAC-SMA)	San Joaquin River basin, California, USA
Ministry of Environment and Forest (2004)	HadRM2	No downscaling	Soil and Water Assessment Tool (SWAT)	Different river basins of India
Rosenzweig et al. (2004)	GCM of GFDL, GISS, MPI, CCC (CGCM2) and HC (HadCM2)	No downscaling	WATBAL, CERES-Maize, SOYGRO, CROPWAT and WEAP	Argentina, Brazil, China, Hungary, Romania and the US
Roy et al. (2004)	HadRM2	No downscaling	Hydrologic Engineering Center's Hydrologic Modeling Systems (HEC-HMS)	Damodar basin (Talaiya, Konar, Maithon, Panchet, and Durgapur sub-catchments), India
Miller et al. (2003)	HadCM2, PCM and hypothetical scenarios	Statistical downscaling	Sacramento Soil Moisture Accounting Model (SAC-SMA)	Smith, Sacramento, Feather, American, Merced and Kings basins of California, USA
Mirza et al. (2003)	CSIRO9, UKTR, GFDL and LLNL	No downscaling	Statistical regression model and MIKE11-GIS hydrodynamic model	Ganges, Brahmaputra, and Meghna (GBM) rivers in Bangladesh
Burlando and Rosso (2002)	HadCM2GHG and HadCM2SUL	Stochastic downscaling	Precipitation-Runoff Modeling System (PRMS)	Arno River basin, Central Italy
Evans and Schreider (2002)	CSIRO9	Stochastic weather generator	Conceptual rainfall-runoff model (CMD-IHACRES)	Major tributaries of Swan River, Perth, Western Australia
Menzel and Burger (2002)	ECHAM4/OPYC3	Expanded downscaling (EDS) method	HBV-D (semi-distributed conceptual rainfall-runoff model)	Mulde catchment, Southern Elbe, Germany
Pilling and Jones (2002)	HadCM2	Statistical downscaling	Lumped runoff simulation model (HYSIM)	Upper Wye Experimental catchment, mid-Wales, U.K.
Prudhomme et al. (2002)	HadCM2	No downscaling	Semi-distributed rainfall-runoff model CLASSIC (Climate and landuse scenario simulation catchment)	Severn at Haw Bridge, Wales, Western England
Fontaine et al. (2001)	Hypothetical	NA	SWAT	Spring Creek basin, USA

Contd...

(Contd.)

Reference	GCM/Scenario	Downscaling	Hydrological Model	Basin/Location
Limaye et al. (2001)	HadCM2	Interpolated using VEMAP	SWAT	Dale Hollow Watershed, Southeastern United States
Mohseni and Stefan (2001)	GISS and CCC	No downscaling	Deterministic monthly runoff model (MINRUN96)	Baptism and Little Washita River of North-central and South-central US
Nijssen et al. (2001)	HCCPR-CM2, HCCPR-CM3, MPI-ECHAM4 and DOE-PCM3	No downscaling	Variable infiltration capacity (VIC) macroscale hydrological model (MHM)	Nine large continental river basins (Amazon, Amur, Mackenzie, Mekong, Mississippi, Severnaya Dvina, Xi, Yellow, and Yenisei)
Stone et al. (2001)	CSIRO GCM	Regional Climate Model (RegCM) nested within CSIRO GCM	SWAT	Missouri River basin, USA
Tung (2001)	CGCM, GFDL and GISS	Weather generation model	Generalized watershed loading function (GWLF) model	Tsengwen Creek watershed, Taiwan
Muller-Wohlfeil et al. (2000)	ECHAM4/OPYC3	Expanded downscaling (EDS)	Distributed hydrological model (ARC/EGMO)	Upper Stör basin, Germany
Stonefelt et al. (2000)	Hypothetical	NA	SWAT	Upper Wind River basin, northwestern Wyoming
Hamlet and Lettenmaier (1999)	CGCM1, GFDL-CGCM, HadCM2 and ECHAM4	No downscaling	Variable Infiltration Capacity (VIC) Hydrology Model, Columbia Simulation (ColSim)	Columbia River basin
Jose and Cruz (1999)	CCCM, UKMO, GFDL and Hypothetical	No downscaling	Lumped conceptual model (WATBAL)	Angat and Lake Lanao reservoirs, Philippines
Mehrotra (1999)	Hypothetical	NA	Conceptual rainfall-runoff model	Damanaganga, Sher, and Kolar sub-basins of central India
Yates and Strzepek (1998)	GFDL (steady and transient state), GISS, UKMO, MPI and CCC	GCM outputs are interpolated using IDRISI	Monthly water balance model (WBNILE)	Nile basin
McCabe and Hay (1995)	Hypothetical	NA	Precipitation-Runoff Modeling System (PRMS)	East River basin, Colorado, USA
Duell (1994)	Hypothetical	NA	Regression model	American, Carson and Truckee River basins, USA
Sikka (1993)	Hypothetical	NA	Distributed Climate Vegetation Hydrologic Model (CVHM) based on Precipitation Runoff Modeling Systems (PRMS)	Causey watershed in the Weber basin, USA
Wolock et al. (1993)	Hypothetical	NA	Modified Thornthwaite water balance model	Delaware River basin, USA
Nash and Gleick (1991)	Hypothetical and scenarios from GISS, GFDL and UKMO GCM	No downscaling	Conceptual model of NWSRFS	Upper Colorado River basin, USA

Note: NA = Not Applicable.

5. HYDROLOGICAL MODELING FOR CLIMATE CHANGE IMPACT ASSESSMENT IN THE BRAHMANI RIVER BASIN: A CASE STUDY

This section presents a case study for assessing water resources availability in the Brahmani river basin under HadCM3 and PRECIS (Providing REgional Climates for Impact Studies) generated climate scenarios.

5.1 Study Area Description

Brahmani river basin (longitude: 83°52′55″-87°00′38″E, latitude: 20°30′10″-23°36′42″N) is located in the eastern part of India and spreads over the states of Orissa, Jharkhand and Chattisgarh. It has a total catchment area of 39,313.50 km^2 and is composed of four distinct sub-basins, namely Tilga, Jaraikela, Gomali and Jenapur (Fig. 3). The basin has a sub-humid tropical climate with an average annual rainfall of 1305 mm, most of which is concentrated during southwest monsoon season. The Brahmani river rises near Nagri village in Ranchi district of Jharkhand at an elevation of about 600 m and travels a total length of 799 km before it outfalls into the Bay of Bengal.

5.2 Methodology

The impact of climate change on the hydrology and water availability in the Brahmani river basin has been studied using USGS (United States Geological Survey) MMS/PRMS (Modular Modeling Systems/Precipitation-Runoff Modeling Systems) model. MMS (Leavesley et al., 1996) is an integrated system

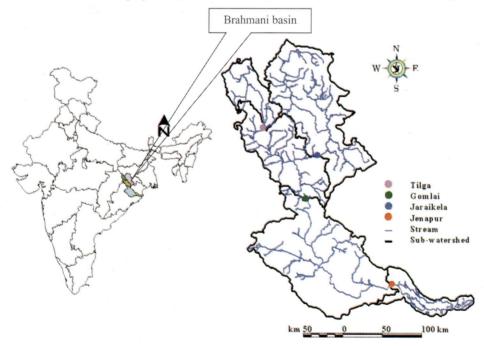

Fig. 3 Location map of Brahmani basin.

of computer software developed to provide a framework for the development and application of models to simulate various hydrological processes. Existing models can be modularized and brought into MMS.

The PRMS (Leavesley et al., 1983) is a deterministic, process-based, distributed parameter modeling system designed to analyze the effects of precipitation, climate, and landuse on streamflow and other general basin hydrology. Distributed parameter capabilities of the model are provided by partitioning the basin into homogenous units, using characteristics such as slope, elevation, aspect, vegetation type, soil type, and precipitation distribution. Each unit, termed as hydrologic response unit (HRU), is assumed to be homogenous with respect to its hydrological response and to the above listed characteristics. In PRMS, the basin is conceptualized as a series of reservoirs (Fig. 4). These reservoirs include interception storage in the vegetation canopy, storage in the soil zone, subsurface storage between surface of watershed and the water table, and groundwater storage. Model considers subsurface flow as relatively rapid movement of water from the unsaturated zone to a stream channel. The subsurface reservoir routes soil-water excess to the ground-water reservoir. Groundwater reservoir is considered as the source of all base flows.

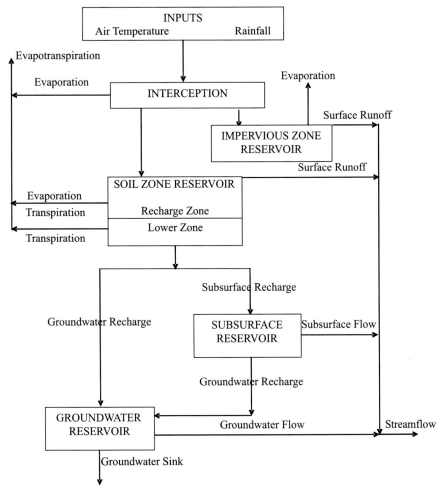

Fig. 4 Conceptual diagram of Precipitation-Runoff Modeling System (Leavesley et al., 1983).

Streamflow is the sum of the various reservoir contributions. Systems inputs included are daily precipitation, and daily minimum and maximum temperature. The model can be operated in two modes—daily and storm mode.

For distributed hydrological modeling using MMS/PRMS, the delineation of basin into different HRUs was done using Digital Elevation Model (DEM) of 30 m resolution. By overlaying elevation, landuse and soil layers, nineteen different classes of hydrological response units were generated (Fig. 5) and basin was divided into 66 HRUs. Different characteristics (such as area, mean and median elevation, slope, land use and soil type etc.) of each HRUs were extracted and used as input to the PRMS model. The model inputs were daily rainfall, maximum and minimum temperature.

For assessing the sensitivity of streamflow to climate change scenarios, hypothetical scenarios were generated by varying the temperature from 0 to 4 °C and rainfall from ± 10 to 30%. For assessing water resources availability in the basin under the climate change scenarios derived from the output of HadCM3 climate model, the HadCM3 predicted mean monthly rainfall and temperature from six nearest grid boxes were extracted. They were then interpolated using inverse distance interpolation method for different periods (i.e., 1980, 2020, 2050 and 2080) and changes were estimated for generation of climate change scenarios. Historical time series of rainfall and temperature were then adjusted by adding monthly changes in temperature (ΔT) to historic temperature series, and by multiplying by monthly changes in rainfall $(1+\Delta P/100)$ to precipitation series.

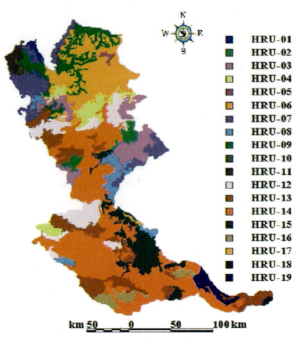

Fig. 5 HRU of Brahmani basin.

5.3 RESULTS

Calibration of MMS/PRMS model by matching the observed and simulated streamflow on annual, monthly and daily basis at four different gauging stations simultaneously (namely, Tilga, Jaraikela, Gomlai and Jenapur) showed a good agreement between observed and simulated streamflow. Calibration (1980-1982) and validation (1983-1985) results of the model at Jenapur gauging station is shown in Fig. 6. The coefficient of determination and modeling efficiency (Nash-Sutcliffe coefficient) were found to vary from 0.96 to 0.98 and 0.81 to 0.69, respectively during calibration phase, and 0.94 to 0.99 and 0.85 to 0.93, respectively during validation phase.

Sensitivity of streamflow to different hypothetical climate change scenarios indicated 76% increase in annual streamflow with a 30% increase in rainfall and no change in temperature (T0P30). If temperature increases by 4 °C (T4P30), increase in streamflow reduces to 62%. A maximum decrease of 33% in annual streamflow is observed with 4°C increase in temperature and 10% decrease in rainfall (Fig. 7). Correlation between changes in temperature and rainfall, and changes in streamflow indicated that rainfall changes had a large effect on monthly, seasonal and annual streamflow. This could be attributed

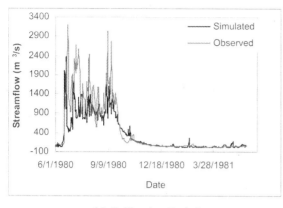 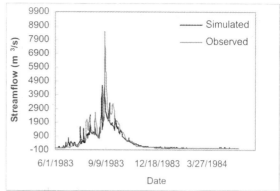

(a) Calibration Period   (b) Validation Period

Fig. 6 Observed and simulated discharges during calibration and validation periods.

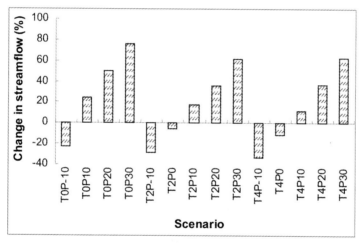

Fig. 7 Changes in the streamflow under hypothetical climate change scenarios.

to sub-humid climatic condition in the basin with lower part of the basin being located in the coastal region.

Simulation using HadCM3 derived climate change scenarios indicated increase in annual as well as seasonal streamflow under both A2a and B2a emission scenarios (Fig. 8). Under A2a emission scenario 15, 9 and 26% increase in annual streamflow was estimated during 2020, 2050, and 2080 respectively. Though, there is increase in streamflow during different periods, the magnitude of increase during 2050 is low as compared to 2020 and 2080. This could be due to changes in monthly rainfall pattern, with no uniform trend during different periods (Fig. 9). There is 6.2, 6.5 and 13.0% increase in annual rainfall during 2020, 2050, and 2080 respectively. Though the annual increase in rainfall during 2020 and 2050 is almost equal, the increase in monsoon rainfall in 2050 (5.41%) is less as compared to 2020 (8.76%) and 2080 (13.09%). Further, increase in rainfall in a particular month not only results in increased streamflow during that month, but also contributes to streamflow as subsurface and groundwater flow

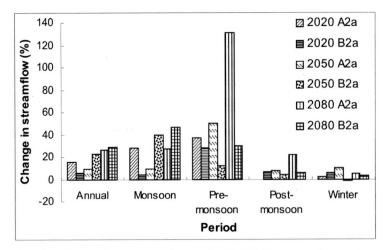

Fig. 8 Changes in the streamflow under HadCM3 generated scenarios.

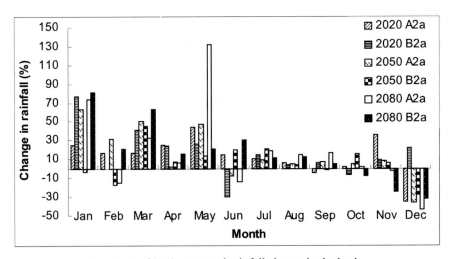

Fig. 9 HadCM3 generated rainfall change in the basin.

during the subsequent months. There is substantial subsurface and groundwater storage during monsoon months and contributes as streamflow during post-monsoon period. For example, increase of 131% rainfall in May 2080 resulted in a substantial amount of subsurface flow during June 2080, and hence higher streamflow (even though there is a decrease in rainfall) as compared to June 2050. Under B2a emission scenario, 23 and 28% increase in annual streamflow is estimated during 2050 and 2080, respectively, which is higher than the estimated streamflow under A2a emission scenario during the same period. This could be attributed to higher monsoon rainfall during 2050 (12.89%) and 2080 (13.87%). Though there is increase in streamflow during all the seasons, the increase is maximum during pre-monsoon season (March-May) under A2a emission scenario, whereas under B2a emission scenario maximum increase in streamflow is estimated in the monsoon season during 2050 and 2080, and pre-monsoon season during 2020.

Analysis of monthly results revealed that the percentage increase in monthly streamflow is maximum in May during all the periods (2020, 2050, and 2080) under A2a emission scenario. During post-monsoon (September-November) and winter (December-February) seasons, increase in streamflow is less than 10%, except for the post-monsoon period of 2080 under A2a emission scenario. Under A2a emission scenario, decrease in monthly streamflow is estimated during October and December in 2020, June in 2050, and June and February in 2080. In case of B2a emission scenario, decrease in monthly streamflow is observed during June and February in 2020, February in 2050, and October to December in 2080.

The results of simulation under PRECIS scenario also indicated an increase in annual as well as seasonal streamflows. An increase of 53% is estimated in the annual streamflow during 2080 (2071-2100) under PRECIS RCM scenario. As there is a variation in the results under different emission scenarios (i.e., A2a and B2a), the estimation of water resources availability using other GCM generated scenarios will help to ascertain these changes. The temporal variability in the availability of water resources in the basin under the influence of climate change indicated the need for developing different adaptation strategies, particularly for winter crops.

6. ADAPTATION TO CLIMATE CHANGE AND MITIGATION MEASURES

Climate change is just one of a number of factors influencing hydrologic systems and water resources. Population growth, changes in land use, restructuring of industrial sectors, and the demands for ecosystem protection and restoration are all occurring simultaneously. Adaptation to climate change takes place through adjustments to reduce vulnerability or enhance resilience in response to observed and expected changes in climate and associated extreme weather events. Individuals, organizations, and society as a whole will inevitably adapt to the changing conditions across a number of scales, sometimes successfully and sometimes unsuccessfully (Dessai et al., 2005). As awareness about the potential impacts of human-induced climate change has grown, so has the desire to plan for the impacts of climate change so that negative hazards can be mitigated and benefits enhanced (Scheraga and Grambsch, 1998). To address the impacts of climate change from long- and medium-term perspectives, urgent mitigation measures are needed such as reducing greenhouse gas (GHG) emissions and enhancing GHG removals as well as adaptation measures. Water is primary medium through which climate change will have an impact on people, ecosystems and economy. Therefore, water resources management should be an early focus for adaptation to climate change (Sadoff and Muller, 2007). Water managers typically rely on well established planning methods and hydrologic estimation tools which assume that the future climate will have same statistical properties of precipitation and temperature that has been experienced in the past. Climate change is likely to result in hydrologic conditions and extremes that will be different from those for which the existing projects were designed.

In the absence of explicit efforts to address the issues of climate variability and climate change, the societal impacts of water scarcity will rise as the competition for water use grows and supply and demand conditions change. As such it may not be safe enough to mention that there are specific adaptation strategies meant to tackle the impact of climate change in India. However, a number of programs and measures are available, which may be capable to alleviate such impacts to a limited extent in water and food sectors. These include supply and demand management measures either meant at conserving or enhancing or improving the water supply. The greatest potential for short-term adaptation is in demand control and more efficient and integrated management of surface and groundwater supplies.

In the agriculture sector, the adaptive measures to counter detrimental impacts of climate variations include changing planting/sowing dates, changing crops and crop varieties that are more tolerant to climatic variations. For irrigated agriculture, increased irrigation efficiency and adoption of drip/sprinkler irrigation systems could reduce water needs. For dryland farming, the implementation of water conservation practices may increase soil moisture, and changes in tillage operation may reduce water losses, and decrease soil erosion. Thus, improving the traditional and community-based irrigation systems, equitable water distribution, rainwater harvesting, groundwater recharge and the development and adoption of efficient irrigation methods such as pressurized irrigation systems could form some of the adaptation strategies. Watershed management could be another example of adaptation strategy especially in the rainfed and dryland areas. Use of seasonal climate forecast in planning and management could possibly reduce the losses due to weather variability and provide opportunities for diversification. Forecasting systems for floods and droughts for people's preparedness may be another example to alleviate the effects of these extreme events. Adaptation measures such as integrated water resources management (IWRM)—"an approach to water management that explicitly recognizes the need to structure and manage the trade-offs required, recognizing that one use affects others and that all depend upon the integrity of the resource base" (Sadoff and Muller, 2007)—will help to achieve water security and sustainability.

The following broad adaptive mechanisms/measures should be planned to reduce the impact of climate change on hydrology and water resources of a basin (Ragab and Prudhomme, 2002; Kumar et al., 2005):

- A strong national climate and water monitoring and research program should be developed, decisions about future water planning and management be flexible, and expensive and irreversible actions be avoided in climate-sensitive areas.
- Improved methods of accounting of climate-related uncertainty should be developed and made part of the decision making process.
- Decision makers at all levels should re-evaluate technical and economic approaches for managing water resources in view of climate variability and climate change.
- Improvements in the efficiency of end uses and the management of water demands must be considered major tools for meeting future water needs, particularly in water scarce regions.
- Water managers should begin a systematic re-examination of engineering design assumptions, operating rules, contingency plans, and water allocation policies under a wider range of climate conditions and extremes than are traditionally used.
- There should be proper coordination and cooperation between water agencies and leading scientific organizations so as to facilitate the exchange of information on the state-of-the-art knowledge about climate change and impacts on water resources.
- There should be timely flow of information among the climate change scientists and the water management community.
- Traditional and alternative forms of water supply can play a significant role in addressing changes in both demands and supplies caused by possible climate change and variability.

7. CONCLUDING REMARKS

It is widely accepted that increasing concentration of greenhouse gases (GHGs) in the atmosphere are causing climate change, but there still exists uncertainty in magnitude, timing and spatial distribution of these changes. With the increasing concern of global climate change, possible impacts of climate change have been widely investigated throughout the world. This chapter focused on the climate change and climate variability in India based on past studies, the role of simulation modeling in assessing climatic change impacts on hydrology and water resources, and a case study on climate change impact assessment in the Brahmani river basin of eastern India.

Continuous increasing trend in mean surface air temperature and changes in rainfall pattern in most parts of the Indian subcontinent has been reported in several studies. An increase in severity of drought and intensity of floods in different river basins of the country has also been reported. Hydrologic simulation models together with the output from GCMs/RCMs are the primary tool for assessing the impacts of climate change on hydrology and water resources. The results of hydrological modeling for assessing water resources availability in the Brahmani river basin using HadCM3 and PRECIS RCM generated scenarios indicated an increase in annual streamflow during different periods.

The majority of the climate change impact assessment studies concentrated on determining the effects of changes in average climate; however emphasis is needed on climate variability and particularly frequency and magnitude of extreme events. Although advances have been made in downscaling approaches, sensitivity analysis using hypothetical scenarios and the simple perturbation approach of altering historical time series with mean monthly changes produced by GCMs are still the most common approach adopted in most impact assessment studies on hydrology and water resources. Impact assessment using the outputs of one or more GCMs provides an estimate of plausible changes, but contains no information about their likelihood. Therefore, tools/techniques are needed to analyze and manage the uncertainty in climate change impact assessment, which is a major concern for climate change studies. Further improvements in spatial and temporal resolution of climate models and development of regional scale climate models with improved hydrologic parameterization at a basin level will enhance the ability of policy makers to use this information in real-world decision making. Integrated modeling frameworks coupling hydrologic, irrigation and crop models is needed to effectively investigate the effects of climate change on water availability, water distribution over space and time, and crop production as well as to prepare adaptation and mitigation strategies.

REFERENCES

Aggarwal, P.K., Joshi, P.K., Ingram, J.S.I. and Gupta, R.K. (2004). Adapting food systems of the Indo-Gangetic plains to global environmental change: Key information needs to improve policy formulation. *Environmental Science & Policy*, **7**: 487-498.

Akhtar, M., Ahmad, N. and Booij, M.J. (2008). The impact of climate change on the water resources of Hindukush-Karakorum-Himalaya region under different glacier coverage scenarios. *Journal of Hydrology*, **355(1-4)**: 148-163.

Andersson, L., Wilk, J., Todd, M.C., Hughes, D.A., Earle, A., Kniveton, D., Layberry, R. and Savenije, H.H.G. (2006). Impact of climate change and development scenarios on flow pattern in the Okavango River. *Journal of Hydrology*, **331**: 43-57.

Arnell, N.W. (1992). Factors controlling the effects of climate change on river flow regimes in a humid temperate environment. *Journal of Hydrology*, **132**: 321-342.

Arnell, N.W. (2004). Climate change and global water resources: SRES emissions and socio-economic scenarios. *Global Environmental Change*, **14**: 31-52.

Arnell, N.W., Hudson, D.A. and Jones, R.G. (2003). Climate change scenarios from a regional climate model: Estimating change in runoff in southern Africa. *Journal of Geophysical Research*, **108(D16)**: 4519.

Arora, M., Goel, N.K. and Singh, P. (2005). Evaluation of temperature trend over India. *Hydrological Sciences Journal*, **50(1)**: 81-93.

Bass, B. (1996). Interim Report on 'The Weather Generator Project'. Focus 4 of IGBP Biospheric Aspects of the Hydrological Cycle (BAHC), Environmental Adaptation Research Group, Atmospheric Environmental Service, Ontario, Canada.

Bathurst, J.C. and O'Connell, P.E. (1992). Future of distributed modeling: The Systeme Hydrologique Europeen. *Hydrological Processes*, **6**: 265-277.

Benioff, R., Guill, S. and Lee, J. (1996). Vulnerability and Adaptation Assessment: An International Guidebook. Kluwer Academic Publisher, Dordrecht, 564 pp.

Beven, K. (1989). Changing ideas in hydrology: The case of physically based models. *Journal of Hydrology*, **105**: 157-172.

Bhutiyani, M.R., Kale, V.S. and Pawar, N.J. (2007). Long-term trends in maximum, minimum and mean annual air temperatures across the northwestern Himalaya during the twentith century. *Climatic Change*, **85**: 159-177.

Booij, M.J. (2005). Impact of climate change on river flooding assessed with different spatial model resolutions. *Journal of Hydrology*, **303**: 176-198.

Brekke, L.D., Miller, N.L., Bashford, K.E., Quinn, N.W.T. and Dracup, J.A. (2004). Climate change impacts uncertainty for water resources in the San Joaquin River Basin, California. *Journal of American Water Resources Association*, **40(1)**: 149-164.

Brumbelow, K. and Georgakakos, A. (2007). Consideration of climate variability and change in agricultural water resources planning. *Journal of Water Resources Planning and Management, ASCE*, **133(3)**: 275-285.

Burlando, P. and Rosso, R. (2002). Effects of transient climate change on basin hydrology. 2. Impacts on runoff variability in the Arno River, central Italy. *Hydrological Processes*, **16**: 1177-1199.

Charlton, R., Fealy, R., Moore, S., Sweeney, J. and Murphy, C. (2006). Assessing the impact of climate change on water supply and flood hazard in Ireland using statistical downscaling and hydrological modeling techniques. *Climatic Change*, **74**: 475-491.

Chattopadhyay, N. and Hulme, M. (1997). Evaporation and potential evapotranspiration in India under conditions of recent and future climate change. *Agricultural and Forest Meteorology*, **87**: 55-73.

Dash, S.K., Jenamani, R.K., Kalsi, S.R. and Panda, S.K. (2007). Some evidence of climate change in twentieth-century India. *Climatic Change*, **85**: 299-321.

Dessai, S., Lu, X. and Risbey, J.S. (2005). On the role of climate scenarios for adaptation planning. *Global Environmental Change*, **15**: 87-97.

Dibike, Y.B. and Coulibaly, P. (2005). Hydrologic impact of climate change in the Saguenay watershed: comparison of downscaling methods and hydrologic models. *Journal of Hydrology*, **307**: 145-163.

Duell, L.F.W. (1994). The sensitivity of northern Sierra Neveda streamflow to climate change. *Water Resources Bulletin*, **30(5)**: 841-859.

Evans, J. and Schreider, S. (2002). Hydrological impacts of climate change on inflows to Perth, Australia. *Climatic Change*, **55**: 361-393.

Fontaine, T.A., Klassen, J.F., Cruickshank, T.S. and Hotchkiss, R.H. (2001). Hydrological response to climate change in the Black Hills of South Dakota, USA. *Hydrological Sciences Journal*, **46(1)**: 27-40.

Gates, W.L., Rowntree, P.R. and Zeng, Q.C. (1990). Validation of climate models. *In:* J.T. Houghton, G.J. Jenkins and J.J. Ephraums (editors), Climate Change: The IPCC Scientific Assessment. Cambridge University Press, New York, pp. 93-130.

Gibson, C.A., Meyer, J.L., Poff, N.L., Hay, L.E. and Georgakakos, A. (2005). Flow regime alterations under changing climate in two river basins: implications for freshwater ecosystems. *River Research and Applications*, **21**: 849-864.

Giorgi, F. and Mearns, L.O. (1999). Introduction to special section: Regional climate modeling revisited. *Journal of Geophysical Research*, **104(D6)**: 6335-6352.

Glantz, M.H. and Krenz, J.H. (1992). Human Components of the Climate System. *In:* K.E. Trenberth (editor), Climate System Modeling. University Press, Cambridge, pp. 27-49.

Gleick, P.H. (1986). Methods for evaluating the regional hydrologic impacts of global climatic changes. *Journal of Hydrology*, **88**: 97-116.

Gleick, P.H. (1987). The development and testing of a water balance model for climate impact assessment: Modeling the Sacramento basin. *Water Resources Research*, **23(6)**: 1049-1061.

Govind Rao, P. (1993). Climate changes and trend over a major river basin in India. *Climate Research*, **2**: 215-233.

Goyal, R.K. (2004). Sensitivity of evapotranspiration to global warming: A case study of arid zone of Rajasthan (India). *Agricultural Water Management*, **69(1)**: 1-11.

Grotch, S.L. and MacCarcken, M.C. (1991). The use of general circulation models to predict regional climate change. *Journal of Climate*, **4(3)**: 286-303.

Hamlet, A.F. and Lettenmaier, D.P. (1999). Effects of climate change on hydrology and water resources in the Columbia river basin. *Journal of American Water Resources Association*, **35(6)**: 1597-1623.

Hewitson, B.C. and Crane, R.G. (1996). Climate downscaling: Techniques and application. *Climate Research*, **7**: 85-95.

Hingane, L.S., Rupa Kumar, K. and Ramana Murty, B.V. (1985). Long-term trends of surface air temperature in India. *International Journal of Climatology*, **5**: 521-528.

Hostetler, S.W. (1994). Hydrologic and atmospheric models: The (continuing) problem of discordant scales. *Climatic Change*, **27(4)**: 345-350.

IPCC (1994). IPCC Technical Guidelines for Assessing Climate Change Impacts and Adaptations. Prepared by Working Group II and WMO/UNEP, University College, London, U.K. and Centre for Global Environmental Research, National Institute for Environmental Studies, Tsukuba, Japan, 59 pp.

IPCC (2001). Climate Change 2001: The Scientific Basis. Contribution of Working Group I to the Third Assessment Report of the Intergovernmental Panel on Climate Change (IPCC). Cambridge University Press, Cambridge, U.K., 1032 pp.

IPCC (2007a). Climate Change 2007: The Physical Science Basis. Contribution of Working Group I to the Fourth Assessment Report of the Intergovernmental Panel on Climate Change (IPCC). Cambridge University Press, Cambridge, U.K., 881 pp.

IPCC (2007b). Climate Change 2007: Synthesis Report. Contribution of Working Groups I, II and III to the Fourth Assessment Report of the Intergovernmental Panel on Climate Change (IPCC), Geneva, 102 pp.

IPCC-TGICA (2007). General Guideline on the Use of Scenario Data for Climate Impact and Adaptation Assessment, Version 2. Prepared by T.R Carter on behalf of the Intergovernmental Panel on Climate Change Task Group on Data and Scenario Support for Impact and Climate Assessment (TGICA), 66 pp. (http://www.ipcc-data.org/guidelines/index.html).

Jager, J. and Feruguson, H.L. (editors) (1991). Climate change: Science, impacts and policy. Proceedings of the Second World Climate Conference, Geneva, Switzerland. Cambridge University Press, Cambridge, U.K., 578 pp.

Jiang, T., Chen, Y.D., Xu, C., Chen, X., Chen, X. and Singh, V.P. (2007). Comparison of hydrological impacts of climate change simulated by six hydrological models in the Dongjiang Basin, South China. *Journal of Hydrology*, **336**: 316-333.

Jose, A.M. and Cruz, N.A. (1999). Climate change impacts and responses in the Philippines: Water resources. *Climate Research*, **12**: 77-84.

Kim, J.W., Chang, J.T., Baker, N.L., Wilks, D.S. and Gates, W.L (1984). The statistical problem of climate inversion: Determination of the relationship between local and large-scale climate. *Monthly Weather Review*, **112(10)**: 2069-2077.

Kouwen, N., Soulis, E.D., Pietroniro, A., Donald, J. and Harrington, R.A. (1993). Grouped response units for distributed hydrologic modeling. *Journal of Water Resources Planning and Management, ASCE*, **119(3)**: 289-305.

Kumar, R., Singh, R.D. and Sharma, K.D. (2005). Water resources of India. *Current Science*, **89(5)**: 794-811.

Lal, M., Cubasch, U., Voss, R. and Waszkewitz, J. (1995). Effect of transient increase of greenhouse gases and sulphate aerosols on monsoon climate. *Current Science*, **69(9)**: 752-763.

Larson, C.L., Onstad, C.A., Richardson, H.H. and Brooks, K.N. (1982). Some particular watershed models. *In:* C.T. Haan, H.P. Johnson and D.L. Brakensiek (editors), Hydrologic Modeling of Small Watersheds. ASAE Monograph No-5, American Society of Agricultural Engineering, St. Joseph, Michigan, pp. 410-436.

Leavesley, G.H. and Stannard, L.G. (1990). Application of remotely sensed data in distributed parameter watershed model. Proceedings of the Workshop on Application of Remote Sensing in Hydrology, Saskatoon, Saskatchewan, Canada, pp. 47-68.

Leavesley, G.H., Lichty, R.W., Troutman, B.M. and Saindon, L.G. (1983). Precipitation-Runoff Modeling System: User's Manual. U.S. Geological Survey Investigation Report 83-4238, 207 pp.

Leavesley, G.H., Restrepo, P.J., Markstrom, S.L., Dixon, M. and Stannard, L.G. (1996). The Modular Modeling System (MMS): User's Manual. U.S. Geological Survey Open-file Report 96-151, 142 pp.

Lenderink, G., Buishand, A. and van Deursen, W. (2007). Estimates of future discharges of the river Rhine using two scenario methodologies: Direct versus delta approach. *Hydrology and Earth System Sciences*, **11(3):** 1145-1159.

Lettenmaier, D.P. and Gan, T.Y. (1990). Hydrologic sensitivity of Sacramento-San Joaquin River Basin, California to global warming. *Water Resources Research*, **26(1):** 69-86.

Limaye, A.S., Boyington, T.M., Cruise, J.F., Bulusu, A. and Brown, E. (2001). Macroscale hydrologic modeling for regional assessment studies in the southeastern United States. *Journal of American Water Resources Association*, **37(3):** 709-722.

Mall, R.K., Gupta, A., Singh, R., Singh, R.S. and Rathore, L.S. (2006). Water resources and climate change: An Indian perspective. *Current Science*, **90(12):** 1610-1626.

McCabe, G.J. and Hay, L.E. (1995). Hydrological effects of hypothetical climate change in the East River basin, Colorado, USA. *Hydrological Sciences Journal*, **40(3):** 303-318.

Mehrota, R. (2000). Hydrologic sensitivity of some Indian basins to expected climate change and its effect on water availability using disaggregated GCMs output. Proceedings of the International Conference on Integrated Water Resources Management for Sustainable Development, Vol. II, pp. 1063-1080.

Mehrotra, R. (1999). Sensitivity of runoff, soil moisture and reservoir design to climate changes in central Indian River basins. *Climatic Change*, **42:** 725-757.

Menzel, L. and Burger, G. (2002). Climate change scenarios and runoff response in the Mulde catchment (Southern Elbe, Germany). *Journal of Hydrology*, **267:** 53-64.

Miller, N.L., Bashford, K.E. and Strem, E. (2003). Potential impacts of climate change on California hydrology. *Journal of American Water Resources Association*, **39(4):** 771-784.

Mimikou, M., Kouvopoulos, Y., Cavadias, G. and Vayianos, N. (1991). Regional hydrological effects of climate change. *Journal of Hydrology*, **123:** 119-146.

Minville, M., Brissette, F. and Leconte, R. (2008). Uncertainty of the impact of climate change on the hydrology of a Nordic watershed. *Journal of Hydrology*, **358:** 70-83.

Mirza, M.M.Q., Warrick, R.A. and Ericksen, N.J. (2003). The implications of climate change on floods of the Ganges, Brahmaputra, and Meghna rivers in Bangladesh. *Climatic Change*, **57:** 287-318.

Mirza, M.M.Q. (1997). The runoff sensitivity of the Ganges river basin to climate change and its implications. *Journal of Environmental Hydrology*, **5:** 1-13.

MOEF (2004). India's Initial National Communication to the United Nations Framework Convention on Climate Change. Ministry of Environment and Forests (MOEF), Government of India, New Delhi.

Mohseni, O. and Stefan, H.G. (2001). Water budgets of two watersheds in different climatic zones under projected climate warming. *Climatic Change*, **49:** 77-104.

Muller-Wohlfeil, D., Burger, G. and Lahmer, W. (2000). Response of a river catchment to climatic change: Application of expanded downscaling to northern Germany. *Climatic Change*, **47:** 61-89.

Nash, L.L. and Gleick, P.H. (1991). Sensitivity of streamflow in the Colorado basin to climate changes. *Journal of Hydrology*, **125:** 221-241.

Nemec, J. and Schaake, J. (1982). Sensitivity of water resource systems to climate variation. *Hydrological Sciences Journal*, **27:** 327-343.

Nijssen, B., O'donnell, G.M., Hamlet, A.F. and Lettenmaier, D.P. (2001). Hydrologic sensitivity of global rivers to climate change. *Climatic Change*, **50:** 143-175.

Notter, B., MacMillan, L., Viviroli, D., Weingartner, R. and Liniger, H. (2007). Impacts of environmental change on water resources in the Mt. Kenya region. *Journal of Hydrology*, **343**: 266-278.

Pal, I. and Al-Tabbaa, A. (2009). Trends in seasonal precipitation extremes: An indicator of 'climate change' in Kerala, India. *Journal of Hydrology*, **367**: 62-69.

Pant, G.B. and Rupa Kumar, K. (1997). Climate of South Asia. John Wiley & Sons Ltd., Chichester, U.K., 320 pp.

Passcheir, R.H. (1996). Evaluation of Hydrologic Model Packages. Technical Report Q2044, WL/Delft Hydraulics, Delft (www.wldelft.nl).

Pilling, C.G. and Jones, J.A.A. (2002). The impact of future climate change on seasonal discharge, hydrological processes and extreme flows in the Upper Wye experimental catchment, mid-Wales. *Hydrological Processes*, **16**: 1201-1213.

Prudhomme, C., Reynard, N. and Crooks, S. (2002). Downscaling of global climate models for flood frequency analysis: Where are we now? *Hydrological Processes*, **16**: 1137-1150.

Ragab, R. and Prudhomme, C. (2002). Climate change and water resources management in arid and semi-arid regions: Prospective and challenges for 21st century. *Biosystems Engineering*, **81(1)**: 3-34.

Richardson, C.W. (1981). Stochastic simulation of daily precipitation, temperature and solar radiation. *Water Resources Research*, **17(1)**: 182-190.

Rosenthal, W.D., Srinivasan, R. and Arnold, J.G. (1995). Alternative river management using linked GIS-hydrology model. *Transaction of American Society of Agricultural Engineering (ASAE)*, **38(3)**: 783-790.

Rosenzweig, C., Strzepek, K.M., Major, D.C., Iglesias, A., Yates, D.N., McCluskey, A. and Hillel, D. (2004). Water resources for agriculture in a changing climate: International case studies. *Global Environmental Change*, **14**: 345-360.

Roy, P.K., Roy, D. and Mazumdar, A. (2004). An impact assessment of climate change and water resources availability of Damodar river basin. *Hydrology Journal* (Indian Association of Hydrologists, Roorkee, India), **27(3-4)**: 53-70.

Rummukainen, M. (1997). Methods for Statistical Downscaling of GCM Simulation. Swedish Regional Climate Modeling Programme (SWECLIM), Reports Meteorology and Climatology No. 80, Swedish Meteorological and Hydrological Institute (SMHI), Norrköping, Sweden, 29 pp.

Running, S.W. and Nemani, R.R. (1991). Regional hydrologic and carbon balance response of forests resulting from potential climate change. *Climatic Change*, **19**: 349-368.

Rupa Kumar, K. and Hingane, L.S. (1988). Long-term variations of surface air temperature at major industrial cities of India. *Climatic Change*, **13**: 287-307.

Rupa Kumar, K., Krishna Kumar, K. and Pant, G.B. (1994). Diurnal asymmetry of surface temperature trends over India. *Geophysical Research Letters*, **21**: 677-680.

Sadoff, C.W. and Muller, M. (2007). Better water resources management — greater resilience, more effective adaptation tomorrow. A perspective paper contributed by Global Water Partnership through its Technical Committee (www.gwpforum.org), pp. 1-14.

Scheraga, J.D. and Grambsch, A.E. (1998). Risks, opportunities, and adaptation to climate change. *Climate Research*, **11**: 85-95.

Seidel, K., Martinec, J. and Baumgartner, M.F. (2000). Modeling runoff and impact of climate change in large Himalayan basins. Proceedings of the International Conference on Integrated Water Resources Management for Sustainable Development, 19-21 December 2000, New Delhi, India, pp. 1020-1028.

Semenov, M.A. and Barrow, E.A. (1997). Use of stochastic weather generator in the development of climate change scenarios. *Climatic Change*, **35**: 397-414.

Sharma, K.P., Vorosmarty, C.J. and Moore III, B. (2000). Sensitivity of the Himalayan hydrology to land-use and climatic changes. *Climatic Change*, **47**: 117-139.

Sikka, A.K. (1993). Modeling the influence of climate change on evapotranspiration and water yield. Unpublished Ph.D. Thesis, Utah State University, Logan, Utah.

Sikka, A.K. and Dhruva Narayana, V.V. (1988). Climatic impacts on water resources and impact mitigation strategies. Proceedings of the UNEP/INSA Sponsored Workshop on Climate Impacts Study, 16-18 March 1988, New Delhi, India.

Singh, N. and Sontakke, N.A. (2002). On climatic fluctuations and environmental changes of the Indo-Gangetic plains, India. *Climatic Change*, **52:** 287-313.

Smithers, J. and Smit, B. (1997). Human adaptation to climate variability. *Global Environmental Change*, **7(2):** 129-146.

Srinivasan, R. and Arnold, J.G. (1994). Integration of a basin scale water quality model with GIS. *Water Resources Bulletin*, **30(3):** 453-462.

Stone, M.C., Hotchkiss, R.H., Hubbard, C.M., Fontaine, T.A., Mearns, L.O. and Arnold, J.G. (2001). Impacts of climate change on Missouri river basin water yield. *Journal of American Water Resources Association*, **37(5):** 1119-1129.

Stonefelt, M.D., Fontaine, T.A. and Hotchkiss, R.H. (2000). Impacts of climate change on water yield in the Upper Wind river basin. *Journal of American Water Resources Association*, **36(2):** 321-336.

Toth, B., Pietroniro, A., Conly, F.M. and Kouwen, N. (2006). Modeling climate change impacts in the Peace and Athabasca catchment and delta: I. hydrological model application. *Hydrological Processes*, **20:** 4197-4214.

Tung, C.-P. (2001). Climate change impacts on water resources of the Tsengwen creek watershed in Taiwan. *Journal of American Water Resources Association*, **37(1):** 167-176.

Vicuna, S. and Dracup, J.A. (2007). The evolution of climate change impact studies on hydrology and water resources in California. *Climatic Change*, **82:** 327-350.

Walker, B.H. and Steffen, W.L. (1999). The nature of global change. *In:* B. Walker, W. Steffen, J. Canadell, and J. Ingram (editors), The Territorial Biosphere and Global Change. Cambridge University Press, U.K. pp. 1-18.

Wilby, R.L. and Wigley, T.M.L. (1997). Downscaling general circulation model output: A review of methods and limitations. *Progress in Physical Geography*, **21(4):** 530-548.

Wilby, R.L., Wigley, T.M.L., Conway, D., Jones, P.D., Hewitson, B.C., Main, J. and Wilks, D.S. (1998). Statistical downscaling of general circulation model output: A comparison of methods. *Water Resources Research*, **34(11):** 2995-3008.

Wilks, D.S. (1989). Conditioning stochastic daily precipitation models on total monthly precipitation. *Water Resources Research*, **25:** 1429-1439.

WMO (1997). A Comprehensive Assessment of Freshwater Resources of the World. World Meteorological Organization (WMO), Geneva.

Wolock, D.M., McCabe Jr., G.J., Tasker, G.D. and Moss, M.E. (1993). Effects of climate change on water resources in the Delaware River basin. *Water Resources Bulletin*, **29(3):** 475-486.

Wood, A.W., Leung, L.R., Sridhar, V. and Lettenmaier, D.P. (2004). Hydrologic implications of dynamical and statistical approaches to downscaling climate model outputs. *Climatic Change*, **62:** 189-216.

Wood, E.F., Sivapalan, M., Beven, K. and Band, L. (1988). Effects of spatial variability and scale with implications to hydrologic modeling. *Journal of Hydrology*, **102:** 29-47.

Wurbs, R.A., Muttiah, R.S. and Felden, F. (2005). Incorporation of climate change in water availability modeling. *Journal of Hydrologic Engineering, ASCE*, **10(5):** 375-385.

Xu, C.-Y. (1999). Climate change and hydrologic models: A review of existing gaps and recent research developments. *Water Resources Management*, **13:** 369-382.

Xu, C.-Y. and Singh, V.P. (1998). A review of monthly water balance models for water resources investigations. *Water Resources Management*, **12:** 31-50.

Yates, D.N. and Strzepek, K.M. (1998). Modeling the Nile basin under climatic change. *Journal of Hydrologic Engineering, ASCE*, **3(2):** 98-108.

18 Global Climate Change vis-à-vis Crop Productivity

Prabhjyot-Kaur* and S.S. Hundal

1. INTRODUCTION

It is now well known that climate is changing worldwide. The past two decades have witnessed globally a rapid increase in the awareness of climatic change and triggered widespread apprehension amongst scientists and governments about its global implications (Gadgil, 1996; Malone and Brenkert, 2008; Baer and Risbey, 2009). These climatic changes are the result of changes in atmospheric gaseous constituents, vegetation/crop cover and biodiversity (Fig. 1).

The Inter-Governmental Panel on Climate Change (IPCC) in its recently released report has reconfirmed that the global atmospheric concentrations of greenhouse gases (GHGs) have increased markedly as a result of human activities (IPCC, 2007). Between 1000 and 1750 AD, the CO_2, methane

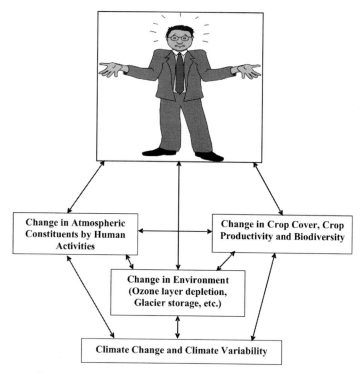

Fig. 1 Climate change as induced by anthropogenic activities and its effect on humans.

---

*Corresponding Author

and nitrous oxide concentrations were 280 ppm, 700 ppb and 270 ppb, respectively. In 2005, these values have increased to 379 ppm, 1774 ppb and 319 ppb, respectively (IPCC, 2007). The global increases in CO_2 concentrations are primarily due to fossil fuel use and land use change, while those of methane and nitrous oxide are primarily due to agriculture. These increases in GHGs have resulted in the warming of climate system by 0.74 °C between 1906 and 2005 (Sathaye et al., 2006; Ghude et al., 2009). Eleven of the last twelve years (1995-2006) rank amongst the 12 warmest years in the instrumental record of global surface temperature since 1850 (IPCC, 2007). The rate of warming has been much higher in the recent decades and the night time minimum temperatures have been increasing at twice the rate of daytime maximum temperatures. The quantity of rainfall and its distribution has also become more uncertain. In some places, climatic extremes such as droughts, floods, timing of rainfall and snowmelt have also increased. The sea level has risen by 10-20 cm with regional variations (Gosain et al., 2006; Unnikrishnan et al., 2006). Similarly, snow cover is also believed to be gradually decreasing (De la Mare, 2009).

CO_2 is vital for photosynthesis and hence for plant growth. An increase in the atmospheric CO_2 concentration affects agricultural production by climate change and changes in photosynthesis and transpiration rate. The rise in atmospheric CO_2 concentration from pre-industrial level of about 280 ppm to about 377 ppm currently is well documented (Keeling and Whorf, 2005). It is therefore important to assess the combined effects of elevated atmospheric concentration and climate change on the productivity of a region's dominant crops (Haskett et al., 1997). The direct effects of increased concentrations of CO_2 are generally beneficial to vegetation (Farquhar, 1997), especially for C_3 plants (wheat, rice, barley, oats, peanut, cotton, sugar beet, tobacco, spinach, soybean and most trees), as elevated levels lead to higher assimilation rates and to an increase in stomatal resistance resulting in a decline in transpiration and improved water use efficiency of crops.

The rising temperature and carbon dioxide and uncertainties in rainfall associated with global climate change may have serious direct and indirect consequences on crop production and hence on food security (Sinha and Swaminathan, 1991; Kumar, 1998). It is, therefore, important to have an assessment of the direct and indirect consequences of global warming on different crops contributing to our food security. Future agricultural planning has to formulate a holistic approach on productivity, stability, sustainability, profitability and equity in Indian agriculture in coming decades. Simulation techniques are easy, time saving, economical and widely used for studying the influence of climatic variability on the growth and yield of crops. The effects of changes in temperature, precipitation and CO_2 concentrations on crop productivity have been studied extensively using crop simulation models (e.g., Bachelet and Gay, 1993; Olszyk et al., 1999; Timsina and Humphreys, 2003; Attri and Rathore, 2003; Parry et al., 2004; Hundal and Prabhjyot-Kaur, 2007; Pathak and Wassmann, 2009). Simulation studies have been conducted for predicting the plausible effects of elevated levels of CO_2 on the yield of crops (Olszyk et al., 1999; Tubeillo and Ewert, 2002). Lal et al. (1998) through sensitivity experiments found that under elevated CO_2 levels, the yields of rice and wheat increased by 15% and 28%, respectively for a doubling of CO_2 in Northwest India. Soybean yields have been reported to increase by nearly 50% for a doubling of atmospheric CO_2 from its current level of 330 ppm (Mall et al., 2004).

The combined effects of climate change have been found to have implications for dryland and irrigated crop yields (Rosenzweig and Iglesias, 1994). However, the effect on production is expected to vary by crop and location as well as by the magnitude of warming, and the direction and magnitude of precipitation change (Adams et al., 1998). Several such attempts have been made for predicting the yield of wheat (e.g., Mearns et al., 1996; Attri and Rathore, 2003; Pathak and Wassmann, 2009), rice (e.g., Bachelet and Gay, 1993; Olszyk et al., 1999) and wheat, rice, maize and groundnut (e.g., Hundal and Prabhjyot-Kaur, 1996) under changing climatic conditions.

The DSSAT-family of models have been used most extensively in predicting the effect of various climate change scenarios on crop yields as these models contain subroutines for examining the impact of climate change on crop yields (e.g., Lal et al., 1998; Attri and Rathore, 2003; Timsina and Humphreys, 2003; Mall et al., 2004; Hundal and Prabhjyot-Kaur, 2007). Keeping into account the anticipated regional climatic changes in India, the effects of changes in temperature, solar radiation and carbon dioxide and their interactions on the growth and yield of salient crops under the agro-climatic conditions of Punjab State were studied by using the CERES- and GRO-family of crop simulation models. The results of these simulation studies, together with an overview of climate change/variability in India are discussed in this chapter.

2. CLIMATE CHANGE AND CLIMATE VARIABILITY IN INDIA

An analysis of the mean annual surface air temperature over India indicates a significant warming of about 0.3-0.6 °C since the 1860's (Hingane et al., 1985; Pant, 2003). This warming trend in general is comparable to the global mean trend of 0.5 °C in last 100 years (Sathaye et al., 2006). Analysis of a representative rainfall series from different locations in India over past more than a century indicate a highly variable but trendless behavior of the rainfall with a prominent epochal nature of variability. However, it has been reported that there do exist some smaller sub-regions with statistically significant increasing and decreasing trends (Sontakke, 1990; Pant and Rupa Kumar, 1997). Some studies have indicated that the rainfall patterns in India are also set to change and the western and central areas could have up to 15 more dry days each year, while in contrast, the north and northeast are predicted to have 5 to 10 more days of rain annually. In other words, the dry areas will get drier and the wet areas will get wetter (Pant and Rupa Kumar, 1997; Pant, 2003).

Since there is a large spatial and temporal variability in weather factors in a region, it is desirable that more detailed scenarios are made available for different agro-climatic zones. There is also greater consensus now that in future climatic variability in India will increase leading to more frequent extremes of weather in the form of uncertain onsets of monsoons, and frequency and intensity of drought, flooding, etc. (Goswami et al., 2006).

3. CLIMATE CHANGE/VARIABILITY: A CASE STUDY IN PUNJAB

Agriculture is sensitive to short-term changes in weather as well as to seasonal, annual and longer-term variations in climate. The variations in the meteorological parameters, in combination with other parameters such as soil characteristic, cultivar, pest and diseases, etc., have paramount influence on the agricultural productivity (Pathak and Wassmann, 2009). In the northern India, though the geographical area of Punjab State is only 1.53 % of the country, it contributes nearly 70% of wheat and 55% of rice to the central pool of foodgrains and hence, is referred to as "Bread Basket" of the country. It is, therefore, of great importance to have an assessment of the variability in these climatic factors in Punjab State.

The climate variability analysis was carried out by analyzing historical data of maximum and minimum temperatures and rainfall for five locations in three agro-climatic zones of the state, i.e., Zone I (Ballowal Saunkhri), Zone III (Amritsar, Ludhiana and Patiala) and Zone IV (Bathinda) (Fig. 2). The daily temperature and rainfall data of past three decades for Ballowal Saunkhri (1984-2005), Amritsar (1970-2005), Ludhiana (1970-2005), Patiala (1970-2005) and Bathinda (1977-2005) were analyzed for annual, *kharif* (1 May to 31 October) and *rabi* (1 November-30 April) crop growing seasons. The results of the study are discussed below.

3.1 Temperature Variability Trends

The trend line obtained by regressing the five-yearly moving averages against time for annual, *kharif* and *rabi* season maximum and minimum temperatures are shown in Tables 1 and 2, respectively for the locations of Ballowal Saunkhri, Amritsar, Ludhiana, Patiala and Bathinda. In general, the maximum temperature has decreased from the normal at Ballowal Saunkhri and Bathinda; however, no trend could be established for the other locations. The *kharif* maximum temperature decreased at the rate of 0.04 °C/year at Ballowal Saunkhri and Bathinda.

The annual and seasonal minimum temperature has increased at the rate of 0.07 °C/year over the past three decades at Ludhiana. At Patiala, the annual and *kharif* minimum temperatures have increased at the rate of 0.02 °C/year and at Bathinda the annual, *kharif* and *rabi* minimum temperatures have increased at the rate of about 0.03, 0.02 and 0.05 °C/year, respectively. However no trend of change in the minimum temperature was observed at Ballowal Saunkhri and Amritsar.

Table 1. Time trend equations for maximum temperature (slope of regression in °C/calendar year) over the past three decades at different locations for annual, *kharif* and *rabi* seasons in Punjab

Station (Latitude, longitude and elevation above MSL)	Annual	Kharif	Rabi
Ballowal Saunkhri (31° 60′ N, 76° 23′ E, 355 m)	$y = 0.058x - 86.08$ $R^2 = 0.46$	$y = 0.045x - 56.17$ $R^2 = 0.46$	$y = 0.089x - 152.30$ $R^2 = 0.56$
Amritsar (31° 37′ N, 74° 53′ E, 231 m)	$y = -0.010x + 50.97$ $R^2 = 0.13$	$y = -0.018x + 73.07$ $R^2 = 0.32$	$y = 0.007x + 10.39$ $R^2 = 0.05$
Ludhiana (30° 56′ N 75° 48′ E 247 m)	$y = -0.0001x + 30.90$ $R^2 = 0.00$	$y = -0.014x + 62.67$ $R^2 = 0.21$	$y = 0.017x - 98.40$ $R^2 = 0.15$
Patiala (30° 20′ N 76° 28′ E 251 m)	$y = 0.004x + 21.10$ $R^2 = 0.04$	$y = -0.007x + 50.33$ $R^2 = 0.17$	$y = 0.020x - 16.39$ $R^2 = 0.26$
Bathinda (30° 12′ N 74° 57′ E 211 m)	$y = -0.023x + 77.42$ $R^2 = 0.21$	$y = -0.040x + 117.00$ $R^2 = 0.31$	$y = -0.001x + 29.41$ $R^2 = 0.00$

Table 2. Time trend equations for minimum temperature (slope of regression in °C/calendar year) over the past three decades at different locations for annual, *kharif* and *rabi* seasons in Punjab

Station (Latitude, longitude and elevation above MSL)	Annual	Kharif	Rabi
Ballowal Saunkhri (31° 60′ N, 76° 23′ E, 355m)	$y = -0.022x + 61.80$ $R^2 = 0.06$	$y = -0.011x + 45.87$ $R^2 = 0.02$	$y = -0.008x + 27.25$ $R^2 = 0.006$
Amritsar (31° 37′ N, 74° 53′ E, 231 m)	$y = -0.004x + 25.17$ $R^2 = 0.017$	$y = -0.002x + 28.13$ $R^2 = 0.004$	$y = -0.006x + 21.01$ $R^2 = 0.02$
Ludhiana (30° 56′ N 75° 48′ E 247 m)	$y = 0.071x - 125.00$ $R^2 = 0.92$	$y = 0.076x - 129.00$ $R^2 = 0.94$	$y = 0.067x - 125.00$ $R^2 = 0.86$
Patiala (30° 20′ N 76° 28′ E 251 m)	$y = 0.015x - 13.34$ $R^2 = 0.43$	$y = 0.022x - 20.81$ $R^2 = 0.59$	$y = 0.010x - 9.803$ $R^2 = 0.15$
Bathinda (30° 12′ N 74° 57′ E 211 m)	$y = 0.038x - 59.57$ $R^2 = 0.59$	$y = 0.025x - 27.30$ $R^2 = 0.25$	$y = 0.053x - 97.18$ $R^2 = 0.71$

3.2 Rainfall Variability Trends

The trend lines obtained by regressing the five-yearly moving averages against time for annual, *kharif* and *rabi* season rainfalls are shown in Table 3 for the locations of Ballowal Saunkhri, Amritsar, Ludhiana, Patiala and Bathinda. In general, no significant change was noted for the annual and seasonal rainfalls at different locations over the past three decades. At Ballowal Saunkhri, the annual, *kharif* and *rabi* rainfalls have decreased at a rate of 16, 12 and 3 mm/year, respectively. At Bathinda, the *rabi* season rainfall has decreased at a rate of 2 mm/year over the past three decades.

Table 3. Time trend equations for rainfall (slope of regression in mm/calendar year) over the past three decades at different locations for annual, *kharif* and *rabi* seasons in Punjab

Station (Latitude, longitude and elevation above MSL)	Annual	Kharif	Rabi
Ballowal Saunkhri (31° 60′ N, 76° 23′ E, 355m)	$y = -16.11x + 3314$ $R^2 = 0.39$	$y = -12.50x + 25948$ $R^2 = 0.34$	$y = -3.26x + 6675$ $R^2 = 0.36$
Amritsar (31° 37′ N, 74° 53′ E, 231 m)	$y = -1.94x + 4579$ $R^2 = 0.06$	$y = -1.32x + 3210$ $R^2 = 0.04$	$y = -1.18x + 2493$ $R^2 = 0.06$
Ludhiana (30° 56′ N 75° 48′ E 247 m)	$y = 3.193x - 5602$ $R^2 = 0.12$	$y = 3.26x - 5860$ $R^2 = 0.13$	$y = -0.476x + 1077$ $R^2 = 0.02$
Patiala (30° 20′ N 76° 28′ E 251 m)	$y = 1.08x - 1358$ $R^2 = 0.007$	$y = 2.25x - 3824$ $R^2 = 0.038$	$y = -1.462x + 3044$ $R^2 = 0.054$
Bathinda (30° 12′ N 74° 57′ E 211 m)	$y = -3.015x + 6562$ $R^2 = 0.034$	$y = -0.276x + 1009$ $R^2 = 0.00$	$y = -2.976x + 6025$ $R^2 = 0.54$

The earlier study conducted by Hundal and Prabhjyot-Kaur (2002) revealed an overall increase in rainfall over a period of 1970-1998 at different locations (Amritsar, Ludhiana, Patiala and Bathinda). However, during the period from 1999 to 2005, below normal rainfall was received at all the five locations during the years 1999, 2002, 2004 and 2005. During the year 2000, below normal rainfall was recorded at Ludhiana and Patiala and during year 2001, below normal rainfall was recorded at all the four locations except Ludhiana. This resulted in arresting the increasing trend of rainfall at different locations in the state. Hence, no significant trend in increase/decrease of rainfall was observed at all the locations except Ballowal Saunkhri where a significant decreasing trend was observed.

4. EFFECT OF CLIMATE CHANGE ON CROP PHENOLOGY, GROWTH AND YIELD: A CASE STUDY

Crop growth and yield were simulated with DSSAT (Decision Support System for Agrotechnology Transfer) models of CERES-Rice (Ritchie et al., 1986), CERES-Wheat (Godwin et al., 1990), PNUTGRO (Boote et al., 1989), SOYGRO (Jones et al., 1988) and CHICKPGRO (Singh and Virmani, 1996) for rice, wheat, groundnut, soybean and gram, respectively under anticipated synthetic climatic scenarios. These models require weather, soil and crop data for simulating the crop phenology, growth and various yield characteristics. The crop models used in this study were: CERES-Rice (Prabhjyot-Kaur and Hundal, 2001), CERES-Wheat (Hundal and Prabhjyot-Kaur, 1997), PNUTGRO (Prabhjyot-Kaur and Hundal, 1999) and SOYGRO (Prabhjyot-Kaur and Hundal, 2002). These models have been validated under Ludhiana (Punjab) environmental conditions for the commonly sown cultivars.

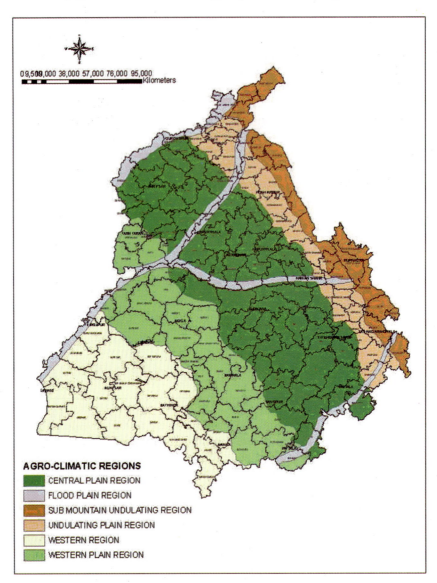

Fig. 2 Agro-climatic map of Punjab State, northern India.

The soil and weather data used in the study were collected from Punjab Agricultural University, Ludhiana, India. It is located at 30° 54′N latitude and 75° 48′E longitude at an elevation of 247 m above mean sea level. This location represents the central irrigated plains of the Indian Punjab in which crops are grown under assured irrigation conditions and hence optimum (non-limiting) moisture conditions were assumed. The simulations were made with the assumption that nutrition was non-limiting and there were no losses from insect-pests.

On the basis of climatic variability trends observed in the state, anticipated synthetic scenarios of increase or decrease from normal temperature, interactions of maximum and minimum temperatures, solar radiation, carbon dioxide levels and intra-seasonal temperature change were generated for the simulation study. One variable at a time was modified and its effect on crop growth and yield was simulated while taking all the other climate variables to be normal. The major reason for using incremental variable scenarios is that they capture a wide range of potential changes. Subsequently, the combination of two variables was interactively modified to assess their combined effect on crop growth and yield.

4.1 Effect of Changes in Temperature

The phenological development, growth and yield attributes of crops were simulated when both maximum and minimum temperatures were increased or decreased by 0.5, 1.0, 2.0 and 3.0 °C from normal while keeping the other climate variables constant. Phenological development of *kharif* crops, i.e., rice, groundnut and soybean was not much affected by increase or decrease in temperature of 1.0 °C from normal (Table 4). On the other hand, phenological development of wheat and gram (*rabi* crops) revealed more drastic changes as the phenology was significantly advanced by increasing temperature but was delayed by decreasing temperature.

The growth and yield of crops was reduced by an increase in temperature, but increased with a decrease in temperature from normal (Table 5). Both the reduction and the increase were more for *rabi* crops (wheat and gram) than for *kharif* crops (rice, groundnut and soybean). With an increase in temperature by 1.0 to 2.0 °C, the simulated maximum leaf area index (LAI) in rice decreased by 3.5 to 9.2%, in wheat by 18.4 to 29.2%, in groundnut by 3.4 to 5.8% and in soybean by 0.3 to 3.0% from normal; biomass yield in rice decreased by 2.3 to 5.0%, in wheat by 13.7 to 22.9%, in groundnut by 2.7 to 5.4%, in soybean by

Table 4. Effect of increase or decrease in temperature from normal on deviations in the phenology (days) of crops

Phenological stages	Temperature level								
	−3.0 °C	−2.0 °C	−1.0 °C	−0.5 °C	Normal*	+0.5 °C	+1.0 °C	+2.0 °C	+3.0 °C
Rice									
Heading date	+5	+2	0	0	223	0	0	+1	+4
Maturity date	+12	+6	+2	0	263	+1	+1	+1	+5
Wheat									
Anthesis date	+25	+17	+8	+3	41	−3	−6	−12	−16
Maturity date	+22	+15	+8	+4	81	−3	−6	−12	−17
Groundnut									
Flowering date	0	0	−1	0	197	0	0	+3	+4
Podding date	0	0	−1	0	218	0	+1	+5	+4
Maturity date	+6	+2	−1	0	285	0	+2	+5	+9
Soybean									
Flowering date	0	0	−1	−1	239	+1	+2	+3	+4
Podding date	0	0	0	0	260	+1	+2	+3	+3
Maturity date	+2	+1	0	0	294	+1	+1	+2	+2
Gram									
Flowering date	+35	+22	+11	+4	08	−4	−7	−19	−23
Podding date	+34	+23	+12	+5	30	−4	−9	−16	−22
Maturity date	+27	+18	+9	+4	99	−5	−8	−16	−24

*Julian day.

Table 5. Effect of increase or decrease in temperature from normal on deviations (%) in the growth and yield attributes of crops

Growth/Yield attributes	Temperature level								
	−3.0 °C	−2.0 °C	−1.0 °C	−0.5 °C	Normal	+0.5 °C	+1.0 °C	+2.0 °C	+3.0 °C
Rice									
Maximum LAI	+7.69	+5.24	+3.15	+1.22	5.72•	−1.75	−3.50	−9.26	−12.94
Grain yield	+26.81	+15.15	+8.07	+6.56	6692♦	−0.16	−2.82	−9.59	−10.14
Biomass yield	+19.93	+9.92	+4.35	+2.93	11717♦	−0.94	−2.35	−5.02	−6.06
Wheat									
Maximum LAI	+41.08	+27.84	+11.62	+5.14	3.70•	−5.94	−18.38	−29.19	−38.90
Grain yield	+9.85	+7.38	+7.16	+6.26	4932♦	−2.75	−9.87	−18.02	−27.03
Biomass yield	+20.84	+16.07	+9.12	+4.11	13304♦	−4.60	−13.76	−22.87	−32.35
Groundnut									
Maximum LAI	+17.10	+11.65	+6.02	+3.76	5.32•	−1.13	−3.38	−5.82	−7.70
Seed yield	+27.75	+16.53	+6.21	+5.31	998♦	−4.01	−4.51	−10.62	−13.13
Biomass yield	+20.50	+12.33	+4.98	+4.37	11467♦	−2.19	−2.79	−5.42	−7.91
Soybean									
Maximum LAI	+3.05	+3.21	+0.76	−0.15	6.55•	−0.31	−0.31	−3.05	−7.63
Seed yield	−11.60	−4.38	−1.61	+0.02	1982♦	+3.98	+2.37	+2.42	−5.60
Biomass yield	−1.93	+1.14	+0.89	+0.80	7369♦	+0.09	−1.25	−3.42	−8.12
Gram									
Maximum LAI	−30.30	−21.30	−10.11	−7.86	0.89•	+7.86	+17.97	+37.08	+47.19
Seed yield	−17.87	−13.94	−7.24	−7.04	1449♦	+0.21	+8.14	+16.01	+17.87
Biomass yield	−23.50	−16.17	−7.51	−6.14	3808♦	+2.25	+8.85	+15.15	+14.20

• Maximum LAI (dimensionless); ♦ Grain/Seed/Biomass Yield (kg/ha).

1.2 to 3.4% and in gram by 8.8 to 15.1% from normal; grain/seed yield in rice decreased by 2.8 to 9.6%, in wheat by 9.8 to 18.0% and in groundnut by 4.5 to 10.6% from normal. However, in gram crop, with an increase in temperature by 1.0 to 2.0 °C, the simulated maximum leaf area index (LAI) increased by 17.9 to 37.0%, biomass yield increased by 8.8 to 15.1%, seed yield increased by 8.1 to 16.0%, while seed yield in soybean also increased by 2.3 to 2.4% from normal.

A decrease in temperature by 1.0 to 2.0 °C led to an increase in the simulated maximum leaf area index (LAI) in rice by 3.1 to 5.2% and in wheat by 11.6 to 27.8% from normal; simulated biomass yield in rice by 4.3 to 9.9%; in wheat by 9.1 to 16.1%, in groundnut by 4.9 to 12.3% and in soybean by 0.8 to 1.1%; grain/seed yield in rice by 8.0 to 15.1%, in wheat by 7.2 to 7.4% and in groundnut by 6.2 to 16.5% from normal. On the other hand, with a decrease in temperature by 1.0 to 2.0 °C, the simulated maximum LAI in gram decreased by 10.1 to 21.3%, biomass yield decreased by 7.5 to 16.1%, seed yield decreased by 7.2 to 13.9%, while seed yield in soybean also decreased by 1.6 to 4.3% from normal.

4.2 Effect of Changes in Solar Radiation

The effects of increase or decrease in solar radiation on growth and yield of crops are shown in Table 6. In general, increase in solar radiation favored the growth and yield of crops, whereas the decrease in solar radiation favored reduction in growth and yield of crops. With an increase in solar radiation by 5.0%, the simulated maximum LAI increased in rice by 2.6%, in wheat by 4.0%, in groundnut by 5.8%

and in gram by 2.2% from normal; biomass yield increased in rice by 4.6%, in wheat by 3.9%, in groundnut by 7.4%, in soybean by 1.0% and in gram by 3.6% from normal; grain/seed yield increased in rice by 6.2%, in wheat by 3.6%, in groundnut by 7.7%, in soybean by 2.3% and in gram by 3.5% from normal. On the other hand, with a decrease in solar radiation by 5.0%, the simulated maximum LAI decreased in rice by 2.7%, in wheat by 4.5% and in groundnut by 1.1%, from normal; biomass yield decreased in rice by 4.4 %, in wheat by 4.3%, in groundnut by 2.7%, in soybean by 2.8% and in gram by 4.0% from normal; grain/seed yield decreased in rice by 6.0%, in wheat by 3.8%, in groundnut by 3.1%, in soybean by 2.8% and in gram by 4.0% from normal.

Table 6. Effect of increase or decrease in radiation from normal on deviations (%) in the growth and yield attributes of crops

Growth/Yield attributes	Radiation level (Percent deviation from normal)						
	−5.0	−2.5	−1.0	Normal	+1.0	+2.5	+5.0
Rice							
Maximum LAI	−2.79	−1.39	−0.52	5.72•	+0.52	+1.57	+2.62
Grain yield	−6.00	−3.00	−1.19	6692♦	+1.19	+3.19	+6.20
Biomass yield	−4.44	−2.19	−0.86	11717♦	+0.86	+3.03	+4.63
Wheat							
Maximum LAI	−4.59	−2.16	−0.81	3.70•	+0.81	+2.16	+4.05
Grain yield	−3.85	−1.90	−0.75	4932♦	+0.75	+1.84	+3.65
Biomass yield	−4.38	−2.12	−0.83	13304♦	+0.83	+2.02	+3.95
Groundnut							
Maximum LAI	−1.13	+0.56	+1.69	5.32•	+3.01	+4.13	+5.82
Seed yield	−3.11	−0.40	+1.30	998♦	+3.41	+5.01	+7.72
Biomass yield	−2.78	−0.20	+1.33	11467♦	+3.37	+4.89	+7.40
Soybean							
Maximum LAI	0.00	0.00	0.00	6.55•	−0.15	−0.31	−0.61
Seed yield	−2.87	−1.36	−0.50	1982♦	+0.55	+1.26	+2.32
Biomass yield	−2.89	−1.39	−0.05	7369♦	+0.53	+0.91	+1.03
Gram							
Maximum LAI	0.00	−1.12	0.00	0.89•	0.00	0.00	+2.25
Seed yield	−4.00	−1.93	−0.76	1449♦	+0.69	+1.79	+3.59
Biomass yield	−4.02	−1.94	−0.76	3808♦	+0.76	+1.89	+3.68

• Maximum LAI (dimensionless); ♦ Grain/Seed/Biomass Yield (kg/ha).

4.3 Effect of Interactions between Maximum and Minimum Temperatures

When the maximum temperature decreased by 0.25 to 1.0 °C from normal and minimum temperature increased simultaneously from 1 to 3 °C from normal keeping the other climate variables constant, the phenology of rice, wheat and gram were advanced by as much as 1 to 15 days (Table 7). When the minimum temperature increased by 1.0 to 3.0 °C and maximum temperature decreased by 0.25 to 1.0 °C from normal, the heading in rice was advanced by 1 to 4 days, while the physiological maturity was advanced by 2 to 8 days from normal; in groundnut physiological maturity was advanced by 1 to 3 days from normal; in wheat both the anthesis and maturity were advanced by up to eight days from normal; and in gram flowering was advanced by 2 to 11 days while podding and physiological maturity were

Table 7. Effect of increasing minimum temperature and decreasing maximum temperature on deviations (days) in the phenology of crops

Phenological stages	Minimum temperature								
	At +1.0 °C			At +2.0 °C			At +3.0 °C		
	Maximum temperature			Maximum temperature			Maximum temperature		
	−0.25 °C	−0.5 °C	−1.0 °C	−0.25 °C	−0.5 °C	−1.0 °C	−0.25 °C	−0.5 °C	−1.0 °C
Rice									
Heading date	−1	−1	−2	−2	−3	−3	−4	−4	−4
Maturity date	−2	−2	−3	−4	−5	−4	−7	−8	−8
Wheat									
Anthesis date	−2	−2	0	−6	−4	−3	−8	−8	−6
Maturity date	−1	−1	+1	−5	−4	−3	−8	−7	−6
Groundnut									
Flowering date	0	0	−1	0	0	0	+1	0	0
Podding date	0	0	−1	0	0	0	+1	0	0
Maturity date	−2	−2	−3	−2	−2	−3	−1	−2	−3
Soybean									
Flowering date	+1	+1	+1	+1	+1	+1	+3	+3	+2
Podding date	+1	+1	+1	+1	+1	+1	+2	+2	+2
Maturity date	0	0	0	0	0	0	+1	+1	+1
Gram									
Flowering date	−3	−3	−2	−5	−5	−6	−11	−10	−9
Podding date	−4	−4	−2	−10	−10	−7	−15	−14	−13
Maturity date	−4	−4	−2	−10	−9	−8	−15	−14	−13

advanced by 2 to 15 days from normal. On the other hand, under similar interactive scenario of decreasing maximum temperature and increasing minimum temperature, the phenological development of soybean was either not affected or was delayed by 1 to 3 days from normal.

The effect of increasing minimum temperature and decreasing maximum temperature on simulated maximum LAI, biomass yield and grain/seed yield for crops are shown in Table 8. It is apparent from this table that generally, the maximum LAI, biomass yield and grain/seed yield of crops were adversely affected by increasing the minimum temperature from normal. However, these adverse effects were partially counteracted by decreasing maximum temperature from normal. When minimum temperature increased by 1.0 °C and maximum temperature decreased by 0.25 to 1.0 °C from normal, the deviations in the growth and yield attributes were low and the yields were not affected significantly. At further higher levels of increase in minimum temperature, reductions in growth and yield were greater and more so in *rabi* season crops (wheat and gram) than in *kharif* season crops (rice, groundnut and soybean).

4.4 Effect of Interactions between Temperature and Solar Radiation

The effect of increasing temperature (by 1, 2 and 3 °C from normal), and decreasing radiation levels (by 1, 2 and 5%) on maximum LAI, biomass and grain/seed yield of crops are shown in Table 9. When temperature increased by 1.0 °C and radiation levels decreased by 1, 2.5 and 5% from normal, the maximum LAI decreased respectively in rice by 4.0, 4.9 and 6.6%, in wheat by 19.2, 20.8 and 23.2% and

Table 8. Effect of increasing minimum temperature and decreasing maximum temperature on deviations (per cent) in the growth and yield attributes of crops

Growth/Yield attributes	Minimum temperature								
	At +1.0 °C			At +2.0 °C			At +3.0 °C		
	Maximum temperature			Maximum temperature			Maximum temperature		
	−0.25 °C	−0.5 °C	−1.0 °C	−0.25 °C	−0.5 °C	−1.0 °C	−0.25 °C	−0.5 °C	−1.0 °C
Rice									
Maximum LAI	−0.52	−0.52	−0.52	−1.39	−2.45	−0.87	−5.24	−3.84	−2.45
Grain yield	+1.45	+1.45	+1.40	−3.55	−4.38	−2.25	−9.71	−7.94	−7.78
Biomass yield	−0.82	−0.82	−0.58	−2.94	−3.36	−2.29	−6.28	−5.60	−3.35
Wheat									
Maximum LAI	−2.97	−2.97	−3.51	−16.21	−15.40	−12.16	−40.81	−20.27	−17.84
Grain yield	+2.67	+2.67	+5.61	−3.22	−2.27	+0.06	−6.33	−3.38	−3.22
Biomass yield	−2.02	−2.02	+0.28	−12.19	−10.68	−9.18	−16.02	−15.59	−12.38
Groundnut									
Maximum LAI	+0.18	+0.18	+0.94	−0.93	−1.13	−1.13	−2.25	−1.32	−1.50
Seed yield	−1.90	−1.90	−0.90	−4.20	−3.11	−4.41	−6.8	−6.81	−6.51
Biomass yield	−0.18	−0.18	−0.29	−2.27	−1.70	−1.58	−2.44	−2.87	−2.87
Soybean									
Maximum LAI	+0.15	+0.15	+0.61	−1.22	−0.92	−0.61	−0.92	−0.61	−0.76
Seed yield	+2.22	+2.22	+3.33	+2.37	+2.97	+4.18	+5.09	+5.8	+7.87
Biomass yield	−0.23	−0.03	+0.35	−1.64	−1.32	−0.62	−1.97	−1.52	−0.76
Gram									
Maximum LAI	+3.37	+7.87	+7.87	+16.85	+17.97	+2.25	+30.34	+34.83	+33.7
Seed yield	−4.62	−0.21	−0.21	+0.48	+5.11	+5.79	+7.18	+9.87	+9.32
Biomass yield	−1.71	+18.64	+18.64	+3.72	+5.93	+7.58	+8.69	+11.24	+10.5

Table 9. Effect of increasing temperature above normal and decreasing radiation below normal on deviations (percent) in the growth and yield attributes of crops

Growth and yield attributes	Temperature change +1 °C			Temperature change +2 °C			Temperature change +3 °C		
	Radiation change (%)			Radiation change (%)			Radiation change (%)		
	−1.0	−2.5	−5.0	−1.0	−2.5	−5.0	−1.0	−2.5	−5.0
Rice									
Maximum LAI	−4.02	−4.90	−6.64	−9.97	−11.19	−13.29	−13.81	−15.73	−18.88
Grain yield	−4.05	−5.89	−8.97	−10.82	−12.64	−15.69	−11.39	−13.28	−16.42
Biomass yield	−3.27	−4.69	−7.08	−6.03	−7.58	−10.25	−7.14	−8.84	−11.73
Wheat									
Maximum LAI	−19.19	−20.81	−23.24	−30.00	−31.35	−33.78	39.73	−41.08	−43.24
Grain yield	−10.69	−11.94	−14.05	−18.86	−20.11	−22.26	−27.86	−29.70	−31.22
Biomass yield	−14.67	−16.06	−18.40	−23.76	−25.10	−27.36	−33.19	−34.48	−36.64

Contd...

(Contd.)

Growth and yield attributes	Temperature change +1 °C			Temperature change +2 °C			Temperature change +3 °C		
	Radiation change (%)			Radiation change (%)			Radiation change (%)		
	−1.0	−2.5	−5.0	−1.0	−2.5	−5.0	−1.0	−2.5	−5.0
Groundnut									
Maximum LAI	−1.31	−2.25	−3.94	−3.20	−4.32	−6.01	−5.63	−6.57	−8.27
Seed yield	−2.71	−4.25	−6.89	−8.71	−10.22	−12.7	−11.22	−12.6	−15.03
Biomass yield	−0.91	−2.41	−4.96	−3.16	−4.64	−7.13	−5.26	−6.7	−9.12
Soybean									
Maximum LAI	0.00	−0.15	−0.31	−3.05	−3.05	−3.05	−7.63	−7.63	−7.79
Seed yield	+1.82	+0.95	−0.55	+1.87	+1.01	−0.55	−6.1	−6.91	−8.37
Biomass yield	−1.79	−2.63	−4.13	−3.95	−4.77	−6.24	−8.63	−9.43	−10.83
Gram									
Maximum LAI	+16.85	+15.73	+14.61	+35.96	+34.83	+32.50	+46.06	+43.82	+41.57
Seed yield	+7.25	+5.79	+3.38	+14.91	+13.25	+10.35	+16.63	+14.83	+11.73
Biomass yield	+7.9	+6.49	+3.99	+14.08	+12.39	+9.53	+13.02	+11.26	+8.22

in groundnut by 1.3, 2.2 and 3.9% from normal. Under same levels of temperature and radiation, the grain/seed yield decreased respectively in rice by 4.0, 5.8 and 8.9%, in wheat by 10.7, 11.9 and 14.1% and in groundnut by 2.7, 4.2 and 6.8% from normal, whereas the biomass yield in rice decreased by 3.3, 4.7 and 7.1%, in wheat by 14.7, 16.1 and 18.4% and in groundnut by 0.9, 2.4 and 4.9% from normal. The interactive effects of increasing temperature and decreasing radiation revealed a cumulative adverse effect on growth and yield of rice, wheat, groundnut and soybean. However, similar response was not observed for the gram crop.

4.5 Effect of Interactions between CO_2 and Temperature

The direct effects of increased concentrations of CO_2 are generally beneficial to vegetation as elevated levels lead to higher assimilation rates. The interactive effects of increasing CO_2 concentration and increasing temperature on crop growth and yield are shown in Table 10. The results of the simulation study revealed that increasing CO_2 levels were able to counteract the adverse effects of temperature increase on growth and yield of crops to some extent. A temperature increase of 2.0 °C from normal and doubled CO_2 concentration of 600 ppm in rice crop reduced the maximum LAI by 5.5%, biomass yield by 2.6% and grain yield by 2.8% from normal. However, a temperature increase of 2.0 °C from normal and doubled CO_2 concentration of 600 ppm increased the maximum LAI of wheat, groundnut, soybean and gram by 2.8, 35.6, 25.6 and 64.4% from normal, respectively; biomass yield of wheat, groundnut, soybean and gram by 3.9, 31.4, 28.6 and 58.4% from normal, respectively and grain yield of wheat, groundnut, soybean and gram by 5.6, 30.1, 35.2 and 55.7% from normal, respectively.

4.6 Effect of Intra-Seasonal Temperature Change

Wheat is a major winter cereal crop in northern India and it requires cool climate during its early growth stages for potential productivity. Any abrupt changes in weather parameters, especially an increase in

Table 10. Effect of increasing temperature and CO_2 above normal on deviations (percent) in the maximum LAI, grain yield and biomass yield of crops

Growth and yield attributes	Temperature change from normal +1 °C				Temperature change from normal +2 °C			
	CO_2 concentration (ppm)				CO_2 concentration (ppm)			
	330 (Normal)	400	500	600	330 (Normal)	400	500	600
Rice								
Maximum LAI	−9.3	−6.1	−4.0	+0.8	−12.3	−11.9	−7.8	−5.5
Grain yield	−6.6	−4.3	−2.8	+0.5	−7.5	−7.2	−4.4	−2.8
Biomass yield	−6.0	−4.0	−2.9	+0.8	−7.3	−7.1	−4.0	−2.6
Wheat								
Maximum LAI	−18.3	−11.2	−2.3	+7.8	−29.1	−17.6	−4.5	+2.8
Grain yield	−9.9	−5.6	+2.1	+10.4	−18.0	−10H.4	−1.4	+5.6
Biomass yield	−13.7	−10.7	−1.4	+8.6	−22.9	−12.5	−3.3	+3.9
Groundnut								
Maximum LAI	−3.4	+17.3	+34.5	+47.2	−5.8	+15.4	+28.3	+35.6
Seed yield	−4.5	+14.1	+29.0	+38.4	−10.6	+12.0	+25.7	+30.1
Biomass yield	−2.8	+14.9	+29.9	+40.6	−5.4	+12.1	+25.6	+31.4
Soybean								
Maximum LAI	−0.3	+12.3	+23.3	+27.8	−3.0	+11.4	+19.6	+25.6
Seed yield	+2.4	+12.5	+25.9	+35.0	+2.4	+12.6	+25.9	+35.2
Biomass yield	−1.2	+11.6	+23.4	+31.6	−3.4	+11.4	+22.9	+28.6
Gram								
Maximum LAI	+17.9	+24.4	+51.2	+64.5	+27.1	+33.9	+51.2	+64.4
Seed yield	+8.1	+21.1	+43.3	+58.8	+16.0	+22.5	+44.2	+55.7
Biomass yield	+8.8	+21.8	+45.0	+57.9	+15.1	23.5	+45.7	+58.4

maximum/minimum temperature from normal at any growth stage of crop adversely affects the growth and ultimately the potential yield of wheat. A simulation study was conducted using CERES-Wheat model to assess the effect of intra-seasonal increase of temperature from normal on yield of wheat sown on different dates (Prabhjyot-Kaur et al., 2007). The simulation study was carried out with the assumption that weather remained normal in rest of the crop growth period, and the crop remained free from water and nutrient stress and pest infestation. The simulation results revealed that in general, an increase in temperature from mid-February to mid-March severely affected the yield of early, normal and late sown wheat (Table 11). A further scrutiny revealed that the temperature increase mostly affected the early (October) sown crop during 4th week of January, February and up to 1st fortnight of March; the timely (November) sown crop during February and March; the late (4th week of November) sown crop during March; and very late (December) sown crop during March and 1st week of April.

The analysis revealed that an increase of temperature from normal decreased the grain yield of wheat at the following rates (Table 12):

- Temperature increase in the 4th week of January decreased the grain yield by 0.99, 0.66 and 0.70% per degree Celsius for wheat sown in the 4th week of October, 1st week of November, and 2nd week of November, respectively.
- Temperature increase in the 1st fortnight of February decreased the grain yield by 2.88 and 1.87% per degree Celsius for wheat sown in the 4th week of October, and 1st week of November, respectively.

Table 11. Effect of intra-seasonal temperature increase from normal on the grain yield (% deviation) of wheat sown on different dates

Month	Time period	Temperature increase from normal					
		+1.0 °C	+2.0 °C	+3.0 °C	+4.0 °C	+5.0 °C	+6.0 °C
Early sown (28th October)							
January	Last week	−3.0	−3.1	−3.1	−6.3	−6.8	−7.1
February	1st fortnight	−3.4	−3.7	−7.6	−11.5	−13.0	−17.2
	2nd fortnight	−2.4	−2.8	−5.2	−8.1	−10.9	−13.8
March	1st fortnight	−2.3	−4.6	−6.8	−13.8	−8.2	−10.4
Normal sown (8th November)							
January	Last week	+1.9	+0.1	+0.4	−1.5	−2.0	−1.1
February	1st fortnight	+1.7	−1.6	−1.8	−3.9	−7.7	−7.3
	2nd fortnight	−0.4	−4.1	−5.1	−9.9	−14.2	−16.4
March	1st fortnight	−2.7	−3.3	−6.0	−9.5	−9.5	−13.0
	2nd fortnight	+1.1	−1.5	−0.5	−0.1	−1.9	−1.5
Normal sown (15th November)							
January	Last week	+1.8	+1.7	+1.4	+0.5	−0.2	−1.8
February	1st fortnight	−0.5	−2.7	−1.5	−2.0	−1.3	−1.9
	2nd fortnight	−2.0	−5.8	−6.0	−8.7	−9.7	−14.2
March	1st fortnight	−4.8	−9.3	−10.1	−14.2	−16.0	−20.8
	2nd fortnight	−2.5	−1.6	−4.3	−6.9	−5.9	−8.1
Late sown (25th November)							
January	Last week	+0.1	+3.7	+3.5	+3.4	+3.4	+7.2
February	1st fortnight	+0.5	+2.4	+2.8	+4.7	+4.9	+7.0
	2nd fortnight	+2.5	+1.1	+3.4	−0.6	−2.6	−3.3
March	1st fortnight	−0.5	−5.4	−6.7	−3.3	−16.0	−19.4
	2nd fortnight	−0.1	−4.7	−5.6	−9.2	−10.1	−11.2
Late Sown (2nd December)							
January	Last week	0.0	+0.6	+0.6	+0.5	+0.6	+3.4
February	1st fortnight	+0.7	+0.6	+0.6	+3.4	+3.6	+3.7
	2nd fortnight	−0.5	−0.4	−1.7	−2.3	−3.1	−3.6
March	1st fortnight	−2.3	−1.6	−6.8	−7.6	−12.5	−17.7
	2nd fortnight	−5.5	−6.6	−12.3	−14.5	−19.1	−21.4
April	1st week	−1.4	−1.8	−2.6	−2.2	−3.5	−3.1

Table 12. Rate of change (increase/decrease) from normal in the grain yield of wheat sown on different dates due to intra-seasonal temperature increase from normal

Time period of temperature increase	Date of sowing	Rate of change in grain yield (Percent/°C)
Fourth week of January	Early sown (28th October)	−0.99
	Normal sown (8th November)	−0.66
	Normal sown (15th November)	−0.70
	Late sown (25th November)	+0.98
	Late sown (2nd December)	+0.48

Contd...

(*Contd.*)

First fortnight of February	Early sown (28th October)	−2.88
	Normal sown (8th November)	−1.87
	Normal sown (15th November)	–
	Late sown (25th November)	+1.19
	Late sown (2nd December)	+0.76
Second fortnight of February	Early sown (28th October)	−2.40
	Normal sown (8th November)	−3.30
	Normal sown (15th November)	−2.15
	Late sown (25th November)	−1.26
	Late sown (2nd December)	−0.69
First fortnight of March	Early sown (28th October)	−2.40
	Normal sown (8th November)	−2.10
	Normal sown (15th November)	−2.98
	Late sown (25th November)	−3.51
	Late sown (2nd December)	−3.15
Second fortnight of March	Normal sown (15th November)	−1.24
	Late sown (25th November)	−2.15
	Late sown (2nd December)	−3.40
First week of April	Late sown (2nd December)	−0.38

- Temperature increase in the 2nd fortnight of February decreased the grain yield by 2.40, 3.30, 2.15, 1.26 and 0.69% per degree Celsius for wheat sown in the 4th week of October, 1st week, 2nd week, 4th week of November, and 1st week of December, respectively.
- Temperature increase in the 1st fortnight of March decreased the grain yield by 2.40, 2.10, 2.98, 3.51 and 3.15% per degree Celsius for wheat sown in the 4th week of October, 1st week, 2nd week, 4th week of November, and 1st week of December, respectively.
- Temperature increase in the 2nd fortnight of March decreased the grain yield by 1.24, 2.15 and 3.40% per degree Celsius for wheat sown in the 2nd week, 4th week of November, and 1st week of December, respectively.

5. CONCLUSIONS

Crop productivity is constrained by the inter- and intra-seasonal variation of the weather parameters through their direct effect as well as indirect effects such as weather induced changes in incidence of pests and diseases, and requirement or availability of water for irrigation (Timsina and Humphreys, 2003). The given uncertainties in regional climates are even aggravated by global warming, which may have serious direct and indirect consequences on crop production and hence on food security. It is, therefore, important to have an assessment of the consequences of climatic variability on crops especially cereals and possible adaptation strategies (Pathak and Wassmann, 2009). This chapter has highlighted the climate change/variability in India and as well as assessed the effect of probabilistic climate change scenarios on the growth duration and yield of crops using simulation models for rice, wheat, groundnut, soybean and gram by presenting a case study in Punjab State of northern India.

Crop simulation models are able to analyze how weather and genetic traits can affect the potential productivity under a given set of management practices. They are very useful tools to evaluate the impacts of climate change/variability in a region or country on the direction of changes in phenological development, growth and yield of crops, which in turn help address the food security issue in the face of global climate change. However, the results of such simulation should be viewed in light of the model limitations. The major limitations of such simulation studies are as follows:

- The effects of nutrients other than nitrogen are not simulated.
- The temperatures are increasing as a consequence of greenhouse gases including carbon dioxide. The positive role of carbon dioxide in enhancing photosynthesis and yield of C_3 crops is expected to counteract the negative impacts of increase in temperature and decrease in solar radiation. Such interactive effects of carbon dioxide increase are not accounted for in some simulation studies (Attri and Rathore, 2003; Mall et al., 2004; Pathak and Wassmann, 2009).
- The adverse effects of extreme weather hazards, weeds, insect-pests and disease damage are not considered in the existing crop simulation models.
- Moreover, rise in ambient air temperature coupled with enhanced precipitation levels may create favorable conditions for pest and disease infestation in tropical countries like India. Such interactions are presently not accounted for in simulation studies as the available dynamic simulation models do not simultaneously simulate the interactive effect of the crop and pest.

The results of the case study presented in this chapter revealed that with an increase in temperature above normal, the phenological development in *rabi* season crops (wheat and gram) was advanced, but that of *kharif* season crops (rice, groundnut and soybean) was not much affected. On the other hand, the growth and yield of all the crops was reduced by increase in temperature and decrease in radiation and vice-versa. The interaction effects of simultaneous increase/decrease in parameters were also simulated. When the maximum temperature decreased by 0.25 to 1.0 °C while minimum temperature increased by 1.0 to 3.0 °C from normal, reductions in growth and yield were greater and more so in *rabi* season crops (wheat and gram) than in *kharif* season crops (rice, groundnut and soybean). The enhanced concentrations of CO_2 were able to counteract the adverse effects of temperature increase on growth and yield of crops to some extent. The simulation study conducted to assess the effect of intra-seasonal increase of temperature from normal on yield of wheat sown on different dates revealed that in general, an increase in temperature from mid-February to mid-March severely affected the yield of early, normal and late sown wheat.

India is a large developing country with nearly two-thirds of the population depending directly on the climate sensitive sectors such as agriculture, fisheries and forests. The projected climate change under various greenhouse gas emission scenarios is likely to have implications on food production, water supply, biodiversity and livelihoods (Sinha and Swaminathan, 1991; Sathaye et al., 2006). It is, therefore, important to assess direct and indirect consequences of global warming on different crops (Attri and Rathore, 2003; Pathak and Wassmann, 2009). This necessitates improved scientific understanding, capacity building, networking and broad consultation processes. Thus, future agricultural planning has to take note of the overall goal of attaining congruence in productivity, stability, sustainability, profitability and equity in Indian agriculture in coming decades in order to ensure food security in the country.

REFERENCES

Adams, R.M., Hurd, B., Lenhart, S. and Leary, N. (1998). The effects of global warming on agriculture: An interpretative review. *Journal of Climate Research*, **11**: 19-30.

Attri, S.D. and Rathore, L.S. (2003). Simulation of impact of projected climate change on wheat in India. *International Journal Climatology*, **23**: 693-705.

Bachelet, D. and Gay, C.A. (1993). The impact of climate change on rice yield: A comparison of four models performances. *Ecological Modelling*, **65**: 71-93.

Baer, P. and Risbey, J.S. (2009). Uncertainty and assessment of issues posed by urgent climate change. An editorial comment. *Climatic Change*, **92**: 31-36.

Boote, K.J., Jones, J.W., Hoogenboom, G., Wilkerson, G.G. and Jagtap, S.S. (1989). PNUTGRO V1.02: Peanut Crop Growth and Simulation Model. User's Guide, Florida Agricultural Experiment Station Journal No. 8420, University of Florida, Gainsville, Florida, 76 pp.

De la Mare, W.K. (2009). Changes in Antarctic sea-ice extent from direct historical observations and whaling records. *Climatic Change*, **92**: 461-493.

Farquhar, G.D. (1997). Carbon dioxide and vegetation. *Science*, **278**: 1411.

Gadgil, S. (1996). Climate change and agriculture: An Indian perspective. *In:* Y.P. Abrol, S. Gadgil and G.B. Pant (editors), Climate Variability and Agriculture. Narosa Publishing House, New Delhi, pp. 1-18.

Ghude, S.D., Jain, S.L. and Arya, B.C. (2009). Temporal evolution of measured climate forcing agents at South Pole, Antartica. *Current Science*, **96(1)**: 49-57.

Godwin, D., Ritchie, J.T., Singh, U. and Hunt, L. (1990). A User's Guide to CERES-Wheat V2.10. International Fertilizer Development Center, Muscle Shoals, Alabama, 94 pp.

Gosain, A.K., Rao, S. and Debajit, B.D. (2006). Climate change impact assessment on hydrology of Indian river basins. *Current Science*, **90(3)**: 346-353.

Goswami, B.N., Venugopal, V., Sengupta, D., Madhusoodan, M.S. and Xavier, P.K. (2006). Increasing trend of extreme rain events over India in a warming environment. *Science*, **314**: 1442-1445.

Haskett, J.D., Pachepsky, Y.A. and Acock, B. (1997). Increase of CO_2 and climate change effects on Iowa soybean yield simulation using GLYCIM. *Agronomy Journal*, **89**: 167-176.

Hingane, L.S., Rupa Kumar, K. and Ramana Murty, B.V. (1985). Long-term trends of surface air temperature in India. *Journal of Climatology*, **5**: 521-528.

Hundal, S.S. and Prabhjyot-Kaur (1996). Climatic change and its impacts on crop productivity in Punjab, India. *In:* Y.P. Abrol, S. Gadgil and G.B. Pant (editors), Climate Variability and Agriculture, Narosa Publishing House, New Delhi, pp. 377-393.

Hundal, S.S. and Prabhjyot-Kaur (1997). Application of the CERES-Wheat model to yield prediction in the irrigated plains of the Indian Punjab. *Journal of Agricultural Science* (Cambridge, U.K.), **129**: 13-18.

Hundal, S.S. and Prabhjyot-Kaur (2002). Annual and seasonal climatic variability at different locations of Punjab state. *Journal of Agrometeorology*, **4(2)**: 113-125.

Hundal, S.S. and Prabhjyot-Kaur (2007). Climatic variability and its impact on cereal productivity in Indian Punjab: A simulation study. *Current Science*, **92(4)**: 506-511.

IPCC (2007). Climate Change 2007: Climate Change Impacts, Adaptation and Vulnerability. Working Group II Contribution to Intergovernmental Panel on Climate Change (IPCC) Fourth Assessment Report. Summary for Policy Makers, 23 pp.

Jones, J., Boote, K., Jagtap, S., Hoogenboom, G. and Wilkerson, G. (1988). SOYGRO V5.41 Soybean crop growth simulation model. User's Guide, Florida Agricultural Experiment Station Journal No. 8304, IFAS, University of Florida, Gainesville, FL.

Keeling, C.D. and Whorf, T.P. (2005). Atmospheric CO_2 records from sites in the SIO air sampling network. *In:* Trends: A Compendium of Data on Global Change. Carbon Dioxide Information Analysis Center, Oak Ridge National Laboratory, U.S. Department of Energy, Oak Ridge, TN.

Kumar, K. (1998). Modeling and analysis of global climate change impacts on Indian agriculture. Ph.D. Dissertation, Indira Gandhi Institute of Development Research, Mumbai, India.

Lal, M., Singh, K.K., Rathore, L.S., Srinivasan, G. and Saseendran, S.A. (1998). Vulnerability of rice and wheat yields in NW India to future changes in climate. *Agricultural and Forest Meteorology*, **89**: 101-114.

Mall, R.K., Lal, M., Bhatia, V.S., Rathore, L.S. and Singh, R. (2004). Mitigating climate change impact on soybean productivity in India: A simulation study. *Agricultural and Forest Meteorology*, **121**: 113-125.

Malone, E.L. and Brenkert, A.L. (2008). Uncertainty in resilience to climate change in India and Indian states. *Climatic Change*, **91**: 451-456.

Mearns, L.O., Rosenzweig, C. and Goldberg, R. (1996). The effects of changes in daily and Inter-annual climate variability on CERES-Wheat: A sensitivity study. *Climate Change*, **32**: 257-292.

Olszyk, D.M., Centeno, H.G.S., Ziska, L.H., Kern, J.S. and Matthews, R.B. (1999). Global climate change, rice productivity and methane emissions: Comparison of simulated and experimental results. *Agricultural and Forest Meteorology*, **97**: 87-101.

Pant, G.B. (2003). Long-term climate variability and change over monsoon Asia. *Journal Indian Geophysical Union*, **7(3)**: 125-134.

Pant, G.B. and Rupa Kumar, K. (1997). Climates of South Asia. John Wiley & Sons, Chichester, U.K., 320 pp.

Parry, M.L., Rosenzweig, C., Iglesias, A., Livermore, M. and Fischer, G. (2004). Effects of climate change on global food production under SRES emissions and socio-economic scenarios. *Global Environmental Change*, **14**: 53-67.

Pathak, H. and Wassmann, R. (2009). Quantitative evaluation of climatic variability and risks for wheat yield in India. *Climatic Change*, **93**: 157-175.

Prabhjyot-Kaur and Hundal, S.S. (1999). Forecasting growth and yield of groundnut (*Arachis hypogaea*) with a dynamic simulation model "PNUTGRO" under Punjab conditions. *Journal of Agricultural Science* (Cambridge, U.K.), **133**: 167-173.

Prabhjyot-Kaur and Hundal, S.S. (2001). Forecasting growth and yield of rice with a dynamic simulation model CERES-Rice under Punjab conditions. Proceedings of the National Seminar on Agrometeorological Research for Sustainable Agricultural Production, 27-28 September 2001, Gujarat Agricultural University, Anand, Gujarat, India.

Prabhjyot-Kaur and Hundal, S.S. (2002). Forecasting growth and yield of soybean (*Glycine max. L*) with a dynamic simulation model SOYGRO under Punjab conditions. Proceedings of the Second International Agronomy Congress, 26-30 November 2002, IARI, New Delhi, India, pp. 1385-1387.

Prabhjyot-Kaur, Singh, H. and Hundal, S.S. (2007). Application of CERES-Wheat model in evaluating the impact of within-season temperature rise on wheat yield in Punjab. Proceedings of the National Conference on Impacts of Climate Change with Particular Reference to Agriculture, 22-24 August 2007, Tamil Nadu Agricultural University, Coimbatore, Tamil Nadu, India, pp. 124-125.

Rao, G.D. and Sinha, S.K. (1994). Impacts of climate change on simulated wheat production in India. *In:* C. Rosenzweig and I. Iglesias (editors), Implications of Climate Change for International Agriculture: Crop Modeling Study. Environmental Protection Agency (EPA), USA, pp. 1-10.

Ritchie, J.T., Alocilja, E.C., Singh, V. and Uehara, G. (1986). IBSNAT and CERES-Rice model: Weather and rice. Proceedings of the International Workshop on the Impact of Weather Parameters on Growth and Yield of Rice, 7-10 April 1986, pp. 271-281.

Rosenzweig, C. and Iglesias, A. (1994). Implications of Climate Change for International Agriculture: Crop Modeling Study. Environmental Protection Agency (EPA) 230-B-94-003, http://www.gcrio.org (accessed in March 2009).

Sathaye, J., Shukla, P.R. and Ravindranath, N.H. (2006). Climate change, sustainable development and India: Global and national concerns. *Current Science*, **90(3)**: 314-325.

Singh, P. and Virmani, S.M. (1996). Modeling growth and yield of chickpea (*Cicer arietinum* L.). *Field Crops Research*, **46**: 41-59.

Sinha, S.K. and Swaminathan, M.S. (1991). Deforestation, climate change and sustainable nutrition security: A case study of India. *Climate Change*, **19**: 201-209.

Sontakke, N.A. (1990). Indian Summer Monsoon Rainfall Variability during the Longest Instrumental Period 1813-1988. M.Sc. Thesis, University of Poona, Pune, India.

Timsina, J. and Humphreys, E. (2003). Performance and Application of CERES and SWAGMAN Destiny Models for Rice-Wheat Cropping Systems in Asia and Australia: A Review. CSIRO Land and Water Technical Report 16/03. CSIRO Land and Water, Griffith, NSW 2680, Australia, 57 pp.

Tubeillo, F.N. and Ewert, F. (2002). Simulating the effect of elevated CO_2 on crops: Approaches and applications for climate change. *European Journal of Agronomy*, **18**: 57-74.

Unnikrishnan, A.S., Kumar, K.R., Fernandes, S.E., Michael, G.S. and Patwardhan, S.K. (2006). Sea level changes along the Indian coast: Observations and projections. *Current Science*, **90(3)**: 314-325.

19

Adapting Smallholder Dairy Production System to Climate Change

Smita Sirohi[*], S.K. Sirohi and Poonam Pandey

1. INTRODUCTION

Livestock sector is socially and economically very significant in developing countries like India due to the multi-functionality of livestock performing output, input, asset and socio-cultural functions. Rapid population growth, urbanization and income growth in developing countries is fuelling a massive increase in global demand for food from animal origin. Driven by the drivers of demand, the world agriculture is slated to witness livestock revolution in the next 20 years or so (Delgado et al., 1999). However, with looming threat of climate change posing formidable development challenge to biological production systems, concerns have emerged regarding the ability of the livestock system to sustain increase in supply for keeping pace with the burgeoning demand of livestock products. As the issues of vulnerability and adaptation of livestock production to climate change have begun to occupy the center stage for the future course of development of the sector, this chapter focuses on the coping strategies that will have to be put in place for countering the sensitivity of livestock to changing climate.

The discussion in this chapter deliberates around dairy production as dairying has predominant share in livestock production and population. The dairy sector in India produces output worth Rs. 1245.2 billion (2005-2006) that is 67% of the value of output from livestock sector and highest among all the agricultural commodities. Among the various species of livestock, cattle and buffalo account for 61% of the livestock population in the country. India possesses about 105 million dairy animals (2003 livestock census) producing 100 million tonnes of milk. With 15% of the world milk production, 16 and 58% of world population of cattle and buffaloes, respectively, it is the top ranking country in the world in terms of milk production and number of dairy animals. In this backdrop, this chapter throws light on the sensitivity of livestock production to climate change, particularly in the context of dairy production in India, and focuses on the need to target the climate change adaptation responses. A detailed discussion on the various adaptation strategies that can insulate the smallholder dairy production to climate change vulnerability is also presented.

2. VULNERABILITY OF DAIRY PRODUCTION TO CLIMATE CHANGE

Climate strongly influences the growth, production, reproduction, health and well-being of the livestock through affecting animal physiology; incidence of diseases; feed, fodder and water availability etc.

---

*Corresponding Author

2.1 Heat Stress

Livestock must regulate their body temperature within a relatively narrow range to remain healthy and productive. The ambient temperature below or above the thermoneutral range creates stress conditions in animals. The approximate thermal-comfort zone for optimum performance of adult cattle is reported to be 5 to 15 °C (Hahn, 1999), however, significant changes in feed intake or in numerous physiological processes will not occur within the range of 5 to 25 °C (McDowell, 1972).

Increase in ambient temperature decreases the difference between the temperature of the animal's surroundings and its body, hence, increasing reliance on evaporative cooling (sweating and panting) to dissipate body heat. In the situation of high relative humidity the effectiveness of evaporative cooling is reduced. Thus, during hot, humid weather conditions the cow cannot eliminate sufficient body heat and suffers from heat stress. The critical values for minimum, mean and maximum Temperature Humidity Index (THI), which incorporates the combined effects of temperature and relative humidity, are determined to be 64, 72 and 76 respectively (Igono et al., 1992).

Net effect of heat stress is increase in heat loss by evaporation and decrease in heat production by metabolism. Heat stress induces physiological changes in cattle, which include reduced feed intake and metabolic activity and thereby declining their productivity (NRC, 2001). The estimated milk yield reduction per unit increase in THI was reported to range from 0.20 to 0.32 kg (Ingraham et al., 1979; Ravagnolo et al., 2000). A few studies give a much higher magnitude of decline; for instance, the milk yield for Holestins was observed to decline by 0.88 kg per THI unit increase for the two-day lag of mean THI (West et al., 2003).

Systematic studies of similar nature for the Indian dairy animals are not available; albeit the experimental studies have shown milk yield of crossbred cows in India (e.g., Karan Fries, Karan Swiss and other Holstein and Jersey crosses) to be negatively correlated with temperature-humidity index (Shinde et al., 1990; Kulkarni et al., 1998; Mandal et al., 2002a). The influence of climatic conditions on milk production is also observed for local cows which are more adapted to the tropical climate of India. The rising temperature decreased the total dry matter intake and milk yield in Haryana cows (Lal et al., 1987). The productivity of Sahiwal cows also showed a decline due to increase in temperature and relative humidity (Mandal et al., 2002b). In case of buffaloes also, heat stress has detrimental effect on the reproduction of buffaloes (Kaur and Arora, 1982; Tailor and Nagda, 2005) even though the morphological and anatomical characteristics of buffaloes make them well-suited to hot and humid climates.

Some preliminary estimates of economic losses from heat stress in dairy animals, at the national and sub-national level, work out to be whopping Rs. 2661.62 crores (at 2005-06 prices), about 2% of the value of output from milk group. The economic losses were highest in UP (> Rs. 350 crores) followed by Tamil Nadu, Rajasthan and West Bengal (Fig. 1). With likely increase in temperature due to climate change the heat stress in dairy animals would accentuate, thereby, further increasing the magnitude of economic losses attributable to heat stress. The high resolution climate change scenarios and projections for India, based on regional climate modeling system, known as PRECIS (Providing REgional Climates for Impacts Studies) developed by Hadley Center for Climate Prediction and Research, shows that by the end of the century, the annual mean surface temperature is expected to rise by 2.5 to 5 °C, with warming more pronounced in the northern parts of the country (Kumar et al., 2006).

2.2 Susceptibility to Extreme Events

Besides being susceptible to increased heat stress from climate change, the cattle in India are also exposed to the increased risk of extreme events. UNEP (1989) identifies India among the 27 countries that are

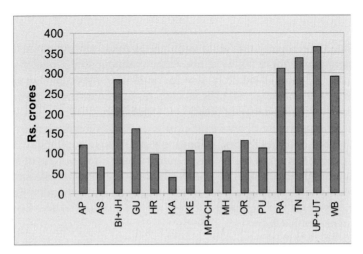

Fig. 1 Statewise value of milk production losses due to heat stress (Sirohi, 2007).

most vulnerable to increase in the frequency and intensity of extreme events, such as heat waves, storm surges, droughts, floods, etc. Simulation models show an increase in frequencies of tropical cyclones in the Bay of Bengal; particularly intense events are projected during the post-monsoon period (Sathaye et al., 2006). In the recent past, there has been an increase in the occurrence of extreme temperature events together with changes in its intensity and pattern (Dash et al., 2007). In 2003, during the 23-day heat wave period from 19 May to 10 June, the highest maximum temperature varied between 45 and 50 °C at four stations on the east coast, breaking their 100 years record in maximum temperatures. During this period, Andhra recorded the highest daily maximum temperature in the country instead of the northwest and central plains of India where such temperature peaks occur normally. Similar type of unusual severe heat waves took place in Orissa during May-June 1998.

Animal functions can become impaired when the intensity and duration of adverse environmental conditions exceed threshold limits with little or no opportunity for relief (Hahn and Becker, 1984; Hahn, 1999). The short-term extreme events (e.g., summer heat waves, winter storms) can result in the death of vulnerable animals (Balling, 1982; Hahn and Mader, 1997). In India, every year thousands of cattle are lost due to heavy rains, floods and cyclones in various parts of the country. During 1953-97, on an average about 93.7 thousand cattle were lost each year due to floods. In 2000, heavy rains and flooding during the Southwest monsoon caused the death of nearly 93 thousand cattle, of which 83.6 thousand died in the state of West Bengal (CSO, 2000). The extreme significance of impact related to climate variability was demonstrated in the 1999 tropical cyclone that hit the state of Orissa, which resulted in a death toll of about 55,000 cattle (CSO, 2000). Severe drought conditions in 1987 affected over 168 million cattle in India, due to decline in feed and fodder availability and serious water shortages. In one of the worst drought affected state of Gujarat, 18 million cattle out of 34 million were reported to have died before it rained the next year.

2.3 Incidence of Diseases

Weather has critical effect on the timing and intensity of disease outbreaks. Warm and moist conditions are conducive to the growth of insects. Also, the rates of insect biting and maturation of the microorganisms within them are temperature dependent and increase when the air warms. The potential impact of climate change on livestock diseases would be primarily on vector-borne diseases. Changes in rainfall pattern can influence an expansion of vectors during wetter years. Also, increasing temperatures have supported the expansion of vector populations into cooler areas, either into higher altitude systems or into more temperate zones. For instance, the rapid northward spread of bluetongue disease from southern Europe has been linked to changing climate patterns. Until fairly recently considered an exotic disease, the bluetongue virus (BTV) has spread to northern European countries in endemic proportions with more than 50,000 reported cases during 2007 in the EU.

Climate-driven models of the temporal and spatial distribution of pests, diseases and weeds have been developed for some key species e.g. the temperate livestock tick Haemaphysalis longicornis and the tropical cattle tick Boophilus microplus (Ralph, 1987). Potential climate change impacts on buffalo fly and sheep blowfly have also been inferred (Sutherst et al., 1996). Climate scenarios in New Zealand and Australia have suggested increased incidence of epidemics of animal diseases as vectors spread and extension of cattle tick infestations, both of which are directly related to changes in temperature and rainfall (Sutherst, 1995).

In India, the incidence of livestock diseases and epidemics is very high. The magnitude of the economic losses due to animal diseases in India is not available, although it is generally agreed that recurring epidemic and other diseases cause phenomenal production losses. The estimated loss in milk and meat production from Foot and Mouth Disease (FMD) was estimated to be around 40-45 billion per annum during 1990-2001 (GoI, 2002). The meteorological parameters like temperature, humidity and rainfall have bearing on this livestock disease and have been found to explain 52 and 84 percent variations in the seasonality of FMD in cattle in hyper-endemic division of Andhra and meso-endemic region of Maharashtra states, respectively (Ramarao, 1988). Research studies indicate that the outbreak of the FMD is correlated with the mass movement of animals which in turn is dependent on the climatic factors (Sharma et al., 1991). Mastitis is another cattle ailment that inflicts heavy losses on the livestock producers. The production losses from clinical and sub-clinical mastitis were calculated as Rs 28 billion at 1994 prices (Sirohi and Sirohi, 2001), with sub-clinical cases accounting for 78% of the loss. The incidence of clinical mastitis in dairy animals is higher during hot and humid weather due to increased heat stress and greater fly population associated with hot-humid conditions (Singh et al., 1996). In addition, the hot-humid weather conditions were found to aggravate the infestation of cattle ticks like Boophilus microplus, Haemaphysalis bispinosa and Hyalomma anatolicum (Singh et al., 2000; Basu and Bandhyopadhyay, 2004; Kumar et al., 2004). Global warming will create favorable climatic conditions for the growth of causative organisms during most part of the year that will increase the probability of the spread of diseases in any season, causing heavy losses to livestock holders due to decline in milk and meat production, reduced work capacity, increase in abortions, subsequent infertility and sterility of animals.

2.4 Feed and Fodder Shortages

Dairy animals in India either subsist on poor quality grasses available in the pastures and non-pasture lands or are stall-fed, chiefly on crop residues. There is deficit of feed and fodder in the country to the tune of 22% for dry fodder, 62% for green fodder and 64% for concentrates (GoI, 2002). The low productivity of dairy animals in the country is largely attributable to the poor quality and lack of adequate availability of feed resources.

The predicted negative impact of climate change on crop production implies that such shortages would aggravate, further constraining the economic viability of dairy production. Simulations of the impact of climate change on rice and wheat yields for several stations in India using dynamic crop growth models (e.g., WTGROWS, INFOCROP, CERES) indicated that in north India, a 2 °C rise in mean temperature reduced potential grain yields of both the crops by about 15-17% (Aggarwal and Sinha, 1993; Hundal and Kaur, 2007). In Tamil Nadu, during the kharif season, the rice yields are anticipated to reduce by 10-15 percent by 2020 due to temperature and precipitation changes (Geethalakshmi and Dheebakaran, 2008). The magnitude of yield decline would aggravate further to 30-35% by 2050. Similarly, projections of climate induced decrease in yields of coarse cereals have been made (Chatterjee, 1998; Ramakrishna et al., 2000). Notably, wheat straw in the Northern India and paddy straw in the rice dominant regions constitute bulk of dry fodder fed to dairy animals. Nearly 44% of the animal feed produced in India is estimated to come from crop residues, such as rice and wheat straw, stovers of coarse cereals and about one-third comes from cultivated green fodder (NIANP, 2005). The potential decline in production of these cereal crops would decrease the dry fodder availability.

Additionally, the projected adverse effects of rising temperatures on productivity of foodgrains (Rao and Sinha, 1994; Aggarwal, 2000, 2003; Aggarwal and Mall, 2002) could mean bringing in more area under food crops to compensate for decrease in production due to yield effect. The area expansion towards food crops at the expense of fodder crops would further impinge on the availability of feed resources for livestock.

Besides being susceptible to a general temperature increase and changing rainfall pattern, feed and fodder production is also exposed to an increased risk of drought due to climate change. The 1999-2000 drought in the arid state of Rajasthan damaged 7.8 million ha of cropped area in the state and fodder availability fell from 144 to 127 million tons thus affecting about 40 million cattle (CSO, 2000). During the recent all-India drought in 2002, the total fodder production (dry + green) fell to 880 million tonnes, about 8.5% lower than the previous year.

3. TARGETING ADAPTATION RESPONSES

While the overall prognosis for climate change impact on crop and livestock agriculture in tropical regions is not good, an even greater worry is the more substantial impact that will occur on pastoralists and smallholders due to their limited adaptive capacity. The dairy production in India is characterized by predominance of smallholders with the average herd size of 2-3 animals. Not only are the herd size small, 71% of the in-milk dairy animals are owned by small and marginal farmers with weak resource position (Fig. 2).

These farmers face number of constraints in dairy production, such as lack of resource for maintaining animals of high genetic potential, feeding high concentrate diet, providing advanced veterinary health care etc. However, as livestock are vital assets held by poor people and are crucial coping mechanism in variable environments, particularly in the event of weather induced shocks, it is imperative that the adaptation responses to climate change are in accordance with the wherewithal of the smallholders and pastoral families.

4. ADAPTATION STRATEGIES AND OPTIONS

Unlike the well-developed models for adaptation assessment for crops, there is a general lack of simulations of livestock adaptation to climate change. Nevertheless, a wide range of possible adaptation strategies exists—ranging from technological and management interventions for sustaining productivity of animals to market responses, institutional and policy changes for reducing the vulnerability of dairy farmers to climate change. As brought out in the previous section that most of the animals are owned by small producers, the adaptation options discussed in this section focus specifically on measures that are directly

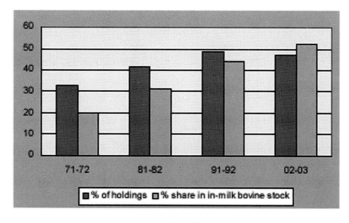

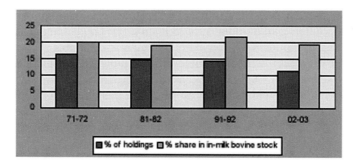

Fig. 2 Share of marginal and small holdings in dairy stock (Source: NSSO, 2006).

relevant for these livestock keepers, while also making a mention of desired macro strategies that would have far reaching implications for adapting small holder dairy production to climate change.

4.1 Shelter Management

In the hot environment, energy exchanges by radiation are dominant, while convective energy exchanges tend to dominate in cold environments. Therefore, the first step to moderate the stressful effects of a hot climate is to protect the dairy animals from direct and indirect solar radiation. Shade against solar radiation could be provided by either trees or constructions made of straw and other locally available material. Trees are considered to be the most effective shades as their leaves are cooled by vaporization, but there are differences among the species with respect to the protection given. For instance, a research study from Brazil observes that best shade was given by the mango tree (*Mangifera indica*), with the least radiant heat load; while the worst tree type was *Pinus*, which presented high heat loads (Waldige, 1994). However, in summers the mango tree is usually avoided as a shade for cattle, because if its fruit is swallowed by a cow, it closes the oesophagus tightly, and may even cause death of the animal.

The appropriate height of the shading structure and the area of shade available per animal are also important in offering adequate protection to the animals. In areas with clear, sunny afternoons, shades should be 3 to 4.5 m high in order to permit maximum exposure to the relatively cool sky, which acts as an efficient radiation sink (Bond et al., 1967). On the other hand, in areas with cloudy afternoons, shades of 2 to 2.5 m in height are better, in order to limit the diffuse radiation received from the clouds by animals beneath the shade (Hahn, 1981). For ascertaining the area of shade that is adequate for a given location/environment, the best way is to observe the behavior of the animals in the range and record the average distance between them. The observed values can then be used in the planning of corrals and housings.

For tropical climates, generally enclosed shelters are not recommended because they decrease natural air velocity. Experimental studies have shown that thick walled, all-brick stall with a black exterior and white-washed under-surface adds to the thermal stress of sheltered crossbreds during Indian summer. Instead, straw thatched sheds or tall, simple, asbestos shades with reflective white outside and absorptive black on the under-surface of roof with plenty of shady trees in the surroundings to act as heat sinks are suitable for crossbred animals (Thomas, 1966). Under semi-arid conditions of Gujarat the thatched roof shed was found to be more conducive for buffaloes as compared to the brick walled shed (Patel et al., 1995).

In Indian conditions, generally loose housing system is presumed to be appropriate except in heavy precipitation areas. However, the loose housing system does not offer much protection during extremes of climate prevalent in northern part of India, such as in cold and chilly winter nights when temperature drops to below 10 °C and in summers when maximum temperature shoots above 40 °C. Aggarwal and Singh (2005) observed the effect of microclimate modification against extreme cold conditions on buffaloes. The feed intake and milk yield of buffaloes under open loose housing system was significantly lower than those housed in sheds with paddy straw bedding. In such areas, semi-loose housing—with sand bedding on the concrete floor in the covered area during summers and straw bedding during winters, together with thatched roofing—provides effective buffer during the extreme climatic conditions to reduce the peak stress on the animals housed.

The basic principle of shelter management to minimize the heat load of animals is construction of farm building to reduce heat gain and promote heat loss from the structures of the animal house by radiation and conduction during summers. In addition, the use of water as a cooling agent has been reported as excellent technique for reducing heat stress (Lin et al., 1996; Frazzi et al., 1997). Sprinklers, evaporative cooling foggers and misters are very effective in evaporating the moisture and cooling the air surrounding the cow, but such proactive measures are beyond the means of smallholder livestock producers. Therefore, alternate cost effective ways of water usage are required for alleviating heat stress. For instance, water application on buffaloes for 15-20 minutes before milking increases the milk yield (Verma and Hussain, 1988). Also, wallowing in buffaloes was reported to be more effective than water showers for increasing milk yield in hot season (Gangwar, 1985; Chauhan, 2004). Management of village ponds is essential for providing clean water for the wallowing buffaloes. Further, low cost, renewable energy operated evaporative cooling systems need to be fabricated as important adaptive devices catering to the needs of rural India.

4.2 Nutritional Management

Reduction in voluntary feed intake in heat stressed dairy animals is the major reason for decrease in milk productivity. In such a situation, the practical approaches to increase the dry matter intake (DMI), which is the key to good performance, include more frequent feeding, improved forage quality, use of palatable feeds, good nutrient balance and greater nutrient (including energy) density. Feed intake declines with hot conditions and rations must be reformulated in an attempt to deliver an adequate quantity of nutrients for sustaining the productivity.

Water is the most important nutrient for the heat stressed dairy animals. Water is lost from the animal body via urine, respiration and sweat. The lactating animal loose additional water through milk as water constitutes 87% of milk. Body of adult animal normally contains 55- 65% of water and a loss of one-fifth of body water is considered to be fatal. Drinking water is the major source of water, and satisfies 80-90% of dairy animals total water needs. Inclusion of green fodders in the diet also helps in catering to the water requirement of the animals. Water consumption is variable, and depends on ambient temperature, DMI, milk yield, sodium intake, physiological stage, minerals present in water etc. In lactating dairy animals the water consumption may be 45-75 liters per day (lpd) per head, while in the dry animals it ranges from 25-40 lpd. Under heat stress, water intake could significantly increase by 120-200%. For *Bos indicus*, for example, water intake increases from about 3 kg per kg DM intake at 10 °C ambient temperature, to 5 kg at 30 °C, and to about 10 kg at 35 °C (NRC, 1981). This increased water intake helps to dissipate heat through the lungs (respiration) and by sweating. The drinking behavior of cows is interesting. Cows spend about six hours a day eating, but only five to ten minutes in drinking. They drink mainly after being milked and when fresh feed is offered. Since it is difficult to define how much water is adequate, it is crucial to supply abundant, clean and easily accessible drinking water to animals all the time.

Managing the feeding schedule of animals is another simple and effective adaptation option. The feeding behavior of animals changes when it is hot. Animals consume more feed during cooler evening hours (West, 1999). Thus, the quantity of feed and the feeding schedule should be adjusted to accommodate this behavior. Having fresh feed in the mangers after milking is a good way to encourage DMI. When the weather is very hot, at least 70% of the daily feed should be given fresh at night. More frequent feeding could keep feed fresher, and encourage cows to eat more frequently, thus stimulating DMI. Theoretically, more frequent feeding might decrease the diurnal fluctuations in metabolites and increase feed utilization efficiency in the rumen (Robinson, 1989). Commercial dairy producers usually believe that frequent feeding is crucial in achieving and maintaining high productivity. This practice could be introduced very easily in our country by smallholders to achieve maximum dry matter intake under heat stress. This practice may be even more important during hot weather, because feed is fermented faster after preparation when air temperatures are high.

The fall in the DMI during hot weather conditions affects the availability of adequate amounts of nutrients viz. energy, protein, and fats. Digestion and metabolism of feed create heat, and this heat production should be cut down as much as possible to provide relief to the animals exposed to environmental heat stress. Each kind of feed has its own heat increment (HI) value that is the value of energy expenditure associated with the digestion and assimilation of food. A diet with a higher nutrient density and low HI (higher energy conversion efficiency) for lactating animals under heat stress is desirable. This is achieved by inclusion of more or highly digestible nutrients rather poor quality roughages like straws.

The dairy rations during hot weather conditions should have low fiber content as there is greater heat production associated with metabolism of acetate compared with propionate. Feeding more concentrates at the expense of fibrous ingredients increases the energy density of rations, and should reduce HI (West, 1999). Also increasing the level of grain in animal ration would reduce the fiber content but a decrease in the fiber level has to be monitored carefully. In some experiments results have indicated that giving cows more grain in their feed leads to a lower rumen pH, especially in hot summer (Mishra et al., 1970), and sorting the feed to remove fiber could make the rumen pH even lower. The level of Acid Detergent Fiber (ADF) should be maintained at a minimum of 18-19%, or alternatively the Neutral Detergent Fiber (NDF) should be at least 25-28% of diet DM. Feeding very high-quality forage to lactating cows in hot summer is also recommended, because it reduces heat build-up and supplies necessary fiber content.

The addition of fat to the diets of lactating dairy animals is another practice that has the potential to reduce the heat increment due to the greater energy density and high energy conversion efficiency of high-fat diet. Although, the results of the research on the positive effects of dietary fat during hot weather are not robust (Skaar et al., 1989; Huber et al., 1994; Chan et al., 1997), biological principles argue in favor of fat supplementation under conditions of heat stress. Extension nutritionists still suggest fat supplements to give a final fat content of 6-7% of diet DM, especially for high milk producing animals. Sources of fat supplements include whole oilseed, crushed oil seeds and/or protected fat products.

The quantity and quality of protein in the diet needs to be considered when feed is being provided for heat-stressed animals as animals suffering from heat stress often have a negative nitrogen (N) balance, because of reduced feed intake. However, simply increasing the level of crude protein (CP) may increase energy requirements and cause problems of environmental pollution as excess dietary protein is converted into urea and excreted. Based on research, it was suggested that during heat stress, the level of crude protein (CP) in the diet should not exceed 18%, while the level of rumen degradable protein should not exceed 61% of CP or 100 grams of N/day (Huber et al., 1994). Cows fed diets with a low level of degradable protein had a higher percentage of milk fat and milk lactose, and a lower level of urea nitrogen in their blood.

Electrolyte minerals, sodium (Na) and potassium (K) are important in the maintenance of water balance, ion balance and the acid-base status of heat-stressed cows. The mineral requirements recommended by the National Research Council (NRC), 1989 do not seem high enough for cows suffering from heat stress. When heat-stressed cows sweat, they lose a considerable amount of K. Increasing the concentration of dietary K to 1.2% or more result in a 3-9% increase in milk yield, and also an increased DMI. Increasing the concentration of sodium in the diet from the NRC recommended level of 0.18% to 0.45% or more improved milk yield by 7-18% (Sanchez et al., 1994). If magnesium oxide (MgO) was added, thus increasing the Mg concentration from 0.25% to 0.44%, the milk yield of heat-stressed cows increased by 9.8% (Teh et al., 1985). In hot weather, the level of milk fat is usually lower. Supplementation with buffers such as sodium bicarbonate ($NaHCO_3$) and magnesium oxide (MgO) is common practice to provide these minerals which also help in maintaining the rumen pH.

Some feed additives were also available in terms of their ability to provide relief animals suffering from heat stress. Most research with lactating cows concerned with direct fed microbials or "probiotic" products deals with either *Aspergillus oryzae* (a mold classified as a fungus) or *Saccharomyces cervisiae* (a yeast). Both *A. oryzae* and *S. cervisiae* may influence the fermentation pattern and microbial population in the rumen (Yoon and Stern, 1996). Results indicated that three grams of *A. oryzae* supplementation had little effect on rectal temperature, respiration rate, or milk composition, but gave a 4% increase in milk yield (1 kg/day) (Huber et al., 1994). On the other hand, trials with niacin feed additive showed that under moderate to severe heat stress conditions, niacin at a rate of 12-36 grams a day lowered skin temperature by about 0.3 °C but it did not improve milk yield (DiCostanzo et al., 1997). In Indian conditions, however, feed additives for managing heat stress may not be economical for low producing animals but if supplemented in high producing animals positive results are expected.

4.3 Health Management

As brought out earlier, air temperature, relative humidity, intensity and duration of sunshine, precipitation etc. have important bearing on the occurrence of livestock diseases and spread of parasites. Successful control of disease requires timely and accurate diagnosis and adequate availability of veterinary medicines and vaccines. To counter the effect of climate change on the incidence of disease, the existing network of veterinary health support services will have to be strengthened in a big way, with particular emphasis on preventive health care services.

Presently, most of the government dispensaries and hospitals are stationary and are primarily engaged in providing curative health cover and breeding services. Only a meager 3.5% of the total staff engaged in livestock health institutions are taking care of disease investigation and control (Ahuja et al., 2003). The state machinery for providing preventive health care services is supported by about 250 disease diagnostic laboratories, 26 veterinary vaccine production units (seven in the private sector), one National Veterinary Biological Products Quality Control Centre and animal quarantine stations at the four metropolitan cities. Disease control programs and vaccinations are, however, sporadic, unsystematic and have limited coverage. Over 25 million vaccinations against FMD are carried out each year, as against the 420 million animals at risk and this does not confer adequate protection against the disease. FMD is a contagious disease and until more than 85% of the animal population in an area is vaccinated the herd immunity cannot be established (Chawla et al., 2004). In light of the fact that immunisation is one of the most economical means of preventing specific diseases, providing long-lasting immunity, the present approach of veterinary health management, needs to undergo a sea change. In general, vaccines offer a substantial benefit for comparatively low cost, a primary consideration for developing countries.

Apart from stepping up the capacities to cater to the infrastructure and pharmaceutical needs for health management, the trained human resources would also be required to meet the enhanced demand for veterinary services. The Tenth Plan Working Group on Animal Husbandry suggested that a minimum of 3,000 veterinarians will be required in the country annually in order to meet the growth rate of 10% by the year 2020. In addition, an input of 3,560 dairy graduates is required in order to sustain a growth rate of 10% in the sector. The current student intake of veterinary colleges may not be enough to fulfill this requirement and hence, an expansion is imperative.

Dealing with emerging challenges of climate change would also require new initiatives aimed at making veterinary services to smallholders more efficient through decision making and planning aids. An initiative in this direction has been proposed by IIT, Mumbai, for taking 'Technology to Masses' through utilization of ICT to provide efficient veterinary services to rural people and their cattle. The design ideas include creating a digitized history of the animals, a system for interaction between cattle owner, veterinary service center and doctor and automatic visit scheduling. Such initiatives need to be up-scaled with additional provision of weather based livestock disease advisory services.

4.4 Managing Common Property Resources

The poor livestock keepers depend heavily on common property resources for their survival. The land holding of these farmers is not sufficient even for crop cultivation. In the absence of CPR lands, they would be unable to maintain their ruminant livestock and shall lose the subsistence source of dairy income as well as face further degradation of their crop land due to deprivation of the only source of manure (dung) and draught power. Unfortunately, there has been rapid deterioration of CPRs on account of physical loss of resources due to infrastructure development, degradation of their physical productivity and re-assignment of usage and property rights. The brunt of developmental pressures may further fall disproportionately on the CPR, more so, as there are few organized efforts for the development of common lands and its sustainable management.

As the change in the status and productivity of common property resources, like pastures and grazing land, village ponds and rivulets etc. will directly influence the economy of the smallholders, particularly the climate change coping options related to nutrition and shelter; the key elements of an approach to regulate the use and enhance regeneration of CPR need to include: (1) introduction of technological investments and creation of economic incentives to conserve such resources while raising their productivity and (2) regulation of common resource use with the involvement of user groups and mobilization of a community strategy that complements state interventions with the essential participation of local people (Jodha, 1995). Available experiences of successful participatory natural resource management initiatives can offer useful lessons for replication.

4.5 Genetic Selection and Conservation

The genotype environment interactions have adapted the Zebu non-descript cattle of India to thermally stressful conditions by reducing the metabolic and heart rate and increasing the sweating capacity. However, as in the process of this adaptation, the low metabolism characters have been selected, it has resulted in low milk production potential of animals. High milk production requires high metabolic activity and more energy expenditure, reducing the capability to withstand high heat loads—a characteristic of *Bos taurus* cattle prevalent in temperate countries. Although the selection of animals in *Bos indicus*, the tropical cattle has been an adaptive one, the requirement of higher milk production led to their crossbreeding with exotic semen from *Bos taurus* cattle. One of the challenges associated with managing high producing cattle in a hot environment is that selection for increased performance is often in conflict with maintaining homeothermy. As genetic variation exists for traits important to thermoregulation, the adaptation challenge is to improve productivity traits while maintaining adaptive traits.

Unfortunately, although the developing countries like India are endowed with vast domestic animal diversity that have high adaptive potential to biotic and abiotic stresses, most of these animal genetic resources are still not characterized and the structured breeding programs are limited. Local breeds are often defined on the basis of subjective data and information obtained from local communities. Reliance on these criteria as the basis for classification for utilisation and/or conservation may be misleading. Hence, application of biotechnology and other advanced techniques for characterization of local breeds and subsequently building inventories, including spatial information, of breeds and valuable breeding stocks can be instrumental in the long run to exploit a genetic approach to heat tolerance while selecting for high milk yield potential. This would also be an adaptation response in the event of loss of local and rare breeds due to increased incidence of droughts, floods, or disease epidemics etc. resulting from climate change.

Besides, characterization of animal genetic resources, its conservation should be one of the priority livestock development activities for developing countries. Experience has indicated that in Indian conditions, "conservation through use" is insufficient due to widespread situation of indiscriminate crossbreeding. Hence, establishment of genebanks for local breeds and ex-situ especially, in vitro conservation needs to be considered as an important component of a broad-based strategy to conserve critical adaptive genes and genetic traits. The ex-situ approaches include cryopreservation of semen, ova and embryos for which the technology is sufficiently developed to be applied in developing countries. There is ardent requirement of financial support for implementing animal genetic resources conservation programs.

Further, development in genetic engineering, cryobiology, cell biology and embryology will provide techniques that may enhance our ability to preserve germplasm in vitro. Techniques such as transfer of DNA within and between species and the production of viable transgenic animals are far from practical application. However, biotechnology will certainly contribute newer and cheaper methods for preservation such as storage of catalogued DNA. At present, other than live animal and embryo preservation, the other techniques do not allow preservation of genomes in a form which can be reactivated in toto at a later stage, but they permit the preservation of individual genes or gene combinations for possible future regeneration (Kannaiyan, 2007).

4.6 Livestock Insurance

The increased risk resulting from climate change faced by the livestock farmers can also be mitigated to certain extent through management of climate perils using the instrument of insurance. In India, the livestock insurance programs are still in nascent stage with extremely low coverage of animals other than commercial poultry. For the dairy animals, the insurance policies only indemnify death due to disease and accident. In the changing circumstances, innovative insurance products will have to be designed aiming at: (a) livestock production insurance that would protect farmers from loss and business interruption due to illness or death as well as recovery of veterinary costs due to on-farm diseases; (b) net revenue insurance for protecting farmers against losses from the market place; and (c) catastrophe insurance, that would protect farmers against extreme price losses due to the emergence of a disease that correlates with rapid decreases in market prices.

The Index Based Livestock Insurance approach adopted in Mongolia to assist the herders in the management of the losses caused by weather induced calamity called *dzuds* can also provide useful learning experience for insuring the smallholders in India against potential threat of climate change to dairy production system.

4.7 Extension Strategy

One important prerequisite for farmers in adapting to the negative effects of climate change is to understand and know its impacts. Although many farmers already use strategies to cope with varying conditions, but as weather becomes less predictable, some of these strategies may no longer work or require additional information in order to remain of value. Hence, effective communication approaches are critical to help farmers adapt to climate change.

Electronic media like radio and television are very effective ways to reach farmers. The available information on climate change is, by and large, not aimed at a farming audience. The challenge for media is to ensure that their clientele understands climate change messages and finds them relevant. Media can also encourage communities to assess local problems, identify local solutions to climate change and establish collective action plans to reduce their vulnerability.

5. CONCLUDING REMARKS

As climate change poses formidable challenge to the development of livestock sector in India, this chapter discussed the vulnerability of dairy animals to climatic conditions and the coping mechanisms that could be instrumental in mitigating the negative effects of changing climate. The anticipated rise in temperature between 2.5 °C and 5 °C over the entire country, together with increased precipitation and occurrence of extreme events resulting from climate change is likely to adversely affect productive and reproductive performance of dairy animals by way of aggravated heat stress, susceptibility to diseases and feed and fodder shortages.

The animals employ physiological mechanisms to counter the heat stress. The adaptation to higher temperature is also complemented by the behavioral process, such as buffaloes prefer wallowing during summer to reduce thermal loads and maintain thermal equilibrium. While new knowledge about animal responses to adverse weather continues to be developed, there is a need for additional knowledge on managing livestock to reduce the impact of climatic changes. Management intervention is needed to ameliorate the constraints on production set by the climate, the physical environment and the health hazards in a region.

Responding to the challenge of climate change requires formulation of a wide range of adaptation options for the smallholder livestock producers, encompassing human intervention for physical modification of the environment, improvement in nutritional management practices, improved animal health technology, genetic selection for stress tolerant high yielding dairy breeds, market responses that are potentially effective adaptation measures to climate change, such as insurance schemes, income diversification opportunities, and institutional and policy changes. The on-farm decisions need to involve selection, design and management of production facilities, while the collective impacts would guide regional or national policy and determine responses to potential large-scale changes. Another critical requirement is the development of collaborative learning processes to support the adaptation of livestock systems to better cope with the impacts of climate change. Research cannot hope to contribute to improving adaptive capacity without a comprehensive understanding of the context in which decisions about adaptation are made and of the capacity of decision makers to change (Thorton et al., 2007). Farmers possess a wealth of indigenous knowledge for dealing with climate variability and risk that should be assessed for its efficacy and subsequently documented for wider dissemination. This exercise is to be done in conjunction with well-targeted capacity building efforts to help farmers deal with changes in their systems that go beyond what they have experienced in the past.

Despite the vital importance of the livestock sector in providing livelihood to the millions of people in India, unfortunately, the sector has by-passed the attention of the planners for reducing its vulnerability to climate change. The National Action Plan on Climate Change (GoI, 2008), which aims to provide the roadmap for the sustained development of various sectors in the future, is virtually silent about the strategies to adapt livestock production to climate change. Its vision for the National Mission for Sustainable Agriculture is essentially crop-centric with only lip-service to the livestock sector by including the use of biotechnology for the development of nutritional strategies to manage heat stress in dairy animals as one of the priority areas under the Mission. The research and development (R&D) efforts to fine-tune and upscale the adaptation measures discussed in this chapter would be instrumental in developing multi-pronged, long-term, integrated and inclusive strategies for insulating the livestock producers from the hazards associated with changing climate. The policy support from the Government through appropriate institutional mechanisms suited for effective delivery of various adaptation options is urgently required for achieving key goals in the context of climate change and livestock sector.

REFERENCES

Aggarwal, P.K. (2000). Application of systems simulation for understanding and increasing yield potential of wheat and rice. Ph.D. Thesis, Wageningen University, The Netherlands, 176 pp.

Aggarwal, P.K. (2003). Impact of climate change on Indian agriculture. *Journal of Plant Biology*, **30**: 189-198.

Aggarwal, P.K. and Mall, R.K. (2002). Climate change and rice yields in diverse agro-environments of India. II. Effects of uncertainties in scenarios and crop models on impact assessment. *Climatic Change*, **52**: 331-343.

Aggarwal, P.K. and Sinha, S.K. (1993). Effects of probable increase in carbon dioxide and temperature on wheat yields in India. *Journal of Agricultural Meteorology*, **48**: 811-814.

Aggarwal, A. and Singh, M. (2005). Impact of microclimate modification on milk production, composition and physiological responses in Murrah buffaloes during winter. *Indian Journal of Dairy Science*, **58(4)**: 269-274.

Ahuja, V., McConnell, K.E., Umali D. and Haan, C. (2003). Are the Poor Willing to Pay for Livestock Services? Evidence from Rural India. *Indian Journal of Agricultural Economics*, **58(1)**: 84-100.

Balling, R.C. Jr. (1982). Weight gain and mortality in feedlot cattle as influenced by weather conditions: refinement and verification of statistical models. *In:* Center for Agricultural Meteorology and Climatology Report, 82-1, University of Nebraska-Lincoln, Lincoln, NE, USA, 52 pp.

Basu, A.K. and Bandhyopadhyay, P.K. (2004). The effect of season on the incidence of ticks. *Bulletin of Animal Health and Production in Africa*, **52(1):** 39-42.

Bond, T.E., Kelly, C.F., Morrison, S.R. and Pereira, N. (1967). Solar, atmospheric, and terrestrial radiation received by shaded and unshaded animals. Transactions of the American Society of Agricultural Engineers, 10, pp. 622.

Chan, S.C., Huber, J.T., Chen, K.H., Simas, J.M. and Wu, Z. (1997). Effects of ruminally inert fat and evaporative cooling on dairy cows in hot environmental temperatures. *Journal of Dairy Science,* **80:** 1172-1178.

Chatterjee, A. (1998). Simulating the increasing of CO_2 and temperature on growth and yield of maize and sorghum. M.Sc. Thesis, Division of Environmental Sciences, Indian Agricultural Research Institute, New Delhi.

Chauhan, T.R. (2004). Feeding strategies for sustainable buffalo production. Proceedings of the XI Animal Nutrition Conference, Jabalpur, M.P., India, pp. 74-83.

Chawla, N.K., Kurup, M.G.P. and Sharma, V.P. (2004). Animal Husbandry. State of the Indian Farmer: A Millennium Study, Vol. XII. Academic Foundation, New Delhi, India.

CSO (2000). Compendium of Environment Statistics. Central Statistical Organisation, Ministry of Statistics and Programme Implementation, Government of India, New Delhi, 306 pp.

Dash, S.K., Jenamani, R.K., Kalsi, S.R. and Panda, S.K. (2007). Some evidence of climate change in twentieth-century India. *Climatic Change*, **85:** 299-321.

Delgado, C., Rosegrant M., Steinfeld, H., Ehui, S. and Courbois, C. (1999). Livestock to 2010: The Next Food Revolution. Food, Agriculture and the Environment Discussion Paper No. 28. International Food Policy Research Institute, Washington D.C., USA.

Di Costanzo, A., Spain, J.N. and Spiers, D.E. (1997). Supplementation of nicotinic acid for lactating Holstein cows under heat stress condition. *Journal of Dairy Science,* **80:** 1200-1206.

Frazzi, E., Calamari, L., Calagari, F., Maianti, M.G., Cappa, V., Bottcher, R.W. and Hoff, S.J. (1997). The aeration with and without misting: effects on heat stress in dairy cows. *Livestock Environment,* **5(2),** Proceedings of the 5th International Symposium, Minnesota, USA, pp. 907-914.

Ganagwar, P.C. (1985). Importance of photoperiod and wallowing in buffalo production. *Indian Journal of Dairy Science*, **38:** 150-155.

Geethalakshmi, V. and Dheebakaran, G. (2008). Impact of climate change on agriculture over Tamil Nadu. In: G.S.L.H.V. Rao Prasada, G.G.S.N. Rao, V.U.M. Rao and Y.S. Ramakrishna (editors), Climate Change and Agriculture Over India. Central Research Institute of Dryland Agriculture (CRIDA), Hyderabad, A.P., pp. 80-93.

GoI (2002). Report of the Working Group on Animal Husbandry & Dairying for the Tenth Five Year Plan (2002-2007). Working Group Sr. No. 42/2001, Government of India, Planning Commission, 214 pp.

GoI (2008). National Action Plan on Climate Change. Prime Minister's Office, Government of India New Delhi.

Hahn, G.L. (1981). Housing and management to reduce climatic impacts on livestock. *Journal of Animal Science,* **52(1):** 175-186.

Hahn, G.L. (1999). Dynamic responses of cattle to thermal heat loads. *Journal of Animal Science*, **77(2):** 10-20.

Hahn, G.L. and Becker, B.A. (1984). Assessing livestock stress. *Agricultural Engineering*, **65:** 15-17.

Hahn, G.L. and Mader, T.L. (1997). Heat waves in relation to thermoregulation, feeding behavior and mortality of feedlot cattle. Proceedings of the 5th International Livestock Environment Symposium, Minneapolis, MN, USA, pp. 563-571.

Huber, J.T., Higginbotham, G., Gomez-Alarcon, R.A., Taylor, R.B., Chen, K.H., Chan, S.C. and Wu, Z. (1994). Heat stress interactions with protein, supplemental fat, and fungal cultures. *Journal of Dairy Science,* **77:** 2080-2090.

Hundal, S.S. and Kaur, P. (2007). Climatic variability and its impact on cereal productivity in Indian Punjab: a simulation study. *Current Science*, **92(4):** 506-511.

Igono, M.O., Bjotvedt, G. and Sanford-Crane, S.T. (1992). Environmental profile and critical temperature effects on milk production of Holstein cows in desert climate. *International Journal of Biometeorology*, **36:** 77-87.

Ingraham, R.H., Stanley, R.W. and Wagner, W.C. (1979). Seasonal effects of tropical climate on shaded and nonshaded cows as measured by rectal temperature, adrenal cortex hormones, thyroid hormone, and milk production. *American Journal of Veterinary Research*, **40**: 1792-1797.

Jodha, N.S. (1995). Common property resources and the dynamics of rural poverty in India's dry regions. *Unasylva*, **46(1)**: 23-29.

Kannaiyan, S. (2007). Conservation of animal genetic resources. Keynote address delivered in the Kerala State Veterinarians Annual Convention: Travancore Meet, Kottayam, Kerala, India.

Kaur, H. and Arora, S.P. (1982). Influence of level of nutrition and season on the oestrus cycle rhythm and on fertility in buffaloes. *Tropical Agriculture*, **59(4)**: 274-278.

Kulkarni, A.A., Pingle, S.S., Atakare, V.G. and Deshmukh, A.B. (1998). Effect of climatic factors on milk production in crossbred cows. *Indian Veterinary Journal*, **75(9)**: 846-847.

Kumar, R.K., Sahai, A.K., Kumar, K.K., Patwardhan, S.K., Mishra, P.K., Revadekar, J.V., Kamala, K. and Pant, G.B. (2006). High-resolution climate change scenarios for India for the 21st century. *Current Science*, **90(3)**: 334-345.

Kumar, S., Prasad, K.D. and Deb, A.R. (2004). Seasonal prevalence of different ectoparasites infecting cattle and buffaloes. *Journal of Research, Birsa Agricultural University*, **16(1)**: 159-163.

Lal, S.N., Verma, D.N. and Husain, K.Q. (1987). Effect of air temperature and humidity on the feed consumption, cardio respiratory response and milk production in Haryana cows. *Indian Veterinary Journal*, **64(2)**: 115-121.

Lin, J.C., Moss, B.R., Koon, J.C., Floed, C.A. and Smita, R.C. (1996). Cows in the mist: Misting system—an efficient way to keep dairy cattle cool. *Highlights of Agricultural Research*, Albama-Agricultural Experimental Station, **43**: 23-24.

Mandal, D.K., Rao, A.V.M.S., Singh, K. and Singh, S.P. (2002a). Effects of macroclimatic factors on milk production in a Frieswal herd. *Indian Journal of Dairy Science*, **55(3)**: 166-170.

Mandal, D.K., Rao, A.V.M.S., Singh, K. and Singh, S.P. (2002b). Comfortable macroclimatic conditions for optimum milk production in Sahiwal cows. *Journal of Applied Zoological Researches*, **13(2/3)**: 228-230.

McDowell, R.E. (1972). Improvement of Livestock Production in Warm Climates. W.H. Freeman & Co., San Francisco, 711 pp.

Mishra, M., Martz, F.A., Stanley, R.W., Johnson, H.D., Campbell, J.R. and Hilderbrand, E. (1970). Effect of diet and ambient temperature-humidity on ruminal pH, oxidation reduction potential, ammonia and lactic acid in lactating cows. *Journal of Animal Science*, **30**: 1023-1028.

NIANP (2005). Feed Base. Animal and Feed Resources Database, National Institute for Animal Nutrition and Physiology (NIANP), Bangalore, Karnataka, India.

NRC (2001). Nutrient Requirement of Dairy Cattle. National Research Council (NRC), National Academy Press, Washington DC, USA.

NRC (1981). Effect of Environment on Nutrient Requirements of Domestic Animals. National Academy Press, Washington DC, USA.

NSSO (2006). Livestock Ownership across Operational Land Holding Classes in India, 2002-03. Report No. 493, National Sample Survey Organization (NSSO), Ministry of Statistics and Programmes Implementation, Government of India, New Delhi, India.

Patel, J.B., Patel, J.P., Pandey, M.P. and Prajapathi, K.B. (1995). Effects of three housing systems and microclimate on growth performance of Mehsana buffalo heifers in semi-arid region of North Gujarat. *Indian Journal of Animal Production and Management*, **11**: 156-160.

Ralph, W. (1987). A simple model predicts an insect's distribution. *Rural Research*, **136**: 14-16.

Ramakrishna, Y.S., Rao, A.S., Rao, G.G.S.N. and Kesava Rao, A.V.R. (2000). Climatic constraints and their management in the Indian arid zone. Proceedings of the Symposium on Impact of Human Activities on the Desertification in the Thar Desert, 14 February 2000, Jodhpur, Rajasthan, India.

Ramarao, D. (1988). Seasonal indices and meteorological correlates in the incidence of foot and mouth disease in Andhra Pradesh and Maharashtra. *Indian Journal of Animal Sciences*, **58(4)**: 432-434.

Rao, D. and Sinha, S.K. (1994). Impact of climate change on simulated wheat production in India. *In:* C. Rosenzweig and A. Iglesias (editors), Implications of Climate Change for International Agriculture: Crop Modelling Study. US Environmental Protection Agency, EPA 230-B-94-003, Washington DC, USA.

Ravagnolo, O., Misztal, I. and Hoogenboom, G. (2000). Genetic component of heat stress in dairy cattle, development of heat index function. *Journal of Dairy Science*, **83:** 2120-2125.

Robinson, P.H. (1989). Dynamic aspects of feeding management for dairy cows. *Journal of Dairy Science*, **72:** 1197-1209.

Sanchez, W.K., McGuire, M.A. and Beede, B.K. (1994). Macromineral nutrition by heat stress interactions in dairy cattle: Review and original research. *Journal of Dairy Science*, **77:** 2051-2079.

Sathaye, J., Shukla, P.R. and Ravindranath, N.H. (2006). Climate change, sustainable development and India: Global and national concerns. *Current Science*, **90(3):** 314-325.

Sharma, S.K., Singh, G.R. and Pathak, R.C. (1991). Seasonal contours of foot-and-mouth disease in India. *Indian Journal of Animal Sciences*, **61(12):** 1259-1261.

Shinde, S., Taneja, V.K. and Singh, A. (1990). Association of climatic variables and production and reproduction traits in crossbreds. *Indian Journal of Animal Sciences*, **60(1):** 81-85.

Singh, A.P., Singla, L.D. and Singh, A. (2000). A study on the effects of macroclimatic factors on the seasonal population dynamics of *Boophilus microplus* infesting the crossbred cattle of Ludhiana district. *International Journal of Animal Sciences*, **15(1):** 29-31.

Singh, K.B., Nauriyal, D.C., Oberoi, M.S. and Baxi, K.K. (1996). Studies on occurrence of clinical mastitis in relation to climatic factors. *Indian Journal of Dairy Science*, **49(8):** 534-536.

Sirohi, S. (2007). Estimates of Economic Losses form Heat Stress in Dairy Animals. Annual Report, ICAR Network Project on Impact, Adaptation and Vulnerability of Indian Agriculture to Climate Change. National Dairy Research Institute, Karnal, Haryana, India.

Sirohi, S. and Sirohi, S.K. (2001). Cost of Bovine Mastitis to Indian dairy farmers. Proceedings of the Round Table on Mastitis, VIII Annual Conference of Indian Association for the Advancement of Veterinary Research, Intas Polivet, Vol. IV, No. 2, pp. 164-170.

Skaar, T.C., Grummer, R.R., Dentine, M.R. and Stauffacher, R.H. (1989). Seasonal effects of prepartum and postpartum fat and niacin feeding on lactation performance and lipid metabolism. *Journal of Dairy Science*, **72:** 2028-2038.

Sutherst, R.W. (1995). The potential advance of pest in natural ecosystems under climate change: implications for planning and management. *In:* J. Pernetta, C. Leemans, D. Elder and S. Humphrey (editors), Impacts of Climate Change on Ecosystems and Species: Terrestrial Ecosystems. IUCN, Gland, Switzerland, pp. 83-98.

Sutherst, R.W., Yonow, T., Chakraborty, S., O'Donnell, C. and White, N. (1996). A generic approach to defining impacts of climate change on pests, weeds and diseases in Australia. *In:* W.J. Bouma, G.I. Pearman and M.R. Manning (editors), Greenhouse: Coping with Climate Change CSIRO, Melbourne, Australia, pp. 281-307.

Tailor, S.P. and Nagda, R.K. (2005). Conception rate in buffaloes maintained under subhumid climate of Rajasthan. *Indian Journal of Dairy Science*, **58(1):** 69-70.

Teh, T.H., Hemken, R.W. and Harmon, R.J. (1985). Dietary magnesium oxide interactions with sodium bicarbonate on cows in early lactation. *Journal of Dairy Science*, **68:** 881-890.

Thomas, C.K. (1966). Feed Utilization by Dairy Cattle as Influenced by Season and Shelter. Unpublished M.Sc. Thesis, Punjab University, Chandigarh, India.

Thornton, P., Herrero, M., Freeman, A., Mwai, O., Rege, E., Jones, P. and McDermott, J. (2007). Vulnerability, climate change and livestock: Research opportunities and challenges for poverty alleviation. *SAT eJournal*, **4(1),** ejournal.icrisat.org.

UNEP (1989). Criteria for Assessing Vulnerability to Sea Level Rise: A Global Inventory to High Risk Area. Delft Hydraulics, Delft, The Netherlands, 51 pp.

Verma, D.N. and Hussain, K.G. (1988). Effect of shower on physiological parameters, nutrient utilization and milk production in buffaloes. Proceedings of the II World Buffalo Congress, India.

Waldige, V. (1994). Avaliação do sombreamento proporcionado por alguns tipos de árvores em pastagens [Evaluation of the Shade given in the Range by Some Tree Types]. Monograph, Faculty of Agricultural and Veterinary Sciences, Jaboticabal, SP, Brazil, 65 pp.

Wathes, C.M., Jones, C.D.R. and Webester, A.J.F. (1983). Ventilation, air hygiene and animal health. *The Veterinary Record*, **113(24):** 554-559.

West, J.W. (1999). Nutritional strategies for managing the heat stressed dairy cow. *Journal of Animal Science*, **77(Suppl. 2/J):** 21-35.

West, J.W., Mullinix, B.G. and Bernard, J.K. (2003). Effects of hot, humid weather on milk temperature, dry matter intake, and milk yield of lactating dairy cows. *Journal of Dairy Science*, **86:** 232-242.

Yoon, I.K. and Stern, M.D. (1996). Effects of Saccharomyces cerevisiae and Aspergillus oryzae cultures on ruminal fermentation in dairy cows. *Journal of Dairy Science*, **79:** 411-417.

20 Climate Change-Proof Disaster Risk Reduction: Prospects and Challenges for Developing Countries

S.V.R.K. Prabhakar

1. INTRODUCTION

Natural disasters are resultant of hazards meeting the vulnerable population and vulnerable infrastructure. Hazards have been thought to be aberration of the weather systems while there has been an increasingly dominant thought that the hazards are natural part of the weather systems and that the hazards are increasingly becoming disasters due to the human systems coming under their way. While this hypothesis doesn't tell much about how hazards have changed over the years, the thought that one location's hazard profile could change in terms of the intensity and magnitude of hazard or even from one kind of hazard to the other kind is more of recent in nature. This thought has emerged much in line with the increasing consensus on how our climate has been changing over the years. During the early years of the topic of climate change coming into vogue, not many researchers and policy makers heed to it as real. However, climate change today is perceived to be real and the issue is only that whether it is a man-made or natural process. The role of Intergovernmental Panel on Climate Change (IPCC) and national and international media is paramount in raising the awareness and action on climate change.

While the discussion on the origin of climate change is out of the purview of this chapter, what this chapter intends to do is to impress upon the reader that the climate change has implications for disasters and disaster management personnel because it influences both the hazard and vulnerability parts of disaster and hence the growing risks. This chapter also presents climate change in the context of disaster risk mitigation, offers some suggestions and identifies some important challenges to be overcome. This chapter thus can serve as a primer on the intriguing field of interface between disasters and climate change which is progressively becoming complex.

2. DEVELOPMENTAL AND DISASTER VULNERABILITY INTERACTION IN ASIAN COUNTRIES

Climate change has important implications for the Asia-Pacific region because the region is undergoing dramatic changes in population and economic status. The Asia-Pacific region hosts a population density 1.5 times the global average where India and China together host 2.426 billion population with an average population density of 197.2 persons/km^2 (U.S. Census Bureau, 2007). In comparison, during the same period, the world population density stood at 49.9 persons/km^2. In addition, the growth rate of population in Asia-Pacific has been one of the concerns. In comparison with the world population growth rate of 1.17%, the population in India grew at a rate of 1.63% though the growth rate in China has come down to 0.59% during recent years.

The Asia-Pacific region is also characterized by rapid economic changes. A glimpse at the economic growth rate of some of the countries provides us a good picture of the state of affairs. The GDP growth rate of India and China was put at 9.3 and 10.7%, respectively by the end of the first quarter of 2007 (Ministry of Statistics and Programme Implementation, 2007). This growth rate is not restricted to China and India alone in the region. Economies of countries such as Vietnam too recorded a growth rate of about 8.7% during 2006-2007. Concerns have been raised on the sustainability of these growth rates and the environmental consequences these rates may have (York et al., 2003). Concerns have also been raised on the economic path these countries are taking with growing disparity between haves and have-nots. Despite the economic growth, much of the population in these countries continues to depend on agriculture for their livelihood. Hence, climate change is a concern for the countries in Asia and the Pacific as a large proportion of their population depends on the climate-influenced sectors such as agriculture, animal husbandry and forestry for their livelihoods. For example, nearly 17.7% of the GDP in India comes from agriculture, fisheries and forestry, while this figure is around 15.5% for China though more than 50% of the population in both the countries depends on agriculture for their livelihoods (Prabhakar, 2007).

Since these economies are growing at a greater pace and since a majority of population is still dependent on agriculture and allied sectors for their livelihoods, it is very likely that any aberrations owing to climate change could mean a doom for these economies. Despite the recent economic growth in Asia and the Pacific, nearly 2/3rd of the world's poor people live here demanding more need for economic growth (Kainuma, 2006). While the economic status of the countries in Asia make them most vulnerable to climate change related impacts, the same vulnerability factors are also responsible for the increasing losses due to disasters in the region. Figure 1, as an example, depicts a growing number of disasters, mostly hydro-meteorological in nature over the years in Vietnam and India. The growing number of disasters could be interpreted in different ways. One interpretation could be that there has been a steady increase in the number and intensity of meteorological events which are responsible for disasters. The other interpretation could be that more and more population and man-made structures (e.g., cities, infrastructure, economic activity, etc.) are coming under the way of more or less same number of natural hazards over the years.

3. CLIMATE CHANGE VIS-À-VIS SUSTAINABLE DEVELOPMENT

The United Nations Conference on Environment and Development (UNCED) which met at Rio de Janeiro from 3 to 14th June 1992 proclaimed that, "In order to achieve sustainable development, environmental protection shall constitute an integral part of the development process and cannot be considered in isolation from it" (Prabhakar, 2007). The well-publicized Brundtland Commission Report (World Commission on Environment and Development, 1987) gave a significant thrust to a new kind of development termed "sustainable development" which was defined as *"development which meets the needs of the present without compromising the ability of future generations to meet their own needs"*. Since its advent, the phrase "sustainable development" became the center stage of development discourse across the globe. It has been well debated in every major forum on development and has become a mother's prescription for all the evils that humanity is facing today.

Since the inception of "sustainable development" concept, though the world has made some progress, environmental problems continue to plague human development and sustainable development seems to demand even more attention than it was expected. Many threats were identified for realizing sustainable development. Some of them include climate, biodiversity, forests, savannas, deserts and semi-arid areas, freshwaters and oceans, toxic and nuclear wastes, energy, new technologies, communication, poverty,

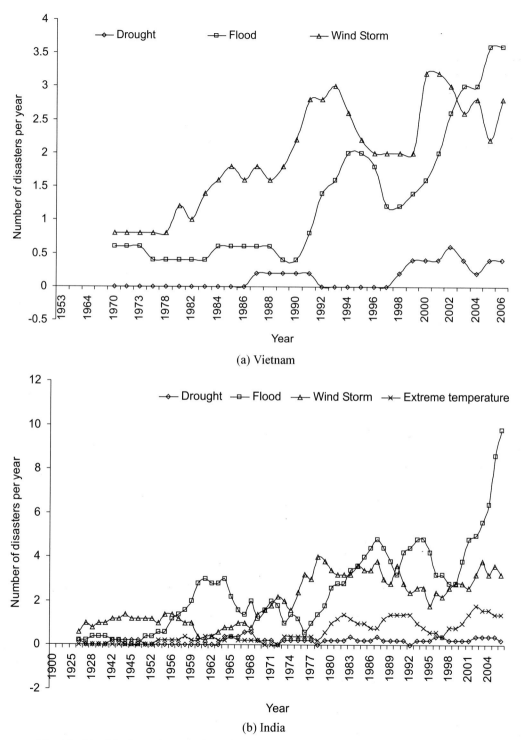

Fig. 1(a, b) Moving averages of number of disasters in Vietnam and India (CRED, 2007).

urban and rural disparities, violence, racism, militarization, population growth, foreign debt, debilitating diseases such as malaria and AIDS (Earth Council, 1994). Climate risks have been increasing over the past 20th century. One of the most significant indicators has been the rise of global average temperatures by 0.6 °C. Greenhouse gas concentrations are higher now than in the past 450,000 years and are projected to keep on rising. The 'climate change', defined by the increasing global temperatures and associated impacts, has been threatening the very existence of mankind on the earth. One of the alarming implications of climate change is to threaten sustainable development through a change in the disaster profile of countries (Helmer and Hilhorst, 2006; Prabhakar et al., 2008; Prabhakar and Shaw, 2008).

Since most people argue that there is a long-drawn debate over environment and sustainable development, one may ask what is the new threat posed by climate change? A recent survey of more than 250 experts and practitioners from 71 countries rated climate change as the second-most important issue (after poverty eradication) in terms of achieving sustainable development (Najam et al., 2002). Therefore, any discourse on sustainable development will inevitably lead to the climate and its change for the same reason that the sustainable development and environmental sustainability are inseparable. The similarity between climate change and any other environmental problem is that both of them can drastically undermine the sustainable development, complexly interwoven and comparable in their magnitude. In addition to its magnitude, the climate change deserves attention due to its ability to question the global future. Hence, climate change needed a global action which was made possible through the United Nations Framework Convention on Climate Change (UNFCCC).

The climate convention, an outcome of the UNCED, aims at stabilizing the greenhouse gases at a safer level and prescribes precautionary measures to achieve the targeted reduction in greenhouse gases (UNEP, 1992). The subsequently constituted Intergovernmental Panel on Climate Change (IPCC), established by the World Meteorological Organization (WMO) and United Nations Environment Program (UNEP), was mandated to assess the available scientific and socioeconomic evidence on climate change and its impact and options of mitigating climate change and adapting it; and to provide, on request, scientific/technical/socioeconomic advice to the Conference of Parties to the United Nations Framework Convention on Climate Change (UNFCCC). Since its inception, IPCC has brought out four comprehensive assessment reports on the status of climate change and its impacts. All the reports to date have brought out the evidence that the climate change is emerging as a significant threat to the human development.

The Fourth Assessment Report of IPCC has clearly mentioned that *"climate change is projected to impinge on sustainable development of most developing countries of Asia, as it compounds the pressures on natural resources and the environment associated with rapid urbanization, industrialization, and economic development"* (IPCC, 2007a). The observations made by the Working Group I to the Fourth Assessment of IPCC summarize the current state of affairs on climate change (IPCC, 2007b). The anthropogenic influence on climate change has also become more evident from the report. Further, the report has identified several impacts of climate change across the globe, viz., changes in precipitation amounts, ocean salinity, wind patterns, droughts, heavy precipitation, heat waves and tropical cyclones.

4. RISKS OF CLIMATE CHANGE

4.1 Defining Climate Risk

The concept of risk is worth elaborating here before we arrive at full understanding of what constitutes climate risk. In the disaster management terminology '*risk*' is defined as 'the probability that a hazard will turn into a disaster'. Einstein (1988) defined the risk as 'the probability of an event multiplied by the consequences if the event occurs'. Chapman (1994) defined the risk as 'a function of the probability of the specified natural hazard event and vulnerability of cultural entities'. In all these definitions, it is apparent that the risk involves probability and loss factors. The probability factor denotes the chances of the loss to occur for a given intensity of the natural hazard. '*Natural hazard*' here is defined as *"those elements of the physical environment, harmful to man and caused by forces extraneous to him"* (Burton et al., 1978). The definition of '*disaster*' has been subjected to a large debate in the literature. According to the World Health Organization (WHO), *"disaster is an occurrence disrupting the normal conditions of existence and causing a level of suffering that exceeds the capacity of adjustment of the affected community"* (WHO, 2003). Disaster can also be defined as 'an event, natural or man-made, sudden or progressive, which impacts with such severity that the affected community has to respond by taking exceptional measures' (Carter, 1992). In both the definitions, it is apparent that a natural event turns into a disaster when the impacts far exceed the coping capacity of local communities requiring an external assistance.

It can be said that the *climate risk* constitutes sum of all the risks posed by the climatic change in a given region at a given point of time. However, it is to be understood that the risk still exists in a non-changing climate because climate inherently has instability that could still give rise to extreme events such as hurricanes and extreme rainfall events. Hence, in the context of climate change, 'climate risk constitutes enhanced risk due to the climate change in addition to the inherent risk due to the nature of climate'.

4.2 Characteristics of Climate Risk

The risk posed by the climate and its change is unique in itself. The most distinguishing characteristics of climate risks are as follows (Prabhakar and Srinivasan, 2008):

- They span across long periods of time beyond the scales that human systems use in their planning.
- They are global but not uniform throughout the globe as there are some geographical areas which may get benefited by climate change while many others may not.
- They are too complex to comprehend as impacts are interrelated and compounded.
- Very little information is available on the exact nature of the full risk because climate risks are evolving in nature as our understanding improves over the time.
- Climate risk is inherently uncertain in its behavior due to the limitation of our understanding about the physical processes of climate and the climate and human interaction.
- Climate change is crescive (i.e., cumulative, incremental, no definitive beginning and no specific convenient location to act upon) (Ungar, 1998).
- It is a future oriented problem with immense costs to incur by the current generation.

According to Ionescue et al. (2009), the vulnerability to climate change is different in different places due to three main reasons. Firstly, climate change vulnerability is different in higher latitudes due to greater warming there than near the equator, and sea-level rise will not be uniform around the globe; secondly there are differences between different groups and sectors of the society which determine the relative importance of impacts; and thirdly there would be differences in the way different regions and social groups can be prepared for climate change impacts. Chichilnisky and Heal (1993) have put forward that the climate risk is unique in two ways. One is the uncertainty in basic scientific principles and the other is the uncertainty about the relationship between global mean temperatures and climate. They proclaimed that the classical formulations of uncertainty in economics no longer suffice to explain and deal with the climate change related uncertainties. The incomplete information and inherent uncertainty of climate risk makes it unique and distinguishing from other forms of risks such as environmental degradation being faced by humanity today. Willows and Connell (2003) opined that the uncertainty is a result of lack of knowledge of either the probability of the event or its consequences (Fig. 2). In Fig. 2, the top-right quadrant shows risk. The other three quadrants show different kinds of uncertainty. The uncertainty in climatic risk is due to *data uncertainty*, *knowledge uncertainty* and *model uncertainty*. The *data uncertainty* arises from measurement errors, incomplete or insufficient data and extrapolation based on uncertain data. The *knowledge uncertainty* arises from partial understanding of the problem which includes uncertainty about the future. *Model uncertainty*, from the simulation models used in producing future scenarios, could be due to choices made in selection of model parameters and model input values which in turn are due to partial understanding of the future which is imperfectly modeled (it should be noted that models are often said as GIGOs which is a well-known acronym for *garbage in and garbage out*). However, decisions must be made despite the uncertainties, because the price of not taking a decision could be costlier than taking a decision. The knowledge that the climate has changed in the past, and is now changing as a result of elevated atmospheric concentrations of greenhouse gases, requires that decisions be taken to exploit potential benefits and reduce deleterious impacts (DETR, 2000). In this context, a decision to do nothing should be viewed as an appropriate and positive risk management option. It is advised that the climate risk management professionals and policy makers acknowledge the uncertainty while making decisions and keep the decision making process transparent so that the society is well informed about the choices made.

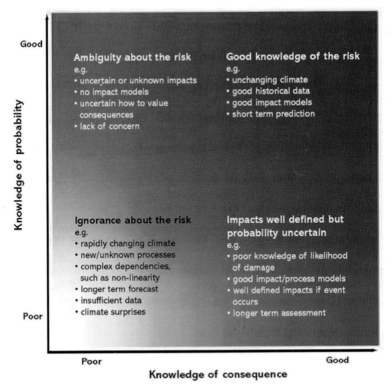

Fig. 2 Uncertainty as a result of lack of knowledge of consequence and probability of an event (Willows and Connell, 2003).

5. CLIMATE CHANGE AND DISASTER MANAGEMENT

5.1 Disasters as Climate Change Impacts

As the 2001 World Disaster Report indicates, 97% of all disaster-related deaths occurred in the poorest of the developing countries, while only 2% took place in the industrialized societies (International Federation of Red Cross and Red Crescent Societies, 2001). This differential impact of disasters is a cause of concern to the developmental prospects of developing countries. Therefore, the linkage between development and disasters is undisputable (UNDP, 2004). Climate change has brought another dimension to the development (McCarthy et al., 2001). The Brundtland Report identified climate change as one of the three problems having important say on our survival way back in 1987 (World Commission on Environment and Development, 1987). Subsequent reports by the Intergovernmental Panel on Climate Change (IPCC) have reaffirmed the role of climate change behind increasing disaster vulnerabilities. The Fourth Assessment Report of IPCC recorded an increase of global atmospheric temperature by 0.74 °C in the past 100 years (IPCC, 2007b). The Fourth Assessment Report of IPCC also indicated a growing evidence of climate change influenced changes in precipitation. There are evidences for longer droughts in tropics and subtropics, increasing frequency of heavy rainfall events on most land areas and for increasing intensity of tropical cyclones in North Atlantic. These changes are expected to have multi-fold impacts in the form of floods and droughts in various parts of the world. The extreme events can be devastating for the developing countries which have less capacity to adapt (Winkler, 2005). This establishes an undeniable unholy alliance between *climate change*, *disasters*, and *development*. This calls for better understanding of the impacts of climate change in terms of disasters and what it means to disaster risk management professionals and policy makers.

Mitigation and adaptation approaches were devised to address the problem of climate change (McCarthy et al., 2001). While mitigation aims at reducing greenhouse gas emissions, adaptation aims at reducing current and future impacts of climate change. Adaptation gained importance due to the fact that the already caused damage to the global environment would continue to show impacts long time into the future, irrespective of the mitigation practices taken up and implemented at present (Smithers and Smit, 1997). Adaptation enhances the capacity of people and governments to reduce climate change impacts (Tompkins and Adger, 2003). There have been cautions for not to invest heavily in adaptation because such investments may lead to mal-adaptations and unsustainable development (Adger et al., 2003). In the context of climate change, disaster risk management is considered as an adaptation option (Smit and Pilifosova, 2001; Vulnerability and Adaptation Resource Group, 2006). Disaster risk management has also been seen as a way to sustainable development (Smit and Pilifosova, 2001; Yodmani, 2001). The linkage between climate change and disaster risk reduction was subject of intensive formal and informal debates at the World Conference on Disaster Reduction (WCDR), Kobe, Japan (Vulnerability and Adaptation Resource Group, 2006). In addition, the '*Hyogo Framework for Action 2005-2015: Building the Resilience of Nations and Communities to Disasters*' identified climate change as one of the threats posing the world future and identified disaster risk management planning as one of the key points of entry to tackle the climate change threats (United Nations International Strategy for Disaster Reduction, 2005). In the words of the Hyogo Framework:

> "*Promote the integration of risk reduction associated with existing climate variability and future climate change into strategies for the reduction of disaster risk and adaptation to climate change, which would include the clear identification of climate related disaster risks, the design of specific risk reduction measures and an improved and routine use of climate risk information by planners, engineers and other decision-makers.*"

There is a clear evidence for growing trend of disasters undermining the disaster management capacities of countries. The data available from the Center for Research on Epidemiology of Disasters reveals a staggering increase in number of hydro-meteorological disasters during the period of 1900 to 2006 (CRED, 2007). During this period, the number of hydro-meteorological disasters had risen from single digit number to nearly 343 per year with corresponding increase in the number of people affected. Though the number of lives lost does not follow the similar trend, the economic losses out of these disasters had risen to nearly US$ 16,338 millions with a peak in 2004 (Fig. 3). According to MunichRe, the frequency of natural disasters have more than doubled between 1960 and 2005 out of which more than 55% were caused by earthquakes and volcanic eruptions (MunichRe, 2007). A closer look at the developed, developing, and under-developed countries reveals a more disappointing performance of developing and under-developed nations in terms of disaster losses. While the economic losses were higher in developed countries, the deaths are concentrated in less developed countries (O'Brien et al., 2006). While considering the role of combination of growing population, expanding infrastructure, and propensity to locate new development in areas of high hazard, Burton et al. (1997) raised the apprehension that they may be indicators for the onset of climate change. There are also apprehensions that the impacts felt till-to-date are not yet severe and that the consequences are likely to be incremental and cumulative (Burton et al., 2002). Examining some of the highly disaster-prone countries gives an indication of changes happening in their disaster profiles (Fig. 1). For example, among other disasters, the number of drought events had increased during recent times in Vietnam. Similar rise could be seen in the number of extreme temperature events in India. There was a steep increase in the number of floods in both the countries.

Surprises in terms of extreme events have become common during recent years. The year 2004 proved to be most devastating for Japan because 10 intense typhoons landed in the same year, while the earlier record used to be landing of six typhoons in 1990 and 1993 (Government of Ehime Prefecture,

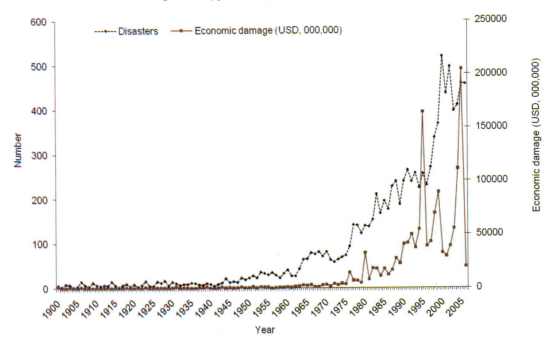

Fig. 3 Number of natural disasters and the economic losses since 1900 (CRED, 2007).

2005). Climate change was attributed to this behavior (Japan Meteorological Agency, 2004). The year 2004 also saw many other natural disasters including drought-like conditions in the Indian subcontinent, devastating floods in South Asian countries of Bangladesh, India, Nepal and Pakistan; typhoon in Philippines and a series of hurricanes in Florida reflecting impacts of changing climate (Asian Economic News, 2005). The hurricane *Katrina* and *Rita* which occurred in August and September 2005 respectively further reaffirm the debate on their linkage with the global climate change (Anthes et al., 2006). In the case of hurricane *Katrina*, the risk being known couldn't make much difference in terms of how quickly the local residents and governments could react and reduce the impacts (Travis, 2005). This shows how human designed physical and social protection systems could fail when a catastrophic event has to occur (Bohannon and Enserink, 2005).

5.2 Existing Disaster Risk Management: Shortcomings and Issues

Considerable amount of efforts have gone into understanding disaster risks; thanks to the major disasters that struck humanity from time to time acting as reminders (Pelling, 2003; Wisner et al., 2004). One of the major approaches of disaster risk reduction is through pragmatic disaster risk management planning (Salter, 1997; Christoplos et al., 2001). The disaster risk management plans are developed by identifying local hazards, risks, vulnerabilities, and capacities leading to planned interventions by the governments, corporations and communities to reduce disaster vulnerabilities and risks while enhancing the capacities. The present day disaster risk management planning largely aim at reducing the current disaster risks, i.e., those risks emanating out of current hazards and vulnerabilities. Often, these risk assessments heavily rely on the historical data of hazards at a given location (Ferrier and Haque, 2003; Dilley, 2005). However, the future is not always the repetition of the past (Quarantelli, 1996). Moreover, the assessments from historical data often fail to look into the future vulnerabilities and risks, and hence cannot incorporate them in terms of added strength in the plan. Many times, the hazard assessments fail to consider the changing frequencies and magnitudes of disasters in their fine details. We are also limited by our understanding on what proportion of our current vulnerabilities and risks are contributed by the climate change, though a broad conclusion is possible that the risks assessed at a given point of time are the results of interaction between past climate change impacts and vulnerabilities. It should be noted that the current and future risks are equally important for the risk management professionals as they aim at the welfare of the society from the angle of risks and sustainability. Thomalla and her colleagues tried to compare the contexts of disaster risk reduction and climate change adaptation and emphasized that the disaster risk management community focuses more on the current risks, while the climate change experts look more into the future risks (Thomalla et al., 2006). Independent working of these two communities has largely resulted in continuous increase in vulnerability of communities to natural hazards. It is important that these two communities should properly interact with each other and arrive at a functional plan of disaster risk reduction that reasonably considers the future risks as well.

One of the important questions to be asked is what makes a disaster risk management plan to work even in a climate change scenario or what is termed as abrupt climate change (Alley et al., 2003). It has been agreed that the current responses to disasters will no longer be sufficient in a changed climate (Sperling and Szekely, 2005). Adaptation to current climate variability has also been suggested as an additional way to approach adaptation to long-term climate change (Burton, 1997). Furthermore, it was also suggested to identify 'win-win' or 'no regrets' measures that address the current vulnerabilities (Schipper and Pelling, 2001). The possible best 'win-win measures' could be to tighten the disaster risk management systems by identifying loopholes and improving upon them, while continuously being in touch with the developments in risk projection methodologies.

Since uncertainty is an important aspect that is limiting in taking advanced decisions on making changes in disaster risk reduction systems, increasing redundancy in disaster risk reduction systems could be one important approach (Prabhakar et al., 2008). The redundancy in disaster risk reduction could be illustrated as in Fig. 4. In this figure, the X-axis denotes the successive periods of disaster-risk reduction planning (in practice, disaster management planning or preparing disaster management plans is done at a certain time when governments have funds or projects to do so and these plans are implemented until a review of the plan is carried out). The Y-axis denotes disaster risks. While disaster risks could be declining at some places due to successful risk mitigation efforts, one can generally assume that risks could also increase continuously due to the changing pressures and stresses including climate change, unsustainable development and population

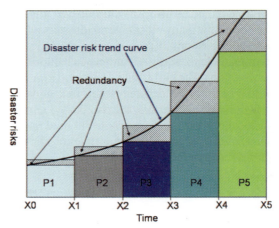

Fig. 4 Disaster management plans with built-in redundancy. The shaded areas show the redundancy built in a disaster risk management plan (Prabhakar et al., 2008).

growth. The growth trend depicted here should only be considered as 'indicative' and it may be possible that the trend in disaster risks could also be linear, logarithmic or polynomial. When a local disaster management authority undertakes disaster risk assessment at a given point of X1, that assessment only considers the vulnerabilities prevailing at that particular point of time and often would not consider the progression of vulnerabilities in future nor is any effort made to understand such progression from the historic data. As a result, the plan prepared at X1 point of time soon becomes negligent of growing risks, and hence is ineffective to deal with the growing risks. Though it is expected that the local authorities review the risks at regular intervals and review the disaster management plans and mitigation systems, such an activity is seldom carried out due to limitations related to funds and personnel, and even due to changing focus of the governance. Such limitations could be overcome if a plan is prepared and implemented with more than necessary 'strength' in it. This added strength could mean additional personnel to be involved in disaster response, or added strength in the bridges and drainage culverts constructed. The added strength has been depicted as shaded areas in Fig. 4 and marked as redundancy.

However, added strength and redundancy in disaster management systems means additional financial burden for governments at various scales. Such decisions would also have to be justified to the tax payers whose money is being used for decisions whose impact may be felt down the line or not at all depending on how effective they will turn out to be. Therefore, decision makers in disaster management domain would be faced with important questions that not only hinder the process of decision making, but would also question their credibility in taking such decisions. In the subsequent section, a process has been elaborated to overcome such limitations.

6. INCORPORATING CLIMATE CHANGE CONCERNS IN DISASTER RISK MANAGEMENT PLANS

One of the best ways to mainstream climate change concerns in disaster risk management planning is to understand current and future impacts and address them in developmental and risk reduction planning. For this, one need to look at climate scenarios generated and overlay them with future socio-economic scenarios to obtain future risks (Jones and Mearns, 2005). However, there are limitations such as the lack of availability of dependable high-resolution climate change scenarios and unaddressed uncertainties even in best available scenarios and projections (Burton, 1997).

It has been suggested that, for any mainstreaming to happen, it is important for the local disaster risk managers and other stakeholders to understand what national and regional climate change assessments mean for the scales at which these personnel operate. Since these personnel work at a local scale (e.g., city, group of villages, etc.) and often lack the perspective of climate and long-term implications of climate change, it is essential that a local Climate Task Group (CTG) is established (Prabhakar et al., 2008). The CTG should consist of personnel from disaster risk management, climate, and atmospheric and policy making domains (O'Brien et al., 2006). Such a group is necessary because the disaster management personnel alone cannot obtain and infer the often challenging climate information available from global and regional climate change studies and reports. However, the personnel required for CTG may not always be available at the administrative scale under consideration (e.g., small and medium cities, and institutions). Under such circumstances, cross-scale collaboration becomes necessary. There may be similar groups existing for monitoring drought or flood conditions; similar to the 'Drought Monitoring Center' established in Karnataka State of India or 'Flood Management Boards' in Vietnam, which are provided with capacities to assess and monitor local drought and flood situations (Samra, 2004). These Centers and Boards could form a good beginning as they have capacities such as data processing that are relevant for the operation of CTG. In addition to its role in disaster risk reduction planning, the CTG can also play a vital role in integrating disaster and climate risk reduction aspects in developmental planning as well. For example, the representative of CTG could be a member in local-level development committees similar to the 'District Development Committees' and 'Village Watershed Committees' established in India. Such integration would enable a free flow of information.

The question is how these CTGs can help in designing better disaster risk management plans and mitigation programs. Since CTGs are better equipped with the skills and knowledge on how to interpret and translate the climate information more into actionable points, these groups would be able to make a difference in short- and long-term decisions taken at the local level.

7. CONSIDERATIONS FOR MAINSTREAMING CLIMATE CHANGE IN DISASTER MANAGEMENT

Implementation of the scheme presented in the preceding section could face certain hurdles. These hurdles include uncertainty in future climate change trends and disaster-related impacts, limited capacity of the people and institutional systems to deal with such kind of impacts, and limitations related to people's perceptions and attitudes, and most importantly the limited resources to deal with the climate change impacts. One of the important impediments in implementing the above suggested scheme is uncertainty in projected climate impacts itself. Global Circulation Models (GCMs) are considered to be too coarse in resolution, and hence are not sufficient for decision making, for adaptation (Prudhomme et al., 2003). Therefore, the employment of techniques such as Regional Circulation Models has come into vogue. However, there are limitations with such downscaling too (Wilby, 2002). One of the major bottlenecks limiting decision making based on these techniques is the unaddressed uncertainty of climate change (Trenberth, 2005). Predictions are uncertain because of unknown future concentrations of greenhouse gases and other anthropogenic and natural forcing agents (e.g., injections of stratospheric aerosol from explosive volcanic eruptions), because of natural (unforced) climate variations and because our models (which we use for prediction) are imperfect (Collins et al., 2006). Heal and Kristrom (2002) provided a comprehensive review on the sources of uncertainty in climate change. Questions are also raised on the

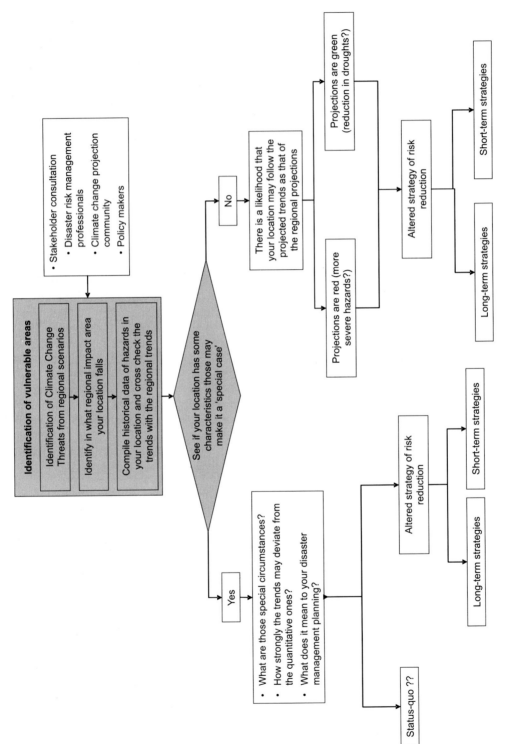

Fig. 5 Flowchart showing steps to be followed by the governments at various levels to mainstream climate change considerations in their developmental and risk reduction policies and plans (Prabhakar et al., 2008).

amount of certainty required in climate change projections to justify investments in adaptation measures and whether such certainty can be delivered (Dessai and Hulme, 2007). Various techniques have been identified to quantify the uncertainty and probabilistic climate scenarios were tried with a certain degree of success (Hulme et al., 2002). Employing probabilistic climate scenarios were too found to misrepresent the uncertainty (Hall, 2007).

There are several reports stating the uncertain nature of climate change. It is important to set the record straight that there is no uncertainty in climate change (IPCC, 2007). IPCC has unequivocally concluded that the climate change is real and happening. Then the question arises on the issue of uncertainty. From where does the question of uncertainty arise and how it impacts the decision making in disaster risk reduction. Uncertainty in climate change context could arise in several contexts. Some of them include:

(1) *Extent of impacts:* How severe the impacts would be as against whether or not there will be impacts.
(2) *When:* It is not the question of whether or not, but how soon. There are several projections from different groups of scientists showing how quick the impacts could escalate to the extent that the human resilience power is undermined to a critical level.
(3) *Where:* The question here is where the impacts will be severe and how soon rather than whether or not a particular region is climate change proof. All the regions are vulnerable to climate change, however with differences in the magnitude and nature of impacts.
(4) *A combination of when, where and to what extent:* When we combine the uncertainty in when, where and to what extent, the uncertainty level even goes up making it difficult to take decisions those matter on the ground.

Hence, uncertainty here is concerned with the need for quantified probability statements as a basis for policy decisions (Baer and Risbey, 2009). Swart et al. (2009) have presented the topology of uncertainty for better understanding this issue. In this topology, they argue that the uncertainty can be better dissected in terms of location uncertainty, level of uncertainty (in terms of probability), and nature of uncertainty (epistemic and variability). They suggest that these kinds of uncertainties could be easily understood to a level with tools such as expert judgement and analyzing the data with innovative approaches such that useful decisions could be taken. They also suggest that, depending on the level of agreement on a particular finding (e.g., finding of severity of climate change) and the amount of evidence available (e.g., in terms of real impacts discernable to climate change), the decision makers should be able to identify the areas and events of climate change which have high agreement and high evidence so that dependable policy interventions could be taken, while in the areas with less agreement and less evidence no regret options could be identified.

The uncertain nature of climate change has led to a greater need for capacity enhancement. In the context of the climate change, capacity considerations include adaptive capacity to increasing threats of climate change. This refers to potential and capacity to improve to a higher state in order to face the impacts of climate change (Brooks and Adger, 2004). Efforts have been made to explain the concept of adaptation and adaptive capacity (Burton et al., 2002). In general, the improvement in adaptive capacity refers to either increment in financial resources, reduction in poverty, provision of diversified income sources, better governance, social and political capital and even equitable flow of resources, etc. (Yohe and Tol, 2001).

Capacity building has been the integral part of various disaster risk management programs worldwide (Rocha and Christoplos, 2001). The usual topics covered in such programs are disaster risk management planning, rescue and evacuation planning, relief planning, emergency communication, fire fighting, conducting risk and vulnerability assessments, hazard and vulnerability mapping, disaster risk mitigation systems, etc. Sometimes, these programs also include the role of different stakeholders and achieving coordination among them in disaster risk reduction (Seth and Jain, 2002). Involvement of communities in disaster risk management and planning has been considered a necessity owing to the capacities and knowledge that the communities possess, which could be of use in effective disaster risk reduction (Allen, 2006). There is a need to enhance the existing capacities in order to deal with the future disasters and increasing uncertainty (Coutney et al., 1997). Table 1 provides a list of capacities suggested by the available literature for an effective adaptation.

Perception can be viewed as a process of transforming inputs (e.g., flood warning) to output (e.g., public mitigation response) (Burn, 1999). People who perceive that they are vulnerable are more likely to respond to warnings and undertake protective measures (Michael and Fasil, 2001). Thus, understanding how people will perceive the risks communicated to them will influence how effective a risk management measure will be. Creation of appropriate perception was found to be important for devising and implementing suitable policy interventions. The importance of risk perception in shaping people's behavior and disaster management planning has been affirmed in several studies.

The nature of climate change is such that it is invisible to many as an entity because it can only be identified through some 'proxy' indicators such as 'change in temperatures' and 'change in rainfall intensities' or 'increase in extreme events'. Often the degree of change could have important bearing on how perceptions are formed. For example, the change in a given geographical location may be perceived insignificant in the short run, but such changes may have compounding impacts which are significant in the long run. This very nature of climate change makes it difficult to create uniform perception across all communities. In a study conducted by the author (unpublished), it was revealed that the old members of a community are more likely to perceive changes than the young members. The difficulty to form appropriate perception of climate change may also be due to the lack of uniformity in the impacts across geographical and time scales. Added to this is the unpredictable nature of climate change. The unpredictability of climate has led to the lack of trust among respondents to forecasts leading to poor response in many situations. Such a poor response to weather forecasts has been cited in literature (Patt and Gwata, 2002). Patt and Gwata (2002), while examining the constraints in effective seasonal climate forecast applications, identified six constraints limiting the usefulness of climate forecasts. They are credibility, legitimacy, scale, cognitive capacity, procedural and institutional barriers, and available choices.

A recent study by the Oxfam-Vietnam and the Kyoto University revealed that the communities have difficulty to express what climate constitutes and that it is difficult for them to explain the concept of climate as different from weather phenomenon which has been used for planning agricultural operations on a daily basis (Oxfam-Vietnam, 2007). It appears that communities didn't have a suitable word in their local languages that readily represents climate or its change. However, the respondents were able to identify changes in climate as more attributed to changes in weather parameters such as temperatures and rainfall over the period that they could remember. They attributed such changes to reduction in forest cover than anything else. The respondents believed that the actions of 'other communities' at 'other location' were more responsible for changes and they couldn't identify any bad management practice within their community which might have contributed to the global problem. These findings are in agreement with the findings elsewhere that the communities are aware about changes in climate, but are flawed in terms of why (causes) such a change is happening (Kempton, 1997).

From the perspective of economists, climate change is sometimes viewed as a new and untested hypothesis (Howarth and Hall, 2001). Economists showed interest on the economics of global climate change and its impacts with a major thrust on the estimates of costs and benefits of GHG emissions and abatements. Application of welfare economics, as a rule rather than exception, and employing complex

Table 1. Nature of capacities required to be built by various stakeholders for mainstreaming climate change concerns in the disaster risk management

Stakeholders	Capacity needs
1. Communities	Human capital (Yohe and Schlesinger, 2002); Social capital of societies (Adger et al., 2003); Underlying health of the communities under question to deal with the climate change threats (Adger, 1999); Knowledge on climate change and its implications for the disaster profile of their region (Yohe and Schlesinger, 2002); Enhanced response capacity (Kelly and Adger, 2000; Tompkins and Adger, 2003); Functional social networks (Yohe and Schlesinger, 2002; Tompkins and Adger, 2003); and Empowerment and enfranchisement (Ribot, 1995).
2. Government and Non-governmental Disaster Management Personnel	Consideration of uncertainty in planning (Willows and Connell, 2003); Flexibility and innovation in the institutions (Kelly and Adger, 2000; Adger et al., 2003); Policies and regulations (Adger et al., 2005); Strengthening early warning systems (Klein, 2002); Spatial planning (Nichollas, 1995); Finances (Kandlikar and Sagar, 1999); Analytical skills to identify climate change impacts and related disaster dynamics (Kandlikar and Sagar, 1999); Respond to the developmental pressures and resource crises (Downing et al., 1997); Risk spreading instruments (Yohe and Schlesinger, 2002); and Governance (Denton et al., 2002).
3. Research and Educational Institutions	Resources (manpower and funds) (Kandlikar and Sagar, 1999); Proactive participation in relevant policy research (Kandlikar and Sagar, 1999); and Integrating with the local disaster management community (Alast, 2006).

models that integrate climate and economy have been used in these assessments (Nodhuas, 1994). However, in the context of adaptation in general and disaster risk management in particular, the economic assessments are still in a nascent phase. The crux of the problem seems to lie in understanding the uncertainty of future impacts on human and natural systems, which form the basis for investments on the adaptation programs. Schneider and Kuntz-Duriseti (2001) argue that the questions of uncertainty are poorly answered in the current generation works of climate change.

Although the 'Kyoto Protocol' is heavily based on a *mitigation approach*, it has been accepted globally that the *adaptation* also need to be considered along with the mitigation, and policy makers do understand and agree that the adaptation efforts would affect the mitigation targets (Kane and Shogren, 2000). Kane and Shogren (2000) also suggested that the society can select any combination of risk avoidance systems among mitigation and adaptation, and such decisions affect the level of risks and the costs of risk reduction. Cost-effectiveness has been found to be crucial factor in climate change. They further emphasized that the risk reduction efforts of people are affected by the economic circumstances, and hence the economic circumstances must be taken into consideration while making risk reduction recommendations.

8. CONCLUDING REMARKS

The science of climate change in itself is full of challenges and uncertainties. Nations (both developed and developing) have only partly been successful in dealing with the disasters that are part of the normal climate process. Dealing with the added risks due to climate change is still an added challenge. However, there are opportunities to deal with the emerging disaster threats. Some of them, as discussed in this chapter, include instituting self-learning and evolving organizations, networks and people which will help in dealing with the uncertainties. Establishing cross-institutional and cross-sectoral linkages is a challenging issue, especially when governance systems in many countries are still in nascent stage. In addition, participatory approaches that educate people on disaster risks and help institutions learn the underlying causes of their vulnerability could establish a cycle of learning-oriented disaster risk reduction. However, there are challenges too. Challenges such as people's perceptions, larger uncertainties in the future climate change impacts and economic limitations of taking advanced actions such as enhancing the redundancy in disaster risk mitigation would have to be overcome. Undoubtedly, much more dedicated and visionary efforts are indispensable by the researchers, planners, decision makers and donor agencies in order to bring a paradigm shift in disaster management in the face of climate change and other environmental changes.

ACKNOWLEDGEMENTS

Author would like to acknowledge that much of the content in this chapter came from his work and understanding from working at Kyoto University, Japan and UNDP-GoI DRM Program in India.

REFERENCES

Adger, W.N. (1999). Social vulnerability to climate change and extremes in coastal Vietnam. *World Development*, 27: 249-269.

Adger, W.N., Huq, S., Brown, K., Conway, D. and Hulme, M. (2003). Adaptation to climate change in the developing world. *Progress in Development Studies*, 3(3): 179-195.

Alast, M.K. (2006). The impacts of climate change on the risk of natural disasters. *Disasters*, 30(1): 5-18.

Allen, K.M. (2006). Community-based disaster preparedness and climate adaptation: Local capacity building in the Philippines. *Disasters*, 30(1): 81-101.

Alley, R.B., Marotzke, J., Nordhaus, W.D., Overpeck, J.T., Peteet, D.M., Pielke, Jr. R.A., Pierrehumbert, R.T., Rhines, P.B., Stocker, T.F., Talley, L.D. and Wallace, J.M. (2003). Abrupt climate change. *Science*, 299(5615): 2005-2010.

Anthes, R.A., Corell, R.W., Holland, G. and Hurrell, J.W. (2006). Hurricanes and global warming-Potential linkages and consequences. *Bulletin of American Meteorological Society*, 87: 623-628.

Asian Economic News (2005). Major natural disasters of 2004. Asian Economic News, http://www.findarticles.com/p/articles/mi_m0WDP/is_2005_Jan_18/ai_n8704072 (accessed on March 8, 2008).

Baer, P. and Risbey, J.S. (2009). Uncertainty and assessment of the issues posed by urgent climate change: An editorial comment. *Climatic Change*, 92: 31-36.

Bohannon, J. and Enserink, M. (2005). Scientists weigh options for rebuilding New Orleans. *Science*, 309 (5742): 1808-1809.

Brooks, N. and Adger, W.N. (2004). Assessing and enhancing adaptive capacity. *In:* L.B. Lim (editor), Adaptation Policy Framework. United Nations Development Program, New York.

Burn, D.H. (1999). Perception of flood risk: A case study of the Red River flood of 1997. *Water Resources Research*, 35(11): 3451-3458.

Burton, I. (1997). Vulnerability and adaptive response in the context of climate and climate change. *Climatic Change*, **36(1-2)**: 185-196.
Burton, I., Huq, S., Lim, B., Pilifosova, O. and Schipper, E.L. (2002). From impacts assessment to adaptation priorities: The shaping of adaptation policy. *Climate Policy*, **2**: 145-159.
Burton, I., Kates, R.W. and White, G.F. (1978) The Environment as Hazard. Oxford University Press, New York.
Carter, W.N. (1992). Disaster Management: A Disaster Manager's Handbook. Asian Development Bank (ADB), Manila, Philippines, 417 pp.
Chapman, D. (1994). Natural Hazards. Oxford University Press, Melbourne, 174 pp.
Christoplos, I., Mitchell, J. and Liljelund, A. (2001). Re-framing risk: The changing context of disaster mitigation and preparedness. *Disaster*, **25(3)**: 185-198.
Collins, M., Booth, B.B.B., Harris, G.R., Murphy, J.M., Sexton, D.M.H. and Webb, M.J. (2006). Towards quantifying uncertainty in transient climate change. *Climate Dynamics*, **27**: 127-147.
Courtney, H., Kirkland, J. and Viguerie, P. (1997). Strategy under uncertainty. *Harvard Business Review*, **75(6)**: 67-79.
CRED (2007). Country profiles. EM-DAT Emergency Disasters Database, http://www.em-dat.net/disasters/Visualisation/profiles/countryprofile.php (accessed on May 15, 2008).
Denton, F., Sokona, Y. and Thomas, J.P. (2002). Climate Change and Sustainable Development Strategies in the Making: What Should West African Countries Expect? Organization for Economic Cooperation and Development, Paris, 27 pp.
Dessai, S. and Hulme, M. (2007). Assessing the robustness of adaptation decisions to climate change uncertainties: A case study on water resources management in the East of England. Global *Environmental Change*, **17**: 59-72.
DETR (2000). Guidelines for Environmental Risk Assessment and Management: Revised Departmental Guidance. Institute for Environment and Health, Stationary Office, London, U.K.
Dilley, M. (2005). Natural disaster hotspots: A global risk analysis. Risk Identification for Disaster Risk Management. Proceedings of the World Conference on Disaster Reduction, 18-22 January 2005, Kobe, Hyogo, Japan, http://www.unisdr.org/wcdr/thematic-sessions/presentations/session2-5/hotspots.pdf (accessed on April 15, 2008).
Downing, T.E., Ringius, L., Hulme, M. and Waughray, D. (1997). Adapting to climate change in Africa. *Mitigation and Adaptation Strategies for Global Change*, **2**: 19-44.
Earth Council (1994). The Earth Summit-Eco 92: Different Visions. Earth Council and Inter American Institute for Cooperation on Agriculture, San Jose, Costa Rica.
Einstein, H.H. (1988). Landslide risk assessment procedure. Proceedings of the Fifth International Symposium on Landslides, Lausanne, Switzerland, pp. 1075-1090.
Ferrier, N. and Haque, C.E. (2003). Hazard risk assessment methodology for emergency managers: A standardized framework for application. *Natural Hazards*, **28**: 271-290.
Government of Ehime Prefecture (2005). Road Memory of Disasters in 2004. Department of Civil Engineering and Urban Road Maintenance Section, Government of Ehime Prefecture, Japan, 21 pp.
Hall, J. (2007). Probabilistic climate scenarios may misrepresent uncertainty and lead to bad adaptation decisions. *Hydrological Processes*, **21**: 1127-1129.
Heal, G. and Kristrom, B. (2002). Uncertainty and climate change. *Environmental and Resource Economics*, **22**: 3-39.
Helmer, M. and Hilhorst, D. (2006). Natural disasters and climate change. *Disasters*, **30(1)**: 1-4.
Howarth, R.B. and Hall, D.C. (2001). Beyond a doubling: Issues in the long-term economics of climate change. *In:* C.H. Darwin and B.H. Richard (editors), The Long-term Economics of Climate Change: Beyond a Doubling of Greenhouse Gas Concentrations. Elsevier Science, The Netherlands, pp. 1-9.
Hulme, M., Jenkins, G.J., Lu, X., Turnpenny, J.R., Mitchell, T.D., Jones, R.G., Lowe, J., Murphy, J.M., Hassell, D., Boorman, P., McDonald, R. and Hill, S. (2002). Climate Change Scenarios for the United Kingdom: The UKCIP02 Scientific Report. Tyndall Centre for Climate Change Research, Norwich, U.K.
International Federation of Red Cross and Red Crescent Societies (2001). World Disasters Report. International Federation of Red Cross and Red Crescent Societies, Geneva.

Ionescue, C., Klein, R.J.T., Hinkel, J., Kumar, K.S.K. and Klein, R. (2009). Towards a formal framework of vulnerability to climate change. *Environmental Modeling & Assessment*, **14:** 1-16.

IPCC (2007a). Summary for Policymakers. *In:* M.L. Parry, O.F. Canziani, J.P. Palutikof, P.J. van der Linden and C.E. Hanson (editors), Climate Change 2007: Impacts, Adaptation and Vulnerability. Contribution of Working Group II to the Fourth Assessment Report of the Intergovernmental Panel on Climate Change (IPCC). Cambridge University Press, Cambridge, U.K., pp. 7-22.

IPCC (2007b). Summary for Policymakers. *In:* S. Solomon, D. Qin, M. Manning, Z. Chen, M. Marquis, K.B. Averyt, M. Tignor and H.L. Miller (editors), Climate Change 2007: The Physical Science Basis. Contribution of Working Group I to the Fourth Assessment Report of the Intergovernmental Panel on Climate Change (IPCC). Cambridge University Press, Cambridge, USA.

Japan Meteorological Agency (2004). Heavy Rains and Typhoons Occurred in 2004. Japan Meteorological Agency, Tokyo, Japan, 39 pp.

Jones, R. and Mearns, L. (2005). Assessing future climate risks. *In:* B. Lim and E. Spanger-Siefried (editors), Adaptation Policy Frameworks for Climate Change: Developing Strategies, Policies and Measures. Cambridge University Press, Cambridge, U.K., pp. 119-144.

Kainuma, M. (2006). Low carbon scenario towards 2050 for Japan. Proceedings of the China-Korea-U.S. Economic and Environmental Modeling Workshop, 20-21 April 2006, Xia Da Du International Conference Center, Beijing, China.

Kandlikar, M. and Sagar, A. (1999). Climate change research and analysis in India: An integrated assessment of a South-North divide. *Global Environmental Change*, **9:** 119-138.

Kane, S. and Shogren, J.F. (2000). Linking adaptation and mitigation in climate change policy. *Climate Change*, **45:** 75-102.

Kelly, P.M. and Adger, W.N. (2000). Theory and practice in assessing vulnerability to climate change and facilitating adaptation. *Climatic Change*, **47:** 325-352.

Kempton, W. (1997). How the public views climate change. *Environment*, **39(9):** 12-21.

Klein, R.J.T. (2002). Climate change, adaptive capacity and sustainable development. Proceedings of the OECD Informal Expert Meeting on Development and Climate Change. 13-14 March 2004, Paris, France, http://www.oecd.org/dataoecd/1/37/1933851.pdf (accessed on March 10, 2008).

McCarthy, J.J., Canziani, O.F., Leary, N.A., Dokken, D.J. and White, K.S. (2001). Climate Change 2001: Impacts, Adaptation, and Vulnerability: Contribution of Working Group II to the Third Assessment Report of the Intergovernmental Panel on Climate Change, Intergovernmental Panel on Climate Change (IPCC). Cambridge University Press, Cambridge, U.K., 1008 pp.

Michael, B. and Fasil, A.G. (2001). Worldwide public perception of flood risk in urban areas and its consequences for hydrological design in Ireland. Proceedings of the National Hydrology Seminar on Flood Risk Management: Impacts and Development, Ireland, pp. 6-24.

Ministry of Statistics and Programme Implementation (2007). Estimates of gross domestic product for the first quarter (April-June) of 2007-08. Government of India, New Delhi, http://www.mospi.nic.in/t1_31august07.htm (accessed on March 20, 2008).

Munich Re (2007). Natural Catastrophes 2006: Analyses, Assessments, Positions. Knowledge Series, Munich Re, Germany, 50 pp.

Najam, A., Poling, J.M., Yamagishi, N., Straub, D.G., Sarno, J., DeRitter, S.M. and Kim, E.M. (2002). From Rio to Johannesburg: Progress and prospects. *Environment*, **44(7):** 26-38.

Nichollas, R.J. (1995). Coastal mega cities and climate change. *GeoJournal*, **37(3):** 369-379.

Nordhaus, W.D. (1994). Managing the Global Commons: The Economics of Climate Change. MIT Press, Cambridge, Massachusetts.

O'Brien, G., OKeefe, P., Rose, J. and Wisner, B. (2006). Climate change and disaster management. *Disasters*, **30(1):** 64-80.

Oxfam-Vietnam (2007). Drought Management Considerations for Climate Change Adaptation: Focus on Mekong Region. Oxfam-Vietnam and Graduate School of Global Environmental Studies, Kyoto University, Kyoto, Japan, 56 pp.

Patt, A.G. and Gwata, C. (2002). Effective seasonal climate forecast applications: Examining constraints for subsistence farmers in Zimbabwe. *Global Environmental Change*, **12**: 185-195.

Pelling, M. (editor) (2003). Natural Disasters and Development in a Globalizing World. Routledge, New York.

Prabhakar, S.V.R.K. (2007). Climate change implications for sustainable development: Need for holistic and inclusive policies in agriculture, land, rural development, desertification, and drought. Background Paper prepared for the 16th Commission on Sustainable Development, Review of the Implementation Status of the Outcomes of the World Summit on Sustainable Development: An Asia-Pacific Perspective. United Nations Economic and Social Commission for Asia and the Pacific, 26-27 November 2007, Jakarta, Indonesia.

Prabhakar, S.V.R.K. and Srinivasan, A. (2008). Climate change risk reduction: Decision making in uncertainty. *In:* R. Shaw and R.R. Krishnamurthy (editors), Disaster Management: Global Challenges and Local Solutions. Universities Press (India) Private Limited, New Delhi, India, pp. 85-101.

Prabhakar, S.V.R.K., Srinivasan, A. and Shaw, R. (2008). Climate change and local level disaster risk reduction planning: Need, opportunities and challenges. Mitigation and Adaptation Strategies to Global Change, DOI 10.1007/s11027-008-9147-4.

Prudhomme, C., Jakob, D. and Svensson, C. (2003). Uncertainty and climate change impact on the flood regime of small UK catchments. *Journal of Hydrology*, **277**: 1-23.

Quarantelli, E.L. (1996). The future is not the past repeated: Projecting disasters in the 21st century from current trends. *Journal of Contingencies and Crisis Management*, **4(4)**: 228-240.

Ribot, J.C. (1995). The casual structure of vulnerability: Its application to climate impacts analysis. *GeoJournal*, **35(2)**: 119-122.

Rocha, J.L. and Christoplos, I. (2001). Disaster mitigation and preparedness on the Nicaraguan Post-Mitch agenda. *Disasters*, **25(3)**: 240-250.

Salter, J. (1997). Risk management in a disaster management context. *Journal of Contingencies and Crisis Management*, **5(1)**: 60-65.

Samra, S.J. (2004). Review and Analysis of Drought Monitoring, Declaration and Management in India. Working Paper 84, International Water Management Institute (IWMI), Colombo, Sri Lanka.

Schipper, L. and Pelling, M. (2006). Disaster risk, climate change and international development: Scope for, and challenges to, integration. *Disasters*, **30(1)**: 19-38.

Schneider, S.H. and Kuntz-Duriseti, K. (2001). Integrated assessment models of climate change: Beyond a doubling of CO_2. *In:* C.H. Darwin and B.H. Richard (editors), The Long-term Economics of Climate Change: Beyond a Doubling of Greenhouse Gas Concentrations. Elsevier Science, The Netherlands, pp.11-64.

Seth, A. and Jain, S.K. (2002). Training of teachers for capacity building towards earthquake safety in India. *The Indian Concrete Journal*, **10**: 629-632.

Smit, B. and Pilifosova, O. (2001). Adaptation to climate change in the context of sustainable development and equity. *In:* J.J. McCarthy, O.F. Canziani, N.A. Leary, D.J. Dokken and K.S. White (editors), Climate Change 2001: Impacts, Adaptation, and Vulnerability. Contribution of Working Group II to the Third Assessment Report of the Intergovernmental Panel on Climate Change, Intergovernmental Panel on Climate Change (IPCC). Cambridge University Press, Cambridge, U.K., pp. 877-912.

Smithers, J. and Smit, B. (1997). Human adaptation to climatic variability and change. *Global Environmental Change*, **7(2)**: 129-146.

Sperling, F. and Szekely, F. (2005). Disaster risk management in changing climate. Discussion paper for the World Conference on Disaster Reduction, 18-22 January 2005, Kobe, Japan.

Swart, R., Bernstein, L., Ha-Duong, M. and Peterson, A. (2009). Agreeing to disagree: Uncertainty management in assessing climate change, impacts and responses by the IPCC. *Climatic Change*, **92**: 1-29.

Thomalla, F., Downing, T., Siegfried, E.S., Han, G. and Rockstrom, R. (2006). Reducing hazard vulnerability: Towards a common approach between disaster risk reduction and climate adaptation. *Disasters*, **30(1)**: 39-48.

Tompkins, E.L. and Adger, W.N. (2003). Building resilience to climate change through adaptive management of natural resources. Working Paper 27, Tyndall Center for Climate Change Research, U.K., 19 pp.

Travis, J. (2005). Hurricane Katrina: Scientists' fears come true as hurricane floods New Orleans. *Science*, **309(5741)**: 1656-1659.

Trenberth, K. (2005). Uncertainty in hurricanes and global warming. *Science*, **308**: 1753-1754.

U.S. Census Bureau (2007). International Database. http://www.census.gov/ipc/www/idb/ind ex.html (accessed on March 20, 2008).

UNDP (2004). Reducing Disaster Risk, a Challenge for Development: A Global Report. United Nations Development Program, New York, 169 pp.

UNEP (1992). The Rio Declaration on Environment and Development. UNCED Secretariat, Rio de Janeiro, Brazil. http://www.unep.org/Documents.Multilingual/Default.asp?DocumentID=78&ArticleID=1163 (accessed on July 17, 2008).

Ungar, S. (1998). Bringing the issue back *In:* Comparing the marketability of the ozone hole and global warming. *Social Problems*, **45**: 510-527.

United Nations International Strategy for Disaster Reduction (2005). Hyogo Framework for Action 2005-2015: Building the resilience of nations and communities to disasters. World Conference on Disaster Reduction, 18-22 January 2005, Kobe, Hyogo, Japan, 22 pp.

Vulnerability and Adaptation Resource Group (2006). Linking Climate Change Adaptation and Disaster Risk Management for Sustainable Poverty Reduction: A Synthesis Report. Vulnerability and Adaptation Resource Group, The World Bank, Washington DC, 30 pp.

WHO (2003). Disasters and Emergencies. Definitions Training Package, WHO/EHA, Panafrican Emergency Training Center, Addis Ababa, Ethiopia. http://www.who.int/disasters/repo/7656.pdf (accessed on July 17, 2008).

Wilby, R.L., Dawson, C.W. and Barrow, E.M. (2002). A decision support tool for the assessment of regional climate change impacts. *Environmental Modeling and Software*, **17(2)**: 145-157.

Willows, R. and Connell, R. (editors) (2003). Climate Adaptation: Risk, Uncertainty and Decision-making. U.K. Climate Impacts Programme Technical Report. UNCIP, Department for Environment, Food and Rural Affairs, Oxford, U.K.

Winkler, H. (2005). Climate change and developing countries. *South African Journal of Science*, **101**: 355-364.

Wisner, B., Blaikie, P., Cannon, T. and Davis, I. (2004). At Risk: Natural Hazards, People's Vulnerabilities and Disasters. Second Edition, Routledge, New York.

World Commission on Environment and Development (1987). Our Common Future. Oxford University Press, Oxford, U.K., 416 pp.

Yohe, G. and Schlessinger, M. (2002). The economic geography of the impacts of climate change. *Journal of Economic Geography*, **2**: 311-341.

Yohe, G. and Tol, R.S.J. (2001). Indicators for social and economic coping capacity: Moving toward a working definition of adaptive capacity. *Global Environmental Change*, **12**: 25-40.

York, R., Rosa, E.A. and Dietz, T. (2003). Footprints on the earth: The environmental consequences of modernity. *American Sociological Review*, **68**: 279-300.

Potential of Geospatial Technologies for Mitigating Land and Water Related Disasters

S. Bandyopadhyay* and Madan Kumar Jha

1. INTRODUCTION

Land and water are primary ingredients in generating rural livelihoods, growing food, producing energy, supporting industrial and service sector growth, and ensuring the integrity of ecosystems and the goods and services they provide. Water also poses its own development challenges. The occurrence of disasters such as floods, droughts and water-related diseases often produce a huge impact on communities as well as on national economies (GWP, 2004; NDMD, 2004). The ever-increasing human and livestock population is creating tremendous pressure on the land and water to get food, fodder, fuel and shelter. To feed the enormous population, cultivation of land has been intensified and even expanded into unsuitable and unprotected lands, and fallow cycles have been shortened (Sombroek and Sene, 1993; Sommer et al., 2001). However, soil resources, highly diversified over different geographic locations, are vulnerable, fragile and prone to degradation due to improper management practices. In addition, growing water scarcity is a serious concern in the 21st century (Postel, 1993; Loucks, 2000; Evans and Sadler, 2008).

India with about 2.5% land area supports 16% of global population and about 4% of world's water resources area. While India has a wide variation of climate, with striking contrasts of meteorological conditions, the climate is mainly tropical. About 60% of the net cultivable area (142 Mha) of the country falls in the dryland region which produces low and unstable yield accompanied with the problem of soil erosion. According to the Ministry of Agriculture estimate, degraded land area is nearly 107 Mha and soil degradation is the most serious problem with maximum degradation by soil erosion (MOA, 1994). This is supported by other estimates available on soil degradation in India from different sources (Table 1). The average annual soil loss is estimated at about 5300 Mha (Dhubanarayana and Babu, 1983). Around 40% of total agricultural production comes from dryland agriculture. Thus, the highly skewed production gains, limited largely to well-endowed irrigated areas, are major concerns to feed the ever-increasing population of the country.

The availability of water in India is highly uneven with space and time. Moreover, the pressure on water resources is mounting with fast growing population and industrialization. The per capita availability of water resources is reducing day by day—it was 2309 and 1902 m³ in the years 1991 and 2001, respectively and these figures are projected to reduce to 1401 m³ in 2025 and 1191 m³ in 2050 (Kumar et al., 2005). The annual water requirements are rising continuously as shown in Table 2 (CWC, 2007). At present, about 85% of the total available water is being utilized as irrigation for food production. The floods and droughts are regular phenomena in the country due to vagaries of monsoon. Over-exploitation

---

*Corresponding Author

Table 1. Soil degradation trend in India

Type of degradation	1980*	1985*	1994[#]
Soil erosion (Mha)	150.0	141.2	162.4
Saline and alkaline soil (Mha)	8.0	9.4	10.1
Waterlogged soil (Mha)	6.0	8.5	11.6
Shifting cultivation (Mha)	4.4	4.9	
Total	168.4	175.1	175.0

* MOA (1994); [#] Sehgal and Abrol (1994).

of groundwater is leading to the reduction of low flows in rivers/streams or drying of rivers, depletion of groundwater resources, and seawater intrusion in the coastal aquifers. Excess use of irrigation water in some of the command areas has resulted in severe waterlogging and salinity. The quality of surface and groundwater resources is also deteriorating because of increasing pollutant loads from point and non-point sources. On the top of it, the climate change is expected to significantly affect precipitation and water availability in the future (Mall et al., 2006). Undoubtedly, the efficient management of land and water resources is a major challenge for the scientists in the 21st century to ensure food security, water security and environmental security for the present as well as future generations (Gleick, 1993; Loucks, 2000; Saheb and Singh, 2002). Consequently, holistic approaches for sustainable development and management of land and water resources are greatly stressed upon throughout the world (e.g., Bouwer, 2000; Gupta and Despande, 2004; GWP, 2004; Rakesh et al., 2005; Falkenmark, 2007).

Table 2. Trend of annual water requirements in India

Purpose	Annual water requirement (10^9 m^3)				
	1990	2000	2010	2025	2050
Domestic	32	42	56	73	102
Irrigation	437	541	688	910	1072
Industry	-	8	12	23	63
Energy	-	2	5	15	130
Others	33	41	52	72	80
Total	502	634	813	1093	1447

In view of the above-mentioned problems, geospatial technologies such as Remote Sensing (RS), Global Positioning System (GPS), and Geographical Information System (GIS) as such offer immense potential for land and water resources development and management (e.g., Schultz, 1994, 2000; Bastiaanssen et al., 1999; Pietroniro and Prowse, 2002; Jha and Peiffer, 2006; Akbari et al., 2007; Karatas et al., 2009). Utilizing the physical properties of earth's surface in electromagnetic spectrum ranging from visible to microwave through thermal range provides a means of estimating hydrological state variables such as land surface temperature, soil moisture, snow cover, vegetation status, etc. over large areas (Engman, 1991; Laymonn et al., 1998; Ray and Dadhwal, 2001). These state variables are used in estimating the dynamic land surface processes such as hydrometeorological fluxes, evapotranspiration, snowmelt runoff, etc. (Schmugge et al., 2002). Besides the geologic framework, there are various parameters such as relief, slope, ruggedness, depth and nature of weathering, thickness and nature of deposited materials, distribution of surface water bodies, river/stream network, precipitation, percolation from canals, land use/land cover, etc. which influence the groundwater regime. Therefore, the use of remote sensing as a tool to address earth's surface mapping and in many geological applications has been successfully developed in past few decades (e.g., Erdelyi and Galfi, 1988; Edet et al., 1998; Schultz, 2000; Jackson, 2002; Jha and Peiffer, 2006).

Geographical Information System (GIS) is an integrated system of hardware, software, and procedures designed to support capture, management, manipulation, analysis, modeling and display of spatially referenced data in a map form for solving complex planning and management problems. The unique advantage of using RS data for hydrological monitoring and modeling is its ability to generate information in spatial and temporal domain, which is very crucial for effective model development and prediction that in turn can ensure efficient decision making. As the use of RS technology involves a large amount of spatio-temporal data management, it requires an efficient system to handle such data. The GIS technology provides a suitable alternative for the efficient management of large and complex databases. GIS helps in integrating the satellite derived spatial information with the non-spatial databases for generating comprehensive output. RS data and GIS play a rapidly increasing role in water resources development and management in terms of mapping and process monitoring (e.g., Engman and Gurney, 1991; Schultz and Engman, 2000; Chen et al., 2004; Bandyopadhyay et al., 2007; Jha et al., 2007) and also in the field of mesoscale hydrological modeling (e.g., Schultz, 1994; Su, 2000; Vieux, 2004). Remote sensing in combination with the Global Positioning System (GPS) and GIS can produce terrain maps with high accuracy and containing detailed information of the variables under study (e.g., Rao and Bandyopadhyay, 2004).

The goal of this chapter is to highlight the land and water degradation problem in India and to demonstrate the potential of geospatial technologies (RS, GPS and GIS) in managing land and water resources so as to minimize/avoid land and water related disasters and ensure sustainable management of these vital resources. In addition, the current and future missions of Indian remote sensing satellites are discussed from the viewpoint of their hydrological applications, together with technical and practical impediments in using geospatial technologies.

2. LAND AND WATER DEGRADATION IN INDIA: AN OVERVIEW

Increasing pressure on land and water resources to raise the agricultural production has cost us dearly in terms of resource degradation. Arable land resources are continually being depleted by several processes of land degradation such as erosion by wind and water, anaerobiosis, salinization, alkalinization, compaction and hard setting, depletion of soil organic matter and nutrient imbalance. Such large-scale ecological losses are reported in cropland, grassland, and forestland. There are six major causes of land degradation in the region: deforestation, shortage of land due to increased populations, poor land use, insecure land tenure, inappropriate land management practices and poverty (FAO, 1995). Non-sustainable land management practices such as intensive agriculture is being practiced in various parts of the country for growing wheat and rice crops without introducing legumes in the crop rotation. Main thrusts have been given to produce maximum quantity per unit area. To meet the objective and keeping in view the economic returns, cropping intensity has been increased by more than 300% in many regions. High yielding varieties are being cultivated to get maximum return. Crop inputs such as water, fertilizers, manures etc. are not being adjusted for the reversal of soil fertility status by replenishment of nutrients. As the balance between input and output is low, serious consequences in form of soil degradation is being manifested. Moreover, groundwater is getting contaminated due to heavy use of nitrogen fertilizers.

While soil erosion is the critical problem in the dryland, waterlogging and soil salinity are important issues in the irrigated lands. Providing canal irrigation in arid and semi-arid regions resulted widespread soil salinization and waterlogging. If such exploitations are continued for long time, they would turn the situation disastrous as the reckless exploitation of land and water resources leads to severe physical, chemical and biological degradation along with various eroding processes (Table 3), which in turn have serious socio-economic and environmental consequences.

Perturbations in hydrologic conditions due to land degradation are also of great concern. The removal of vegetative cover causes increased surface runoff, less recharge of groundwater, unusual fluctuation or depletion of streamflows, and extensive flooding in the downstream area. These hydrologic conditions also promote soil erosion. As a result, sediment loads in rivers are increasing, dams are filling with silt, hydroelectric schemes are being damaged, navigable waterways are being blocked, and surface water quality is deteriorating. In several parts of India, the potential life of reservoirs has been reduced by more than 50% due to enhanced sedimentation rate which is ten times higher than that envisaged by the designers.

Das (1987) estimated loss of land productivity due to soil erosion as shown in Table 4. Brandon et al. (1995) estimated total annual loss in the productivity of major crops due to soil erosion as 7.2 million tonnes. However, there is no steadfast information available on type, intensity and severity of land degradation for India. A rough estimate of soil erosion and sedimentation for India reveals that about 5300 million tonnes of topsoil are eroded annually and 24% of this quantity is carried by rivers as sediments and deposited in the sea, and nearly 10% is deposited in reservoirs reducing their storage capacity. As for waterlogging and salinisation, the available estimates indicate that canal command area constitutes 48% of the total waterlogged area, and 45% of the total salt affected area in India. In fact, for a few states like Andhra Pradesh, Tamil Nadu, Orissa, Punjab and Gujarat, canal irrigated area occupies 100% of the total waterlogged area. The rate of soil loss without affecting land productivity is considered as less than 12 t/ha/year.

Table 3. Distribution of areas under different soil erosion, land degradation and land utilization problem (Dandapani, 1990)

Categories	Area (Mha)	
1. Geographical Area	329.00	
2. Area Subjected to Water and Wind Erosion	150.00	
2.1. Area subjected to wind erosion and aridity		38.74
2.2. Area affected by sand dunes out of 38.74 Mha		7.0
3. Area Degraded through Special Problems	25.00	
3.1. Waterlogged area		6.00
3.2. Alkali soils		2.50
3.3. Saline soils including coastal sandy areas		5.50
3.4. Ravines and gullies		3.97
3.5. Area subjected to shifting cultivation		4.36
3.6. Riverine and torrents		2.73
4. Total Problem Area (2 + 3)	175.00	
5. Average Annual Rate of Encroachment of Table Lands by Ravines	8.00	
6. Average Area Annually Subjected to Damage by Shifting Cultivation	1.00	
7. Total Flood-prone Area	40.00	
7.1. Annual Average Area affected by Flood		9.00
8. Total Drought-prone Area	260.00	

Table 4. Rate of soil erosion and long-term decline in land productivity (Das, 1987)

Rate of soil loss (t/ha/year)	Anticipated long-term productivity losses
<12	No change in land productivity
12-50	50% of the area of very productive land down-grades to productive lands: the remainder remains unchanged
51-100	100% of all productive lands downgrades by one productivity class
101-200	50% of the area of all productive land down-grades to not suitable (non-productive land): the remainder downgraded by one productivity class
>201	All extents of productive land downgrades to not suitable (non-productive land)

Various types of wastelands are the output of the land degradation process. To map the entire country in 1:50,000 scale, National Remote Sensing Agency, Department of Space (DOS) carried out National Wasteland Mapping Mission during 1986-2000 and subsequently updated the maps through National Wasteland Updating Mission in year 2003 using IRS LISS III satellite data. The project was sponsored by the Ministry of Rural Development (MRD). About 55.27 million hectares, equivalent to 17.46% of total geographical area of the country, are reported in the Wasteland Atlas (NRSA, 2005a). The information on wasteland has been categorized into 13 broad groups as shown in Table 5, which was further subdivided into 28 sub-classes.

Table 5. Category-wise wasteland in India

Sl. No.	Class	Wasteland area (km^2)	Percentage of total geographical area
1.	Gullied/Ravinous land	19039.34	0.60
2.	Land with or without scrub	187949.5	5.94
3.	Waterlogging and marshy land	9744.97	0.31
4.	Land affected by salinity/alkalinity	12024.05	0.38
5.	Degraded forest land	18765.86	0.60
6.	Degraded pasture/grazing land	126551.8	3.99
7.	Shifting cultivation area	19344.3	0.61
8.	Degraded pasture/grazing lands	2138.24	0.07
9.	Sands	33984.2	1.07
10.	Mining/industrial wasteland	1977.35	0.06
11.	Barren rocky	57747.11	1.82
12.	Steep sloping area	9097.38	0.29
13.	Snow/glacial covered area	54328.16	1.72
	Total	552692.26	17.46

3. HISTORICAL PERSPECTIVE OF INDIAN REMOTE SENSING (IRS) SATELLITES

The Indian space program aimed at developing, launching and operating satellites to cater to the data requirement in support of national development. Being a developing country, the Indian space program was driven by a vision of reaching the benefits of the space technology program to the society at a grassroot level, even as remarkable achievements were made in mastering the complex technologies in a self-reliant manner. The Indian space program started in the early 1960s with the scientific investigation of upper atmosphere and ionosphere over the magnetic equator that passes over Thumba near Thiruvananthapuram using small sounding rockets. Later, the space program embarked on developing indigenous operational satellite systems and launch vehicles. It was done through a well orchestrated planned strategy. Initial learning efforts on spacecraft development culminated in the successful development of Aryabhatta in 1975, a scientific satellite carrying celestial X-ray, solar neutron, gamma and ionosphere experiments and weighing about 360 kg (Katti, 2007). The demonstration of the efficacy of the space systems was performed using the available international missions as done during Satellite Instructional Television Experiment (SITE) in 1975-1976 and Satellite Telecommunication Experimental Project (STEP) in 1977-1979, followed by developing experimental satellites such as Bhaskara and Apple missions; and later transgressing to operational IRS and INSAT systems (Bhaskaranarayana, 2007). Both SITE and STEP paved the way for Indian Space Research Organization (ISRO) to take implementation of operational Indian National Satellite System (INSAT) in the early 1980s, which has now become the largest domestic communication satellite system. The development of operational launch vehicles also followed through a planned sequence of development efforts through SLV and ASLV to reach the current operational PSLV and GSLV systems.

Indian remote sensing era started with Bhaskara, the first experimental EO (Earth Observation) satellite launched in 1979 and is continuing with the launch of Cartosat-2a and IMS-1 in 2008 as well as RISAT-2 in 2009. India has achieved a range of spatial resolution ability from 1 km to better than 1 m. Presently, the Indian earth observation program consists of theme-specific remote sensing satellites (Table 6), with current constellation of seven satellites in operation (Oceansat-1, Resourcesat-1, Cartosat-1, 2 & 2A and IMS-1). India has also launched geostationary satellites in INSAT series (viz., INSAT 1, 2 & 3) and Metsat (Kalpana-1) for higher repetitivity mapping and meteorological applications. INSAT series deployed in orbit comprises Very High Resolution Radiometer (VHRR) with imaging capability in visible (0.55-0.75 micron), thermal infrared (10.5-12.5 micron) and water vapor channel (5.7-7.1 micron) and provides 2×2, 8×8 and 8×8 km ground resolution, respectively. Metsat (Kalpana-1), which carries VHRR and Data Relay Transponder (DRT) payload, provides meteorological services in India (DOS, 2007, 2009). Aerial remote sensing capability with high-resolution digital camera and laser terrain mapper (for detailed surveys of local area) further strengthens the earth observation program. Satellite/aerial remote sensing payloads and the INSAT meteorological payloads together provide an immense imaging capability to the national and global communities.

The IRS series of satellites was operationalized with the commissioning of IRS-1A in March 1988. An identical satellite, IRS-1B, was launched in August 1991 to continue the image data from the satellite. IRS-1C, IRS-P2, IRS-P3, IRS-1D, Technology Experiment Satellite (TES), Resourcesat-1 and Cartosat-1 have further enhanced the imaging capability. All these satellites have been launched by indigenous launch vehicle called Polar Satellite Launch Vehicle (PSLV) (Chandrasekhar et al., 1996).

Oceansat-1, launched in May 1999, carries an Ocean Color Monitor (OCM) and a Multi-frequency Scanning Microwave Radiometer (MSMR)—both of which are useful to study physical and biological aspects of oceanography. IRS-1C and IRS-1D, launched in December 1995 and September 1997 respectively, were identical satellites carrying three cameras, viz., Panchromatic Camera (PAN), Linear Imaging Self Scanner (LISS-III) and Wide Field Sensor (WiFS). The spatial resolution of PAN is 5.8 m; LISS-III 23.8 m in VNIR bands, 70 m in SWIR and 188 m in WiFS. These two satellites are considered as the workhorse of Indian remote sensing applications.

The state-of-the-art satellite Resourcesat-1 (IRS-P6), which was launched on board PSLV-C5 in October 2003, carries three cameras: (i) A high resolution Linear Imaging Self Scanner (LISS-4) operating

Table 6. Salient specifications of IRS satellites launched so far (Navalgund et al., 2007; DOS, 2008)

Satellite (Year)	Sensor	Spectral bands (μm)	Spatial resolution(m)	Swath (km)	Radiometric resolution (bits)	Repeat cycle (days)
IRS-1A/1B (1988, 1991)	LISS I	0.45–0.52 (B) 0.52–0.59 (G) 0.62–0.68 (R) 0.77–0.86 (NIR)	72.5	148	7	22
	LISS-II	Same as LISS-I	36.25	74	7	22
IRS-P2 (1994)	LISS-II	Same as LISS-I	36.25	74	7	24
IRS-1C/1D (1995, 1997)	LISS-III	0.52–0.59 (G), 0.62–0.68 (R) 0.77–0.86 (NIR)	23.5	141	7	24
		1.55–1.70 (SWIR)	70.5 (SWIR)	148		
	WiFS	0.62–0.68 (R) 0.77–0.86 (NIR)	1	810	7	24 (5)
	PAN	0.50–0.75	5.8	70	6	24 (5)
IRS-P3 (1996)	MOS-A	0.755–0.768 (4 bands)	1570 x 1400	195	16	24
	MOS-B	0.408–1.010 (13 bands)	520 x 520	200	16	24
	MOS-C	1.6 (1 band)	520 x 640	192	16	24
	WiFS	0.62–0.68 (R) 0.77–0.86 (NIR) 1.55–1.70 (SWIR)	188	810	7	5
IRS-P4 (1999)	OCM	0.402–0.885 (8 bands)	360 x 236	1420	12	2
	MSMR	6.6, 10.65, 18, 21 GHz (V & H)	150, 75, 50 and 50 km, respectively	1360	–	2
IRS-P6 (2003)	LISS-IV	0.52–0.59 (G) 0.52–0.59 (G) 0.62–0.68 (R) 0.77–0.86 (NIR)	5.8	70	10 (7)	24 (5)
	LISS-III	0.52–0.59 (G), 0.62–0.68 (R) 0.77–0.86 (NIR) 1.55–1.70 (SWIR)	23.5	141	7	24
	AWiFS	0.52–0.59 (G), 0.62–0.68 (R) 0.77–0.86 (NIR) 1.55–1.70 (SWIR)	56	737	10	24(5)
IRS-P5 (Cartosat-1), 2005	PAN (Fore (+26°) & Aft (-5°)	0.50–0.85	2.5	30	10	5
Cartosat-2 (2007)	PAN	0.50–0.85	0.8	9.6	10	5
IMS-1 (Indian mini satellite) (2008)	MxT	0.45-0.52, 0.52-0.59 0.62-0.68, 0.77-0.86	37	151	10	-
	HySI	0.4-0.95 (64 bands)	550	130	11	
Cartosat-2A (2008)	PAN	0.50–0.85	0.8	9.6	10	5

in three spectral bands in the Visible and Near Infrared Region (VNIR) with 5.8 m spatial resolution and steerable up to ± 26 deg across track to obtain stereoscopic images and achieve five-day revisit capability; (ii) A medium resolution LISS-3 operating in three spectral bands in VNIR and one in Short Wave Infrared (SWIR) band with 23.5 m spatial resolution and 142 km swath; and (iii) An Advanced Wide Field Sensor (AWiFS) operating in three spectral bands in VNIR and one band in SWIR with 56 m spatial resolution and a combined swath of 730 km achieved through two AWiFS cameras. Resourcesat-1 also carries a Solid State Recorder to store the images taken by its cameras, which can be received later by the ground stations. These sensors of Resourcesat-1 satellite possess enormous application potential. LISS-4 data help to assess in detail the alignment of canals, site suitability for dams/reservoirs, tunnels, roads, railway, etc. apart from the land and water resources information on detailed scales, i.e., up to village level on 1:10,000/12,500 scale. LISS-3 is useful in preparing thematic maps like land use/land cover, geomorphology, soil, drainage, water bodies, etc. on larger scales (up to 1:50,000 scale), which can be used to arrive at land and water resources management plans up to *taluk*/block level. This data can also be used for the detailed assessment of damages due to disasters like flood, drought, earthquake, tsunami, etc., crop acreage and production estimation, forest type/cover mapping, command area studies, reservoir capacity evaluation, alignment of canals, selection of sites for reservoirs, urban sprawl mapping, and so on. AWiFS with a wide swath of 740 km, resolution of 56 m and repetitivity of five days is best suited for observing rapidly changing features/phenomena like crop condition/acreage, forest cover/fires, large-scale flood monitoring, snow cover, agricultural drought, coastal features, land degradation or desertification, damage assessment due to disasters, command area studies, etc.

Cartosat-1 satellite, launched on May 05, 2005, is primarily intended for advanced mapping applications. Cartosat-1 has two panchromatic cameras with a spatial resolution of 2.5 m and a swath of 30 km each. The cameras are mounted with a tilt of +26 deg and −5 deg along track with respect to nadir to provide stereo pairs of images for generation of Digital Terrain Models (DTM)/Digital Elevation Models (DEM). The satellite has a revisit capability of five days, which can be realized by steering the spacecraft about its roll axis by +26 degrees. The stereoscopic images from Cartosat-1 data will be useful in making three-dimensional maps and generation of contour/slope information which is required for the alignment of canals, command area studies, site suitability assessment of dams/reservoirs, various water conservation/recharge measures, etc. Cartosat-1 data could also be used effectively for updating information on roads, drainage/streams, water bodies, etc. as well as for studying urban areas and rural settlements. The Cartosat-2 (launched on January 10, 2007) and Cartosat-A (launched on April 28, 2008) have been designed to provide much higher resolution stereo data (1 m spatial resolution) for cartographic mapping. The satellites can be steered up to 45 degree along as well as across tract. The Indian Mini Satellite (IMS-1), launched along with Cartosat-2A, carries two payloads – Multispectral Camera (MxT) and Hyperspectral Camera (HySI). In April, 2009, ISRO has launched radar imaging satellite (RISAT-2) with X-band SAR and having all-weather capability to take images of the earth. This satellite will enhance national capability for disaster management aplications. Today, many ground stations around the world are receiving Indian remote sensing data.

Since 1970s, the satellite data are being used for multifold applications, especially for natural resources mapping and inventory. In India, hydrogeomorphological maps have been prepared on 1:250,000 and 1:50,000 scales for groundwater exploration during late 1980s and early 1990s using the imagery obtained from IRS-LISS-I and II (Linear Imaging Self-scanning Sensor) of IRS 1A and 1B (Sahai et al., 1991; Krishnamurthy and Srinivas, 1995). With the availability of data from IRS 1C and 1D in the latter half of 1990s, the mapping scale improved to 1:12,500. The Panchromatic (PAN) data (5.8 m spatial resolution) merged with LISS-III data (23 m spatial resolution) generate hybrid false color composite (FCC) which has an advantage for both spectral and spatial resolution and was used for hydrogeologic mapping for groundwater development and utilization (Reddy et al., 1996; Bhattacharya, 1999). The Resourcesat-1 (IRS P6) was launched in 2003 with Advanced Wide Field Sensors (AWiFS) as one of the sensors to provide high-resolution multi-spectral data having 10 bits radiometry. The data derived from AWiFS, LISS-III and LISS-IV sensors of Resourcesat-1 are being utilized for geological and geomorphological mapping (Vinod Kumar and Martha, 2004). IRS P6 LISS-III data were analyzed to study the lithology and landform of Ramtek area near Nagpur, Central India (Joshi et al., 2004). The false color composite was generated using SWIR, IR and green bands, which showed light greenish tone in the areas identified as shallow alluvium soils underlain with surrounding rocks.

Thus, the Indian Earth Observation (EO) satellite constellation provides data at various spatial and temporal resolutions, and is operationally used in India for many applications, which are of direct social relevance such as water resources management (including groundwater), environmental degradation (desertification, deforestation, and soil erosion), agricultural applications (e.g., estimation of crop acreage and yield, crop suitability analysis, etc.), and land and water developmental applications (e.g., watershed development, irrigation water management, and reservoir sedimentation).

4. ROLE OF RS AND GIS IN LAND & WATER RESOURCES DEVELOPMENT AND MANAGEMENT

Land evaluation is carried out to prepare a management plan for efficient use of land in rural or urban areas. The rural economy mainly depends on productivity of agricultural land, which provides the means to raise income and ensure food security. To turn agriculture profitable, the dependence on marginal lands has to be brought down by providing alternate means of livelihoods and the scope for sustainable sources for irrigation and drinking water must be expanded. In this context, remote sensing provides ample information in addressing these issues. High-resolution EO images provide community-centric, geo-referenced spatial information relevant to the management of natural resources like land use/land cover, terrain morphology, surface water and groundwater, soil characteristics, environment, forest, and infrastructure.

The RS data, being digital in nature, can be efficiently interpreted and analyzed using various kinds of image processing software packages (e.g., ERDAS IMAGINE, PCI geomatics, and IGIS). The raster information, thus generated, can be brought into the GIS environment for integration with various spatial and non-spatial data. A wide range of image analysis systems with GIS facilities are also commercially available such as ArcInfo GRID, IDRISI, GEOMATICA and GRASS. In addition, GPS measurements provide positional coordinates of various features under study in the map layer and can be directly integrated to the GIS database. Thus, the remote sensing and GPS along with GIS aid to collect, analyze and interpret the data rapidly on a large scale, and are very helpful for the planning and management of land and water resources.

5. APPLICATIONS OF IRS DATA AND GIS TO WATERSHED DEVELOPMENT AND MANAGEMENT

5.1 Development of Watersheds

Watershed development in general consists of integrated development of land and water resources. The treatments include soil and water conservation, afforestation, horticulture and other vegetative treatment. However, non-land based activities, viz., income generating activities, training, capacity building, skill upgradation, agro-processing, transport, marketing of agricultural produce, animal husbandry activities, etc. are built into the programs of watershed development at present. These activities would ensure that there is all round development of the villages that are part of the micro watershed. Satellite data greatly facilitate periodic monitoring and mapping of forest, vegetation cover, land cover, geology and soils over watersheds, which are helpful for the study of land use changes, watershed potential, degradation, etc. (e.g., Kudrat and Saha, 1993).

Remote sensing based Integrated Mission for Sustainable Development (IMSD) project was carried out by ISRO, based on the integration of various themes and collateral information, in about 84 million ha in 175 districts of the country. In this project, space-borne multi-spectral data have been operationally used for generating information on various themes and integrated on a watershed basis for generating location-specific action plans for land and water resources development and for assessment of the impact of implementation. The project has been successful in improving groundwater level, increasing cropping intensity and finally conserving land and water resources (NRSA, 2002).

Under the National Watershed Development Program for Rainfed Areas (NWDPRA), 84 desert/drought-prone watersheds of the country were monitored to assess the impacts of various developmental programs undertaken in these watersheds (NNRMS-TR-98, 1998; Chowdary et al., 2001). Indian Remote Sensing Satellites (IRS-1A & 1B) LISS-II data with a spatial resolution of 36 m of pre- and post-treatment of the watersheds have been analyzed. The classified land use/land cover maps and Normalized Vegetation Index (NDVI) outputs have been compared to derive the information on changes that occurred over a given time period. The area statistics with respect to increase or decrease in land use/land cover categories in the watershed and changes with respect to vegetation status were derived. Under NWDPRA, several watershed treatment activities were identified which are broadly categorized as conservation and production measures, drainage-line treatments, creation of infrastructure, livestock development, and people's participation. These studies demonstrated community participation right from the planning stage, which helped in implementing the framed measures for improving watersheds in terms of productivity.

Participatory watershed development taken up through multi-stakeholder involvement across five districts of Karnataka under Sujala Watershed Development Program (Diwakar et al., 2008) is an excellent example of watershed development and management using remote sensing and participatory approaches. The major goal of the project is towards improving the productive potential of degraded watersheds in dryland areas and poverty alleviation of rural community. Remote sensing based products on natural resources are effectively used for integrated land and water resources developmental planning to enhance the productivity levels in dryland areas. Remote sensing inputs along with GIS and Management Information System (MIS) have helped in characterization, prioritization, developmental planning, implementation strategies generation with peoples' participation, and concurrent monitoring and impact assessment at various stages of implementation. The uniqueness of this study lies in: (i) infusion of modern technologies in the form of remote sensing inputs with simple MIS/GIS tools in local language and simple client-server solutions; and (ii) participation of people, experts, local government, NGOs, and other stakeholders with one clear objective of achieving sustainable development. Multi-seasonal and high spatial resolution remote products are used discretely at various stages of the project implementation for effective project monitoring. The mid-term assessment of the project has indicated encouraging trends. The average crop yields have increased by 24% over the baseline. Annual household income from employment, income generating activities and improvements in agricultural productivity has increased by 30% from a baseline of Rs. 10,300 to 13,000. The average groundwater level has increased by 3 to 5 feet. Shift to agro-forestry and horticulture, and reduction in non-arable lands can also be seen.

5.2 Impact Assessment of Climate Change and Land use

Addressing climate change requires a good scientific understanding as well as coordinated action at national and global levels. The Fourth Assessment Report of the Intergovernmental Panel on Climate Change (IPCC, 2007) predicts the surface air warming of 1.8 to 4.0 °C (under different greenhouse gas emission scenarios), sea level rise of 0.18 to 0.59 m, high frequency in heat waves and heavy precipitation events. Increases in the amount of precipitation are very likely in high latitudes, while decreases in precipitation are likely in most sub-tropical regions (by as much as about 20% under A1B scenario in 2100). Although there are uncertainties in the projections of climate change into 2100 and beyond, most Global Circulation Models (GCMs) are robust in predicting global warming. The projected climate change under different greenhouse gas emission scenarios is likely to have adverse implications on freshwater supply, food production, coastal settlements, livelihoods and biodiversity. The climate change issue along with salient case studies from India on impact assessment is discussed in detail in other chapters.

Reliable scientific and economic studies are required at national and regional levels within India to help policy makers, industry, and climate-sensitive communities and sectors (e.g., agriculture, forest and fisheries) in developing strategies for mitigating and adapting to climate change. Space-based remote sensing data help in mapping earth resources, monitoring their changes, and deriving bio-geophysical parameters. These pieces of information are useful for identifying the indicators and agents of climate change. The space-based inputs can also be integrated with physical simulation models to predict the impact of climate change and climate variability. Space-based inputs provide information related to three aspects of climate change: (i) indicators of climate change such as change in polar ice cover, glacial retreat in the Himalayas, upward shift in timberline and vegetation in the alpine zone, desertification, bleaching of coral reefs, etc.; (ii) assessment of agents of climate change such as greenhouse gases and aerosol, their sources and distribution pattern; and (iii) modeling impact of climate change in various fields, including natural resources. Thus, satellite data are of great help in planning adaptation, preparedness and mitigation measures for climate change and climate variability.

Land use/land cover mapping on 1:250,000 scale covering entire country using multi-temporal IRS P6 AWiFS dataset and digital analysis technique was initiated in December 2004 with the main objective to get the net sown area of crops on an annual basis for different cropping seasons starting from 2004-2007. Further, under Natural Resource Census (NRC) project, land use/land cover mapping and land degradation mapping at 1:50,000 scale using IRS data has been initiated and being executed with the support drawn from various partner institutions (DOS, 2008). However, the status of land use needs to be evaluated under various climate change scenarios. Desertification is one of the important indicators of climate change. SAC (2007) carried out a national project on desertification and land degradation status mapping on 1:500,000 scale using multi-date AWiFS data from Resourcesat (IRS-P6) satellite. It was reported that 81.45 Mha area of the country is under desertification. More and more studies are required in this direction to evaluate the impacts of environmental changes under different hydrologic and hydrogeologic conditions.

5.3 Monitoring and Management of Drought

Drought is a perennial and recurring feature in many parts of India. According to NDMD (2004), about 68% of the country is prone to drought in varying degrees. The 'chronically drought-prone' areas (around 33%) receive less than 750 mm of rainfall, and 'drought-prone' areas (35%) receive a rainfall of 750-

1200 mm. Such areas of the country are confined to peninsular and western India, primarily in arid, semi-arid and sub-humid regions (Singh and Rao, 1988).

Drought management involves development of both short-term and long-term strategies. Short-term strategies include early warning, monitoring, and assessment of droughts. However, long-term strategies aim at drought mitigation measures through proper irrigation scheduling, soil and water conservation measures, cropping pattern optimization, etc. Early warning of drought is useful for on-farm operations and to arrive at an optimal local water utilization pattern. Rainfall anomalies as observed from geostationary/meteorological satellites are being used for the early warning of drought, which is yet to be fully operationalized. Studies have indicated that certain large-scale meteorological patterns are associated with the failure of the summer southwest monsoon, which is the main cause of droughts in the Indian subcontinent (Sikka, 1980). Factors that can provide early indication of possible droughts include upper air winds over India, development of hot low-pressure areas over southern Asia, and the El Nino/Southern oscillation phenomena in the Pacific Ocean. Other factors that can be observed by satellites and which are related to rainfall patterns are sea surface temperature, snow cover, cloud patterns, wind velocity and direction, and atmospheric temperature/humidity profiles (Pandey and Ramasastri, 2001; Dilley and Heyman, 2007).

Satellite derived vegetation index (VI) which is sensitive to moisture stress is now being used continuously to monitor drought conditions on a real-time basis often helping the decision makers initiate strategies for recovery by changing cropping patterns and practices (Seiler, 1998; Kogan, 2001). In India, a RS-based National Agricultural Drought Assessment and Monitoring System (NADAMS) has been developed for countrywide drought monitoring. Monthly drought assessment reports are being generated under NADAMS by National Remote Sensing Centre (NRSC). With the operationalization of IRS-1C WiFS and IRS-P3 WiFS and SWIR bands, in-season agricultural drought monitoring capability has been further improvised. Seasonal drought occurs when rainfall is deficient by more than twice the mean deviation. In rainfed areas, as long as the risk associated with the adoption of a technology appears to be high for the farmers, that technology tends to remain unaccepted. Therefore, sustainable dryland farming systems should include arable cropping, livestock management, alternative land uses, and management of land and water resources to stabilize crop production over the years (Singh et al., 2000). The drought and risk resistant farming systems have been developed for dryland farming taking into account in situ rainwater conservation, harvesting of surface runoff, and recycling.

IRS-1C/1D LISS III geocoded FCC on a 1:50,000 scale was used for the preparation of land and water resources management/utilization plans to aid state and district/block-level officials in planning development works on desertification control and drought mitigation (IRISDA Atlas, 2004). The thematic information on land use/land cover; land forms/geology, groundwater, drainage and surface water bodies, soil, land degradation, etc. were obtained through the interpretation of satellite data. Survey of India topographical maps on 1:50,000 scale were used to derive information on slope. Collateral data (maps and non-spatial data) were collected and integrated suitably with the image-derived information. Field studies were carried out to assess the interpretation accuracy and to collect data on rocks, crops, wells, reservoirs, soils, etc. Socio-economic/demographic data were also collected from the Census 2001 and State/District sources. Maps were finalized by integrating all the information. A digital database of both spatial and non-spatial data had been created in the GIS environment and was used to assess desertification status. A set of decision rules, based on the existing land use/land cover, soil, slope, groundwater prospects (hydro-geomorphology), had been prepared taking into consideration terrain condition, ground observations and interaction with resource scientists. The decision rules were implemented on the integrated layer derived from GIS analysis for the generation of suitable land and water resources development plans. The action plans contained recommendations to amend the present land use system. The action plan maps thus generated depict the spatial categorization of land suitability under updated land use systems such as agro-forestry, double cropping, horti-pastoral, silvipastoral, agro-horticulture, etc. for mitigating desertification (e.g., Fig. 1).

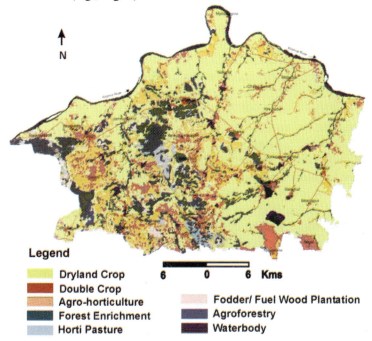

Fig. 1 Land resources development plan of Devedurga Taluka, Raichur District, Karnataka, India (IRISDA Atlas, 2004).

5.4 Monitoring and Assessment of Forest Cover

Forest ecosystems over the time have been subjected to severe biotic and abiotic pressures. In this context, Earth Observation Systems constituting diverse satellite and airborne sensors are contributing substantially in assessing the global, regional and local impacts on forest ecosystems and develop different means for sustainable development. Currently several countries (India, USA, Brazil, Sri Lanka, etc.) are preparing inputs to monitor forest cover at regional and local scales, biodiversity assessment, forest degradation, forest fire mapping, biomass potential/carbon sequestration, inventories, forest hydrology, etc. using satellite data.

The Forest Survey of India (FSI), Ministry of Environment and Forest, is conducting forest mapping in the country using IRS data. State of Forest Cover Report 2003, prepared using IRS 1C/1D data, was released by FSI in 2005. According to the estimate, total forest cover as of 2003 is 678,333 km^2 (20.64% of total geographical area) as against 20.55% in 2001. A new class, Very Dense Forest (VDF) with a canopy density of above 70% has been added in the 2003 mapping. The very dense forest and dense forest together as of 2003 is 11.88% as against 12.68% in 2001, open forest as of 2003 is 8.76% as against 7.87% in 2001, and mangroves as of 2003 are 0.140% as against 0.141% in 2001. Overall, the forest cover has increased by 2,795 km^2 from 2001 to 2003, but the dense forest cover in the country has decreased by 26,246 km^2 (FSI, 2005). The open forest, on the other hand, has increased by 29,040 km^2.

5.5 Landslide Hazard Assessment

It has been reported that almost all the landslides have resulted from the combined effects of two or more of the several causative factors such as steep slopes, deep weathering, loose overburden with clayey material, heavy rains, slope disturbance by construction of roads and buildings, toe and headward erosion by streams, etc. (Sarkar et al., 1995). The hilly terrains of India, particularly in the Himalayas and the Western Ghats, are most vulnerable to landslides. The Himalayan mountain belt consists of tectonically unstable younger geological formations and often the landslides are huge and massive and in most cases the overburden along with the underlying lithology is displaced during sliding, such as the Malpa landslide of 1998 when one entire village was buried by a huge landslide (Paul et al., 2000). In contrast, the Western Ghats and Nilgiri Hills are geologically stable but have uplifted plateau margins influenced by neo-tectonic activity and the landslides are usually confined to the overburden without affecting the bedrock. The landslides are generally in the form of debris flows occurring mainly during monsoon seasons, but the effects are felt more acutely due to higher density of population in this region.

Remotely-sensed data provide a synoptical view of large areas on a temporal basis for evaluating the surface conditions before and after a landslide event. Terrain information such as land cover, geology, geomorphology and drainage could also be derived from satellite imagery and the existing thematic information can be updated to enable the quantification of human interference on the earth's surface. The spatial and temporal thematic information derived from remote sensing, thematic maps, and ground-based information need to be integrated for landslide hazard zoning (Nagarajan et al., 1998). This requires voluminous data storage, manipulation and analytical capabilities, which are offered by a Geographical Information System even on a PC-based platform.

A landslide hazard zone map of Aizawl was generated for district using the thematic maps prepared on 1:50.000 scale (Anonymous, 1991). In this study, the thematic maps, viz., geomorpholigical map with major structures and landslides locations, soil, land use/land cover, slope, drainage and watershed, transport networks and settlement locations on 1:50,000 scale for Aizawl district in Mizoram were prepared using IRS-1B LISS-II data. Each of the existing landslides was evaluated in terms of its setting with reference to landform, lithology, transport network and settlement locations and the main factors (parameters) that have contributed to slope failures at each of these locations were identified. Only the causative factors/parameters were considered for integration to generate a landslide hazard zone map. In Mizoram, the lithology, geomorphology, slope, transport network, and settlements have been found to influence slope failures. Integration of these themes was carried out toposheet-wise by overlay method and the entire district was broadly demarcated into three categories of landslide hazard zones namely 'high hazard zone', 'medium hazard zone' and 'low hazard zone'.

Moreover, a study was carried out (Kimothi et al., 1999) to ascertain the causes of the 1998 landslides in Madhyamaheshwar and the Kaliganga sub-watersheds in Okhimath Tehsil of Rudraprayag district, Uttaranchal. The objective of the project was to identify the forcing factors using high resolution IRS-1C/1D (pre- and post-damage LISS III and PAN) data in conjunction with archived multi-temporal satellite and ancillary data. The temporal change monitoring indicated marginal increase in landslides (156 ha) during 1966 to April 1998 (pre-damage). The August 1998 landslide devastated nearly 718 ha (16.7%) in Madhyamaheshwar and 101 ha (4.4%) in Kaliganga. The severe damages were observed in pine and oak mixed village forest, agriculture, built-up and grassland by comparing the satellite images of April 1998 (pre-damage) and October 1998 (post-damage period). The study identified about 21 villages in Madhyamaheshwar watershed for immediate rehabilitation before the onset of monsoon.

5.6 Wetland Monitoring and Mapping

Wetland is a complete hydrological system and is susceptible to changes in hydrological conditions with various geomorphic-geographic settings. Because of their diverse origin, geographic locations, geomorphic settlings, variable watershed areas and water regimes, sediment character, and biotic (both plants and animals) components, wetlands exhibit wide variations in their nature and sustenance (Maltby, 1991). Wetlands are very intricate and their ecological and physico-chemical functions are influenced and affected by a variety of environmental factors (anthropogenic and natural). It is well recognized that wetlands are rich in biodiversity and various flora and fauna populate in wetland ecosystems. As a result, wetlands perform many socio-economic, ecological and environmental functions based on hydrological, chemical and biological properties, out of which society may gain different category of values such as fish and wildlife values, environmental quality values and socio-economic values (Maltby, 1991). Despite the tremendous importance for protecting inland and coastal environments, and as potential sources of a number of goods and services, wetlands are fragile ecosystems, susceptible to change and the easiest to destroy permanently. The wetlands are under constant threats due to siltation, over-exploitation, aquaculture development, encroachment, pollution, etc. Consequently, the health of wetland ecosystems is gradually deteriorating because of increasing population and industrialization (Moss, 1988), especially in developing nations. In addition, global environmental changes such as climate change and socio-economic changes are posing additional challenges for the sustainable management of wetland ecosystems. Thus, there is an urgent need to conserve and protect wetland ecosystems in a sustainable way using modern management tools and techniques so as to avoid catastrophes in the future. In fact, proper management of wetlands is a serious issue these days, which has received a global recognition.

For protecting wetland ecosystems and developing conservation measures, monitoring, mapping and inventory of wetlands are of prime importance; geospatial technologies can play an important role in accomplishing these tasks effectively. For example, a national project on wetland mapping was accomplished in India during 1992-1993 (Garg et al., 1998). IRS 1A/1B (LISS I/II) false color composite (FCC) data were visually interpreted for the delineation of wetlands, assigning turbidity levels and distribution of vegetation in inland wetlands. Based on the hue evident in FCC, the quantitative turbidity ratings were assigned to wetlands such as 'low', 'medium' and 'high' for dark blue, medium blue and light blue, respectively. More and more studies in this direction are necessary in order to ensure sustainable management of wetland ecosystems.

6. APPLICATIONS OF IRS DATA AND GIS TO WATER RESOURCES MANAGEMENT

6.1 Monitoring and Assessment of Glaciers

The Himalayas, the youngest and fragile mountain system of the earth, has direct influence on climate control, regional hydrology and environment of Indian subcontinent. The Himalayas has one of the largest concentrations of glacier-stored water outside the Polar Regions (Bandyopadhyay and Gyawali, 1994; Thompson et al., 2000). About 17% of its mountain area is covered by glaciers. These glaciers are receding faster than those in other parts of the world (Kumar, 2005). Melt water from these glaciers forms an important source of runoff into the North Indian rivers during critical summer months.

The Himalayan glaciers are located in remote and rugged mountainous terrain. Therefore, normally conventional methods are difficult to use and satellite remote sensing can provide vital information about the Himalayan glaciers (Kulkarni, 1991, 1992; Philip and Ravindran, 1998). This technique has been extensively used for mapping various glacier features such as accumulation area, ablation area, transient snow line, moraines, moraine-dammed lakes and glacier boundary. An investigation in the Satluj basin indicated a presence of 2321 glaciers and permanent snowfields covering an area of 2697 km² (Kulkarni et al., 1999). Moreover, satellite-imagery derived glacier surface topographies obtained at intervals of a few years were adjusted and compared (Anonymous, 2007a). Calculations indicated that 915 km² of Himalayan glaciers of the test region, Spiti/Lahaul (Himachal Pradesh, India), has been thinned by an annual average of 0.85 m between 1999 and 2004. Though the technique used in this investigation is still experimental, it has been validated in the Alps and could prove effective for watching over all the Himalayan glacier systems. However, there is a need to have a better technique for reliable estimates of glaciers, which can overcome several sources of error and approximation inherent in satellite-based observations (Anonymous, 2007a).

6.2 Reservoir Capacity Evaluation

Conventional hydrographic surveys on reservoir sedimentations are tedious, labor intensive and costly as it involves boat-level surveying and subsequent data analysis. Satellite remote sensing method provides a time- and cost-effective approach for the estimation of water spread area. The assessment of reservoir capacity is based on the fact that the water spread area of a reservoir decreases with sediment deposition. The quantity of sediment deposited in a reservoir is determined by the reduction in reservoir water-spread area at a specified elevation over a time period. Using this approach, the storage area-capacity curve was determined for assessing the loss of reservoir capacity (Agarwal et al., 2000; Jeyakanthan et al., 2002). The reservoir capacity at various elevations (derived from multi-date satellite data) is computed using following trapezoidal formula:

$$C = \frac{1}{3}(H_{i2} - H_{i1})(A_{i1} + A_{i2}) + \sqrt{(A_{i1} \times A_{i2})} \tag{1}$$

where C = computed storage capacity [L³], and A_{i1} and A_{i2} = water-spread areas obtained from satellite imagery corresponding to reservoir stages H_{i1} and H_{i2}, respectively [L²].

Optimum reservoir operation schedules are determined based on the realistic assessment of available storage. Reduction in the storage capacity beyond a limit prevents the reservoir from fulfillment of the purpose for which it is designed. Periodical capacity surveys of reservoir help in assessing the rate of sedimentation and reduction in storage capacity. Operational methodologies, integrating satellite and ground data, for sediment yield estimation were developed and applied in many locations for the realistic assessment of reservoir capacity, appropriate catchment treatment plan and efficient reservoir operation (Khan et al., 2001; Sekhar and Rao, 2002; Jain et al., 2002; Beguería, 2006).

In India, reservoir capacity evaluation with remote sensing data was started using Landsat data during 1980s. Rao et al. (1985) applied visual interpretation techniques to estimate water-spread area of Sriramsagar reservoir from Landsat MSS imagery and concluded that the results were comparable with the hydrographic survey. Mohanty et al. (1986) employed area-capacity curve to estimate the sedimentation of Hirakud reservoir in Orissa, India, using multi-date Landsat MSS satellite data. Subsequently, Manavalan et al. (1990) used IRS-1A satellite data for evaluating the capacity of the Malaprava Reservoir, Karnataka, India. Jain et al. (2002) estimated reservoir sedimentation in the Bhakra reservoir located on the Satluj river basin in the foothills of the Himalayas. Multi-temporal IRS-1B LISS-II (36.5 m resolution) data of the years 1996 and 1997 were analyzed for determining the water-spread area of the reservoir and for computing the rate of sedimentation. The estimated average rate of sedimentation through remote sensing data was found to be 25.23 Mm^{3}yr^{-1} compared to 20.84 Mm^{3}yr^{-1} measured by hydrographic survey. The error has been attributed to the accuracy in determination of water-spread area using satellite data. Sekhar and Rao (2002) used satellite remote sensing technique to identify soil erosion zones of Sriramsagar catchment area in Andhra Pradesh, India, and to suggest appropriate measures for controlling soil erosion. IRS-1C LISS-III data, Survey of India toposheets (1:50,000 scale), and precipitation data were used to estimate the sediment yield from the Sriramsagar catchment area (called 'Phulang Vagu watershed'). The soil conservation measures such as vegetation cover and 12 check dam sites were proposed by integrating drainage map, slope map and land use/land cover map. In another study, Shanker (2004) assessed the water-spread area of Panshet reservoir in Maharashtra using both IRS-P6 multispectral LISS-IV (5 m resolution) and LISS-III (24 m resolution) images. The reservoir water-spread areas obtained from the LISS-IV and LISS-III images were comparable. However, the error associated with border (mixed) pixels while delineating water spread was significantly lower in the case of LISS-IV image. Therefore, the use of higher resolution imagery such as LISS-IV improved the accuracy of reservoir water-spread estimation, especially for small to medium size reservoirs having small 'shape factor' (area divided by perimeter).

6.3 Monitoring and Management of Flood

About 40 Mha area (i.e., nearly one-eighth of the country's geographical area) is flood prone. The total area affected annually on an average is about 7.7 Mha. The cropped area affected annually is about 3.5 Mha and was as high as 10 Mha in the worst year (Planning Commission, 2002). On an average, as many as 1439 lives are lost every year due to floods. One of the most important elements in flood management is the availability of timely information for taking decisions. Hence, an identified system should be developed to address various information needs and to provide an operational service with its framework. The main goal is to provide timely and accurate information about inundation extent on operational basis to the user departments so as to assist in organizing relief operations and making a quick assessment of damages. The information has to be made available within 24 hours to the user community for effective utilization of the information.

For making quick damage assessment, the most important aspect is the detailed database development for the chronic flood-prone areas. This is an essential element towards generating information required by the users at Central and State levels. The database consisting of various thematic layers such as land use/cover, topography, infrastructure, etc. needs to be created which can be used to overlay disaster related information collected from various sources to assess the extent and impact of flood disaster. Furthermore, most of the flood-prone rivers in India change their courses frequently, and hence it is necessary to understand the river behavior and its latest configuration so as to plan flood control measures effectively. At the same time, it is equally important to monitor existing flood control structures from time to time to avoid breaches bearing in mind frequent changes in river configuration. In order to provide bank protection works, vulnerable areas subjected to bank erosion along the rivers need to be monitored as well. Flood hazard zonation maps at a large scale are required for planning non-structural measures. Thus, the data requirement for flood disaster management is quite complex and an extensive effort is needed to collect necessary data, especially by conventional methods. To obtain flood-related information by conventional means is time consuming and laborious; the flooded areas may not be accessible. In this regard, satellite remote sensing along with GIS provides an excellent source of flood-related information, even for inaccessible areas.

In order to provide space-based inputs concerning flood inundation to the users, a constant watch was kept by Disaster Support Centre (DSC), National Remote Sensing Centre (NRSC) on the flood situation in the country and the satellite data from Indian Remote Sensing Satellites (P4/P6), Modis-Terra and Radarsat along with pre-flood satellite dataset were procured. These satellite data were analyzed for the delineation of flood inundation thematic layer. The flood inundation map was then integrated with district boundaries and district-wise flood inundation area statistics were generated. The flood inundation maps along with the affected area statistics were furnished to the National Disaster Management Authority (NDMA), Ministry of Home Affairs, Government of India, New Delhi; Chairman, Central Water Commission (CWC), Government of India, New Delhi; Relief Commissioner of the concerned States and other departments associated with flood management in the country. Apart from providing flood inundation information, the historic flood inundation information obtained from satellite imagery was used to generate flood hazard zone maps for planning non-structural flood control measures. Using multi-date satellite data maps pertaining to river configuration, bank erosion and flood control works were prepared and furnished to the users for planning structural flood control measures. Moreover, the flood-risk zone maps are updated with high spatial resolution satellite data and digital elevation models based on close contour information by DSC, NRSC. Flood damage vulnerability analysis requires integration of the information on satellite derived physical damage and socio-economic data. Since 1987, all major flood events of the country have been mapped in near real-time and statistics on crop area affected and number of marooned villages generated by NRSC. Near real-time flood monitoring is being carried operationally in Brahmaputra (Assam), Kosi/Ganga (Bihar), Indus (J&K), Godavari (AP) and Mahanadi (Orissa) river basins using optical and microwave data (Venkatachary et al., 2001; DOS, 2007). Studies have also been carried out for the delineation of flood risk zones in Bramhaputra and Kosi river basins (Bhanumurthy and Behera, 2008).

6.4 Groundwater Evaluation and Management

Remote sensing data coupled with GIS help in evaluating groundwater potential in a basin in less time and cost-effective manner than the conventional methods (Jha et al., 2007 and references therein). Remote sensing captures the key geological, geomorphological, structural and hydrological surface features, and thus enables qualitative groundwater prospecting (e.g., Fig. 2). The thematic layers on geology, geomorphology, drainage, soil and land use generated from satellite data are essential for the preparation of hydro-geomorphological maps and subsequently identifying groundwater prospective zones in an area (Das, 1990; Krishnamurthy et al., 1996; Saraf and Choudhury, 1998; Edet et al., 1998; Chowdhury et al., 2009). Further follow-up investigation using hydrogeologic and geophysical methods can help select suitable sites for well drilling.

Moreover, RS and GIS can be helpful in delineating artificial recharge zones in basin and/or selecting suitable sites for artificial recharge or rainwater harvesting (Jha and Peiffer, 2006; Jha et al., 2007). For example, one study was carried out for the entire state of Tamil Nadu, India, and thematic maps were prepared on 1:50,000 scale indicating the hydrological features and favorable areas for artificial recharge (Ramalingam and Santhakumar, 2000). Considering the terrain conditions and favorable zonation, suitable artificial recharge structures such as percolation ponds, check dams, recharge pits/shafts, contour bunds, contour trenches, *nallah* or stream bunds, and subsurface dykes were recommended. Recently, Chowdhury et al. (2009) delineated artificial recharge zones and identified possible sites for artificial recharge in the West Medinipur district of West Bengal, India, using RS, GIS and multi-criteria decision making (MCDM) techniques. Besides these studies, several studies have been conducted in different parts of India as well as in other countries dealing with groundwater evaluation and artificial recharge using RS and GIS techniques, and a review of such studies could be found in Jha et al. (2007). For the details about the application potentials of RS and GIS techniques in groundwater hydrology, current status and research needs in this area, the readers are referred to Jha and Peiffer (2006).

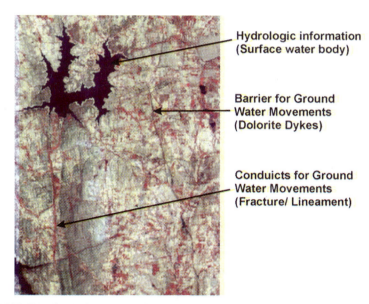

Fig. 2 IRS LISS-III False Color Composite image depicting the hydrogeomorphological features for groundwater potential assessment (Reddy, 2003).

In fact, the use of remote sensing and GIS technologies for groundwater evaluation was started in India for the preparation of hydrogeomorphological maps of the entire country on 1:250,000 scale using satellite images under the National Drinking Water Mission (NDWM), launched by the Government of India in 1986 as an effort to provide sources of safe drinking water to 160,000 problematic villages. Under this mission, broad groundwater potential zones were delineated for further follow-up through hydrogeologic and geophysical methods to provide potable drinking water to the villages (Sahai et al., 1991; Reddy et al., 1996). Subsequently, the Drinking Water Mission was reinitiated in 1999 on 1:50,000 scale for the preparation of groundwater potential maps using high-resolution satellite data (IRS LISS-III data). Under this project, groundwater potential maps on 1:50,000 scales are prepared by incorporating geological (lithological and structural), geomorphological and hydrological information. Such integrated information is used to derive hydro-geomorphological maps for assessing the probable groundwater potential as well as providing information for selecting sites for constructing recharge structures to augment drinking water sources wherever required. These maps are then evaluated critically in terms of the lithology, associated structural features, drainage, soil type and depth of weathering, and type of land use/land cover prevailing in the area. Thereafter, suitable sites for well drilling are selected using hydrogeological and geophysical methods. With the help of these maps, a number of wells have been drilled with a success rate of about 90% and artificial recharge structures have also been successfully constructed (NRSA, 2005b).

6.5 Monitoring and Mapping of Water Pollution

Remote sensing of water quality can complement ground efforts in mapping and monitoring point and non-point pollution sources. Point source identification calls for high-resolution satellite data. Regional models of non-point source pollution loading is arrived based on remote sensing derived inputs on land use/land cover, supported by sample ground data collection (Chowdary et al., 2005; Ouyang et al., 2008). Salinity intrusion into the Hooghly estuary (located in the east coast of India) has been aerially surveyed and studied (INCOIS, 2002). Growths of aquatic weeds and algal blooms, as an indication of eutrophication, have also been mapped from satellite and aircraft data. In general, remote sensing techniques can be successfully applied in all environments where there is a change in color, temperature or turbidity. Care needs to be taken to support statistical models of remote sensing for water quality with proper understanding of physics (Mertes, 2002). Ground truth requirements are also more stringent in this case compared to land remote sensing. GIS technology provides enhanced capability for water quality modeling. Further, satellite data have been used in River Action Plan such as in the Yamuna River (New Delhi) to identify sites for sewage treatment plants. Thus, remote sensing and GIS are useful tools for the monitoring and mapping of water pollution.

7. APPLICATIONS OF IRS DATA AND GIS TO COMMAND AREA DEVELOPMENT AND MANAGEMENT

7.1 Performance Evaluation of Irrigation Systems

Rainfall in India, as in other tropical countries, is confined mainly to the southwest monsoon months of June to September. The rainfall is not uniform and has significant spatial and temporal variation causing droughts in some parts of the country and floods in other parts. Since time immemorial, irrigation has been recognized as a vital input for agriculture in India. About 55.0 Mha (40%) out of 142 Mha of net sown area is under irrigation and the remaining area is under rainfed condition. The average yield of foodgrain is 1.0 t/ha from the rainfed areas compared to 2.3 t/ha from the irrigated areas. Thus, the irrigated areas account for 59% of total foodgrain production in India (MOA, 2006).

The irrigation scenario in the developing country is generally characterized by low level of performance (Bos et al., 1994). The major problems in command areas are associated with excess water applications, inappropriate delivery methods, poor infrastructure, deviation from recommended cropping patterns, unauthorized irrigation, lack of irrigation schedules, absence of water pricing, and lack of awareness about water saving, etc. In India, the command area development program was launched during 1974-1975 with a major objective of bridging the gap between the irrigation potential created and that utilized through micro-level infrastructure development and efficient on-farm water management in order to enhance agricultural production and productivity (Asawa, 1993). However, most canal commands of the country are still performing poorly and the key concern is to enhance water use efficiency and to ensure sustainable agricultural production and food security (Asawa, 1993; Jha et al., 2001). Most command areas are suffering from the problems of waterlogging and salinity/alkalinity. The extent of command area affected by waterlogging and salinity/alkalinity has been monitored and mapped using remote sensing technique to aid reclamation measures (Choubey, 1998; Dwivedi et al., 1999).

In a study conducted by Chari et al. (1994), multi-date LISS-I data (72 m resolution) of IRS 1A satellite for the three winter seasons prior (1986-1987) and after (1992-1993 and 1993-1994) the implementation of National Water Management Project (NWMP) were analyzed to provide information about system performance and diagnostic analysis of canal irrigation projects. The system performance indicators were depth of water applied, irrigation intensity, major crop yield and water use efficiency. The diagnostic analysis of distributaries was carried out in temporal and spatial domains to find out water distribution problem. The study revealed anomaly in water use in the head reach against the tail reach. Subsequently, under NWMP ensuring the equitable water distribution and improved water management as corrective measures helped in increasing irrigation intensity, depth of water application, winter crop yield and area under rice crop. Furthermore, IRS-1A and 1B data (LISS-II) have helped in the development of methodologies for disaggregated inventory of irrigation system with regard to area, cropping pattern, crop condition and productivity (Prasad et al., 1996; Thiruvengadachari and Sakthivadivel, 1997; Murthy et al., 1998). The launch of IRS-1C and 1D satellites has facilitated better spatial resolution satellite data, which have been used for the performance evaluation of irrigation projects (Chakraborti et al., 2002).

A study on performance evaluation of Nagarjun Sagar Irrigation Command (4200 km^2) in Andhra Pradesh, India, using remote sensing, GPS and GIS techniques demonstrated the potential utilization of IRS-1A LISS II and IRS-1C/1D LISS III satellite data (Chakraborti et al., 2002). The objectives of the study was to estimate the utilization of the created irrigation potential, performance of canals and distributaries in terms of irrigation intensity, percentage area under different crops, crop condition and productivity and delineation of waterlogged and salt-affected soils. The methodology involved acquisition of multi-date satellite data of winter crop season (November to March) during 1990-1991 and 1998-1999 and crop classification using ground information to determine the changes occurred over the years. The NDVI from satellite data was also computed to assess spatial distribution of crop vigor and development of crop yield model. The other indices such as irrigation intensity and water utilization index were also worked out from the data collected from command area authority. These indices helped in assessing water management scenario of the command area. The progressive development of waterlogging and extent of salt-affected soils were obtained from interpretation of satellite data and subsequent field verification. The study revealed anomaly in water utilization in the canal command. This is due to the fact that there was much higher water application at the head reach of the canal system for irrigating the crops like rice and pulses during both the years resulting in low irrigation water availability for the tail enders. The higher water utilization index during 1998-1999 corresponding with lower productivity of rice over 1990-1991 indicated extensive irrigation application during 1998-1999 without considering optimal water management practices (Fig. 3). Further, the extent and nature of salt-affected soils were also mapped to take up suitable reclamation measures.

7.2 Monitoring and Mapping of Problematic Soils

Waterlogging and salinity are the major forms of land degradations and generally associated either with the landforms or with the irrigated agriculture. The global extent of primary salt-affected soils is about 955 Mha, while secondary salinization affects some 77 Mha, with 58% of these in irrigated areas (Metternicht and Zinck, 2003). Nearly 20% of all irrigated land is salt-affected, and this proportion tends to increase in spite of considerable efforts dedicated to land reclamation. There is an urgent need for careful monitoring of soil salinity status and variation in order to curb degradation trends, and ensure sustainable land use and management. In southwestern Australia, satellite remote sensing data were used for the broad spatial assessment of saline and waterlogged soils which are developed due to soil type and terrain characteristics (McFarlane and Williamson, 2002). Irrigation induced soil salinity was characterized effectively using field-derived spectra of saline soils and related vegetation in the Murray-Darling Basin of Australia (Dehaan and Taylor, 2002). Spaceborne, airborne and ground-based electromagnetic induction meters, combined with ground data, have shown potential to extract surface information about soil salinity (Howari, 2003; Metternicht and Zinck, 2003). Shrestha (2006) carried out a study in northeast Thailand to develop the relationship of spectral reflectance and physico-chemical soil properties with electrical conductivity (EC) using remote sensing data (Landsat ETM+) and result of soil sample analysis. Multiple

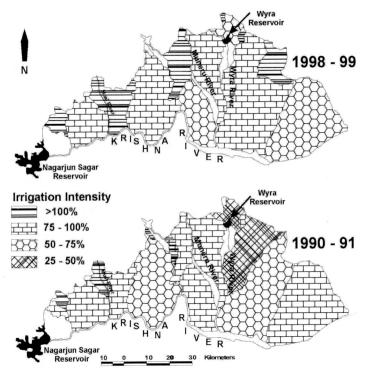

Fig. 3 Irrigation intensity in Nagarjun Sagar Irrigation Command for two time periods (Chakraborti et al., 2002).

regression analysis indicated that the mid-infrared band (band 7) and the near-infrared band (band 4) correlated well with the observed EC values of the surface soil layer.

India is no exception and about 7.6 Mha of India's land area is afflicted with the twin problems of waterlogging and salinity, which have appreciably reduced agricultural productivity and have grave implications for nation's food security (Anonymous, 2007b). The mapping of degraded land and salt-affected soil started with using Landsat TM data (Sharma and Bhargava, 1988; Saha et al., 1990). Dwivedi and Rao (1992) adopted a quantitative approach to identify the most appropriate three-band combination of Landsat TM reflective-band data for delineating salt-affected soils of the Indo-Gangetic alluvial plain. They used the standard deviation and correlation coefficient values of TM data to compute a statistical parameter called 'optimum index factor' (OIF), which is indicative of the variance of the data. Of the 20 possible 3-band combinations, the combination of 1, 3, and 5 was found to be the best in terms of information content. Dwivedi et al. (2001) adopted an approach of merging high spatial resolution (5.8 m) Panchromatic (PAN) sensor data and the Linear Imaging Self-scanning Sensor (LISS-III) data (23.6 m resolution) from the Indian IRS-1C using Intensity, Hue, and Saturation (IHS) transformation and a subsequent supervised classification using Gaussian maximum-likelihood classification algorithm to evaluate the potential for detection and delineation of salt-affected soils in a portion of the Indo-Gangetic alluvial plains of northern India. Results indicated deterioration in the overall accuracy of salt-affected soils derived from the LISS-III data as compared to the IRS-1B LISS-II data owing to an improvement in the spatial resolution (23.5 m for LISS-III versus 36.5 m for LISS-II), leading to higher intra-class spectral variability. The PAN and LISS-III hybrid data without any transformation ranked last in terms of overall accuracy. Overall accuracy figures for the LISS-II, LISS-III, and PAN and LISS-III hybrid data with the IHS transformation have been on the order of 89.6, 85.9, and 81.5%, respectively. Sharma et al. (2000) mapped and characterized the salt affected soils of Etah District, Uttar Pradesh, India, using multi-date IRS-1A and 1B LISS-II false color composite images. By integrating visual image interpretation, physiographic analysis, ground data and laboratory analysis of soil samples, a legend for mapping salt-affected soils (SAS) was formulated. Based on the variations in physicochemical properties, i.e., nature, intensity and depth-wise distribution of salts, five categories of SAS requiring specific reclamation measures were identified. An example of sodic soil mapping using IRS-1C PAN and LISS-III merged false color composite data and classified output for Bhind Village, Pratapgarh District, Uttar Pradesh, India, is shown in Fig. 4 (UPRSAC, 2004).

Moreover, development of waterlogging and subsequent salinization and/or alkalinization is a major land degradation problem associated with the irrigated agriculture. They are among the principal causes of decreasing production on many irrigated projects (Rhoades et al., 1992). Dwivedi et al. (1999) used satellite multispectral data for deriving information on the nature, extent, spatial distribution and magnitude of various degraded lands. Indian Remote Sensing Satellite (IRS-1B) LISS-I and Landsat Thematic Mapper (TM) data in the form of standard false color composite (FCC) at 1:100,000 scale covering Nagarjunsagar Left Bank canal command in Andhra Pradesh, southern India, were used to delineate waterlogged areas and salt-affected soils through a visual interpretation approach. Waterlogged areas were estimated at 1380 ha, whereas salt-affected soils of mostly saline-sodic nature were estimated at 6830 ha out of created irrigation potential of 35.5 thousand hectares. In another study, Choubey (1998) made an assessment of waterlogged areas in the Sriram Sagar command using remote sensing and field data. The Indian Remote Sensing Satellite (IRS-1A-LISS-II) data of 12 April and 6 October 1989 were analyzed to assess the areas affected by waterlogging and the areas prone to waterlogging. The results obtained from this study indicated that in April, 1989 about 388 km^2 area was waterlogged and 689 km^2

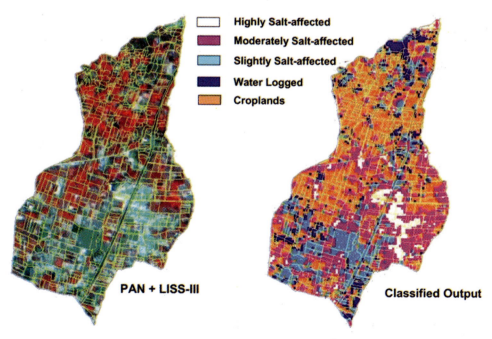

Fig. 4 Sodic soil mapping of Bhind Village, Pr atapgarh District, U.P., India (UPRSAC, 2004).

area was prone to waterlogging, while during October 1989, the extent were about 540 km^2 (waterlogged) and 802 km^2 (prone to waterlogging), respectively. The IRS data proved to be very useful for the assessment of waterlogging in the canal command. It was concluded that the density slicing and principal component analysis are useful techniques, as far as image processing is concerned, for assessing waterlogged areas in canal commands.

8. CONSTRAINTS OF USING GEOSPATIAL TECHNOLOGIES

There are some inherent limitations of satellite remote sensing, which restrict its utilization. The disadvantages of satellite remote sensing include the inability of many sensors working in visible region of electromagnetic spectrum to obtain data and information through cloud cover (although microwave sensors can take the earth's image through clouds) and the relatively low spatial resolution achievable with many satellite-borne sensor payloads. In addition, the satellite sensors provide information restricted to a few centimeters below the earth's surface (Lillesand and Kiefer, 1987; Beven, 2001). These datasets do not provide the details of subsurface features/processes and also obscure the dynamic and minute information of the earth's surface due to the resolution (both spatial and temporal) limitations. Another major issue is the influence of atmosphere while sensing the earth due to the presence of water vapor, aerosol, ozone, etc. (Lillesand and Kiefer, 1987). It requires suitable corrections for atmospheric absorption and scattering and also for the absorption of radiation through water on the ground to obtain desired data and information on particular variables. Also, satellite remote sensing generates large volumes of digital data that typically require extensive processing as well as storage and analysis. Therefore, there is an urgent need to improve the accuracy and reliability of remote sensing estimates, which are uncertain until now (Beven, 2001). In addition, the data from satellite remote sensing are often costly if purchased from private vendors or value-adding resellers, and this initial cost, together with intellectual property restrictions, can limit the dissemination of products from such sources. Finally, the end users look for the required information from RS data without understanding much of the technicalities. To overcome this problem, remote sensing data should be transformed into usable information to enable the users to apply RS technology effectively. Moreover, the cost associated with commercially available GIS software packages and the complications in software handling are bottlenecks for the general users. The uncertainty involved in the GIS analysis is another important issue, which needs to be addressed properly in order to improve the accuracy of GIS results in the future (Jha and Peiffer, 2006). Further details about the constraints and challenges of using geospatial technologies could be found in Jha and Chowdary (2007) and Jha et al. (2007).

9. FUTURE INDIAN REMOTE SENSING MISSIONS

Periodic inventory of natural resources, generation and updation of large-scale maps, disaster monitoring and mitigation, improved weather forecasting at better spatial and temporal scales, ocean-state forecasting, facilitating infrastructure development and providing information services at the community level for better management of natural resources continue to be the thrust areas of applications for the Indian Earth Observation (EO) Program. These thrust areas call for the development of new remote sensors some of which related to hydrologic application areas are presented in Table 7 (Navalgund et al., 2007). A number of Indian EO missions have been planned to meet future observational requirements, and they include operational polar orbiting Resourcesat, Cartosat, Oceansat, and Risat; experimental polar orbiting Altika-Argos and low-inclination orbit Megha Tropiques in cooperation with CNES (Centre National d'Etudes Spatiale, France); and augmented with the geostationary INSAT systems with Imagers and

Sounders (Table 8). It is also planned to use the complementary and supplementary data from other international missions to augment the data sources to meet the increasing demands of user communities in India.

Table 7. Future requirements of earth observations sensors for land and water resources applications (Navalgund et al., 2007)

Sensor requirement	Sensor specifications	Possible applications
High Spatial and Temporal Resolution Multispectral Sensor	Four bands in VNIR-SWIR; resolution: 20-30 m, 5-day repetitivity, >700 km swath.	Multiple crop forecasting
High Temporal and Moderate Spatial Resolution Sensor	VNIR, SWIR, MIR and TIR bands, Resolution: 50-100 m (VNIR), 250-500 m (TIR), Daily repetitivity, wide swath ~1000 km.	Regional vegetation monitoring; Parameter retrieval.
Very High Spatial Resolution Pan	Panchromatic, Resolution: <40 cm, ~8 km swath, Spot imaging Stereo pairs.	1:1000 scale mapping, urban and local area planning, facility management, City 3-D modeling.
Hyper-Spectral Sensors	Large number of narrow bands (0.4-2.5 µm), >10 nm bandwidth, 20-30 m resolution, 20-25 day repetitivity.	Applications in forestry, agriculture, coastal zone and inland waters, soil, geology
Atmospheric Corrector	Coarse resolution (500 m); Four bands (485, 940, 1625, 2100 nm).	Need for simultaneous measurement along with other multi-spectral sensors for atmospheric correction
Synthetic Aperture Radar (SAR)	Dual frequency (L and C band) polarimetric system	Crop monitoring, soil moisture, tree canopy

Radar Imaging Satellite (RISAT-1), planned for launch in 2010, will have night and day imaging capability as well as imaging under cloudy conditions. The proposed RISAT SAR system will work at C-band with improved remote sensing capabilities like coarse to fine resolution, multi-polarization and multi-incidence angle. RISAT is the first indigenously developed SAR system, and hence it happens to be the first operational SAR system to which easy access will be available to the Indian user community. This will enhance the scope of many applications like disaster monitoring and management, *kharif* crop production forecasting, etc. (Chakraborty et al., 2005; Anonymous, 2007c). Further, the mission will extend the scope of applications in areas like estimation of soil moisture, snow wetness and various oceanographic parameters. The soil moisture estimation and monitoring in rainfed crop production systems and the impact of larger agricultural heterogeneity could be addressed in the future. RISAT will complement the band of electro-optical sensors on board Indian Remote Sensing satellites launched so far.

Oceansat-2 is envisaged for providing continuity to Oceansat-1 and is scheduled to be launched in 2009. It will carry Ocean Color Monitor (OCM), Ku-band Scattterometer and a piggy-back payload, viz., Radio Occultation Sounder for Atmospheric studies developed by the Italian Space Agency. OCM has eight narrow spectral bands operating in visible and near-infrared bands. It will help in investigating the impact of agricultural practices on coastal water productivity, algal bloom detection, coastal zone management, etc. (Navalgund et al., 2007; DOS, 2008). This satellite will also help in locating potential fishing zones, forecasting sea condition, and in the studies related to coastal zones, climate and weather.

Table 8. Capabilities of some of future Indian satellites for hydrological studies (DOS, 2009)

Satellite (Probable launch date)	Sensor Spatial resolution (M)/Swath (km)	Application areas
Resourcesat-2 (2009)	AWiFS - 56/740 4Xs LISS III - 23/140; 4Xs LISS IV - 5.8/70; 3Xs *Repetitivity: 22 Days*	Agriculture, Geology, Forestry, Urban, Land use, Soil, Terrain, Water resources, DEMs, Environment, Disasters (damage assessment/relief).
RISAT -1 (2010)	C band SAR - 5.35 GHz Five imaging modes; single, dual and quad polarization modes; 18.3° to 48.6° look angle *Repetitivity: 13 Days*	Soil moisture, agriculture, flood & water spread, geology, snow & glaciers, forest, topography and oceanography.
Megha-Tropiques (2010)	SAPHIR: Sounder SCARAB: Four channels Radiometer MADRAS: Microwave imaging radiometer (10, 18, 37, 89 and 157 GHz) [*Frequent simultaneous observations of tropics*]	Studying water cycle and energy exchanges to better understand the life cycles of the tropical convective system.
Oceansat-2 (2009)	OCM: 360/1420 Scatterometer: Ku Band ROSA (Radio Occultation Sounder for atmospheric studies), *Repetitivity: two days*	Coastal zone mapping, retrieval of oceanic parameters
INSAT-3D (2010)	Visible, TIR and WV Continuous in geo-stationary orbit	Clouds, ST, WV image, mesoscale temperature and humidity profiles.

Note: X: Number of bands; MX: Multispectral Mode; ST: Surface Temperature; WV: Water Vapor.

Moreover, the future series of Cartosat satellites will provide the sub-meter resolution panchromatic image with stereo capability. It would be suitable for the cadastral and infrastructure mapping, development of detailed land and water resources management plans, site selection for hydrological structures like dams and the generation of seamless mosaic base image of a District/State on which the natural resources information can be overlaid to retrieve the current status of information. The forthcoming Resourcesat-2 will provide the continuity of services of Resourcesat-1, which is planned to be launched by 2009. The satellite will continue to study crops and vegetation dynamics, natural resources census, disaster management and large-scale mapping of them. Beyond 2011, ISRO has envisaged a disaster management satellite DM-SAR-1. The satellite will have C-/X-band radar mainly to overcome the problem of cloud during observation. It will be most useful for the monitoring and management of floods and cyclones.

In order to derive maximum benefits from the above planned missions, the Indian EO Program is also planning to address corresponding improvements on the ground segment. The emphasis will be towards multi-mission acquisition and processing; effective delivery mechanisms; web-based services; mission-oriented outreach activities; development of freeware tools for data products access, etc. (Navalgund et al., 2007). Thus, it is expected that the Indian remote sensing satellites (IRS) will create better avenues in the future, which in turn will ensure efficient management of natural resources, especially land and water resources of the country.

10. CONCLUSIONS

It is clear from the examples presented in this chapter that several critical developmental issues in India have driven the applications of geospatial technologies (i.e., RS, GPS and GIS) to natural resources management in general and land and water resources management in particular. In fact, there are a variety of applications, especially those related to real-life situations where geospatial technologies could be put to use for solving real-world problems. The analysis of thematic layers such as land cover, drainage, geology, soil, land degradation, etc. using geospatial technologies (RS, GPS and GIS) has helped in generating action plans, especially in amending existing land use plans which were economically and/or environmentally non-viable. These action plan maps suggest improved land use practices such as agro-forestry, double cropping, horti-pastoral, silvipastoral, agro-horticulture, etc. for combating land degradation and water scarcity. On the other hand, the assessment of groundwater potential and vulnerability as well as the selection of suitable sites for artificial recharge and rainwater harvesting is considerably aided by the use of thematic layers such as geology, geomorphology, lineament, drainage density, land use/land cover, slope, soil, recharge, etc. generated in a GIS environment using satellite imagery and other ancillary data. Geospatial technologies are also helpful in addressing the emerging issues due to environmental changes (i.e., climate change, climate variability and socio-economic changes).

While there are operational applications facilitating the decision-making process in support of more efficient land and water resources management, there are semi-operational and research areas wherein the conjunctive use of thermal, micro-wave and hyperspectral capabilities have been envisaged. Remote sensing retrieval of land surface parameters and hydrological variables using hyperspectral, thermal and microwave data are mostly in research domain in India. However, it is expected that several areas of research today will move into operational arena in the near future. For instance, hyperspectral imaging will address some of the critical issues related to precision agriculture and crop pest and diseases detection; thermal imaging will help in crop water stress detection and drought condition assessment; multi-polarimetric SAR will enable improving crop statistics during rainy seasons and flood monitoring within small turn around time. In this regard, land surface process and hydrological models, using remote sensing derived parameters as inputs, have a major role to play. These models offer an efficient tool for improving our understanding about spatial processes and patterns affecting the distribution and movement of water in different landscapes as well as for evaluating the long-term impacts of environmental changes on land and water resources.

In India, the initiatives have already gained momentum for resource planning by bringing into focus to setup a well-developed information system. A standardized national information system is being planned with a network of data banks and databases, integrating and strengthening the existing Central and State agencies, and improving the quality and quantity of data and processing capabilities. It will not only overcome the current impediments of remote sensing and GIS application and boost RS- and GIS-based studies but will also create new avenues for better research and efficient planning and management of land and water resources. Further, with the availability of high resolution satellite data and the gradual advancement in GIS, information systems and modeling techniques, it is expected that geospatial technologies will play an increasing role in improving our knowledge about hydrologic and hydrogeologic processes as well as in effectively quantifying the spatio-temporal variations of hydrologic parameters and in facilitating efficient transmission, collection, storage, processing and visualization of hydrologic data which in turn can ensure adequate and rapid analyses of various facets of land and water systems in the face of environmental changes. The improved management of land and water systems will certainly help minimize or avoid land and water related disasters and thereby ensure sustainable development on the earth.

REFERENCES

Agarwal, K.K., Agarwal, C.K. and Kumar, P. (2000). Remote Sensing in reservoir studies: A case study of Sriram Sagar reservoir. Proceedings of the Tenth National Symposium on Hydrology, New Delhi, India, pp. 509-522.

Akbari, M., Toomanian, N., Droogers, P., Bastiaanssen, W. and Gieske, A. (2007). Monitoring irrigation performance in Esfahan, Iran using NOAA satellite imagery. *Agricultural Water Management*, **88:** 99-109.

Anonymous (1991). Natural Resource Database Generation of Mizoram State and Landslide Hazard Zonation of Aizawal District. Project Report, State Land Use Board, Government of Mizoram and Indian Institute of Remote Sensing (IIRS), Department of Space, Dehradun.

Anonymous (2007a). Satellites confirm fear of glaciers melting. *Deccan Herald*, March 29, 2007.

Anonymous (2007b). www.Indiastat.com (accessed in May 2007).

Anonymous (2007c). Flood Inundation Mapping: Service under Disaster Management Support Programme—Manual. DSC, National Remote Sensing Agency (NRSA), Hyderabad, A.P., India.

Asawa, G.L. (1993). Irrigation Engineering. Wiley Eastern Ltd., New Delhi, India.

Bandyopadhyay, J. and Gyawali, D. (1994). Himalayan water resources: Ecological and political aspects of management. *Mountain Research and Development*, **14(1):** 1-24.

Bandyopadhyay, S., Srivastava, S.K., Jha, M.K., Hegde, V.S. and Jayaraman, V. (2007). Harnessing Earth Observation (EO) capabilities in hydrogeology: An Indian perspective. *Hydrogeology Journal*, **15(1):** 155-158.

Bastiaanssen, W.G.M., Thiruvengadachari, S., Sakthivadivel, R. and Molden, D.J. (1999). Satellite remote sensing for estimating productivities of land and water. *International Journal of Water Resources Development*, **15(1-2):** 181-196.

Beguería, S. (2006). Identifying erosion areas at basin scale using remote sensing data and GIS: a case study in a geologically complex mountain basin in the Spanish Pyrenees. *International Journal of Remote Sensing*, **27(20):** 4585-4598.

Beven, K.J. (2001). Rainfall-Runoff Modeling: The Primer. John Wiley & Sons Ltd., Chichester, pp. 297-306.

Bhanumurthy, V. and Behera, G. (2008). Deliverables from space data sets for disaster management—present and future trends. Proceedings of the International Archives of the Photogrammetry, Remote Sensing and Spatial Information Sciences. ISPRS Congress Beijing 2008, Volume XXXVII, Part B8, Commission VIII, http://www.isprs.org/congresses/beijing2008/proceedings/8_pdf/2_WG-VIII-2/15a.pdf (accessed in March 2009).

Bhaskaranarayana, A., Bhatia, B., Bandyopadhyay, S.K. and Jain, P.K. (2007). Applications of space communication. *Current Science*, **193(12):** 1737-1746.

Bhattacharya, A. (1999). Remote sensing applications for geology and geomorphology: Retrospective and perspective. Proceedings of the National Symposium on Natural Resources: Retrospective and Perspective, 19-21 January 1999, Bangalore, Karnataka, India, pp. 456-459.

Bos, M.G., Murray-Rust, D.H., Merrey, D.J., Johnson, H.G. and Snellen, W.B. (1994). Methodologies for assessing performance of irrigation and drainage management. *Irrigation and Drainage Systems*, **7:** 231-261.

Bouwer, H. (2000). Integrated water management: Emerging issues and challenges. *Agricultural Water Management*, **45:** 217-228.

Brandon, C., Hommann, K. and Kishore, N.M. (1995). The cost of inaction: Valuing the economy wide cost of environmental degradation in India. Proceedings of the UN University Conference on the Sustainable Future of the Global System, 16-18 October 1995, Tokyo, Japan.

Chakraborti, A.K., Rao, V.V., Shanker, M. and Suresh Babu, A.V. (2002). Performance evaluation of an irrigation project using satellite Remote Sensing, GIS and GPS. www.gisdevelopment.net/application/agriculture/irrigation/agriir001.htm (accessed in April 2007).

Chakraborty, M., Manjunath, K.R., Panigrahy, S., Kundu, N. and Parihar J.S. (2005). Rice crop parameter retrieval using multi-temporal, multi-incidence angle Radarsat SAR data. *ISPRS Journal of Photogrammetry and Remote Sensing*, **59(5):** 310-322.

Chandrasekhar, M.G., Jayaraman, V. and Rao, M. (1996). Indian remote sensing satellites: Planned missions and future applications. *Acta Astronautica*, **38**: 647-658.

Chari, S.T., Jonna, S., Raju, P.V., Murthy, C.S. and Hakeem, K.A. (1994). System performance evaluation and diagnostic analysis of canal irrigation projects. Proceedings of the Fifteenth Asian Conference on Remote Sensing, 17-23 November 1994, Bangalore, India.

Chen, Y., Takara, K., Cluckie, I.D. and Smedt, F.H.D. (editors) (2004). GIS and Remote Sensing in Hydrology, Water Resources and Environment. IAHS Publication No. 289, IAHS Press, Wallingford.

Choubey, V.K. (1998). Assessment of waterlogging in Sriram Sagar command area, India by remote sensing. *Water Resources Management*, **12(5)**: 343-357.

Chowdary, V.M., Paul, S., Srinivas, K.T., Sudhakar, S., Adiga, S. and Rao, P.P.N. (2001). Remote sensing and GIS approach for watershed monitoring and evaluation: A case study in Orissa State, India. Proceedings of the 22nd Asian Conference on Remote Sensing, 5-9 November 2001, Singapore.

Chowdary, V.M., Rao, N.H. and Sarma, P.B.S. (2005). Decision support framework for assessment of non-point-source pollution of groundwater in large irrigation projects. *Agricultural Water Management*, **75(3)**: 194-225.

Chowdhury, A., Jha, M.K. and Chowdary, V.M. (2009). Delineation of groundwater recharge zones and identification of artificial recharge sites in West Medinipur District, West Bengal using RS, GIS and MCDM techniques. *Environmental Geology* (in press).

Chowdhury, A., Jha, M.K., Chowdary, V.M. and Mal, B.C. (2009). Integrated remote sensing and GIS-based approach for assessing groundwater potential in West Medinipur district, West Bengal, India. *International Journal of Remote Sensing*, **30(1)**: 231-250.

CWC (2007). Central Water Commission, New Delhi, http://cwc.gov.in/ (accessed in December 2007).

Dandapani, K.R. (1990) Degrading land as a constraint to national food security system. *Journal of Soil and Water Conservation*, **34**: 67-95.

Das, D. (1990). Satellite remote sensing in subsurface water targeting. Proceedings of the ACSM-ASPRS Annual Convention, 18-23 March 1990, Denver, CO, pp. 99-103.

Das, D.C. (1987). Soil and water conservation in the perspective of national planning for security of productivity food livelihood and environment. *Journal of Soil and Water Conservation*, **31**: 198-208.

Dehaan, R. and Taylor, G.R. (2002). Field-derived spectra of salinized soils and vegetation as indicators of irrigation induced soil salinization. *Remote Sensing of Environment*, **80**: 406-417.

Dhubanarayana, V.V. and Babu, R. (1983). Estimation of soil erosion in India. *Journal of Irrigation and Drainage Engineering, ASCE*, **109(4)**: 419-434.

Dilley, M. and Heyman, B.N. (2007). ENSO and disaster: droughts, floods and El Niño/Southern Oscillation warm events. *Disasters*, **19(3)**: 181-193.

Diwakar, P.G., Ranganath, B.K., Gowrisankar, D. and Jayaraman, V. (2008). Empowering the rural poor through EO products and services—An impact assessment. *Acta Astronautica*, **63**: 551-559.

DOS (2007). Annual Report (2006-07). Department of Space, Government of India, Bangalore, India.

DOS (2008). Annual Report (2007-08). Department of Space, Government of India, Bangalore, India.

DOS (2009). Annual Report (2008-09). Department of Space, Government of India, Bangalore, India.

Dwivedi, R.S. and Rao, B.R.M. (1992). The selection of the best possible Landsat TM band combination for delineating salt-affected soils. *International Journal of Remote Sensing*, **13(11)**: 2051-2058.

Dwivedi, R.S., Ramana, K.V., Thammappa, S.S. and Singh, A.N. (2001). The utility of IRS-1C LISS-III and PAN-merged data for mapping salt-affected soils. *Photogrammetric Engineering and Remote Sensing*, **67(10)**: 1167-1176.

Dwivedi, R.S., Sreenivas, K. and Ramana, K.V. (1999). Inventory of salt-affected soils and waterlogged areas: A remote sensing approach. *International Journal of Remote Sensing*, **20(8)**: 1589-1599.

Edet, A.E., Okereke, C.S., Teme, S.C. and Esu, E.O. (1998). Application of remote-sensing data to groundwater exploration: A case study of the Cross River State, southeastern Nigeria. *Hydrogeology Journal*, **6(3)**: 394-404.

Engman, E.T. (1991). Applications of microwave remote sensing of soil moisture for water resources and agriculture. *Remote Sensing of Environment*, **35(2-3):** 213-226.

Engman, E.T. and Gurney, R.J. (1991). Remote Sensing in Hydrology. Chapman and Hall, London.

Erdelyi, M. and Galfi, J. (1988). Surface and Subsurface Mapping in Hydrogeology. John Wiley and Sons, New York, 384 pp.

Evans, R.G. and Sadler, E.J. (2008). Methods and technologies to improve efficiency of water use. *Water Resources Research*, **44,** W00E04, doi:10.1029/2007WR006200.

Falkenmark, M. (2007). Shift in thinking to address the 21st century hunger gap: Moving focus from blue to green water management. Water Resources Management, **21:** 3-18.

FAO (1995). The conservation of lands in Asia and the Pacific (CLASP): A framework for action. Food and Agriculture Organization, Rome, Italy, http://www.fao.org/docrep/v9909e/v9909e00.htm (accessed in March 2008).

FSI (2005). Annual Report. Forest Survey of India (FSI), Ministry of Environment and Forest, Government of India, Dehradun.

Garg, J.K., Singh, T.S. and Murthy, T.V.R. (1998). Wetlands of India. Project Report RSAM/SAC/RESA/PR/01/98, Space Applications Centre, Ahmedabad, India.

Gleick, P.H. (1993). Water and Conflict: Fresh Water Resources and International Security. *International Security*, **18(1):** 79-112.

Gupta, S.K. and Despande, R.D. (2004). Water for India in 2050: First order assessment of available water. *Current Science*, **86(9):** 1216-1224.

GWP (2004). Catalyzing Change: A Handbook for Developing Integrated Water Resources Management (IWRM) and Water Efficiency Strategies. GWP Secretariat, Stockholm, Sweden, 52 pp.

Howari, F.M. (2003). The use of remote sensing data to extract information from agricultural land with emphasis on soil salinity. *Australian Journal of Soil Research*, **41(7):** 1243-1253.

INCOIS (2002).www.incois.gov.in/documents/2001-2002_English.pdf (accessed in March 2009).

Indiastat (2007). www.indiastat.com (accessed in May 2007).

IPCC (2007). Fourth Assessment Report: Climate Change 2007. Intergovernmental Panel on Climate Change (IPCC), Geneva, Switzerland, http://www.ipcc.ch (accessed in June 2008).

IRISDA Atlas (2004). Integrated Natural Resources Management in Raichur District, Karnataka State Using Remote Sensing and GIS Techniques — Atlas. National Remote Sensing Agency (NRSA), Hyderabad, A.P., India.

Jackson, T.J. (2002). Remote sensing of soil moisture: Implications for groundwater recharge. *Hydrogeology Journal*, **10(1):** 40-51.

Jain, S.K., Singh, P. and Seth, S.M. (2002). Assessing the vulnerability to soil erosion of the Ukai Dam catchments using remote sensing and GIS. *Hydrological Sciences Journal*, **47(2):** 31-40.

Jeyakanthan, V.S., Sreenivasalu, V., Rao, Y.R.S. and Ramasastri, K.S. (2002). Reservoir capacity estimation using satellite data. Proceedings of the ISPRS Commission VII: Symposium on Resource and Environmental Monitoring, 3-6 December 2002, Vol. 34, Part 7, Hyderabad, A.P., India, pp. 863-866.

Jha, M.K. and Chowdary, V.M. (2007). Challenges of using remote sensing and GIS in developing nations. *Hydrogeology Journal*, **15(1):** 197-200.

Jha, M.K. and Peiffer, S. (2006). Applications of Remote Sensing and GIS Technologies in Groundwater Hydrology: Past, Present and Future. BayCEER, Bayreuth, Germany.

Jha, M.K., Chowdhury, A., Chowdary, V.M. and Peiffer, S. (2007). Groundwater management and development by integrated remote sensing and geographic information systems: Prospects and constraints. *Water Resources Management*, **21(2):** 427-467.

Jha, M.K., Kamii, Y. and Chikamori, K. (2001). Irrigation and water management: An Indian perspective. *Rural and Environmental Engineering*, **40(2):** 46-66.

Joshi, A.K., Krishanamurthy, Y.V.N. and Jayaraman, V. (2004). Resourcesat-1 Data in Geological Applications. Bulletin of the National Natural Resources Management System, NNRMS(B)-29, Department of Space, Bangalore, India.

Karatas, B.S., Akkuzu, E., Unal, H.B., Asik, S. and Avci, M. (2009). Using satellite remote sensing to assess irrigation performance in Water User Associations in the Lower Gediz Basin, Turkey. *Agricultural Water Management*, **96**: 982-990.

Katti, V.R., Thyagarajan, K., Shankara, K.N. and Kiran Kumar, A.S. (2007). Spacecraft technology. *Current Science*, **193(12)**: 1715-1736.

Khan, M.A., Gupta, V.P. and Moharana, P.C. (2001). Watershed prioritization using remote sensing and geographical information system: A case study from Guhiya, India. *Journal of Arid Environments*, **49**: 465-475.

Kimothi, M.M., Garg, J.K., Joshi, V., Semwal, R.L., Pahari, R. and Juyal, N. (1999). Slope Activation and its Impact on the Madhyamaheshwar and the Kaliganga Sub-watersheds Okhimath Using IRS 1C/1D Data. Scientific Report, SAC/RESA/FLPG-FED/SR/04/99, SAC (ISRO), Ahmedabad, Gujarat.

Kogan, F.N. (2001). Operational space technology for global vegetation assessment. *Bulletin of the American Meteorological Society*, **82(9)**: 1949-1964.

Krishnamurthy, J. and Srinivas, G. (1995). Role of geological and geomorphological factors in ground water exploration: A study using IRS-LISS-II data. *International Journal of Remote Sensing*, **16**: 2595-2618.

Krishnamurthy, J., Venkatesa, K.N., Jayaraman, V. and Manivel, M. (1996). Approach to demarcate groundwater potential zones through remote sensing and a geographical information system. *International Journal of Remote Sensing*, **17(10)**: 1867-1884.

Kudrat, M. and Saha, S.K. (1993). Land productivity assessment and mapping through integration of satellite and terrain (slope) data. *Indian Journal of Remote Sensing*, **21(3)**: 151-166.

Kulkarni A.V., Philip, G., Thakur, V.C., Sood, R.K., Randhawa, S.S. and Chandra, R. (1999). Glacial inventory of the Satluj basin using remote sensing technique. *Himalayan Geology*, **20(2)**: 45-52.

Kulkarni, A.V. (1991). Glacier inventory in Himachal Pradesh using satellite data. *Journal of Indian Society of Remote Sensing*, **19(3)**: 195-203.

Kulkarni, A.V. (1992). Mass balance of Himalayan glaciers using AAR and ELA methods. *Journal of Glaciology*, **38(128)**: 101-104.

Kumar, K. (2005). Receding glaciers in the Indian Himalayan region. *Current Science*, **88(3)**: 342-343.

Kumar, R., Singh, R.D. and Sharma, K.D. (2005). Water resources of India. *Current Science*, **89(5)**: 794-811.

Laymonn, C., Quattrochi, D., Malek, E., Hipps, L., Boettinger, J. and McCurdy, G. (1998). Remotely sensed regional-scale evapotranspiration of a semi-arid Great Basin desert and its relationship to geomorphology, soils, and vegetation. *Geomorphology*, **21(3-4)**: 329-349.

Lillesand, T.M. and Kiefer, R.W. (1987). Remote Sensing and Image Interpretation. 2nd edition, John Wiley & Sons, New York.

Loucks, D.P. (2000). Sustainable water resources management. *Water International*, **25(1)**: 3-10.

Mall, R.K., Gupta, A., Singh, R., Singh, R.S. and Rathore, L.S. (2006). Water resources and climate change: An Indian perspective. *Current Science*, **90(12)**: 1610-1626.

Maltby, E. (1991). Wetlands and their values. *In:* M. Finlayson and M. Moser (editors), Wetlands. International Waterfowl and Wetlands Research Bureau, Facts On File, New York.

Manavalan, P., Sathyanath, P., Sathyanarayn, M. and Raje Gowda, G.L. (1990). Capacity Evaluation of the Malaprabha Reservoir Using Digital Analysis of Satellite Data. Technical Report, Regional Remote Sensing Service Centre, Bangalore and Karnataka Engineering Research Station, Krishnarajsagar, Karnataka, India.

McFarlane, D.J. and Williamson, D.R. (2002). An overview of waterlogging and salinity in southwestern Australia as related to the 'Ucarro' experimental catchment. *Agricultural Water Management*, **53(1-3)**: 5-29.

Mertes, L.A.K. (2002). Remote sensing of riverine landscapes. *Freshwater Biology*, **47**: 799-816.

Metternicht, G.I. and Zinck, J.A. (2003). Remote sensing of soil salinity: Potentials and constraints. *Remote Sensing of Environment*, **85(1)**: 1-20.

MOA (1994). Draft report on Status of Land Degradation in India. Department of Agriculture and Cooperation, Ministry of Agriculture, Government of India, New Delhi.

MOA (2006). Annual Report, 2005-06. Ministry of Agriculture (MOA), Government of India, New Delhi, www.agricoop.nic.in (accessed in February 2006).

Mohanty, R.B., Mahapatra, G., Mishra, D. and Mahapatra, S.S. (1986). Report on application of remote sensing to sedimentation studies in Hirakud reservoir. Orissa Remote Sensing Application Centre, Bhubaneswar and Hirakud Research Station, Hirakud, Orissa, India.

Nagarajan, R., Mukherjee, A., Roy, A. and Khire, M.V. (1998). Temporal remote sensing data and GIS application in landslide hazard zonation of part of Western Ghat, India. *International Journal of Remote Sensing*, **19(4)**: 573-585.

Navalgund, R.R., Jayaraman, V. and Roy, P.S. (2007). Remote sensing applications: An overview. *Current Science*, **193(12)**: 1747-1766.

NDMD (2004). Disaster Management in India: A Status Report. National Disaster Management Division (NDMD), Ministry of Home Affairs, Government of India, New Delhi, India.

NNRMS-TR-98 (1998). Monitoring and Evaluation of Watersheds in Karnataka Using Satellite Remote Sensing. RRSSC-NNRMS Technical Report, ISRO, Bangalore, Karnataka, India, 151 pp.

NRSA (2002). Integrated Mission of Sustainable Development. National Remote Sensing Agency (NRSA), Department of Space, Hyderabad, A.P., India.

NRSA (2005a). Wasteland Atlas of India. National Remote Sensing Agency (NRSA), Department of Space (DOS), Hyderabad, A.P., India

NRSA (2005b). Atlas: Ground water prospects maps of Jharkhand State. Vol. 1, National Remote Sensing Agency (NRSA), Department of Space (DOS), Hyderabad, A.P., India.

Ouyang W., Hao, F. and Wang, X. (2008). Regional nonpoint source organic pollution modeling and critical area identification for watershed best environmental management. *Water, Air, & Soil Pollution*, **187(1-4)**: 251-261.

Pandey, R.P. and Ramasastri, K.S. (2001). Relationship between the common climatic parameters and average drought frequency. *Hydrological Processes*, **15(6)**: 1019-1032.

Paul, S.K., Bartarya, S.K., Rautela, P. and Mahajan, A.K. (2000). Catastrophic mass movement of 1998 monsoons at Malpa in Kali Valley, Kumaun Himalaya (India). *Geomorphology*, **35(3-4)**: 169-180.

Philip, G. and Ravindran, K.V. (1998). Glacial mapping using Landsat Thematic Mapper data: A case study in parts of Gangotri glacier, NW Himalaya, India. *Journal of Indian Society of Remote Sensing*, **26(112)**: 29-34.

Pietroniro, A. and Prowse, T.D. (2002). Applications of remote sensing in hydrology. *Hydrological Processes*, **16(8)**: 1537-1541.

Planning Commission (2002). http://planningcommission.nic.in/plans/planrel/fiveyr/10th/default.htm (accessed in March 2009).

Postel, S. (1993). Water and Agriculture. *In:* P.H. Gleick (editor), Water in Crisis: A Guide to the World's Fresh Water Resources. Oxford University Press, New York, pp. 56-66.

Prasad, V.H., Chakraborti, A.K. and Nayak, T.R. (1996). Irrigation command area inventory and assessment of water requirements using IRS-1B satellite data. *Journal of Indian Society of Remote Sensing*, **24(2)**: 85-96.

Rakesh, K., Singh, R.D. and Sharma, K.D. (2005). Water resources of India. *Current Science*, **89(5)**: 794-811.

Ramalingam, M. and Santhakumar, A.R. (2000). Case Study on Artificial Recharge using Remote Sensing and GIS. National Natural Resources Management System (NNRMS) Bulletin, Department of Space (DOS), Government of India, Bangalore, Karnataka, India.

Rao, H.G., Rao, R. and Viswanatham, R. (1985). Project Report on Capacity Evaluation of Sriramsagar Reservoir Using Remote Sensing Techniques. Andhra Pradesh Engineering Research Laboratory, Hyderabad, A.P., India.

Rao, V.V. and Bandyopadhyay, S. (2004). Application of remote sensing and GIS in interlinking of rivers in India. *ISG Newsletter*, **10(1-2)**: 7-18.

Ray, S.S. and Dadhwal, V.K. (2001). Estimation of crop evapotranspiration of irrigation command area using remote sensing and GIS. *Agricultural Water Management*, **49**: 239-249.

Reddy, P.R., Vinod Kumar, K. and Sheshadri, K. (1996). Use of IRS-1C data for groundwater studies. *Current Science*, **70(7)**: 600-605.

Reddy, S. (2003). Geological features from satellite data. Personal Communication, National Remote Sensing Agency (NRSA), Hyderabad, India.

Rhoades, J.D., Kandiah, A. and Mashali, A.M. (1992). The Use of Saline Waters for Crop Production. FAO Irrigation and Drainage Paper 48, Food and Agriculture Organization of the United Nations, Rome, Italy.

SAC (2007). Desertification and land degradation atlas of India. Space Application Centre, ISRO, Ahmedabad.

Saha, S.K., Kudrat, M. and Bhan, S.K. (1990). Digital processing of Landsat TM data for wasteland mapping in parts of Aligarh district, Uttar Pradesh, India. *International Journal of Remote Sensing*, **11**: 485-492.

Sahai, B., Bhattacharya, A. and Hegde, V.S. (1991). IRS-1A application for groundwater targeting. *Current Science*, **61(3-4)**: 172-179.

Saheb, P.S. and Singh, R.N. (2002). Water resource and climate change. *Science and Culture*, **68(9-12)**: 244-256.

Saraf, A.K. and Choudhury, P.R. (1998). Integrated remote sensing and GIS for groundwater exploration and identification of artificial recharge sites. *International Journal of Remote Sensing*, **19(10)**: 1825-1841.

Sarkar, S., Kanungo, D.P. and Mehrotra, G. S. (1995). Landslide hazard zonation: A case study in Garhwal Himalaya, India. *Mountain Research and Development*, **15(4)**: 301-309.

Schmugge, T.J., Kustas, W.P., Ritchie, J.C., Jackson, T.J. and Rango, A. (2002). Remote sensing in hydrology. *Advances in Water Resources*, **25(8)**: 1367-1385.

Schultz, G.A. and Engman, E.T. (editors) (2000). Remote Sensing in Hydrology and Water Management. Springer, Berlin, 483 pp.

Schultz, G.A. (1994). Mesoscale modelling of runoff and water balances using remote sensing and other GIS data. *Hydrological Sciences Journal*, **39(2)**: 121-142.

Schultz, G.A. (2000). Potential of modern data types for future water resources management. *Water International*, **25(1)**: 96-109.

Sehgal, J.L. and Abrol, I.P. (1994). Soil Degradation in India: Status and Impact. Oxford and IBH Publishing, New Delhi, India.

Seiler, R.A., Kogan, F. and Sullivan, J. (1998). AVHRR-based vegetation and temperature condition indices for drought detection in Argentina. *Advances in Space Research*, **21(3)**: 481-484.

Sekhar, K.R. and Rao, B.V. (2002). Evaluation of sediment yield by using remote sensing and GIS: A case study from the Phulang Vagu watershed, Nizamabad District (AP), India. *International Journal of Remote Sensing*, **23(20)**: 4499-4509.

Shanker, M. (2004). Comparative Evaluation of the Water Spread Area Estimation using Resourcesat-1 LISS IV and LISS III Data for Reservoir Sedimentation Studies. *Bulletin of the National Natural Resources Management System*, NNRMS(B) **29**, pp. 55-58.

Sharma, R.C. and Bhargava, G.P. (1988). Landsat imagery for mapping saline soils and wetlands in north-west India. *International Journal of Remote Sensing*, **9**: 39-44.

Sharma, R.C., Saxena, R.K. and Verma, K.S. (2000). Reconnaissance mapping and management of salt-affected soils using satellite images. *International Journal of Remote Sensing*, **21(17)**: 3209-3218.

Shrestha, R.P. (2006). Relating soil electrical conductivity to remote sensing and other soil properties for assessing soil salinity in Northeast Thailand. *Land Degradation and Development*, **17(6)**: 677-689.

Sikka, D.R. (1980). Some aspects of the large scale fluctuations of summer monsoon rainfall over India in relation to fluctuations in the planetary and regional scale circulation parameters. *Journal of Earth System Science*, **89(2)**: 179-195.

Singh, H.P., Venkateswarlu, B., Vittal, K.P.R. and Ramachandran, K. (2000). Management of rainfed agroecosystem. Proceedings of the International Conference on Managing Natural Resource for Sustainable Agricultural Production in 21st Century, 14-18 February 2000, New Delhi, India.

Singh, R.P. and Rao, R. (1988). Agricultural Drought Management in India: Principle and Practices. Technical Bulletin No. 1, Central Research Institute for Dryland Agriculture (CRIDA), Hyderabad, A.P., India.

Sombroek, W. and Sene, E.H. (1993). Land degradation in arid, semi-arid and dry sub-humid areas: rainfed and irrigated lands, rangelands and woodlands. FAO presentation at INCD, Nairobi, 24-28 May 1993, Food and Agriculture Organization (FAO) of the United Nations, Rome, Italy.

Sommer, R., de Sá, T.D.A., Vielhauer, K., Vlek, P.L.G. and Fölster, H. (2001). Water and nutrient balance under slash-and-burn agriculture in the Eastern Amazon, Brasil: The role of a deep rooting fallow vegetation. *In:* W.J. Horst et al. (editors), Plant Nutrition: Food Security and Sustainability of Agro-ecosystems, Kluwer Academic Publishers, pp. 1014-1015.

Su, Z. (2000). Remote sensing of land use and vegetation for mesoscale hydrological studies. *International Journal of Remote Sensing*, **21(2):** 213-233.

Thiruvengadachari, S. and Sakthivadivel, R. (1997). Satellite Remote Sensing for Assessment of Irrigation System Performance: A Case Study in India. Research Report 9, International Irrigation Management Institute (IIMI), Colombo, Sri Lanka.

Thompson, L.G., Yao, T., Mosley-Thompson, E., Davis, M.E., Henderson, K.A. and Lin, P.N. (2000). A high resolution millennial record of south Asian monsoon from Himalayan ice cores. *Science*, **298:** 1916-1919.

UPRSAC (2004). Personal communication. Director, Uttar Pradesh State Remote Sensing Application Centre, U.P., India.

Venkatachary, K.V., Bandyopadhyay, K., Bhanumurthy, V., Rao, G.S., Sudhakar, S., Pal, D.K., Das, R.K., Sarma, U., Manikiam, B., Meena Rani, H.C. and Srivastava, S.K. (2001). Defining a space-based disaster management system for floods: A case study for damage assessment due to 1998 Brahmaputra floods. *Current Science*, **80(3):** 369-377.

Vieux, B.E. (2004). Distributed Hydrologic Modeling Using GIS. Springer, Dordrecht.

Vinod Kumar, K. and Martha, T.R. (2004). Evaluation of Resourcesat-1 Data for Geological Studies. *Bulletin of the National Natural Resources Management System*, NNRMS(B) **29**, Department of Space, Bangalore, India.

Decision Support System: Concept and Potential for Integrated Water Resources Management

P. Tiwary*, Madan Kumar Jha and M.V. Venugopalan

1. FRESHWATER SCARCITY: GLOBAL AND INDIAN PERSPECTIVES

Water is the lifeblood of the biosphere. It makes life itself possible, and nourishes our ecosystems, grows our food and powers our industry. Hardly any economic activity can be sustained without water (Haddadin, 2001). Thus, water plays a vital role in our life. Different dimensions of water functions in society and nature are (Iyer, 2004; Falkenmark and Rockstrom, 2004): water as *life-support* and hence, as a basic need and as a human and animal right; water as an *economic* commodity in some uses; water as an *integral part of the ecosystem* (sustaining it and being sustained by it); water as a *sacred resource*; and water as *an ineluctable component of cultures and civilizations*.

Although 70% of the earth's surface is covered by water, the amount of water that can be consumed and used for different purposes (known as freshwater) is very limited. Out of a total global storage of some 1.4 billion km^3 in solid, liquid and gaseous forms, about 97.5% is salty water in the seas and oceans and as such unsuitable for human use. The remaining 2.5% is freshwater, of which 85% is not readily available—it is either locked up in ice sheets and glaciers, stored in deeper aquifers, or is in the soil, atmosphere and in living things (Gupta, 1992). Further, about 70% of the total freshwater resources are snow, ice, permafrost, etc., with only 30% available in the liquid form. Of the liquid freshwater, approximately 98.7% exists as groundwater. Thus, the amount of water readily available for human development is only a small part—less than 0.01% of the total water in the hydrosphere (or, about 1% of all liquid freshwater), which is renewed annually (Ayibotele, 1992).

Global climatic change, socio-economic changes and growing anthropogenic pollution have significant impacts on freshwater resources and their availability. Therefore, there is an urgent need for widespread realization that water is a finite and vulnerable resource, which must be used efficiently, equitably and in an ecologically sound manner for present and future generations. To convey this idea, the term "sustainable development" was coined around 1980, but it became popular and entered the political mainstream in 1987 when the Brundtland report "Our Common Future" (WCED, 1987) was published. There are many definitions of sustainable development. However, in different ways all the views are building on the Brundtland definition: "Sustainable development is the development that meets the needs of the present without compromising the ability of future generations to meet their own needs". Specifically, Gleick (1996) provided a broad definition of sustainable water use as: "The use of water that supports the ability of human society to endure and flourish into the indefinite future without undermining the integrity of the hydrological cycle or the ecological systems that depend on it". It should be noted that despite widespread recognition and acceptance of sustainable development as an appropriate goal for human kind, the application of sustainable development concept has proved difficult.

---

*Corresponding Author

At present, about 10% of the world's freshwater supplies are used for maintaining health and sanitation, whereas agriculture accounts for about 70% of the world's freshwater supplies. Industries use about 20% of the world's freshwater supplies, often as a vital part of the production process (Shiklomanov, 1997). Food production is indeed one of the water-intensive sectors. It has been estimated that about one liter of liquid water gets converted to water vapor to produce one calorie of food. Every person is responsible for consuming 2000 to 5000 liters of water every day depending on one's diet and the method of food production, which is far more than 2 to 5 liters we drink every day (Rodriguez and Molden, 2007). A heavy meat diet undoubtedly requires much more water than a vegetarian diet. Unfortunately, the demand for water is gradually increasing with the growing population as well as rapid urbanization and industrialization in different parts of the world. As a result, water demand is outstripping the limited freshwater resource and ecosystem degradation is rising at a greater pace. Thus, water-related development prospects are threatened at the dawn of the 21st century. On the top of it, in future, more people will need more water for food, fiber, livestock, fish, and industrial crops (for producing bio-energy).

Water experts predict that the global food demand will double over next 50 years, and if the current practices don't get changed, water use will also double (Rodriguez and Molden, 2007). Water problems will be further complicated by the impacts of climate change (Falkenmark, 2007a; Harris and Baveye, 2008). This situation will lead to widening of the gap between water supply and water demand, which in turn will result in water stress or chronic water scarcity worldwide. Various studies have been conducted to characterize and quantify the severity of water scarcity or water stress. Falkenmark attracted the world's attention to the growing problem of relative water scarcity by introducing an empirical water competition scale that focuses on the number of persons sharing one flow unit of blue water. Later, this empirical scale was inverted as a relative indicator which measures water stress/scarcity in terms of volume of freshwater availability per person, instead of number of persons sharing a flow unit. This water scarcity indicator has been referred to as '*Falkenmark indicator*' or '*water stress index*' and was renamed later by Population Action International as '*standard indicator*' (Engelman and LeRoy, 1995). The original and adapted Falkenmark indicators are summarized in Table 1. In this table, the 600 persons/flow unit corresponds to 1667 m^3/capita/annum which was rounded to 1700 m^3/capita/annum. Based on the Falkenmark indicator, nearly 1.4 billion people will experience water scarcity (i.e., water supply less than 1000 m^3/capita/annum) within the first 25 years of this century, mostly in semiarid regions of Asia, North Africa and Sub-Saharan Africa. By 2025, 1.8 billion people will be living in countries or regions with '*absolute water scarcity*', and two thirds of the world's population could be under '*water stress*' conditions (UN Water, 2007). It is worth mentioning that from a water management point of view, the Falkenmark indicator has little use as it is based on water supply and population and does not consider the actual water demand of a country/region. To overcome this limitation, the UN Study on Comprehensive Assessment of the World's Freshwater Resources (CFWA, 1997) introduced an indicator based on total annual water withdrawals (i.e., the amount of water withdrawn from surface and groundwater sources to satisfy human needs) instead of water supply, which is known as '*Water Resources Vulnerability Index*' (Table 2). This indicator is defined as a ratio of total annual water withdrawals to the annual available water resources in a country/region and is expressed in percentage.

Some water experts such as Frank Rijsberman and David Molden have reported that above mentioned two indicators become complex if a greater accuracy is desired. However, they may be helpful in estimating overall water demand and supply relationship at global, regional and national levels (i.e., at large scales). For water management decision making, it is more important to understand the variations in water availability and demand that occur within a country. To address this concern, the International Water Management Institute (IWMI) developed an exhaustive water scarcity indicator considering two important factors: (a) current withdrawals as percentage of available water resources, and (b) future withdrawals as percentage of current withdrawals. Based on this indicator, the water scarcity has been categorized into three types (Seckler et al., 1998): (i) *physical water scarcity* (i.e., where the available water resources cannot meet the demands of the population), (ii) *economic water scarcity* (i.e., where countries lack the necessary infrastructure to harness water from rivers and aquifers), and (iii) *institutional/political water scarcity* (i.e., people not having access to available water resources due to various political or social reasons). Unlike other indicators, the IWMI indicator can be used even at a small scale (e.g., district level). At present, around 1.2 billion people (almost one-fifth of the world's population) live in the areas of *physical scarcity* and 500 million people are approaching this situation. Another 1.6 billion people (almost one quarter of the world's current population) face *economic water scarcity* (Molden, 2007). Undoubtedly, water scarcity is a major concern for sustainable development in the 21st century. In order to correctly address this issue, a better understanding of the links between *water security, food security, energy security* and *environmental security* is needed.

Table 1. 'Falkenmark Indicator' for characterizing water stress (Falkenmark and Rockstrom, 2004)

Original Falkenmark Indicator (persons per flow unit)	*Adopted Falkenmark Indicator (m^3/capita/annum)*	*Water stress implication*
>600	<1700	Water stress
>1000	<1000	Chronic water scarcity
>2000	<500	Absolute water scarcity

Table 2. 'Water Resources Vulnerability Index' and 'IWMI Indicator' for characterizing water stress (Amarsinghe et al., 1999)

Water stress condition	UN Indicator	IWMI Indicator
	\multicolumn{2}{c}{*Extent of annual withdrawal*}	
Low water stress	<10%	I_{IWMI1} <50% and I_{IWMI2} <125%
Medium low water stress	10-20%	I_{IWMI1} <50% and 125% < I_{IWMI2} < 200%
Medium high water stress	20-40%	I_{IWMI1} <50% and I_{IWMI2} > 200%
High water stress	>40%	I_{IWMI1} > 50%

Note: I_{IWMI1} = Current withdrawals as % of Available water resources; I_{IWMI2} = Future withdrawals as % of Current withdrawals.

India is no exception and is suffering from water scarcity (Garg and Hassan, 2007). In India, per capita surface water availability in the years 1991 and 2001 was 2309 and 1902 m^3, respectively and it is expected to decline to 1401 and 1191 m^3 by the years 2025 and 2050, respectively (Kumar et al., 2005b). Thus, the average water availability in India is likely to fall below the water-stress level in the near future and given the wide variations across the country, water stress conditions already exist in several parts of the country. The total annual water requirement for different sectors in India was about 634 km^3 in 2000, which will increase to 1093 km^3 in 2025 and 1447 km^3 in 2050 (Table 3). It is apparent from Table 3 that by 2050, the annual water demand in all the sectors would be more than two times the water requirement in 2000. In the sectors like industry and energy, the increase in water demand would be about 8 and 65 folds, respectively due to rapid growth in industrial activities and increased power demand.

Thus, the limited freshwater resources to meet the ever-increasing competitive demands in different sectors on one hand, and the widespread water scarcity, growing pollution of freshwater resources, and climate change impacts on future water supply coupled with the progressive encroachment of incompatible activities on the other hand, have compelled the water managers and policy makers to think over holistic water resources planning and management. As a result, water resources management has been undergoing a change worldwide—moving from a mainly supply-oriented and engineering-biased approach towards a demand-oriented and multi-sectoral approach, which has been termed integrated water resources management (IWRM). The adoption of modern water management concepts such as IWRM in practice is an urgent need throughout the world, especially in developing nations in order to ensure holistic planning and management of water resources.

Table 3. Sectoral and temporal variations of annual water requirements in India (CWC, 2000)

Sector	Annual water requirement (km^3)		
	2000	2025	2050
1. Domestic	42	73	102
2. Irrigation	541	910	1072
3. Industry	8	23	63
4. Energy	2	15	130
5. Other	41	72	80
Total	634	1093	1447

The aim of this chapter is to highlight the gravity of freshwater crisis in the world in general and in India in particular and to present IWRM concept and its implication for solving real-world water problems. In addition, the fundamentals of decision support systems are described, together with a critical review of DSS applications in water management for combating water-related disasters.

2. IWRM: BASIC CONCEPT, IMPLICATIONS AND CHALLENGES

The concept of integrated water resources management (IWRM) has been developing since the beginning of eighties. However, it has attracted a lot of attention in recent years, which is a way to deliberately move away from the prevalent fragmented approaches followed in the planning and management of water resources. IWRM is defined by the Global Water Partnership as follows (GWP, 2000):

"IWRM is a process which promotes the coordinated development and management of water, land and related resources in order to maximize the resultant economic and social welfare in an equitable manner without compromising the sustainability of vital ecosystems".

It is clear from the above definition that IWRM considers the use of resources in relation to social and economic activities and functions. They also determine the need for laws and regulations for the sustainable use of water resources. Infrastructure made available, in relation to regulatory measures and mechanisms, will allow for effective use of resources taking due account of the environmental carrying capacity. Thus, the IWRM approach is anchored on three basic pillars, viz., *'enabling environment'*, *'institutional framework'* and *'management instruments'* (Fig. 1) (GWP, 2004) and it explicitly aims at avoiding the classical fragmented approach for water resources management. Consequently, the implementation of IWRM approach requires enabling environment of appropriate policies, strategies and legislative framework for sustainable water resources development and management. These policies and legislation can be implemented through institutional framework with the help of different management instruments. Overall, IWRM is about change in water governance, i.e., the change of political, social, economic and administrative systems which are responsible for developing and managing water resources and delivering water services at different levels of society.

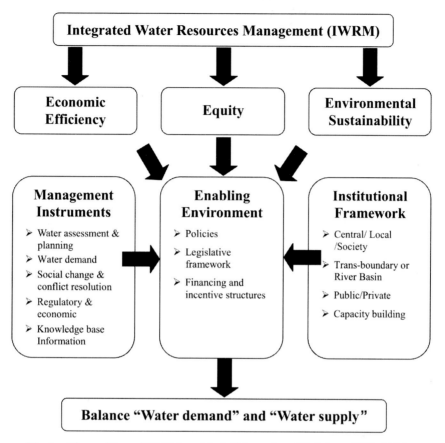

Fig. 1 Three pillars of IWRM and their linkage (modified from GWP, 2004).

The above concept of IWRM has received a widespread support because it addresses the interconnections among water, development and sustainability. However, integration needs to be interpreted broadly—it is about integrating within human as well as natural systems (Lenton, 2004). The concept has been discussed in major international conferences throughout the 1990s and several countries are preparing themselves for its implementation. Integrated water resources management considers the viewpoints of water management agencies with specific purposes, governmental and stakeholder groups, geographic regions, and knowledge of multi-disciplines. These viewpoints have been described in a variety of ways. For example, according to Mitchell (1990), integrated water management considers three broad aspects: *dimensions of water* (surface water and groundwater, and quantity and quality); *interactions with land and environment*; and *interrelationships with social and economic development*. Lenton (2004) emphasizes that to really develop and manage the water resources of a basin in ways that ensure sustainable development, an IWRM approach must be viewed as a process of change in political, social, economic and administrative systems. However, major initial reforms are not essential, rather easily implementable first steps are often sufficient to catalyze change. Further, Falkenmark and Rockstrom (2004) recommend that while human livelihood security and pure drinking water, household water, water-dependent food and income generation have to be safeguarded, ecological (or environmental) security must also be ensured with healthy terrestrial and aquatic ecosystems and long-term resilience to unavoidable changes. To achieve this goal, we must take an integrated approach to land, water, and ecosystems as well as broaden the perception of water. We must take into account the involvement of water in terrestrial and aquatic ecosystems. We must also increase our understanding of the multiple water functions in the planet's life-support system, of the value of ecosystem water, and of the value of water for socio-economic development. Thus, besides the paradigm shifts in water management, a new type of water expertise with a capacity to focus on cross-disciplinary issues is required, which can help in solving the future complex challenges of dealing with various functions of water in nature and society (Falkenmark and Rockstrom, 2004).

According to Falkenmark and Rockstrom (2004), a successful water resources management will have to incorporate the following three basic efforts:

(i) to *secure* water and water-dependent services (e.g., food production);
(ii) to *avoid* degradation of water and land resources, and of ecosystem integrity; and
(iii) to *foresee* and *assess* changes in climate, population, socio-economic conditions, etc.

The above efforts call for an integrated approach considering both blue and green water as well as water quantity and quality. According to Falkenmark (2007b), the best way will be to focus on integrated water resources management (IWRM) on a catchment or river basin scale (*blue water* approach) in conjunction with *green water* through its interaction with *blue water* as illustrated in Fig. 2. This approach would lead to an integrated land and water resources management (ILWRM), which is being practiced in several transnational rivers supported by the Global Environment Facility (Duda, 2003).

The IWRM approach offers several important benefits such as:

- Greater utility from a given amount of water through adjusted allocations and enhanced water supply through water conservation strategies.
- Reduced groundwater mining through conjunctive management of groundwater and surface water.
- More intensive reuse/recycle of water through planned sequencing of various uses.
- Improved water quality through more comprehensive data collection, monitoring and enforcement.
- Incorporation of current social as well as environmental values into the decision making related to water allocation, supply and management.
- Inclusion of a wider range of basin stakeholders into decision making.
- Reduced conflict among water users.

It is evident from the above discussions that although the concept of IWRM is sound and has many benefits, it involves a highly challenging and complex approach wherein decisions need to be taken under many conflicting interests/issues in order to fulfill the needs and demands of different stakeholders. Even taking a decision about water resources allocation and management is very tedious because of the different stakeholders pursuing multiple and conflicting objectives, the influence of spatial distribution of land use, socio-economic activities and the environmental constraints. Therefore, there is a need to have a tool or a support system which can help in data/information gathering, analysis of physico-chemical, biological and socio-economic processes, weighing of interests and decision-making related to the sustainable development and management of vital water resources. Nowadays, an ever-increasing role is played by complex scientific models, computers and tools to assist water resources management in order to satisfy today's water demands of competing sectors (McKinney et al., 1999; Horlitz, 2007). With a vast development in the computer technology, the computer-based decision support system (DSS)

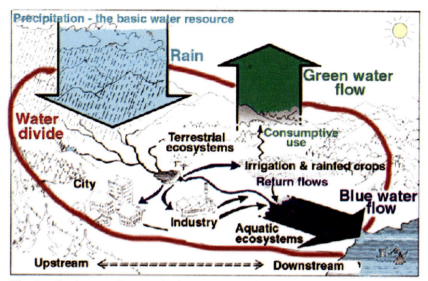

Fig. 2 Catchment allows an integrated approach to all water-related phenomena within the water divide. All the rain falling in the catchment gets partitioned between humans and ecosystems (terrestrial as well as aquatic), between land use and water, and between upstream and downstream (after Falkenmark, 2007b).

is emerging as an important tool for decision making related to real-world problems involving different layers of information and conflicting interests/issues such as land, air and water problems (Watkins and McKinney, 1995; Giupponi et al., 2002). Decision support systems are very efficient and fast in data/information storing and retrieval, analysis, querying and decision process. Thus, DSS is a promising computer tool, which can play a central role in solving real-world land and water problems as well as in mitigating natural and anthropogenic disasters.

3. WHAT IS A DSS?

Decision support system (DSS) is a computer-based system, which is used to assist and aid decision makers in their decision-making processes. From the inception of DSSs, it has become clear that DSSs aid and assist decision makers but do not replace them. This feature distinguishes a DSS from other information systems (IS). In practice, the term DSS is used for many different kinds of computer-based information and modeling systems. Several attempts have been made to define DSS since 1970, which are succinctly described below:

- Little (1970), in one of the earliest works on computer-based decision support, proposed that a DSS be "a model-based set of procedures for processing data and judgments to assist a manager in his decision making".
- According to Keen and Scott-Morton (1978), "decision support systems couple the intellectual resources of individuals with the capabilities of computers to improve the quality of decisions. It is a computer-based support for management decision makers who deal with semi-structured problems."

- Moore and Chang (1980) defined a DSS in terms of its features and use: "DSS is a system that is extendable, capable of supporting ad hoc analysis and decision modeling, oriented towards future planning, and of being used at irregular, unplanned intervals".
- Bonczek et al. (1980) defined DSS in terms of its components: "a generic DSS consists of a language system for communication between the user and the DSS, a knowledge system containing problem domain knowledge consisting of data and procedures, and a problem processing system consisting of programs capable of solving decision problems".
- Sprague and Carlson (1982) defined DSS as "interactive computer-based support systems that help decision makers utilize data and models to solve unstructured problems."

Key terms in above definitions are: interactive, data, and models, which are a recurring theme among the developers of water management DSSs.

- Adelman (1992) defined DSS as "interactive computer programs that utilize analytical method, such as decision analysis, optimization algorithms, program scheduling routines, and so on for developing models to help decision makers formulate alternatives, analyze their impacts, and interpret and select appropriate options for implementation."
- Recently, Poch et al. (2003) defined DSS as "an intelligent information system that reduces the time in which decisions are made, and improves the consistency and quality of those decisions."

Thus, it can be inferred from the various definitions of DSS that although a single, unambiguous and universally accepted definition of DSS is lacking, to date there is some consensus on the purpose of DSS, i.e., to support decision making in more or less complex situations (i.e., multi-objective or semi-structured or unstructured problems).

4. HISTORICAL PERSPECTIVE OF DSS

Computerized decision support systems became practical with the development of minicomputers, timeshare operating systems and distributed computing. The history of the implementation of such systems begins in the mid-1960s (Power, 2007). Initially, the concept of DSS evolved from two main areas of research: the theoretical studies of organizational decision making done at the Carnegie Institute of Technology during the late 1950s and early 1960s, and the technical work on interactive computer systems, mainly carried out at the Massachusetts Institute of Technology (MIT), USA in the 1960s (Keen and Scott Morton, 1978). The original DSS concept, the most clearly defined, was given by Gorry and Scott Morton (1971). They combined Anthony's (1965) categories of management activity and Simon's (1960) description of decision types, using the terms *structured*, *semi-structured*, and *unstructured*. They emphasized that Management Information Systems (MIS) primarily focus on structured decisions and hence, the supporting systems for semi-structured and unstructured decisions should be better termed "Decision Support Systems".

With the development of modern technologies, DSS technology and its applications have evolved significantly since its early development in the 1970s and over the last two decades or so, several additional concepts and views have been evolved through DSS research (Fig. 3). DSS once utilized more limited database, modeling and user interface functionality, and supported individual decision-makers, but technological innovations have enabled far more powerful DSS functionality and are being used by organizations, workgroups or teams in solving more complex problems and in decision making process. Around 1985, '*Group Decision Support Systems*' (GDSS), or simply '*Group Support Systems*' (GSS) evolved to provide brainstorming, idea evaluation, and communication facilities to support team problem

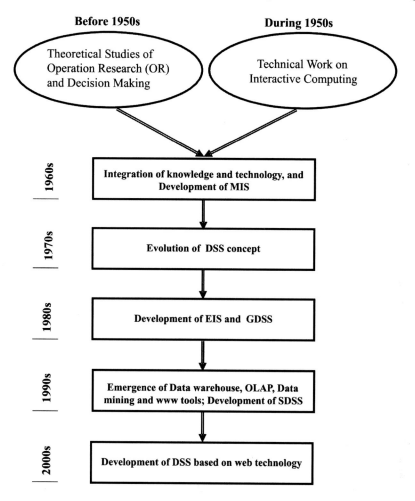

Legend: MIS = Management Information System; DSS = Decision Support System; EIS = Executive Information System; GDSS = Group Decision Support System; SDSS = Spatial Decision Support System; OLAP = On-Line Analytical Processing.

Fig. 3 Chronology of the evolution of Decision Support Systems.

solving. It is an interactive computer-based system which facilitates solution of unstructured problems by a set of decision makers working together as a group (DeSanctis and Gallupe, 1985). A GSS must have a communications base as well as the model base, database, and interface as required in conventional DSS (Bui and Jarke, 1986). The communication and coordination activities of group members are facilitated by technologies that can be characterized in time, space, and level of group support (DeSanctis and Gallupe, 1987; Johansen, 1988; Alavi and Keen, 1989). Thus, GSS facilitates more effective group interaction and thereby leading to greater effectiveness of decision-making (Warkentin et al., 1997).

Executive information systems (EIS) have extended the scope of DSS from personal or small group use to the corporate level. An EIS is a type of management information system used to support the information and decision making needs of senior executives by providing easy access to both internal and external information relevant to meeting the strategic goals of the organization. It is commonly considered as an enterprise-wide DSS. These days, Business Intelligence (with the sub-areas of reporting, analysis, and digital dash boards) is getting more popularity in place of EIS. Model management systems and knowledge-based decision support systems have used techniques from artificial intelligence and expert systems to provide smarter support for the decision-maker (Bonczek et al., 1981; Courtney and Paradice, 1993). In the early 1990s, four powerful tools emerged for building a DSS: (i) data warehouse, (ii) on-line analytical processing (OLAP), (iii) data mining, and (iv) World Wide Web (WWW). Data warehousing and on-line analytical processing (OLAP) have become essential elements of decision support. A *'data warehouse'* is a subject-oriented, integrated, time-variant, nonvolatile collection of data (Inmon, 1992). It is a complex system which has unique design and operational requirements that are significantly different to other kinds of databases. A large data warehouse often leads to an increased interest in analyzing and using historical data. *'On-line analytical processing'* (OLAP) tools, on the other hand, are used to analyze the historical data in a data warehouse. On-line analytical processing is a category of "software technology that enables analysts, managers, and executives to gain insight into data through fast, consistent, interactive access to a wide variety of possible views of information that has been transformed from raw data to reflect the real dimensionality of the enterprise as understood by the use" (Power, 1997). Although OLAP tools have become more powerful in recent years, a set of artificial intelligence and statistical tools collectively called *'data mining tools'* (Edelstein, 1996) has been proposed for more sophisticated data analysis. The rapidly expanding volume of real-time data, due to advanced monitoring techniques such as remote sensing, GPS and in situ sensors, has also contributed to the demand for data mining tools.

The traditional decision support systems, used in managerial decision making, have a major limitation of inability to exploit spatial and temporal data. Therefore, in recent years, a new type of DSS, known as the *'Spatial Decision Support System'* (SDSS), has emerged. Spatial Decision Support Systems are designed to help decision makers in solving complex problems such as site selection, urban planning, and routing that have a strong spatial component. An SDSS incorporates both geographic information systems (GIS) functionalities such as spatial data management, cartographic display, etc. as well as analytical modeling capabilities, a flexible user interface, and complex spatial data structures (Goodchild, 2000). Thus, SDSS provides a framework for integrating (a) analytical and spatial modeling capabilities, (b) spatial and non-spatial data management, (c) domain knowledge, (d) spatial display capabilities, and (e) reporting capabilities. Currently, there is a growing interest in developing both spatial models and SDSS for managerial decision making (Sikder and Gangopadhyay, 2002).

At the beginning of the 21st century, the web environment is emerging as a very important DSS development and delivery platform and has been the center of activity in developing DSS (Eom, 1999). The primary web tools are 'web servers' based on 'Hypertext Transfer Protocol' (HTTP) containing web pages that are developed using Hypertext Mark-up Language (HTML) and JavaScript.

5. COMPONENTS OF DSS FOR WATER MANAGEMENT

Three main subsystems must be integrated in an interactive manner in a DSS (Orlob, 1992): (1) a user-interface for dialog generation and managing the interface between the user and the system; (2) a model management subsystem; and (3) an information management subsystem. Considering this in more detail, a water management DSS should consist of the following five components (Fig. 4) (McKinney, 2004):

- *Data input System* to receive various data (e.g., water level and temperature, precipitation, air temperature, concentrations, etc.) from hydrological stations over the river basins or watersheds being managed as well as weather data and forecasts.

Legend: IWRM = Integrated Water Resources Management; GW = Groundwater; ANN = Artificial Neural Network; MOP = Multiobjective Programming; MCDA = Multicriteria Decision Analysis; AI = Artificial Intelligence.

Fig. 4 General framework of a decision support system for water resources management (McKinney, 2004).

- *Data Processing System* to store the data related to the processes of interest in the river basins or watersheds, both spatial and feature related as well as time series data.
- *Analytical System* of models and tools designed to predict watershed response and provide river forecasts, using data from the Data Input System, and time-series data of river basin needed to calibrate hydrologic models.
- *Decision Formulation and Selection System* for gathering and merging conclusions from knowledge-based and numerical techniques, and the interaction of users with computer through an interactive and graphical user interface.
- *Decision Implementation System* for disseminating decisions regarding water use under normal conditions, and flood/drought warnings, river forecasts, weather forecasts, and disaster response in affected areas.

6. TYPES OF DSS

DSS can be classified into five major categories as follows (Power, 2007):
- Data-Driven DSS
- Model-Driven DSS

- Communications-Driven DSS
- Document-Driven DSS, and
- Knowledge-Driven DSS.

A recent addition appears to be the Web-based DSS category, though it is, in a way, an implementation of all other types of DSS through the widely used World-Wide Web (WWW).

6.1 Data-Driven DSS or Data-Oriented DSS

It is a type of DSS that emphasizes access to and manipulation of a time series data (historical data). A simple data-driven DSS is based on file system with the most elementary level of functionality wherein data are accessed by query and retrieval tools. Data-driven DSSs with data warehouse systems offer additional functionality for data manipulation whereas with On-line Analytical Processing (OLAP) or data mining tools, it provides the highest level of functionality for retrieval and analysis of large collections of historical data. Early versions of data-driven DSSs were called Retrieval-Only DSSs (Bonczek et al., 1981). Executive Information Systems are examples of data-driven DSS (Power, 2002). The main components of a data-driven DSS are shown in Fig. 5.

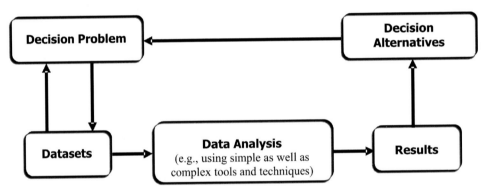

Fig. 5 Main components of a data-driven DSS (Kersten et al., 2002).

6.2 Model-Driven DSS or Model-Oriented DSS

In a model-driven DSS, a model (e.g., statistical, financial, optimization and/or simulation model) is used for decision making and the type of model to be used depends on the decision making process. It uses limited data and parameters to aid decision makers in analyzing a situation. Simple statistical and analytical tools provide the most elementary level of functionality. Some OLAP systems that allow complex analysis of data may be classified as hybrid DSS systems providing both modeling and data retrieval and data summarization functionality. Data mining is also a hybrid approach to DSS. Early versions of model-driven DSSs were called Computationally Oriented DSSs (Bonczek et al., 1981). The sequence of main activities in a model-driven DSS is shown in Fig. 6.

6.3 Communications-Driven DSS

It is a type of DSS that emphasizes communications, collaboration and shared decision-making support. It enables two or more people to communicate with each other, share information and coordinate their activities. In this system, communication tools/technologies are the dominant architectural components. The tools used in a Communications-Driven DSS are: audio conferencing, bulletin boards and web conferencing, document sharing, electronic mail, computer supported face-to-face meeting software, and interactive video. The WWW and intranet infrastructures have expanded the functionality of Communications-Driven DSSs and make them more powerful. A Group Decision Support System (GDSS) is a Communication-Driven DSS that allows multiple users to work collaboratively in a group. Communications-Driven DSS has at least one of the following characteristics:

- Enabling communication between groups of people
- Facilitating the sharing of information
- Supporting collaboration and coordination between people
- Supporting group decision tasks

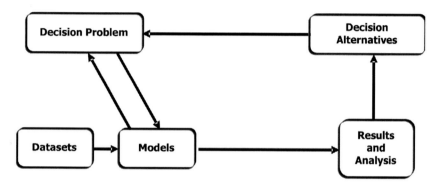

Fig. 6 Main components of a model-driven DSS (Kersten et al., 2002).

6.4 Document-Driven DSS

It manages, retrieves, and manipulates unstructured information using storage and processing technologies. This type of DSS assists in knowledge categorization, deployment, enquiry, discovery and communication. The most elementary level of document-driven DSS is a hyperlinked collection of documents such as web pages.

6.5 Knowledge-Driven DSS

Knowledge-driven decision support systems have specialized problem-solving expertise stored as facts, rules, and procedures or in similar structures. The expertise consists of knowledge about a particular domain, and skill in solving some specific problems. Knowledge-driven DSSs are also called 'expert systems'.

7. TOOLS AND TECHNIQUES FOR DECISION MAKING IN LAND AND WATER MANAGEMENT

Recent advances in information technology have brought a significant shift in the evolution of DSS. Advanced simulation and optimization computer models, and modern tools/techniques such as 'geographic information system (GIS)', 'remote sensing (RS)' 'data warehouse', 'on-line analytical processing

'OLAP)', 'artificial intelligence (AI)', fuzzy logic', 'artificial neural network (ANN)', genetic algorithm (GA), etc. have enhanced the capabilities and functionality of DSS to help in decision making for more complex and higher level tasks. Salient tools and techniques often used in a DSS are briefly discussed below.

7.1 Simulation and Optimization Models

Over the past decades, the planning and management of water resources have been increasingly challenging because of intense competition between water demands for drinking and other purposes such as manufacturing, agriculture, recreation, and hydroelectricity generation. In addition, the intensive uses of land and aquifer have caused the degradation of water quality and the contamination of water supply sources. Therefore, simulation and optimization techniques are used to analyze complex water resource systems with an ultimate goal of efficient water resources management.

There are various types of simulation models used in the field of water resources engineering, which range from deterministic to stochastic, dimensionless to one-dimension, two-dimensions and three-dimensions, and static to dynamic models (Table 4). Simulation models are used for analyzing hydrologic and hydrogeologic processes occurring at varying scales. These models are developed and applied to different water problems such as flood management, urban water management, reservoir operation, groundwater use, conjunctive use of surface water and groundwater, quality of surface and groundwater, and irrigation and drainage at both subsystem and river basin levels (Loucks et al., 1981; Maidment, 1993; McKinney et al., 1999; Loucks and van Beek, 2005).

Although simulation models can accurately represent system operations and are useful in examining long-term reliability of operating systems, they are not well suited to determining the 'best' or optimum strategies when flexibility exists in coordinated system operations (McCartney, 2007). Instead, prescriptive optimization models are often used to systematically derive optimal solutions, or families of solutions, under specified objectives and constraints. The application of optimization models in water resources management is extensive. The most commonly used optimization techniques are Linear Programming (LP) and Dynamic Programming (DP) (Loucks et al., 1981; Willis and Yeh, 1987; Loucks and van Beek, 2005). Nonlinear Programming (NLP) is also used. In recent years, these techniques have been combined with new approaches, viz., 'optimal control theory', 'fuzzy logic', 'genetic algorithm' or 'genetic programming' (Goldberg, 1989; McCartney, 2007).

7.2 Geographical Information System (GIS) and Remote Sensing (RS)

The physiographic properties of a region or river basin/watershed greatly influence the development and management of water resources. Geographical Information System (GIS), Remote Sensing (RS) and Global Positioning System (GPS) techniques have emerged as efficient and powerful tools for data collection and analysis. The RS technology is used to collect detailed information in space and time even from inaccessible areas and is advancing gradually. These two promising techniques complement each other. GIS is a set of computer-based systems for capturing, storing, querying, analyzing, and displaying geographic data and using these data to solve a variety of spatial problems (Lo and Yeung, 2003). It provides comprehensive facilities for storing, retrieving, displaying and manipulating data essential to the decision-making process. Two common data manipulation and storage systems or tools are: the relational database, which relates information in a tabular form so that the rules of relational algebra can be applied, and the geographic database which relates information pertaining to fundamental spatial

Table 4. Summary of the models needed for integrated water resources management

Model type	Description
1. Watershed Models	They relate meteorological phenomena to the runoff quantity and/or quality from a watershed. They vary from simple empirical relationships like Rational formula to complex physically deterministic models that take into account of land and vegetation types, antecedent moisture, evapotranspiration, infiltration, solar radiation, and other details.
2. Surface Water Quantity Models	They describe the movement of water on the earth's surface, including rivers, lakes and artificial channels. They include representations of structures such as dams, power plants, and control structures, as well as laws or policies such as operating procedures and water rights. In mathematical complexity, they range from simple mass balance to three-dimensional partial differential equations of fully dynamic flow. They may also include boundary conditions with the atmosphere and/or the groundwater. In addition, some models address the problem of sediment transport.
3. Surface Water Quality Models	They include water quantity as required, but focus on the fate of physical, chemical and biological constituents carried by the water. Modeled variables include temperature, dissolved oxygen, plant biomass, metals, toxic organics, sediments, and dissolved solids. The importance of water quality models have increased over past two decades in managing land and water resources because of serious environmental concerns.
4. Subsurface Water Quantity and Quality Models	They simulate quantity and/or quality of water in the unsaturated, saturated and variably saturated porous media. They include infiltration, evapotranspiration, pumping, flow and solute movement through geological formations, aquifer management, contaminant plume tracking, conjunctive management of surface water and groundwater, etc.
5. Economic Models	They do not explicitly model physical processes, but consider the economic aspects of water resources development and management. Economic aspects of land and water resources projects include cost-benefit analysis of projects or policies, regional economic impacts of water and/or land resources projects, costs of pollution control measures, and risk-benefit analysis.
6. Social Models	They consider the social impacts of land and water resources projects and policies, including migration, displacement, farming practices, education, and the uneven distribution of benefits among various stakeholders.

features such as points, lines, and polygons (Lo and Yeung, 2003; Chang, 2006). GIS not only brings spatial dimensions into the traditional water resources database, but also, more significantly, has the ability to better integrate the various social, economic, and environmental factors related to water resources planning and management for use in a decision-making process. GIS offers a spatial representation of water resources systems, but currently few predictive and related analytical capabilities are available in GIS for solving complex water resources planning and management problems (Chang, 2006).

An excellent review of GIS applications in water resources, in general, is presented by Tsihrintzis et al. (1996), while Jha et al. (2007) present a state-of-the-art review of remote sensing and GIS applications in groundwater hydrology.

In order to create a truly useful DSS for water management, a data model with geometric representations and spatial referencing is needed that has an open architecture to facilitate the integration of GIS and models. There are several strategies for coupling an environmental model to a GIS, ranging from a *loose coupling* where data are transferred between models and GIS, and each has separate database management capabilities and systems; to a *tight coupling* where data management in the GIS and model are integrated and they share the same database (McKinney, 2004). For instance, ArcHydro is a data model that can facilitate tight coupling between water resources models and GIS. However, more and more research on integration of GIS and expert systems is needed.

7.3 Artificial Intelligence

Artificial intelligence (AI) is a state-of-the-art technology that resembles the human thinking process in decision making and strategy learning. It has been well recognized for its outstanding ability to handle complex systems (Lin and Su, 2000; Kurosh, 2005). AI techniques such as artificial neural network (ANN), genetic algorithm (GA) and fuzzy logic have been used for the development of intelligent systems (IS) and decision support systems (DSS) in water resources management (Westphal et al., 2003; Karbowski et al., 2005; Chaves and Chang, 2008).

In last decade, ANNs have been widely applied with success to various water resources problems such as rainfall-runoff modeling (e.g., Chang et al., 2004, Kumar et al., 2005a), rainfall forecasting (e.g., Olsson et al., 2004), groundwater problems (e.g., Daliakopoulos et al., 2005; Lin and Chen, 2006) and reservoir operation problems (e.g., Chaves et al., 2004; Chaves and Chang, 2008). An excellent review of the application of ANNs in hydrology can be found in the ASCE Task Committee (2000) report. In addition, GA has been used to solve several water resources related problems (e.g., Chang et al., 2005; Cheng et al., 2008). In recent years, some hybrid techniques such as GA-DP, ANN-GA and neuro-fuzzy are also being used for analyzing water resources systems (Huang et al., 2002; Chaves and Chang, 2008).

Moreover, '*expert system*' is one of the important application-oriented branches of Artificial Intelligence. It helps in decision making processes by mimicking human reasoning processes, relying on logic, belief, rules of thumb opinion and field experiences (Regan, 2008). It has been used in reservoir operation, urban water management, wastewater treatment, sewerage operation and maintenance, etc. (Fedra, 1993).

7.4 Multiobjective and Multicriteria Decision Analysis

Water resource management decisions are complex and multifaceted involving many different stakeholders, often with conflicting priorities or objectives. In these cases, tradeoffs exist among stakeholders in terms of social, economic and environmental which must be considered when searching for the best compromise decisions (Loucks and van Beek, 2005). Multiobjective (MO) and Multicriteria Decision Analysis (MCDA) techniques are considered to be very useful in resolving conflicts related to water resources management. The MO technique can help provide valuable trade-off information among conflicting objectives, whereas the MCDA provides a systematic procedure to help decision makers to compare alternative course of action on the basis of multiple factors and to identify the best performance solution (Massam, 1988). According to RAC (1992) and Voogd (1983) as cited in Hajkowicz and Higgins (2008), MCDA can be defined as "a grouping of techniques for evaluating decision options against multiple criteria measured in different units". Fundamentally, MCDA has inherent properties that make it appealing and practically useful.

The MCDA or MCDM (multicriteria decision making) process generally involves the following steps:

1. To identify final decision makers and stakeholders (people affected by the decision making process);
2. To select criteria for decision making;
3. To define alternatives;
4. To select suitable MCDM technique(s);
5. To assign weights to the criteria;
6. To assess the performance of the alternatives against the criteria;
7. To transform the criteria performance values to commensurable units and normalized values;
8. To apply the selected MCDM technique(s);
9. To perform sensitivity analysis; and
10. To make final decision.

Weighting the criteria and assessing the performance of alternatives against the criteria are two of the most important and difficult aspects of applying the MCDM methodology and are potential sources of considerable uncertainty (Roy and Vincke, 1981; Larichev and Moshkovich, 1995). Therefore, the use of fuzzy decision analysis has been suggested by some researchers to address the uncertainties involved in MCDM process (e.g., Bender and Simonovic, 2000; Gemitzi et al., 2007).

In recent years, with the advancements in modeling tools for water resources planning and management, MCDM techniques have become a major component of decision support systems (Goicoechea et al., 1992; Qureshi and Harrison, 2001; Fassio et al., 2005) and have been applied to an array of problems in water resources management (Karamouz et al., 2003; Lee and Chung, 2007; Van Cauwenbergh et al., 2008), collaborative environmental planning (Yurdusev and O'Connel, 2005; Hermans et al., 2007), land use planning and optimal spatial distribution (Aerts et al., 2005; Geneletti and van Duren, 2008), and group decision making (Cai et al., 2004). An excellent review on the application of MCDA techniques in water resources planning and management has been presented by Hajkowicz and Collins (2007).

8. DESIGN AND DEVELOPMENT OF DSS FOR WATER MANAGEMENT

There are many factors that must be considered in designing and developing a DSS for water resources management. Some of these factors related to the basic structure or components of DSS and others are related to the target users/organizations, and prevailing political, financial and socio-psychological conditions within the organizations. Hence, designing a DSS is a very complex process and involves several development stages/phases right from its inception to the final stages of fielding, testing, training and long-term maintenance.

Several researchers (e.g., Bennett, 1983; Sauter, 1997) have investigated the theory and practice of development phases of a DSS. A comprehensive set of phases for DSS design and development as suggested by Loucks (1995) consists of: identification of DSS model structure, identification of DSS algorithms and interactions, identification of input data requirements and assembly of data, design of the interface incorporating data input and output, code development and testing, DSS calibration and verification, and DSS documentation and training. However, Loucks (1995) has emphasized more on the technical aspects of development phases and an important phase namely, needs identification and analysis is missing, which is essential to make a DSS successful and adaptive. Reitsma et al. (1996) have suggested eight phases of DSS design and development including needs identification and analysis as illustrated in Fig. 7.

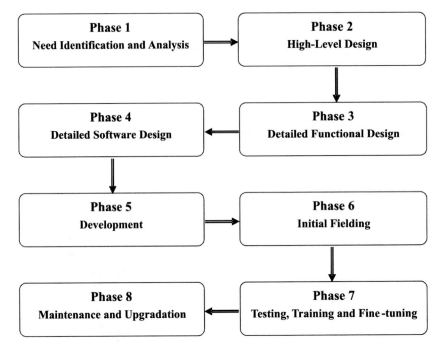

Fig. 7 Flowchart showing design and development phases of a DSS.

8.1 Phase 1: Needs Identification and Analysis

This phase is very critical for any DSS to be successful and having wide acceptance. The needs for and uses of the system have to be assessed and clearly defined. The specific aspects of water management to be addressed by the DSS must be defined i.e., a DSS for flood management cannot be used as water allocation DSS and vice versa. Also, the types of users have to be clearly identified and their needs for system functions need to be assessed.

The primary goal of a needs analysis is to identify the functions supported by the system. It is a systematic inventory and based on the analyzed results, a high-level system design focusing on the overall system architecture and its various components can take place. There are various techniques such as problem-centered, task-centered, user-centered, etc. for arriving at this initial inventory of needs and requirement (Lewis et al., 1991; Polsen et al., 1992; Lewis and Riemen, 1993). In the problem-centered design, the needs analysis is done by means of problem scenarios rather than through asking future users for a list of functional specifications. In this design, the needs analysis also yields as comprehensive a set of scenarios as possible. These scenarios are used in almost all subsequent phases of design and development processes for the reference and evaluation of system components, user interface, and overall capability of the system.

8.2 Phase 2: High-level Design

During this phase, two tasks need to be performed: (i) development of system architecture, and (ii) selection/design of models. DSS developers identify the database-management system, the hardware and software platforms for DSS implementation, and the software tools such as programming languages, programming tools, graphic libraries, etc., to be used in the design and development of a DSS. All required models, their availability, source codes, and compatibilities are also listed in this phase. It should also give an indication of the duration and costs of the project.

8.2.1 Development of System Architecture

In this part of the high-level design phase, decision about how to integrate different components of a DSS is taken, which ultimately defines the system functionality. The first and foremost decision to be made during this phase is the selection of appropriate system architecture. In an environmental DSS, there are four types of architecture depending on the integration of various functions: (i) Dedicated models, (ii) Dedicated DSS, (iii) Data-centered DSS, and (iv) GIS-based DSS. In recent years, web-based architecture has become very popular. It should be mentioned that the choice of architecture has profound consequences on the modality, long-term maintenance, extensibility, and performance of a DSS (Table 5).

Table 5. Evaluation criteria for DSS architectures (Reitsma et al., 1996)

Criteria	Description
Extensibility State	Ease with which architecture can accommodate an extension of its database or can be applied to another case such as another watershed, stream, or river basin.
Extensibility Process	Extensibility with respect to models. Concerns the effort needed to extend an existing system with new versions of already integrated models or addition of new models.
Maintainability	Operation and maintenance of a particular architecture. Systems mainly based on third-party software are easier to maintain than the custom-developed software.
Performance	Execution speed of a system.
Modularity of Implementation	Availability of intermediate or by-products during system development. Some architecture allow for parts of the system/functionality to become available during development, whereas others allow products to be brought on-line after complete development is finished.
Development Costs	Costs to implement the architecture.
Extensibility Costs	Costs to extend an existing system.

8.2.2 Selection or Design of Models

In this part of the high-level design phase, decisions on the models to be integrated in the system need to be made. This means the developer has to decide whether to integrate existing models or to develop new ones. This depends on the type of models that are available and their ownership, acceptability of the models, costs of modifying an existing model to fit the requirements of a particular DSS, and the development cost of new models. The criteria to be employed for the selection of models to be integrated in a DSS are: (i) representation, (ii) accuracy and flexibility, (iii) type of model, (iv) robustness, (v) propriety, and (iv) acceptability, which can also be considered as design criteria for the new models to be developed (Reitsma et al., 1996).

8.3 Phase 3: Detailed Functional Design

The detailed functional design describes in detail all components and all infrastructures of a DSS from a functional viewpoint. In this phase, a detailed description of all user interfaces is made wherein all generic system functions need to be specified in detail and mapped onto a set of user interfaces which support these functions in an efficient, consistent, and user-friendly manner. User interface design has two aspects: *control flow* and *dialog elements*. The control flow specifies the order in which user interface actions need to occur and thus order-dependencies between various interface elements. The user interface dialogs contain the actual interface components such as prompts (command line interfaces), forms (form-based interfaces), or graphical user interface (GUI) elements. Similar designs must be developed for the models, their data, their interconnections, the various visualization tools and all system utilities. An important component of the detailed functional design is a development schedule and an associated budget.

8.4 Phase 4: Detailed Software Design

The detailed software design takes the specifications from the detailed functional design and translates them into software representations (i.e., data structures, algorithms, program functions, function calls and parameters, and so on). It also outlines other complementary issues and conventions needed during the actual software development process. Examples of these are coding standards, in-line documentation standards, conventions for function naming, variable and function declarations, specification of libraries, etc. If databases are to be part of the DSS, then this phase also includes the specification of database schema or data model.

8.5 Phase 5: Software Development

During this phase, the software design and actual coding are performed. There are different groups of developers who follow different approaches in conducting the software design and the actual coding. Some groups choose to do a comprehensive software design before actual coding, while others follow a more liberal approach where coding of certain components can start before the software design is completed. Naturally, the systems determine to a large extent whether such a freedom is allowed or not. Generally, the more modular a system, the more freedom developers have to accept a liberal design and coding methodology. However, even the most modular architectures require a fair amount of software design before actual coding can begin. Rapid prototyping is a technique commonly used in software design and actual coding. It refers to a design and development method where the developers, after gaining some insight into the organization's needs for a DSS, quickly produce a first, initial version of a system (Budde and Bacon, 1992).

8.6 Phase 6: Initial Fielding

Once the software development is almost finished, the DSS should be subjected to an *alpha-testing* or an initial fielding phase. This encompasses to installing the DSS on the sponsor's machine and making sure that all the data and functional components as well as the connecting infrastructure work properly. Initially fielding implies that all the functions of a DSS are tested by the developers by running it on the sponsor's machine or network.

8.7 Phase 7: Testing, Training and Fine-tuning

When all the basic functions of the DSS are integrated and the initial fielding is completed, *beta-testing* and user training are conducted. Training sessions offer not only an opportunity to make the users familiar with the features and functionalities of the DSS, but also for software-testing purposes. Proper system design and appropriate coding styles, however, facilitate the fixing of the bugs and the realization of small changes. During this phase, many fixes and/or flaws/bugs in the DSS can be incorporated or corrected.

8.8 Phase 8: Maintenance and Further Development

Once a DSS is installed and starts functioning, both immediate and long-term system maintenance and version control policies such as version control strategies, database update and archival, new releases of supporting software, newly emerging hardware, and the extension of the DSS with new capabilities should be decided. Unfortunately, this consideration is often left out of the initial design phases of a DSS. The long-term system maintenance is an issue which should be dealt with during the early phases of DSS design and development.

9. APPLICATION DOMAINS OF DSS

Since 1970s, DSS research has mainly concentrated on financial (banking, insurance) and medical systems. In recent years, the following areas have been explored wherein the DSS technology is being used in decision-making (Kersten et al., 2002):

- Land and water management;
- Food production and distribution;
- Poverty alleviation;
- Primary health services;
- Public services and administration, governance;
- Education;
- Pollution control;
- Environmental management;
- Urban planning and management;
- Recovery from a natural disaster;
- Population growth control; and
- Economic planning.

For the details about DSS applications in other areas, interested readers are referred to journals such as *Interfaces, Decision Support System, Environmental Modeling and Software, Journal of Hydrology, Computers and Electronics in Agriculture*, etc., and trade publications like *Information Week* (http://www.informationweek.com). Many case studies on application of DSS are also available on World-Wide Web which can be accessed using search engines like 'Google'.

The decision support system (DSS) has been applied for water-related problems since mid-1970s (Loucks et al., 1985; Labadie and Sullivan, 1986; Loucks, 1995; Georgakakos and Martin, 1996; McKinney, 2003), but most of the DSSs have been developed for specific domains of water management, viz., reservoir operation, flood management, etc., and hence they have very limited applications. To date, very few DSSs have been developed that can be used for integrated water resources management; as such DSS requires the consideration of various social, economic and environmental aspects of resource use and protection.

There are several areas of water management problems wherein decisions need to be taken to meet increasingly competing water demands and protect our environment. Depending on the type of decisions, these areas can be categorized into two principal areas as follows (McKinney, 2004):

- *Emergency water management*: It involves floods, droughts, tsunamis, cyclones, earthquakes or chemical spills; and
- *Water regulation and allocation*: It involves water supply for municipalities, agriculture, industry, hydropower production, and environmental protection.

However, there are no DSSs which can handle a wide variety of water management problems involving various physical, chemical and biological processes that need to be handled in both temporal and spatial scales and require different data sets. Flooding problems often need models that use time steps of tens of minutes, whereas water allocation and groundwater models require time steps of one month (McKinney, 2004).

9.1 DSS for Emergency Water Management

Occurrence of natural disasters or accidents viz., floods, droughts, tsunamis, cyclones/hurricanes, earthquakes, chemical spills, etc. cause loss of life and bring considerable damage to property and infrastructures, which have cascading effect on the development of society and economy (UNDP, 2004). Despite advancements in IT, communication technologies and computer simulation techniques, we are still struggling to mitigate, prepare for and respond to catastrophic events. This is because most disasters are difficult to predict or even well predicted events (e.g., hurricanes) pose a wide range of unprecedented challenges. To address these challenges, DSSs or expert systems are being developed and used to collect, organize and interpret information in order to ensure effective emergency management. DSSs can also be used as early warning systems for forecasting natural disasters. Early warning systems are information systems designed to send automated data concerning disasters to decision makers, who combine them with meteorological data and simulation models to disseminate hazard forecasts and formulate strategies for reducing economic damage and loss of life.

Various local, national and international organizations are engaged in disaster management or emergency management operations. These operations are generally carried out in three stages: pre-event, crisis-event and post-event (Harrald, 2005) which involve the following processes (van der Walle and Turoff, 2008):

- Preparedness (analysis, planning, and evaluation),
- Training,
- Mitigation,
- Detection,
- Response, and
- Recovery/normalization.

DSSs or expert systems can help in coordination and collaboration, and to disseminate gathered or analyzed information among the organizations for successful and effective emergency management.

9.2 DSS for Water Regulation and Allocation

9.2.1 River Basin Management

In the area of general river basin management, DSSs can help decision makers/planners in tackling or solving numerous problems such as:

- Operation of reservoirs for the supply of water for various purposes including recreation, municipal and industrial water uses, in-stream flows, irrigation, and hydropower production;
- Examination of the effects of land-use and land-management policies on the quality of surface water and groundwater;
- Assessment of eutrophication in surface water bodies;
- Development of pollution control plans for river basins and estuaries, including hydrodynamic and water quality impacts of alternative control strategies;
- Design and operation of wastewater treatment plants, i.e., what level of treatment is necessary to meet water quality goals under specific flow conditions; and
- Management of river basins, including the evaluation of interrelationships between economic productivity and environmental degradation in a basin.

9.2.2 Lake and Reservoir Management

In the area of lake and reservoir management, support is required to make decisions for the operation of multiple reservoir systems, water supply, hydropower operation, pollution control, mitigation of climate change effects, etc. Depending on the purpose, different types of models are required to provide decision support such as water allocation models to determine the distribution of water for economic production and environmental protection in a basin; or two- and three-dimensional models to analyze water quality and surface water-groundwater interactions.

9.2.3 Conjunctive Use Management

DSSs have great potential for improving the planning and management of conjunctive use (surface water and groundwater) systems by integrating simulation and optimization models with graphic user interface capabilities to provide an adequate framework to resolve water allocation conflicts in a river basin or canal commands. Conjunctive use models and multi-objective decision methods can be combined to provide decision support for inter-basin water transfer planning (e.g., interlinking of rivers) and thereby allowing decision makers to analyze the social, economic, and environmental impacts of water transfers.

9.2.4 Pollution Management

Decision support is needed to make plans for agricultural chemical use or protection of vulnerable water bodies, streams and aquifers. Modeling and managing agricultural non-point source pollution typically requires the use of a distributed parameter watershed model. The data management and visualization capabilities are required to allow decision makers to identify and analyze problem areas easily. DSS can play an important role in managing the pollution of land and water resources.

9.2.5 Water Treatment and Distribution Systems

Design and operation of water treatment and distribution systems are also complex tasks in which the experience of the designer or operator is critical. The physical accuracy of models is generally sacrificed to get timely solutions of the problems related to water treatment and distribution systems. Trade-offs between model solvability and accuracy in the design of water supply and distribution systems, investment options, and interaction between water quantity and quality need to be evaluated. General network simulation and optimization models can be used in scheduling and operating water distribution systems in urban areas to determine proper structural changes to the system that minimize disruption in water supply. Recently, evolutionary methods have been used to solve realistic models of large urban water distribution systems which are difficult to manage with more traditional methods.

10. AVAILABLE DECISION SUPPORT SYSTEMS FOR WATER MANAGEMENT: AN OVERVIEW

Several DSSs have been developed to date and applied for different aspects of water management all over the world (Table 6). However, in spite of recent rapid advances in computer technology and the proliferation of computer software for decision support, there are few DSSs available that can help to solve water problems covering more than one areas of water management (McKinney, 2004) and very few DSSs such as mDSS (Giupponi et al., 2004), OPTIMA (Fedra et al., 2007) and NRWS (Olsen, 2005) have the capability to include the concept of integrated water resources management. TERRA (Reitsma et al., 1994), AQUATOOL (Andreu et al., 1996), WaterWare (Jamison and Fedra, 1996), **ToDSS** (Arumugam and Mohan, 1997), MODSIM DSS (Labadie et al., 2000), RiverWare (Zagona et al., 2001), and CTIWM (Ito et al., 2001) are the DSSs meant for reservoir operation and river basin planning and management. A succinct description of some of the DSSs is presented below.

AQUATOOL (Andreu et al., 1996) is a generalized DSS developed at the Universidad Politecnica de Valencia (UPV), Valencia, Spain. Initially, the DSS was designed for the planning stage of decision-making in complex basins comprising multiple reservoirs, aquifers and demand centers. Later, the system was expanded to incorporate modules for the operational management stage of decision-making. It is used by several River Basin agencies in Spain to develop the basin hydrological plans and to effectively manage water resources on a short-term or medium-term basis. The DSS consists of a simulation module, an optimization module, an aquifer flow modeling module, two modules for risk assessment, six modules for the analysis and reporting of results, and a general utility module.

mDSS (Giupponi et al., 2004) is a prototype DSS developed under MULINO (**Mu**lti-sectoral **In**tegrated and **O**perational) project financed by the European Commission. It is used for the sustainable management of water resources at a catchment scale according to the EU Water Framework Directive. The software integrates socio-economic and environmental modeling, along with geo-spatial information and multi-criteria techniques. The integrated assessment modeling provides the values of quantitative indicators to be used for transparent and participated decisions, through the application of value functions, weights and decision rules chosen by the end user. Simple routines for the sensitivity analysis and comparison of alternative weight vectors provides effective decision support by exploring and finding compromises between conflicting interests/perspectives in a multi-stakeholder context.

Table 6. Summary of available DSSs for water resources management

Name of DSS	Tools/Techniques used	Application
1. RiverWare (Zagona et al., 2001)	Simulation and optimization models	It is used for reservoir and river systems operation and planning and is capable of modeling short-term (hourly to daily) operations and scheduling, mid-term (weekly) operations and planning, and long-term (monthly or annually) policy and planning.
2. WaterWare (Jamison and Fedra, 1996)	Simulation and optimization models, GIS and Expert System	It is used for water-resource assessment, reservoir site selection, decontamination of groundwater, estimation of sustainable irrigation abstractions and derivation of required effluent-quality standards in a river basin.

Contd...

(*Contd.*)

3.	MODSIM (Labadie et al., 2000)	Simulation and optimization models, and GIS	It is a generic river basin management DSS for the analysis of long-term planning, medium-term management, and short-term operations and planning. It can also be used for water quality issues in stream-aquifer systems.
4.	CALSIM (DWR, 2004)	Simulation and optimization models	It is used to simulate and optimize existing and potential water allocation and reservoir operating policies and constraints.
5.	RIBASIM (Delft Hydraulics, 2006)	Simulation model	It generates water distribution patterns and provides a basis for detailed water quality and sedimentation analyses in river reaches and reservoirs.
6.	Mike-Basin (DHI, 2004)	Simulation and optimization models, GIS	It is used for hydrological simulation, routing river flows, water quality simulation, diversions and extractions from multiple reservoirs based on priority and water allocation.
7.	WEAP (SEI, 2005)	Simulation and optimization models, GIS	It is used for integrated water resources planning and management through sectoral demand analyses, water conservation, water rights and allocation priorities, streamflow simulation, reservoir operation, ecosystem requirements, and project cost-benefit analyses.
8.	Nile DST (Georgakakos, 2006)	Simulation and optimization models, GIS and RS	It is used to assess the benefits and trade-offs associated with various water development and sharing strategies. It can be used for reservoir flow regulation for irrigation projects, hydropower generation and wetland protection.
9.	CRSS (Schuster, 1987)	Simulation model	It is used to schedule, forecast and plan reservoir operations for water allocation.
10.	AQUARIUS (Diaz et al., 1997)	Simulation and optimization models	It is used to simulate the allocation of water using any time interval, including days, weeks, months, and time intervals of nonuniform lengths and to identify tradeoffs between water uses by examining the feasibility of reallocating water to alternative uses.
11.	WSM (Manoli et al., 2001)	Simulation and optimization models, and GIS	It is used for water demand calculation and water allocation. Alternative scenarios can be generated and trends and interactions of a complex water system can be analyzed to suggest strategies to solve water allocation conflicts and to plan necessary infrastructure interventions in order to meet water needs.
12.	BASINS (USEPA, 2004)	Simulation model and GIS	It used to simulate and analyze water quality by integrating environmental data, analytical tools, and modeling tools to support the development of cost-effective approaches to environmental protection.
13.	Elbe-DSS (Matthies et al., 2006)	Simulation models, GIS and artificial intelligence (AI)	It is used for the estimation of water quantity, chemical quality and ecological status of surface water for integrated river basin management.
14.	GLOWA Volta DSS (Rodgers et al., 2007)	Simulation and optimization models, GIS and AI	Knowledge-based decision support system for the planning, management and use of water resources in the Volta river basin, West Africa.

Olsen (2005) developed a DSS called NRWS (**N**ilgiris **R**ural **W**ater **S**upply) for identifying key issues in selecting sustainable water sources, and systematically guides the user through various methodologies to quantify potential water sources. The shell of NRWS has been developed through Microsoft® Excel using Visual Basic for applications programming language. A user-friendly interface directs the user through the program functions by a network of links and forms. The NRWS DSS consists of six modules that represent different criteria used to evaluate potential water sources: (i) water source yield, (ii) capital costs, (iii) cost and ease of operation and maintenance, (iv) impact of development, (v) political and legal constraints, and (vi) water quality. The criteria are organized in a decision matrix that provides a total score and rank for each potential water source. There are many different sources that can be used to supply water for domestic use, but only five water sources (rooftop rainwater harvesting, check dams, reservoirs, springs, and dug wells) sources considered in NRWS due to their popularity in Nilgiris District.

11. CONCLUSIONS

In view of freshwater crisis (both physical and economic), IWRM (Integrated Water Resources Management) has emerged as a new management paradigm which promotes a holistic approach to the planning and management of water resources, and thereby ensures sustainability of water resources, long-term benefits and environmental protection. However, there are many constraints such as lack of quantitative information about water use and a wide range of institutional and other socio-economic issues which create barriers to the effective application of IWRM in practice. In recent years, however, with the development of advance computer models and DSS, IWRM is getting popularity and its wider application is being advocated in various forums to combat freshwater crisis. The full implementation of IWRM in practice requires international cooperation, transfer of technology and expertise, the adaptation of technologies to local situations and requirements. Financial resource should also be made available for promoting institutional reform and building capacity as well as for acquiring more data and knowledge of the systems to be managed.

Despite the considerable technical progress in DSS development and design, which has made it an important tool for IWRM implementation, there is limited use of DSS for real-world decision making processes which can be partly due to lack of awareness and/or reluctance among the concerned agencies and policy makers in using modern/emerging tools; they rely more on traditional modes of decision making. Although a couple of DSSs have been developed for water resources management, the need to further develop decision-support tools in this field is worldwide recognized because of location-specific problems and the limitations of existing DSSs. There is also an urgent need to develop DSSs which can help in assessing vulnerability to natural disasters as well as in developing preparedness and mitigation plans so as ensure security against disasters in the future.

Integration of advanced computer models with modern tools/techniques such as data warehouse, OLAP, data mining tools, artificial intelligence, GIS, ANN, fuzzy logic, GA, etc. is expected to expand the functionality of a DSS and make it more powerful in the future for analyzing risks and uncertainty in decision-making processes, which is a serious concern in the 21st century. Integration of expert systems (ES) and decision support systems (DSS) with GIS has not been much applied in water resources engineering. This is an area of potential research to aid the effective and timely decision making concerning planning, design, operation, and maintenance of land and water resources systems which in turn can greatly reduce time and efforts required in traditional approaches. Such a decision support system can automate the process of solving water problems on local, regional or national scales as well as can help in selecting cost-effective and efficient management alternatives. Much of real-world decision making can also be automated through the use of an expert GIS.

Overall it can be concluded that DSS development and its implementation should be encouraged to improve water governance and to overcome the existing obstacles to IWRM implementation. Besides DSS design and development, there is a need to formulate '*Best Practice Guidelines (BPG)*' for policy/decision makers highlighting the potential of DSS to improve land and water management as well as the critical knowledge gaps that still exist in land and water resources management and DSS design. These guidelines can ensure effective DSS implementation for solving real-world water problems.

REFERENCES

Adelman, L. (1992). Evaluating Decision Support and Expert Systems. John Wiley and Sons, New York.

Aerts, J.C.J.H., Herwijnen, M., Janssen, R. and Stewart, T.J. (2005). Evaluating spatial design techniques for solving land-use allocation problems. *Journal of Environmental Planning and Management*, **48(1)**: 121-142.

Alavi, M. and Keen, P.G.W. (1989). Business teams in an information age. *The Information Society*, **6(4)**: 179-195.

Amarasinghe, U.A., Mutuwatta, L. and Sakthivadivel, R. (1999). Water Scarcity Variations within a Country: A Case Study of Sri Lanka. Research Report 32, International Water Management Institute (IWMI), Colombo, Sri Lanka.

Andreu, J., Capilla, J. and Sanchis, E. (1996). Aquatool: A generalized decision support system for water resources planning and operational management. *Journal of Hydrology*, **177(3-4)**: 269-291.

Anthony, R.N. (1965). Planning and Control Systems: A Framework for Analysis. Harvard University Press, Cambridge, MA.

Arumugam, N. and Mohan, S. (1997). Integrated decision support system for tank irrigation system operation. *Journal of Water Resources Planning and Management, ASCE*, **123(5)**: 266-273.

ASCE Task Committee (2000). Application of artificial neural networks in hydrology. I: Preliminary concepts. *Journal of Hydrologic Engineering, ASCE*, **5(2)**: 115-123.

Ayibotele, N.B. (1992). The world's water: Assessing the resource. Proceedings of the International Conference on Water and the Environment: Development Issues for the 21st Century, 26-31 January 1992, Dublin, Ireland, pp. 1.1-1.25.

Bender, M.J. and Simonovic, S.P. (2000). A fuzzy compromise approach to water resource systems planning under uncertainty. *Fuzzy Sets and Systems*, **115**: 35-44.

Bennett, J.L. (1983). Building Decision Support Systems. Addison-Wesley Publishing Company, Boston.

Bonczek, R.H., Holsapple, C.W. and Whinston, A.B. (1980). Evolving roles of models in decision support systems. *Decision Sciences*, **11(2)**: 337-356.

Bonczek, R.H., Holsapple, C.W. and Whinston, A.B. (1981). Foundations of Decision Support Systems. Academic Press, New York.

Budde, R. and Bacon, P. (1992). Prototyping: An Approach to Evolutionary System Development. Springer-Verlag, Berlin.

Bui, T. and Jarke, M. (1986). Communications requirements for group decision support systems. Proceedings of the 19th Hawaii International Conference on Systems Science, January 1986, Honolulu, Hawaii, pp. 524-533.

Cai, X., Lasdon, L. and Michelsen, A.M. (2004). Group decision making in water resources planning using multiple objective analysis. *Journal of Water Resources Planning and Management, ASCE*, **130(1)**: 4-14.

CFWA (1997). Comprehensive Assessment of the Freshwater Resources of the World. World Meteorological Organization (WMO), Geneva, Switzerland.

Chang, F.-J., Chen, L. and Chang, L.-C. (2005). Optimizing the reservoir operating rule curves by genetic algorithms. *Hydrological Processes*, **19**: 2277-2289.

Chang, K.T. (2006). Introduction to Geographic Information Systems. 3rd Edition, Tata McGraw-Hill Publishing, New Delhi.

Chang, Y.-M., Chang, L.-C. and Chang, F.-J. (2004). Comparison of static-feedforward and dynamic-feedback neural networks for rainfall-runoff modelling. *Journal of Hydrology*, **290(3-4)**: 297-311.

Chaves, P. and Chang, F.-J. (2008). Intelligent reservoir operation system based on evolving artificial neural networks. *Advances in Water Resources*, **31**: 926-936.

Chaves, P., Tsukatani, T. and Kojiri, T. (2004). Operation of storage reservoir for water quality by using optimization and artificial intelligence techniques. *Mathematics and Computers in Simulation*, **67(4-5)**: 419-432.

Cheng, C.-T., Wang, W.-C., Xu, D.-M. and Chau, K.W. (2008). Optimizing hydropower reservoir operation using hybrid genetic algorithm and chaos. *Water Resources Management*, **22**: 895-909.

Courtney, F. and Paradice, D.B. (1993). Studies in managerial problem formulation systems. *Decision Support Systems*, **9**: 413-423.

CWC (2000). Assessment of Availability and Requirement of Water for Diverse Uses in India. Standing Sub-Committee Report, Central Water Commission, Government of India, New Delhi.

Daliakopoulos, I.N., Coulibaly, P. and Tsanis, I.K. (2005). Groundwater level forecasting using artificial neural networks. *Journal of Hydrology*, **309(1-4)**: 229-240.

Delft Hydraulics (2006). Software: RIBASIM (River Basin Planning and Management). http://www.wldelft.nl/soft/ribasim/ int/index.html (accessed in October 2006).

DeSanctis, G. and Gallupe, B. (1985). Group decision support systems: A new frontier. *Data Base*, **(Winter 1985)**: 3-10.

DeSanctis, G. and Gallupe, B. (1987). A foundation for the study of group decision support systems. *Management Science*, **33(12)**: 1589-1609.

DHI (2004). Mike-Basin Description. Danish Hydraulics Institute (DHI), http://www.dhisoftware.com/mikebasin/Description/ (accessed in April 2004).

Diaz, G., Brown, T. and Stevens, O. (1997). Aquarius: A Modeling System for River Basin Water Allocation. General Technical Report RM-GTR-299, U.S. Department of Agriculture, Forest Service, Rocky Mountain Forest and Range Experiment Station, Fort Collins, CO., http://www.fs.fed.us/rm/value/aquariusdwnld.html (accessed in April 2004).

Duda, A.M. (2003). Integrated management of land and water resources based on collective approach to fragmented international conventions. *Philosophical Transactions, Royal Society, London*, **358**: 2051-2062.

DWR (2004). CALSIM Water Resources Simulation Model. Department of Water Resources (DWR), Bay-Delta Office, Sacramento, CA, http://modeling.water.ca.gov/hydro/model/index.html (accessed in April 2004).

Edelstein, H. (1996). Mining data warehouses. *Information Week*, January 8, 1996, http://www.informationweek.com/561/61oldat.htm (accessed in September 2007).

Engelman, R. and LeRoy, P. (1995). Sustaining Water. An Update. Population Action International, Washington, DC.

Eom, S.B. (1999). Decision support systems research: Current state and trends. *Industrial Management and Data Systems*, **99(5)**: 213-220.

Falkenmark, M. (2007a). Global warming: Water the main mediator. *Water Front*, Stockholm International Water Institute (SIWI), **No. 2**, pp. 6-7.

Falkenmark, M. (2007b). Shift in thinking to address the 21st century hunger gap: Moving focus from blue to green water management. *Water Resources Management*, **21**: 3-18.

Falkenmark, M. and Rockstrom, J. (2004). Balancing Water for Humans and Nature: The New Approach in Ecohydrology. Earthscan, London.

Fassio, A., Giupponi, C., Hiederer, R. and Simota, C. (2005). A decision support tool for simulating the effects of alternative policies affecting water resources: An application at the European scale. *Journal of Hydrology*, **304**: 462-476.

Fedra, K. (1993). Expert Systems in Water Resources Simulation and Optimization. *In:* J.B. Marco et al. (editors), Stochastic Hydrology and Its Use in Water Resources Systems Simulation and Optimization. Kluwer Academic Publishers, Dordrecht, The Netherlands, pp. 397-412.

Fedra, K., Kubat, M. and Aloise, M.Z. (2007). Water resources management: Economic valuation and participatory multi-criteria optimization. Proceedings of the 2nd IASTED International Conference on Water Resources Management, 20-22 August 2007, Honolulu, Hawaii.

Garg, N.K. and Hassan, Q. (2007). Alarming scarcity of water in India. *Current Science,* **93:** 932-941.

Gemitzi, A., Tsihrintzis, V.A., Vondrias, E., Petalas, C. and Stravodimos, G. (2007). Combining geographic information system, multicriteria evaluation techniques and fuzzy logic in siting MSW landfills. *Environmental Geology,* **51:** 797-811.

Geneletti, D. and van Duren, I. (2008). Protected area zoning for conservation and use: A combination of spatial multicriteria and multiobjective evaluation. *Landscape and Urban Planning,* **85:** 97-110.

Georgakakos, A. (2006). Decision Support Systems for Integrated Water Resources Management with an Application to the Nile Basin. *In:* A. Castelletti and R. Soncini-Sessa (editors), Topics on System Analysis and Integrated Water Resources Management. Elsevier, Amsterdam, The Netherlands.

Georgakakos, A. and Martin, Q. (1996). An international review of decision support systems in river basin operation. Proceedings of the 5th Water Resources Operations Management Workshop, ASCE, March 1996, Arlington, VA.

Giupponi, C., Mysiak, J., Fassio, A. and Cogan, V. (2004). MULINO-DSS: A computer tool for sustainable use of water resources at the catchment scale. *Mathematics and Computers in Simulation,* **64:** 13-24.

Giupponi, C., Mysiak, J., Fassio, A. and Cogan, V. (2002). Towards a spatial decision support system for water resource management - MULINO-DSS 1st release. Proceedings of 5th AGILE Conference on Geographic Information Science, 25-27 April 2002, Palma, Spain, pp. 1-6.

Gleick, P.H. (1996). Global water resources in the 21st century: Where should we go and how should we get there? Proceedings of 5th Stockholm Water Symposium on Water Quality Management: Heading for a New Epoch, 13-18 August 1995, Stockholm, Publication No. 5, pp. 73-78.

Goicoechea, A., Stakhiv, E.Z. and Li, F. (1992). Experimental evaluation of multiple criteria decision models for application to water resources planning. *Water Resources Bulletin,* **28:** 89-102.

Goldberg, D. E. (1989). Genetic Algorithms in Search, Optimization and Machine Learning. Addison-Wesley, Reading, Mass.

Goodchild, M.F. (2000). The current status of GIS and spatial analysis. *Journal of Geographical Systems,* **2(1):** 5-10.

Gorry, G.A. and Scott Morton, M.S. (1971). A framework for management information systems. *Sloan Management Review,* **13(1):** 55-70.

Gupta, D.B. (1992). The importance of water resources for urban socio-economic development. Proceedings of the International Conference on Water and the Environment: Development Issues for the 21st Century, 26-31 January 1992, Dublin, Ireland, pp. 5.1-5.19.

GWP (2000). Integrated Water Resources Management. TAC (Technical Advisory Committee) Background Papers No. 4, Global Water Partnership (GWP), Stockholm.

GWP (2004). Integrated Water Resources Management and Water Efficiency Plans by 2005. TEC (Technical Committee) Background Papers No. 10, Global Water Partnership (GWP), Stockholm.

Haddadin, M.J. (2001). Water scarcity impacts and potential conflicts in the MENA region. *Water International,* **26(4):** 460-470.

Hajkowicz, S. and Collins, K. (2007). A review of multiple criteria analysis for water resource planning and management. *Water Resources Management,* **21:** 1553-1566.

Hajkowicz, S. and Higgins, A. (2008). A comparison of multiple criteria analysis techniques for water resource management. *European Journal of Operational Research,* **184:** 255-265.

Harrald, J. (2005). Supporting agility and discipline when preparing and responding to extreme events. Proceedings of the 2nd International ISCRAM, April 2005, Brussels, Belgium.

Harris, R. and Baveye, P.C. (2008). Water on the table: Sigma Xi's Year of Water affords unique opportunities to share hydrological information. *Journal of Hydrology*, **354**: v-vii.

Hermans, C., Erickson, J., Noordewier, T., Sheldon, A. and Kline, M. (2007). Collaborative environmental planning in river management: An application of multicriteria decision analysis in the White River watershed in Vermont. *Journal of Environmental Management*, **84**: 534-546.

Horlitz, T. (2007). The role of model interfaces for participation in water management. *Water Resources Management*, **21(7)**: 1091-1102.

Huang, W.-C., Yuan, L.-C. and Lee, C.-M. (2002). Linking genetic algorithm with stochastic dynamic programming to the long-term operation of multireservoir system. *Water Resources Research*, **38(12)**: 1304, doi:10.1029/2001WR001122.

Inmon, W.H. (1992). Building the Data Warehouse. QED Information Sciences, Wellesley, MA.

Ito, K., Xu, Z.X., Jinno, K., Kojiri, T. and Kawamura, A. (2001). Decision support system for surface water planning in river basins. *Journal of Water Resources Planning and Management, ASCE*, **127(4)**: 272-276.

Iyer, R.R. (2004). A critical view in integrated water resources management: IWRM carries the seeds of centralization and gigantism. *Water Front*, Stockholm International Water Institute (SIWI), **No. 4**, pp. 10-11.

Jamison, D. and Fedra, K. (1996). The WaterWare decision support system for river basin planning. 1. Conceptual design. *Journal of Hydrology*, **177(3-4)**: 163-175.

Jha, M.K., Chowdhury, A., Chowdary, V.M. and Peiffer, S. (2007). Groundwater management and development by integrated remote sensing and geographic information systems: Prospects and constraints. *Water Resources Management*, **21(2)**: 427-467.

Johansen, R. (1988). Groupware: Computer Support for Business Teams. The Free Press, New York.

Karamouz, M., Zahraie, B. and Kerachian, R. (2003). Development of a master plan for water pollution control using MCDM techniques: a case study. *Water International*, **28(4)**: 478-490.

Karbowski, A., Malinowski, K. and Niewiadomska-Szynkiewicz, E. (2005). A hybrid analytic/rule-based approach to reservoir system management during flood. *Decision Support Systems*, **38**: 599-610.

Keen, P.G.W. and Scott Morton, M.S. (1978). Decision Support Systems: An Organizational Perspective. Addison-Wesley Publishing, Reading, MA.

Kersten, G.E., Mikolajuk, Z. and Yeh, A.G. (2002). Decision Support System for Sustainable Development: A Resource Book of Methods and Applications. Kluwer Academic Publishers, Dordrecht, The Netherlands.

Kumar, A.R.S., Sudheer, K.P., Jain, S.K. and Agarwal, P.K. (2005a). Rainfall-runoff modeling using artificial neural networks: Comparison of network types. *Hydrological Processes*, **19(6)**: 1277-1291.

Kumar, R., Singh, R.D. and Sharma, K.D. (2005b). Water resources of India. *Current Science*, **89(5)**: 794-811.

Kurosh, M. (editor) (2005). Artificial neural networks and intelligent information processing. Proceedings of the 1st International Workshop on Artificial Neural Networks and Intelligent Information Processing (ANNIIP), September 2005, Barcelona, Spain.

Labadie, J.W. and Sullivan, C. (1986). Computerized decision support systems for water managers. *Journal of Water Resources Planning and Management, ASCE*, **112(3)**: 299-307.

Labadie, J.W., Baldo, M.L. and Larson, R. (2000). ModSim: Decision Support System for River Basin Management, Documentation and User Manual. http://modsim.engr.colostate.edu (accessed in April 2004).

Larichev, O.I. and Moshkovich, H.M. (1995). ZAPROS-LM: A method and system for ordering multiattribute alternatives. *European Journal of Operational Research*, **82**: 503-521.

Lee, K.S. and Chung, E.-S. (2007). Development of integrated watershed management schemes for intensively urbanized region in Korea. *Journal of Hydro-Environmental Research*, **1(2)**: 95-109.

Lenton, R. (2004). A critical view in integrated water resources management: IWRM integration needs broad interpretation. *Water Front*, Stockholm International Water Institute (SIWI), **No. 4**, pp. 10-11.

Lewis, C.H. and Riemen, J. (1993). Task-centered user interface design: A practical introduction. http://uow.ico5.janison.com/ed/subjects/EDGI957/resources/HCIComplete.pdf (accessed in August 2008).

Lewis, C.H., Riemen, J. and Bell, B. (1991). Problem-centered design for expressiveness and facility in a graphical programming system. *Human Computer Interaction*, **6**: 319-355.

Lin, C.-L. and Su, H.-W. (2000). Intelligent control theory in guidance and control system design: An overview. Proceedings of the National Science Council, Republic of China, Part A: *Physical Science and Engineering*, **24(1)**: 15-30.

Lin, G.-F. and Chen, G.-R. (2006). An improved neural network approach to the determination of aquifer parameters. *Journal of Hydrology*, **316(1-4)**: 281-289.

Little, J.D.C. (1970). Models and managers: The concept of a decision calculus. *Management Science*, **16(8)**: 35-43.

Lo, C.P. and Yeung, A.K.W. (2003). Concepts and Techniques of Geographic Information Systems. Prentice-Hall of India, New Delhi.

Loucks, D.P. (1995). Developing and implementing decision support systems: A critique and a challenge. *Water Resources Bulletin*, **31(4)**: 571-582.

Loucks, D.P. and van Beek, E. (2005). Water Resources Systems Planning and Management: An Introduction to Methods, Models and Applications. Studies and Reports in Hydrology, UNESCO Publishing, UNESCO, Paris, 680 pp.

Loucks, D.P., Kindler, J. and Fedra, K. (1985). Interactive water resources modeling and model use: An overview. *Water Resources Research*, **21(2)**: 95-102.

Loucks, D.P., Stedinger, J.R. and Douglas, A.H. (1981). Water Resource Systems Planning and Analysis. Prentice-Hall, Inc., Engelwood Cliffs, New Jersey, 559 pp.

Maidment, D.R. (editor-in-chief) (1993). Handbook of Hydrology. McGraw-Hill, Inc., 1400 pp.

Manoli, E., Arampatzis, G., Pissias, E., Xenos, D. and Assimacopoulos, D. (2001). Water demand and supply analysis using a spatial decision support system. *Global Nest*, **3(3)**: 199-209.

Massam, B.H. (1988). Multicriteria decision making techniques in planning. *Programme Planning*, **30**: 1-84.

Matthies, M., Berlekamp, J., Lautenbach, S., Graf, N. and Reimer, S. (2006). System analysis of water quality management for the Elbe river basin. *Environmental Modelling and Software*, **21**: 1309-1318.

McCartney, M.P. (2007). Decision Support Systems for Large Dam Planning and Operation in Africa. IWMI Working Paper 119, International Water Management Institute (IWMI), Colombo, Sri Lanka, 47 pp.

McKinney, D.C. (2003). Basin scale integrated water resources management in Central Asia. Third World Water Forum on Regional Cooperation in Shared Water Resources in Central Asia, 18 March 2003, Kyoto, Japan, http://www.adb.org/Documents/Events/2003/3WWF/RC_McKinney.pdf (accessed in April 2004).

McKinney, D.C. (2004). International Survey of Decision Support Systems for Integrated Water Management. Support to Enhance Privatization, Investment, and Competitiveness in the Water Sector of the Romanian Economy (SEPIC), Technical Report, Bucharest, 67 pp.

McKinney, D.C., Cai, X., Rosegrant, M.W., Ringler, C. and Scott, C.A. (1999). Modeling Water Resources Management at Basin Level: Review and Future Directions. System-Wide Initiative on Water Management (SWIM), SWIM Paper 6, International Water Management Institute (IWMI), Colombo, Sri Lanka.

Mitchell, B. (1990). Integrated Water Management. *In:* B. Mitchell (editor), Integrated Water Management: International Experiences and Perspectives. Belhaven Press, London, U.K.

Molden, D. (editor) (2007). Summary of Water for Food, Water for Life: A Comprehensive Assessment of Water Management in Agriculture. Earthscan, London, U.K., 40 pp.

Moore, J.H. and Chang, M.G. (1980). Design of decision support systems. *Data Base*, **12(1-2)**: 8-14.

Orlob, G. (1992). Water quality modeling for decision making. *Journal of Water Resources Planning and Management, ASCE*, **118(3)**: 295-307.

Olsen, D.R. (2005). Decision Support System for Rural Water Supply in the Nilgiris District of South India. Unpublished Master Thesis (Civil Engineering), McMaster University, Hamilton.

Olsson, J., Uvo, C.B., Jinno, K., Kawamura, A., Nishiyama, K., Koreeda, N., Nakashima, T. and Morita, O. (2004). Neural networks for rainfall forecasting by atmospheric downscaling. *Journal of Hydrologic Engineering*, **9(1)**: 1-12.

Poch, M., Comas, J., Rodríguez-Roda, I., Sànchez-Marrè, M. and Cortés, U. (2003). Designing and building real environmental decision support systems. *Environmental Modelling and Software*, **19(9)**: 857-873.

Polsen, P.G., Lewis, C.H., Riemen, J. and Wharton, C. (1992). Cognitive walkthroughs: A method theory-based evaluation of user interfaces. *International Journal of Man-Machine Studies*, **36(5)**: 741-773.

Power, D.J. (1997). DSS Glossary. http://www.dssresources.com/glossary/olaptrms.html (accessed in September 2007).

Power, D.J. (2002). Decision Support Systems: Concepts and Resources for Managers. Greenwood Publishing Group, Westport, CT, 251 pp.

Power, D.J. (2007). A Brief History of Decision Support Systems, version 4.0. http://www.DSSResources.com/history/dsshistory.htm (accessed in September 2007).

Qureshi, M.E. and Harrison, S.R. (2001). A decision support process to compare riparian revegetation options in Scheu Creek catchment in north Queensland. *Journal of Environmental Management*, **62**: 101-112.

Regan, G.O. (2008). A Brief History of Computing. Springer London Limited, U.K., 243 pp.

Reitsma, R.F., Ostrowski, P. and Wehrend, S.C. (1994). Geographically distributed decision support: The Tennessee valley authority (TVA) TERRA system. Proceedings of the 21st Annual Conference on Water Policy and Management: Solving the Problems, ASCE, 23-26 May 1994, Denver, Colorado, pp. 311-314.

Reitsma, R.F., Zagona, E.A., Chapra, S.C. and Strzepek, K.M. (1996). Decision Support Systems for Water Resources Management. *In:* L.W. Mays (editor), Water Resources Handbook. McGraw-Hill, New York, pp. 33.1-33.35.

Rodriguez, D. and Molden, D. (2007). Water for food, water for life: Influencing what happens next. *Water Front*, Stockholm International Water Institute (SIWI), **No. 2**, pp. 12-13.

Rodgers, C., van de Giesen, N., Laube, W., Vlek, P.L.G. and Youkhana, E. (2007). The GLOWA Volta project: A framework for water resources decision making and scientific capacity building in a transnational West African basin. *Water Resources Management*, **21**: 295-313.

Roy, B. and Vincke, P. (1981). Multicriteria analysis: Survey and new directions. *European Journal of Operational Research*, **8(3)**: 207-218.

Sauter, V. (1997). Decision Support Systems: An Applied Managerial Approach. John Wiley and Sons, New York.

Schuster, R. (1987). Colorado River Simulation System: System Overview. U.S. Department of Interior Bureau of Reclamation, Denver, Colorado.

Seckler, D., Amarasinghe, U.A., Molden, D., de Silva, R. and Barker, R. (1998). World Water Demand and Supply, 1990 to 2025: Scenarios and Issues. Research Report 19, International Water Management Institute (IWMI), Colombo, Sri Lanka.

SEI (2005). WEAP: Water Evaluation and Planning System-User Guide for WEAP21. Stockholm Environment Institute (SEI), Boston.

Shiklomanov, I.A. (1997). Comprehensive Assessment of the Freshwater Resources of the World: Assessment of Water Resources and Water Availability in the World. Comprehensive Assessment of the Freshwater Resources (CFWA), World Meteorological Organization (WMO), Geneva, Switzerland, 88 pp.

Sikder, I. and Gangopadhyay, A. (2002). Design and implementation of a web-based collaborative spatial decision support system: Organizational and managerial implications. *Information Resources Management Journal*, **15(4)**: 33-47.

Simon, H.A. (1960). The New Science of Management Decision. Harper Brothers, New York.

Sprague, R.H. and Carlson, E.D. (1982). Building Effective Decision Support System. Prentice-Hall, Inc., Engelwood Cliffs, New Jersey.

Tsihrintzis, V.A., Hamid, R. and Fuentes, H.R. (1996). Use of geographic information systems (GIS) in water resources: A review. *Water Resources Management*, **10(4)**: 251-277.

UN Water (2007). Coping with Water Scarcity: Challenge of the Twenty-First Century. Report for World Water Day 2007, http://www.unwater.org/wwd07/downloads/documents/escarcity.pdf (accessed on 23 March 2007).

USEPA (2004). Better Assessment Science Integrating Point and Nonpoint Sources (BASINS). http://www.epa.gov/docs/ostwater/BASINS/ (accessed in April 2004).

Van Cauwenbergh, N., Ointe, D., Tilmant, A., Frances, I., Pulido-Bosch, A. and Vanclooster, M. (2008). Multiobjective multiple participant decision support for water management in Andarax catchment, Almeria. *Environmental Geology*, **54**: 479-489.

Van de Walle, B. and Turoff, M. (2008). Decision support for emergency situations. *Information Systems and E-Business Management*, **6**: 295-316.

Warkentin, M.E., Sayeed, L. and Hightower, R. (1997). Virtual teams versus face-to-face teams: An exploratory study of a web-based conference system. *Decision Sciences*, **28(4)**: 975-996.

Watkins, D. Jr. and McKinney, D.C. (1995). Recent Developments in decision support systems for water resources. U.S. national contributions in hydrology 1991-1994. Reviews of Geophysics, **Supplement**, pp. 941-948.

WCED (1987). Our Common Future. World Commission on Environment and Development (WCED), Oxford University Press, Oxford, UK.

Westphal, K.S., Vogel, R.M., Kirshen, P. and Chapra, S.C. (2003). Decision support system for adaptive water supply management. *Journal of Water Resources Planning and Management, ASCE*, **129(3)**: 165-177.

Willis, R. and Yeh, W.W-G. (1987). Groundwater Systems Planning and Management. Prentice-Hall, Inc., Engelwood Cliffs, New Jersey, 415 pp.

Yurdusev, M.A. and O'Connel, P.E. (2005). Environmentally-sensitive water resources planning: 1. Methodology. *Water Resources Management*, **19**: 375-397.

Zagona, E.A., Fulp, T.J., Shane, R., Magee, T. and Goranflo, H.M. (2001). RiverWare; A generalized tool for complex reservoir system modeling. *Journal of the American Water Resources Association*, **37(4)**: 913-929.

Sustainable Forest Management: Key to Disaster Preparedness and Mitigation

Tajbar S. Rawat*, D. Dugaya, B.K. Prasad and Savita Bisht

1. INTRODUCTION

Sustainable forest management has been considered as an integral component of sustainable development since the UNCED Conference at Rio de Janeiro in 1992, also called the Earth Summit (UN, 1992). After the summit, where international 'Forest Principles' were formulated for the first time by world leaders and the first global policy on sustainable forest management was adopted, the notion of sustainable forest management rapidly gained interest. Sustainable management of natural and managed forests has been a worldwide issue after the summit. Accordingly, the forest resources and lands should be managed sustainably to meet the social, economic, ecological, cultural and spiritual functions and for the maintenance and enhancement of biological diversity (FAO, 1995).

The movement got wide acceptance in later half of nineteen century. Among the first international recognition addressing growing concern "about the accelerating deterioration of the human environment and natural resources and the consequences of that deterioration for economic and social development" was the World Commission on Environment and Development now called the Brundtland Commission (WCED, 1987). The need of sustainable forest management approach for sustainable development was further emphasized in the World Summit on Sustainable Development at Johannesburg in 2002. Moreover, for the first of its kind, UN General Assembly in 2007 adopted the "Non-Legally Binding Instrument on All Types of Forests" (UN, 2007) resolution which is a reflection of the strong international commitment to promote the implementation of sustainable forest management through a new and more holistic approach that brings all stakeholders together.

Some indicators of sustainable forest management framework, specifically the forest area change, have also been included among 48 indicators of the Millennium Development Goals (MDGs) of the United Nations. The Goal 7 of MDGs is to "ensure environmental stability", and Target 9 is to "integrate the principles of sustainable development in country policies and programs and reverse the loss of environmental resources". The indicators for the achievement of this Goal of MDG are: Indicator 25 (proportion of land area covered by forest) and Indicator 26 (ratio of area protected to maintain biological diversity to surface area). The UN Millennium Declaration adopted in 2000 also declares to "adopt in all our environmental action new ethics of conservation and stewardship and as first steps .... to intensify cooperation to reduce the number and effects of natural and man-made disasters" (UN, 2000, 2008). International Strategy for Disaster Reduction, a global framework established within the United Nations for the promotion of action to reduce social vulnerability and risks of natural hazards and related technological and environmental disasters, has also identified environment as one of the vulnerability factors. Disaster and risk reduction is, therefore, emerging as an important prerequisite for sustainable development (ISDR, 2002).

*Corresponding Author

The concept got support and recognition in various international forums for the management, conservation and sustainable development of all types of forests. As the sustainable forest management frameworks are increasingly accepted and used all over the world, many countries now produce national reports assessing their progress towards sustainable forest management. However, the international initiatives towards sustainable development offered little in terms of measurable objectives. Also, it is not easy to measure the natural resources. Conceptual difficulties, unavailability of measuring techniques coupled with lack of adequate data pose additional challenges. In this context, there have been numerous initiatives and processes in the world to streamline the efforts towards sustainable forest management. The criteria and indicators approach developed as a potent tool for assessment, monitoring and reporting of sustainability of forest resources. This approach is now considered as a tool for monitoring, assessment and reporting of sustainable forest management (ITTO, 2005). The periodic monitoring of the indicators provides a trend scenario whether we are moving towards sustainable forest management or not.

Although naturally occurring events have greatly affected societies from the earliest history of man, significant interest and research of natural hazards and disasters has been recent development. As we move more towards economic, social development, we are forced to compromise on ecological values making us more vulnerable to disasters. There have been many debates on the definitions of terms hazard and disaster (Smith, 1996). To begin with, the natural calamities having potential to cause harm to life property and environment in an area are termed *hazards*. Many authors would not limit hazards to naturally occurring event or phenomenon and it has been viewed as a naturally occurring or human-induced process or event with the potential to create losses (Smith, 1996). When such a hazard is realized involving an area with human population, it is termed *disaster*. A disaster is the impact of a natural or man-made hazard that negatively affects society or environment. A disaster generally results from the interaction, in time and space, between the physical exposure to a hazardous process and a vulnerable human population (Smith, 1996). On an average, 104 million hectares of forest were reported to be significantly affected each year by forest fire, pests (insects and disease) or climatic events such as droughts, wind, snow, ice and floods (FAO, 2006). The damage to forests and natural systems can be great. For example, the Indian Ocean tsunami in December 2004 damaged 51 to 100% mangrove ecosystems, 41 to 100% coral reef ecosystems and 6.5 to 27% forest ecosystems in the four islands of Nicobar Islands, India (Ramachandran et al., 2005). However, the area of forest affected by disturbances is often severely under-reported, with information missing from many countries, especially for forest fires in Africa (FAO, 2006). India's population is slated to grow to become most populous country in the world overcoming China by 2050. Such a tremendous demographic pressure shall mean much more forest losses and more land degradation in the future. Such a scenario is most likely in many other developing nations as well.

The natural environment is an important component to providing stability and increasing our resilience against disasters. The forest ecosystems are one of most important components of this natural environment. In the recent past there have been an increased number of natural disasters. Some documented studies show that there has been a significant increase in natural disasters over the few last decades (UNEP, 2005; IPCC, 2007). Some studies also document the increase in the degree of losses due to these disasters in recent times (ISDR, 2002). The loss of life in the under-developed and developing countries is also much higher than in the industrialized countries. This might be due to lack of facilities for disaster preparedness and management. However, that has more to do with effective preparedness for disasters, poor infrastructure, mitigation and response systems. More of the developing countries are in tropical and subtropical regions which expose them to more of storms, heavy rains and landslides. The same is

> **Environment and Disaster: Terms**
>
> *Ecosystem:* A functional unit consisting of all the living organisms (plants, animals and microbes) in a given area, as well as the non-living physical and chemical factors of their environment, linked together through nutrient cycling and energy flow. An ecosystem can be of any size – a log, a pond, a field, a forest or the Earth's biosphere – but it always functions as a whole unit.
>
> *Ecosystem services:* The benefits people derive from ecosystems. These include provisioning services such as food and water; regulating services such as flood and disease control; cultural services such as spiritual, recreational, and cultural benefits; and supporting services such as nutrient cycling that maintain the conditions for life on Earth. The concept "ecosystem goods and services" is synonymous with ecosystem services.
>
> *Environment:* All of the external factors, conditions, and influences that affect an organism or a community. Also, everything that surrounds an organism or organisms, including both natural and human-built elements.
>
> *Disaster:* A serious disruption of the functioning of a community or a society causing widespread human, material, economic or environmental losses which exceed the ability of the affected community or society to cope using its own resources. A disaster is a function of the risk process. It results from the combination of hazards, conditions of vulnerability and insufficient capacity or measures to reduce the potential negative consequences of risk.
>
> *Disaster risk reduction:* The conceptual framework of elements considered with the possibilities to minimize vulnerabilities and disaster risks throughout a society, to avoid (prevention) or to limit (mitigation and preparedness) the adverse impacts of hazards, within the broad context of sustainable development.
>
> *Environmental degradation:* The reduction of the capacity of the environment to meet social and ecological objectives, and needs. Potential effects are varied and may contribute to an increase in vulnerability and the frequency and intensity of natural hazards. Some examples include: land degradation, deforestation, desertification, wildland fires, loss of biodiversity, land, water and air pollution, climate change, sea level rise and ozone depletion.
>
> (UNEP, 1996; ISDR, 2004)

true for disasters relating to environment and forests. The concentration of world's forest resources in developing countries is immense. Simultaneously, many looming social and economic challenges in terms of poverty, literacy, and low resource base put additional pressure on their natural resource base making them more vulnerable to disasters.

If we look at the attainment of greater aim of sustainable development, we have to consider risk of natural hazards, disasters and their impacts effecting forests and environment. To achieve sustainable development, we need to look deep into the risks of natural disasters and their lasting effects. The forests are renewable natural resource which undoubtedly represent the wealth of a country and also provide a basis for the existence of the society. Today, sometimes economic and social roles of forest get priority over ecological functions. However, these are closely interrelated which is reflected in the assessment frameworks of sustainable forest management.

> **Flooding in Dominican Republic and Haiti**
>
> The May 2004 debris flow in Jimani, Dominican Republic, killed an estimated 400 of the 11,000 Jimani residents, displaced up to 3,000 individuals, and destroyed at least 300 homes, or approximately 10 to 15 percent of the town's housing stock.
>
> Primary causes were an intense rainfall event driving flash flooding (a combined total of over 500 mm of rainfall to the Haitian/Dominican Republic border regions—as much as the usual annual precipitation in the region), the location of the town on an alluvial fan, and deforestation in the upper catchment—the Haitian portion of the catchment is "virtually treeless", with some estimates suggesting that up to 97% of the original forest cover has been removed, most within the last 20 years. Secondary causes included the geomorphology of the area (poorly consolidated sediments, gravels and boulders), the hydrology of the watershed, and poor local capacity for weather forecasting, river monitoring, and communication systems.
>
> (Brent, 2006 cited in UNEP, 2008)

This chapter emphasizes the importance of sustainable forest management in relation to disaster preparedness. Although a good number of literatures are available on sustainable forest and natural resources management, the goal of the present chapter is to highlight the need of shift towards sustainable forest management to increase our capacities in disaster preparedness, response, and reporting mechanisms for disasters in forestry and other natural ecosystems. The chapter discusses issues and challenges about the role of sustainable forest and other natural resource management in improving our understanding, which in turn can help minimize our vulnerability to disasters. We start with the genesis and evolution of the concept of sustainable forest management, and its historical context. This is followed by the operational framework for sustainable forest management, the international initiatives and the initiatives at the national level in India. Thereafter, we discuss about the significance and application of sustainable forest management in monitoring and evaluation framework and disaster preparedness and mitigation. Lastly, we draw attention to some important issues about sustainable forest management.

2. SUSTAINABLE FOREST MANAGEMENT

2.1 Conceptual Background

Sustainable forest management (SFM) is the process of management of forests in accordance with the principles of sustainable development. The sustainable development encompasses environmental, economical and socio-cultural aspects. The sustainable forest management applies very broad socio-cultural, economic and ecological goals and a wide range of forestry institutions now practicing various forms of sustainable forest management and a broad range of methods and tools have been developed and being tested over time. Sustainable forest management ensures that the goods and services derived from the forests meet present-day needs while at the same time securing their continued availability and contribution to development in the long-term. In its broadest sense, sustainable forest management encompasses all the three components of sustainability, viz., ecological, economic and socio-cultural well-being encompassing the administrative, legal and technical aspects of the conservation and use of forests as shown in Fig. 1 (Prasad, 1999, 2000; IIFM, 2005; Kotwal et al., 2006). Sustainable forest management has been defined by the International Tropical Timber Organization (ITTO) as 'the process of managing permanent forest land to achieve one or more clearly specified objectives of forest management with regard to the production of a continuous flow of desirable forest products and services without undue reduction of its inherent values and future productivity and without undue undesirable effects on the physical and social environment' (ITTO, 1998). FAO has adopted the definition sustainable forest management as defined by the Ministerial Conferences on the Protection of Forests in Europe, as 'the stewardship and use of forests and forest lands in a way, and at a rate, that maintains their biodiversity, productivity, regeneration capacity, vitality and their potential to fulfill, now and in the future, relevant ecological, economic and social functions, at local, national, and global levels, and that does not cause damage to other ecosystems' (MCPFE, 1993; FAO, 1995).

Sustainability is not an absolute, independent conceptual framework. Rather, it is always set in the context of decisions about what type of system is to be sustained and over what spatio-temporal scale (Allen and Hoekstra, 1994). Given the abstract nature of sustainability, the criteria and indicators approach provides a framework to define the parameters and goals of socio-cultural, economic and ecological aspects relating to sustainability and assess progress towards them (Higman et al., 2005).

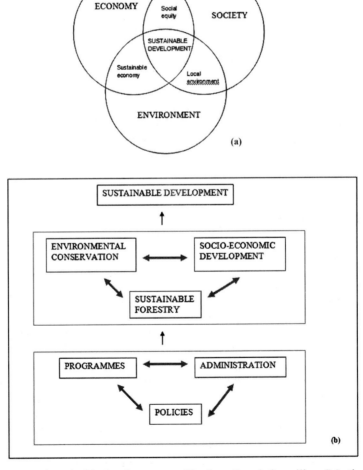

Fig. 1 (a, b) Sustainable development and its three foundation pillars (Martin, 1994).

2.2 Why Sustainable Forest Management?

Increased pressure on forest resources of the country over the last few decades has threatened the livelihoods of millions of forest-dwellers and other poor people living in the vicinity of the forests. Forest resources have been important for the prosperity of any nation and its communities. They are an essential natural resource providing multiple benefits to people besides other important functions such as biodiversity conservation, global carbon storage and a storehouse for future option values. The pressure on existing forest resources is immense in India. Having only 2.5% of the world's geographic area and 1.85% of the world's forest area, we have 17% of the world's population and 18% of livestock population (NFAP, 1999). In this background it becomes imperative to preserve the forests and manage it sustainably, to ensure livelihood security of the forest-dependent communities as well as conserving our biological diversity and environmental resilience.

According to FAO (2006), the deforestation and forest degradation of world's forest continues at an alarming rate, about 13 million hectares net per year mainly due to conversion of forests to agricultural land. As it may take care of area added due to afforestation, reforestation and natural expansion of forests, the loss of natural forest areas might be much higher. Under such situation of degradation of forest resources it was felt necessary by the global community that a management system for forests should be endured so that they not only meet the needs of the communities now, but also is perpetual to satiate the requirements of the future generations.

The implementation of sustainable forest management in a diverse country like India is a challenging task. To be more effective, criteria and indicators approach should be incorporated into national forestry legislations and regulation; not only as voluntary application. Being analogous with sustainable development, sustainable forest management also has important implications in the global economic scenario (Kotwal et al., 2007; Rawat et al., 2008). Besides contributing to environmental, social and economic well-being of the communities, it also facilitates market-oriented tools like certification and eco-labeling. This requires active participation and coordination among the stakeholders for proper implementation. A wider application of criteria and indicators shall require a long maturity process.

2.3 Sustainable Development and Management of Forests

Sustainable forest management has been described as forestry's contribution to 'sustainable development'. According to the World Commission on Environment and Development held in 1987, "sustainable development is development that meets the needs of the current generation without compromising the needs of future generations" (WCED, 1987). There have been many deliberations for the definition of the term 'sustainable development' since then. There have been even whole books devoted to defining sustainability (Reid, 1995). However, it has been widely accepted that the development which does not compromise the future needs and which is economically viable, environmentally sound is socially beneficial and acceptable. Rightly, these three aspects form the basis of sustainability in sustainable forest management.

Increased focus on development in the last two centuries has been well known and documented. All these years, forestry has emerged as best development option in some rural areas (Kennedy et al., 2001). There are many areas which are best suited to forestry than any other form of productive land use like agriculture. There may be such areas where forestry based enterprises contribute to livelihoods and local economy much more than any other development activity, but the forestry resources are important to the people of all type of areas enhancing their quality of life. It is estimated that about 400 million people are dependent on forests for at least part of their livelihood (World Bank, 2002). Forestry sector is widely recognized as having advantages when it comes to ensuring and enhancing livelihood security and provide resource safety nets to people (Mayers and Vermenlen, 2002).

Forest resources have been important for the prosperity of any nation and its communities. In India, increased pressure on forest resources of the country over the last few decades has threatened the livelihoods of millions of forest-dwellers and other poor people living in the vicinity of the forests. In India, poverty, as well as large and expanding human and livestock populations puts inexorable pressure on the forests. The consequence is severe degradation of the country's forest resources. The dependency of majority of populace on forest resources for their livelihood support systems is immense; also forestry in many developing countries, including India, is seen as a means for eradicating rural poverty and achieving sustainable development (WBOED, 2000).

We cannot proceed towards our goals of sustainable development without having an integrated framework to deal with disasters keeping in view all aspects. Without such a framework, we not only miss out the benefits of sustainable forest management to deal with disasters but we also jeopardize the resources itself. The natural disasters like the tsunami on 26 December 2004 caused extensive damage to life and property in South and Southeast Asia. In India, it is estimated that a total of 12,565 ha of forest area including mangroves were damaged in this disaster which caused extensive damages to people's livelihood support systems including fisheries, farmlands and forest lands (FSI, 2006). The affected areas included Andaman and Nicobar, Tamil Nadu, Andhra Pradesh, Kerala and Pondicherry.

3. SFM FRAMEWORK IN INDIA

3.1 Status of Forests in India

The State of Forest Report 2005 published by Forest Survey of India (FSI) shows that the area under forest is 67.71 Mha, which is 20.60% of its total geographic area. This contains 1.66% as very dense area (crown density more than 70%), 10.12% as moderately dense (crown density between 40% and 70%), and 8.82% open (crown density between 10% and 40%) including 0.44 Mha of mangroves (Fig. 2). The tree cover (tree outside forest) is estimated at 9.17 million ha constituting about 2.8% of the geographic area of the country. The forest and tree cover together is estimated at 76.88 Mha covering 23.4% of forest area which comes to 24.76% if area unavailable for planting is excluded (FSI, 2007). The forest cover in hilly areas is estimated to be 38.85%. However, by excluding area unavailable for planting, like area under permanent snow/glaciers/rocky terrain, the percentage comes to 52.4% (FSI, 2007). Ravindranath et al. (2008) have noted that forest cover is projected to reach 72 Mha by 2030. The population density of our country is 363 persons/km^2 as per the 2001 Census. The forest area per capita is only 0.06 ha in India, compared to the world average of 0.062 ha/capita and Asian average of 0.15 ha/capita (FSI, 2007). The per capita forest and other wooded land is less not only in developing countries, where forests and other wooded lands per capita is higher by 3 to 40 times, but it is less even when compared to developed countries like Germany and France, nearly two and five times that of India. Table 1 presents a comparison of per capita forest and other wooded area of key developing countries and some developed countries (Ravindranath et al., 2008).

Concerning the afforested area, the cumulative area afforested in India during the period 1980–2005 is about 34 million ha, at an average annual rate of 1.32 Mha (MoEF, 2008). This includes community woodlots, farm forestry, avenue plantations and agro-forestry. Afforestation and reforestation in India are being carried out under various programs, namely social forestry initiated in the early 1980s, Joint Forest Management Programme initiated in 1990, afforestation under National Afforestation and Eco-development Board (NAEB) programs since 1992, and private farmer and industry initiated plantation forestry.

Sustainable Forest Management: Key to Disaster Preparedness and Mitigation 543

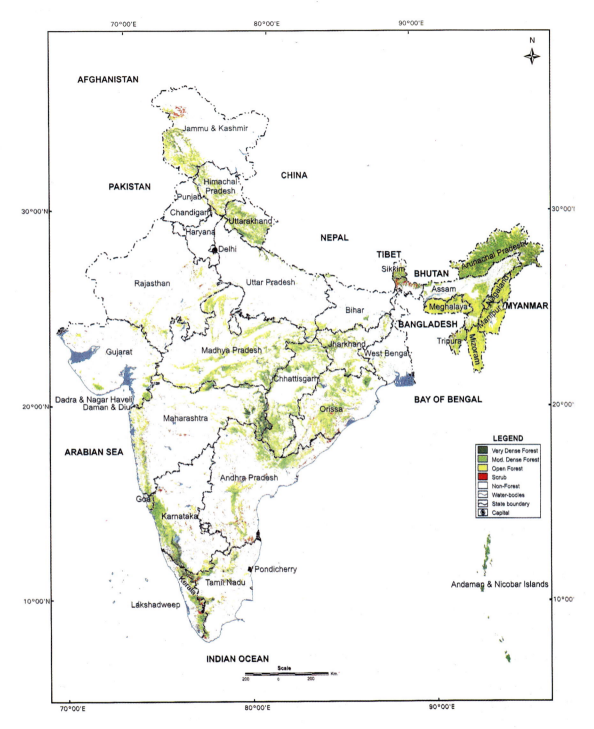

Fig. 2 Forest cover of India (FSI, 2007).

Table 1. Comparison of total forest area and forest area/1000 population (FAO, 2005)

Country	Population (million)	Forest area ('000 ha)	Other wooded land ('000 ha)	Total area under forest and wooded land ('000 ha)	Forest and wooded land (ha/1000 population)
India	1079	67,701	4110	71,811	66
China	1326	197,290	87,615	284,905	215
Brazil	178	477,698	0	477,698	2673
Indonesia	217	88,495	0	88,495	406
Germany	82	11,076	0	11,076	134
United Kingdom	59	2,845	20	2865	48
France	59	15,554	1708	17,262	287

The forestry sector increasingly faces many challenges like the growing human and cattle population, lack of infrastructure and diversion of forest ecosystems and increased pressure due to economic activities. The forestry sector in India has seen many changes in the recent past. It is managed on modern scientific management for more than a century now which was previously managed by communities based on their traditional ecological and indigenous knowledge. Moreover, there is a clear shift from non-participatory productive forestry policies during British period to participatory, conservation oriented forestry policies after independence. This shift is more evident from a holistic sustainable forest management approach encompassing economic, environmental, social and cultural dimensions in cognizance with National Forest Policy 1988 and National Forestry Action Programme 1999 (MoEF 1988; NFAP, 1999).

The system of criteria and indicators can help monitor the direction of change, whether towards or away from sustainable forest management. There have also been many efforts for institutionalization of the criteria and indicators approach. The forests in India are managed according to a scientifically sound, written management plan known as the 'Working Plan', and every division has a working plan which is revised after every ten years. Incorporation of monitoring and evaluation frameworks for sustainable forest management in working plans itself is a first step towards institutionalization of criteria and indicator (C&I) approach for sustainable forest management (SFM) (IIFM, 1999, 2000, 2005). With the tropical climate and unstable landforms, coupled with high population density, poverty, illiteracy and lack of adequate infrastructure, India is one of the most vulnerable developing countries to suffer very often from various natural disasters such as drought, flood, cyclone, earthquake, landslide, forest fire, hail storm, locust, volcanic eruption, etc.

3.2 Initiatives for Sustainable Forest Management

There have been many internationally recognised initiatives with potential application to define and assess sustainable forest management, such as criteria and indicators, life cycle assessment, cost-benefit analysis, knowledge-based systems and environmental impact assessment (Baelemans and Muys, 1998). The criteria and indicator approach has been widely accepted and immense work has been done towards its refinement and practical application. Over the years, it has developed as a potent tool for assessment, monitoring and reporting of sustainability of forest resources. Currently, about 160 countries are participating in nine regional and international processes of sustainable forest management following the criteria and indicator approach, mostly within the framework of an international initiative, which are specific to various forestry conditions (Castañeda, 2000a; FAO, 2003). The criteria and indicators approach presents a tool for assessing the magnitude and direction of change in given forestry situations, and this provides critical information to the managers and other stakeholders for forest-related decision-making. It is an important framework to assist countries to collect, store and disseminate reliable science based forest information needed to monitor and assess forest conditions (FAO, 2003). Criteria define and characterize the essential elements, as well as a set of conditions or processes, by which sustainable forest management may be assessed. The criteria and indicators provide a robust framework not only to define sustainability in the context of individual countries, but also provide a mechanism for understanding, monitoring and analysing national and global trends (IIFM, 2000). These are instruments through which progress towards sustainable forest management may be evaluated and reported. Castañeda (2000b) defines criteria as the range of forest values to be addressed and the essential elements or principles of forest management against which the sustainability of forests may be assessed. Each criterion relates to a key element of sustainability and may be described by one or more associated indicators. Indicators are parameters that measure specific quantitative and qualitative attributes and help monitor trends in the sustainability of forest management over a given time.

3.2.1 Global Initiatives

The criteria and indicators approach for sustainable forest management was conceptualized and initiated by the ITTO even prior to the Earth Summit in 1992 (ITTO, 1992). At present, it appears that the international consensus is getting impetus on the key elements of sustainable forest management. Currently, there are nine on-going international and/or regional criteria and indicators initiatives, involving approximately 160 countries with some member-countries participating in more than one process. Table 2 summarizes these nine processes (Castañeda et al., 2001). SFM has been a leading concept in international deliberations and work. The result today is a broad consensus on principles, guidelines, criteria and indicators for SFM on international level. Seven common thematic areas of sustainable forest management have emerged based on the criteria of the nine ongoing regional and international sustainable forest management initiatives. These were acknowledged by the international forest community at the fourth session of the United Nations Forum on Forests (2004) and the 16th session of the Committee on Forestry (2003). These seven thematic areas include: (i) extent of forest resources; (ii) biological diversity; (iii) forest health and vitality; (iv) productive functions of forest resources; (v) protective functions of forest resources; (vi) socio-economic functions, and (vii) legal, policy and institutional framework.

3.2.2 Indian Initiatives

The forestry sector in India is among the first in the world to be managed on the lines of modern scientific management. Establishment of forest management from the middle of the eighteenth century incidentally coincided with the industrial revolution in the West. The forests emerged as important resources during the pre-independence period, as the demand for raw materials increased, and a need was felt to expand the railway network (Tucker, 1993; Saxena, 1997). Forestry was thus production-oriented at that time. However, the basic change in perception was brought by the National Forest Policy of 1952, from production forestry to focus on meeting objectives of maintaining ecological balance on the one hand and meeting the needs of stakeholders in the best possible way on the other. The 1988 National Forest Policy focused on the maintenance of environmental stability, conservation of natural heritage by preserving the natural forests and meeting the basic needs of people, and also maintaining the relationship between the tribals and other dependent people, thus encompassing ecological, economic and social aspects of forest management (MoEF, 1988). There is however an urgent need to monitor and ensure

Table 2. Brief description of major internationally recognized processes on criteria and indicators and number of participating countries (Castañeda et al., 2001)

Process	No. of criteria	No. of indicators	Place of adoption	Date of adoption	No. of countries	Reference
ITTO Initiative on criteria and indicators	7	57	Yokohama, Japan	March 1992	59	ITTO, 1998, 2005
Dry-Zone Africa Process	7	47	Nairobi, Kenya	November 1995	30	FAO, 1996
Pan-European Forest Process	6	27 quantitative 101 descriptive	Helsinki, Finland, Lisbon, Portugal	June 1993 June 1998	37	Pan-European Process on Forests, 1995
Montreal Process	7 (non-legally binding)	67	Santiago, Chile	February 1995	12	Montreal Process 1995
Tarapoto Proposal	1 global 7 national 4 forest management unit (FMU)	7 (global) 47 (national) 22 forest management unit (FMU)	Tarapoto, Peru	February 1995	8	Tarapoto Proposal, 1995
Near East Process	7	65	Cairo, Egypt	October 1996	30	FAO, 1997a
Lepaterique Process of Central America	4 regional 8 national	40 (regional) 53 (national) 50 (FMU)	Tegucigalpa, Honduras	January 1997	8	FAO, 1997b
African Timber Organization	28	60	Libreville, Gabon	January 1993	13	ITTO, 2003
Regional Initiative for Dry Forests in Asia	8	49	Bhopal, India	December 1999	9	FAO, 2000

proper implementation of these policy implications. The quantifiable approach like criteria and indicators to monitor and implement these objectives of sustainability is imperative for which one pioneer attempt has been initiated at Indian Institute of Forest Management, Bhopal (IIFM, 2000, 2005; Rawat et al., 2008). Table 3 shows the criteria and indicators of the Indian Initiative, known as "National Set of Criteria and Indicators" (IIFM, 2005; MoEF, 2008).

Considering the importance of criteria and indicator approach for sustainable forest management, Ministry of Environment and Forests, Government of India, constituted a Task Force on sustainable forest management in 1999 (MoEF, 1999). Based on the recommendations of the National Task Force and various recommendations made by other authorities and institutions at national level regarding the adaptation of SFM initiatives (NFC, 2006) a "SFM Cell" has been created in the Ministry of Environment and Forest, Government of India. Under the aegis of SFM Cell, a "National Set of criteria and indicators"

Table 3. Criteria and Indicators of pioneering Indian initiative on SFM (IIFM, 2005; MoEF, 2008)

Criteria	Indicators
Criterion 1: Maintenance/increase in the extent of forest and tree cover	1.1 Area of forest under different forest Acts or management Plans 1.2 Percentage of forest with secured boundaries 1.3 Change in area of forest cover—dense, open, scrub forests, pastures, deserts, etc. 1.4 Change in tree cover outside forest area
Criterion 2: Maintenance, conservation and enhancement of biodiversity	2.1 Protected area network 2.2 Species diversity 2.3 Genetic diversity 2.4 Status of biodiversity conservation in forests 2.5 Status of species prone to over-exploitation 2.6 (a) Status of non-destructive harvest of wood 2.6 (b) Status of non-destructive harvest of Non-Timber Forest Produce (NTFP)
Criterion 3: Maintenance and enhancement of forest health and vitality	3.1 Status of regeneration 3.2 (a) Area affected by forest fire 3.2 (b) Area damaged by natural calamities 3.3 Area protected from grazing 3.4 Area infested by invasive weeds in forests 3.5 Incidences of pest and diseases
Criterion 4: Conservation and maintenance of soil and water resources	4.1 Area treated under soil and water conservation measures 4.2 (a) Duration of water flow in the selected seasonal streams 4.2 (b) Water bodies and wetlands in forest areas 4.2 (c) Water level in the wells in the vicinity (up to 5 kms) of forest area
Criterion 5: Maintenance and enhancement of forest resource productivity	5.1 Growing stock of wood 5.2 Increment in volume of identified species of wood 5.3 (a) Efforts towards enhancement of forest productivity area brought under high-tech plantations 5.3 (b) Extent of seed production area and seed orchards
Criterion 6: Optimization of forest resource utilization	6.1 (a) Recorded removal of timber 6.1 (b) Recorded removal of fuel wood 6.1 (c) Recorded removal of bamboo 6.2 Recorded removal of locally important NTFP. 6.3 Direct employment in forestry activities 6.4 Demand and Supply of Timber and important Non-Timber Forest Produce 6.5 (a) Import and Export of wood and wood products 6.5 (b) Import and Export of NTFPs 6.6 Value and percentage contribution of forestry sector to Gross Domestic Products (GDP)
Criterion 7: Maintenance and enhancement of social, cultural and spiritual benefits	7.1 (a) Number of JFM committees and area(s) protected by them 7.1 (b) Status of people's participation in management and benefit-sharing 7.2 Use of indigenous knowledge 7.3 Extent of cultural/sacred groves

Contd...

Table 3. (*Contd.*)

Criteria	Indicators
Criterion 8: Adequacy of Policy, Legal and Institutional framework	8.1 (a) Existence of policy and legal framework 8.1 (b) Status of approved working plan 8.2 Number of forest related offences 8.3 Status of Research and Development 8.4 Human resource capacity building efforts 8.5 (a) Forest Resource Accounting (FRA) 8.5 (b) Budgetary allocations to the forestry sector 8.6 Monitoring and Evaluation mechanisms of forestry development activities 8.7 Status of data collection, information dissemination and utilization 8.8 Adequate manpower in FMU

for sustainable forest management has been developed (MoEF, 2008). The National Set of eight criteria and 37 indicators provides a framework to monitor, assess and report progress of sustainable forest management activities at the national, regional/state and forest management unit level.

In 1999, a workshop on 'Development of National Level Criteria and Indicators for Sustainable Management of Dry Forests in Asia' was held at the Indian Institute of Forest Management (IIFM), Bhopal, with support from the Food and Agriculture Organization of the United Nations and the United Nations Environment Programme in collaboration with the ITTO, the United States Department of Agriculture Forest Service and the IIFM. Now referred to as the 'Dry Forest in Asia Process', ten Asian countries jointly developed a regionally applicable set of national-level criteria and indicators relevant for dry forests in the region (FAO, 2000). The National Task Force on Sustainable Forest Management appointed by the Ministry of Environment and Forests, Govt. of India, endorsed this regional initiative. Thus, the Indian initiative of criteria and indicators approach for sustainable forest management was spearheaded by the IIFM in collaboration with ITTO and the Ministry of Environment and Forests, Government of India (IIFM, 1999, 2000, 2005). A total of eight criteria and 51 indicators specific to Indian forestry conditions were evolved after a consultative process involving a gamut of stakeholders. The criteria and indicators of the Bhopal-India process have evolved after a lot of deliberations and field-testing over the years. The indicators were revisited in 2008 after discussion with the stakeholders on the line of Bhopal- India process and are given in Fig. 3 (IIFM, 2005; MoEF 2008).

4. MONITORING AND EVALUATION MECHANISMS FOR SUSTAINABLE FOREST MANAGEMENT

The monitoring and evaluation mechanism is vital to success of any program and attainment of desired goals. The monitoring and evaluation becomes more important for sustainable forest management owing to the long gestation period. The effective monitoring and evaluation mechanism will ensure the desired visible impact on ground. The National Forest Commission, in its report released in 2006, has recommended creating an enabling environment to facilitate assessment, monitoring and reporting on national-level criteria and indicators for sustainable forest management.

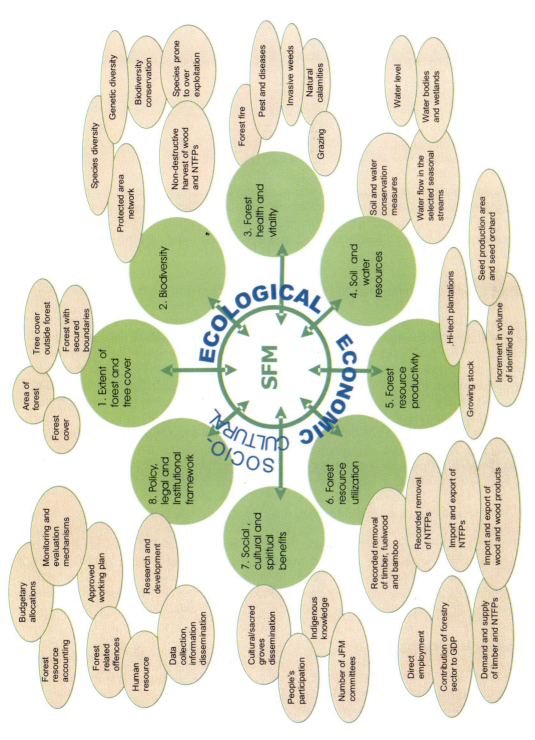

Fig. 3 An illustrated representation of the three pillars of sustainable forest management approach showing criteria and indicators (without adjectives) (IIFM, 2005; MoEF, 2008).

4.1 Forest Working Plans—Potential for all Monitoring and Evaluation Activities

In India, most of the forest management units, i.e., territorial forest divisions have forest Working Plans. These Working Plans are prepared for a period of 10 years or so and are within the framework of which detailed management activities of the Forest Divisions are drawn up. As per the present legal and policy framework, no forest can be worked without an approved Working Plan. These Working Plans in turn are prepared according to the guidelines prescribed in the Working Plan Code. The national set of criteria and indicators for sustainable forest management has semblance with National Working Plan Code (2004) and 83% of data on criteria and indicators are already being collected at the time of preparation of Working Plan (IIFM, 2005; Kotwal et al., 2005). It is worthwhile to mention here that the working plan code forms the basis and provides guidelines to prepare forest working plans at forest management units. Incorporating all monitoring and evaluation activities in the Working Plan, it has potential to take care of all our M&E needs. The incorporation of criteria and indicators approach in Working Plan Code will help in formulation of Working Plans incorporating effective monitoring and evaluation mechanism (Kotwal et al., 2005).

4.2 Market Links to Sustainable Forest Management

Recently, as a result of increasing public awareness and various treaties and conventions all over the world, there is a movement towards accepting only those forest products which have originated from sustainably managed forests (Rametsteiner, 2003). The overall aim of such programs is to orient production system towards sustainable processes that ensure their availability in perpetuity. It has emerged as a market-based mechanism in support of sustainable forest management. Certification and eco-labeling are such new mechanisms enhancing forest-product positioning for a premium price on one hand, and ensuring better managing practices for forests on the other.

Forest certification is a process to determine whether a forest is being sustainably managed or not. Over the years certification and eco-labeling have emerged as a market based tool towards sustainable forest management. It acts as a market-based system for verification that assesses forest management operations against a standard set of principles as elucidated by the criteria. This approach ensures that management of forests and its ecosystem is carried out in accordance to accepted principles. It links market with the three aspects of sustainability viz., ecological, social and economical.

The concept of *eco-label* is derived from the word "eco" which means natural environment, and "label" which differentiates it from other products. Eco-label helps consumers in selecting environment friendly products as well as tools for producers to inform the consumers of the environment friendly products. Moreover, the concept of certification can be attributed to the society's concern for the social and environmental significance of forests, furthered by increased environmental awareness in 1960s and early 1970s (Granholm et al., 1996; Hansen, 1997).

4.3 Ecosystem Services

Forests provide many valuable ecological services like climate stabilization, carbon sequestration, protection of hydrological function, watershed protection, soil conservation and improvement of agro-ecological system, biodiversity conservation (Creedy and Wurzbacher, 2001). Ecosystem services are generated due to the interaction and exchange between biotic and abiotic components of the ecosystem

(Singh, 2002). There are many academicians, scientists and researchers who are studying the inter-relationship of natural calamities like forest fires, floods, extreme storms, droughts and forest destruction including the costs and benefits of forest degradation. Ground level ozone, due to its steadily increasing concentrations and high phyto-toxic potential, is the most important pollutant affecting forests worldwide. Information from a range of sources have suggested that mangroves and other coastal forests mitigated the effects of the tsunami, which led to tree planting programs and calls to establish coastal buffer zones in a number of tsunami-affected countries (Forbes and Broadhead, 2007).

Market based development of environmental services from forests has recently attracted attention as a potential tool for promoting sustainable forest management by providing new sources of financing and providing incentives to adopt sustainable practices (Katila and Puustjärvi, 2004). The services of a healthy ecosystem have earlier been taken for granted as being available in close vicinity for free and also indefinitely. However, recently, as demands grew drastically and natural resources became scarce, there have been new mechanisms evolving for financing forest conservation and payments to those who bear the cost of forest degradation (Pagiola et al., 2002).

Recognition to this role of forests has given rise to the Reduced Emissions from Deforestation and Degradation (REDD) concept, creating a carbon market mechanism to pay local communities in developing countries for protecting their forests, an alternative income stream to what would be earned by logging or farming the same land. This mechanism aims to provide incentives for developing countries to cut emissions by preserving forests or having better forest management practices—all of this as an effort to bring down emission of carbon dioxide from this source and to ensure that there are sufficient forests remaining for the uptake of carbon dioxide (and provide other crucial global ecological services). This mechanism can achieve its goals if sustainable forest management is integrated into broader carbon emission reduction strategies.

Valuing Ecosystem Services

While the degree of protection provided by ecosystems depends on a number of factors, social and natural scientists have been working to calculate value for providing ecosystem services. Some examples are:

The forests of Indian state of Himachal Pradesh, which are 66% of total geographical area, play a pivotal role in the regional and global economy. When the Gross State Domestic Product (GSDP) of the state is corrected for total economic value including the contribution of the forestry sector for all the products and services including ecosystem services, increases from 5.26% of GSDP to 92.40% of corrected GSDP. The generated economic valuation of the forest derived is to the tune of Rs. 7.45 lakh/hectare. The maximum per hectare value is generated by watershed function followed by carbon sink, biodiversity, ecotourism (all non-marketed values) (Verma, 2000).

Sri Lanka's Muthurajawela marsh, a coastal peat bog covering some 3,100 ha, plays an important part in local flood control. The marsh significantly buffers floodwaters from the Dandugam Oya, Kala Oya and Kelani Ganga rivers and discharges them slowly into the sea. The annual value of these services was estimated at more than USD 5 million, or USD 1,750 per hectare of wetland area (Emerton and Bos, 2004).

Shoreline stabilization is important for inland rivers. In the eastern United Kingdom, the cost of the loss of vegetation along riverbanks was estimated at USD 425 per meter of bank. This is the cost of maintaining artificial bank reinforcement to prevent erosion (Ramsar Convention, 2005).

(ISDR, 2005)

5. DISASTER PREPAREDNESS AND MITIGATION

Integrated natural resource management is a critical strategy to prevent disasters and reduce risks/vulnerabilities of disasters to the prone countries and communities. Vulnerability can be considerably reduced through effective and long-term environmental and natural resource management practices. The sustainable forest management ensuring economic, ecological and socio-cultural well-being through range of forest goods and services does not only enhance ecosystem resilience to climate change but also helps forest dependent communities to cope with impacts of disasters.

Sustainable forest management approach aims to ensure that the goods and services derived from forests meet present needs while at the same time securing their continued availability and contribution to future generations and long-term development. It may be noted that forest management encompasses the administrative, legal, technical, economic, social and environmental aspects of conservation and use of forests. The forest ecosystems are known to act positively towards increasing resistance to certain disasters. If not managed properly, the immediate environment within which we have to operate also poses many hazards and potentials of causing disasters. The response of forest management systems towards mitigating the impacts of disaster shall be imperative in shaping effective integrated response systems to deal with adverse effects of disasters.

With growing population and infrastructures, our exposure to natural hazards is inevitably increasing. This is particularly true as the maximum population growth is located in coastal areas where there are greater chances of floods, cyclones and tidal waves. To make matters worse any land remaining available for urban growth is generally risk-prone, for instance flood plains or steep slopes are subject to landslides (UNEP, 2005). The effects of disasters fall most heavily on developing nations, as the developing countries are more prone to the damage caused due to their lack of disaster response mechanism and limited infrastructure. Natural resource-based communities are often theoretically assumed to be vulnerable to the negative effects of environmental and social change. Such communities are also seen as being especially vulnerable to risks and disasters. Disaster research perspectives echo this in their emphasis on environmental and social vulnerability.

The forest ecosystems play a vital role in reducing/mitigating certain disaster hazards and act positively increasing resistance to them. It is, for example, well known that coastal forests and trees can, under certain conditions, act as bio-shields to protect lives and valuable assets against coastal hazards, including tsunamis, cyclones, wind and salt spray and coastal (tidal) erosion. However, coastal forests and trees are not able to provide complete protection against all hazards (e.g., extremely large tsunami waves, flooding from cyclones and super-cyclones, and certain types of erosion in coastal areas). The development of bio-shields may not be possible in all situations due to biological limitations, space constraints, incompatibility with priority land uses, prohibitive costs, etc. However, taking this aspect while implementing sustainable forest management approach, more so at the macro level, shall be important in effecting our response to disaster management. It shall also be helpful in organized information system to address the challenges of disaster management.

In the recent times, there are new tools available for monitoring and providing working frameworks to fight the disasters. We have advanced in science, information and technology which were unimaginable a few decades ago. Latest tools like remote sensing and GIS applications are helpful in planning, disaster forewarning, risk zonation, development of communication net-works, telemedicine related to disasters effecting forests and other natural resources like forest fires, cyclones, coastal erosions etc. Integrating disaster management related data with respect to forests with monitoring and evaluation frameworks like sustainable forest management approach shall go a long way ensuring long term sustainability of such resources. This approach shall be helpful in generating reliable data regarding disasters and forests as we increasingly feel the need to accurate, up-to-date and unbiased data relating to forests and environment. World over, as concerns grow about the potential causes and effects of global climate change, scientific researchers are seeking reliable data to assess the conditions of natural ecosystems including forests.

Disasters are serious disruption of the functioning of a ecological system, community or a society causing widespread human, material, economic and environmental losses which exceed the ability of the affected community/society to cope using its own resources. Disaster risk management is systematic process of using administrative decisions, organization's operational skills and capacities to implement policies, strategies and coping capacities of the society and communities to minimize the impacts of natural hazards and related environmental and technological disasters. This comprises all forms of activities, including structural and non-structural measures to avoid (prevention) or to limit (mitigation and preparedness) adverse effects of hazards. Disaster reduction is the systematic development and application of policies, strategies and practices to minimize vulnerability and disaster risks throughout a society, to avoid (prevention) or to limit (mitigation and preparedness) adverse impact of hazards, within the broad context of sustainable development (ISDR).

In January 2005, countries at the World Conference on Disaster Reduction adopted the Hyogo Framework for Action (HFA), which guides disaster risk reduction activities globally (ISDR, 2005). The Hyogo Framework recognizes that environmental degradation contributes to disaster risk, and that disasters occur when hazards interact with, among other things, environmental vulnerability. The document urges governments to pursue the "substantial reduction of disaster losses, in lives and in the social, economic and environmental assets of communities and countries" (UNEP, 2008). The Framework recognizes both the role of environment as a trigger of disaster risk, and the sensitivity of the environment to the forces of hazards.

The human societies are closely related to the environment system. They shape the environment with their actions and are influenced by the environment in development. The less degraded the environmental component of this system, the lower its overall vulnerability and the higher its coping capacity (UNEP, 2008).

Reducing the Underlying Risk Factors—Priority for Action

The Hyogo Framework defines "reducing the underlying risk factors" as a Priority for Action and specifically recommends environmental and natural resource management and other efforts that:

(a) Encourage the sustainable use and management of ecosystems, including through better land-use planning and development activities to reduce risk and vulnerabilities;

(b) Implement integrated environmental and natural resource management approaches that incorporate disaster risk reduction, including structural and non-structural measures, such as integrated flood management and appropriate management of fragile ecosystems; and

(c) Promote the integration of risk reduction associated with existing climate variability and future climate change into strategies for the reduction of disaster risk and adaptation to climate change, which would include the clear identification of climate-related disaster risks, the design of specific risk reduction measures and an improved and routine use of climate risk information by planners, engineers and other decision makers.

(ISDR, 2005)

The cause of the increase in natural disasters can cause widespread human intervention in the equilibrium of fragile ecosystems like forest clearance, soil erosion and single cropping practices. In India, it is observed that we are lacking the preparatory arrangements to combat/mitigate the effects of natural disasters. It is important to identify the disaster prone areas or the vulnerable situations. It is also suggested to arrange disaster fighting mechanism to undertake timely measures in case of occurrence of disasters.

6. CLIMATE CHANGE VIS-À-VIS FOREST MANAGEMENT

Among various potential hazards climate change can bring to our lives, a crucial one is stressing forests and other natural resources through increase in mean annual temperatures, altered precipitation patterns and more frequent and extreme weather events. According to the Intergovernmental Panel on Climate Change (IPCC), climate change will stress critical ecosystems and pose a threat to life and livelihoods support systems of the millions of poor people who are dependent on them (IPCC, 2001, 2007). It is an irony that the changes we may see, as the climate warms with increased storms of this magnitude, could be accelerating the source of emissions that create the change, and hence the change could be accelerating itself.

According to FAO (2006), forests store enormous amounts of carbon—in total, the world's forests and forest soils currently store 1.2 trillion tons of carbon that is twice the amount found floating free in the atmosphere. Thus, together with carbon stored in dead wood, litter and soil as humus, forests are one of the most important storehouses of carbon. The forestry sector is very critically placed in this battle of survival for mankind. The climate change has emerged among the topmost action priority in whole development sector. The present challenges in forestry sector, particularly in India, may make it very difficult to adopt and suitably respond to overcome the threats due to climate change. The climate change causes flooding, drought and big forest fire all over the world, mentioned as the natural disasters.

According to FAO (2005), total forest carbon stock in India is 10.01 GtC. According to some other estimates the present carbon stock in Indian forests in both soil and vegetation range from 8.58 to 9.57 GtC (Ravindranath et al., 1997; Ravindranath et al., 2008). With present policy and administrative framework, the carbon stock in the forests is projected to be nearly stable over the next 25-year period at 8.79 GtC. Given the challenges faced by forests in India, stabilization or increase of forest carbon stock is a big achievement. Ensuring sustainable forest management in the years to come shall not only ensure saving our forests from adverse effects of climate change but also holds a great promise to mitigate them by storing additional carbon stocks.

7. ADAPTATION AND MITIGATION STRATEGIES TOWARDS SUSTAINABLE FOREST MANAGEMENT

It has been commonly acknowledged that natural forest ecosystems are great reservoirs for carbon storage and this carbon remains vital in maintenance of carbon balance in globe. One of most striking effects of climate change is going to be on natural resources including forests and water with extensive complications for agri-ecological systems affecting livelihoods of millions. The adverse effects due to loss of natural ecosystems shall lead to the 'multiplier' effect of adverse consequences which shall be a greater challenge to deal with. These consequences have compelled almost all the nations of the world, irrespective of their geo-physical conditions, to think about the sustainable management of their forest resources (Spittlehouse and Stewart, 2004). The climate change is expected to bring more extreme weather events.

The first step towards climate change adaptation is to address existing vulnerabilities to these extremes. Many of the required climate change adaptation measures, such as early warning systems, risk assessment, and the use of sustainable natural resources, are disaster risk reduction activities in practice (UNEP, 2008).

Climate change and forests are closely linked. Trees absorb carbon dioxide, the primary gas causing global climate change and retain the carbon from the molecule and release oxygen into the atmosphere. The carbon makes up half the dry weight of a tree and forests are the world's second largest carbon reservoirs (oceans are the largest). Unlike oceans, however, we can grow new forests. Planting new trees and sustainably managing resources remains one of the cheapest and most effective means of drawing excess carbon from the atmosphere. One acre of forestland will sequester between 150-200 tons of carbon in its first 40 years. Forest ecosystems are important components of the global carbon cycle in at least two ways. First, terrestrial ecosystems remove nearly three billion tons of anthropogenic carbon every year (3 Pg C per year) through net growth, absorbing about 30% of all CO_2 emissions from fossil fuel burning and net deforestation (Canadell and Raupach, 2008). The forests can play an important role in this not just by preventing this carbon sink to maintain its carbon, but through *afforestation* (new plantings) and *reforestation* (replanting of deforested areas) of non-forested lands (Rowe et al., 1992). It is estimated that the global carbon retention resulting from reduced deforestation, increased forest re-growth and more agro-forestry and plantations could mitigate about 15 percent of carbon emissions from burning of fossil fuels over the next 50 years.

To discourage greenhouse gases from one of its important source, the policies on mitigation encourage the replacement of fossil-based energy with biomass energy. It has also been modeled by some workers that though the green house gases mitigation policies would generally reduce their atmospheric emissions by using more energy from biomass, part of the endeavors would be counteracted by the land-use conversion including forests as a result of large-scale production of biomass energy (Wang, 2008).

With its fragile ecosystems, diverse terrain, rich biodiversity and long coastline, India is also vulnerable to climate change variations. Studies have projected that India is likely to suffer from long term adverse impacts of climate change, such as rise in mean winter temperature, decline in summer rainfall leading to unfavorable consequences for agriculture, drinking water supply and hydropower generation; melting of glacial ice that can drastically reduce water flows in the rivers of the northern plains; reduction in the duration of crop cycles shortening of the grain fill period that could substantially reduce agricultural productivity and output; sea level rise that can affect biodiversity, rich coastal wetlands, increased flooding, erosion and salt intrusion in the deltas, increase in vector-borne diseases due to the rise in temperature and humidity levels.

Sustainable forest management contributes significantly in mitigation of harmful effects of greenhouse gases. In general, the term mitigation refers to all activities aimed at reducing green house gas emissions and/or removal of CO_2 from the atmosphere with the aim of stabilizing its concentrations (Ravindranath, 2007). A summary of climate change mitigation options and its relevance with respect to adaptation strategies through forestry as applicable under sustainable forest management framework is discussed in Table 4.

Sustainable forest management shall be a critical component of any strategy to address the global concerns about the impacts of climate change. Sustainable forest management has to play a vital role in both adaptation and mitigation to climate change. Almost all of these options can be addressed through sustainable forest management (Table 5). In India, the criteria and indicators approach has over the years endeavored to formulate a working framework for the achievement of the goals of sustainable forest management, specific to the national forestry conditions.

Table 4. Climate change adaptation and possible mitigation mechanisms applicable to the eight criteria of Bhopal-India process

Criteria	Annotations with respect to possible applicability in adaptation	Possible mitigation mechanism
1. Maintenance or increase in the extent of forest and tree cover	Maintenance of extent of permanent forest estate shall be a big challenge to climate change, like all the ecosystems on Earth, forests are also vulnerable and they have further implications—forest ecosystems store one trillion tonnes of carbon—50% more carbon than found in atmosphere (FAO, 2006).	Committing forests as carbon reservoirs, Forest Restoration (not defined in Climate Change methodologies—restore degraded carbon stocks)
2. Maintenance, conservation and enhancement of biodiversity	Climate and land use are the two major factors controlling biological diversity. Critically threatened biodiversity is facing new threat of climate change. The major threat shall be habitat loss accompanied with disturbances in the dynamic ecosystem relationships.	Substitution of carbon
3. Maintenance and enhancement of forest health and vitality	Increase in temperatures and low moisture content can enhance fire incidences and loss. Overall, these disturbances shall enhance loss of forest cover due to natural calamities, pests and diseases.	Reforestation (CDM), reducing deforestation and forest degradation (REDD)
4. Conservation and maintenance of soil and water resources	The erratic weather events has already started happening. Incidences of floods and droughts have increased. With global warming happening faster than expected we need to take additional measures to address this important criterion.	Afforestation/reforestation (CDM), forest restoration
5. Maintenance and enhancement of forest resource productivity	Though increased carbon concentrations in atmosphere may experience slight to moderate increases in forest productivity, certain regions may experience significant reductions in forest productivity.	Substitution of carbon, afforestation/reforestation (CDM)
6. Optimization of forest resource utilization	The bio-energy from sustainably managed forests is renewable and it is only cycling carbon through ecosystems unlike fossil fuels which is releasing the captive carbon stored millions of years ago. The recent crisis in fossil fuels has highlighted forests role as a bio-energy resource. Many people, particularly tribal and indigenous groups, are also dependent on forests for their livelihood. We need to strengthen effective utilization mechanism for forest produce.	Substitution of carbon, afforestation (CDM)
7. Maintenance and enhancement of social, cultural and spiritual benefits	People's participation is vital in success of any program. People's institutions can ensure effective implementation of conservation programs. Involving institutions like JFM has been found helpful in participatory plantation and protection.	REDD

Contd...

(Contd.)		
8. Adequacy of Policy, Legal and Institutional framework	It is important to create and operationalise legal and institutional frameworks to respond to the needs of climate change. Institutions are a critical dimension of sustainability and sustainable forest management. Similarly, continuous research and development, budgetary allocation is important for improvisation.	REDD

Table 5. A summary of climate change mitigation options in the forestry sector

Climate change mitigation	Mitigation options in the forestry sector
1. Carbon sequestration	Afforestation (CDM: on land not forested since at least 50 years) Reforestation (CDM: on land not forested on/after 01/01/1990) Forest Restoration (Not defined in CC: restore degraded carbon stocks)
2. Emission reduction of Green House Gases	Reducing deforestation and forest degradation (REDD)
3. Substitution of carbon	Use of wood products

8. CONCLUSIONS

Nature provides enormous goods and services as natural resources which are often over-exploited. Uncontrolled anthropogenic activities have changed the ecosystems to a greater extent during past couple of centuries, and there has been much debate on their adverse effects on environment in the past. The unwarranted depletion of resources to meet the increasing demands, rising consumption lifestyle and ever-growing human population cannot be continued for ever. Whether we work for it or not, there shall be equilibrium to be achieved; it is up to us to make it favorable for all the life forms existing on the earth. Responsible environment behavior in all our roles is greatly needed at this juncture for the conservation of natural resources and the protection of environment.

One of the biggest challenges towards the outlook of forests in the recent times has been concerns about the 'sustainability' of our resources. In India, the progressive conservation-oriented forest policies and afforestation programs are contributing to the conservation of biodiversity, stabilization of watersheds, ecosystems and carbon stocks in forests, and the forest sector is projected to keep making positive contributions to global change and sustainable development (Ravindranath et al., 2008). The sustainability of people-oriented management initiatives like 'joint forest management' can be enhanced by involving the communities in applying and monitoring the sustainability using criteria and indicators approach. For the application and monitoring of criteria and indicators by the communities, it is imperative that we take care of the institutionalization mechanism and capacity-building needs of the communities. The criteria and indicators offer an opportunity to monitor and assess the state of forest management. This approach provides a powerful and user-friendly tool for forest managers. However, as with other monitoring and assessment frameworks, it ultimately rests with the forest managers to implement and analyze the framework to make sustainable forestry decisions. The criteria and indicators approach besides measuring sustainability of forests at a national level, envisages to monitoring it effectively. Close international cooperation in forest science and related disciplines is required to enable forests to satisfy the manifold human needs in a sustainable way on a long-term basis. Though the evolution of regional initiatives for criteria and indicators has been possible because of such cooperation in the first place, we need to strengthen them further for ensuring our goals of sustainability in the future.

The working frameworks for sustainable forestry have to address social well-being, economic prosperity and environmental conservation to ensure sustainable development. Addressing disaster management within such frameworks is important from social, economic and environmental viewpoints. With increasing threats due to a number of incidents, hazards, climate change and increasing population, it is important for policy makers, administrators and citizens to help make informed decisions about natural resources keeping in mind a long-term perspective. A sustainable forest management framework aims to answer this aspect with respect to an important aspect of natural resource, i.e., the forests. We also need to emphasize that sustainable development along with the international policy framework aiming at poverty reduction and environmental protection cannot be successful without taking into account the risks of natural hazards and their impacts. We simply cannot afford the increasing costs and losses to our delicate ecological and human systems due to natural and anthropogenic disasters. Conversely, having an effective monitoring and evaluation framework for sustainable natural resource, management shall be imperative in order to protect these resources from the adverse effects of different disasters and shape our response mechanisms to deal with them efficiently.

The sustainable forest management operationalised through the criteria and indicators approach gives an opportunity to monitor and assess the state of forests. It can be of great help in disaster preparedness, enhancing the resilience of ecosystems to face disasters and preparing our responses in rehabilitation work after the disasters occur. It is recommended that the national platforms for disaster risk reduction should integrate environmental concerns, which in turn should be supported by environment-related institutions. The environmental regulatory frameworks should include risk reduction criteria. Greater interaction between policymakers, environmental managers and the scientific community can breed familiarity with technological alternatives and innovations, and can stimulate targeted research towards reducing disaster risk. Strengthening new technologies and processes for managing natural resources and taking into consideration the ecological, social and cultural dimensions of resource management, presents many opportunities for reducing disaster risk (ISDR, 2004). The environmental and natural resource management professionals should be an integral part of any development, analysis and planning team, who should advocate for greater awareness to disaster risk reduction. Finally, the recovery, reconstruction and risk reduction efforts must be carried out with proper environmental guidance and safeguard to avoid devastating short- and long-term impacts on the environment.

REFERENCES

Allen, T.F.H. and Hoekstra, T.W. (1994). Toward a definition of sustainability. *In:* W.W. Covington and L. De Bano (editors), Sustainable Ecological Systems: Implementing an Ecological Approach to Land Management, Rocky Mountain Forest and Range Experiment Station. US Department of Agriculture (USDA).

Baelemans, A. and Muys, B. (1998). A critical evaluation of environmental assessment tools for sustainable forest management. *In:* D. Ceuterick (editor), Proceedings of the International Conference on Life Cycle Assessment in Agriculture, 3-4 December 1998, Agro-industry and Forestry, Brussels, pp. 65-75.

Canadell, J.G. and Raupach, M.R. (2008). Managing forests for climate change mitigation. *Science*, **320**: 1456-1457.
Castañeda, F. (2000a). Criteria and indicators for sustainable forest management: International processes, current status and the way ahead. *Unasylva*, **203**: 34-40.
Castañeda, F. (2000b). Why national and forest management unit level criteria and indicators for sustainable management of dry forests in Asia? *In:* T.L. Cheng and P.B. Durst (editors), Development of National-level Criteria and Indicators for the Sustainable Management of Dry Forests in Asia: Background Papers. Food and Agriculture Organization of the United Nations (FAO), Rome, Italy.
Castañeda, F., Palmberg-Lerche, C. and Castañeda, P.V. (2001). Criteria and Indicators for Sustainable Forest Management: A Compendium. Working Paper FM/5, FAO, Rome, Italy.
Creedy, J. and Wurzbacher, A.D. (2001). The economic value of forested catchment with timber, water and carbon sequestration benefits. *Ecological Economics*, **38**: 71-83.
Emerton, L. and Bos, E. (2004). Value: Counting Ecosystems as Water Infrastructure. Water and Nature Initiative, IUCN, Gland.
FAO (1995). The Challenge of Sustainable Forest Management: What Future for the World's Forests? Food and Agriculture Organization of the United Nations (FAO), Rome, Italy, http://www.fao.org/ (accessed on 22 October 2008).
FAO (1996). Criteria and Indicators for Sustainable Forest Management in Dry-Zone Africa. UNEP/FAO Experts' Meeting, 21-24 November 1995, Nairobi, Kenya. Food and Agriculture Organization of the United Nations (FAO), Rome, Italy.
FAO (1997a). Workshop on Criteria and Indicators for Sustainable Forest Management in Near East Process. 30 June-3 July 1997, Cairo, Egypt.
FAO (1997b). Criteria and Indicators for Sustainable Forest Management in Central America. FAO/CCAD/CCAB-AP Experts' Meeting, Lepaterique Process of Central America, Tegucigalpa, Honduras.
FAO (2000). Report of the FAO/UNEP/ITTO/IIFM/USFS Workshop on Regional Initiative for the Development and Implementation of National Level Criteria and Indicators for the Sustainable Management of Dry Forests in Asia. 30 November-03 December 1999, Bhopal, India. FAO Regional Office for Asia and the Pacific, Bangkok, Thailand.
FAO (2003). State of the World's Forests. Food and Agriculture Organization of the United Nations (FAO), Rome, Italy.
FAO (2005). State of the World's Forests 2005. Food and Agriculture Organization of the United Nations (FAO), Rome, Italy.
FAO (2006). Global Forest Resource Assessment 2005. Food and Agriculture Organization of the United Nations (FAO), Rome, Italy, http://www.fao.org/forestry/fra2005/en/ (accessed on 23 November 2008).
Forbes, K. and Broadhead, J. (2007). The Role of Coastal Forests in the Mitigation of Tsunami Impacts. FAO Regional Office for Asia and the Pacific Publication 2007, http://www.fao.org/forestry/media/14561/1/0/ (accessed on 23 November 2008).
FSI (2006). An Overview of Projects in Forest Survey of India. Forest Survey of India (FSI), Dehradun, Ministry of Environment and Forests (MoEF), Government of India, New Delhi.
FSI (2007). State of Forest Report 2005. Forest Survey of India (FSI), Dehradun, Ministry of Environment and Forests (MoEF), Government of India, New Delhi.
Granholm, H., Vähänen, T. and Sahlberg, S. (editors) (1996). Background document. Intergovernmental Seminar on Criteria and Indicators, Ministry of Agriculture and Forestry, Helsinki, Finland, 131 pp.
Hansen, A.J. (1997). Sustainable forestry in concept and reality. *In:* C. Freese (editor), The Consumptive Use of Wild Species: A Global Study of Conservation Benefits. Johns Hopkins Press, Baltimore, MD, pp. 217-245.
Higman, S., James, M., Bass, S., Neil, R. and Nussbaum, R. (2005). The Sustainable Forestry Handbook: A Practical Guide for Tropical Forest Management. Earthscan, London, U.K.

IIFM (1999). Proceedings of the National Technical Workshop for Evolving Criteria and Indicators for Sustainable Forest Management in India. 21-23 January 1999, Indian Institute of Forest Management (IIFM), Bhopal, India.

IIFM (2000). Bhopal-India Process for Sustainable Management of Indian Forests. Indian Institute of Forest Management (IIFM), Bhopal, India.

IIFM (2001). Trees Resources Outside Forest in India. SFM Series 2/01, Indian Institute of Forest Management (IIFM), Bhopal, India.

IIFM (2005). Proceedings of the National Workshop on Refining Indicators of Bhopal-India Process and Implementation Strategy of C&I for SFM in India. Indian Institute of Forest Management (IIFM), Bhopal, India.

IPCC (2001). Climate Change 2001: Synthesis Report. Contribution of Working Groups I, II and III to the Third Assessment Report of the Intergovernmental Panel on Climate Change, http://www.ipcc.ch/ipccreports/ar4-wg2.htm (accessed on 18 June 2008).

IPCC (2007). Impacts, Adaptation and Vulnerability. Contribution of IPCC Working Group II to the Intergovernmental Panel on Climate Change Fourth Assessment Report. Cambridge University Press, Cambridge, http://www.ipcc.ch/ (accessed on 24 June 2008).

ISDR (2002). Disaster Reduction and Sustainable Development: Understanding the Links between Vulnerability and Risk related to Development and Environment. Background Document for the World Summit on Sustainable Development (WSSD), International Strategy for Disaster Reduction (ISDR), http://www.unisdr.org (accessed on 16 June 2008).

ISDR (2004). Terminology of Disaster Risk Reduction. International Strategy for Disaster Reduction (ISDR), http://www.unisdr.org (accessed on 16 June 2008).

ISDR (2005). Hyogo Framework for Action 2005-2015: Building the Resilience of Nations and Communities to Disasters. World Conference on Disaster Reduction, January 2005, Kobe, Hyogo, Japan. International Strategy for Disaster Reduction (ISDR).

ITTO (1992). Criteria for the Measurement of Sustainable Forest Management. ITTO Policy Development Series No. 3. International Tropical Timber Organization (ITTO), Yokohama, Japan.

ITTO (1998). Criteria and Indicators for Sustainable Forest Management of Natural Tropical Forests. ITTO Policy Development Series No. 7. International Tropical Timber Organization (ITTO), Yokohama, Japan.

ITTO (2003). ATO/ITTO Principles, Criteria and Indicators for the Sustainable Management of African Natural Tropical Forests. ITTO Policy Development Series No. 14. International Tropical Timber Organization (ITTO), Yokohama, Japan.

ITTO (2005). Revised ITTO Criteria and Indicators for the Sustainable Management of Tropical Forests including Reporting Format. International Tropical Timber Organization, Yokohama, Japan.

Katila, M. and Puustjärvi, E. (2004). Markets for forests environmental services: Reality and potential. *Unasylva*, **219(55):** 53-58.

Kennedy, J.J., Thomasb, J.W. and Glueckc, P. (2001). Evolving forestry and rural development beliefs at midpoint and close of the 20th century. *Forest Policy and Economics*, **3:** 8-95.

Kotwal, P.C., Omprakash, M.D., Gairola, S. and Dugaya, D. (2007). Ecological indicators: Imperative to sustainable forest management. *Ecological Indicators*, **8(1):** 104-107.

Kotwal, P.C., Omprakash, M.D., Bhomia, R.K. and Dugaya, D. (2005). Incorporation of criteria and indicators in Forest Working Plans. *C&I India Update*, **4(3):** 2-12.

Kotwal, P.C., Rawat, T.S. and Dugaya, D. (2006). Standardising indicators and their values for measuring sustainability of forests. *In:* N.P. Todaria, B.P. Chamola and D.S. Chauhan (editors), Concepts in Forestry Research, Proceedings of the International Seminar on Forest, Forest Products and Services: Research, Development and Challenges Ahead, 1-3 November, HNB Garhwal University, Srinagar. International Book Distributors, Dehradun, pp. 349-359.

Martin, R.B. (1994). Alternative approaches to sustainable use: What does and doesn't work. Proceedings of the Conference on Conservation through Sustainable Use of Wildlife, 8-11 February 1994, University of Queensland, Brisbane, Australia.

Mayers, J. and Varmenlon, S. (2002). Power from trees: How good forest governance can help reduce poverty. World Summit on Sustainable Development Opinion, International Institute of Environment and Development, London, www.iied.org (accessed on 08 July 2008).

MCPFE (1993). Ministerial Conferences on the Protection of Forests in Europe. http://www.mcpfe.org/ (accessed on 23 December 2008)

MoEF (1988). National Forest Policy 1988. Ministry of Environment and Forests (MoEF), Government of India, New Delhi.

MoEF (1999). Report of the National Task Force on SFM. Ministry of Environment and Forests (MoEF), Government of India, New Delhi.

MoEF (2004). National Working Plan Code 2004. Ministry of Environment and Forest (MoEF), Government of India, New Delhi.

MoEF (2008). Annotations for Modified/Improved C&I for Sustainable Forest Management (SFM). Ministry of Environment and Forests (MoEF), Government of India, New Delhi.

Montreal Process (1995). The Montreal Process on Criteria and Indicators for the Conservation and Sustainable Management of Temperate and Boreal Forests. Montreal Process Working Group, www.mpci.org/ (accessed on 16 June 2008).

NFAP (1999). National Forestry Action Programme India. Ministry of Environment and Forests (MoEF), Government of India, New Delhi.

NFC (2006). Report of the National Forest Commission. Ministry of Environment and Forests, Government of India, New Delhi, http://www.envfor.nic.in/divisions/nfr.html (accessed on 01 July 2008).

Pagiola, S., Bishop, J. and Landell-Mills, N. (2002). Selling Forest Environmental Services: Market based Mechanism for Conservation and Development. Earthscan, London, U.K.

Pan-European Process on Forests (1995). Criteria and indicators for the conservation and sustainable forest management. Proceedings of the Ministerial Conference on the Protection of Forests in Europe, Antalya, Turkey.

Prasad, R. (1999). National Forest Policy imperatives: Criteria and indicators of sustainable forest management in India. *In:* R. Prasad (editor), Proceedings of the National Technical Workshop on Evolving Criteria and Indicators for Sustainable Forest Management in India. Indian Institute of Forest Management (IIFM), Bhopal, India.

Prasad, R. (2000). Sustainable Forest Management for Dry Forests of South Asia. *In:* T.L. Cheng and P.B. Durst (editors), Development of National Level Criteria and Indicators for the Sustainable Management of Dry Forests in Asia: Background Papers, FAO, Bangkok, Thailand.

Ramachandran, S., Anitha, S., Balamurugan, V., Dharanirajan, K., Ezhil Vendhan, K., Divien, M.I.P., Senthil Vel, A., Hussain, I.S. and Udayaraj, A. (2005). Ecological impact of tsunami on Nicobar Islands (Camorta, Katchal, Nancowry and Trinkat). *Current Science*, **89(1)**: 195-200.

Rametsteiner, E. and Simula, M. (2003). Forest certification: An instrument to promote sustainable forest management. *Journal of Environmental Management*, **67**: 87-98.

Ramsar Convention (2005). Ramsar Convention on Wetlands, 2005. www.ramsar.org (accessed on 25 October 2008).

Ravindranath, N.H. (2007). Mitigation and adaptation synergy in forest sector. *Mitigation and Adaptation Strategies for Global Change*, **12(5)**: 843-853.

Ravindranath, N.H., Chaturvedi, R.K. and Murthy, I.K. (2008). Forest conservation, afforestation and reforestation in India: Implications for forest carbon stocks. *Current Science*, **95(2)**: 216-222.

Ravindranath, N.H., Somashekhar, B.S. and Gadgil, M. (1997). Carbon flows in Indian forests. *Climate Change*, **35**: 297-320.

Rawat, T.S., Menaria, B.L., Dugaya, D. and Kotwal, P.C. (2008). Sustainable forest management in India. *Current Science*, **94(8)**: 996-1002.

Reid, D. (1995). Sustainable Development: An Introductory Guide. Earthscan, London, U.K.
Rowe, R., Sharma, N.P. and Browder, J. (1992). Deforestation: Problems, Causes and Concerns. *In:* Managing the World's Forests: Looking for Balance between Conservation and Development. Dubuque, Iowa.
Saxena, N.C. (1997). The Saga of Participatory Forestry Management in India. CIFOR, Jakarta, Indonesia.
Singh, S.P. (2002). Balancing the approaches of environmental conservation by considering ecosystem services as well as biodiversity. *Current Science*, **82(11)**: 1331-1335.
Smith, K. (1996). Environmental Hazards: Assessing Risk and Reducing Disaster. Routledge, London, U.K.
Spittlehouse, D.L. and Stewart, R.B. (2003). Adaptation to climate change in forest management. *Journal of Ecosystems and Management*, **4:** 1-11.
Tarapoto Proposal (1995). Proposal of criteria and indicators for sustainability of the Amazon forests. Results of the Regional Workshop on the Definition of Criteria and Indicators for Sustainability of Amazonian Forests, 25 February 1995, Tarapoto, Peru. Pro Tempore Secretariat, Amazon Cooperation Treaty, Lima, Peru.
Tucker, R.P. (1993). Forests of the Western Himalaya and the British Colonial System 1815-1914. *In:* A.S. Rawat (editor), Indian Forestry: A Perspective. Indus Publishing, New Delhi.
UN (1992). Non-legally binding authoritative statement of principles for a global consensus on the management, conservation and sustainable development of all types of forests. Report of the UN Conference on Environment and Development, 3-14 June 1992, Rio de Janeiro, United Nations, Vol. III.
UN (2000). United Nations Millennium Declaration. United Nations General Assembly, A/RES/55/2, September 2000, http://www.unisdr.org/eng/mdgs-drr/link-mdg-drr.htm (accessed on 08 July 2008).
UN (2007). Non-legally binding instrument on all types of forests. United Nations General Assembly, Document A/C.2/62/L.5.
UN (2008). The Official list of MDG Indicators. United Nations, http://unstats.un.org/ (accessed on 20 May 2008).
UNEP (1996). Glossary of Environmental Terms. United Nations Environment Program (UNEP), http://www.unep.org/ (accessed on 24 October 2008).
UNEP (2005). UNEP/GRID-Arendal. Trends in Natural Disasters, UNEP/GRID-Arendal Maps and Graphics Library, http://maps.grida.no/go/graphic/trends-in-natural-disasters (accessed on 08 July 2008).
UNEP (2008). Environment and Vulnerability: Emerging Perspectives. UN ISDR Environment and Disaster Working Group, http://postconflict.unep.ch/publications/ (accessed on 22 October 2008).
Verma, M. (2000). Economic Valuation of Forests of Himachal Pradesh. Indian Institute of Forest Management (IIFM), Bhopal, India and International Institute for Environmental Development (IIED), London, U.K.
Wang, X. (2008). GHG Mitigation Policies and Land Use Interactions. *Leadership and Management in Engineering*, **8(3):** 148-152.
WBOED (2000). India: Alleviating Poverty through Forest Development. Evaluation Country Case Study Series. World Bank Operations Evaluations Department, World Bank, http://lnweb18.worldbank.org (accessed on 08 July 2008).
WCED (1987). Our Common Future: The Brundtland Report. World Commission on Environment and Development (WCED). Oxford University Press, New York.
World Bank (2002). A Revised Forest Strategy for the World Bank Group. World Bank, Washington DC, http://lnweb18.worldbank.org (accessed on 20 May 2008).

Participatory Information Management for Sustainable Disaster Risk Reduction

S.V.R.K. Prabhakar* and Md. Shahid Parwez

1. INTRODUCTION

Disasters can result if the risk information, for example in the form of early warning, does not reach the people at risk in time (Pelling, 2003). This is true for both natural and anthropogenic disasters. One example to highlight the importance of communicating risk information in time is the Indian Ocean tsunami of 2004 where a lapse in communicating the tsunami risk warning to India and Sri Lanka allegedly occurred despite the technology and knowhow existed with the U.S. Pacific Tsunami Warning Center and the Japan Meteorological Agency to give a 2-hour advance warning (Goliath, 2005; Zubair, 2005). In this case, getting the advanced early warning could have saved half of the lives lost. Another example comes from even more recent Mumbai, India terror attacks where the available intelligence could not stop the terror attacks from happening in November 2008 (*Deccan Herald*, 2005). While these two examples clearly show the importance of disseminating right information to right people at the larger scale, there also exist examples of such failures at the local level. Such failures were often reported in communicating the cyclone early warning to the remote coastal areas (*The Hindu*, 2000). The contiguous information continuum from the national level to individual level is essential for an efficient transfer of information and risk reduction. Sometimes the information failure occurs due to the lack of trust on the information source on the part of those who receive the information leading to poor response (Leon et al., 2006). This shows that there is a need for a paradigm shift in the way information is generated and disseminated among various stakeholders such that the credibility of the risk information is high leading to effective response and risk reduction.

This chapter proposes a paradigm shift in the disaster risk reduction by emphasizing the need for a participatory process and proposes a conceptual model for an effective information flow among various stakeholders involved in disaster risk reduction. This chapter also provides salient examples of networks which have been successful in disseminating information among the stakeholders with examples from international and national levels. The chapter concludes by emphasizing the need for participatory information management and proper information networking to ensure sustainable disaster risk reduction.

2. PARTICIPATORY DEVELOPMENT PLANNING AND DISASTER RISK REDUCTION

Since time immemorial, the decision making process in development planning has been confined to the domain of researchers and development planning agencies. The target groups, communities, have often

---

*Corresponding Author

been used as 'study material' or 'mere targets' without their involvement in the process of developing and implementing the development programs. Such interventions could not bring out solutions tailored to the needs of the target communities and often resulted in poor adoption. The costs of such interventions were high, could not reach the larger number of needy ones, and their sustainability was often questioned (Platteau, 2002). This has eventually made the developmental institutions to seek community participation in their programs.

Participation has been a basic nature of human activity. However, participation of all the stakeholders in social development has been a recent phenomenon. Participation is the act of sharing by taking part in a group action. Participation is said to have happened when one tries to share physical, emotional, and/or psychological material with others in the group for a common cause. One can look at the participation as a process of information generation and dissemination ensuring the development of a sense of ownership of information on the part of the participants and make them feel accountable for the outcome.

The advent of the concept of participatory development dates back to 1970s when Freire (1972) advocated for alternative form of development different from top-down approach. The idea was that the top-down approaches have failed as they were not inclusive in nature. There are numerous advantages of participatory development which are well documented in the literature (e.g., Jennings, 2000; Platteau, 2002; Khampha, 2008). Some advantages of participatory development are given below:

(1) Enhanced project and program success.
(2) Empowerment of those involved in the development process.
(3) Long-term sustainability of efforts since practices and policies are designed based on the better understanding of local needs leading to better acceptance by the society.
(4) Self-sustaining of developmental efforts.
(5) Better poverty reduction and enhanced and sustainable development.
(6) Better integration and synergism among stakeholders.
(7) Avoids repetition of efforts and waste of resources since everybody knows what others are doing in a participatory setup and hence, projects often have lower costs and higher efficiency.[1]
(8) Emphasis on local knowledge while designing technologies and policy solutions.
(9) Helps in understanding the factors governing the spread of pilot projects into larger areas.

Because of above-mentioned advantages, the participatory development principles have increasingly, but slowly, been employed in social and public medical research and development programs. Subsequently, the concept has spread to other fields such as agriculture where the technology transfer has been known to be hindered due to the absence of location-specific information making it difficult to design tailor-made solutions. The use of participatory development principles and practices in disaster risk reduction is of very recent development as the long history of top-down approaches has failed to produce fruitful results. Traditionally, the disaster risk reduction programs were reactive and non-participatory. Governments have often given an emphasis on relief and rehabilitation with little or no involvement of local people. Even, these relief programs were often not designed keeping in view the specific needs and perceptions of the people who are affected by the disasters. For example, until very recently the drought relief programs in India were confined to the supply of food and fodder with food for work programs, which are decided arbitrarily with little or no consideration of local needs (Prabhakar and Shaw, 2008).

---

[1] There are some reports of higher upfront costs due to the need for training many participants on the program approaches as against the non-participatory approaches. However, the overall costs have always been lower and cost-benefit ratios were reported to be higher due to higher benefits and success rate.

The very reason for a higher emphasis on community participation in disaster risk reduction has been that the communities are the first among those impacted by the disasters, first among those who respond to the disasters, and are highly vulnerable to disasters due to various vulnerability factors such as poverty, lack of appropriate knowledge in risk reduction, lack of capacity to deal with the disasters, and are isolated from governments and other developmental organizations who tend to look at them as 'receivers' rather than equal players in the process. The need for participatory planning, development and information sharing in disaster risk reduction is highlighted by the fact that disaster risks are often spatially varied sometimes governed by the variables such as income, perception of risks and vulnerabilities and hence requires deeper understanding of these location specific factors which cannot be known through non-participatory methods.

3. PARTICIPATORY RISK REDUCTION AS A PROCESS OF INFORMATION FLOW

Disasters are often linked to the failure in information flow from expert or non-expert sources (Pelling, 2003). This emphasizes the need to understand what kind of information is needed and how it should flow so that the risks are reduced by using that information. In this context, participatory information development and sharing assumes importance.

Participatory risk reduction should work by defining the stakeholders (people who affect and or are affected by the actions of an intervention) and seek their active participation in designing and implementing risk reduction processes, interpreting the information generated, and deciding how to use the generated information and end results. Hence, the process of participatory risk reduction can also be viewed as the process of information generation and information flow from one stakeholder to another stakeholder in a multidirectional way which in turn results in desirable outputs.

The overall success of a risk reduction process depends on how the gradient of information flow is set and how the information flow is controlled such that the ultimate sinks (i.e., stakeholders) get benefited. The information flow is enabled through the 'information gradient' at the community end and this is represented by the extent of knowledge that is available with the communities and putting to use of such knowledge for beneficial purposes. This information gradient is mostly formed by the believers and potential believers who often tend to act upon the information received and put them to use thereby increasing the efficiency of the information. The misbelievers often reduce the efficiency of information. Some of the believers and potential believers become champions and pioneers as they accept and act early on new ideas. Here, there are also some classes of people who are satisfied by the information and its results but become more conservative in terms of adoption and enthusiasm generated. Nevertheless, this philosophy of participatory risk reduction information flow assumes that a major portion of target community use the information as they are involved in the process holistically and get benefited gradually. The participants who accumulate more positive information, oriented towards the direction set in the risk reduction process, will become leaders and innovators.

3.1 Types of Information

Broadly, risk reduction information sources can be classified as *electronic* and *non-electronic* types, which are also known as *soft* and *hard* sources, respectively. The velocity and effectiveness of these sources depend mostly on how the user is familiar with them and his perception about the sources. The electronic sources are considered highly skill oriented and hence are often preferred by the educated and often at higher levels of the user hierarchy. Electronic information sources have high velocity of spread as they could be transmitted through internet and other means which work at higher speeds (Table 1).

Table 1. Sources and velocity of information

Information type	Information source	Velocity of information
Non-electronic	Publications	Low
	Demonstrations	High
	Local leaders	Very high
	Communities	High
Electronic	Websites/internet-based databases	Very high
	Information kiosks	High

Digital information sources have become more prevalent and effective media for dissemination. However, such sources are also blamed for creating a digital divide in the society, bigger than the economic divide in some instances. In this age of revolution in information and communication technologies (ICTs), the prosperity lies in using ICTs for the well-being of all the sections of a society. Since ICTs may limit the way the information is spread among the users, they should be chosen with proper attention. Depending on the education levels and access to electronic sources, local communities may predominantly depend on non-electronic form of information including tacit knowledge that is available among the members (Table 2). Electronic type of information becomes more significant as one move up the ladder of education and economic levels and from local to national and international levels.

Table 2. Predominant information types for different stakeholders in risk reduction

Hierarchy of stakeholders	Predominant information type
International	Mostly electronic, non-electronic
National	Mostly electronic, non-electronic
Local governments	Mostly non-electronic
Communities	Mostly Non-electronic, tacit knowledge, electronic.

3.2 Information Needs

For the proper information flow, one must analyze the information needs of various stakeholders in a risk reduction process. Information needs of the target community and its relevance sets the information gradient and speed of information flow, and this varies with the factors like awareness and education levels, socioeconomic and cultural conditions and geographic location of a participant and the information gain is dependent on the fact that whether the participant is active or passive in the process. An effective risk reduction strategy should properly consider this fact and should design and implement its awareness generation programs after assessing the information needs of the communities. Sometimes it may happen that the information being sent out may not be the one needed by the communities and may reduce the overall efficiency.

3.3 Models of Information Flow

Different models of information flow could be observed in the risk reduction process. Figure 1 shows a traditional model of information flow. Here communities and other stakeholders tend to 'react' to the stimuli of disaster rather than work in anticipation and reduce risk. Literature suggests that the memory of a disaster event makes people to be more prepared and the preparedness and awareness levels go down after certain time and attain a peak after the next disaster only (Paton, 2003; Davis et al., 2005; Becker et al., 2001). This emphasizes that the disaster events often compete with social information systems in terms of modifying people's behavior.

In the conventional scheme of information flow (Fig. 1), the arrows emanating from the disaster event are uni-directional rather than bi-directional signifying that there is little or no influence of stakeholders on the disaster as they always tend to respond to it with little or no emphasis on the mitigation of a disaster and its impacts. The figure also shows that the relationship between communities and other intervening agencies is, often, unidirectional in nature.

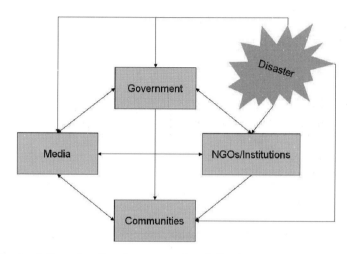

Fig. 1 Information flow in a conventional disaster management system.

3.4 A Conceptual Model for Participatory Information Flow

In advanced and better conceptual model for participatory information flow the stakeholders participating in risk reduction process could be visualized in different spheres of activity influencing the information generation, dissemination and utilization representing sources, sinks and carriers of information respectively. While all the participants in the information chain are the users of the information in varying degrees, they also act as information carriers and sinks at various stages. However, it is difficult to conclude which stakeholder is an absolute sink or source as they also involve in other activities of information flow. Due to this reason, one can only classify them as potential sources, potential carriers and potential sinks. Figure 2 classifies the participants in the risk reduction process into three spheres. It can be seen that the disaster, donors and researchers are potential information sources while the legal, managerial and service providing community tend to be potential carriers. The important target community acts as potential sinks. Within the target community, the lead adopters are active sources as well as sinks too and the rest of the community acts either as potential sinks or as leakage (the unbelievers). However, unlike in the conventional model, here all the linkages are bidirectional which means communities would also act as source of information for all the other actors leading to risk reduction. This very vital aspect of being a source of information could make a significant difference in disaster risk reduction as it

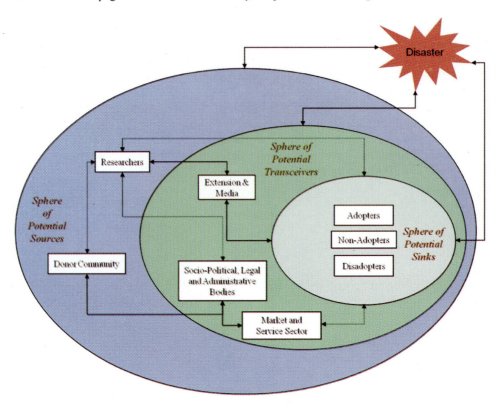

Fig. 2 Conceptual model for participatory information flow: Position of different stakeholders in three information spheres.

enables the other players to better understand the underlying vulnerabilities and risks and hence help in better planning (Pearce, 2004).

4. PRESENTING RISK INFORMATION

Presenting the risk information is the key aspect in a participatory risk reduction process (Fessenden-Raden et al., 1987). Individuals are receptive to information which is relevant to them (Fischhoff et al., 1993). In order to make the risk information suitable to the ultimate stakeholders, i.e., communities, one needs to ask a set of questions in the process of risk information communication. Table 3 provides a list of questions which should often be asked by various stakeholders for the effective information generation and dissemination. Broadly, the presentation of information should be in such a way that it fulfills the objective of information receivers. In addition, the credibility of the information source also plays a major role. This brings us to the concept of '*brand strength*' which can serve as a measure of trust that the end user has on the vehicle through which the information flows in a society (Farr and Hollis, 1997). This market research terminology has potential application in information dissemination for risk reduction. Higher is the brand strength better the impact because information communicated through the most trusted vehicle could initiate response in many end users.

Table 3. Frequently asked questions in information generation and flow

Stakeholder	Key questions to be asked
Donors	• What information is relevant to the clients I am catering to? • How this information sensitizes the target community? • Is this information understandable to the target community? • How my funds were utilized in line with my objectives? • How the information could impact the livelihoods of the target community? • What is the transaction cost?
Researchers	• Is the information relevant to the target community? • Is the information feasible socio-economically? • What would be the output and its worth? • What would be the extent of impact? • What would be the transaction cost involved? • What should be the vehicle of information flow to community? • Does our capacity allow the information to be transmitted effectively? • In what time frame the community would reap the benefits from the information? • What is the impact of information? • What are the externalities?
Legal/Managerial/ Service providers	• Does the information lead to conflict? • How the target community will get benefited? • What is the opportunity cost? • What is the transaction cost? • Is there enough networking?
Target community	• Is the information and its source authentic (social distance and brand distance)? • Is the information relevant for me? • What are the opportunity costs? • Is it feasible for me to act on the information received? • Where do we stand in the process? • Whether to continue participation or not?

5. TOOLS FOR PARTICIPATORY INFORMATION SHARING

There are several tools for collecting and disseminating the risk information in a participatory setup. They are traditionally and collectively called '*participatory appraisal techniques*'. If these techniques are used in a rural background, they are called '*participatory rural appraisal*' (PRA) techniques. These techniques could be effectively used both for information collection as well as for dissemination. For example, the focus group discussion (FGD) could be used to obtain the collective opinion of the community on a particular vulnerability or for knowing their preference of locating a cyclone shelter (for example). At the same time, the FGD can also be used to enhance the awareness of participants on subjects such as gender sensitivity and disaster relief efforts.

Taking the advantage of advanced information technology tools would help in effective generation and dissemination of the disaster risk information. One of such advanced tools which can be effectively combined with the participatory techniques is geographical information system (GIS) (Rambaldi et al., 2006). Participatory GIS can be used for a variety of purposes. For example, participatory tools can be combined with GIS for obtaining the opinion of local communities on the vulnerabilities mapped and risks assessed in a GIS. Participatory tools also help in collecting the information on local vulnerabilities, which can subsequently be input into a GIS. Another example for the combination of advanced information technology and PRA tools is information kiosks. These kiosks have often been used as information hubs offering various services related to e-governance, land records, market prices, weather forecasts, etc. Some noteworthy examples to quote are The World Bank Kiosk project in Maharashtra, India, for distributing the information to sugarcane growers (World Bank, 2009) and M.S. Swaminathan Research Foundation (MSSRF) initiative which has grown into a big alliance called "National Alliance for Mission 2007" and has an objective of establishing information kiosks in every village in India (MSSRF, 2004). These kiosks form a very important resource for the delivery of disaster risk information to the village level.

6. INFORMATION NETWORKING: SALIENT EXAMPLES

It is possible that the effectiveness of the information disseminated is determined by the brand strength of the information vehicle itself. Hence, in order to enhance the strength of the information vehicles, it is advised that many sources of information could form into a network utilizing the synergy among the participating agencies. In such a scenario, the effectiveness of information flow is in direct proportion with the effectiveness of coordination among all the stakeholders. Communities and others confront a plethora of information and knowledge providers in the risk reduction process. If unmanaged, this might lead to confusion (Hussain, 2002). The ineffective information coordination could lead to distortion, misinterpretation, and loss in the process of information communication and transformation from one form to other form. Potential stages of information loss could be when a highly technical material is being converted into semi-technical and non-technical information suitable to common man's understanding. Here the networking and coordinated efforts help making available the relevant information to the target community (Fig. 3). Such networking makes sure that the utility of information is not lost by preserving the essential message as different stakeholders try to put quality control measures at each stage of information generation and transmission.

Fig. 3 Coordinated information flow: Networking of right information sources for better action and empowerment.

6.1 Examples at International Level

Some of the successful examples of networks for risk reduction are International Federation of the Red Cross and Red Crescent Societies (IFRC), ProVention Consortium, and Global Forum for Disaster Risk Reduction (GFDRR) of the International Strategy for Disaster Reduction (ISDR) (IFRC, 2006; ProVention Consortium, 2006; La RED, 2006; ISDR, 2006a). While IFRC is the body that coordinates various national level Red Cross societies for bringing in the disaster risk reduction assistance to the local communities and governments, the ProVention Consortium is a global coalition of international organizations, governments, private sector, civil society organizations and academic institutions for increasing the safety of vulnerable societies and reducing the impacts of disasters. These bodies were largely successful in their objective by effective networking enabling efficient sharing of risk reduction information available with the stakeholders. La RED has been another successful entity formed as a forum in Costa Rica with an objective of information exchange. La RED has also been identified as one of the successful lobbying entities reducing the vulnerability and disaster risks of millions of people in Latin America and Caribbean countries.

The Global Forum for Disaster Risk Reduction (GFDRR), under the International Strategy for Risk Reduction (ISDR), forms a global partnership between governments, UN agencies, regional organizations and civil society. It has an objective of supporting the implementation of Hyogo Framework of Action and guiding various associated networks and platforms. The GFDRR envisages strengthening the partnership between universities, academic institutions and research organizations for disaster risk reduction through knowledge generation, dissemination, and implementation which is also supported by the sustained cooperation in research, innovation, and education at a global scale (ISDR, 2006b). The initiative builds upon the existing knowledge sharing networks among universities, developmental organizations, and civil society and replicates the same at global scale.

6.2 Examples at National Level

Several national examples could be cited from Asia (Table 4). The Indian example for successful and large-scale use of participatory risk reduction principles could be found in the disaster risk reduction initiatives of Government of India (GoI, 2002). The Government of India and the United Nations Development Program (UNDP) collaborative project on disaster risk management (often called GoI-UNDP DRM Program) has put in place different community-based disaster risk management teams (DMTs) at various levels starting from the village to the district, ward, and city levels. DMTs at all levels are linked through the planning process. Thousands of disaster management plans are being prepared by DMTs at village, block, district, ward, and city levels. Each team is made by selecting the members from the concerned administrative level, the members are trained, and plans are prepared in a participatory manner. The Program also envisages developing a GIS-based hazard and vulnerability mapping by using participatory rural appraisal techniques. Participatory approaches are also used for disseminating risk information and for training in emergency evacuation, relief, and recovery skills. The risk information is delivered through DMTs. Disaster information kiosks are being established and one of the members of the DMT is trained to operate these kiosks and disseminate the information to the members of the community in easy to understand manner.

Community driven early warning and communication systems were promoted as a part of the GoI-UNDP DRM program. While there are numerous examples to cite, noteworthy to mention is the community-based early warning system established at Jodya Taluka of Jamnagar district in Gujarat, India (Fig. 4 shows a snapshot of participatory approaches used in the DRM program in India). "The initiative was undertaken in Jodya Taluka of Jamnagar District, Gujarat where 52 DMT members were trained. They were introduced to the key institutions and organizations from national to taluka level involved in early warning, operation of control rooms and different communication equipment. The trainees learn to assemble battery terminals, antennas, coaxial cables, hand-held and base station VHF, HF and UHF sets. During the training exercise and with hands-on practice, the DMT members appreciate how the instruments are used for effective dissemination of information that would enable community to take precautionary measures in the event of any disaster" (Ministry of Home Affairs, 2004).

Table 4. Disaster information networks in the Asian countries

Country	Disaster information system	Remarks
India	India Disaster Knowledge Network (IDKN)	A knowledge portal providing information on different aspects of natural disasters.
	India Disaster Resource Network (IDRN)	Provides information on available resources for disaster response, relief, and reconstruction. Currently open only to the government agencies.
Bangladesh	Disaster Risk Information Network	Established with the help of UNOPS, the network provides information on various disaster risks and connects Disaster Management Information Centers across the country.
Philippines	Disaster Information Database	Provides historical record of different disasters in the country.
Vietnam	Disaster Communication System	Sends early warning messages about flood levels to the provincial dyke management officers.
Sri Lanka	Social Management Information System	Broadly caters to regular governance as well as disaster response time.
Japan	Central Disaster Prevention Radio Network, Prefectural Disaster Prevention Radio Network, Municipal Disaster Wireless Network, etc.	Connects various disaster management officers for the effective management of disasters.
China	Integrated Risk Information Network (iRiskNet)	Established as a part of Integrated Risk Governance of China, the network is supported by the Ministry of Science and Technology and provides information to government and public users.

Fig. 4 Villagers taking part in a participatory village vulnerability mapping process with the help of government officials in one of the Indian villages as a part of the GoI-UNDP DRM program (Photo copyright@Tom Pietrasik/UNDP, 2008).

Some participatory examples can also be derived from developed countries. Noteworthy to mention is the '*Jishu-bosai-soshiki*' system, or '*Jishubo*' in short, meaning 'community-based autonomous organization for disaster risk reduction', established in Japan as per the law stipulated by the 'Basic Law on Natural Disasters' (Matsuda and Bajek, 2007; Bajek et al., 2008). The system comprises a group of people formed into a team at community level for the purpose of participatory disaster management planning, information sharing, and capacity building. The *Jishubo* system is promoted by the government through community-based units called '*Chonai-kai*' or '*Jichi-kai*' (Ajisaka, 2000). *Jichi-kai* means 'neighborhood association'. These associations form the basic community governance unit in Japan. *Jichi-kai* maintains community bulletin board and disseminates the information provided by the local governments, provides avenues for training on various social activities and disaster risk reduction, and organizes festivals. These groups play an active role in disaster risk management planning, regular updating of plans, and conducting mock drills (Fig. 5). These groups often take the opportunity of festivals for community gathering and generate awareness about various developmental programs and disaster risk reduction.

Transfer of risk information to individuals is an area where latest developments in information and communication technology could be effectively used. The advent of mobile communication technologies has opened avenues for sending disaster risk information on to the individual cellular mobile handsets. One such example could be found in Japan where flood, cyclone, and earthquake related information has been effectively disseminated to the people at risk (Hartmann et al., 2004). Tools such as multilingual disaster information systems have also been tried for their applicability in disseminating various kinds of disaster risk information (Miyao et al., 2007). Many mobile phone service providers in Japan provide disaster information services called '*Saigaiji-Navi*'. These services enable the subscribers to locate nearest evacuation area and find major disaster risk management related contact numbers and addresses, etc. on their handsets by connecting to the location maps through Internet (Obi, 2007). In another initiative, the

Fig. 5 Members of *jichi-kai* discussing disaster management plans in a meeting in one of the cities in Japan (Matsuda and Bajek, 2006).

Japan Meteorological Agency (JMA) has established an 'Area Mail Disaster Information Service' which consists of sending earthquake early warning and evacuation messages to mobile handsets. Other services include "My Rescue" Crisis Management Information Mobile Delivery Service which integrates the disaster risk information generated and disseminated by various hazard monitoring and early warning agencies in the country (Government of Japan, 2006).

7. CONCLUSIONS

The field of disaster risk reduction has largely been confined to the specialized agencies and governments who often have ignored to involve the communities at risk in the process of disaster risk reduction planning. In the recent past, there has been a paradigm shift in this approach with an evolving emphasis on the need to involve communities at various stages of disaster risk reduction planning and implementation, which is known as 'participatory risk reduction'. Participatory information management is key to any participatory risk reduction process.

Participatory risk reduction, in fact, is an information generation, dissemination, and assimilation process which needs to be understood properly for achieving sustainable risk reduction. This includes the identification of potential sources, carriers, and sinks, and understanding their strengths and weaknesses in a particular setup. When the information flow is implemented properly, it could lead to desired results what is called 'sustainable risk reduction'. In addition, proper information management through networking and quality control at appropriate stages of information flow could considerably enhance the efficiency of risk information shared among stakeholders. Coordination, networking, and adoption of a holistic approach are the key features of effective information flow in a participatory process. Networks at the local, regional, and global scales have come to stay and are the integral parts of the risk reduction. In this chapter, some examples were highlighted. However, there is a need to look at the functionality of these networks at a greater detail on how they could succeed in transferring the context and location specific information to different stakeholders involved in disaster risk reduction.

While focusing on the participatory risk reduction in this chapter, disasters are considered as a consequence of information loss. The chapter presented the conventional form of risk reduction information generation and sharing and offered a conceptual model for better risk information generation and dissemination. While presenting the risk information in a participatory setup, this chapter also described various forms of information and spheres of information flow and provided a list of important questions to be asked by various stakeholders. It is emphasized that the information networking has become an important aspect of sustainable risk reduction and there is a need to bring in more players to contribute substantially in the information chain.

There are certain challenges to be overcome for promoting the participatory risk reduction and information sharing explained in this chapter. There is a need for greater efforts to network various stakeholders at the beginning of the intervention rather than towards the end. Disaster risk reduction efforts could fail entirely due to improper planning at takeoff stage. Bringing all the stakeholders together is a big challenge especially if the individual agenda is not reflected in the collective action. Some challenges are inherent with any participatory process. Therefore, involving the stakeholders at the project design phase itself is suggested so that they adhere to the agreed objectives, build trust, and ownership. Several sustainability and scaling-up issues were identified during the review of the GoI-UNDP DRM program in India. These include issues such as long-term funding for community-based risk reduction initiatives; mainstreaming the participatory practices in the governance procedures so that things happen 'automatically' and 'continually' than only when funds are available or when there is an external stimulus.

An important limitation that needs to be kept in mind while emphasizing the participatory information management is that all information may not be rightly communicated in a participatory setup. Information that has high technical jargon may lose its value and content if too much simplification is done. Hence, efforts should be made such that alternative paths are identified for the effective dissemination of information which cannot be easily understandable. Another limitation in employing participatory methods is that they are often considered as time consuming and hence may not be effectively used during some phases of disaster management especially when a rapid response is required. However, these limitations could be overcome by planning and collecting information prior to the occurrence of a disaster. For instance, the preferences of different gender classes could be identified during the phase of disaster management planning rather than when the actual relief is to be distributed.

Overall, it can be concluded that there is a greater need on the part of practitioners working with various disaster risk reduction agencies to use their imagination for the innovative use of participatory methodologies. These ideas need not be suggested by governments who implement programs or by donor agencies that fund these interventions. Instead, there are opportunities for practitioners to act within the flexibility available to them while fulfilling their duties leading to a greater impact. Then only a paradigm shift in disaster risk reduction could be realized.

ACKNOWLEDGEMENTS

We acknowledge that the great deal of understanding on this subject came from our working with the Rice-Wheat Consortium for the Indo-Gangetic Plains, an ecoregional program of CGIAR, and Disaster Risk Management Program of United Nations Development Program in India.

REFERENCES

Ajisaka, M. (2000). Local organizations and groups in Japan: Chiefly in neighborhood associations (*Chonan-kai* and *Jichi-kai*). *Hyoron Hsakaikagaku* (Bulletin of Universities and Institutes), **63**: 1-13.

Bajek, R., Matsuda, Y. and Okada, N. (2008). Japan's Jishu-bosai-soshiki community activities: Analysis of its role in participatory community disaster risk management. *Natural Hazards*, **44(2)**: 281-292.

Becker, J., Smith, R., Johnston, D. and Munro, A. (2001). Effects of the 1995-1996 Ruapehu eruptions on communities in central North Island, New Zealand, and people's perceptions of volcanic hazards after the event. *Australasian Journal of Disaster and Trauma Studies*, http://www.massey.ac.nz/~trauma/issues/2001-1/becker.htm (accessed on 20 March 2008).

Davis, M.S., Ricci, T. and Mitchell, L.M. (2005). Perceptions of risk for volcanic hazards at Vesuvio and Etna, Italy. *Australasian Journal of Disaster and Trauma Studies*, http://www.massey.ac.nz/~trauma/issues/2005-1/davis.htm (accessed on 15 March 2008).

Deccan Herald (2005). Mumbai attacks: Chidambaram admits security, intelligence failure. *Deccan Herald*, 5 December 2005. http://www.deccanherald.com/Content/Dec52008/national20081205104854.asp?section=updatenews (accessed on 20 January 2009).

Farr, A. and Hollis, N. (1997). What do you want your brand to be when it grows up: Big and strong? *Journal of Advertising Research*, **37**: 23-36.

Fessenden-Raden, J., Fitchen, J.M. and Heath, J.S. (1987). Providing risk information in communities: Factors influencing what is heard and accepted. *Science, Technology & Human Values*, **12(3/4)**: 94-101.

Fischhoff, B., Bostrom, A. and Quadrel, M.J. (1993). Risk perception and communication. *Annual Review of Public Health*, **14**: 183-203.

Freire, P. (1972). Cultural Action for Freedom. Harmondsworth, Penguin, London, 74 pp.

GoI (2002). Disaster Risk Management Programme: Community Based Disaster Reduction and Recovery through Participation of Communities and Local Governments. Ministry of Home Affairs, Government of India and United Nations Development Program (UNDP), New Delhi, India, http://www.ndmindia.nic.in/EQProjects/goiundp2.0.pdf (accessed on 20 December 2008).

Goliath (2005). Preparing for natural disasters: Indian Ocean Tsunami Warning System to become operational in 2006. Goliath Internet Information Service, http://goliath.ecnext.com/coms2/gi_0199-4972103/Preparing-for-natural-disasters-Indian.htm (accessed on 15 December 2008).

Government of Japan (2006). Japan's Natural Disaster Early Warning Systems and International Cooperative Efforts. Early Warning Subcommittee of Inter-Ministerial Committee on International Cooperation for Disaster Reduction, Government of Japan, Tokyo, March 2006, 27 pp.

Hartmann, J., Okada, N. and Levy, J.K. (2004). Integrated disaster risk management strategy to prevent exposure to hazardous substances due to inundation triggered releases: A concept for Japan. *Journal of Natural Disaster Science*, **26(2)**: 87-93.

Hussain, S. (2002). Challenges of reducing hunger and poverty. The Hindu Survey of Indian Agriculture. *The Hindu*, India, pp. 27-31.

IFRC (2006). Who we are. International Federation of Red Cross and Red Crescent Societies, http://www.ifrc.org/who/index.asp? (accessed on 18 December 2008).

ISDR (2006a). Global Partnership of Universities, Academic Institutions and Research Organizations for Disaster Risk Reduction under GFDRR/ISDR. Video Conference, 13 December 2006. International Environmental House 1, Geneva, Switzerland.

ISDR (2006b). Preparations for the first global platform for disaster risk reduction. International Strategy for Disaster Reduction (ISDR), http://www.unisdr.org/eng/isdr-system/In-a-nutshell.htm (accessed on 18 December 2008).

Jennings, R. (2000). Participatory development as new paradigm: The transition of new development professionalism. Presented at the Conference on Community-based Reintegration and Rehabilitation in Post-Conflict Settings, 5-8 October 2000, Washington DC.

Khampha (2008). Implementing gross national happiness: A participatory development communication approach. http://www.bhutanstudies.org.bt/main/index.php (accessed on 14 November 2008).

La RED (2006). About La RED. http://www.desenredando.org (accessed on 20 November 2008).

Leon, J.C.V., Bogardi, J., Dannenmann, S. and Basher, R. (2006). Early warning systems in the context of disaster risk management. *Entwicklung & Ländlicher Raum* (Germany), **2**: 23-25.

Matsuda, Y. and Bajek, R. (2006). Discussion on the factors to be identified for the sustainability of community based disaster management. Presented at The Sixth DPRI-IIASA Forum on Integrated Disaster Risk Management, 13-17 August 2006, Istanbul, Turkey, http://www.iiasa.ac.at/Research/RAV/conf/IDRiM06/ (accessed on 19 January 2009).

Ministry of Home Affairs (2004). Local level risk management: Indian experience. An Initiative under the GoI-UNDP Disaster Risk Management Program, Ministry of Home Affairs, Government of India, New Delhi, 22 pp.

Miyao, M., Sato, K., Hasegawa, S., Fujikake, K., Tanaka, S. and Okamoto, K. (2007). Multilingual disaster information system for mobiles in Japan. *Lecture Notes in Computer Science*, **4558**: 592-599.

MSSRF (2004). National Alliance for Mission 2007: Every Village a Knowledge Center. M.S. Swaminathan Research Foundation (MSSRF), Chennai, Tamil Nadu, India, 21 pp.

Obi, T. (2007). The role of CIO for disaster management. Presented at the Global CIO Leadership Roundtable: Creating Better Government and Society through Enhanced CIO Leadership, 27-29 October 2007, George Mason University, Virginia, USA.

Paton, D. (2003). Disaster preparedness: A social cognitive perspective. *Disaster Prevention and Mitigation*, **12(3)**: 210-216.

Pearce, L. (2004). Disaster management and community planning, and public participation: How to achieve sustainable hazard mitigation. *Natural Hazards*, **28(2-3)**: 211-228.

Pelling, M. (2003). The Vulnerability of Cities: Natural Disasters and Social Resilience. Earthscan, London, U.K.

Platteau, J.P. (2002). Pitfalls of Participatory Development. Report Submitted to the United Nations Center for Research in the Economics of Development, University of Namur, Belgium, 47 pp.

Prabhakar, S.V.R.K. and Shaw, R. (2008). Climate change adaptation implications for drought risk mitigation: A perspective for India. *Climatic Change*, **88(2)**: 113-130.

ProVention Consortium (2006). ProVention at a glance. http://www.proventionconsortium.org (accessed on 15 August 2008).

Rambaldi, G., Kyem, P.A.K., McCall, M. and Weiner, D. (2006). Participatory spatial information management and developing countries. *The Electronic Journal on Information Systems in Developing Countries*, **25(1)**: 1-9.

The Hindu (2000). DoS admits failure in cyclone early warning. 26 April 2000, *The Hindu*, http://www.hinduonnet.com/2000/04/26/stories/0226000j.htm (accessed on 21 January 2009).

World Bank (2009). Village information kiosks for the Warana Cooperatives in India. The World Bank, Washington DC, http://go.worldbank.org/QV3KOF7VV0 (accessed on 20 January 2008).

Zubair, L. (2005). The tsunami, science and disaster management, http://www.recoverlanka.net/docs/summary.html (accessed on 15 November 2008).

Cyclone Emergency Preparedness in DAE Coastal Installations, India

D.V. Gopinath

1. INTRODUCTION

It may not be out of place here to present a brief account of the origin and development of cyclone emergency management planning in the Indian Department of Atomic Energy (DAE) installations. Reactor Research Centre [the present Indira Gandhi Centre for Atomic Research (IGCAR)] was set up in early seventies on the charming and serene coast of Bay of Bengal, close to Mahabalipuram. It was intended to be a center for the development of advanced reactor technology with Fast Breeder Test Reactor (FBTR) as its core activity. On a heavily overcast day in mid seventies, while a few of us in the Health and Safety group along with experts from other disciplines were intensely discussing about the safety evaluation of FBTR, the then Director of the Centre Mr. N. Srinivasan walked in and said 'You people are working so hard to protect us against the reactor accidents which may never happen. But what have you done, or doing, about an emergency which is highly probable for this area and which may occur any day now? I am talking about the cyclone emergency; there is a warning that the cyclone may cross the coast in this region in the next 48 hours.' He was dead right! We were discussing about the Maximum Credible Accident (MCA) in the reactors with the probability of the order of 10^{-6} per year, while it is well known that every year 2-3 severe cyclones cross the peninsular coast with devastating effects (see Section 2). The Centre, still being in the formative stage with low-lying areas, temporary structures and shacks, asbestos roofing, improvised and inadequate drainage, etc., was highly vulnerable for cyclone damage. Obviously, our priority should have been the planning for management of cyclone emergency at site and it was taken up in right earnest.

There was not much of published literature to help us in our task. State Government and District administration did have some plan, but that did not particularly address the issues involved in the cyclone emergency preparedness of the major industrial installations on the coast. It was essentially confined to warning the vulnerable section of the population and in the extreme cases moving out people from the areas likely to be severely affected. Even that was in a developmental stage. It was recognized that whatever plans the State Governments have or may evolve, it is imperative that major industrial installations located in the coastal area should have their own emergency preparedness/protection plans for internal action as well as for effective coordination with the State Government agencies. This led to the development of IGCAR Cyclone Emergency Preparedness manual in 1978 (KAL, 1978) and it is being periodically updated to take into account the technological developments and changes in the Centre. Around that period, the Indian Rare Earths Limited (IREL), a constituent unit of DAE, was setting up a major industrial installation named Orissa Sands Complex (OSCOM) on the sea coast near Chatrapur in Orissa. OSCOM's main objective was seashore mining of minerals like Ilmenite, Rutile, Zircon, Monazite, Garnet, etc. and upgrading them to value-added products. Though the complex is located in a highly cyclone-prone region, in the initial stages the complex did not have any systematic cyclone emergency management plan. During 17-18 October 1999, a severe cyclone hit the Orissa coast close to OSCOM and this was followed by another super cyclone, with wind speeds of about 250 km/h, hitting the coast in the Paradeep port area. The storm of 17th/18th October 1999 caused significant damage at the OSCAM installation as may be seen from Fig. 1. Even though the emergency at site was managed very well without any loss of life by the dedicated efforts of the plant personnel (Mukherjee et al., 2002), absence of a well thought of emergency preparedness and management plan was acutely felt. Immediate action was initiated to develop such a plan, which resulted in an exemplary cyclone protection and relief manual (IREL, 2001). IREL also operates shore-mining units, similar to OSCOM, at Chavara and Manavalakurichi on the West Coast of the peninsula. Though the west coast is less prone to severe cyclones, protection manuals

Synthetic Rutile Godown Shattered

Damaged Dredge Plant

Damaged Upgradation Plant

Damaged Synthetic Rutile Plant

Production Godown Blown Off

Damaged Acid Regeneration Plant

Fig. 1 Damage caused at OSCOM due to Orissa 1999 Super Cyclone.
(Source: Head, OSCOM, IREL, Orissa)

similar to that of OSCOM were developed for these installations also (IREL, 2003a,b). By then, several institutes and installations in the country, such as SHAR at Sriharikota, SERC at Chennai, Naval base at Visakahpattanam, had programs related to cyclone emergency management. In order to obtain state-of-the-art picture in the field and provide a forum for discussion amongst various teams leading to coordinated efforts, a National Symposium on Cyclone Emergency Preparedness was organized in 2002 at Kalpakkam, Tamil Nadu, India (KAL, 2002). An important outcome of this symposium was identification of the critical need for developing national guidelines in this area. This task was taken up by the National Safety Council, India, resulting in the preparation and issue of guidelines for cyclone emergency management in coast-based industrial installations in 2003 (NSC, 2003).

This chapter, drawing significantly from IREL (2001) and NSC (2003), presents an overview of the requirements and measures to minimize the damage likely to be caused to the coast-based major industrial installations by severe cyclonic storms in India. It begins with a brief description of the Indian cyclones and their damage potential. This is followed by an account of the essential components of a cyclone emergency management plan for industrial installations and basic requirements for such a plan. Finally, a typical response-action plan in cyclone emergency is given. Even though the plan described here is essentially in line with the one developed at OSCOM, Orissa, there is nothing in it preventing its applications to other coast-based industrial installations in general.

2. TROPICAL CYCLONES: CHARACTERISTICS AND IMPACT

A cyclonic storm is a huge rotating mass of moist air in the form of a disc with its horizontal dimensions extending up to 1000 km. The disturbance can extend vertically up to 15 km with peak wind velocities going up to 200-250 km/h (IMD, 2002). Cyclones invariably start as low pressure zones over the ocean bodies and under favorable conditions, sequentially develop into very severe or super cyclonic storms. The different stages in the development of severe cyclonic storms and corresponding peak wind velocities are given in Table 1.

Severe cyclones have a small central region, called the *Eye* of the cyclone, with a diameter of 20 to 30 km. The intensity or the damaging potential of the cyclone is closely related to the pressure deficiency in this region. For very severe cyclonic storms the pressure drop is in the range of 30 to 50 hectoPascals (hPa) going up to 80 hPa for super cyclonic storms. The eye region, which is calm, cloud free, rainless and relatively warm, is surrounded by a region of reducing pressure drop with very tall, convective clouds. This region, known as *eye wall*, produces torrential rain which can be as much as 50 cm in 24 hours (IMD, 2002). Beyond the eye wall region, major convective clouds of the cyclone appear as spiral bands, which are sometimes hundreds of kilometers long and a few kilometers wide. Satellite images of two typical severe cyclones are shown in Fig. 2. These spirals continuously change their orientation and produce heavy rainfall at about 3 cm/h, which can go up to 10 cm/h in extreme cases. At the time of cyclone eye crossing the coast, because of its large pressure deficiency, there will be a rapid rise in sea level close to the place of landfall. This phenomenon, known as the "storm surge", resulting in large-scale inundation of low-lying coastal areas with saline water, is the single major cause of devastation from cyclones.

Table 1. Evolution of tropical cyclones (IMD, 2002)

Grading in the evolution	Associated wind speed	
	Knot	km/h
1. Low pressure area	<17	<31
2. Depression	17-27	31-49
3. Deep depression	28-33	50-61
4. Cyclonic storm	34-47	62-88
5. Severe cyclonic storm	48-63	89-118
6. Very severe cyclonic storm	64-119	119-221
7. Super cyclone	≥120	≥222

Fig. 2(a) Satellite image of the super cyclone crossing the coast at Orissa, India in October 1999. (Source: http:///www.en.wikipedia.org/wiki/1999_orissa_cyclone)

Fig. (2b) Satellite image of Cyclone Leon-Eline crossing the coast in Madagascar, Africa in February 2000 (Cyclone eye is clearly seen in the picture). (Source: www.en.wikipedia.org/wiki/cyclone_leon_eline)

It is well known that the east coast of India and, to a less extent, the west coast are prone to severe cyclonic storms. On an average, 6 to 7 cyclonic disturbances per year originate in the Bay of Bengal and Arabian Sea of which 2 or 3 develop into very severe or super cyclonic storms and cross the peninsular coast. A list of major cyclonic storms crossing the East and West Coasts of India is given in Tables 2(a) and 2(b), respectively.

Risk due to the cyclone is a composite function of its frequency, hazard potential and vulnerability of the community. Vulnerability to cyclone damage depends upon factors like coastal topography, population density and the type of infrastructure. The impact due to storm surge is not uniform all along the coastline. There are certain segments of Indian coast, such as North Orissa and West Bengal coasts, Andhra coast between Ongole and Machilipatnam, Tamilnadu coast south of Nagapattinam and Gujarat coast on the western side, which are more vulnerable to surge effects. In other areas the damage due to wind and rain may be more significant. The characteristics of cyclones of different categories and their impact are given in Table 3. A detailed account of the damage potential of the cyclones is given in the document issued by the India Meteorological Department (IMD, 2002).

As a result of all the devastating potential described above, cyclonic storms, particularly the very severe and super ones, can cause death and untold sufferings to the population besides severe damage to the public and private property in the affected areas. While one can exercise very little control over the occurrence of the cyclones with the associated storm surges, gales and flooding, it is possible to minimize their impact in terms of loss of life and damage to the property by a well-prepared plan for pre- and post-incident actions. Over a period of time, the well-developed emergency management systems based on gigantic strides in the technology for tracking and predicting the course of cyclonic storms and phenomenal improvement in the mass communication systems, have considerably reduced the disastrous consequences of cyclones.

Table 2(a). 60 years-record of devastating cyclonic storms formed in the Bay of Bengal and crossed the East Coast of India (IMD, 2002)

Sl. No.	Date and year	Type of disturbance	Landfall and relevant information
1.	8-11 October, 1967	Severe cyclonic storm	Crossed Orissa coast between Puri and Paradip on the morning of October 9 and crossed Bangladesh coast during the night of 10-11 October. Lives Lost: People 1,000; Cattle 50,000.
2.	26-30 October, 1971	Severe cyclonic storm	Crossed Orissa coast near Paradip, morning of October 30. Maximum Wind Speed: 150-170 km/h. Surge Height: 4-5 m, north of Chandbali. Lives Lost: People 10,000; Cattle 50,000.
3.	14-20 November, 1977	Super cyclone	Crossed Andhra coast near Nizampatnam at 17:30 IST on November 19. Wind Speed: Max. estimated 260 km/h. Surge Height: 5 m. Lives Lost: People 10,000; Cattle 27,000. Damage to Crops and Properties: ~350 crores.
4.	4-11 May, 1990	Super cyclone	Crossed Andhra coast, about 40 km south west of Machilipatnam on May 9. Wind Speed: Max. estimated 235 km/h. Surge Height: 4-5 m. Lives Lost: People 967. Damage to Crops and Properties: ~Rs. 2,248 crores.
5.	5-6 November, 1996	Very severe cyclonic storm	Crossed Andhra coast near Kakinada at midnight of November 6. Maximum Wind Speed: 200 km/h. Surge Height: 3-4 m. Lives Lost: People 2,000 (missing 900). Crops in 3,20,000 hectares damaged
6.	25-31 October, 1999	Super cyclone	Crossed Orissa coast near Paradip at noon of October 29. Wind Speed: Max. estimated 260 km/h. Surge Height: 6-7 m. Lives Lost: People 9,885; Cattle 3,70,297. Crops in 16,50,000 hectares damaged.

Table 2(b). 60 years-record of devastating cyclonic storms formed in the Arabian Sea and crossed the West Coast of India (IMD, 2002)

Sl. No.	Date and year	Type of disturbance	Landfall and relevant information
1.	23-25 May, 1961	Severe cyclonic storm	Crossed coast near Devgad on the night of May 24. Loss and Damage: 5 lakh fruit trees razed to the ground. 1,700 houses completely and 25,000 houses partially damaged.

Contd...

(*Contd.*)

2.	9-13 June, 1964	Severe cyclonic storm	Crossed coast just west of Naliya on June 12. Recorded Wind Speed: Naliya 135 km/h. Lives lost: People 27.
3.	19-24 October 1975	Very severe cyclonic storm	Crossed Saurashtra coast about 15 km to the northwest of Porbandar on October 22. Recorded Wind Speed: Jamnagar 160-180 km/h Surge Height: 4-6 m at Porbandar and Okha. Lives Lost: People 85. Damage: Estimated to be Rs. 75 crores.
4.	31 May-5 June, 1976	Sever cyclonic storm	Crossed coast near Bhavnagar on June 3. Wind Speed: Reported speed 167 km/h. Lives Lost: People 70; Cattle 4,500. Damage: Estimated to be Rs. 3 crores.
5.	13-23 Nov., 1977	Very severe cyclonic storm	Crossed between Mangalore and Honavar in the morning of 22 November. Lives Lost: People 72. Damage: 8,400 houses total and 19,000 partial damage. Estimated loss: Rs. 10 crores.
6.	4-9 November, 1982	Very severe cyclonic storm	Crossed Saurashta coast, about 45 km east of Veraval on November 8. Lives Lost: People 507; Livestock 1.5 lakh. Damage: Thousands of houses collapsed.
7.	17-20 June, 1996	Severe cyclonic storm	Crossed Gujarat coast between Veraval and Diu in the early morning of June 19. Maximum Wind Speed: Veraval recorded 86 km/h. Storm Surge: 5-6 m near Bharuch. Lives Lost: People 47; Cattle 2113. Damage: Houses damaged 29,595. Loss of Property: Rs.1,805 lakhs.
8.	4-10 June, 1998	Very severe cyclonic storm	Crossed Gujarat coast near Porbandar in the morning of June 9. Maximum Wind Speed: Jamnagar recorded 182 km/h. Surge Height: 2-3 m above the tide of 3.2 m. Lives lost: People killed 1,173; missing 1,774. Loss of Property: ~ Rs.1,865 crores.

3. CYCLONE MONITORING AND WARNING SYSTEM

The most important factor contributing to the success of cyclone emergency management is the cyclone warning system, that is, availability of advance information on the movement of the cyclone. IMD provides such cyclone warnings in India and the tropical cyclone warning service is one of its most important functions. It was the first service undertaken by the Department in as early as 1865 and with time, IMD has vastly improved it by continuous updating.

Table 3. Impact of cyclonic storms

Category and wind speed	Structures	Communication and power	Road/Rail	Marine interests	Coastal zone	Overall damage
Deep depression, 52–61 km/h	Minor damage to loose/unsecured structures.	Not significant	Some breaches in *Kutcha* road due to flooding.	Very rough seas. Sea waves about 4–6 m high.	Minor damage to *Kutcha* embankments.	**Minor**
Cyclonic storm, 62–87 km/h	Damage to thatched huts.	Minor damage to power and communication lines due to falling trees.	Major damage to *Kutcha* and minor damage to *Pucca* roads.	High to very high sea waves about 6–9 m.	Seawater inundation in low-lying areas.	**Minor to moderate**
Severe cyclonic storm, 88–117 km/h	Damage to thatched houses/huts. Rooftops may blow off. Unattached metal sheets may fly.	Minor damage to power and communication lines.	Major damage to *Kutcha* and some damage to *Pucca* roads. Flooding of escape routes	Phenomenal seas with wave height 9-14 m. Movement in motor boats unsafe.	Storm surge up to 1.5 m (area specific); damage to embankments/salt pans. Inundation up to 5 km in specific areas.	**Moderate**
(a) Very severe cyclonic storm, 118–167 km/h	Destruction of thatched houses and extensive damage to *Kutcha* houses. Some damage to *Pucca* houses. Threat from flying objects.	Bending or uprooting of power and communication poles.	Major damage to *Kutcha* and *Pucca* roads. Flooding of escape routes. Minor disruption of railways, overhead power lines and signaling system.	Phenomenal seas with wave heights more than 14 m. Visibility severely affected. Movements in motor boats and small ships unsafe.	Storm surge up to 2 m. Inundation up to 10 km in specific areas. Small boats, country crafts may get detached from moorings.	**Large**

Contd...

(Contd.)

(b) Very severe cyclonic storm, 168–221 km/h	Extensive damage to all types of *Kutcha* houses, some damage to old badly managed *Pucca* structures. Potential threat from flying objects.	Extensive uprooting of communication and power poles.	Disruption of rail and road link at several places.	Phenomenal seas with wave heights of more than 14m. Movement in motor boats and small ships not advisable.	Storm surge of 2-5 m, inundation may extend up to 10-15 km over specific areas. Large boats and ships may get torn from their moorings.	**Extensive**
Super cyclonic storm, 222 km/h and above	Extensive damage to non-concrete residential and industrial buildings. Structural damage to concrete structures. Air full of large projectiles.	Uprooting of communication and power poles. Total disruption of communication and power supply.	Extensive damage to *Kutcha* roads and some damage to poorly repaired *Pucca* roads. Large-scale submerging of coastal roads. Total disruption of railway and road traffic due to major damages to bridges, signals and railway tracks. Washing away of rail/road links at several places.	Phenomenal seas with wave heights of more than 14 m. All shipping activity unsafe.	Extensive damage to port installations. Storm surge more than 5 m, inundation up to 40 km in specific areas and extensive beach erosion. All ships torn from their moorings. Flooding of escape routes.	**Catastrophic**

Important requirements of an efficient warning system are: (i) accurate forecasting ability, and (ii) dependable and extensive communication system. Accurate and detailed forecasts of dangerous conditions in sufficient advance time calls for a dependable and powerful cyclone tracking system. IMD's cyclone tracking system is an integrated one consisting of about 560 observatories for taking meteorological data from the earth's surface, 100 observatories making measurements of wind in the upper atmosphere up to altitudes of 20 to 25 km, 35 observations for making measurements of temperature and humidity up to an altitude of 225 km, ships' observations, 10 cyclone detection radars along the coasts and geostationary INSAT satellites (IMD, 2001).

With the help of satellite imagery, a constant watch is kept on the Arabian Sea and Bay of Bengal for the likely genesis of tropical cyclones. Once the cyclonic disturbance approaches the coastline, its subsequent development and movement are monitored by a chain of Cyclone Detection Radars set up by IMD to cover the entire coastal belt. The likely movement of the storms is predicted with the help of track prediction models and by reference to past climatology, which has been built up using 125 years of cyclone data.

Information concerning tropical cyclones and warnings are included in the cyclone advisory or warning bulletins. Types of advisories issued are (IMD, 2001): (i) Bulletins for the high seas, (ii) Coastal bulletins, (iii) Tropical Cyclone Bulletins to All India Radio (AIR) for broadcast, (iv) Port Warnings, and (v) Fisheries Warnings. Cyclone warnings are provided by the IMD from the Area Cyclone Warning Centres (ACWCs) at Calcutta, Chennai and Mumbai and Cyclone Warning Centres (CWCs) at Visakhapatnam, Bhubaneswar and Ahmedabad. The cyclone warning process is coordinated by the IMD's Weather Centre at Pune and the Northern Hemispheric Analysis Centre at New Delhi. Doordarshan and AIR stations at New Delhi are provided Cyclone Warning Bulletins for inclusion in the national telecast/broadcast.

The cyclone warnings are issued at two stages. The first stage warning known as "Cyclone Alert" is issued 48 hours in advance of the expected commencement of adverse weather over the coastal areas. The second stage warning known as "Cyclone Warning" is issued 24 hours in advance. When the cyclone is close to the coast, advisories are issued at frequent intervals. A pre-cyclone watch may be instituted prior to the cyclone alert and a post-landfall outlook is issued for areas in the interior, which may be affected by the cyclone as it continues to move inland and dissipate.

4. POSSIBLE IMPACT OF CYCLONIC STORM ON THE INSTALLATION

In case an industrial installation is exposed to a cyclonic storm and/or severe flooding, the plant operations as well as the infrastructure facilities are quite likely to be affected. To be well prepared for the management of such situations, it is necessary to have a good assessment of the likely damage at site. Such an assessment, based on the site and installation data and the impact data given in Table 3, is presented below.

Power Supply: In the event of a severe cyclonic storm, snapping up of power supply lines is a distinct possibility. Even if the grid supply is not affected, it may become necessary to cut-off the mains power supply to the plant and associated facilities to avoid the possibility of short-circuiting and consequent danger to the personnel or damage to the equipment.

Telecommunications: The cyclonic storm may snap all the cable connections and make the land-based communication system defunct. Mobile telephones could be of use but their relay towers being knocked off cannot be ruled out.

Water Supply: Generally, water requirements of an industrial installation and the associated township are met from a major reservoir or dam across a nearby river, drawn through adequately sized pipelines. If the pipelines are on the surface there is a possibility that the storm surges and severe winds may damage them and disrupt the water supply. If they are sub-surface lines, while the possibility of their breakage is remote, water supply can still be disrupted due to non-availability of pumping power. Further, one has to consider the possibility of impairment of even the captive water supply from the in-situ bore holes as some bore holes located in low lying areas may not be available due to flooding of the area.

Transport Facilities: This has to be considered in two parts: (i) availability of vehicles and (ii) operability of roads. Roads may become inoperable due to (i) blockage of roads by the uprooted trees and muck, (ii) flooding of roads, and (iii) wash out of roads. Probability of non-availability of vehicles may be minimized with on-site maintenance units and provision for parking them at strategic and protected locations.

Tall Structures: High-rise structures such as power transmission towers, chimneys, high-rise structural buildings are likely to be affected by severe winds. The towers and light poles may get uprooted, chimneys may fall and high-rise buildings may collapse. They also pose the danger of hitting the nearby objects during their fall and cause extensive damage.

Fall of Trees: Trees are the most affected ones during the cyclonic storm. They can get uprooted and fall on nearby objects, leading to road blockage, damage to the structures, disruption of the power supply etc.

Some of the above occurrences, particularly the snapping of power lines, flooding and falling objects may lead to casualties and even loss of life.

5. COMPONENTS OF EMERGENCY MANAGEMENT

A complete Cyclone Emergency Management plan has the following four major components: (i) Mitigation, (ii) Preparedness, (iii) Response, and (iv) Relief and Restoration (NSC, 2003).

Mitigation actions aim at lasting reduction of exposure to cyclone hazards. They involve site selection, zoning and compliance with the building code requirements of the cyclone-prone area.

Preparedness involves identifying the problems likely to be faced in a given area under cyclonic conditions, building up all the facilities and equipment necessary to handle such situations, establishing authority and responsibility structures for emergency action and ensuring all the necessary support to them. It also involves laying down a clear-cut action plan for different agencies at different phases from the beginning of the cyclone season.

Response involves phased and time bound action during the cyclone emergency to minimize the loss of life and property. The action sequence starts with the first information about the impending cyclone and lasts till the cyclone dissipates after crossing over the coast.

Relief and Restoration involve ensuring shelter, food, drinking water, medical facilities, etc. to the affected sector. This phase also involves restoring the infrastructure such as power, communication, water supply, road transport and expeditiously bringing back the affected zone to normalcy.

Of the components mentioned above, the mitigation measures are to be incorporated right at the design, site selection and construction stages. Operating organization has little role in this part of emergency management except that in the extreme cases backfitting the existing layout and construction may have to be resorted to. Detailed information on the design criteria for buildings in the cyclone prone area is given in the Guideline document issued by the Structural Engineering Research Centre, Chennai, (SERC, 1998). The remaining emergency management components are briefly described in subsequent sections.

6. FACTORS INFLUENCING EMERGENCY MANAGEMENT

Vulnerability of the installation for cyclone damage depends on several site, plant and infrastructure details; they become important inputs for emergency planning. Some of such important factors are briefly described below.

6.1 Site Data

(1) Geographic location of the site including the *Tahasil*, District and the State. Proximity to the sea and other large water bodies such as backwaters, large lakes, rivers, dams, etc.
(2) Land characteristics such as elevations and gradients at the site. Green belt in and around the site. Drainage characteristics of the soil. Subsurface data such as water table and flow pattern.
(3) Meteorological data including maximum wind speed experienced and maximum precipitation occurring in a spell in the region. It also includes cyclonic and flood history of the region.
(4) Distance of the installation from major urban centers and railway stations. Type of road linkages to these centers, such as highways, all weather roads etc. Waterways, if any, assume importance.
(5) Neighborhood details including settlements and major industrial installations.

In case the installation encompasses more than one non-contiguous site and/or has an associated township, the above data are required for each one of the sites and the township.

6.2 Plant Data

(1) Concise description of the plant operations, including the type and magnitude of the input materials and final products.
(2) Concise description of special and/or hazardous materials, if any, being handled — their type, magnitude, storage and handling facilities.
(3) Number of persons employed in different shifts and their location details.
(4) Plant layout with description of the major structures, temporary buildings, low-lying structures, large structures and structures with sheet roofing.
(5) Auxiliary facilities and utilities such as boilers, captive power supply, air handling systems, waste treatment and waste disposal systems, etc.

6.3 Infrastructure Details

Invariably, industrial installations are supported by several infrastructure facilities and in case of a cyclonic storm and/or flooding, these facilities are quite likely to be affected. Hence, an important aspect of cyclone emergency preparedness is to protect them to the extent possible and have contingency plans. Important infrastructure facilities and the required information about them for emergency planning are listed below.

6.3.1 Power Supply

(1) Full-load power requirement.
(2) Power source and the distance of the installation from the point at which power is drawn from the grid.
(3) Details of the distribution system within the installation, with appropriate line drawings.
(4) Details of the captive/standby/emergency power supply system.

6.3.2 Water Supply

(1) Total requirements of water for process and drinking purposes.
(2) Primary source of water supply, its distance from the installation and mode of water transport (surface pipeline, underground pipeline, canal, etc.).

(3) Power requirements for water supply and the provision for the same during normal as well as abnormal situations.
(4) Water storage facilities within the installation, their capacity and distribution.
(5) Captive water sources such as bore wells and wells—their capacity, location with respect to different units within the installation and elevation. Means of lifting and supplying water from such sources (provision of pumps and DG power supply, etc.).

6.3.3 Drainage System
(1) Routing of the storm water, sanitary and plant effluent drainage lines with line drawings.
(2) Type of plant effluent drainage system; underground or surface line; if surface line, open channel or piped system.
(3) Drainage pumps if any and their power supply system.
(4) Ultimate destination of the drainage lines.

6.3.4 Telecommunication
Normally, the installation will have multi-tier telecommunication system such as:
- Internal PABX
- External PABX (possibly with DID facility)
- Direct P&T lines
- Wireless communication system
- Emergency communication system

With advancing technology, the installation may also have a network of mobile phones. Data required about these systems are:

- Type and capacity of the PABX exchanges and whom they cater.
- Location and elevation of the exchange housing in the context of their flood-prone nature.
- Backup time of the power supplies for communication systems.
- Arrangements for contacting the outside Liaison Office, Corporate Office and District authorities in case of an emergency.

6.3.5 Medical and First-Aid Facilities
Information required on the medical organization and facilities in the installation, and the township includes
- Number of medical and paramedical staff available.
- Ambulance services and location details of medical centers (including those in the township, if any) and dispensaries.
- First-aid system operating in the installation including the details of first-aid training program, number of trained staff members available at any time in the plant and distribution of the first-aid facilities within the installation.
- Major medical centers and facilities available in the neighborhood and contact details for the same.

6.3.6 Transport Facilities
Generally, installations have three types of transport facilities as follows:
- Mass transport system for bringing the staff from their residential area.
- Light vehicle fleet for personnel transport.

- Heavy vehicles for material transport.

Data required on these facilities are:
- Type and number of vehicles and their capacities.
- Type of fuel used and extent of the fuel storage at site.
- Operation and maintenance, plant controlled or controlled by outside agencies.
- Parking arrangements and availability of their services on emergency.

6.3.7 Residential Colony
Major industrial installations would generally have residential colonies managed by the installation for its work force, for which the following information is required:
  (i) Distance from the Plant and the type of linking road.
  (ii) No. of residential units and the type of construction.
  (iii) Approximate size of the colony population.
  (iv) Infrastructure facilities such as hospital, community centre, school and college buildings (which can serve as emergency shelters) and shopping centre (which can serve to provide essential commodities during emergencies).
  (v) Water supply; details about the source, storage and distribution.
  (vi) Power supply; requirements, sources and distribution, and emergency arrangements.

7. EMERGENCY RELATED FACILITIES AND EQUIPMENT

Protection measures in the emergency management involve the following actions:
- Safe shutdown or scale down of operations.
- Safe storage of hazardous materials.
- Providing emergency power supply.
- Providing emergency water supply.
- Providing emergency food and fuel.
- Providing emergency transport.
- Providing emergency Medical and First-aid Rescue operations.
- Evacuation and providing safe shelters for evacuees.

These are described in detail in the Relief Action Plan (Section 9). Preparedness for all these actions calls for special facilities, special equipment, emergency organization with well-established hierarchy and clear-cut responsibilities and an unambiguous action plan. Some major facilities related to the emergency preparedness are described below and the emergency organization is described in Section 8.

7.1 Emergency Control Center

During the emergency period (commencing from the receipt of the cyclone alert message till the crossover and dissipation of the cyclone), all the activities need to be coordinated from a central, well-identified place and for this purpose it is necessary to establish a control center. The control center should be at a location, which is easily accessible and least likely to be affected by the cyclone storm. It should be provided with facilities such as:

(i) Emergency power supply
(ii) P&T lines, wireless telephones and Cell phones
(iii) Radio and TV sets
(iv) First Aid kit
(v) Emergency lights
(vi) Cyclone Emergency Manual
(vii) Prominent display of vital phone numbers and shelter details

It is essential that all the above equipment is always maintained in trouble-free, operating condition. The control center should have facility for meeting the Cyclone Action group with provision for displaying the status of the cyclone storm, protective tasks and task forces. The control center should be manned round the clock during the emergency period.

7.2 Shelters

In a cyclonic storm, it is likely that the personnel from their operating areas and buildings (in extreme cases, even the families in the township from their dwellings) may have to be moved to safer places. Hence, it is essential that there should be well-identified places to house such personnel. The shelters should satisfy the following criteria:

(i) Spacious enough for the number of persons expected to be accommodated from the area.
(ii) Strong enough to withstand the winds with the anticipated velocity.
(iii) Not prone for flooding.
(iv) Easily accessible.
(v) Have adequate drinking water supply and toilet facilities.
(vi) Provided with adequate quantity of dry food such as biscuits and bread.
(vii) Equipped with emergency lighting and first aid facilities.

Adequate number of shelters, complying with the above criteria, should be identified in the plant site as well as township. The emergency plan should clearly bring out the distribution of the shelters, their capacity and the area to which they cater. It should also have information on the location and other details of the shelters provided by the district authorities in the neighborhood.

7.3 Emergency Equipment

During the cyclone emergency, normal life would be disturbed due to the failure of power and water supply, flooding, road blockage, personnel injury etc. To manage such situations it is necessary to have ready access to special equipment such as:

- Portable generators.
- Portable pumps.
- Mobile PA system.
- Walkie-Talkie sets or Mobile phones.
- Petrol operated saw.
- Life Buoys/Belts.
- Resuscitators.
- Rain coats and gum boots.

- Torch lights/Search lights.
- PVC hose and manila rope.
- Shovels, *gamelas* and spades.
- Personal protective equipment (if the plant is handling hazardous material).

The emergency plan should make provision for adequate number/quantity of all the above equipments in good and operable condition. The plan should have information on their availability, number and distribution.

8. EMERGENCY ORGANIZATION

To effectively implement all the protection measures for minimizing the damage by the cyclone and provide relief measures for expeditiously bringing back the conditions to normalcy after the cyclone, it is essential that the installation should have a *Cyclone Protection and Relief Organization* with a clear-cut hierarchical structure and responsibilities. A general scheme for such an organization is given below (what is given is only a general scheme. Installations, depending on their conditions, may have different but effective organizational schemes).

8.1 Installation Cyclone Protection Committee (ICPC)

The ICPC is the apex committee at site for coordinating all protection and relief measures. It will activate all the protection-related action groups in the installation and endeavor to continuously keep in touch with the Corporate Office and the District authorities. It will be headed by the Unit Head or a very senior officer of the installation commanding all the authority at the time of emergency. It will have as its members, heads of important groups, one amongst whom would be the convener of the Committee. In the absence of Chairman, the Convener will take over and carry out all the functions of the Committee.

Zonal Action Committees (ZAC): In case the installation encompasses more than one non-contiguous zone, each zone will have a Zonal Action Committee (ZAC). The respective ZAC would be responsible for the implementation of the protection and relief measures within the zone. All the Zonal Committees report to ICPC and act under its overall guidance. One of the Zonal committees, designated as ZAC(T), is responsible for all the protection measures in the township.

8.2 Service Groups (SGs)

For timely and efficient implementation of the protection and relief measures, the Action Committees would require dedicated services of several technical groups. To meet this requirement, the emergency organization will have several Service Groups (SGs). Some of the important SGs are:
- Electrical Services (ESG)
- Civil Engineering Services (CESG)
- Mechanical Services (MSG)
- Water Supply and Utilities Services (W&USG)
- Telecommunication Services (TSG)
- Road Restoration and Transport Services (RTSG)
- First-Aid and Medical Services (F&MSG)
- Rescue and Security Services (RSSG)

Depending on its own need, the installation may identify additional SGs. Further, based on the anticipated workload, each SG may have appropriate number of persons, headed by a senior person in the respective area. The Group should have clearly specified alternate Head to implement the tasks of the Group in the absence of the designated Head. The SGs are activated by ICPC and they work in tandem with the Zonal Action Committees (ZACs). Functions of these committees prior to, during and after the emergency are described in Section 9.

9. PROTECTION AND RELIEF ACTION PLAN

9.1 Primary Emergencies

Protection and Relief Action Plan is the most important operative part of emergency management. To avoid too many false alarms inevitably leading to complacency on subsequent occasions, the prescribed precautionary measures and the action taken should be commensurate with the gravity of the situation. Therefore, the action plan is drawn up based on the warning sequence available to the installation. For instance, warnings may be available at 48, 24 and 12 hours in advance of the impending cyclone storm. Accordingly, the actions to be taken by different components of the emergency organization would be in the following sequence:

(i) Action at the start of the cyclone season.
(ii) Action to be taken at 48 hours warning.
(iii) Action to be taken at 24 hours warning.
(iv) Action to be taken at 12 hours warning.
(v) Restoration action.
(vi) Post-emergency review.

Actions to be taken during the different phases mentioned above by different components of emergency organization are given in Tables 4 to 8. In order to facilitate identifying various tasks for different groups and drawing up the detailed action plan, these tables are in line with a plan in existence in one of the units of DAE (IREL, 2001). Needless to say that this is only an example and each installation has to draw up its plan based on a thorough review of site conditions.

9.2 Secondary Emergencies

The operations in some installations may involve hazardous materials such as radioactivity, chemical toxins, alkali metals, etc., calling for site-storage of such materials in significant quantities. It is likely that the cyclonic storms may damage their containment leading to secondary emergencies. Managing such emergencies would depend on the type and magnitude of material involved and this chapter does not specifically address such secondary emergencies. However, it is mandatory that all the installations handling hazardous material should have emergency plans against the loss of containment of such materials. Those plans need to be integrated with the general plan described in this chapter.

10. AID AND COOPERATION

Quite often, the installation is located away from major urban centers and it would be surrounded by villages with improper dwellings and inadequate infrastructure. In case of a cyclone emergency, the district authorities are expected to conduct rescue and relief operations in such villages. However, because

Table 4. Tasks of different action groups at the start of the cyclone season[*]

Action group	Tasks
ICPC and ZAC(P)	Inspect the plant site and ensure that: (i) Window doors and latches in all the buildings are in proper condition; arrange to repair or strengthen wherever necessary. (ii) Sheet roofing of all the structures is in good condition and wherever necessary arranges for their proper fastening for holding against storms. (iii) Plant surroundings are clear of loose objects. (iv) All the shelter areas in the plant site are properly equipped and their access is well maintained.
ICPC and ZAC(T)	Inspect the township and the surroundings for loosely lying objects and ensure their clearance. Ensure that the shelters in the township and their access are well maintained and all the necessary facilities are available at the shelters.
ZAC(T)	(i) Issue notice to the residents to report to civil section regarding defective doors, windows etc., if any, for rectification. (ii) Inspect the Medical Center and ensure adequate stock of first-aid materials and medicines.
CESG	(i) Inspect all the storm drainage systems at the plant site as well as the township and arrange for their de-clogging and proper maintenance. (ii) Maintain constant surveillance on installations by way of periodic inspection for protection against heavy rains and strong winds.
EESG	Check temporary power lines, if any, in different buildings and ensure their adequate protection against wind and rain.
MSG	(i) Maintain constant surveillance on installations by way of periodic inspection for proper support of pipelines, vessels, equipment, etc. and ensure no leakage through any joints, valves, flanges, etc. (ii) Maintain emergency stock of nuts, bolts, gaskets, valves, slip plates, tools, etc.

[*]In Tables 4 to 8, ZAC(P) refers to Zonal Action Committee for the specific plant site. The other abbreviations are as given in Sections 8.1 and 8.2.

Table 5. Tasks of different action groups at 48 hours warning

Action group	Tasks
ICPC	(i) Activate ZAC(P) and ZAC(T) and alert all Service Groups. (ii) Establish and maintain contact with the concerned District or State Authorities and the Corporate Office.
ZAC(P)	(i) Ensure the operability of DG sets and emergency lighting systems. (ii) Ensure that the bore-well is functional with DG power. (iii) Identify areas, which require manning to operate essential services. (iv) Ensure availability of adequate stock of diesel/petrol for emergency transport including site evacuation. (v) Ensure that all the emergency equipment is at the designated locations and operational. (vi) Ensure that the shelters are properly equipped.
ZAC(T)	(i) Instruct the colony residents to stock food for about a week. Instruct Cooperative Stores and other shops in the township to keep adequate stock of food items and other essentials. (ii) Ensure that the Medical Service Group is fully prepared with enough stock of appropriate medicines and first aid kits. (iii) Ensure that the DG sets are operable. Ensure that the power supply to the drinking water pump, Health Center and shelters is available.

Table 6. Tasks of different action groups at 24 hours warning

Action group	Tasks
ICPC	(i) Inform the Corporate Office about the development. (ii) Inspect the plant site for compliance of all emergency related requirements. (iii) Defer any previously planned activity requiring more than 12 hours.
ZAC(P)	(i) Alert security and transport personnel. (ii) Ensure that a minimum of 2000 m^3 of water is kept in the water reservoir, controlling the water supply to the plant. (iii) Ensure availability of emergency vehicles at the designated places.

Table 7. Tasks of different action groups at 12 hours warning

Action group	Tasks
ICPC	(i) Continuously monitor the situation and maintain contact with the District Authorities and the Corporate Office. (ii) Coordinate rescue operations. (iii) Depending on the severity, advise plant/site evacuation.
ZAC(P)	(i) Coordinate rescue operations with the Service Groups. (ii) On advice from the ICPC, initiate plant/site evacuation.
ZAC(T)	(i) Coordinate the rescue operations with Service Groups. (ii) Depending on the severity, advise the colony residents to remain indoors.
SGs	Provide the required services for rescue and relief operations.

Table 8. Tasks of different action groups immediately after the cyclone event

Action group	Tasks
ICPC	(i) Continuously monitor the relief operations and provide the overall coordination. (ii) Keep the Corporate Office informed of the status. (iii) After the site has reached a reasonable degree of normalcy, along with ZAC(T), ZAC(P) and SGs review all the preparatory, rescue and relief operations. Identify the shortcomings and take corrective measures.
ZAC(P), ZAC(T) and SGs	Continue the relief operations to bring back the site to normalcy expeditiously.

of logistic problems, such aid may not be forthcoming on time and it is quite likely that the villagers look for some relief from the installation. It is prudent to anticipate such an eventuality and make adequate provision for this contingency in the cyclone emergency plan of the installation. Towards this end, the plan should have information on the surrounding villages from which population may look for help. The type of help that is planned to be provided such as food, drinking water, shelter, medical aid, etc. should be clearly brought out in the plan.

There could be other major industrial installations in the neighborhood. While each one of such installations is expected to have its own 'stand-alone' emergency plan, mutual aid and cooperation amongst the installations may be very helpful on certain occasions. Hence, the plan should have information on the nearby major industrial installations, status of their emergency management plan and the protocol to receive and provide aid such as transport, medical facilities, food, etc. in case it is required during the cyclone emergency.

11. EDUCATION, TRAINING AND EXERCISES

For effective emergency management, it is necessary that all the persons concerned should be well acquainted with emergency plan and they should be well aware of their duties and responsibilities during the emergency period. This calls for well-established training program for (i) building up the awareness amongst the installation work force/township residents about the cyclone emergency plan, (ii) preparing the action groups to effectively handle their duties during the emergency, and (iii) periodic mock exercises to test the readiness of the emergency organization.

12. FEEDBACK AND REVIEW

No emergency plan, no matter how well deliberated and written, can be considered perfect and experience may bring out the deficiencies/shortcomings in the plan. Further, the conditions in the installation may change over time; there can be addition of new units/facilities or change in the process conditions. Therefore, the emergency document describing the preparedness and action plan needs to be updated periodically. Revision number and date of revision of the prevalent action plan should be clearly indicated on the emergency document.

13. CONCLUDING REMARKS

Notwithstanding the fact that on an average 6 to 7 cyclonic disturbances per year originate in the Bay of Bengal and Arabian Sea, of which 2 or 3 develop into very severe or super cyclonic storms and cross the peninsular coast, major industrial installations do exist on the peninsular coast. While there is very little one can do to prevent the occurrence of cyclones, a well-defined emergency response plan can greatly help reducing the misery and loss of life in the cyclone-prone areas. An essential requirement for an effective cyclone emergency planning is a good cyclone monitoring and warning system. With continuous upgradation through decades, the India Meteorology Department has made our cyclone monitoring and warning system as one amongst the most modern in the world today. Furthermore, with the advances in space and telecommunication technology, there have been gigantic strides in mass communication systems. Utilizing these developments, district and state administrations have prepared elaborate and effective emergency rescue and relief plans for the general public in the affected areas. However, the industrial installations in the cyclone-prone area, because of their high concentration of equipment, materials and personnel, should have their own elaborate cyclone emergency preparedness and response plans. They serve to efficiently organize their internal emergency response action as well as to coordinate with the public administration. Over a period of time, such elaborate plans have been evolved in the coast-based installations of the Department of Atomic Energy, India. Based on this experience, the National Safety Council of India has generated guidelines for cyclone emergency preparedness for coast-based installations in India in general. With these developments, it is hoped that the coastal industrial installations would be well prepared to manage the cyclone emergencies with a minimum loss of life and property.

ACKNOWLEDGEMENTS

Thanks are due to Dr. S.K. Satpathy for helpful discussions regarding the cyclone emergency plan at Kalpakkam and Sri M.S. Roy for providing photographs of the damage at OSCOM caused by the cyclone that crossed the Orissa coast in 1999.

REFERENCES

IMD (2001). Cyclone Warning Services. India Meteorological Department (IMD), Government of India, New Delhi, www.imd.in/cyclone/cyclone-warning-services (accessed in June 2008).

IMD (2002). Damage Potential of Tropical Cyclones. India Meteorological Department (IMD), Government of India, New Delhi.

IREL (2001). Cyclone Protection and Relief Manual—Orissa Sands Complex. Indian Rare Earths Ltd., Orissa, India.

IREL (2003a). Cyclone and Flood Protection and Relief Manual—Manavalakurichi. Indian Rare Earths Ltd., Manavalakurichi, Tamil Nadu, India.

IREL (2003b). Cyclone and Flood Protection and Relief Manual—Chavara. Indian Rare Earths Ltd., Chavara, Kerala, India.

KAL (1978). Cyclone Protection and Relief Manual for IGCA. Indira Gandhi Centre for Atomic Research, Department of Atomic Energy, Kalpakkam, India.

KAL (2002). Proceedings of the DAE Symposium on Cyclone Emergency Preparedness. Indira Gandhi Centre for Atomic Research, Department of Atomic Energy, Kalpakkam, India.

Mukherjee, T.K., Siddiqui, A.S. and Maharana, L.N. (2002). Experience of cyclonic storm at Indian Rare Earths Ltd., OSCOM and post disaster management. Proceedings of the DAE Symposium on Cyclone Emergency Preparedness, Indira Gandhi Centre for Atomic Research, Department of Atomic Energy, Kalpakkam, India.

NSC (2003). Guidelines for the Preparation of Cyclone Emergency Management Manual for Coast-based Industrial Installations. National Safety Council (sponsored by Ministry of Labour, Government of India), Navi Mumbai, India.

SERC (1998). Guidelines for Design and Construction of Buildings and Structures in Cyclone-prone Areas. Structural Engineering Research Centre (SERC), CSIR, Government of India, Chennai.

Sustainable Management of Disasters: Challenges and Prospects

Madan Kumar Jha

1. CHALLENGES OF SUSTAINABLE DISASTER MANAGEMENT

The major challenges of the 21st century are: how to achieve *water security, food security, energy security* and *environmental security*? Because of growing incidence of natural and anthropogenic disasters in different parts of the world, these crucial challenges become much more complex and daunting, especially for developing and low-income countries. On the top of it, climate change is expected to exacerbate disaster risk and disaster management problems that the world already faces. The warming of climate system is indisputable as is evident from the observations such as increase in global average air and ocean temperatures, shrinking of glaciers and snowfields, rise in global mean sea level, change in rainfall patterns and intensity, gradual loss of biodiversity, etc. The IPCC studies predict that climate change will increase droughts, heat waves and fires in some areas, while in other areas, more intense tropical storms and higher precipitation will increase floods, landslides and mud slides (IPCC, 2007a). Global climate change and socio-economic changes have implications for food production, natural ecosystems, freshwater supply, health and hygiene, etc. Climate change is considered as a major challenge to the efficient management of natural resources and a barrier to the transition from poverty to prosperity (UNDP, 2007). According to GWP (2009), it is arguably the most severe long-term threat to development faced by present generation as well as future generations. Specifically, for the developing and low-income countries, climate change is a foremost threat to sustainable development because they are most vulnerable and worst sufferers. Thus, the gravity of the problem faced by mankind at the dawn of the 21st century is apparent from Kauai Declaration (2007): *"We are losing our cultural heritage at a rate that will seriously diminish our opportunities to achieve sustainability in the future"*.

Apart from the above-mentioned four major challenges, it is worth to identify the challenges of sustainable disaster management that can lead to a disaster-resilient society. In fact, there are multiple challenges of sustainable disaster management and they have direct or indirect links with the major challenges mentioned above; the poorest countries, as always, will face the greatest challenges. These challenges can be summarized as follows (modified after Coppola, 2007):

(1) Combating global climate change and socio-economic changes. Climate change is one of the most important global environmental challenges faced by humanity in the 21st century. Having a large domain of influence on human and environmental systems, the climate change also poses several challenges in different sectors/fields as well as increases the complexity and scale of those challenges.

(2) Generating political willpower and dedication for disaster mitigation.

(3) Strengthening coordination and collaboration between central and state government organizations as well as allied government organizations responsible for natural resources management and/or disaster management.

(4) Creating public awareness about disaster prevention and promoting the culture of pre-disaster preparedness and mitigation.
(5) Improving early warning and forecasting systems as well as knowledge sharing and communication systems.
(6) Building institutional capacity at all levels to develop comprehensive disaster management capacities.
(7) Developing strong and coherent coordination among government and non-government organizations, including international organizations before and during disaster response.
(8) Ensuring effective coordination and information sharing between news media and disaster management officials as well as honest and accurate dissemination of facts and figures by the news media. The news media can play a significant role in disaster preparedness and response.
(9) Curbing increasing and widespread corruption, especially in developing nations. Needless to mention that the corruption not only complicates management problems but also poses a major obstacle to proper human development in most developing nations. The eradication of widespread corruption is one of the biggest challenges of this century.
(10) Maintaining equality in humanitarian assistance and relief distribution as well as ensuring efficient distribution of disaster funds by the donors during emergency situations, particularly during composite disasters.

Except for the first and fifth challenges that have both technical and institutional components, the remaining challenges of sustainable disaster management are solely concerned with the institutional aspect of disaster management which increases the complexity of these challenges. Nevertheless, these challenges should be seriously understood and addressed by the policy makers, resource managers and disaster managers at local, national and international levels in order to ensure sustainable disaster management in both developed and developing countries of the world. The research and development efforts required to overcome some of these challenges are discussed in the subsequent section.

2. FUTURE RESEARCH AND DEVELOPMENT NEEDS

Natural hazards cannot be eliminated. However, anthropogenic hazards can be minimized provided that communities are informed and resilient, and ecosystems are allowed to perform as they should ideally perform as much as we can. Thus, disaster risk reduction is an integral part of sustainable development and global fight against poverty (UNDP 2004, 2007). However, reducing disaster risks has not been on a high priority by many governments because the politicians are too often disinclined to allocate scarce resources to preventing something which may not occur during their period of office and the prevention effort may remain invisible if it is successful (UNESCO, 2007). As a result, relief and rehabilitation constitute the primary form of disaster risk management and account for most of expenditure on disaster-related activities annually, leaving a very low balance for preventive measures. This trend is also obvious from the case studies presented in this book. Apart from this institutional/political problem, there are many other institutional and technical issues which need to be properly addressed in the future to overcome the challenges of disaster management mentioned in the previous section. As to the future research and development needs for sustainable disaster management, I believe that the recommendations of the Hyogo World Conference on Disaster Reduction held in January 2005 should be strictly followed for efficient disaster management, which in turn can ensure our future life safer and sustainable. Besides these recommendations, some other related research and developments needs are also discussed in this section.

2.1 Adherence to the Hyogo Recommendations

As discussed in Chapter 1, there is a global recognition that more concerted efforts are needed by all countries to minimize global vulnerability and ensure safer future. The 'Hyogo Framework for Action' sets out priorities for disaster risk reduction and calls upon the international community to take pragmatic steps to make nations and communities safer by 2015 (UNISDR, 2005). The recommendations of the Hyogo conference, which provide excellent guidelines for comprehensive disaster management, must be strictly followed by the disaster-prone countries with top priority. Furthermore, the future research and development needs for sustainable disaster management are also elaborated in the Report of the 2005 World Conference on Disaster Reduction in Hyogo, Japan (UNISDR, 2005), some of which are highlighted below for the quick benefit of the readers:

(1) "Taking into account the importance of international cooperation and partnerships, each State has the primary responsibility for its own sustainable development and for taking effective measures to reduce disaster risk, including for the protection of people on its territory, infrastructure and other national assets from the impact of disasters. At the same time, in the context of increasing global interdependence, concerted international cooperation and an enabling international environment are required to stimulate and contribute to developing the *knowledge, capacities* and *motivation* needed for disaster risk reduction at all levels".

(2) "An integrated, multi-hazard approach to *disaster risk reduction* should be factored into policies, planning and programming related to sustainable development, relief, rehabilitation, and recovery activities in post-disaster and post-conflict situations in disaster-prone countries".

(3) "A gender perspective should be integrated into all *disaster risk management policies, plans* and *decision-making processes,* including those related to *risk assessment, early warning, information management,* and *education* and *training.* Also, cultural diversity, age, and vulnerable groups should be taken into account when planning for disaster risk reduction, as appropriate".

(4) "Both communities and local authorities should be empowered to manage and reduce disaster risk by having access to the *necessary information, resources* and *authority* to implement actions for disaster risk reduction".

(5) "There is a need to enhance international and regional cooperation and assistance in the field of disaster risk reduction".

(6) "The promotion of a culture of prevention, including through the mobilization of adequate resources for disaster risk reduction, is an investment for the future with substantial returns. Risk assessment and early warning systems are essential investments that protect and save *lives, property* and *livelihoods,* contribute to the *sustainability of development,* and are far more *cost-effective* in strengthening *coping mechanisms* than is primary reliance on post-disaster response and recovery".

(7) "There is also a need for proactive measures, bearing in mind that the phases of relief, rehabilitation and reconstruction following a disaster are windows of opportunity for the rebuilding of livelihoods and for the planning and reconstruction of physical and socio-economic structures, in a way that will build community resilience and reduce vulnerability to future disaster risks".

(8) "Disaster risk reduction is a cross-cutting issue in the context of sustainable development and therefore an important element for the achievement of internationally agreed development goals, including those contained in the Millennium Declaration. In addition, every effort should be made to use humanitarian assistance in such a way that risks and future vulnerabilities will be lessened as much as possible".

2.2 Adoption of Modern Management Approaches

As mentioned above, concerted efforts should be made to reduce the number and effects of natural and anthropogenic disasters so as to save the earth by ensuring sustainable development. Sustainable development leads to a better life for the present generation and survival for future generations (Munier, 2005). The widely used definition of *"sustainable development"* is (WCED, 1987): *"Development that meets the needs of the present without compromising the ability of future generations to meet their own needs"*. In other words, 'we have no moral right to pass on a degraded environment to our future generations'. Thus, the principle of sustainability demands for more serious and comprehensive examination of the long-term *environmental, economic,* and *social* impacts of proposed natural resources development following *holistic* and *multidisciplinary* (also called *interdisciplinary*) approaches (Newson, 1992; Munier, 2005; Marshall and Toffel, 2005). The casual use of "sustainability" simply as a synonym for "good" or for "having reduced environmental impacts" should be discouraged (Marshall and Toffel, 2005). Since the ability of scientists, economists, and social scientists to predict various impacts is presently limited, a **"precautionary principle"** as a companion to the principle of sustainability has been formulated which states (Dovers and Handmer, 1995): *"Where there are threats of serious or irreversible environmental damage, lack of full scientific certainty should not be used as a reason for postponing measures to prevent environmental degradation, [and development] decisions should be guided by (a) careful evaluation to avoid, wherever practicable, serious or irreversible damage to the environment; and (b) an assessment of the risk-weighted consequences of various options"*.

There is a need to continually review and revise management approaches because of the changing and uncertain nature of our socio-economic and natural environments. In fact, considering global environmental changes and uncertain future, an *evolving* and *adaptive* strategy for natural resources development and management is a necessary condition for sustainable development. The concept of **"adaptive management"** as discussed in Chapter 1 appears to be the only feasible approach in coping with the uncertainties in our knowledge and the variability of societal attitudes towards the resource over time. Although recent trends in resource management also focus on the concept of *"adaptive management"* (Ludwig et al., 1993; Maimone, 2004; Loucks and van Beek, 2005; Pahl-Wostl, 2007; van der Keur et al., 2008), its use in practice is highly limited in the world in general and developing countries in particular.

Moreover, sustainable water management has become an issue of major concern in the 21st century. It is predicted that the impact of climate change will be severe on the water sector, which will have large implications for other sectors. Therefore, if the challenges of climate change for the world's water are not understood and addressed, it would be almost impossible to ensure sustainable future (GWPTEC, 2007). In addition, there is danger as well: "If we fail to understand the interaction between climate change and water, other climate change strategies may actually aggravate the problems and increase the vulnerability of communities to both natural and man-made calamities" (GWPTEC, 2007). As a result, the water experts emphasize "improving the way we use and manage our water today will make it easier to address the challenges of tomorrow" (GWPTEC, 2007).

The adoption of modern and structured water management approach known as **"Integrated Water Resources Management"** (IWRM) in practice is an urgent need throughout the world in order to ensure sustainable management of vital water and land resources. IWRM has received a widespread support because it focuses on holistic and multidisciplinary approaches and thereby addresses the interconnections among water, socio-economic development, and sustainability. It involves both 'hard' infrastructural and 'soft' institutional strategies and is advocated as an intelligent approach for climate change adaptation (GWPTEC, 2007). Thus, IWRM can play a pivotal role in achieving water, food, environmental and energy securities, which in turn can ensure sustainable development on the earth. The concept of IWRM and its benefits are discussed in Chapter 22, and for further details, the readers are referred to the Global Water Partnership website (http://www.gwpforum.org). Since uncertainty is an inherent part of managing resources in general, a new paradigm "Adaptive Water Resources Management" (AWRM) has been recently introduced (Pahl-Wostl, 2007; van der Keur et al., 2008). It is an emerging approach to advance the concept of IWRM to a stage where it can deal with uncertainty and indicate the necessity of flexible governance systems and management strategies (Pahl-Wostl, 2007). Therefore, it would be prudent to coin the new term (Adaptive Water Resources Management) as *"Adaptive Integrated Water Resources Management"* (AIWRM). These modern approaches, together with rainwater harvesting and artificial recharge have great potential to foster more efficient and sustainable use of water resources, and thereby achieve the Millennium Development Goals (MDGs). IWRM approach will support not just achievement of the MDGs but also the long-term economic development, poverty reduction and environmental sustainability (GWPTEC, 2006). There is a need for linking IWRM to national development and local action plans for its effective implementation. The efforts needed by national policy makers and managers in the water sector to successfully implement IWRM to achieve the MDGs are highlighted in GWPTEC (2006).

It is our responsibility as scientists and engineers to critically investigate as many of the impacts as possible taking advantage of technological improvements and knowledge enhancement, share the results with stakeholders, and help them make management decisions that minimize adverse impacts and maximize benefits. Through the active participation of all the interested stakeholders, we should arrive at some shared vision of what is best to do, at least, until conditions change or new knowledge or new goals or new requirements emerge—it is the best way to identify a plan which is politically, technically, financially as well as socially and institutionally feasible (Loucks and van Beek, 2005). Also, water resource managers and other natural resource managers should always remember that the fundamental principles must not be overlooked as they seek to demonstrate that the use of natural resources is sustainable. Changing the institutional and social components of environmental systems for sustainable development is the most difficult task because it requires changing the people's behavior (i.e., the way people think and act) and the prevalent institutional practices. Nevertheless, concerted efforts at all levels are essential to gradually achieve this goal for ensuring sustainable development on the earth, and hence better future for the present generation and better world for generations to come.

2.3 Need for Integrated Approach and International Vision to Combat Climate Change

Undoubtedly, climate change is a serious global threat and raises a wide range of concerns which need to be urgently addressed by the national and local water managers and planners, climatologists, hydrologists, policy makers, and public (GWP, 2009; Şen, 2009). It is predicted that the global warming and related climate changes will present significant challenges over the next century. These challenges are increasingly better understood and there is a growing consensus on their likely scale (IPCC, 2007a). There is an increasing need to address the vital issue of climate change so that awareness is created among the people about climate change and its disastrous effects, and technical solutions are found for mitigating climate change impacts as well as for developing management strategies to adapt to climate change. The climate change being part of the larger challenge of sustainable development, the policies on climate change can be more effective when consistently embedded within broader development strategies (UNDP, 2007). Such an approach is necessary to make development more resilient to disasters and thereby ensuring national and regional development paths more sustainable. According to UNDP (2007), "Climate change will be one of the defining forces shaping prospects for human development during the 21st century". Thus, there is an urgent need to integrate disaster management into overall development activities at both international and national levels (Coppola, 2007; UNDP, 2004, 2007; UNESCO, 2007). Besides the increasing support through overseas development assistance, international funding should also support improved information on the regional impacts of climate change, and research on new crop varieties which could be more resilient to droughts and floods (Cooperative Program on Water and Climate, http://www.waterandclimate.org). GWPTEC (2007) suggests that adequate funds must be allocated today for water resources management so as to achieve the goal of climate-proofing future; otherwise climate change could be more dangerous and remedial measures much more expensive in the future. As emphasized in the previous section, the implementation of IWRM worldwide is indispensable to effectively tackle climate change. In conjunction with the climate change, the impacts of socio-economic changes should also be addressed for sustainable human development.

According to IPCC (2007b), "Irrespective of the scale of mitigation measures, adaptation measures are necessary." This demands an integrated approach to combat climate change considering both *mitigation* and *adaptation* measures; the former deals with the drivers of climate change, while the latter focuses on the measures necessary to accommodate such changes. Such an integrated approach is currently lacking, and hence it should be initiated globally without delay in order to effectively manage the impacts of climate change. Particularly, adaptation efforts need to be accelerated in developing and low-income countries which are much more vulnerable to climate change. In the words of GWPTEC (2007), *"Unless we act now, we will miss opportunities to make it easier to ensure a more sustainable long-term future"*.

The international agencies like UNDP and Global Water Partnership also recommend that climate-proofing infrastructure and adaptation are the best ways to reduce the impact of climate change (UNDP, 2007; GWP, 2009). A typology of coping and adaptation strategies is an essential component of any disaster risk reduction process. Existing typologies of adaptation should be reviewed and, if necessary, new typologies should be developed to explore suitable adaptation strategies and options in different sectors of human vulnerability to a hazard in general and climate change in particular. In addition, the commercialization of low emission technologies in various sections of application, the conservation of natural resources by adopting management strategies such as saving, reusing, recycling and augmentation, the efficient management of wastes, and the engineering solutions (e.g., increased use of solar and wind power, second generation biofuels, fourth generation nuclear fuels, efficient power transmission, etc.) for reducing emissions of greenhouse gases (GHGs) are the need of the day. Future studies are required to understand the mechanisms linking large-scale climate variability with regional conditions, which can help reduce uncertainty involved in the assessment of regional impacts of climate change (Şen, 2009). Region-specific studies are also needed to predict the consequences of global environmental change for the efficient planning and management of social, economic, and environmental systems.

Overall, a *multidisciplinary* or *interdisciplinary* approach considering uncertainties in climate change and the role of indigenous and other local peoples is essential for finding efficient solutions which can help society at large to cope with the present and future changes in climate conditions (Hallegatte, 2009; Salick and Ross, 2009). It is now well recognized that climate change being a global problem, there must be a collective response to its impacts by creating a shared international vision of long-term goals and agreement on frameworks which will accelerate action, and by building mutually reinforcing approaches at national, regional and international levels (UNDP, 2007; Cooperative Program on Water and Climate, http://www.waterandclimate.org).

2.4 Need for Improved Understanding of Human-Environment Interactions

We should always remember that without conserving natural resources and maintaining ecosystem services, it is not possible to eliminate poverty by simply economic development. However, our knowledge about human-environment interactions is presently limited. Therefore, an integrated approach taking into account environmental, human and technological factors and their interdependence is required to better understand complex human-environment interactions, which calls for a *multidisciplinary* or *interdisciplinary* team for comprehensive and in-depth research. More and more field-based research should be carried out using both *top-down* (data-driven) and *bottom-up* (process-driven) approaches to advance our knowledge about natural processes and their interaction with human. Intensive research is needed to predict how potential ecosystem perturbations may affect short- and long-term ecosystem functionality. The *"syndrome"* approach, whereby investigators evaluate the symptoms of sustainability similar to a doctor evaluating a patient's symptoms in order to identify a specific disease or syndrome, appears to be promising (Petschel-Held et al., 1999).

Natural scientists and social scientists need to work together and cooperate with each other in an effective way to better understand human-environment interactions (IPCC, 2001). Also, the need for integrative (*interdisciplinary* or *transdisciplinary*) approaches and increased science-society interactions for addressing environmental changes has been emphasized by some researchers (e.g., Tress and Tress, 2009 and references therein). Tress et al. (2009) provide salient suggestions to improve integrative research on environmental and landscape change. While several problems that can harm ecosystem function have been dealt with, comprehensive methods for quantifying ecosystem function have not been developed. Given the ecosystems' non-equilibrium dynamics, a major technical challenge is quantifying whether and how much the ability of ecosystems to meet human needs is changing over time (Marshall and Toffel, 2005). More research following real *holistic* and *interdisciplinary/transdisciplinary* approaches is required in this direction under varying hydrological and ecological conditions. Thus, improved knowledge about human-environment interactions is the key to long-term sustainable development.

2.5 Need for Efficient Early Warning and Communication Systems

Efficient early warning and communication systems are the basic requirements for efficient disaster management. The need for reliable and timely information is as vital as is coherent and prompt coordination among disaster managers at local, national, and international levels. Thus, the real success of a disaster early warning system depends on unrestricted and free access to data/information; timely dissemination of information; prompt information sharing among the stakeholders; strong and coherent coordination among government and non-government agencies at local, state, and national levels; and realistic and workable plans. Lack of awareness and pre-disaster preparedness, coupled with reliable databases and proper information sharing among the stakeholders in different phases of disaster management have been proved to be important challenges in many recent disasters. This is obvious from the statement of UNESCO's Director-General, Koïchiro Matsuura, during the UNESCO's General Conference in 2005 (UNESCO, 2007): "The Indian Ocean tsunami of December 2004 and the devastating impact of Hurricane Katrina and Hurricane Rita in the United States of America have many aspects but one crucial factor is the importance of public awareness, preparedness and information transmission". It is unfortunate that despite the evidence of increasing extreme events as a result of global climate change and population pressures, disaster relief rather than pre-disaster preparedness (risk prevention) is usually considered as much more important by the aid donors, and government and non-government organizations, including the affected countries (UNESCO, 2007). This raises new challenges for the concerned scientists, which need to be addressed properly.

Moreover, forecasting and early warning systems based on short- to medium-term weather forecasts and up-to-date hydrological information, and linked to an effective disaster-preparedness organization are inevitable for effective coping with future climate changes (Cooperative Program on Water and Climate, http://www.waterandclimate.org). Unfortunately, such systems are presently available only in a very few developing countries. They require a high level of local participation and credibility as well as a robust and well-maintained network for data monitoring and dissemination. There is an urgent need, particularly in developing countries, to first create and then maintain such facilities with full dedication and professionalism. Besides developing adequate monitoring facilities to generate necessary field data in data-scarce developing nations, acquiring good-quality data is much more important for the healthy progress of science and humanity. It is rightly said that it is better to have no data than to have poor-quality or spurious data which have no value at all; they merely mislead scientists and planners, and hinder knowledge enhancement or progress of science. Therefore, there is a crucial need that the value of data be seriously recognized by the scientists, practitioners and policy makers, which is usually neglected in most developing countries, and every effort should be made to monitor and process data with utmost accuracy for the benefit of humanity. It is also indispensable that the professional ethics and the commitment for doing good science be never compromised by the researchers/scientists and practitioners of all levels for their self-vested interests. Thus, given the proper hardware for forecasting and early warning, a high level of credibility and devotion for the profession can lead to improved and reliable forecasting and early warning systems in developing countries as well.

In general, it is necessary to continually improve the reliability of forecasting and early warning systems for both geophysical and hydro-meteorological disasters taking advantage of emerging tools/techniques, improved modeling techniques, and new knowledge about natural systems. In this direction, I suggest initiating some innovative research in collaboration with biological scientists to explore the possibility of some kind of biological sensor for the early warning of disasters learning from the response of some animals to an impending danger. For example, the quick response of dogs before the devastating 2004 Tsunami took place and the unusual behavior of some birds when they sense a hazard. I believe such an innovative early warning system has greater potential for disaster risk reduction than the traditional one. Although it may appear to be unfeasible at present, sincere efforts are necessary for a breakthrough in early warning systems. Further, apart from the improvement in the hardware of forecasting and early warning systems, there is an urgent need for developing institutional frameworks for effective information sharing and communication among different stakeholders before, during, and after a disaster.

2.6 Need for Decision Support System and Extensive Use of Modern Tools

Decision Support Systems (DSSs) not only help decision makers in making efficient decisions but also help organizations/agencies to become more watchful and prepared for emergency situations. As highlighted in Chapter 22, there is a limited use of decision support systems for real-world decision making processes which can be partly due to lack of awareness and/or reluctance among the concerned agencies and decision makers in using modern/emerging tools even in the era of information technology. Although a couple of DSSs have been developed for water resources management, the need to further develop decision-support tools in this field is worldwide recognized because of location-specific problems and the limitations of existing DSSs. There is also an urgent need to develop DSSs which can help in assessing vulnerability to natural/anthropogenic disasters as well as in developing preparedness and mitigation plans so as to ensure security against disasters in the future. According to Van de Walle and Turoff (2008), "Often in large-scale disasters, the people who must work together have no history of doing so; they have not developed a trust or understanding of one another's abilities, and the totality of resources they bring to bear has never been exercised before. As a result, the challenges for individual or group decision support systems (DSS) in emergency situations are diverse and immense". The recent advances in DSS for disaster management, an overview of existing DSSs for certain emergency situations, and important challenges are discussed in Van de Walle and Turoff (2008). Furthermore, integration of expert systems and DSS with GIS (geographic information system) is an area of potential research to aid to the effective and timely decision making.

As demonstrated in Chapters 15 and 21, the advances in information and communication technologies (ICTs) as well as in automation technologies in the recent past have enabled them to serve as important tools for efficient natural resources management as well as for mitigating hazards/disasters. The promising tools/techniques are: remote sensing (RS), global positioning system (GPS), geographic information system (GIS), decision support system (DSS), expert system (ES), artificial intelligent system (AIS), modeling, internet, mobile phones, etc. Particularly, rapid and continued advancements in satellite remote sensing, GPS, GIS, and computer technologies as well as in communication technologies are making them more and more powerful tools for managing natural resources, tackling climate change, and mitigating hazards/disasters. The extensive and proper use of these modern and promising tools/techniques is essential for efficient resource management, and effective and timely communication among the stakeholders before, during, and after the disaster incidence. Such a need has also been emphasized for the effective management of different types of disasters discussed in this book. In addition, there is also a need for participatory information management and proper information networking to ensure sustainable disaster management as discussed in Chapter 24.

Finally, it is worth mentioning that the joint efforts of professional associations and cultural heritage institutions have resulted in the development of a variety of tools to help professionals in preparing disaster and recovery plans (http://en.wikipedia.org/wiki/Disaster). In many cases, these tools are made available to external users, including the plan templates of some organizations. Although individual organizations will have to formulate their own plans and tools for their specific requirements, some of these tools and plan templates can serve as a useful starting point in the planning process (http://en.wikipedia.org/wiki/Disaster), especially for new organizations.

2.7 Need for Disaster Education and Training

No management system, no matter how well designed or how impressive data and tools it contains, can be effective unless it improves the decisions and behavior of the people who operate the facility as well as of the stakeholders. Therefore, proper education and comprehensive emergency management training, including mock exercises/drills are the key to efficient disaster management. There is a need to have wide recognition of the fact that proper education (basic as well as advanced) on disaster management is vital and has great potential for reducing disaster losses. It will not only help enhance general awareness about various disasters but will also better equip the people to tackle hazards/disasters, and thereby will strengthen community-based disaster preparedness which in turn will contribute to the long-term goal of disaster-resilient society. The international agencies such as UNESCO, UNDP and UNEP have to play an important role in persuading national governments, especially in disaster-prone developing countries, to take up this task urgently as part of their disaster management strategies. All efforts should be made by the national governments to make each town/city and village of disaster-prone areas disaster-literate through adequate education and training.

An excellent example highlighting the importance of education related to disaster is 10-year-old British girl, "Tilly Smith", who was holidaying with her parents and 7-year-old sister on Maikhao Beach in Phuket, Thailand in December 2004. Two weeks before she had studied tsunami in a Geography class at Danes Hill Preparatory School in Surrey, England. Her knowledge about the signs of an impending tsunami (i.e., receding shoreline and frothing bubbles on the sea surface) and her quick reaction saved her family and 100 other tourists (http://en.wikipedia.org/wiki/Tilly_Smith). Because of her timely warning, the beach could be evacuated before the devastating December 2004 Tsunami hit the shore, and it was one of the few beaches on the Phuket Island with no reported casualties. Thus, proper education and training in the field of disaster management is a prime necessity, which should be initiated at national and local levels without any delay.

3. EPILOGUE

The concept of "**disaster risk reduction**" as emphasized in the 2005 World Conference on Disaster Reduction is sound and promising, which reminds us of the well-known proverb, *"Prevention is better than cure"*. It emphasizes the vital role of human thought and action in minimizing disaster risk, and can provide a practical basis for adapting to climate change. Cost-benefit analyses have clearly indicated that appropriate investments in disaster prevention can substantially reduce the burden of disasters, which falls disproportionately on the world's poor countries. The economic losses are gradually becoming more burdensome. For example, in the 1990s, economic losses were three times higher than in the 1980s, almost nine times higher than in the 1960s and 15 times higher than in the 1950s (UNESCO, 2007). Considering the scale, frequency, human suffering and fatalities, and economic and environmental losses caused by disasters, the investment in disaster risk reduction is of immense significance. Undoubtedly, disaster risk reduction, disaster relief, and sustainable development are intimately linked. Also, vulnerability to disasters is linked to poverty, and vice versa.

In view of the above facts, there is an urgent need for a paradigm shift in disaster management approach away from *"post-disaster reaction"* to *"pre-disaster action"* and thereby improving disaster preparedness, mitigation, and response (UNISDR, 2005; UNESCO, 2007). This new, emerging approach is being advocated by UNESCO which stresses the merit of preventive and mitigative measures through scientific understanding and technical know-how. The key steps needed to follow this novel approach are: creating public awareness, imparting information effectively, involving local communities and other stakeholders, making disaster prevention part of education, and providing adequate financial support and infrastructure. The international organizations like UNESCO, UNDP, UNEP, ISDR, UNICEF, WMO, WHO, FAO, World Bank, etc. have to play a vital role in inculcating a culture of 'disaster preparedness and mitigation' in all disaster-prone countries and its adoption as best disaster management practice in the national development process, though they have been catalyst for international, interdisciplinary cooperation in many aspects of disaster prevention and mitigation. Today's disaster managers have a wealth of information available, which along with modern tools and approaches can help them develop expertise in disaster preparedness, mitigation, and response that in turn can strengthen their decision-making capability. Such an expert disaster management team has large potential to develop a disaster-resilient society in the disaster-prone countries of the world, and thereby contributing to sustainable human development on the earth.

What is much more required, especially in developing countries, is that national and state governments as well as concerned organizations/agencies must have strong willpower, seriousness and dedication for disaster preparedness, mitigation, and response; simply nice documentation of plans/strategies is not enough. Lip service, ad hoc and short-term approaches, and short-sightedness must be avoided, coupled with significant reduction in corruption (if not elimination) in order to ensure efficient disaster management in a real sense. In addition, there is an urgent need to have a general realization that many disasters could be greatly mitigated with adequate prudence and preparation, and that the cost of mitigation is much smaller than that of relief and recovery efforts (UNESCO, 2007). Modern and innovative approaches, together with modern tools and techniques have significant potential to ensure sustainable management of natural resources and thereby avoiding catastrophes in the future resulting from mismanagement of natural resources. Sincere and concerted efforts are needed at all levels to ensure sustainable disaster management which can help build disaster-resilient communities and nations worldwide. Finally, the insightful guidance provided by James Lee Witt (Federal Emergency Management Agency's famous and successful director) for the disaster managers should be followed in practice so as to ensure efficient disaster management: "We need to take a common-sense, practical approach to reducing the risks we face and protecting our citizens and our communities. We need to identify our risks, educate and communicate to our people about those risks, prepare as best we can for the risks, and then, together, form partnerships to take action to reduce those risks" (Coppola, 2007).

REFERENCES

Coppola, D.P. (2007). Introduction to International Disaster Management. Elsevier, Amsterdam, The Netherlands.
Dovers, S.R. and Handmer, J.W. (1995). Ignorance, the precautionary principle, and sustainability. *Ambio*, **24(2):** 92-97.
GWP (2009). GWP Strategy 2009-2013. Global Water Partnership (GWP), Stockholm, Sweden, 23 pp.
GWPTEC (2006). How IWRM will Contribute to Achieving the MDGs. Catalyzing Change Series Policy Brief 4, Global Water Partnership Technical Committee (GWPTEC), Global Water Partnership, Stockholm, Sweden, 8 pp.
GWPTEC (2007). Climate Change Adaptation and Integrated Water Resource Management: An Initial Overview. Catalyzing Change Series Policy Brief 5, Global Water Partnership Technical Committee (GWPTEC), Global Water Partnership, Stockholm, Sweden, 12 pp.
Hallegatte, S. (2009). Strategies to adapt to an uncertain climate change. *Global Environmental Change*, **19:** 240-247.
IPCC (2001). Third Assessment Report: Climate Change 2001. Intergovernmental Panel on Climate Change (IPCC), Geneva, Switzerland, http://www.ipcc.ch (accessed in March 2005).
IPCC (2007a). Climate Change 2007: Impacts, Adaptation and Vulnerability. Contribution of Working Group II to the Fourth Assessment Report of the Intergovernmental Panel on Climate Change (IPCC), Cambridge University Press, Cambridge, U.K.
IPCC (2007b). Climate Change 2007: Mitigation. Contribution of Working Group III to the Fourth Assessment Report of the Intergovernmental Panel on Climate Change (IPCC), Cambridge University Press, Cambridge, U.K. and New York, USA.
Kauai Declaration (2007). Ethnobotany, the science of survival: A declaration from Kauai. *Economic Botany*, **61(1):** 1-2.
Loucks, D.P. and van Beek, E. (2005). Water Resources Systems Planning and Management: An Introduction to Methods, Models and Applications. Studies and Reports in Hydrology, UNESCO Publishing, UNESCO, Paris.
Ludwig, D., Hilborn, R. and Walters, C. (1993). Uncertainty, resource exploitation, and conservation: Lessons from history. *Science*, **260:** 17-18.
Maimone, M. (2004). Defining and managing sustainable yield. *Ground Water*, **42(6):** 809-814.
Marshall, J.D. and Toffel, M.W. (2005). Framing the elusive concept of sustainability: A sustainability hierarchy. *Environmental Science & Technology*, **39(3):** 673-682.
Munier, N. (2005). Introduction to Sustainability: Road to a Better Future. Springer.
Newson, M. (1992). Land, Water, and Development: River Basin Systems and Their Sustainable Development. Routledge, London.

Pahl-Wostl, C. (2007). Transitions towards adaptive management of water facing climate and global change. *Water Resources Management*, **21(1)**: 49-62.

Petschel-Held, G., Block, A., Cassel-Gintz, M., Kropp, J., Lüdeke, M.K.B., Moldenhauer, O., Reusswig, F. and Schellnhuber, H.J. (1999). Syndromes of global change: A qualitative modeling approach to assist global environmental management. *Environmental Modeling & Assessment*, **4**: 295-315.

Salick, J. and Ross, N. (2009). Traditional peoples and climate change. *Global Environmental Change*, **19**: 137-139.

Şen, Z. (2009). Global warming threat on water resources and environment: A review. *Environmental Geology*, **57**: 321-329.

Tress, B. and Tress, G. (2009). Environmental and landscape change: Addressing an interdisciplinary agenda. *Journal of Environmental Management*, **90(9)**: 2849-2850.

Tress, B., Tress, G. and Fry, G. (2009). Integrative research on environmental and landscape change: PhD students' motivations and challenges. *Journal of Environmental Management*, **90(9)**: 2921-2929.

UNDP (2004). Reducing Disaster Risk: A Challenge for Development. United Nations Development Program (UNDP), Bureau for Crisis Prevention and Recovery, United Nations Plaza, New York, 146 pp.

UNDP (2007). Human Development Report 2007/2008. United Nations Development Program (UNDP), United Nations Plaza, New York, 384 pp.

UNESCO (2007). Disaster Preparedness and Mitigation: UNESCO's Role. Section for Disaster Reduction, Natural Sciences Sector, the United Nations Educational, Scientific and Cultural Organization (UNESCO), Paris, France, 48 pp.

UNISDR (2005). Report of the World Conference on Disaster Reduction. Kobe, Hyogo Prefecture, Japan, 18-22 January 2005, United Nations International Strategy for Disaster Reduction (UNISDR), Geneva, Switzerland, 42 pp.

Van de Walle, B. and Turoff, M. (2008). Decision support for emergency situations. *Information Systems and E-Business Management*, **6(3)**: 295-316.

van der Keur, P., Henriksen, H.J., Refsgaard, J.C., Brugnach, M., Pahl-Wostl, C., Dewulf, A. and Buiteveld, H. (2008). Identification of major sources of uncertainty in current IWRM practice: Illustrated for the Rhine basin. *Water Resources Management*, **22**: 1677-1708.

WCED (1987). Our Common Future. World Commission on Environment and Development (WCED), Oxford University Press, Oxford, UK.

Index

3D geological modeling, 332
3D plant design, 332

Acid Detergent Fiber (ADF), 439
Adaptive Integrated Water Resources Management, 602
adaptive management, 14, 15, 601
agricultural drought, 198, 199, 201, 203-207, 224, 225
Agricultural Drought Indicators, 243, 246
agro-climatic zones, 415
agro-forestry, 542, 555
airborne sensors, 481
alkalinization, 471, 491
anaerobiosis, 471
Antecedent Precipitation Index, 173
anthropogenic nitrogen fixation, 386
anticyclone, 324, 326, 328, 330
apparent drought, 199, 203- 206, 212, 213, 215, 221
aquifer contamination, 63
aquifer restoration, 87
ArcGlobe, 379
Aridity Index, 243, 246
Armchair Mining, 361, 362
artificial intelligence, 149, 512, 513, 516, 518, 528
Artificial Neural Network, 102, 109-111, 148, 151
Asian Summer Monsoon Circulation, 330
automation technology, 345, 346
autonomous mining system, 344

backwater, 64, 74, 76
barrier reef ecosystem, 65
biological degradation, 472
biological disasters, 1
biological diversity, 536, 541, 545, 556
bluetongue virus (BTV), 435
bottom pressure recorders, 103

carbon sequestration, 63
carbon stock, 63

categories of tropical cyclones, 262
celerity-discharge relationships, 191
Celerity-Stage Relationship, 179, 180
challenges of the 21st century, 598
Chaman fault region, 17
chlorination, 85, 92, 93
chlorine treatment, 50, 52
chronic drought, 203-209, 216, 219, 220
civil defense, 10, 11
climate change impacts, 453, 455, 457, 459, 463, 464
climate change indices, 279
climate extreme events, 276
climate risk, 452-455, 459
Climate Task Group, 459
coastal aquifer, 55
coastal disaster, 141, 142
coastal flooding, 127, 129, 134, 139, 144
community acts, 567
community governance, 573
composite disasters, 6
comprehensive vulnerability management, 6
conceptual lumped parameter models, 396
continental shelf, 64, 72, 78
coping capacity, 7
crop insurance, 228, 229
crop models, 417
Crop Moisture Index, 245
crop phenology, 417
crop productivity, 414, 427
Crop Water Stress Index, 246
cultivars, 417
cyclogenesis, 257, 258, 264, 265, 267, 269, 270
cyclone damage, 578, 581, 587
cyclone detection radars, 141
cyclone emergency preparedness, 578, 580, 588, 596
cyclone monitoring, 583, 596
cyclone shelter, 569
cyclone warning dissemination systems, 270
cyclone warning system, 583

cyclonic storms, 126-129, 134, 140, 141, 256-258, 260-262, 272, 273

dairy production, 432, 435-437, 443
data uncertainty, 454
Decile Indices, 239
Decision Support System (DSS), 506, 508-513, 515, 518, 519, 523, 526-528, 606
desertification, 198, 199, 224
devastating cyclonic storms, 582
Digital Elevation Model (DEM), 402
digital terrain model, 151
direct waves, 64, 69, 78
disaster management cycle, 13
disaster preparedness and mitigation, 539, 552
disaster risk management, 7, 455, 457-459, 461-463, 573, 575
disaster risk mitigation, 462, 464
disaster risk mitigation systems, 462
disaster-related terms, 3
disastrous forest fires, 382
drought assessment, 237, 242, 247
drought indicators, 237, 238, 242, 243, 246
Drought Indices, 224
Drought Monitoring Center, 459
drought prediction, 228, 231, 232
drought week, 199
dryland agriculture, 469
dryland farming, 406
DSSAT, 415, 417
dust storms, 60
dynamic crop growth models, 435
dynamic downscaling, 394
dynamic risk assessment, 381
dynamic wave model, 192

early warning system, 18, 37, 38, 56, 463
Earth Summit, 536, 545
earthquake alarms systems, 37
eco-centric approach, 77
ecological services, 550, 551
economic divide, 566
e-governance, 570
El Niño, 199, 200, 369, 373, 375
El Niño-Southern Oscillation, 277
electrical conductivity, 52, 59
emergency equipment, 594
emergency management, 10, 14

emergency planning, 587, 588, 596
emergency response, 142-144
emergency water management, 524
Enhanced Vegetation Index, 247
environmental degradation, 553
environmental resilience, 541
ephemeral drought, 202-206, 208, 211- 214, 218, 220
evolution of tropical cyclones, 580
evolutionary algorithm, 149, 152
Executive Information Systems (EIS), 511, 514
expected time of arrival, 107
expert system, 149, 168, 512, 515, 518, 524, 526, 528
extreme flood events, 151
extreme temperature indices, 280, 283

famine, 198, 199, 201, 228
fecal pollution, 48
Fire Advisory Group, 380
Fire Disaster, 367, 369, 373, 382
fire emissions, 368
fire-safe building codes, 382
flash floods, 60, 126, 129, 130, 133
Floating Month Drought Index, 249
flood control measures, 485, 486
flood control, 147, 148, 150-153, 156, 158, 160, 165
flood envelope curve, 148
flood forecasting, 137
flood frequency analysis, 148, 396
flood hazard, 126, 127, 130, 137, 138, 142, 144
flood insurance, 139
flood management software, 174
flood management, 127, 134, 135, 516, 520, 523
flood preparedness, 138, 143
flood risk zone, 151
flood warning systems, 150
Flood Wave Model, 173, 192
flooding waves, 82, 89
floodplain zoning maps, 138
Folded Dynamic Programming, 153-155, 158, 161
Foot and Mouth Disease (FMD), 435
forecast dissemination, 172
forecast error, 171, 184, 188
forecast evaluation, 172
forest cover of India, 543
Forest Principles, 536
frequency indices, 281-283
frequency of tropical cyclones, 256, 257
fresh water zone, 52

freshwater-saltwater interface, 93
Fuzzy logic, 148, 151, 516, 527, 528

genetic algorithm, 516, 518, 529
Geographical Information System (GIS), 332, 355, 470, 471, 516
Geohazard View, 63
geospatial technologies, 64, 78
global circulation models, 459, 479
global climate model, 393, 394
Global Environmental Disaster Information System, 351
Global Fire Monitoring Centre, 382
Global Positioning System (GPS), 332, 344, 352, 470, 471, 516, 606
global warming, 387, 390
GPS geodesy, 36
greenhouse effect, 386
groundwater protection, 84, 86, 89, 96, 97
groundwater quality, 71
groundwater salinity, 52, 53
groundwater sustainability, 84
Group Support Systems (GSS), 510
Grouped Response Unit, 396

Haveli system, 215
Hazara arc, 17
hazard and vulnerability mapping, 462
high-fat diet, 440
Himalayan arc region, 18
Himalayan Seismic Belt, 19, 20, 27, 28, 32
humanitarian crisis, 7
Huygens principle, 106, 110
hydraulic models, 173
hydrogeologic mapping, 476
hydrologic assessment, 394
hydrologic models, 391, 394-398
hydrologic response units, 396
hydrologic simulation models, 148
hydrologic units, 294
hydrological drought, 203
hydrological drought indicators, 242
hydrological forecasting system, 172, 173
hydro-meteorological disasters, 1, 2
Hyogo Framework for Action, 12, 13, 600
Hyogo Recommendations, 600
hyper-endemic division, 435

impact of cyclonic storms, 583
impact of disasters, 7, 8, 10, 13, 14
Index of Areal Representativeness, 300
Indian Ocean Dipole Mode Index, 258
Indian plate, 18, 30, 32
Indian Remote Sensing Missions, 492
Indian Shield Region, 18, 20, 27, 28, 32
Indo-Burmese arc, 17
Indo-Gangetic Plains, 295, 296, 303, 305, 309, 326, 329, 388, 389, 407
information and communication technologies, 566, 606
Information Networking, 563, 570, 574
inland lagoon, 89
Integrated Coastal Zone Management, 77
integrated nutrient management, 218, 222, 227
Integrated Water Resources Management (IWRM), 506-508, 513, 517, 523, 526, 528
integrated watershed management, 214
intelligent DSS, 151
intensity indices, 281, 282, 284, 285, 288
interlinking of rivers, 165
international disaster, 4
interplate earthquake, 18, 20, 21
interseismic deformation, 31, 32
intrinsic permeability, 54
IRS satellites, 475
isochrome charts, 102

Kauai Declaration, 598
kinematic-wave equation, 177
knowledge uncertainty, 454
Kyoto Protocol, 463

La Niña, 277, 278, 280, 284-288
land management units, 225, 226
land production system, 372
land use policy, 229, 230, 232
landslide hazard zones, 482
Laser Terrain Mapping, 139
Leaf Area Index (LAI), 419, 420
limitations of satellite remote sensing, 492
livestock insurance, 443
local disaster, 4

macro strategies, 437
Management Information Systems (MIS), 510

marine ecosystem, 79
meteorological drought indicators, 238
meteorological drought, 198, 199
microbiological contamination, 85
Millennium Development Goals, 3, 13, 536, 602
mine disaster, 333-335, 338-342, 345, 347, 350, 362, 363
mine fires, 343
mine multimedia rescue communication system, 350
mine robots, 347
mine safety, 332, 333, 343-346, 358, 362
mitigation phase, 13
mixed integer linear programming, 151
model uncertainty, 454
modern approaches, 3, 13
modern management approaches, 601
monsoon depressions, 257, 264, 268, 269, 273
monsoon trough position, 293
mudslides, 60
multicriteria decision making, 150
multi-hazard approach, 600
multiple-peak floods, 129
multi-temporal satellite, 482
Muskingum method, 178
Muskingum routing method, 156
Muskingum stage-hydrograph, 171, 175, 192
Muskingum-Cunge method, 172

national disaster, 4
National Emergency Response Force, 143
National Research Council, 440
natural ecosystem, 276, 294, 295
natural flooding, 169
natural flushing, 84
natural remediation processes, 90
network architecture, 118
network learning, 117
non-structural measures, 7, 485
Normalized Difference Vegetation Index (NDVI), 247, 375
North Indian Ocean, 256, 257, 261, 267, 269, 273
North Sub Polar Low, 325, 330
nutritional management, 438, 444

oceanogenic disasters, 103
Oldham Fault, 22
on-line analytical processing, 511, 512, 514, 515
optimal control theory, 516
optimization techniques, 516

Pacific Ocean signatures, 277
Palmer Drought Severity Index, 239
participatory appraisal techniques, 569
participatory land use planning, 225
participatory rural appraisal, 569, 571
pathogenic contamination, 82, 91, 96
physiographic division, 294, 296-300, 303-305, 329
physiographic units, 294
physiological maturity, 421
plate boundary earthquake, 26
power haulage, 336
precautionary principle, 601
preparedness phase, 14
probable maximum flood, 157
probable maximum storm surge, 132
productive farming systems, 222, 223

radiocarbon ages, 27
rainfall sequence, 297, 299-301, 309, 311, 314, 316, 319
rainfed agriculture, 197, 214, 225, 229, 230
rainfed Agro-Economic Zones, 229, 230
rainwater management, 198, 213, 227
rapid-onset disasters, 4
real time risk assessment, 381
real-time flood forecasting, 171-174, 183, 184, 192
Reclamation Drought Index, 242
recovery phase, 14
rehabilitation, 50, 51, 54, 55
Relief Action Plan. 590, 593
Remote Sensing, 332, 347, 352, 354, 357, 358, 470, 471, 473, 474, 476, 477-480, 482-488, 491-495, 512, 515, 516, 517, 606
renewable energy, 438
representative elemental areas, 396
research and development needs, 599, 600
reservoir operation, 516, 518, 523, 526, 527
resonance effects, 45
response phase, 14
river flooding, 129, 130, 134, 140, 141
riverbed aggradation, 129
rockfalls, 17
routing method, 171, 172, 175, 176, 178, 179, 184, 191, 192
rule curves, 155, 158, 161, 164

salinity remediation, 86
salinization, 471, 472, 489, 491
saltwater flooding, 84-89, 93, 96, 97

salty water zone, 52
satellite communication, 332, 352
seawall, 66, 68, 73
seawater intrusion, 71
seismic hazard, 18, 32, 36
seismic intensity, 23
seismic microzonation, 37
seismic monitoring, 345
seismic productivity, 28
seismic zone, 17, 30
seismic-hazard map, 37
seismograph, 21
seismometers, 37
seismotectonic models, 32
severe cyclonic storm, 256-258, 273
shelter management, 437, 438
simulation models, 516, 524, 527
single-peak floods, 129
slow-onset disasters, 4
societal drought, 238, 250
socio-economic drought, 199
sociological disasters, 1
soil conservation measures, 485
soil conservation units, 225, 226
soil fertility management, 217
soil quality units, 225, 226
soil resilience, 225
solid waste management, 54
Southern Africa Fire Network, 382
Southern Oscillation Index, 200
spatial decision support system, 250
stage forecasting, 171, 192
Standardized Precipitation Index, 240, 241, 246
Standardized Water Level Index, 242
static risk assessment, 381
statistical downscaling, 394, 395
statistical routing methods, 172
stochastic weather generator, 394
storm surge, 102, 104, 123
storm tide, 130
streamflow forecast, 151
structural measures, 7
submarine earthquake, 61
submarine seismic activity, 100
super cyclone, 262
Surface Water Supply Index, 242
Sustainable disaster management, 598-600, 606, 607
sustainable forest management, 536, 542, 544-546, 548, 550, 551, 553-555, 557, 558

sustainable resilient community, 7
SWAT (Soil Water Assessment Tool), 389
synchronized floods, 129
syntaxial bends, 17
Synthetic Aperture Radar, 138

technological disasters, 1, 130
tectonic earthquakes, 18
tectonic plates, 17
telemetry system, 172, 173
temperature extremes, 276-278, 280-285, 287, 288
temporal downscaling, 393, 394
terminal drought, 205, 211, 212, 217
thermal imaging, 495
tropical cyclone forecasting, 267, 270
tropical cyclone, 17
tropical depression, 262, 264, 273
tropical disturbances, 261, 264, 266, 273
tropical storm, 262, 266
Tsunami Arrival Time, 109, 110, 120
tsunami influx, 90
tsunami process, 56
tsunami surge, 78
tsunami travel time (TTT), 102, 105-109, 119, 123, 124
tsunami vulnerability, 63
Tsunami Warning Centre, 44
tsunamigenic zones, 101
turbidity, 55, 59
turbine efficiency, 163

underwater earthquakes, 61

Vegetation Drought Response Index, 249
vertical drought, 203

warning dissemination, 144
water scarcity indicator, 504
water use efficiency, 414
waterlogging, 470, 472, 473, 489-492
watershed models, 517
wave energy, 45
waveform analysis, 29
Weather Code, 292, 232
Web Fire Mapper, 378
wildfires, 366, 369, 376
Working Plan Code, 550

zero-order approximation, 102